BASIC PARTIAL DIFFERENTIAL EQUATIONS

BASIC PARTIAL DIFFERENTIAL EQUATIONS

David Bleecker
George Csordas

Department of Mathematics
University of Hawaii
Honolulu, Hawaii

VNR VAN NOSTRAND REINHOLD
New York

Library of Congress Catalog Card Number 92-6226
ISBN 0-442-01253-5

Manufactured in the United States of America

Published by Van Nostrand Reinhold
115 Fifth Avenue
New York, New York 10003

Chapman and Hall
2-6 Boundary Row
London, SE1 8HN, England

Thomas Nelson Australia
102 Dodds Street
South Melbourne 3205
Victoria, Australia

Nelson Canada
1120 Birchmount Road
Scarborough, Ontario M1K 5G4, Canada

16 15 14 13 12 11 10 9 8 7 6 5 4 3 2 1

Library of Congress Cataloging-in-Publication Data

Bleecker, David.
Basic partial differential equations / David Bleecker, George Csordas.
p. cm.
Includes bibliographical references (p.) and indexes.
ISBN 0-442-01253-5
1. Differential equations, Partial. I. Csordas, George.
II. Title.
QA374.B64 1992 92-6226
515'.353–dc20 CIP

TABLE OF CONTENTS

9. PDEs in Higher Dimensions

Dependence of Sections

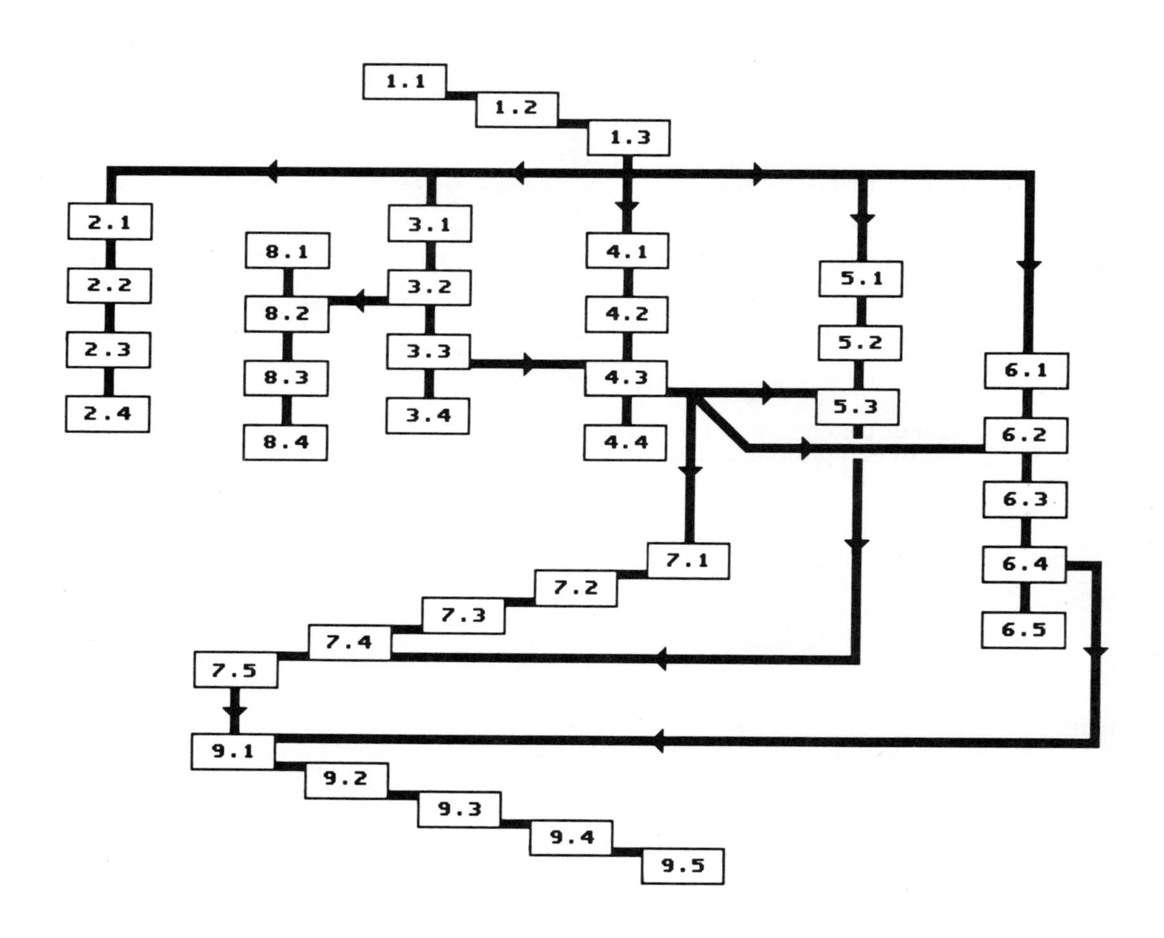

PREFACE

Quantities which depend on space and/or time variables are often governed by differential equations which are based on underlying physical principles. Partial differential equations (PDEs) not only accurately express these principles, but also help to predict the behavior of a system from an initial state of the system and from given external influences. Thus, it is hard to overestimate the relevance of PDEs in all forms of science and engineering, or any endeavor which involves reasonably smooth, predictable changes of measurable quantities.

Having taught from the material in this book for ten years with much feedback from students, we have found that the book serves as a very readable introduction to the subject for undergraduates with a year and a half of calculus, but not necessarily any more. In particular, one need *not* have had a linear algebra course or even a course in ordinary differential equations to understand the material. As the title suggests, we have concentrated only on what we feel are the absolutely essential aspects of the subject, and there are some crucial topics such as systems of PDEs which we only touch on. Yet the book certainly contains more material than can be covered in a single semester, even with an exceptional class. Given the broad relevance of the subject, we suspect that a demand for a second semester surely exists, but has been largely unmet, partly due to the lack of books which take the time and space to be readable by sophomores. A glance at the table of contents or the index reveals some subjects which are regarded as rather advanced (e.g., maximum principles, Fourier transforms, quasi–linear PDEs, spherical harmonics, PDEs on manifolds, complex variable theory, conditions under which Fourier series are uniformly convergent). However, despite general impressions given (perhaps unwittingly) by mathematical gurus, *any* valid mathematical result or concept, regardless of how "advanced" it is, can be broken down into elementary, trivial pieces which are easily understood by all who desire to do so. With regard to the so–called "advanced" topics in this book, we feel that we have accomplished this to a degree which surprised even us. For us it was a constant and worthwhile challenge to make the book completely self–contained for those who have only been through the typical freshman/sophomore calculus sequence, even if they forgot most of it. We have successfully taught students who did not recall how to solve $y'(x) = y(x)$ with $y(0) = 1$ at the beginning of the semester, as was the case with over half of our students according to initial survey tests. However, before the semester's end, these same students could prove and understand the Maximum Principle for the heat equation and could easily deduce the continuous dependence of solutions on initial and boundary data. In essence, "advanced topics" are rarely difficult per se, but they may seem so, if (for the sake of elegance) too little time is spent explaining and motivating them.

We have avoided the temptation to first prove unmotivated results in great generality and then use them to deduce an abundance of particular cases. By and large, we have introduced results and techniques inductively through many solved examples. By the time students have seen enough examples, they can often anticipate, as well as understand, the argument for the general case. In particular, we have found that, in spite of the fact that Sturm–Liouville Theory provides a uniform approach to boundary–value problems, it is not so wise to teach it first to students who are barely familiar with sines and cosines, and then cover the elementary boundary–value problems as special cases. We have proceeded in the opposite manner. After we have handled a variety of simple boundary conditions for the heat equation and treated Fourier series, the student is prepared to study and appreciate Sturm–Liouville Theory as a natural continuation of what has been learned without it. Proceeding from examples to theorems may result in a book which is physically longer, but students learn more rapidly and effectively this way. In short, it is easier to build from the ground up than from the roof down. In the process, we may have sacrificed some degree of elegance, but we have not sacrificed rigor. Nearly every basic result is proved rigorously at some stage, or at least we give a reference (e.g., for the convergence of eigenfunction expansions on manifolds). We certainly do not recommend proving everything in class, since this would severely limit the range of the material covered, but instead the interested student may be directed to the many detailed, thoroughly digestible proofs in the text. On the point of rigor, we mention that many solutions of PDEs are expressible in terms of integrals of Greens functions against boundary and/or initial data. In most PDE texts, such integral formulas are derived (if at all) under the assumption that solutions of the PDEs actually exist. To be honest, one should have the tools to check that the functions defined by such integral formulas actually solve the given problem. This necessarily entails the use of Leibniz's rule for differentiation under an integral, sometimes when the interval of integration is unbounded. One feature of this book, which appears to be absent in other texts, is that there is a complete, elementary proof of Leibniz's rule in the Appendix. To experts, this may be surprising, since many standard proofs entail the use of the Lebesgue Dominated Convergence Theorem. However, in the Appendix, we have proven a suitable version of dominated convergence which avoids the notion of Lebesgue measure and integration. (The idea originated in [Lewin, 1986, 1987].)

Solving problems is the major part of learning any mathematical subject. This book contains many problems which range from the purely routine to those which will challenge even the most brilliant student. Sometimes one finds that although some students can arrive at a solution to a problem through mimicking procedures, they still may not be able to interpret or use the solution or even understand why the expression they have found is actually a solution of the problem. We have tried to counter this tragedy by including many exercises which require the student to think, draw some conclusions, and express themselves, instead of simply implementing purely computational procedures. Since some students will do anything to get the answer in the back of the book, we have been sparing with the answers. However, a solution manual (with complete solutions to all but the most trivial problems) is available to instructors only. We personally worked out each of the problems.

Since the whole book cannot be covered in a single semester, instructors who are limited to a single semester must decide which sections or chapters to cover. Given the demand, instructors might consider the introduction of a second semester of PDEs. Below, we summarize the material covered in the chapters and sections. Following this, some suggestions are given on what sections must, should or could be included in a one–semester or two–quarter course.

Acknowledgements. It is our pleasure to acknowledge the comments and suggestions of our colleagues and students. In particular, we thank Hans Broderson, Karl Heinz Dovermann, Christopher Mawata, Ken Rogers, Mi–Soo Smith, Wayne Smith, David Stegenga, Joel Weiner, George Wilkens, and Les Wilson, who have adopted the notes in their courses. We also acknowledge Paolo Agliano and Paul Kohs who helped us with the typing and the graphics. In addition, a warm *mahalo* is due to the secretarial staff of the Department of Mathematics at the University of Hawaii. A special *mahalo nui loa* is due to Pat Goldstein who cheerfully helped us with much of the clerical work. Last, but not least, we wish to thank our families for their patience and support during the preparation of this work.

Honolulu, 1992 David Bleecker & George Csordas

Chapter–by–chapter synopsis and suggestions for the instructor

Chapter 1 (Review and Introduction): If the students have had a course in ODEs, then Section 1.1 can be skipped, or assigned as reading. Some coverage of Sections 1.2 and 1.3 is necessary for a general overview of PDEs and their applications, and for an introduction to certain topics, such as separation of variables and the superposition principle. These concepts are used often in the sequel.

Chapter 2 (First–Order PDEs): For instructors who regard first–order PDEs as devoid of any real application, we urge them to read the introduction to Chapter 2, before deciding to skip Chapter 2 entirely. Not only are there wide applications to birth and death processes (e.g., the evolution of population densities), continuum mechanics and the development of shocks in traffic flow, but also the student sees how a change of variables can greatly simplify a PDE. Incidentally, we elected not to include examples and drill exercises for putting second–order, linear PDEs (with constant coefficients) into the standard normal forms (e.g, by rotation of axes, etc.), for the simple reason that second–order PDEs which arise in applications are already in a standard form. However, a complete statement of the Classification Theorem is given in Section 1.2, and a complete proof is given in the Appendix A.1. To compensate for lack of practice in change of variables drill for second–order PDEs, there are plenty of change–of–variable problems for first–order PDEs in Chapter 2. First–order PDEs which arise in applications are seldom in the standard form of a parametrized ODE. While Chapters 3–9 do not depend on Chapter 2, instructors should seriously consider doing at least Section 2.1 in which $au_x + bu_y + cu = f(x,y)$ is solved, when a, b, and c are constants. The case of variable coefficients is covered in Section 2.2, and the quasi–linear case is covered in Section 2.3. The fully nonlinear case is covered in the purely optional Section 2.4.

Chapter 3 (The Heat Equation): Section 3.1 begins with a derivation of the heat equation. The simplest initial/boundary–value problems are solved *without* first introducing Fourier series. Here, we use separation of variables to find product solutions of the heat equation which meet the homogeneous boundary conditions B.C. and then find a linear combination which meets the initial condition. In Chapter 3, initial temperatures are chosen so that they are expressible (via trigonometric identities) as finite linear combinations of sines or cosines of the appropriate form. Students then naturally ask what can be done if this is not the case. In other words, they are naturally motivated for the introduction of Fourier series which is the topic of Chapter 4. In Section 3.2, uniqueness of solutions of various initial/boundary–value problems for the heat equation is proved, by showing that for homogeneous B.C. of the first or second kind, the mean–square of the temperature is non–increasing. The Maximum Principle provides a second approach. We first illustrate the Maximum Principle through a number of examples and we show that it easily leads to continuous (uniform) dependence of solutions on initial/boundary data. The proof of the Maximum Principle is then given at the end of Section 3.2. Section 3.3 deals with the case of various simple B.C. which are time–independent, but possibly inhomogeneous. In Section 3.4, the case of time–dependent B.C. and heat sources are handled by means of Duhamel's principle. Section 3.4 can be skipped or covered later if time permits, and Section 3.3 can be covered quickly and lightly. However, Section 3.1 is certainly part of any first PDE course, and we strongly recommend that Section 3.2 be covered in some detail.

Chapter 4 (Fourier Series and Sturm–Liouville Theory): Students see the need for Fourier series in Chapter 3. In Section 4.1, we introduce the notion of functional orthogonality, and the definition of Fourier series of a function as a formal expression which may or may not converge to the function. Many examples are computed, and the question of convergence is motivated. An estimate for the number of terms needed to uniformly approximate a C^2 function is stated (but the proof is deferred until Section 4.2). We provide a technique for obtaining much sharper estimates by means of integral estimates of the tail of a Fourier series. Section 4.2 contains detailed proofs of the convergence of Fourier series under various assumptions. We gently introduce the difference between pointwise convergence and uniform convergence. Pointwise convergence is proved for piecewise C^1 functions and uniform convergence for *continuous* piecewise C^1 functions. Without the luxury of time, we recommend that the lengthier proofs be skipped or assigned for reading. However, certainly one should get across the general idea that the smoother a function is on a circle, the more rapid is the convergence of its Fourier series. In Section 4.3, we introduce Fourier sine and cosine series which are used to handle (at least formally) the case (left dangling in Chapter 3) that the initial temperature was not a *finite* linear combination of the appropriate form. It is emphasized that infinite sums of C^2 functions need not be C^2, and hence the formal solutions obtained need not be strict solutions. However, by truncating the series at a large enough number of terms one can often meet the I.C. within any positive error, which is all that is needed for applications. The validity of formal solutions under certain assumptions is deferred to Chapter 7. Sturm–Liouville Theory is covered in Section 4.4. At this point the student is ready to savor this subject which extends what is known already to the case of inhomogeneous rods and boundary conditions of the third kind. We provide a convincing sketch of a proof of the infinitude of the eigenvalues for Sturm–Liouville problems, by means of the Sturm Comparison Theorem. Practically none of the rest of the book depends on Section 4.4, except the statement found in Chapter 9 (Section 9.5) that Bessel functions have infinitely many zeros. Thus, in the face of time pressures, it is possible to omit Section 4.4 entirely, although one should at least tell students what it is about. We have found that Section 4.3 can and should be covered rapidly, and that one should stress the statements of the theorems in Section 4.2, but not necessarily the details of the proofs. Section 4.1 should be covered in detail, as it is frequently used later.

Chapter 5 (The Wave Equation): In Section 5.1, the wave equation for a transversely vibrating string is derived from Newton's equation. Some care is taken to explain why the assumption of transverse vibrations actually *implies* a linear wave equation instead of an approximately linear equation. The dubious assumption of "small" vibrations is thus eliminated. The simplest initial/boundary–value problems for a finite string are solved. Uniqueness of solutions of these problems is also proved in Section 5.1, using the energy–integral method. In Section 5.2, we cover D'Alembert's solution of wave problems on the infinite string. Consequences of D'Alembert's solution, such as finite propagation speed are covered, and the method of images for semi–infinite strings is explained. For finite strings, the method of images provides an alternative to the Fourier series approach. The continuous dependence of solutions for the finite string on initial conditions is also an easy consequence of D'Alembert's formula and the method of images. In Section 5.3 a variety of boundary conditions for the string are handled. Also, the inhomogeneous wave equation (i.e., with forcing term) is treated via both Duhamel's principle and the Fourier series approach. Section 5.1 should be covered in some detail, with the complete derivation possibly assigned as reading. Section 5.2 is equally crucial, but if time is running short Section 5.3 can simply be summarized, so that students are aware of what is covered in case they need it.

Chapter 6 (Laplace's Equation): In Section 6.1, Laplace's equation is motivated and it is shown that solutions may be interpreted as steady–state temperature distributions. The Dirichlet and Neumann problems are introduced. Section 6.2 concerns the solution of these problems on a rectangle. Since students are familiar with separation of variables and superposition, this material can be done quickly. Uniqueness and the Maximum Principle are motivated and utilized, but proofs are deferred until Section 6.4. In Section 6.3, we solve Dirichlet and Neumann problems on annuli and disks using polar coordinates. The Mean–Value Theorem and Poisson's Integral Formula are carefully proved, and the regularity of harmonic functions is demonstrated. In

Section 6.4, the Maximum Principle for harmonic functions on bounded domains is proved along with continuous dependence of solutions of the Dirichlet problem on boundary data. The importance of these results has been amply demonstrated to students in the previous sections. Section 6.5 is on the application of complex variable theory to Laplace's equation. We assume *no* knowledge of complex–variables. We do not cover Cauchy's theorem, contour integration, or residue theory, for the simple reason that we do not need it. However, the intimate connection between complex analytic functions and harmonic functions is brought out and exploited. Moreover, the concept and use of conformal mapping to solve problems in steady–state temperatures, fluid flow and electrostatics are handled without any difficulty. All of the material in Chapter 6 is important, and if too much time is spent on material in previous chapters, it may not be possible to cover all of Chapter 6. For a class of mostly engineers, it may be wiser to cover Section 6.5 instead of Section 6.4, if a choice must be made, whereas for mathematics majors the reverse choice is desirable.

Chapter 7 (Fourier Transforms): It will take an exceptional class to reach Chapter 7 in one semester, without skipping all but the most essential material in the previous chapters. However, if students are likely to take a full complex variable course in the future, many concepts in Chapter 6 will be treated in that course. Then, skipping much of Chapter 6 and proceeding with Chapter 7 becomes an attractive possibility. Of course, the possibility of introducing a second semester (or more quarters) of PDEs should be contemplated. The demand is there. In Section 7.1, we introduce complex Fourier series and define the Fourier transform. Many examples are computed. In Section 7.2, we develop the basic properties of Fourier transforms which make them a useful tool for finding solutions of PDEs (i.e., differentiation is carried to a multiplication operator, and multiplication of transforms corresponds to convolution). The idea that the regularity of a function increases the rate of decay of its Fourier transform (and vice versa), is brought out. Although, this is typically regarded as an advanced topic, we treat it in an elementary way, and it is a close relative of the idea (covered in Section 3.2) that the smoothness of a function on a circle increases the rate of decay of its Fourier coefficients. Section 7.3 covers use of the Inversion Theorem, inverse Fourier transforms, and Parseval's equality. The proof of the Inversion Theorem is deferred to a supplement at the end of Chapter 7. In Section 7.4, Fourier transforms are applied to solving PDEs. One may wish to cover Sections 7.1 to 7.3 quickly and concentrate on Section 7.4. Here, we solve the heat problem on the infinite rod, and the Dirichlet problem for the half plane. We felt that it was a good idea to emphasize the fact that Fourier transform methods not only presume that a solution of a problem exists, but also that it has certain decay properties. Thus, integral formulas for solutions obtained in this fashion should be checked independently through a careful application of Leibniz's rule for differentiating under the integral. For a class of mostly engineers, this point can be made, without going through the details of the verification. Although a derivation of D'Alembert's formula for the wave equation is given in Chapter 5, we also show how to get it by Fourier transform techniques and the Dirac delta distribution is discussed. In Section 7.5, heat problems for semi–infinite and finite rods are solved via the method of images. The validity of formal infinite–sum solutions, found in Chapter 4, is now handled with ease. Also, Fourier sine and cosine transformations are introduced and applied.

Chapter 8 (Numerical Solutions of PDEs) : While the solution of PDEs by numerical methods could constitute a whole course, we offer an introduction to the subject in Chapter 8. Our aim is not to present, without proof or motivation, a huge number of algorithms. Instead, we have concentrated on the foundations of the numerical approach, and we work mostly with the familiar heat equation to illustrate the nature and possible pitfalls of the numerical approach. To broaden the horizons, we do provide an optional overview of other numerical methods for other PDEs for the interested reader in Section 8.4. In Section 8.1, the "big O" notation is introduced. There is discussion of Taylor's Theorem which is the basis for the approximation of partial derivatives by finite differences. This allows the approximation of PDE problems by a finite system of equations for the values of the unknown function at grid points. For the heat equation, these systems are easily solved by the explicit method in Section 8.2. Moreover, in the case of the heat equation, the discretization error (i.e., the difference of the numerical solution from the actual solution) is proved to approach zero as the grid point separation goes to zero, at least in the absence of round–off errors. In Section 8.3, we obtain exact solutions for a finite grid by means of the theory of difference equations. We then examine how systematic round–off errors lead to the conclusion that best results are not always obtained by taking the grid size as small as possible. Continuing with the simple case of the heat equation, we obtain theoretical estimates for optimal grid sizes, which are born out to be correct in concrete examples. We believe that it is better to discuss in some depth a number of crucial issues for a single equation, than only briefly comment on a lot of PDEs and techniques. Again, Section 8.4 provides some overview and plenty of references for further study.

Chapter 9 (PDEs in Higher Dimensions): In Section 9.1 the fundamental ideas in Chapters 3 though 7 are extended in a straightforward manner to the case of several cartesian spatial coordinates. We solve dynamic heat problems on rectangles and cubes, and consider Laplace's equation on a solid rectangle. Double Fourier transforms and series are easily motivated and introduced. In Section 9.2, it is made clear that the primary objects from which solutions of the heat, wave and potential problems are constructed are the eigenfunctions of the Laplace operator which meet the B.C. . This basic fact is often hidden behind the process of separation of variable and the plethora of special functions which thereby arise in various coordinate systems. A great variety of series expansions for functions all fall into the category of eigenfunction expansions. In Section 9.2, we also prove a uniform convergence result for double Fourier series, and discuss simple properties of double Fourier transforms. In Section 9.3, we begin our study of the standard PDEs in terms of spherical coordinates. The spherical harmonics are defined as eigenfunctions of the Laplace operator on a sphere. They arise as the angular part of eigenfunctions of the Laplace operator on space and can be expressed through associated Legendre functions. We solve a number of heat and wave problems with spherical symmetry. The three–dimensional version of D'Alembert's formula is derived and Huygen's principle is discussed. The determination of all eigenvalues and spherical harmonics, dimensions of eigenspaces, etc. is covered in Section 9.4. There is a complete proof of the uniform convergence of the Laplace series for C^2 functions on a sphere. Moreover, a number of problems with angular dependence (e.g., heat flow in a ball) are solved through the use of spherical harmonics and spherical Bessel functions. In Section 9.5, we consider PDEs in cylindrical coordinate systems and some more PDEs in spherical coordinates, but with nontrivial potentials, such as Schrödinger's equation. The special functions which arise in the process are discussed. While spherical Bessel functions can be expressed in terms of sines and cosines, the cylindrical Bessel functions (of integer order) cannot, which is why we did not handle cylindrical coordinates before spherical ones. We consider a number of applications, ranging from the vibrating circular drum, to the determination of the energy levels and wave functions for the (nonrelativistic) hydrogen atom and the degeneracy of the energy levels which forms the basis for the periodic table. Section 9.6 deals with the standard heat, wave and potential problems on compact submanifolds with boundary in $\mathbb{R}^n$. Laplace operators are defined on these objects in an easily understood way. Although, we do not prove the existence theory for eigenfunctions and eigenvalues in this general setting, some of the more readable references are cited. Admittedly, the eigenfunctions are difficult to concretely compute or approximate, but once the eigenfunctions are given, the solution of the standard heat, wave and potential problems on manifolds proceeds in a way which is quite analogous to the many special cases covered in the rest of the book. This last section essentially unifies and consolidates these special cases into a single framework. Moreover, there is some discussion of Weyl's asymptotic formula for the eigenvalues of the Laplace operator, and the geometric information about the manifold which can be "heard" from the eigenvalues which may be interpreted as frequencies of vibration.

In constructing a one–semester or two–quarter course, we suggest selecting sections from the list below, keeping the indicated priorities in mind. In addition, 1.1 should be covered if your students are weak in ODEs. Sections which are marked with stars can or should be covered in only 2 hours, whereas most instructors will want to spend about 3 hours on the other sections. Leave time for tests and going over some of the homework. Chapters 8 and 9 are probably best left for a second semester or possibly as sources of projects for advanced, gifted and/or highly motivated students. In some schools where students have strong backgrounds or interests in computers one may wish to cover Chapter 8 in lieu of Chapter 7.

crucial sections: 1.2, 1.3*, 3.1, 3.2, 4.1, 4.2, 4.3*, 5.1, 5.2, 6.1*, 6.2*, 6.3

highly desirable sections: 2.1, 3.3*, 5.3*, 6.4, 6.5, 7.1*, 7.2*, 7.3*, 7.4

luxury sections: 2.2, 2.3, 2.4, 3.4, 4.4, 7.5

BASIC PARTIAL DIFFERENTIAL EQUATIONS

CHAPTER 1

REVIEW AND INTRODUCTION

In this chapter, we review those aspects of ordinary differential equations (ODEs) which will be needed in the sequel. We also provide an overview of the applications of partial differential equations (PDEs), and introduce the reader to some elementary techniques, such as separation of variables. The review of ODEs in Section 1.1 is self–contained, since experience dictates that a remedial study of this material is often sorely needed. Even those whose mathematical knowledge of ODEs is sufficient may find the applied examples and problems (dealing with biology, fluid flow, electronics, mechanical vibrations, resonance, etc.) interesting and challenging. Section 1.2 gives the reader a perspective on the uses of PDEs in various scientific applications, such as gravitation, electrostatics, thermodynamics, acoustics, and minimal soap film surfaces. Some of the material (e.g., the use of Green's functions and integral operators), will not be universally appreciated upon a first reading. Indeed, students will find certain aspects of Section 1.2 more illuminating at later stages in their course of study. In Section 1.3, the studies of ODEs and PDEs are contrasted, with regard to the differences in the typical forms for general solutions. We illustrate how side conditions are used to extract particular solutions from general ones. Moreover, the method of separation of variables is also covered in this section.

1.1 A Review of Ordinary Differential Equations

A differential equation is an equation involving an unknown function and its derivatives. If the unknown is a function of more than one variable, then the differential equation is called a **partial differential equation** (henceforth, abbreviated PDE), since the derivatives of the unknown function are partial derivatives. In an ordinary differential equation (ODE), the unknown function depends on a single variable. Before studying PDEs, a review of certain basics of ODEs is desirable, because solutions of PDEs can often be found by solving related ODEs. The following review of first–order ODEs (separable and linear) and homogeneous second–order linear ODEs with constant coefficients will suffice for our purposes.

First–Order ODEs

A first–order ODE is separable, if it can be written in the form

$$f(y)\,\frac{dy}{dx} = g(x)\,, \tag{1}$$

where y is an unknown function of the independent variable x.

One solves such an equation by integrating (if possible) both sides with respect to x. Integrating the left side yields

$$\int f(y)\,\frac{dy}{dx}\,dx = \int f(y)\,dy = F(y) + C_1\,,$$

where F(y) is an antiderivative of f(y) (i.e., $F'(y) = f(y)$) and C_1 is an arbitrary constant. Integrating the right side of (1) also, and letting G(x) denote an antiderivative of g(x), we then obtain

$$F(y) + C_1 = G(x) + C_2 \quad \text{or} \quad F(y) = G(x) + C\;, \tag{2}$$

where we have incorporated the arbitrary constants C_1 and C_2 into the single arbitrary constant $C = C_2 - C_1$. In practice, one can obtain (2) by first rewriting (1) in terms of differentials

$$f(y)dy = g(x)dx\,. \tag{3}$$

Then, integrating both sides of (3) yields (2). Note that in (3) the variables x and y are on different sides of the equation, and hence the term "separable equation" is used. If possible, one solves (2) for y in terms of x. However, there may be more than one value (or possibly no value) for y, given x and C. Observe that for a fixed value of C, equation (2) will usually define a curve in the xy–plane, but there is no guarantee that this curve will be the graph of a function of x. Nevertheless, the family of curves obtained by allowing C to vary in (2), is usually considered to adequately represent the set of solutions of (1) or (3).

Example 1. A certain population has $P(t)$ individuals at time t, and its rate of growth is proportional to its size (i.e., $P'(t) = aP(t)$, for some constant $a > 0$). Find $P(t)$ in terms of the initial population $P(0)$ and a.

Solution. The equation $P'(t) = aP(t)$ is separable, since we can write it in the form

$$\frac{dP}{P} = a\ dt.$$

Integrating, we obtain (assuming $P > 0$) $\log(P) = at + C$ or $P(t) = \exp(at + C) = e^C e^{at}$. Since $P(0) = e^C$, the desired solution is

$$P(t) = P(0)e^{at}\ .$$

Note that the same technique will work in the more general case where $P'(t) = a(t)f(P(t))$ for given functions $a(t)$ and $f(P)$, since this equation is also separable. However, the technique fails for $P'(t) = t + P(t)$ and many other equations which are not separable. □

Example 2. A particle is carried along by a fluid flow in the xy–plane. Suppose that the velocity of the fluid at the arbitrary point (x,y) is $2y\mathbf{i} + 4x\mathbf{j}$ (i.e., the direction and magnitude of the fluid flow varies from point to point). Find the path traced out by the particle, if it is known to pass through the point $(1,3)$.

Solution. The slope of the path of a particle at (x,y) is the ratio $4x/2y$ (assuming that $y \neq 0$) of the components of the fluid velocity vector at (x,y). Assuming that the path is the graph of a function y of x, we then obtain the ODE $y'(x) = 4x/2y$, which is separable $(2y\,dy = 4x\,dx)$. Integrating, we obtain the family of streamlines (cf. Figure 1)

$$y^2 = 2x^2 + C\ , \tag{4}$$

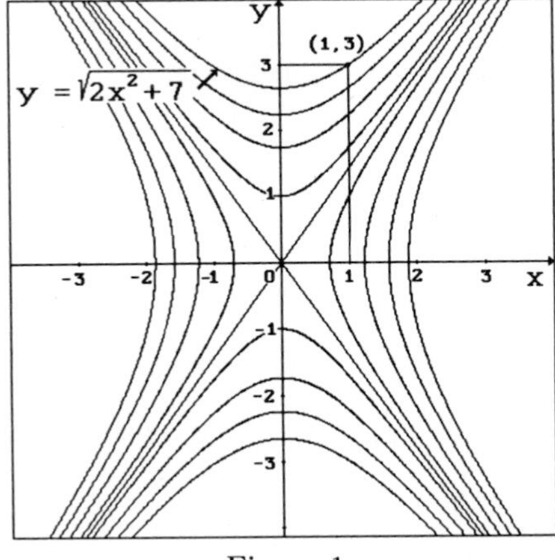

Figure 1

which are hyperbolas. The streamline passing through (1,3) is the upper branch of the hyperbola (4) with $C = 3^2 - 2(1)^2 = 7$, namely $y(x) = (2x^2 + 7)^{\frac{1}{2}}$. □

Another type of first–order ODE which arises in the sequel is the **first–order linear ODE**

$$a(x)y'(x) + b(x)y(x) = c(x) , \tag{5}$$

where $a(x)$, $b(x)$ and $c(x)$ are given continuous functions. Assuming that $a(x) \neq 0$, we may divide (5) by $a(x)$, obtaining an ODE in standard form, in the following sense.

We will say that the first–order linear ODE

$$y'(x) + p(x)y(x) = q(x) \tag{6}$$

is in **standard form.**

If we replace $q(x)$ by 0, the resulting equation

$$y'(x) + p(x)y(x) = 0 \tag{7}$$

is called the **related homogeneous equation** for (6). Unlike (6), (7) is always separable :

$$\frac{dy}{y} = -p(x)\,dx \qquad (y \neq 0) . \tag{8}$$

Thus, by integrating (8), we obtain the following general solution $y_h(x)$ of (7) :

$$y_h(x) = C \cdot \exp[-P(x)] \text{ , where } P(x) \equiv \int p(x)\,dx . \tag{9}$$

The **integrating factor** for equation (6) is defined to be

$$m(x) \equiv \exp[P(x)] = \exp\left[\int p(x)\,dx\right] . \tag{10}$$

Note that by (9) we have $m(x)y_h(x) = C$. Thus,

$$0 = \frac{d}{dx}[m(x)y_h(x)] = m(x)y_h'(x) + m'(x)y_h(x)$$

$$= m(x)y_h'(x) + m(x)p(x)y_h(x) = m(x)[y_h'(x) + p(x)y_h(x)] \tag{11}$$

where we have used the fact that $m'(x) = \exp[P(x)]P'(x) = m(x)p(x)$. For a solution $y(x)$ of (6) with $q(x) \neq 0$, we do not have $m(x)y(x) = C$, but the computation in (11) yields

$$\frac{d}{dx}[m(x)y(x)] = m(x)[y'(x) + p(x)y(x)] = m(x)q(x) \tag{12}$$

Integrating (12), we obtain

$$m(x)y(x) = \int m(x)q(x)\,dx + C$$

or

$$y(x) = [m(x)]^{-1}\left\{\int m(x)q(x)\,dx + C\right\}. \tag{13}$$

Note that (13) reduces to (9) when $q(x) \equiv 0$. While one may simply use formula (13) to write down the solution of (6), it is preferable to remember the steps of the solution process. (For a summary of these steps see the end of this section.)

Example 3. Sclve $(1+x^2)y' + 2xy = 3x^2$.

Solution. First, put the equation into the standard form (6), namely

$$y' + [2x/(1+x^2)]y = 3x^2/(1+x^2). \tag{14}$$

The integrating factor for equation (14) (cf. (10)) is

$$m(x) = \exp\left[\int \{2x/(1+x^2)\}\,dx\right] = \exp[\log(1+x^2)] = 1+x^2.$$

Equation (12) tells us that if we multiply both sides of (14) by $m(x)$, then we will obtain

$$\frac{d}{dx}[m(x)y(x)] = m(x)q(x) = 3x^2. \tag{15}$$

Integrating both sides of (15), we get $m(x)y(x) = x^3 + C$ or $y(x) = (x^3+C)/(1+x^2)$. □

Example 4. Consider two identical cans, A and B. Assume that syrup will leak out of either can at a rate which is proportional to the volume V of the syrup in the can, say $V'(t) = -kV(t)$, where $k > 0$, due to the leakage. Suppose that the initial volume of syrup in can A is $V_A(0)$, while can B is initially empty. If can A begins leaking into can B at $t = 0$, find the volume $V_B(t)$ of syrup in can B at an arbitrary time $t > 0$.

Solution. The rate of change of $V_B(t)$ is

$$V_B'(t) = -kV_B(t) - V_A'(t). \tag{16}$$

Since $V_A'(t) = -kV_A(t)$, we find, as in Example 1, that $V_A(t) = V_A(0)e^{-kt}$. Thus, by (16)

$$V_B'(t) + kV_B(t) = kV_A(0)e^{-kt} .$$

Multiplying by the integrating factor e^{kt} , we obtain (via (12))

$$\frac{d}{dt}[e^{kt}V_B(t)] = kV_A(0) \quad \text{or} \quad V_B(t) = e^{-kt}[ktV_A(0) + C] .$$

Since $V_B(0) = 0$, we know that $C = 0$. So, $V_B(t) = ktV_A(0)e^{-kt}$. □

Example 5. Suppose that tank A contains salt water with 4 pounds of salt per 100 gallons. Tank B is initially filled with 100 gallons of pure water. Over a period of one hour, the water in tank B is drained at the rate of 3 gallons per minute. The water in tank A flows into tank B at the rate of 5 gallons per minute, as tank B is drained. How many pounds of salt are dissolved in tank B at the end of the hour? Assume that tank B is well–mixed at all times and does not overflow.

Solution. Let $S(t)$ denote the number of pounds of salt in tank B at time t. At time t, tank B loses salt (via draining) at the rate of 3 gallons per minute, times the amount of salt per gallon in tank B, namely, $3S(t)/(100 + (5-3)t)$ lbs./min.. The rate at which tank B gains salt from tank A is 5 times 4/100 lbs./min. . Thus, the net rate of salt increase in tank B is given by $S'(t) = [-3S(t)/(100 + 2t)] + 1/5$. Hence,

$$S'(t) + [3S(t)/(100 + 2t)] = 1/5 . \tag{17}$$

The integrating factor is $m(t) = \exp[\frac{3}{2}\log(100 + 2t)] = (100 + 2t)^{\frac{3}{2}}$. Multiplying (17), on both sides, by the integrating factor, we obtain

$$\frac{d}{dt}[m(t)S(t)] = \frac{1}{5}(100 + 2t)^{\frac{3}{2}} \quad \text{and} \quad S(t) = (100 + 2t)^{-\frac{3}{2}}[\tfrac{1}{25}(100 + 2t)^{\frac{5}{2}} + C] .$$

Since $S(0) = 0$, we have $C = -\frac{1}{25}\cdot 10^5$ and $S(t) = \frac{1}{25}(100 + 2t)^{-\frac{3}{2}}((100 + 2t)^{\frac{5}{2}} - 10^5)$. Finally, $S(60) \approx 7.574$ lbs.. □

Second–Order Linear ODEs with Constant Coefficients

We will need a good understanding of the homogeneous second–order linear equation

$$ay''(x) + by'(x) + cy(x) = 0 , \tag{18}$$

where the coefficients a,b, and c are real constants. If $a = 0$, then (18) is either a linear first–order ODE, or (if also $b = 0$) trivial. Thus, we assume that $a \neq 0$. The usual method of solving (18) is to first assume that a solution is of the form $y(x) = e^{rx}$ for some constant r. Substituting this y(x) into (18), we obtain

$$ae^{rx}r^2 + be^{rx}r + ce^{rx} = e^{rx}(ar^2 + br + c) = 0 .$$

Thus, r must satisfy the quadratic equation $ar^2 + br + c = 0$, known as the **auxiliary equation** for (18). Let $d = b^2 - 4ac$. There are three cases : $d > 0$, $d = 0$ and $d < 0$.

If $d > 0$, then there are two distinct real roots, namely

$$r_1 = \frac{-b + \sqrt{d}}{2a} \quad \text{and} \quad r_2 = \frac{-b - \sqrt{d}}{2a} .$$

In this case, the general solution of (18) is the superposition (or linear combination)

$$y(x) = c_1 e^{r_1 x} + c_2 e^{r_2 x} , \tag{19}$$

where c_1 and c_2 are arbitrary constants. Recall that the superposition of two solutions of a homogeneous linear equation is also a solution (cf. Problem 7). Moreover, if the equation is second–order and the ratio of two particular solutions is not constant (i.e., they are linearly independent), then any solution is a superposition of these two solutions (cf. Problem 20).

If $d = 0$, then there is only one solution of $ar^2 + br + c = 0$, namely $r = -b/2a$, which is a root of multiplicity 2. However, we recall that, in addition to e^{rx}, there must be another linearly independent solution of (18). By trying a solution of the form $f(x)e^{rx}$, one finds $f''(x) = 0$ (cf. Problem 9). Thus, choosing $f(x) = x$, we obtain another linearly independent solution, xe^{rx}. Hence, when $d = 0$, the general solution of (18) is

$$y(x) = c_1 e^{rx} + c_2 x e^{rx} , \tag{20}$$

where $r = -b/2a$. If $d < 0$, the roots of $ar^2 + br + c = 0$ are complex, namely

$$r_1 = \frac{-b + i\sqrt{|d|}}{2a} \quad \text{and} \quad r_2 = \frac{-b - i\sqrt{|d|}}{2a} , \tag{21}$$

where i has the property that $i^2 = -1$. (Thus, i cannot be a real number.) Now set

$$r_1 = \alpha + i\beta \quad \text{and} \quad r_2 = \alpha - i\beta \tag{22}$$

where α and β are real numbers. Then it can be shown that

$$y(x) = c_3 e^{(\alpha+i\beta)x} + c_4 e^{(\alpha-i\beta)x} , \tag{23}$$

where c_3 and c_4 are arbitrary constants, satisfies (18). In order to construct a more useful form of the solution (23) (for the details see Problem 11) we use **Euler's formula**

$$e^{iy} = \cos(y) + i\sin(y) . \tag{24}$$

[The Swiss–born mathematician and physicist, Leonhard Euler (1707–1783), made important contributions to many areas of mathematics and celestial mechanics. The number e is named after him.]. Euler's formula can be established by setting $z = iy$ in the power series expansion of the complex exponential e^z :

$$e^z = \sum_{n=0}^{\infty} \frac{z^n}{n!} = 1 + \frac{z}{1!} + \frac{z^2}{2!} + \frac{z^3}{3!} + \dots . \tag{25}$$

Now using the relation $e^{(\alpha+i\beta)x} = e^{\alpha x}e^{i\beta x}$ and Euler's formula, we can express (23) in the form

$$y(x) = e^{\alpha x}[c_3(\cos(\beta x)+i\sin(\beta x)) + c_4(\cos(\beta x)-i\sin(\beta x))]. \tag{26}$$

Setting $c_1 = c_3 + c_4$ and $c_2 = i(c_3 - c_4)$, (26) becomes

$$y(x) = e^{\alpha x}[c_1\cos(\beta x) + c_2\sin(\beta x)] . \tag{27}$$

Finally, with the notation of (22) and (21), we obtain from (27) the following general solution of (18), in the case when $d < 0$:

$$y(x) = e^{-bx/2a}\left[c_1\cos\left[\frac{\sqrt{|d|}\,x}{2a}\right] + c_2\sin\left[\frac{\sqrt{|d|}\,x}{2a}\right]\right] . \tag{28}$$

The foregoing results may be summarized as follows :

Consider the ODE

$$ay''(x) + by'(x) + cy(x) = 0 \ , \tag{29}$$

where a,b and c are real constants and $a \neq 0$. Let r_1 and r_2 be the roots of the associated auxiliary equation $ar^2 + br + c = 0$. Let $d = b^2 - 4ac$, and let c_1, c_2 denote arbitrary constants.

1. If r_1 and r_2 are real and distinct (i.e., $d > 0$), then the general solution of (29) is

$$y(x) = c_1e^{r_1x} + c_2e^{r_2x} \ . \tag{30}$$

2. If $r_1 = r_2 = r$ (i.e. $d = 0$), then the general solution of (29) is

$$y(x) = c_1e^{rx} + c_2xe^{rx} \ . \tag{31}$$

3. If $r_1 = \alpha + i\beta$ and $r_2 = \alpha - i\beta$ (i.e., $d < 0$), then the general solution of (29) is

$$y(x) = e^{\alpha x}[c_1\cos(\beta x) + c_2\sin(\beta x)] \ . \tag{32}$$

Example 6. An object of mass m is attached to a spring which lies along the x–axis, as shown in Figure 2 below. With **Hooke's law** in effect, when the object is displaced to the position x, the spring exerts a force $-kx$ (toward the origin, since the constant k is positive) on the object. Let x(t) be the position of the object at time t. The object is also subject to a force, say due to air resistance, which is $-bx'(t)$, for a constant $b > 0$. If the object is released from the position x_0 at time $t = 0$, find the position of the object at any time $t > 0$, using **Newton's second law** of motion $mx''(t) = F(t)$, where F(t) is the total force on the object at time t. [The English scientist Robert Hooke (1635–1703) and mathematician/physicist Isaac Newton (1642–1727) were often at odds, in particular, over the division of credit for the inverse–square law of gravity.]

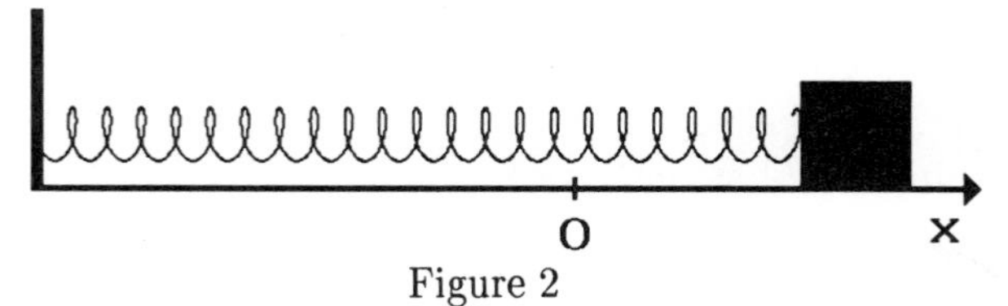

Figure 2

Solution. Since the total force on the object is $F(t) = -kx(t) - bx'(t)$, Newton's second law yields the ODE

$$mx''(t) + bx'(t) + kx(t) = 0 \ . \tag{33}$$

Here $d = b^2 - 4mk$. All three cases $d > 0$, $d = 0$ and $d < 0$ are possible. They are referred to as **over–damped**, **critically–damped** and **under–damped** (or **oscillatory**) respectively. The solutions in these cases are

$$x(t) = e^{-\frac{1}{2}bt/m}[c_1 e^{\frac{1}{2}\sqrt{d}\cdot t/m} + c_2 e^{-\frac{1}{2}\sqrt{d}\cdot t/m}] \quad (d > 0), \tag{34}$$

$$x(t) = e^{-\frac{1}{2}bt/m}[c_1 + tc_2] \quad (d = 0), \tag{35}$$

$$x(t) = e^{-\frac{1}{2}bt/m}[c_1\cos(\tfrac{1}{2}\sqrt{|d|}\cdot t/m) + c_2\sin(\tfrac{1}{2}\sqrt{|d|}\cdot t/m)] \quad (d < 0). \tag{36}$$

The constants c_1 and c_2 are found from the given initial conditions $x(0) = x_0$ and $x'(0) = 0$. For (34), we have that

$$x(0) = c_1 + c_2 = x_0 \quad \text{and} \quad x'(0) = \frac{1}{2m}[(\sqrt{d} - b)c_1 - (\sqrt{d} + b)c_2] = 0$$

imply $c_1 = \frac{1}{2}[1 + (b/\sqrt{d})]x_0$ and $c_2 = \frac{1}{2}[1 - (b/\sqrt{d})]x_0$.

Hence, (34) becomes

$$x(t) = x_0 e^{-\frac{1}{2}bt/m}\Big[\tfrac{1}{2}(e^{\frac{1}{2}\sqrt{d}\cdot t/m} + e^{-\frac{1}{2}\sqrt{d}\cdot t/m}) + (b/\sqrt{d})\cdot\tfrac{1}{2}(e^{\frac{1}{2}\sqrt{d}\cdot t/m} - e^{-\frac{1}{2}\sqrt{d}\cdot t/m})\Big]$$

$$= x_0 e^{-\frac{1}{2}bt/m}[\cosh(\tfrac{1}{2}\sqrt{d}\cdot t/m) + (b/\sqrt{d})\sinh(\tfrac{1}{2}\sqrt{d}\cdot t/m)]. \tag{37}$$

The hyperbolic sine and cosine often occur naturally, when initial conditions are imposed. They are defined by $\sinh(x) = \frac{1}{2}(e^x - e^{-x})$ and $\cosh(x) = \frac{1}{2}(e^x + e^{-x})$. The interested reader who is unfamiliar with these functions and their relation to the usual sine and cosine, should consult Problem 18. The computation of the values for c_1 and c_2 in (35) and (36) is suggested in Problem 12. For certain values of b, m and k, the solutions are graphed in Figure 3 below. □

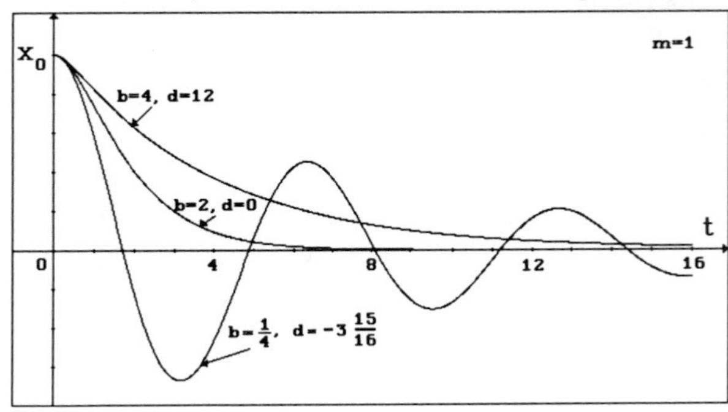

Figure 3

Example 7. Equation (18) also arises in electrical circuit theory. Suppose that a battery of voltage V, a resistor of resistance R, a coil of inductance L and a capacitor of capacitance C are placed in series as shown below in Figure 4. We wish to find the most general expression for the current i(t) in this circuit as a function of time t. [The flow of current in this circuit is governed by the second law of the German physicist Gustav Kirchhoff (1824–1887), who is famous for his contributions in electronics and spectroscopy.]

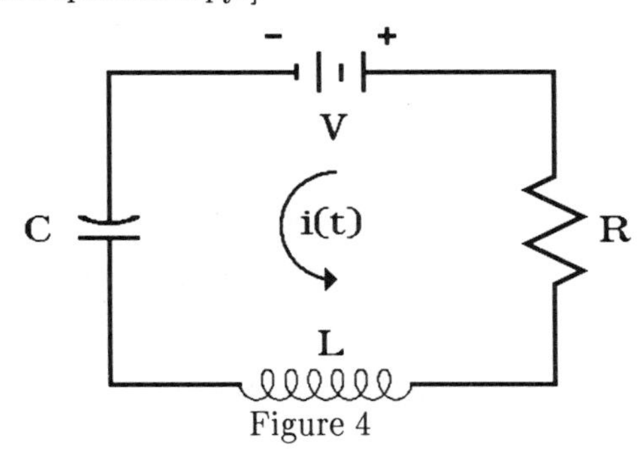

Figure 4

Solution. Kirchhoff's second law asserts that the sum of the voltage drops across the elements of any closed loop in a circuit must be zero. At time t, the voltage drop across the resistor is R times the current i(t). The voltage drop across the coil is $L\,i'(t)$. (This drop is due to the fact that an increasing current in a coil creates a changing magnetic field which induces an opposing electric field and induces a voltage drop across the coil.) The voltage drop across a capacitor is 1/C times the total amount of charge ($\int i(t)\,dt$) which it has accumulated on one of its plates. By Kirchhoff's second law,

$$R\,i(t) + L\,i'(t) + \frac{1}{C}\int i(t)\,dt - V = 0 \ , \tag{38}$$

and differentiating, we obtain

$$L\,i''(t) + R\,i'(t) + \frac{1}{C}\,i(t) = 0 \ . \tag{39}$$

Hence, the current i(t) behaves just as the displacement of an object attached to a spring as in Example 6, with $m = L$, $b = R$ and $k = 1/C$. In particular, with these new values, formulas (34),(35) and (36) give us the general solutions for i(t) in the three cases. □

We will not cover the general case of the inhomogeneous equation

$$ay''(t) + by'(t) + cy(t) = f(t) \tag{40}$$

which would arise in Example 6 when there is an external driving force f(t), or in Example 7 when the voltage source is variable $[f(t) = V'(t)]$. One could solve (40) by adding a particular solution to the general solution of the related homogeneous equation with f(t) replaced by 0. A particular solution can be obtained by the method of variation of parameters which can be found in most ODE books. However, as an illustration of resonance and the utility of the complex approach, we will find a particular solution of (40) in the important case when f(t) is of the form $A\cos(\omega t)$ or $A\sin(\omega t)$, for a constant amplitude A and angular frequency ω.

Example 8. Find a particular solution of (40) with $abc \neq 0$, in the case when $f(t) = A\cos(\omega t)$ or $f(t) = A\sin(\omega t)$ for a real constant ω, by using the following approach. Determine a complex constant C, such that $y(t) = Ce^{i\omega t}$ solves (40) with $f(t) = Ae^{i\omega t}$. Then show that the real and imaginary parts of $y(t)$ will be the desired particular solutions.

Solution. Substituting the trial solution $y(t) = Ce^{i\omega t}$ into (40) with $f(t) = Ae^{i\omega t}$, we obtain

$$Ce^{i\omega t}[a(i\omega)^2 + bi\omega + c] = Ae^{i\omega t} \quad \text{or} \quad C\cdot[(c - a\omega^2) + ib\omega] = A. \tag{41}$$

Using the identity $(r + is)(r - is) = r^2 + s^2$, we see that

$$[r + is]^{-1} = (r - is)/(r^2 + s^2). \tag{42}$$

Thus, multiplying by $[(c-a\omega^2) + ib\omega]^{-1}$ in (41), we obtain

$$C = A[(c - a\omega^2) - ib\omega]/[(c - a\omega^2)^2 + b^2\omega^2] \tag{43}$$

and

$$\begin{aligned} y(t) = Ce^{i\omega t} &= C[\cos(\omega t) + i\sin(\omega t)] \\ &= A[(c - a\omega^2)\cos(\omega t) + b\omega\sin(\omega t)]/[(c - a\omega^2)^2 + b^2\omega^2] \\ &\quad + iA[(c - a\omega^2)\sin(\omega t) - b\omega\cos(\omega t)]/[(c - a\omega^2)^2 + b^2\omega^2] \\ &= y_R(t) + iy_I(t), \end{aligned}$$

where the last equation defines the real and imaginary parts of the solution $y(t)$ of (40) with $f(t) = Ae^{i\omega t} = A\cos(\omega t) + iA\sin(\omega t)$. Since

$$ay'' + by' + cy = (ay_R'' + by_R' + cy_R) + i(ay_I'' + by_I' + cy_I), \tag{44}$$

we see that $y_R(t)$ solves (40) with $f(t) = A\cos(\omega t)$, while $y_I(t)$ solves (40) with $f(t) = A\sin(\omega t)$. If the frequency ω is allowed to vary, then the amplitude $A[(c - a\omega^2)^2 + b^2\omega^2]^{-\frac{1}{2}}$ (cf. Figure 5 below) of y_R and y_I is largest, when ω is chosen so that $h(\omega) = (c - a\omega^2)^2 + b^2\omega^2$ is minimal.

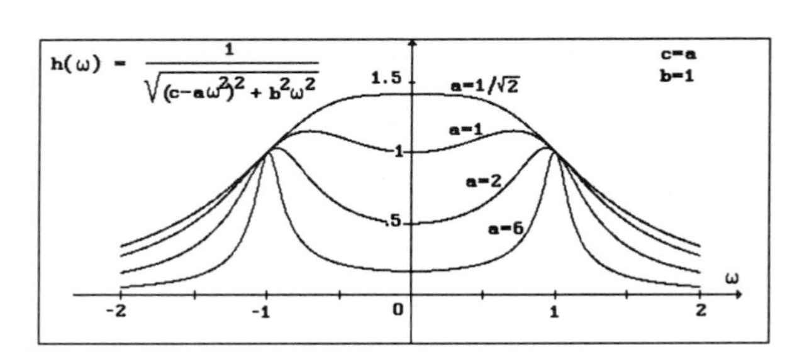

Figure 5

By setting $h'(\omega) = 0$, these resonant frequencies ω_R are found to be

$$\omega_R = \pm\left[\frac{c}{a} - \frac{1}{2}\left[\frac{b}{a}\right]^2\right]^{\frac{1}{2}}, \tag{45}$$

unless $2ac - b^2 < 0$, in which case $\omega_R = 0$ yields the maximum amplitude. Observe that if $b^2 - 4ac < 0$, then $|\omega_R|$ is less than the natural frequency $\nu \equiv |\sqrt{|d|}\,/2a| = [(c/a) - \frac{1}{4}(b/a)^2]^{\frac{1}{2}}$ which occurs in (32). Note that

$$(\omega_R/\nu)^2 = [1 - \tfrac{1}{2}(b^2/ac)]/[1 - \tfrac{1}{4}(b^2/ac)] ,$$

which shows that $|\omega_R|/\nu \longrightarrow 1$ as $(b^2/ac) \longrightarrow 0$. In the above argument, we have assumed that A does not depend on ω. In applications to electronics, A is usually proportional to ω, in which case (45) does not apply. As an illustration see Problem 13. □

Special Systems of ODEs

Occasionally we will meet a system of linear ODEs of the form

$$x'(t) = ax(t) + by(t) \tag{46a}$$

$$y'(t) = cx(t) + dy(t) , \tag{46b}$$

where a,b,c, and d are given real constants and x(t) and y(t) are unknown functions. We are required to solve the system (46a) (46b) for x(t) and y(t), given the initial values x(0) and y(0). If $b = 0$, we can solve the first order ODE (46a) for x(t). Then we substitute the solution x(t) into (46b), and solve the resulting first–order ODE for y(t). If $b \neq 0$, we differentiate both sides of (46a) and use (46b) as follows :

$$\begin{aligned} x''(t) &= ax'(t) + by'(t) = ax'(t) + b(cx(t) + dy(t)) \\ &= ax'(t) + bcx(t) + d\cdot(x'(t) - ax(t)), \end{aligned}$$

where we have used 46(a) for the last equality. Thus, x(t) must satisfy

$$x''(t) - (a + d)x'(t) + (ad - bc)x(t) = 0 . \tag{47}$$

This familiar second–order ODE is solved for x(t), using the initial values x(0) and $x'(0) = ax(0) + by(0)$. There is no need to solve (46b) for y(t), since by (46a)

$$y(t) = (x'(t) - ax(t))/b . \tag{48}$$

The above ideas suffice to solve certain other types of systems of ODEs which arise in the sequel, and there will be no need for differential operator/matrix methods.

Example 9. In Example 2, calculate the position $(x(t),y(t))$ of the particle at any time t, given that $x(0) = 1$ and $y(0) = 3$.

Solution. The velocity vector at time t is $x'(t)\mathbf{i} + y'(t)\mathbf{j}$. Thus, we have the system

$$x'(t) = 2y(t) \tag{49a}$$

$$y'(t) = 4x(t) . \tag{49b}$$

As above, differentiating (49a) and using (49b) we get

$$x''(t) = 2y'(t) = 8x(t) \quad \text{or} \quad x''(t) - 8x(t) = 0 .$$

Since $r^2 - 8 = 0$, we have $r = \pm 2\sqrt{2}$, and the general solution is

$$x(t) = c_1 e^{2\sqrt{2}\cdot t} + c_2 e^{-2\sqrt{2}\cdot t} ,$$

but we need $x(0) = 1$ and $x'(0) = 2y(0) = 6$. These conditions yield

$$\begin{cases} c_1 + c_2 = 1 \\ 2\sqrt{2}\ c_1 - 2\sqrt{2}\ c_2 = 6 \end{cases} \quad \text{or} \quad \begin{cases} c_1 = \frac{1}{2} + \frac{3}{4}\cdot\sqrt{2} \\ c_2 = \frac{1}{2} - \frac{3}{4}\cdot\sqrt{2} \end{cases} .$$

Using (49a),

$$y(t) = \tfrac{1}{2}\, x'(t) = \sqrt{2}\left[c_1 e^{2\sqrt{2}\cdot t} - c_2 e^{-2\sqrt{2}\cdot t}\right] .$$

As t varies, the point $(x(t),y(t))$ traces out a branch of the hyperbola $y^2 - 2x^2 = 7$ (cf. (4) in Example 2 with $C = 7$), because one can verify that $[y(t)]^2 - 2[x(t)]^2 = 7$. The parametric representation $(x(t),y(t))$ for this curve gives us much more information than $y^2 - 2x^2 = 7$ does, since $(x(t),y(t))$ gives us the particle's position at any time t. □

Example 10. The weight $w(t)$ of a certain animal grows at a rate $w'(t) = Cs(t) - K$, where $s(t)$ is the size of the animal's food supply and $K, C > 0$ are constants. We assume that $s(0)$ and $w(0)$ are positive. If $s(t)$ ever becomes 0, then it remains at 0. The animal has starved to death, if $w(t)$ drops to 0. The heavier the animal gets, the more it eats from its food supply which ordinarily would grow at a rate proportional to $s(t)$ in the animal's absence. Thus, while the food supply lasts, $s'(t) = As(t) - Bw(t)$, for constants $A, B > 0$. Show that if $A^2 < 4BC$, then the animal will eventually starve to death (after a number of diet/binge cycles), unless $w(0) = AK/BC$ and $s(0) = K/C$, in which case $w(t)$ and $s(t)$ are constant. (The case where $A^2 > 4BC$ is the subject of Problem 19.)

Solution. We have the system

$$s'(t) = As(t) - Bw(t) \tag{50a}$$

$$w'(t) = Cs(t) - K\ . \tag{50b}$$

Differentiating the first equation and using the second, we have

$$s''(t) - As'(t) + BCs(t) = BK\ . \tag{51}$$

The general solution of (51) is the constant particular solution K/C plus the general solution of the related homogeneous equation

$$s''(t) - As'(t) + BCs(t) = 0\ . \tag{52}$$

Equation (52) is of the form (18), with $a = 1$, $b = -A$ and $c = BC$. Thus, $d = b^2 - 4ac = A^2 - 4BC$. If $A^2 < 4BC$, then $d < 0$ and the general solution of (51) is

$$s(t) = K/C + e^{\frac{1}{2}At}[c_1\cos(\sqrt{|d|}\cdot t/2) + c_2\sin(\sqrt{|d|}\cdot t/2)]\ . \tag{53}$$

The function in brackets can be written as $(c_1^2 + c_2^2)^{\frac{1}{2}}\cos[(\sqrt{|d|}\cdot t/2) + \theta]$ for some constant θ (cf. Problem 10). Thus, if c_1 and c_2 are not both zero, the solution will oscillate about K/C with the growing amplitude $(c_1^2 + c_2^2)^{\frac{1}{2}}e^{\frac{1}{2}At}$, as the animal diets and indulges with greater intensity. Eventually this amplitude will be greater than K/C (provided w(t) remains positive), and s(t) must drop to zero at some time, say t_0, during the next cycle. Thus, if the animal is still alive at time t_0, then after t_0, $w'(t) = -K$, and w(t) drops steadily to zero. If c_1 and c_2 are both zero, then $s(t) = K/C$, and (50a) then says that $w(t) = AK/BC$. Figure 6 shows that w(t) might drop to zero while s(t) is still positive. □

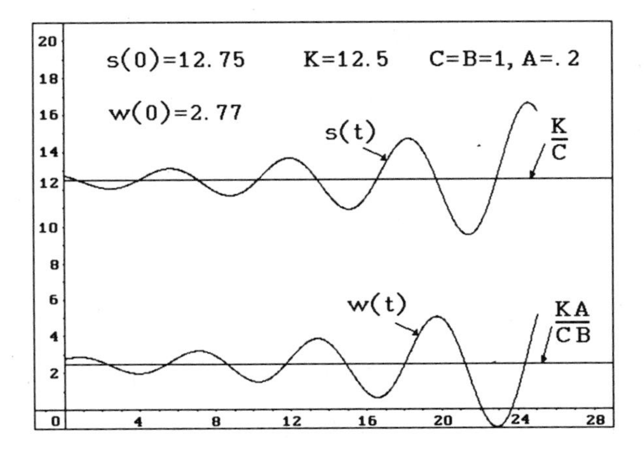

Figure 6

Summary 1.1

1. First–order separable ODEs : To solve the first–order separable ODE $f(y)\,\frac{dy}{dx} = g(x)$, write it in the form $f(y)dy = g(x)dx$ and integrate :

$$\int f(y)\ dy = \int g(x)\ dx + C\ .$$

2. First–order linear ODEs : The general solution of the first–order linear ODE

$$y'(x) + p(x)y(x) = q(x)\ , \qquad \text{(S1)}$$

which is in standard form, can be obtained as follows.

(a) Multiply both sides of (S1) by the integrating factor $m(x) = \exp\left\{\int p(x)\ dx\right\}$ and check that

$$\frac{d}{dx}\,[m(x)y(x)] = m(x)q(x)\ . \qquad \text{(S2)}$$

(b) Integrate both sides of (S2) to obtain $m(x)y(x) = \int m(x)q(x)\ dx + C$, where C is an arbitrary constant.

(c) The general solution of (S1) is then

$$y(x) = [m(x)]^{-1}\left\{\int m(x)q(x)\ dx + C\right\}\ .$$

3. Homogeneous second–order linear ODEs : To determine the general solution of the homogeneous second–order linear ODE

$$ay''(x) + by'(x) + cy(x) = 0\ , \qquad \text{(S3)}$$

where a,b,c are real constants and $a \neq 0$, first find the roots r_1 and r_2 of the associated auxiliary equation $ar^2 + br + c = 0$.

Let c_1 and c_2 denote arbitrary constants and let y(x) denote the general solution of (S4).

(i) If r_1 and r_2 are real and distinct, then $y(x) = c_1e^{r_1x} + c_2e^{r_2x}$.

(ii) If $r_1 = r_2 = r$, then $y(x) = c_1e^{rx} + c_2xe^{rx}$.

(iii) If $r_1 = \alpha + i\beta$ and $r_2 = \alpha - i\beta$, then $y(x) = e^{\alpha x}[c_1\cos(\beta x) + c_2\sin(\beta x)]$.

4. Linear systems : To solve the linear system of ODEs

$$x'(t) = ax(t) + by(t) \quad \text{(A)}$$

$$y'(t) = cx(t) + dy(t) \quad , \quad \text{(B)}$$

where a,b,c, and d are real constants, with given initial values $x(0)$ and $y(0)$, consider the following cases.

Case 1. If $b = 0$, solve (A) for $x(t)$ and substitute the solution into (B). Then solve the resulting ODE for $y(t)$.

Case 2. If $b \neq 0$, differentiate both sides of (A) with respect to t , and then use (B) to get

$$x'' - (a + d)x' + (ad - bc)x = 0 .$$

Using the initial values $x(0)$ and $x'(0) = ax(0) + by(0)$, we first solve the above second–order ODE for $x(t)$, and then set $y(t) = [x'(t) - ax(t)]/b$ by (A).

Exercises 1.1

1. Find the general solutions of the following separable equations :

(a) $\frac{dy}{dx} = xy$ (b) $\frac{dx}{dt} = x(1-x)$ (c) $\frac{dy}{dx} = x^2y^2 + x^2 - y^2 - 1$

(d) $\frac{dy}{dx} = \frac{1+y^2}{1+x^2}$ (e) $\frac{dx}{dt} + x^2\sin(t) = 0$ (f) $\frac{dy}{dx} - \frac{x+e^{-x}}{y+e^y} = 0$

(g) $\frac{dx}{dt} = te^{x+t}$ (h) $x\frac{dy}{dx} = 1 + y^2$ (i) $T'(t) + 3T(t) = 0$.

2. A radioactive substance decays at a rate proportional to the amount of the substance present. If 64% of the substance remains after 10 years, what percentage will remain after 15 years ?

3. Torricelli's law states that (under certain ideal circumstances) fluid will leak out of a hole at the base of a container at a rate proportional to the square root of the height of the fluid's surface from the base. Suppose that a cylindrical container is initially filled to a depth of one foot. If it takes one minute for three quarters of the fluid to leak out, how long will it take for all of the fluid to leak out ? [Italian Evangelista Torricelli (1608–1647) succeeded Galileo as professor of mathematics at the Florentine Academy, and following a suggestion of Galileo, invented the mercury barometer.]

4. Solve the following first–order linear equations, subject to the given conditions :

(a) $y'(x) + 2y(x) = e^x$, $y(0) = 1$

(b) $x'(t) - (2/t)x(t) = 1$, $x(1) = 0$

(c) $\sin(x)y'(x) - \cos(x)y(x) = \sin(2x)$, $y(\pi/2) = 0$

(d) $x'(t) + \frac{x(t)}{t} = t^2$, $x(0) = 0$

(e) $3\frac{dy}{dx} + 6xy = 6e^{-x}$, $y(0) = 1$

(f) $\frac{dy}{dx} = 3y + e^{2x}$, $y(0) = 0$

(g) $x'(t) + x(t)\cos(t) = 0$, $x(\pi) = 100$

(h) $\frac{dy}{dx} + (1+2x+3x^2)y = e^{-x-x^2-x^3}$, $y(0) = 3$

(i) $\frac{dx}{dt} + \frac{3x}{2t+100} = 0$, $x(-49.5) = 1$.

5. A population P of bacteria grows at a rate (say $b \cdot P$, $b > 0$) proportional to its size, but it is destroyed at a steadily increasing rate (say $c \cdot t$, $c > 0$) by a spot of mold which starts growing at $t = 0$. Under what circumstances will the mold completely consume the bacteria ?
Hint. Solve $P'(t) = bP(t) - ct$ in terms of b, c, P(0) and t. Under what condition(s) (on P(0), b and c) will P(t) drop to zero for some $t > 0$?

6. Find the general solution, y(x), of the following second–order homogeneous linear ODEs.

(a) $y'' = 0$

(b) $y'' - 3y = 0$

(c) $y'' + 3y = 0$

(d) $y'' + y' = 0$

(e) $y'' - 3y' = 0$

(f) $4y'' + 3y' + 5y = 0$

(g) $2y'' + 5y' + 2y = 0$

(h) $y'' - 6y' + 13y = 0$

(i) $y'' - 4y' + 4y = 0$

(j) $y'' + 10y' + 25y = 0$.

7. Find the particular solutions y(t), meeting the given initial data, of the following second–order homogeneous linear ODEs.

(a) $y'' - 5y' + 6y = 0$; $y(0) = 1$, $y'(0) = 2$

(b) $y'' - 4y' + 4y = 0$; $y(0) = 0$, $y'(0) = 1$

(c) $y'' + y = 0$; $y(0) = a$, $y'(0) = b$

(d) $y'' - y = 0$; $y(0) = a$, $y'(0) = b$

(e) $5y'' + 8y' + 5y = 0$; $y(0) = 0$, $y'(0) = 1$

(f) $5y'' + 8y' + 5y = 0$; $y(0) = 1$, $y'(0) = 0$.

8. (a) Show that if $y_1(x)$ and $y_2(x)$ are solutions of the homogeneous linear ODE $a(x)y'' + b(x)y' + c(x)y = 0$, then the superposition $c_1y_1(x) + c_2y_2(x)$ is also a solution.

(b) If a(x), b(x) and c(x) are continuous with a(x) never zero, then the ODE in part (a) has a unique solution y(x) with given values for $y(x_0)$ and $y'(x_0)$ (cf. [Simmons, Section 57]). Assuming this, show that no solution of this ODE can have a graph which is tangent to the x–axis at some point, unless the solution is identically zero.

9. (a) If $ar^2 + br + c = 0$ has only one root (of multiplicity 2) $r = -b/2a$, show that $f(x)e^{rx}$ is a solution of $ay'' + by' + cy = 0$, if and only if $f''(x) = 0$.

(b) For distinct numbers r_1 and r_2 observe that $\lim_{r_2 \to r_1} \dfrac{e^{r_2 x} - e^{r_1 x}}{r_2 - r_1} = xe^{r_1 x}$.

How is this observation related to the result in part (a) ?

10. (a) Show that any complex number $z = x + iy$ can be written in "polar form" $re^{i\theta}$, where $r = |z| = [x^2 + y^2]^{\frac{1}{2}}$ and θ are the polar coordinates of the point (x,y) in the Cartesian plane.

(b) For real x, y and ω, note that $x\cos(\omega t) + y\sin(\omega t)$ is the real part of the product $(x + iy)(\cos(\omega t) - i\sin(\omega t)) = re^{i\theta}e^{-i\omega t}$. In view of this show that

$$x\cos(\omega t) + y\sin(\omega t) = r\cos(\omega t - \theta) = r\sin(\omega t - \theta + \pi/2).$$

11. (a) By setting $x = 0$ in the formula $e^z = \sum_{n=0}^{\infty} \frac{z^n}{n!}$, $z = x + iy$, and by using the series expansions for $\cos(y)$ and $\sin(y)$, verify that $e^{iy} = \cos(y) + i\sin(y)$.

(b) If $r_1 = \alpha + i\beta$ and $r_2 = \alpha - i\beta$, verify that $\frac{1}{2}[e^{r_1 x} + e^{r_2 x}] = e^{\alpha x}\cos(\beta x)$ and $-\frac{1}{2}i[e^{r_1 x} - e^{r_2 x}] = e^{\alpha x}\sin(\beta x)$.

(c) Use the definition $\frac{d}{dx}[f(x) + ig(x)] = f'(x) + ig'(x)$ and the formula

$$e^{(\alpha+i\beta)x} = e^{\alpha x}e^{i\beta x} = e^{\alpha x}[\cos(\beta x) + i\sin(\beta x)]$$

to show that $\frac{d}{dx} e^{rx} = re^{rx}$, where $r = \alpha + i\beta$.

(d) Use part (c) to verify that the function $y(x)$ defined by (23) (see also (21) and (22)) satisfies the differential equation (18).

12. Find the constants c_1 and c_2 in (35) and (36) such that $x(0) = x_0$ and $x'(0) = 0$.

13. Suppose that in Example 7 (with $LRC \neq 0$), the voltage source is alternating, say $V(t) = \sin(\omega t)$. For what value of ω is the amplitude of $i(t)$ the greatest, for large t ?

Hint. In the case of variable $V(t)$, the right side of (39) is replaced by $V'(t)$. Show that any solution of the related homogeneous equation approaches 0 as $t \to \infty$ (such a solution is called **transient**). To find a nontransient particular solution, apply Example 8, with $f(t) = V'(t) = \omega \cdot \cos(\omega t)$, noting that A is ω.

Remark. If $V(t) = V_0 e^{i\omega t}$ and $I(t) = I_0 e^{i\omega t}$ for complex constants V_0 and I_0, then the complex number I_0/V_0 is called the **admittance** and it is usually denoted by $Y(\omega)$ since it depends on ω, while $Z(\omega) = [Y(\omega)]^{-1}$ is called the **impedance**. The problem is to determine the "low impedance resonance" ω_0 which makes $|Z(\omega)|$ smallest.

14. For the system (46), we showed that $x(t)$ must satisfy $x'' - (a + d)x' + (ad - bc)x = 0$. Show that $y(t)$ must also satisfy $y'' - (a + d)y' + (ad - bc)y = 0$.

15. Solve the following system subject to the given initial data

$$x'(t) = x(t) + y(t) \qquad x(0) = 1$$

$$y'(t) = -x(t) + y(t) \qquad y(0) = 0 .$$

Draw a rough sketch in the xy–plane of the solution curve $(x(t),y(t))$ as t varies.

16. Consider the system (46). If $(a-d)^2 + 4bc \neq 0$, then show that any complex solution $(x(t),y(t))$ of the system must be of the form

$$x(t) = c_1 e^{r_1 t} + c_2 e^{r_2 t} \quad \text{and} \quad y(t) = d_1 e^{r_1 t} + d_2 e^{r_2 t} ,$$

where c_1, c_2, d_1, d_2 are complex constants and r_1 and r_2 are the (possibly nonreal) roots of $r^2 - (a + d)r + (ad - bc) = 0$. What happens if $(a-d)^2 + 4bc = 0$?

Hint. See Problem 14.

17. Solve each of the following systems subject to the given initial data :

(a) $x'(t) = 3x(t) - 4y(t)$, $y'(t) = x(t) - y(t)$, $x(0) = 1$, $y(0) = 1$

(b) $x'(t) = x(t) - 4y(t)$, $y'(t) = x(t) + y(t)$, $x(0) = 1$, $y(0) = 1$

(c) $x'(t) = x(t) + 2y(t)$, $y'(t) = 3x(t) + 4y(t)$, $x(0) = 0$, $y(0) = 1$.

18. For any complex number z, we define the hyperbolic sine and cosine by

$$\sinh(z) = \tfrac{1}{2}(e^z - e^{-z}) \qquad \cosh(z) = \tfrac{1}{2}(e^z + e^{-z}) .$$

(a) Verify that $\cosh^2(z) - \sinh^2(z) = 1$.

(b) For a real variable x, show that $\frac{d}{dx}\sinh(x) = \cosh(x)$ and $\frac{d}{dx}\cosh(x) = \sinh(x)$.

(c) For a real variable y , check that $\sinh(iy) = i\sin(y)$ and $\cosh(iy) = \cos(y)$.

(d) Define sin(z) and cos(z) for any complex number z, by allowing y to be complex in (c). Check that $\sin^2(z) + \cos^2(z) = 1$.

19. In relation to Example 10 , assume that $A^2 > 4BC$ in each of the following parts.

(a) If the animal does not starve to death first, show that its weight eventually grows at an exponential rate, unless the animal maintains the constant weight AK/BC .

(b) Explain intuitively why it is possible to choose positive initial values for w(0) and s(0) such that the animal starves to death.

(c) Give a concrete example to prove your claim in (b).

20. By completing the following steps, show that the general solution of a second–order homogeneous linear equation $ay'' + by' + cy = 0$ [where a,b and c are constants ($a \neq 0$) ; see, however, the final remark after step (f)] is of the form $c_1y_1(x) + c_2y_2(x)$, where $y_1(x)$ and $y_2(x)$ are any two linearly independent solutions (i.e., neither is a constant multiple of the other). We assume that all functions under consideration here have continuous second derivatives everywhere.

(a) Show that two functions f(x) and g(x) (with g(x)/f(x) or f(x)/g(x) differentiable) are linearly dependent on some open interval I, if and only if their Wronskian function W[f,g](x), defined as $f(x)g'(x) - f'(x)g(x)$, is zero for all x in I. [Jozef M. Hoene–Wronski (1778–1853) was a Polish–born, egocentric mathematician and metaphysician. Wronski became later a French citizen. He is best known for the determinants such as $\begin{vmatrix} f & g \\ f' & g' \end{vmatrix} = fg' - f'g$, which he used in his "highest law" of mathematics. The term "Wronskian" was coined by Thomas Muir around 1882.]

(b) Show that if y(x) and z(x) are any solutions of $ay'' + by' + cy = 0$, then W[y,z](x) is a solution of $aW'(x) + bW(x) = 0$. Thus $W[y,z](x) = C\exp(-bx/a)$, for some constant C which depends on the choice of solutions y and z. (This is **Abel's formula**.)

(c) Conclude from (b) that if $W[y,z](x) = 0$, for some x, then $W[y,z](x) = 0$ for all x .

(d) In (b) and (c), let $z(x) = d_1y_1(x) + d_2y_2(x)$ for constants d_1 and d_2 (a solution, by Problem 8(a)). Show that $W[y, d_1y_1 + d_2y_2](x) = d_1W[y,y_1](x) + d_2W[y,y_2](x)$. Explain why there must be some constants d_1 and d_2 (not both zero), such that $d_1W[y,y_1](x) + d_2W[y,y_2](x) = 0$ for *some* particular x.

(e) Conclude from (c) and (d) that there are constants d_1 and d_2 (not both zero) such that $W[y, d_1y_1 + d_2y_2](x) = 0$ for *all* x.

(f) Conclude from (a) and (e) that $y(x) = c'[d_1y_1(x) + d_2y_2(x)] = c_1y_1(x) + c_2y_2(x)$ for some constants c', c_1 and c_2 on any interval where $y(x)$ is never 0. (We omit the proof of the fact that if $y(x) = c_1y_1(x) + c_2y_2(x)$ on one interval, then the same is true everywhere.)

Remark. The same proof works in the case where a,b and c are replaced by continuous functions $a(x)$, $b(x)$ and $c(x)$, if one assumes that $a(x)$ is never zero. Then, $W[y,z] = C\exp\left[-\int \frac{b(x)}{a(x)}\,dx\right]$.

1.2 Generalities About PDEs

Let $u = u(x,y,z,...)$ be a function of several unrestricted real variables $x, y, z,...$. (In the remark below, we consider the case where $(x,y,z,...)$ is restricted to some region.) Recall that the partial derivative $\frac{\partial u}{\partial x}$ of u, with respect to the variable x, is just the ordinary derivative of u with respect to x, treating the other variables as constants. We use the following convenient notation

$$u_x = \frac{\partial u}{\partial x}, \quad u_y = \frac{\partial u}{\partial y}, \quad u_z = \frac{\partial u}{\partial z}, \quad ...,$$

$$u_{xx} = \frac{\partial^2 u}{\partial x^2}, \quad u_{yx} = \frac{\partial^2 u}{\partial x \partial y}, \quad u_{xy} = \frac{\partial^2 u}{\partial y \partial x}, \quad ... .$$

The order of a partial derivative is then the same as the number of subscripts. The function u is said to be **continuous at a point** $p = (x_0,y_0,z_0,...)$, if the values of the function can be made arbitrarily close to $u(p)$ by allowing the variables $x, y, z,...$ to vary (simultaneously) only within sufficiently small open intervals about $x_0, y_0, z_0, ...,$ respectively. The function u is **continuous** if it is continuous at all points p.

For a nonnegative integer k, a function u is said to be a $\mathbf{C^k}$ **function**, if every k–th order partial derivative of u exists and is continuous.

A function is a C^0 function, if and only if it is continuous. The notation " $u \in C^k$ " is used to indicate that u is a member of the set of the C^k functions. It is a standard fact that $u \in C^k$ implies $u \in C^{k-1}$ for $k > 0$. For a C^2 function u, recall that $u_{xy} = u_{yx}$. More generally, the order in which one takes k or fewer partial derivatives of a C^k function is immaterial.

Remark. We have assumed above that the function u is defined for all values of the independent variables. The function might only be defined for $(x,y,z,...)$ in a certain region D. The regions which we will encounter are rather simple (e.g., rectangles, strips, discs). If such a region includes some point p of its boundary, then technically the notion of partial derivative of u at p is not defined, unless one wishes to deal with one–sided derivatives. Let us simply say that a function u is C^k on a region D with boundary points, if there is a C^k function v, defined on a larger region without boundary points (i.e., an **open** region) such that $u = v$ at all points of D.

Definition 1. A **partial differential equation (PDE) of order** $k > 0$ is an equation involving an unknown function u, such that k is the greatest of the orders of the partial derivatives of u appearing in the equation.

Definition 2. A **solution of a k–th order PDE**, on a prespecified region D, is a C^k function defined on the region D such that the PDE is satisfied at all interior points of D. If no region is specified in advance, then a solution of a k–th order PDE is a C^k function defined on at least some nonempty open region where it satisfies the PDE.

There are many functions of several variables which arise in practice. At a point (x,y,z) at time t, u(x,y,z,t) might be any one of the following quantities : temperature, electrostatic potential energy, gravitational potential energy, pressure, mass density, energy density, concentration of a certain chemical, etc. . The laws of science are frequently stated in terms of PDEs involving such functions as unknowns. One often has the problem of determining the function u(x,y,z,t) for arbitrary t, for given information about u at time $t = 0$ (i.e., **initial conditions**, abbreviated I.C.). Such problems are referred to as **initial–value problems** . In **steady–state problems**, the function u is independent of t. In this case, one is often interested in solving a PDE for u(x,y,z) in a certain region D in space, where information is given about the behavior of u on the boundary of D (i.e., **boundary conditions**, B.C., are given). Such problems are known as **boundary–value problems**. More generally, one often seeks a solution u(x,y,z,t) of some PDE for points (x,y,z) in a region D at arbitrary time $t > 0$, subject to initial conditions at time $t = 0$, as well as boundary conditions specified at each time $t > 0$. Such a problem is aptly called an **initial/boundary–value problem**. It is important that the initial conditions and boundary conditions be chosen in such a way that the PDE has a unique solution satisfying them. Otherwise, one cannot meet the chief goal of predicting the relevant physical quantity represented by u. Mathematicians tend to be more interested in proving the existence, uniqueness and qualitative behavior of the exact solutions of initial/boundary–value problems, while those who apply the theory are concerned with actually finding functions which satisfy the PDE and initial/boundary conditions, at least within experimental error. In this book, we try to adopt an intermediate stance, believing that each camp can benefit from the considerations of the other. Before continuing our general discussion, we will now present some specific examples. Example 1 is lengthy, but it is well worth understanding.

Example 1 (Spherically symmetric gravitational potentials). In the Newtonian (pre–Einstein) theory of gravity, at a fixed time, the gravitational acceleration vector field (force per unit mass) is $-\nabla u$, where $\nabla u \equiv u_x\mathbf{i} + u_y\mathbf{j} + u_z\mathbf{k}$ is the gradient of a function u(x,y,z), called the **gravitational potential**. The function u obeys the second–order PDE

$$u_{xx} + u_{yy} + u_{zz} = 4\pi G\rho \tag{1}$$

where $\rho = \rho(x,y,z)$ is the density (mass per unit volume) of matter at (x,y,z), and G is the gravitational constant, $G \approx 6.668 \times 10^{-11}\ m^3 s^{-2} kg^{-1}$. One can also interpret u in other ways, for example, (i) as a steady–state temperature distribution in a solid with internal heat source density proportional to ρ, or (ii) as an electrostatic potential whose negative gradient is the electric field produced by a charge density proportional to ρ. In any case, equation (1) is known as **Poisson's equation.** In the special case when $\rho = 0$, (1) is better known as **Laplace's equation.** Suppose that we seek a solution u(x,y,z) of Laplace's equation which is spherically symmetric in the sense that u(x,y,z) only depends on the distance $r = [x^2 + y^2 + z^2]^{\frac{1}{2}}$ to the origin (0,0,0). In other words, $u(x,y,z) = f(r)$ for some function f of a single variable $r > 0$. Using the chain rule, we have

$$u_x = \frac{df}{dr}\frac{\partial r}{\partial x} = f'(r)\, r_x\,, \quad \text{where } r_x = \frac{\partial}{\partial x}[x^2 + y^2 + z^2]^{\frac{1}{2}} = \tfrac{1}{2}r^{-1}2x = xr^{-1}\,.$$

Then

$$\begin{aligned} u_{xx} &= f''(r)[r_x]^2 + f'(r)\, r_{xx} \\ &= f''(r)(x^2/r^2) + f'(r)[r^{-1} + x(-r^{-2}r_x)] \\ &= f''(r)(x^2/r^2) + f'(r)[r^{-1} - (x^2/r^3)]\,. \end{aligned}$$

We get similar expressions for u_{yy} and u_{zz}. Adding these results, we obtain

$$\begin{aligned} u_{xx} + u_{yy} + u_{zz} &= f''(r)(x^2 + y^2 + z^2)/r^2 + f'(r)[3r^{-1} - (x^2 + y^2 + z^2)/r^3] \\ &= f''(r) + 2r^{-1}f'(r) = 0\,. \end{aligned} \tag{2}$$

Writing $g(r) = f'(r)$, equation (2) becomes the separable (or linear) ODE $g'(r) + 2r^{-1}g(r) = 0$, whose solution is $g(r) = Cr^{-2}$. Thus, $f(r) = -Cr^{-1} + K$, where C and K are arbitrary constants. Hence the general spherically symmetric solution of Laplace's equation is

$$u(x,y,z) = -C[x^2 + y^2 + z^2]^{-\frac{1}{2}} + K = -Cr^{-1} + K\,. \tag{3}$$

If $C \neq 0$, then this solution is not defined at (0,0,0). Thus, the only spherically symmetric solutions that are defined everywhere are the constant solutions $u = K$, which give rise to a zero gravitational (or electric) field ($-\nabla K = \mathbf{0}$). Of course, one does not expect to find any gravity (or static electrical field) when the density ρ is 0 everywhere. When $C \neq 0$, we obtain a solution defined in any region D which excludes (0,0,0). In the gravitational context, take D to be the exterior, $r > r_0$, of some isolated planet. Suppose that the magnitude $|\nabla u| = f'(r) = Cr^{-2}$ of the gravitational acceleration is known to be g at the planet's surface (e.g., for the earth $g \approx 9.8\ \text{m/sec}^2 \approx 32\ \text{ft/sec}^2$). Then we have the boundary condition $Cr_0^{-2} = g$ or $C = gr_0^2$. Thus,

$$u = -gr_0^2r^{-1} + K \quad \text{and} \quad -\nabla u = -g(r_0/r)^2\, \mathbf{e}_r \quad \text{for } r \geq r_0\,, \tag{4}$$

where $\mathbf{e}_r$ is the unit vector field pointing away from (0,0,0). When $r < r_0$, these formulas do not apply, since $\rho > 0$ inside the planet. (In this case we would have to solve Poisson's equation (1) for $r < r_0$). Since $|\nabla u|$ in (4) is proportional to r^{-2}, we have deduced the inverse–square law for gravity from Laplace's equation. □

Remark 1 (Escape velocity). The potential difference $u(\infty) - u(r_0) = gr_0$ is the energy (per unit mass) required to move an object from the planet's surface to arbitrarily far reaches of space.

Thus, ignoring atmospheric resistance, the kinetic energy per unit mass, namely $\frac{1}{2}v^2$ (v = velocity), which is needed for a projectile to completely escape from the planet is gr_0. In other words, the escape velocity is $\sqrt{2gr_0}$. For the earth, this is about 11.2 km/sec ≈ 7 miles/sec . □

Remark 2 (Spherically symmetric solutions in dimension n). In n–dimensional space, the spherically symmetric solutions $u(x_1,\ldots,x_n)$ of Laplace's equation ($u_{x_1x_1} + \ldots + u_{x_nx_n} = 0$) can be found in the same way as in Example 1 (cf. Problem 6), and are of the form

$$-Cr^{2-n} + K, \quad \text{if } n > 2 \tag{5a}$$

and

$$C\cdot\log(r) + K, \quad \text{if } n = 2, \tag{5b}$$

where $r = [(x_1)^2 + \ldots + (x_n)^2]^{\frac{1}{2}}$. Regardless of the dimension, solutions of Laplace's equation are called **harmonic functions.** The formulas (5a) and (5b) show that in dimension n, the magnitude of the force $-\nabla u$ (per unit mass), associated with a spherically symmetric harmonic potential, is proportional to r^{1-n}. In Problem 19, we prove that when $n \geq 4$, a planet subject to such a force cannot have a closed orbit unless the orbit is a perfect circle, a very unstable possibility. (Of course, a wide variety of closed elliptical orbits are possible when $n = 3$). Thus, perhaps it is not so surprising that the space that we live in has no more than 3 dimensions. □

Remark 3 (Other solutions of Laplace's equation). It should be mentioned that there are actually infinitely many independent solutions (not just depending on r) of Laplace's equation in any dimension $n > 1$. For example, consider $u = x, y, x^2-y^2, 2xy, x^3-3xy^2, e^x\sin(y),\ldots$. Chapter 6 is devoted mainly to Laplace's equation in dimension 2 : $u_{xx} + u_{yy} = 0$. There, we will consider the boundary–value problem (among others) of determining solutions u of Laplace's equation on a plane region D, given the values of u on the boundary of D. This problem has applications to steady fluid flow, electrostatics and steady–state heat theory in which there is no dependence on the spatial variable z. One reason for the appearance of Laplace's equation in so many contexts is that it is the only homogeneous, linear (cf. Definition 3 below) PDE which involves only partial derivatives of orders strictly between 0 and 4 and retains its form under translations and rotations of coordinates. □

Example 2 (Heat problems). Suppose that $u(x,y,z,t)$ is the temperature at time t at the point (x,y,z) in a homogeneous heat conducting solid without heat sources. Under natural assumptions, one can show that u satisfies the following second–order PDE called the **heat equation**:

$$u_t = k(u_{xx} + u_{yy} + u_{zz}), \tag{6}$$

where $k > 0$ is a constant which measures the heat conductivity of the material in the solid. Note that in the case of a steady–state temperature distribution, where u does not depend on t, the left side u_t of (6) vanishes, and thus the steady–state temperature $u(x,y,z)$ satisfies Laplace's equation. In Chapter 3, we will derive and study the heat equation in the simpler one–dimensional setting, where u depends only on x and t. One example of the initial/boundary–value problems which we will consider is

$$\begin{array}{ll} \text{D.E.} & u_t = ku_{xx} \quad 0 \le x \le 1\,,\ t \ge 0 \\ \text{B.C.} & u(0,t) = 0 \quad u(1,t) = 0 \\ \text{I.C.} & u(x,0) = f(x)\ . \end{array} \tag{7}$$

Here $u(x,t)$ is the (uniform) temperature of the cross section at distance x along a solid rod which extends from $x = 0$ to $x = 1$. We assume that the rod is covered with heat insulation except at the end cross sections. The B.C. $u(0,t) = 0$ and $u(1,t) = 0$ state that the ends of the rod are to be maintained at temperature 0 (e.g., the rod is placed in ice water). The I.C., $u(x,0) = f(x)$, tells us that, at $t = 0$, the rod has the given temperature distribution $f(x)$. For example, suppose that $f(x) = \sin(\pi x)$. One can easily verify that

$$u(x,t) = e^{-\pi^2 kt}\sin(\pi x) \tag{8}$$

solves the PDE (7) and the initial condition $u(x,0) = \sin(\pi x)$, as well as the boundary conditions. We expect that the rod's temperature will approach the temperature (zero) of its icy environment. Indeed, the factor $\exp[-\pi^2 kt]$ in (8) tells us that, as $t \to \infty$, the temperature of the rod approaches 0, and it does so more rapidly for larger values of the heat conductivity k. More generally, choosing $f(x) = \sin(n\pi x)$ for an arbitrary positive integer n, we get the solution $u(x,t) = \exp[-n^2\pi^2 kt]\sin(n\pi x)$. Note that the rate at which this solution approaches 0 as $t \to \infty$ is faster for larger values of n. Physically, this is so, because the rate of heat transfer from hot to cold regions is greater when these regions are separated by smaller distances, which is the case when n gets larger. Associated with the heat equation, there are many other types of boundary and initial conditions which will be explored and solved in Chapter 3. □

Example 3 (Wave problems). If $u(x,y,z,t)$ is the deviation of air pressure (from its normal value) at (x,y,z) at time t, then (to a good approximation) u satisfies the **wave equation**

$$u_{tt} = a^2(u_{xx} + u_{yy} + u_{zz})\,, \tag{9}$$

where a is the speed of sound. We assume that the elevation is near sea level, so that a can be taken to be the constant 1087 ft/sec. For another interpretation of (9), the scalar potential (as well as the components of the vector potential) of a possibly time–dependent electromagnetic field in vacuum also satisfies the wave equation, where a is the speed of light in a vacuum ($\approx$ 186,000 mi/sec $\approx 2.998 \times 10^8$ m/sec). Note that when u is time–independent (e.g., when u is an electrostatic potential), (9) reduces to Laplace's equation, since $u_t = 0$. Returning to the case where u measures air pressure deviations, suppose that we wish to find possible sounds (variations of pressure) inside a closed box. As one nears a wall of the box from inside, one finds that the derivative of the pressure in the direction normal (perpendicular) to the wall approaches 0. This is because wind blows in the direction of the negative pressure gradient, $-\nabla u = u_x\mathbf{i} + u_y\mathbf{j} + u_z\mathbf{k}$. Since there can be no wind velocity component normal to the wall, the pressure gradient has no normal component. So, for the box $0 \le x \le A$, $0 \le y \le B$, $0 \le z \le C$, we have the following B.C. :

$$\begin{aligned} &u_x(0,y,z) = 0\,, \quad u_y(x,0,z) = 0\,, \quad u_z(x,y,0) = 0 \\ &u_x(A,y,z) = 0\,, \quad u_y(x,B,z) = 0\,, \quad u_z(x,y,C) = 0\,. \end{aligned} \tag{10}$$

There is a large family of solutions of (9) which satisfies the B.C. in (10). For any triplet (m,n,p) of integers, let $\nu(m,n,p) = \frac{1}{2}a[(m/A)^2 + (n/B)^2 + (p/C)^2]^{\frac{1}{2}}$. Then,

$$u(x,y,z,t) = \sin[2\pi\nu(m,n,p)t]\cdot\cos(m\pi x/A)\cdot\cos(n\pi y/B)\cdot\cos(p\pi z/C) \tag{11}$$

satisfies the PDE (9) and the boundary conditions (10) (cf. Problem 9). Moreover, if in (11) we replace the leading factor by $\cos[2\pi\nu(m,n,p)t]$, then we get another solution. Notice that, at points in the box, the pressure (11) oscillates through $\nu(m,n,p)$ cycles per unit time. Hence, $\nu(m,n,p)$ is called the **frequency** of the solution (11). If $A \le B \le C$, the lowest possible nonzero frequency (called the **fundamental pitch**) is $\nu(0,0,1) = \frac{1}{2}a/C$. Taking the box to be an enclosed shower stall with a 7 ft height (and smaller dimension for the base), we have $\frac{1}{2}a/C = 1087/14 \approx 78$ cycles per second (or 78 Hertz), which is the pitch of a rather low voice. □

Remark. In Chapter 5, we concentrate on the one–dimensional wave equation

$$u_{tt} = a^2u_{xx} \tag{12}$$

for a function $u = u(x,t)$. At a fixed time t, $u(x,t)$ can be interpreted as the transverse displacement at position x of a vibrating string which is stretched along the x–axis when at rest. Here a^2 is T/ρ, where T is the tension at rest and ρ is the mass per unit length of string. In Chapter 5, we provide a derivation of (12) using Newton's second law, and we solve numerous initial/boundary–value problems for the vibrating string. In contrast to the heat equation, in order to determine a unique solution of (12), one needs to know not only the string's initial displacement $u(x,0)$, but also the initial rate of change $u_t(x,0)$. A simple example of an initial/boundary–value problem for the string is

$$\begin{aligned} &\text{D.E.}\quad u_{tt} = a^2u_{xx} \qquad 0 \le x \le 1\,,\ t \ge 0 \\ &\text{B.C.}\quad u(0,t) = 0\,,\ u(1,t) = 0 \\ &\text{I.C.}\quad u(x,0) = f(x)\,,\ u_t(x,0) = g(x)\,. \end{aligned} \tag{13}$$

The B.C. of (13) state that the ends of the string are held fixed on the x–axis at 0 and 1. Intuitively, the motion of the string is determined only if both the initial transverse displacement $f(x)$ and the initial transverse velocity $g(x)$ are given. For example, if $f(x) = \sin(\pi x)$ and $g(x) = \sin(3\pi x)$, the theory of Chapter 5 yields the solution

$$u(x,t) = \cos(\pi at)\sin(\pi x) + (1/3\pi a)\sin(3\pi at)\sin(3\pi x). \quad \square \tag{14}$$

Linear PDEs, Classification, and the Superposition Principle

All of the PDEs in the above examples are linear. The notion of linearity for PDEs is strictly analogous to linearity for ODEs. Recall that the general n–th order linear ODE is an ODE which is expressible in the form

$$a_n(x)y^{(n)} + \ldots + a_2(x)y'' + a_1(x)y' + a_0(x)y = f(x)\ , \tag{15}$$

where $a_0(x)$, $a_1(x)$, ..., $a_n(x)$ and $f(x)$ are given (possibly constant) functions of the independent variable x. In particular, terms involving y^2 or higher powers of y (or more complicated functions of y or its derivatives), which cannot be eliminated, will make an equation nonlinear. For example, the equations $y' + y^2 = 0$, $(y'')^{-1} - x\cdot\log(y) = x$ and $yy' = 1$ are nonlinear. We say that the left side of (15) is a **linear combination** of y, y', y'', ... with coefficients $a_0(x)$, $a_1(x)$, $a_2(x)$, ... , which are given functions of the independent variable x.

Definition 3. A **linear n–th order PDE** is a PDE which can be put in the following form. The left side of the equation is a linear combination of the unknown function u and its partial derivatives (up to order n) with coefficients which are given functions of the independent variables. The right side must be some given function f of the independent variables. If the function f is identically zero, then the linear PDE is called a **homogeneous** PDE.

Example 4. The general second–order linear PDE for an unknown function $u = u(x,t)$ is

$$q(x,t)u_{xx} + r(x,t)u_{xt} + s(x,t)u_{tt} + a(x,t)u_x + b(x,t)u_t + c(x,t)u = f(x,t)\ , \tag{16}$$

where q, r, s, a, b, c and f are given functions (possibly constant) of x and t, with q, r, and s not all zero. If $f \equiv 0$, then (16) is the general second–order homogeneous linear PDE. □

Example 5. The one–dimensional heat equation $u_t = ku_{xx}$ can be put in the form (16) with $q = -k$ and $b = 1$, and with zero values for all other coefficients and f. Thus, the heat equation is a homogeneous linear PDE. When heat sources or sinks are present, they can often be represented by the terms cu and $h(x,t)$ in the inhomogeneous heat equation

$$-ku_{xx} + u_t + cu = h(x,t)\ , \quad k > 0\ , \tag{17}$$

which is again a special case of (16). We call (17) the **generalized heat equation.** □

Example 6. As another instance of (16), in the case of a vibrating string with a transverse applied force density proportional to $-cu(x,t) + F(x,t)$, we obtain the one–dimensional inhomogeneous **Klein–Gordon equation**

$$-a^2u_{xx} + u_{tt} + cu = F(x,t)\ , \quad a > 0\ . \tag{18}$$

If $F \equiv 0$ and $c = 0$, (18) reduces to the (homogeneous) wave equation $-a^2u_{xx} + u_{tt} = 0$ or (12). We refer to (18) as the **generalized wave equation.** □

Example 7. Usually one does not use t as an independent variable in Poisson's equation (cf. (1)), since t usually connotes time, whereas Poisson's equation is used in a steady–state context. However, using t unconventionally, we obtain Poisson's equation $u_{xx} + u_{tt} = g(x,t)$ in dimension 2. More generally (but still as a special case of (16)), we have the equation

$$a^2u_{xx} + u_{tt} + cu = g(x,t)\ , \quad a > 0\ . \tag{19}$$

If t is replaced by y (so that there will be no way of confusing t with time), then equation (19) is known as the **inhomogeneous Helmholtz equation** in dimension 2. Among other things, it is used in the analysis of vibrational modes of the skin of a drum. Roughly speaking, the constant a differs from 1 if the drum has a tension that is higher in one direction than in the other. We refer to (19) as the **generalized Poisson/Laplace equation.** □

It might appear that if we were to concentrate only on the "physical" equations (17),(18) and (19), then we would not make much progress in the study of the more general equation (16). However, in the case where the coefficients in (16) are constants, we have the following result, whose proof is given in Appendix A.1.

The Classification Theorem. Consider the second–order linear PDE

$$aU_{\xi\xi} + bU_{\xi\tau} + cU_{\tau\tau} + dU_{\xi} + eU_{\tau} + kU = F(\xi,\tau) \ , \tag{20}$$

$(a^2+b^2+c^2 \neq 0)$, where the unknown function $U = U(\xi,\tau)$ is C^2 and a, b, c, d, e and k are given real constants and $F(\xi,\tau)$ is a given continuous function. Then there is a change of variables of the form

$$\begin{aligned} x &= \alpha\xi + \beta\tau \quad t = -\beta\xi + \alpha\tau \\ u(x,t) &= \rho^{-1}\exp(\gamma\xi + \delta\tau)U(\xi,\tau) \ , \end{aligned} \tag{21}$$

where α, β, γ, δ and ρ $(\rho \neq 0)$ are real constants with $\alpha^2 + \beta^2 = 1$, such that (20) is transformed into exactly one of the following forms :

1. the form of the generalized wave equation (18), if $b^2 - 4ac > 0$, in which case (20) is called **hyperbolic** ;
2. the form of the generalized Poisson/Laplace equation (19), if $b^2 - 4ac < 0$, in which case (20) is called **elliptic** ;
3. the form of the generalized heat equation (17), if $b^2 - 4ac = 0$, and $2cd \neq be$ or $2ae \neq bd$ in which case (20) is called **parabolic** ;
4. the equation $u_{xx} + cu = g(x,t)$, if $b^2 - 4ac = 0$, $2cd = be$ and $2ae = bd$, in which case (20) is called **degenerate.**

In other words, aside from the degenerate case, equation (20) with constant coefficients is only a disguised version of the generalized wave equation, Poisson/Laplace equation or heat equation, depending on whether (16) is hyperbolic, elliptic or parabolic respectively. While it is good to know the Classification Theorem, it is perhaps not essential to become a virtuoso in performing the required change of variables, because when PDEs are derived from physical considerations in natural coordinates, almost always they are already found to be in a simple standard form. If it is ever needed, the method for the transformation of variables can be gleaned from the proof of the Classification Theorem in the Appendix A.1. Perhaps, the most significant facts to emerge are the following :

(i) Every equation of the form (20) has a physical interpretation, when rewritten in terms of appropriate variables.
(ii) In the general study of (20), there is really no loss of generality in confining one's attention to (17),(18),(19) and the degenerate case which is addressed in Section 1.3 .

The Superposition Principle

A very important fact concerning linear equations is the superposition principle which we will now describe. By definition, a linear PDE can be written in the form $L[u] = f$, where $L[u]$ denotes a linear combination of u and some of its partial derivatives, with coefficients which are given functions of the independent variables. Since $L[u]$ has this form, if we were to replace u by $u_1 + u_2$ the result, namely $L[u_1 + u_2]$, will be the same as $L[u_1] + L[u_2]$. The underlying reason for this is the fact that a partial derivative of the sum of two functions is the sum of their partial derivatives taken separately. More generally, for any constants c_1 and c_2,

$$L[c_1u_1 + c_2u_2] = c_1L[u_1] + c_2L[u_2] . \tag{22}$$

As a direct consequence of (22), we have

The Superposition Principle (or **Property**). **Let u_1 be a solution of the linear PDE $L[u] = f_1$ and let u_2 be a solution of the linear PDE $L[u] = f_2$. Then, for any constants c_1 and c_2, $c_1u_1 + c_2u_2$ is a solution of $L[u] = c_1f_1 + c_2f_2$. In other words,**

$$L[c_1u_1 + c_2u_2] = c_1f_1 + c_2f_2 . \tag{23}$$

In particular, when $f_1 = 0$ and $f_2 = 0$, (23) implies that if u_1 and u_2 are solutions of the homogeneous linear PDE $L[u] = 0$, then $c_1u_1 + c_2u_2$ will also be a solution of $L[u] = 0$.

Proof. By (22), $L[c_1u_1 + c_2u_2] = c_1L[u_1] + c_2L[u_2] = c_1f_1 + c_2f_2$. □

Example 8. Observe that $u_1(x,y) = x^3$ is a solution of the linear PDE $u_{xx} - u_y = 6x$, and $u_2(x,y) = y^2$ is a solution of $u_{xx} - u_y = -2y$. Find a solution of $u_{xx} - u_y = 18x + 8y$.

Solution. Here $L[u] = u_{xx} - u_y$, $f_1(x,y) = 6x$ and $f_2(x,y) = -2y$. Note that $18x + 8y = 3f_1(x,y) - 4f_2(x,y)$, and thus $c_1 = 3$ and $c_2 = -4$. The superposition principle then tells us that $3u_1(x,y) - 4u_2(x,y)$ (or $3x^3 - 4y^2$) will be a solution of $u_{xx} - u_y = 18x + 8y$, as can be easily checked directly. □

Example 9. Observe that $u_1(x,t) = \sin(t)\cos(x)$ and $u_2(x,t) = \cos(3t)\sin(3x)$ are solutions of the wave equation $u_{tt} = u_{xx}$. By applying the superposition principle, find infinitely many other solutions, none of which is a constant multiple of any other.

Solution. Note that $u_{tt} = u_{xx}$ can be written in the form of an homogeneous linear PDE $u_{tt} - u_{xx} = 0$. According to the superposition principle, for any constants c_1 and c_2,

$$c_1\sin(t)\cos(x) + c_2\cos(3t)\sin(3x)$$

is a solution. For each choice for c_1 and c_2, we obtain a different solution (cf. Problem 14). By choosing $c_1 = 1$ and letting c_2 vary, we obtain an infinite family of solutions, none of which is a constant multiple of any other. □

A great difficulty in the study of nonlinear equations is the typical failure of the superposition principle for such equations. This failure makes it difficult to form families of new solutions from an original pair of solutions, as the next example illustrates.

Example 10. Consider the nonlinear first–order PDE $u_x u_y - u(u_x + u_y) + u^2 = 0$ or equivalently $(u_x - u)(u_y - u) = 0$. Note that we have two solutions, namely e^x and e^y. However, show that $c_1e^x + c_2e^y$ will not be a solution, unless $c_1 = 0$ or $c_2 = 0$.

Solution. Defining $N[u] = (u_x - u)(u_y - u)$, observe that for any C^1 functions v and w

$$\begin{aligned} N[v + w] &= (v_x + w_x - v - w)(v_y + w_y - v - w) \\ &= N[v] + N[w] + (v_y - v)(w_x - w) + (v_x - v)(w_y - w) . \end{aligned}$$

This computation shows that $N[v + w] \neq N[v] + N[w]$ in general, due to the nonlinearity of the PDE. Taking $v = c_1e^x$ and $w = c_2e^y$ we obtain $N[c_1e^x + c_2e^y] = N[c_1e^x] + N[c_2e^y] + (-c_1e^x)(-c_2e^y) = c_1c_2e^{x+y}$. Thus, $N[c_1e^x + c_2e^y] = 0$ only if $c_1 = 0$ or $c_2 = 0$. □

Although all of the physically relevant PDEs which we have discussed so far are linear, there are many examples of nonlinear PDEs which are of great importance in physics. For example, Einstein's theory of relativity describes the force of gravity in terms of the curvature in the geometry of space–time. The Einstein field equations form a system of nonlinear PDEs. Because of the nonlinearity, solutions of these field equations are difficult to obtain, except in situations where several degrees of symmetry are assumed. Nonlinearity is also found in the PDEs of fluid mechanics, optics and elasticity theory. Nonlinear equations are often approximated by linear equations which hopefully yield solutions that are close to the corresponding solutions of the nonlinear equations. However, many interesting features of the original equations can be lost in the process, and gross errors can arise. In the next example, we illustrate these issues with the nonlinear minimal surface equation whose solutions yield soap film surfaces.

Example 11. Imagine a soap film surface which remains after a (possibly nonplanar) loop of wire is dipped in a soap solution. Due to the surface tension of the film, it will form a surface of least area spanning the loop (i.e. a minimal surface). If the surface is the graph $z = u(x,y)$ of some function u, defined on a bounded region D, then its area is $\iint_D (1 + u_x^2 + u_y^2)^{\frac{1}{2}}\, dxdy$. In 1760, Joseph Louis Lagrange showed that if $u(x,y)$ minimizes this integral among all functions with the same values on the boundary of D, then u must satisfy the (nonlinear) **minimal surface equation**

$$(1 + u_y^2)u_{xx} + (1 + u_x^2)u_{yy} - 2u_x u_y u_{xy} = 0 \; . \tag{24}$$

If one were to assume that the surface $z = u(x,y)$ is nearly level, then u_x and u_y would be small (say compared with 1), and u_x^2, u_y^2 and $u_x u_y$ would be very small. In this case, it would appear that equation (24) is reasonably approximated by Laplace's equation

$$u_{xx} + u_{yy} = 0 \quad . \tag{25}$$

Indeed, if the wire loop is nearly planar, and is held nearly level, then the minimal surface formed will be close (in a sense which is rather difficult to make precise) to the graph of the corresponding solution of Laplace's equation. Troubles arise when the supposition $u_x^2 + u_y^2 << 1$ turns out to be incorrect. As an illustration, we compare the solutions of (24) and (25) in the case where u is assumed to have the form $u = f(r)$, $r = [x^2 + y^2]^{\frac{1}{2}}$. By the computation done in Example 1, it is found that (24) and (25) become

$$rf''(r) + f'(r)(1 + [f'(r)]^2) = 0 \tag{26}$$

$$rf''(r) + f'(r) = 0 \; , \tag{27}$$

respectively. If we set $g = f'$, then both (26) and (27) are separable first–order ODEs. The corresponding general solutions of (26) and (27) are, respectively,

$$\bar{f}(r) = C\log(\tfrac{1}{2}[r + (r^2 - C^2)^{\frac{1}{2}}]) + K$$
$$f(r) = C\log(r) + K \; .$$

These solutions agree well for large r, where $\bar{f}'(r) \approx f'(r) \approx 0$ (i.e., where $u_x^2 + u_y^2 << 1$). However, $\bar{f}(r)$ and $f(r)$ behave differently as $r \downarrow C$, and $\bar{f}(r)$ is undefined for $0 \le r \le C$, whereas $f(r)$ is defined for all $r > 0$. In Figure 1 below, we have chosen $C > 0$ and $K = -C\log(C/2)$ so that $\bar{f}(C) = 0$. The graph of $u(x,y) = \bar{f}([x^2 + y^2]^{\frac{1}{2}})$ is a minimal surface obtained by revolving the graph of the curve $z = \bar{f}(r)$ about the z–axis. By joining the curve $z = \bar{f}(r)$ with $z = -\bar{f}(r)$, and revolving, we obtain a complete minimal surface running from

$z = -\infty$ to $z = +\infty$. The portion of the surface between the two circles at $z = a$ and $z = b$ is the soap film obtained by dipping those circles (say, wires) in soap solution, provided that $|b - a|$ is not too large (cf. Problem 20). The curve $z = \pm\bar{f}(r)$ is the same as the curve $r = C\cdot\cosh(z/C)$, and the complete minimal surface revolution is called a **catenoid.** The catenoids obtained in this way vary in size but not in shape. The minimum distance of the film to the z–axis is C . If we were to believe the validity of the Laplacian approximation, we might erroneously conclude that the film will continue to approach the z–axis, as $z = f(r)$ does. The reason for the failure of the approximation is that $\bar{f}'(r) = [u_x^2 + u_y^2]^{\frac{1}{2}}$ does not remain small as $r \downarrow C$, but approaches infinity. □

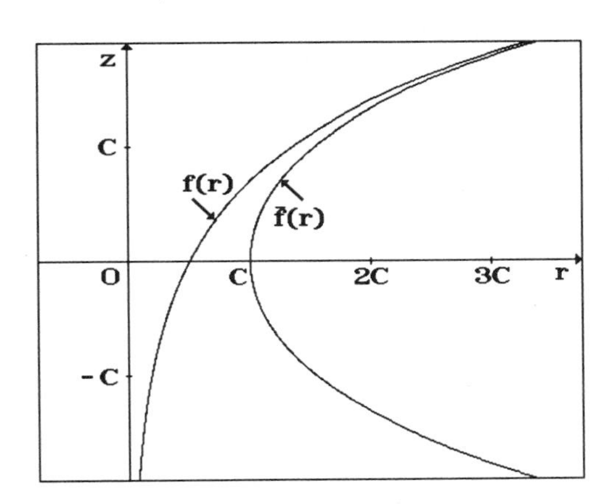

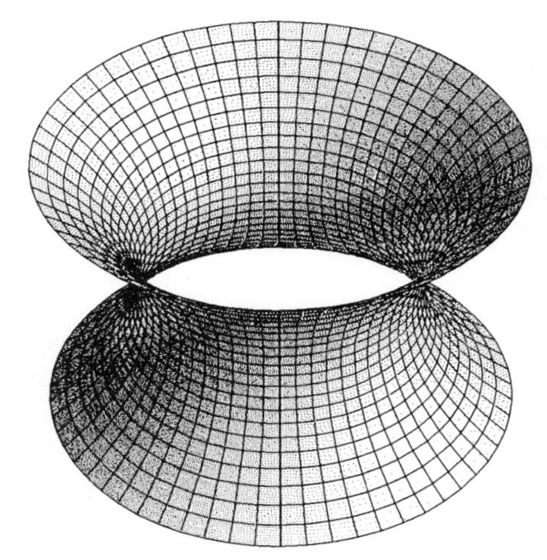

Figure 1

Remark (Black holes). A very similar phenomenon happens when the nonlinear Einstein field equations are used to compute the spherically symmetric geometry for a gravitational field about a ball of mass M and radius r_0. In Example 1, the spherically symmetric Newtonian gravitational potential was found to be $C/r + K$, and it was derived from (the linear) Laplace equation. This formula for the potential is valid as long as $r > r_0$, no matter how small or dense the ball is. However, it is found that Einstein's nonlinear description of the space–time geometry, in terms of the variable r and "time" t , can break down before r reaches r_0. Indeed, if the radius of the ball of mass M is less than the so–called **Schwarzchild radius** $r_M = 2GM/c^2$ (where c is the speed of light and $G = 6.668 \times 10^{-11}\ m^3s^{-2}kg^{-1}$ is the gravitational constant), then the representation of the solution of the field equations becomes undefined as $r \downarrow r_M$. In place of the ball, we then actually have what is known as a **black hole**. As with the soap film, the solution of the Einstein field equations can be mathematically continued, if one changes coordinates from r and t to new variables which can be related to r and t via hyperbolic functions. This continuation of space–time goes inside the throat of the black hole and into "another universe" which is, however, inaccessible by any ordinary means. Indeed, any object that enters the black hole and travels at a speed not greater than c will meet a singular boundary of space–time, never reaching the other universe or returning to our own. These interesting features of Einstein's theory are lost in the linear Newtonian theory which approximates Einstein's theory in less extreme circumstances. □

Operators and Green's Functions

A useful approach to solving linear initial/boundary–value problems for PDEs or ODEs is based on the construction of so–called Green's functions. This concept originated in the memoir, "Essay on the Application of Mathematical Analysis to the Theory of Electricity and Magnetism", published in 1828 by the English mathematician George Green (1793–1841). Green introduced the term "potential" and used what is known now as Green's theorem, to study the properties of electric and magnetic potentials. Here we briefly explain the concepts of linear differential operators, Green's functions and integral operators.

An **operator** is a prescription which assigns to each suitable function some new function. For example, suppose that $L[u] = f$ is a k–th order linear PDE. The operator which assigns, to each C^k function u, the new function $L[u]$, is an example of a differential operator. The concept of such an operator is independent of any particular choice of u, in the same sense that the concept of a certain function, say $\log(x)$, is independent of any particular choice of x. Just as one might prefer to speak of the log function, without any reference to x, it is fashionable to speak of partial differential operators L without any reference to u. For example, the Laplace operator, say in dimension 3, is denoted by

$$\Delta = \frac{\partial^2}{\partial x^2} + \frac{\partial^2}{\partial y^2} + \frac{\partial^2}{\partial z^2} \ . \tag{28}$$

The operator Δ assigns to each C^2 function u, the new continuous function $\Delta[u]$ or simply Δu, which is $u_{xx} + u_{yy} + u_{zz}$. Thus, Poisson's equation is $\Delta u = f$, where $f = f(x,y,z)$ is a given function. To solve an equation such as $\log(x) = C$ for $x > 0$, recall that we simply operate on both sides by the inverse function exp, obtaining $x = \exp[\log(x)] = \exp(C)$. To solve Poisson's equation, one might attempt to find the inverse operator, say Δ^{-1}, of the Laplace operator. Then we would simply apply Δ^{-1} to both sides of $\Delta u = f$, obtaining the solution $u = \Delta^{-1}[f]$. It turns out that the inverse of the Laplace operator (on a certain class of functions) is the operator which operates on the function f to produce the new function $\Delta^{-1}[f]$ defined by

$$\Delta^{-1}[f](x,y,z) = \iiint G(x,y,z;\bar{x},\bar{y},\bar{z})f(\bar{x},\bar{y},\bar{z})\ d\bar{x}d\bar{y}d\bar{z}\ , \tag{29}$$

where

$$G(x,y,z;\bar{x},\bar{y},\bar{z}) = -\frac{1}{4\pi}[(x-\bar{x})^2 + (y-\bar{y})^2 + (z-\bar{z})^2]^{-\frac{1}{2}}. \tag{30}$$

Observe that when the integration with respect to $\bar{x}$, $\bar{y}$ and $\bar{z}$ is performed, we are left with a function of x, y, and z, which is by definition $\Delta^{-1}[f]$. If the function f is C^1 and is zero outside some ball, then armed with the appropriate tools, one could prove that $\Delta^{-1}[f]$ is a solution of $\Delta[u] = f$. The operator Δ^{-1} is an example of an **integral operator**, i.e. an operator B of the form

$$B[f](p) = \int g(p;q)f(q)\ dq\ . \tag{31}$$

Here p and q range over possibly multidimensional domains. When the solution of an initial/boundary–value problem for a PDE (or ODE) is expressed in the form of an integral operator (e.g., as in (29)) the function $g(p,q)$ is called a **Green's function** for the boundary–value problem. In the case of Poisson's equation (roughly speaking, with the boundary condition that solutions tend to zero at infinity), the Green's function is $G(x,y,z;\bar{x},\bar{y},\bar{z})$ in (30). Once the correct Green's function is found, the problem is reduced to computing the integral (31) for arbitrary p. Such a computation can be quite difficult, and numerical methods might be needed. Of course, when possible, one might prefer an algebraic formula, of the solution of a particular initial/boundary–value problem. For the most part, this is what we strive for in this book. Nevertheless, in general circumstances, Green's functions and their associated integral operators provide a tidy way of presenting solutions which we will exploit occasionally. It should be noted that integral operators (31) are linear, in the sense that $B[c_1f_1 + c_2f_2] = c_1B[f_1] + c_2B[f_2]$, and consequently they can only serve as inverses of *linear* operators. In particular, Green's functions and their integral operators *cannot* be used to express solutions of nonlinear PDEs !

Summary 1.2

1. **C^k functions** : For a nonnegative integer k, a function u is said to be a C^k function, if every k–th order partial derivative of u exists and is continuous.

2. **Linear PDEs** : A linear n–th order PDE is a PDE of the form $L[u] = f$, where $L[u]$ is a linear combination of the unknown function u and its partial derivatives (up to order n), where the coefficients and f are given functions of the independent variables. If $f \equiv 0$, the PDE is called homogeneous.

3. **The Classification Theorem** : The Classification Theorem asserts that every second–order linear PDE (cf. (20)) with constant coefficients, where the unknown function has two independent variables, can be transformed (by a change of variables) into exactly one of the following forms (where $u = u(x,t)$) :

(i) the form of the generalized wave equation

$$-a^2u_{xx} + u_{tt} + cu = F(x,t)\ ,\ a > 0\ ,\ \text{(hyperbolic case)}\ ;$$

(ii) the form of the generalized Poisson/Laplace equation

$$a^2u_{xx} + u_{tt} + cu = g(x,t)\ ,\ a > 0\ ,\ \text{(elliptic case)}\ ;$$

(iii) the form of the generalized heat equation

$$-ku_{xx} + u_t + cu = h(x,t)\ ,\ k > 0\ ,\ \text{(parabolic case)}\ ;$$

(iv) the form

$$u_{xx} + cu = g(x,t) \qquad \text{(degenerate case).}$$

4. **The superposition principle** : The superposition principle (or property) asserts that if u_1 and u_2 are solutions of the linear PDEs $L[u] = f_1$ and $L[u] = f_2$, respectively, then for any constants c_1 and c_2, $c_1u_1 + c_2u_2$ is a solution of $L[u] = c_1f_1 + c_2f_2$. In other words, $L[c_1u_1 + c_2u_2] = c_1L[u_1] + c_2L[u_2]$.

5. **Green's functions** : Green's functions and their associated integral operators are used to represent solutions of initial/boundary–value problems for linear PDEs (or ODEs).

Exercises 1.2

1. Show that the given functions satisfy the accompanying PDE.

(a) $u(x,y) = x + y$; $u_{xx} + u_{yy} = 0$

(b) $u(x,y) = f(x) + g(y)$; $u_{xy} = 0$, where the functions f and g are assumed to be C^2 .

(c) $u(x,y) = f(x+y) + g(x-y)$; $u_{xx} - u_{yy} = 0$.

(d) $u(x,t) = x^2 + 2t$; $u_{xx} = u_t$

(e) $u(x,y) = \sin(x)\cosh(y)$; $u_{xx} + u_{yy} = 0$

(f) $u(x,t) = \sin(x-ct)$; $u_{tt} - c^2u_{xx} = 0$, where c is a real constant.

2. Verify that the following functions are solutions of Laplace's equation $u_{xx} + u_{yy} = 0$.

(a) $u(x,y) = e^y\cos(x)$ (b) $u(x,y) = 3x^2y - y^3$

(c) $u(x,y) = \log(x^2 + y^2)$, $x^2 + y^2 \neq 0$

(d) $u(x,y) = e^y\cos(x) + 3x^2y - y^3 + \log(x^2 + y^2)$, $x^2 + y^2 \neq 0$.

3. Show that the following solve the heat equation $u_t - ku_{xx} = 0$.

(a) $u(x,t) = e^{-kt}\sin(x)$

(b) $u(x,t) = e^{-a^2kt}\cos(ax)$, for any real constant a.

(c) $u(x,t) = e^{kt}\cosh(x)$

(d) $u(x,t) = (1/\sqrt{kt})\exp[-x^2/(4kt)]$.

4. Show that the following are solutions of the wave equation $u_{tt} - c^2u_{xx} = 0$, for some c.

(a) $u(x,t) = x^2 + t^2$

(b) $u(x,t) = \cos(ax)\sin(bt)$, for any real constants a, b.

(c) $u(x,t) = \log(x + t) + (x - t)^2$

(d) $u(x,t) = f(x+2t) + g(x-2t)$, for any C^2 functions f and g.

5. Give the orders of the following PDEs, and classify them as linear or nonlinear. If the PDE is linear, specify whether it is homogeneous or inhomogeneous.

(a) $x^2u_{xxy} + y^2u_{yy} - \log(1 + y^2)u = 0$

(b) $u_x + u^3 = 1$

(c) $u_{xxyy} + e^xu_x = y$

(d) $uu_{xx} + u_{yy} - u = 0$

(e) $u_{xx} + u_t = 3u$.

6. Derive formulas (5a) and (5b) for the most general spherically symmetric solution of Laplace's equation in dimension n .

7. (a) Find a solution of Laplace's equation $u_{xx} + u_{yy} = 0$ of the form $u(x,y) = Ax^2 + Bxy + Cy^2$ $(A^2+B^2+C^2 \neq 0)$ which satisfies the boundary condition $u(\cos(\theta),\sin(\theta)) = \cos(2\theta) + \sin(2\theta)$ for all points $(\cos(\theta),\sin(\theta))$ on the unit circle, $x^2 + y^2 = 1$.

(b) Show that the graph of any solution $u(x,y)$ of Laplace's equation of the form in (a), intersects the xy–plane in a pair of perpendicular lines through (0,0).

8. (a) Show that $u(x,t) = \exp[-n^2\pi^2kt]\sin(n\pi x)$ solves the initial/boundary–value problem given in equations (7) with I.C. $f(x) = \sin(n\pi x)$, if and only if n is an integer.

(b) In how many points does the graph of $\sin(n\pi x)$ intersect the x–axis between 0 and 1 ?

(c) Give a physical reason for why the temperature approaches 0 faster if n is larger.

9. Let $u(x,y,z,t)$ be the solution (11) in Example 3 on wave problems.

(a) Show that $u_{tt} = -[2\pi\nu(m,n,p)]^2 u$, $u_{xx} = -(m\pi/A)^2 u$, etc. . Use these facts to deduce that $u(x,y,z,t)$ satisfies the wave equation (9).

(b) Verify that $u(x,y,z,t)$ meets the B.C. (10).

(c) The set of points (x,y,z) inside the box, where $u(x,y,z,t)$ is always zero, is the union of a number of intersecting rectangular surfaces which divide the interior of the box into a number of compartments. How many compartments are there ?

(d) At what points in the box does the pressure experience the greatest changes ?

10. Refer to Example 3, and assume that the box is cubical with $A = B = C = 1$ and $a = 2$.

(a) By giving an example, show that it is possible for two independent solutions of the form of $u(x,y,z,t)$ in (11) to have the same frequency.

(b) List the ten lowest positive *distinct* frequencies for the box.

11. (a) Show that if $f(x) = \sin(\pi x)$ and $g(x) = \sin(3\pi x)$, then $u(x,t)$ in (14) solves the initial/boundary–value problem (13).

(b) Find two solutions $u(x,t)$ of the D.E. and B.C. in (13) such that these two solutions have the same initial profile $u(x,0)$, but have different initial velocity distribution $u_t(x,0)$.

12. For what values of the positive constants m and n will the second–order PDE $u_{xx} + u_{yy} + mu_{xy} + u_x + nu_y = 0$ be (a) hyperbolic, (b) elliptic, (c) parabolic or (d) degenerate ?

13. Observe that $u_1(x,y) = x^3$ solves $u_{xx} + u_{yy} = 2$ and $u_2(x,y) = cx^3 + dy^3$ solves $u_{xx} + u_{yy} = 6cx + 6dy$ for real constants c and d.

(a) Find a solution of $u_{xx} + u_{yy} = Ax + By + C$ for given real constants A, B and C.

(b) How can many more solutions of the problem in (a) be produced ?

14. In relation to Example 9, show that if $c_1\sin(t)\cos(x) + c_2\cos(3t)\sin(3x) = d_1\sin(t)\cos(x) + d_2\cos(3t)\sin(3x)$ for all (x,t) , then $c_1 = d_1$ and $c_2 = d_2$.

15. By direct computation, verify that by revolving the curve $y = \cosh(x)$ about the x–axis, we obtain a solution $u(x,y) = [\cosh^2(x) - y^2]^{\frac{1}{2}}$ of the minimal surface equation (24) on the domain $|y| < \cosh(x)$. In view of the solution $\Upsilon(r)$ found in Example 11, give a purely geometrical reason for why u(x,y) must be a solution.

16. Suppose that u(x,y) is any solution of the minimal surface equation (24), for (x,y) in some open region D in the plane.

(a) Show that it is not always true that cu(x,y) will be a solution for all real c.

(b) Show that if $c \neq 0$, then $cu(x/c,y/c)$ will be a solution on the new region consisting of all (x,y) with (x/c,y/c) in D .

(c) Explain the results of (a) and (b) geometrically in terms of similarity between the shapes of the surfaces.

17. Let u(x) be an arbitrary C^1 function defined for $x \geq 0$, such that $u(0) = 0$. Consider the ordinary differential operator d/dx which assigns to each such function u the new continuous function u′(x). Show that the inverse operator, say B, assigns to each continuous function f(x), defined for $x \geq 0$, the function

$$B[f](x) \equiv \int_0^\infty g(x,z)f(z)\,dz \ , \quad \text{where } g(x,z) = \begin{cases} 1 & 0 \leq z \leq x \\ 0 & z > x \end{cases} .$$

Consequently, the solution of the problem $u'(x) = f(x)$ $(x \geq 0)$ with boundary condition $u(0) = 0$, is given in terms of the integral operator B with Green's function g(x,z).

18. Let p(x) and q(x) be given continuous functions. Show that the solution of the linear ODE $u'(x) + p(x)u(x) = q(x)$ $(x \geq 0)$, with B.C. $u(0) = 0$, is given by

$$A[q](x) = \int_0^\infty G(x,z)q(z)\,dz \ , \ (x \geq 0),$$

where the Green's function is $G(x,z) = \exp[P(z) - P(x)]g(x,z)$, and $P(x)$ is an antiderivative of the function $p(x)$ and $g(x,z)$ is defined in Problem 17. What is the inverse operator of A ? **Hint.** Use Leibniz's rule in Appendix A.3.

19. Here we demonstrate the instability of planetary orbits in dimension greater than 3, assuming a spherically symmetric harmonic potential. In Example 1 (or Problem 6), we showed that such a potential in dimension $n > 2$ is of the form $-Cr^{2-n}$ (where $C > 0$, as we assume that the force is attractive). The angular momentum for the path $(r(t),\theta(t))$ in polar coordinates of a planet of mass m is $mr^2\dot{\theta}$ (where $\dot{\theta} \equiv \frac{d\theta}{dt}$) which is some constant, say A, for a central force. Thus, $\dot{\theta} = A/(mr^2)$. The kinetic energy of the planet is then $\frac{1}{2}m(\dot{r}^2 + r^2\dot{\theta}^2) = \frac{1}{2}m\dot{r}^2 + A^2/(2mr^2)$, where $\dot{r} \equiv \frac{dr}{dt}$. The total energy (kinetic + potential) is a constant

$$E = \tfrac{1}{2}m\dot{r}^2 + [A^2/(2mr^2) - Cmr^{2-n}] \ . \qquad (*)$$

Let $f(r)$ be the function in brackets in (*). Assume that the planet's orbit has at least two consecutive local extrema for r, say r_1 and r_2 (with $r_1 < r_2$). Of course, this assumption is possible when $n = 3$, since then there are elliptical orbits. For $n > 3$, we now show that this assumption leads to a contradiction. At such extreme points on the orbit, we have $\dot{r} = 0$, and thus $f(r_1) = f(r_2) = E$ by (*). Since $\frac{1}{2}m\dot{r}^2 > 0$ while the planet moves between the two consecutive extrema, we must have $f(r) < E$ for $r_1 < r < r_2$ by (*). Hence $f(r)$ must have a local minimum which is strictly less than E at some point r_0 between r_1 and r_2 .

(a) When $n = 4$, show that there is no r_0 such that $f'(r_0) = 0$, unless $f(r) \equiv E = 0$, but then $f(r_0)$ is not strictly less than E .

(b) When $n > 4$, show that there is only one positive value r_0 where $f'(r_0) = 0$, and this value is a local maximum instead of a local minimum, as can be deduced from the fact that $\lim_{r\to 0} f(r) = -\infty$ and $\lim_{r\to\infty} f(r) = 0$.

(c) Show that there is no such contradiction when $n = 3$, since $\lim_{r\to 0} f(r) = +\infty$ and $\lim_{r\to\infty} f(r) = 0$, when $n = 3$.

(d) A circular orbit is possible for $n \geq 4$, but such an orbit is unstable, since the slightest nudge will throw the planet out of a circular orbit. Assume that the orbit is not a perfect circle.
(i) If $n = 4$, show that either $r(t) \to \infty$ as $t \to \infty$, or $r(t) \to 0$ as t approaches some *finite* value.
(ii) If $n \geq 5$, show that in addition to the two possibilities in (i), it can also rarely happen that the orbit will spiral toward a circular orbit. Show that this can only occur if the maximum value of $f(r)$ is E. Why is this a rare occurance ?

20. In the the following, we deduce that a minimal soap film cannot be formed between two coaxial rings of radius R which are separated by a distance of more than $1.3255 \cdot R$.

(a) For $C > 0$, consider the curve $r = C \cdot \cosh(z/C)$ in the zr–plane. Show that the tangent line through the point (z_0, r_0) on this curve passes through the origin only when $\cosh(z_0/C)/\sinh(z_0/C) = z_0/C$ (i.e., $\coth(z_0/C) = z_0/C$).

(b) Show that there is a unique positive solution of $\coth(x) = x$, say $\alpha \approx 1.200$.
Hint. Let $g(x) = \coth(x) - x$. For small $x > 0$, show that $g(x) > 0$, while $g(x) < 0$ for large $x > 0$. Show that $g(x)$ is strictly decreasing for $x > 0$, by computing $g'(x)$.

(c) Show that the tangent lines in part (a) must be of the form $r = \pm\sinh(\alpha) \cdot z$, where α is defined in (b). Hence, regardless of the value of C, these lines are tangent to each of the curves $r = C \cdot \cosh(z/C)$.

(d) From Part (c) and the convexity of the curves $r = C \cdot \cosh(z/C)$ $(C > 0)$, deduce that all of these curves are contained in the wedge $r \geq \sinh(\alpha) \cdot |z|$.

(e) Conclude that there is no minimal surface joining two coaxial rings of radius R, if the rings are separated by a distance of more than $2R/\sinh(\alpha) < 1.3255 \cdot R$.

Remark. If the separation distance is less than $2R/\sinh(\alpha)$, then there are actually two surfaces of the form $r = C \cdot \cosh(z/C)$ that join the rings. The surface with the larger value of C is the one which actually has the minimum area (i.e., the one which arises physically).

1.3 General Solutions and Elementary Techniques

Ideally, one would like to have a general technique that could be used to find all of the solutions of an arbitrary PDE, or at least a relevant solution that satisfies certain initial/boundary conditions. Such a general technique does not exist even for the class of first–order ODEs. Recall that for such equations, there is a variety of techniques which work when the first order ODE is of a particular form (e.g., separable, homogeneous, exact, linear, etc.). Moreover, it is easy to find first–order ODEs that do not have any of these forms. The situation for PDEs is similar. It is easy to find PDEs for which there is no known method which will yield a single solution. Fortunately, the PDEs which arise in practice are not completely arbitrary. Indeed, there are few different kinds of PDEs, or systems of PDEs, which regularly appear in applications. Although there are some procedures that apply to more than one relevant equation, it is better not to develop excessively such procedures apart from the specific PDEs to which they will be applied. Instead, we prefer to handle each relevant equation separately. When a pattern of techniques emerges, we will note it and appreciate it, but we see no advantage in trying to force the solution process into a preconceived mold which could be motivated only with a great deal of hindsight. Also, unlike the theory of ODEs, the methods for solving PDEs often depend more on the form of the imposed boundary conditions than on the PDE itself. This makes it even more difficult to develop a unified theoretical edifice, if that were our goal. Nevertheless, in this section, we discuss some elementary techniques. One technique, known as "separation of variables", is a preliminary step used in solving a wide variety of PDE problems. However, first we shall illustrate some of the differences between PDEs and ODEs. We also explore some of the difficulties in determining the form of the general solution of a PDE, and in finding particular solutions which meet given side conditions.

General Solutions and Particular Solutions of PDEs

Recall that the general solution of an n–th order linear ODE involves n arbitrary constants. These constants are determined when the solution is required to satisfy n initial conditions. For example, the general solution, $y(t)$, of the second–order ODE

$$y'' + y = 0 \tag{1}$$

is

$$y(t) = c_1\cos(t) + c_2\sin(t) , \tag{2}$$

where c_1 and c_2 are arbitrary constants. If we also specify the initial conditions $y(0) = 0$ and $y'(0) = 1$, then the only solution of (1) which meets these conditions is $y(t) = \sin(t)$. Recall (cf. Definition 2 of Section 1.2) that a solution of PDE of order n is required to be a C^n function on the open set (possibly prespecified) where it is defined.

The **general solution** of a PDE is the collection of all solutions of the PDE.

As with ODEs, it is usually not possible to list all of the solutions, but rather one specifies the form of the general solution as in (2). However, the form of the general solution of an n–th order PDE typically involves n arbitrary functions, rather than arbitrary constants. The following Example illustrates this.

Example 1. Find the general solution of the first–order linear PDE for $u = u(x,y)$

$$u_x(x,y) = 2xy , \quad \text{for all } (x,y). \tag{3}$$

Solution. If we hold y fixed and integrate with respect to x, we obtain

$$u(x,y) = x^2y + f(y) . \tag{4}$$

Note that the constant of integration may depend on y, and indeed any function of the form (4) satisfies (3). As a technical point, recall that in Definition 2 of Section 1.2, we require that the function f(y) in (4) be C^1 (i.e., f has a continuous first derivative). If in place of (3) we had the PDE $u_x(x,y,z) = 2xy$, then the form of the general solution would be $u(x,y,z) = x^2y + g(y,z)$ for an arbitrary C^1 function $g(y,z)$. □

> Whenever integrating with respect to one variable, remember to add an arbitrary function of the other variables.

Example 2. Find the general solution of the third–order PDE

$$u_{xyy} = 2\sin(x) , \quad u = u(x,y,z) , \quad \text{for all } (x,y,z). \tag{5}$$

Solution. Integrating (5) once with respect to y, we get $u_{xy}(x,y,z) = 2y\sin(x) + f(x,z)$. Integrating again with respect to y, we obtain $u_x(x,y,z,) = y^2\sin(x) + yf(x,z) + g(x,z)$. Finally, integrating with respect to x, we obtain the general solution

$$u(x,y,z) = -y^2\cos(x) + yF(x,z) + G(x,z) + h(y,z) , \tag{6}$$

where F(x,z) and G(x,z) are antiderivatives (with respect to x) of f(x,z) and g(x,z), respectively. Since we want the solution to be C^3, we require that F, G and h be C^3 functions, and except for this requirement, these functions are arbitrary. □

Of course, as with the ODEs, it is not always possible to find the general solution of a PDE simply by integrating a few times. Nevertheless, the above examples suggest that typically the general solution of an n–th order PDE, for an unknown function u of m independent variables, involves n arbitrary functions of m–1 variables. However, it is easy to find examples which violate this rule. For instance, consider the following example.

Example 3. Find the general solution of

$$(u_{xx})^2 + (u_{yy})^2 = 0 , \quad u = u(x,y) , \quad \text{for all } (x,y). \tag{7}$$

Solution. A C^2 function $u(x,y)$ solves this equation, if and only if $u_{xx} = 0$ and $u_{yy} = 0$. Since $u_{xx} = 0$, u must have the form $u(x,y) = f(y)x + g(y)$. However, since $u_{yy} = 0$, u must also have the form $u(x,y) = h(x)y + k(x)$. The only functions which have both of these forms are of the form

$$u(x,y) = axy + bx + cy + d\ , \tag{8}$$

where a, b, c and d are arbitrary constants. Thus, the general solution of (7) involves four arbitrary constants instead of two arbitrary functions of a single variable. Note also that the superposition of two solutions of the form (8) is also a solution. Hence, (7) also provides us with an example of a nonlinear PDE whose solutions obey a superposition principle. □

In Example 3 of Section 2.2, we show that the homogeneous first–order linear equation $xu_x - yu_y + yu = 0$ has a general solution which depends on *two* arbitrary functions, instead of one. Thus, there are really no precise rules concerning the form of the general solution of (even) linear PDEs. However, it will be convenient to introduce the following notion of a "generic" solution of a PDE. Such a solution has the *expected* form of a general solution, a form which might not be realized for certain PDEs as we have just seen.

Definition 1. A **generic solution** of an n–th order PDE for an unknown function of m independent variables is a solution which involves n arbitrary C^n functions of $m-1$ variables. Moreover, we require that none of these arbitrary functions can be eliminated or combined without losing solutions in the process.

Remark. The last requirement ensures that one cannot simply increase the number of arbitrary functions by replacing some arbitrary function by a sum of two new arbitrary functions, or by some similar artifice. For instance, the solution (6) is generic, but $-y^2\cos(x) + y[k(x,z) - j(x,z)] + g(x,z)$ is not generic (even though there are three arbitrary functions), since $k(x,z) - j(x,z)$ can be replaced by $f(x,z)$. □

While the general solutions (4) and (6) for the PDEs in Examples 1 and 2 above are generic, according to Definition 1, the general solution (8) of the PDE (7) is not generic. It is also possible to have a generic solution which is not a general solution as the following example shows.

Example 4. Find a generic solution of the nonlinear first–order PDE

$$u_x(x,y) = [u(x,y)]^2\ . \tag{9}$$

Solution. By fixing y, we may regard (9) as a first–order separable ODE, namely $u^{-2}du = dx$, assuming that $u \neq 0$. Integrating, we get the solution $-u^{-1} = x + g(y)$, or

$$u(x,y) = -[x + g(y)]^{-1}\ , \tag{10}$$

where g is an arbitrary C^1 function and $u(x,y)$ is defined everywhere except for points (x,y) on the curve $x = -g(y)$. The solution (10) is generic. However, (10) is not the most general solution, because there are solutions of (9) which are not of the form (10). Indeed, $u(x,y) \equiv 0$ is such a solution. One can produce other solutions, by "pasting" two solutions together (see Problem 11). Now suppose that an open region D is given beforehand, and suppose that only solutions which are defined throughout D are allowed, then the function $g(y)$ must satisfy the requirement that the curve $x = -g(y)$ does not intersect D. Since no such region was specified here, we regard all functions of the form (10) as solutions. If we had required that the solution be defined everywhere, then the only solution would be $u(x,y) \equiv 0$. □

Example 5. Consider the first–order linear PDE

$$xu_x - 2xu_y = u \ . \tag{11}$$

Show that

$$u(x,y) = xf(2x + y) \tag{12}$$

is a generic solution of (11), where f is an arbitrary C^1 function.

Solution. First note that despite the involvement of both x and y in $f(2x + y)$, the function f is still really a function of one variable, since f has only one "slot", unlike say $g(x,y)$. Thus, by definition, $u(x,y)$ in (12) will define a generic solution if it satisfies (11). The product and chain rules yield $u_x = f(2x + y) + xf'(2x + y)\cdot 2$ and $u_y = xf'(2x + y)$. Hence, $xu_x - 2xu_y = xf(2x + y) + 2x^2f'(2x + y) - 2x^2f'(2x + y) = xf(2x + y) = u$. Thus, (12) defines a generic solution. Using the theory of Chapter 2, one can prove that the general solution of (11) has the form (12). □

Usually one wants to find a particular solution of a PDE which meets a side condition. The next two examples show how such a solution may be extracted from a generic solution.

Example 6. Find a solution of (11) which satisfies the condition $u(1,y) = y^2$ for all y.

Solution. The condition $u(1,y) = y^2$ specifies the values of the solution $u(x,y)$ for points (x,y) on the line $x = 1$, parallel to the y–axis. Since (12) is a generic solution, it suffices to find a function f such that $u(1,y) = 1\cdot f(2 + y) = y^2$ or $f(2 + y) = y^2$. To find such a function, let $r = 2 + y$. Since $y = r - 2$, we have $f(r) = (r - 2)^2$. Thus, f is the function which takes a number, subtracts 2, and squares the result. In particular, $f(2x + y) = (2x + y - 2)^2$. Hence, $u(x,y) = x(2x + y - 2)^2$. One should check directly that this u satisfies the PDE (11) and $u(1,y) = y^2$. □

Example 7. Show that the wave equation $u_{tt} = c^2u_{xx}$ has a generic solution of the form

$$u(x,t) = f(x + ct) + g(x - ct) \ , \tag{13}$$

where f and g are arbitrary C^2 functions. Find a particular solution meeting the initial conditions

$$\text{I.C.} \quad u(x,0) = h(x) \quad \text{and} \quad u_t(x,0) = 0 \ , \tag{14}$$

where $h(x)$ is a given C^2 function.

Solution. One can directly verify that (13) is a solution of $u_{tt} = c^2u_{xx}$, as in Problem 4(d) of Section 1.2 . Since the wave equation is second–order and there are two arbitrary functions in (13), neither of which can be eliminated without losing solutions (cf. Problem 12), we conclude that (13) is a generic solution. By setting $t = 0$ in (13) and using $u(x,0) = h(x)$, we get that $f(x) + g(x) = h(x)$. By differentiating (13) with respect to t, we obtain $u_t(x,t) = f'(x + ct)c + f'(x - ct)(-c)$, whence $u_t(x,0) = 0$ yields $f'(x) - g'(x) = 0$. Thus, (14) gives us two conditions on the two unknown functions f and g, namely

$$f(x) + g(x) = h(x) \quad \text{and} \quad f(x) - g(x) = K. \tag{15}$$

Adding corresponding sides of the equations (15), we obtain $f(x) = \frac{1}{2}[h(x) + K]$. Similarly $g(x) = \frac{1}{2}[h(x) - K]$. These identities determine the functions f and g in terms of the given function h. Thus, we obtain the following solution of $u_{tt} = c^2u_{xx}$, which meets the initial conditions (14) :

$$u(x,t) = \tfrac{1}{2}[h(x + ct) + K + h(x - ct) - K] = \tfrac{1}{2}[h(x + ct) + h(x - ct)] \ . \tag{16}$$

In Problem 12, the reader is asked to check directly the validity of (16) . □

Elementary Techniques

We have already seen in Example 1 and 2 that PDEs, which simply set a partial derivative of the unknown function equal to a given function, can be solved by direct integration. The PDE $u_x = u^2$ in Example 4 cannot be solved by integrating both sides with respect to x, because the right side involves the unknown function $u(x,y)$. However, we were able to solve this equation by ODE techniques.

> If a PDE involves only partial derivatives with respect to one of the independent variables, then such an equation may be regarded as an ODE for an unknown function of a single variable, where the other variables are held fixed. In the solution, the arbitrary constants are replaced by arbitrary functions of these remaining variables.

By way of illustration, we solve here the homogeneous version of the degenerate equation $u_{xx} + cu = g(x,t)$ which arose in the Classification Theorem of Section 1.2 .

Example 8. Find the general solution of the PDE

$$u_{xx} + cu = 0\ , \quad u = u(x,t) \tag{17}$$

in the three cases $c > 0$, $c = 0$ and $c < 0$.

Solution. For fixed t , (17) is a second–order linear ODE with constant coefficients (discussed in Section 1.1) for u, regarded as a function of x. If $c > 0$, then for each fixed t, the solution is of the form $c_1\sin(\sqrt{c}\cdot x) + c_2\cos(\sqrt{c}\cdot x)$. However, as t varies, the choices for c_1 and c_2 may change (i.e., they may be functions of t). Consequently, the general solution of (17) is

$$u(x,t) = f_1(t)\sin(\sqrt{c}\cdot x) + f_2(t)\cos(\sqrt{c}\cdot x)\ ,$$

where f_1 and f_2 are arbitrary C^2 functions. The general solution in the cases $c = 0$ and $c < 0$ are, respectively,

$$u(x,t) = f_1(t)x + f_2(t) \text{ and } u(x,t) = f_1(t)e^{\sqrt{|c|}\cdot x} + f_2(t)e^{-\sqrt{|c|}\cdot x} \quad \square.$$

Example 9. Find the general solution $u = u(x,y)$ of

$$u_{yy} + u_y = x\ . \tag{18}$$

Solution. By fixing x, we can regard (18) as a linear, inhomogeneous, second–order ODE with y as the independent variable. A particular solution is $u(x,y) = xy$. The auxiliary equation for the related homogeneous equation is $r^2 + r = 0$, which has roots 0 and -1. Remembering that the arbitrary constants may depend on x, we add the general solution of the homogeneous equation to the particular solution, and thus obtain the following general solution of (18) :

$$u(x,y) = xy + f(x) + g(x)e^{-y}\ , \tag{19}$$

where $f(x)$ and $g(x)$ are arbitrary C^2 functions. One can also solve (18) by first integrating with respect to y, obtaining the first–order linear ODE (where x fixed)

$$u_y + u = xy + h(x)\ . \tag{20}$$

We multiply each side of (20) by the integrating factor e^y, obtaining

$$\frac{\partial}{\partial y}[e^y u] = xye^y + e^y h(x) \text{ or } e^y u = x(ye^y - e^y) + e^y h(x) + k(x)\ .$$

Thus, another form of the general solution of (18), is given by

$$u(x,y) = xy - x + h(x) + k(x)e^{-y}, \quad h, k \in C^2 . \tag{21}$$

Note that (19) and (21) appear to be different, but they are actually equivalent. Indeed, adding the function $-x$ to the arbitrary function $h(x)$ simply gives us another arbitrary function which may be identified with $f(x)$ in (19). Often, different methods yield general solutions which appear to be different, but which are actually equivalent in the sense that they generate the same family of solutions as the arbitrary functions vary. □

Separation of Variables

The method of **separation of variables** is used to find those solutions (if any) of a PDE which are products of functions, each of which depends on just one of the independent variables. Such solutions are called **product solutions.**

The following examples illustrate the method of separation of variables.

Example 10. Using separation of variables, find the product solutions of the heat equation with temperature–dependent sink, namely

$$u_t - u_{xx} = -u , \quad u = u(x,t) \tag{22}$$

Solution. Substituting a product solution of the form $u(x,t) = f(x)g(t)$ into (22), we get

$$f(x)g'(t) - f''(x)g(t) = -f(x)g(t) . \tag{23}$$

Then we separate the variables, so that functions in the variable x only appear on one side, and functions in the variable t only appear on the other side. If this is possible, it can usually be accomplished by first dividing by $f(x)g(t)$ and then rearranging :

$$[g'(t)/g(t)] + 1 = f''(x)/f(x) . \tag{24}$$

The only way that a function of x can equal a function of t is for both functions to be the same constant, say λ. Thus, (24) splits into two ODEs, namely

$$[g'(t)/g(t)] + 1 = \lambda \quad \text{or} \quad g'(t) + (1 - \lambda)g(t) = 0 \tag{25}$$

and

$$f''(x)/f(x) = \lambda \quad \text{or} \quad f''(x) - \lambda f(x) = 0 \tag{26}$$

The general solution of (25) is $g(t) = C\exp[(\lambda - 1)t]$. The form of the general solution of (26) depends on whether $\lambda > 0$, $\lambda < 0$ or $\lambda = 0$. If $\lambda < 0$, then $f(x) = c_1\sin(\sqrt{|\lambda|}\cdot x) + c_2\cos(\sqrt{|\lambda|}\cdot x)$, and in this case the product solution $f(x)g(t)$ is

$$u(x,t) = [c_1\sin(\sqrt{|\lambda|}\cdot x) + c_2\cos(\sqrt{|\lambda|}\cdot x)]\exp[(\lambda - 1)t]. \tag{27}$$

Note that the arbitrary constant C in $g(t)$ has been absorbed into c_1 and c_2, without loss of generality. (For the cases where $\lambda > 0$ and $\lambda = 0$, see Problem 5). □

Remark 1. Note that (27) is not a generic solution, since there are no arbitrary functions involved. Thus, solutions obtained by separation of variables are usually far from being general solutions. However, if the PDE is linear and homogeneous, then the linear combinations of product solutions (for various values of λ) will also be solutions according to the superposition principle in Section 1.2 . Often, solutions obtained in this way are sufficiently general for applications, as will be seen repeatedly in Chapter 3 onwards. □

Remark 2. A seasoned separatist, say Dr. XX, will realize in advance that undesirable square roots and absolute value signs will appear in the solution of (27). To avoid this, Dr. XX (by second nature) will write the negative separation constant λ in the form $-\lambda^2$ for some $\lambda > 0$. Then Dr. XX arrives not only at the prettier solution

$$u(x,t) = [c_1\sin(\lambda x) + c_2\cos(\lambda x)]\exp[-(\lambda^2 + 1)t]$$

which is equivalent to (27), but also dazzles fledgling students with her brilliance. We hope that this remark will spare the reader such bewilderment. □

In the case of more than two independent variables, separation of variables involves a number of stages, as we illustrate next.

Example 11. Find some nontrivial product solutions of the following wave equation for the amplitude $u(x,y,t)$ of a transversely vibrating membrane at (x,y) at time t

$$u_{tt} = u_{xx} + u_{yy} . \tag{28}$$

Solution. Let $u(x,y,t)$ be of the form $X(x)Y(y)T(t)$ for functions X, Y and T. This notation for the function is helpful in keeping track of the variables which correspond to the functions. Substituting u into (28), we get $XYT'' = X''YT + XY''T$. Separating t from x and y, we get

$$T''/T = X''/X + Y''/Y .$$

A function of t can only equal a function of x and y when these functions are constant. Thus,

$$T''/T = \lambda \quad \text{or} \quad T'' - \lambda T = 0 \tag{29}$$

and

$$X''/X + Y''/Y = \lambda \quad \text{or} \quad X''/X = \lambda - Y''/Y . \tag{30}$$

Both sides of the last equation in (30) must be a constant, say μ (Why ?). Thus we obtain

$$T'' - \lambda T = 0 \ , \ X'' - \mu X = 0 \ , \ Y'' + (\mu - \lambda)Y = 0 . \tag{31}$$

There are a number of possibilities, depending on the signs of λ, μ and $\mu - \lambda$. Since our aim is not to produce every conceivable product solution, we will make some choices that will produce a popular family of solutions. For constants a, b and c, let $\lambda = -a^2 - b^2$, $\mu = -a^2$, $\mu - \lambda = b^2$. Then selecting some particular solutions of (31), we obtain a nontrivial family of product solutions

$$\cos([a^2 + b^2]^{\frac{1}{2}}t)\sin(ax)\sin(by)\ . \tag{32}$$

Of course, in (32) one can replace the cosine by a sine and any of the sines by cosines; there are eight possibilities. By forming a linear combination of the eight possibilities, we obtain an even larger family of solutions, by the superposition principle. One can even replace all of the sines and cosines by hyperbolic sines and cosines, say in (32), and still get a valid family of product solutions. Indeed, such families would result from setting $\lambda = a^2 + b^2$, $\mu = a^2$ and $\mu - \lambda = -b^2$ (see Problem 18 of Section 1.1). □

Summary 1.3

1. **General solutions** : The general solution of a PDE is the collection of all solutions of the PDE.

2. **Generic solutions** : A generic solution of an n–th order PDE for an unknown function of m independent variables is a solution which involves n arbitrary C^n functions of m–1 variables. Examples 3 and 4 show that a general solution need not be generic, and a generic solution need not be general.

3. **ODE technique** : If a PDE involves only partial derivatives with respect to *one* of the independent variables, then such an equation may be regarded as an ODE for an unknown function of a single variable, where the other variables are held fixed. In the solution, the arbitrary constants are replaced by arbitrary functions of these remaining variables.

4. **Separation of variables** : The method of separation of variables is used to find those solutions $u(x,y)$ (if any) of the form $f(x)g(y)$. Such solutions are called product solutions. Upon substituting the form of the product solution into the PDE, one tries to get expressions involving x on one side of the equation and those involving y on the other (i.e., one tries to separate variables). If this is possible, then both sides can be set equal to a constant, and one obtains an ODE for $f(x)$ and an ODE for $g(y)$. For unknown functions of three or more variables, several stages of the separation process are carried out. Solutions obtained in this way are usually far from being general solutions of the PDE.

Exercises 1.3

1. Find the general solution of each of the following PDEs by means of direct integration.

(a) $u_x = 3x^2 + y^2$, $u = u(x,y)$ (b) $u_{xy} = x^2y$, $u = u(x,y)$

(c) $u_{xyz} = 0$, $u = u(x,y,z)$ (d) $u_{xtt} = \exp[2x + 3t]$, $u = u(x,t)$.

2. Find general solutions of the following PDEs for $u = u(x,y)$ by using ODE techniques.

(a) $u_x - 2u = 0$ (b) $yu_y + u = x$

(c) $u_x + 2xu = 4xy$ (d) $yu_{xy} + 2u_x = x$ (**Hint.** First integrate with respect to x.)

(e) $u_{yy} - x^2u = 0$.

3. For the PDEs (a) through (e) of Problem 2, find a particular solution satisfying the following respective side conditions.

(a) $u(0,y) = y$ (b) $u(x,1) = \sin(x)$

(c) $u(x,x) = 0$ (i.e., $u = 0$ on the line $y = x$) (d) $u(x,1) = 0$ and $u(0,y) = 0$

(e) $u(x,0) = 1$ and $u_y(x,0) = 0$.

4. Find a nontrivial family of solutions of the following PDEs by the method of separation of variables. You need not find the most general solution obtainable in this way.

(a) $u_t = 2u_{xx}$, $u = u(x,t)$ (b) $u_x = 4u_y$, $u = u(x,y)$

(c) $u_{tt} = 16u_{xx}$, $u = u(x,t)$ (d) $u_t = u_{xx} + u_{yy}$, $u = u(x,y,t)$

(e) $u_{xx} + u_{yy} + u_{zz} = 0$, $u = u(x,y,z)$.

5. Find the product solutions of the PDE in Example 10, in the cases where the separation constant (i.e., λ) is positive or zero. When the separation constant is positive, find an equivalent product solution (as in Remark 2) which does not involve square roots.

6. In Section 1.2 we have used trial solutions of the form e^{rx} to find particular solutions of certain ODEs. The higher dimensional analogue of this substitution (as, for example, $u(x,y) = \exp(rx + sy)$, where r and s are constants) is called the **exponential substitution**. Use the exponential substitution to find a nontrivial family of solutions of each of the following PDEs.

(a) $2u_x + 3u_y - 2u = 0$, $u = u(x,y)$ (b) $4u_{xx} - 4u_{xy} + u_{yy} = 0$, $u = u(x,y)$

(c) $u_{xyz} - u = 0$, $u = u(x,y,z)$ (d) $u_{xx} + u_{yy} = 14\exp(2x + y)$, $u = u(x,y)$

(e) $u_{xx} + u_{yy} = u$, $u = u(x,y)$.

7. Consider the problem $u_{xx} + u_{xy} + u_{yy} = 0$, $u = u(x,y)$, and attempt to use the method of separation of variables to arrive at $f''(x)g(y) + f'(x)g'(y) + f(x)g''(y) = 0$.

(a) If $f(x)g(y) \neq 0$, verify that $-f''(x)/f(x) = [g'(y)/g(y)][f'(x)/f(x)] + g''(y)/g(y)$.

(b) Deduce from (a) that if $f'(x)/f(x)$ is not constant, then $g'(y)/g(y)$ is a constant, say s.

(c) Deduce from (b) that $g(y) = ce^{sy}$, and $g''(y)/g(y) = s^2$.

(d) Show that $f''(x) + sf'(x) + s^2f(x) = 0$. Solving this ODE for f(x), obtain the solution

$$u(x,y) = [c_1\cos(\tfrac{1}{2}\sqrt{3}\cdot sx) + c_2\sin(\tfrac{1}{2}\sqrt{3}\cdot sx)]\exp[s(y - \tfrac{1}{2}x)] .$$

8. For each of the following PDEs, find some constants a and b (not both zero), such that $u(x,y) = f(ax + by)$ is a generic solution, where f is an arbitrary C^1 function.

(a) $u_x + 2u_y = 0$ (b) $5u_x + 6u_y = 0$ (c) $cu_x + du_y = 0$, for any constants c and d.

9. Use the technique of Problem 8 to solve the following PDEs, subject to the given side conditions. Explain why one cannot obtain the solutions by using separation of variables.

(a) $u_x + 2u_y = 0$, $u(x,0) = x$ (b) $u_x + 3u_y = 0$, $u(x,2x + 1) = x^2$

(c) $3u_x - 4u_y = 0$, $u(x,x) = x^2 - x$ (d) $u_x + 2u_y = 2x + 4y$, $u(0,y) = y^2 + 1$.

Hint. For (d), first find a particular solution $u_p(x,y)$ of the form $ax^2 + by^2$.

10. For given real constants A, B and C, consider the second–order PDE $Au_{xx} + Bu_{xy} + Cu_{yy} = 0$. Show that if $B^2 - 4AC > 0$ (i.e., the PDE is hyperbolic), then this PDE has a generic solution of the form $u(x,y) = f(ax + by) + g(cx + dy)$, where a, b, c and d are real constants, and where f and g are C^2 functions.

Hint. Assume $u(x,y) = h(rx + sy)$, obtain $Ar^2 + Brs + Cs^2 = 0$, fix r and solve for s.

11. In relation to Example 4, where the PDE $u_x = u^2$ was considered, define

$$u(x,y) = \begin{cases} -[x + g(y)]^{-1} & \text{for } y > 0,\ x \neq -g(y) \\ 0 & \text{for } y \leq 0 \end{cases}$$

(a) Show that if $g(y) = y^{-2}$, then $u(x,y)$, $u_x(x,y)$ and $u_y(x,y)$ are continuous at points of the x–axis. Deduce that u is C^1 (and a solution of $u_x = u^2$), except at points on the curve $x = -y^{-2}$, $y > 0$.

(b) Let $g(y) = y^{-1}$. Show that $u_y(x,y)$ is not continuous at points on the x–axis, because in this case $u_y(x,y)$ jumps as y passes through 0. Why does this imply that $u(x,y)$ is not a solution of the PDE $u_x = u^2$ in the region consisting of the whole plane except for points on the curve $x = -y^{-1}$, $y > 0$?

12. The following parts concern the solution $u(x,t) = f(x + ct) + g(x - ct)$ of the wave equation $u_{tt} = c^2u_{xx}$, where f and g are C^2 functions.

(a) Let $u(x,t) = f(x + ct)$. Suppose that for each fixed time t we graph u as a function of x. Show that as t advances, the graph moves to the left with velocity c. What about $u(x,t) = g(x - ct)$?

(b) Show that if $f(x + ct) = g(x - ct)$ for all x and t, then f and g must be constant.

(c) Deduce from (b) that neither $f(x + ct)$ nor $g(x - ct)$ can be eliminated from the solution $u(x,t) = f(x + ct) + g(x - ct)$ without losing solutions in the process.

(d) Check directly that (16) solves the PDE $u_{tt} = c^2 u_{xx}$ with the I.C. given by (14).

CHAPTER 2

FIRST – ORDER PDEs

In most PDE textbooks, first–order PDEs usually receive only a brief treatment. One reason for this is that the PDEs which have the most obvious applications are the standard second–order PDEs, namely the heat, wave and Laplace equations. Moreover, the theory of first–order PDEs locally reduces to the study of systems of first–order ODEs, which is presumably a subject of another course. Here we will find that first–order PDEs have a variety of applications. Also, there are certain global topological considerations which arise in the study of first–order PDEs which make the theory more than just a study of systems of ODEs.

In Section 2.1, we solve first–order, linear PDEs with constant coefficients by introducing a linear change of variables, which converts the PDE into a family of ODEs depending on a parameter. We apply this theory to population and inventory analysis. In Section 2.2, we handle the case of first–order, linear PDEs with nonconstant coefficients. This is done by making a nonlinear change of variables, so that when all but one of the new variables is held fixed, one obtains a characteristic curve along which the PDE becomes an ODE in the remaining new variable. By piecing together the solutions of the ODEs on these curves, we indicate how certain global considerations may arise. Applications to gas flow and differential geometry are provided. In Section 2.3, we show how this method of characteristics extends to first–order linear PDEs in three dimensions, which we use to solve related first–order quasi–linear PDEs in two dimensions. Among many possible applications, we show how quasi–linear PDEs arise in the study of traffic flow and nonlinear continuum mechanics, particularly with regard to the phenomenon of shock waves. In the optional Section 2.4, the more involved theory of arbitrary nonlinear first–order PDEs is introduced, and there is an application to the study of the motion of wave fronts in an inhomogeneous medium with a variable wave propagation speed. Moreover, in this application, we see the wave/particle duality in the Hamilton–Jacobi theory which foreshadows the analogous duality which lies at the foundations of quantum mechanics.

2.1 First–Order Linear PDEs (Constant Coefficients)

Perhaps the simplest nontrivial type of PDE is the first–order linear PDE

$$au_x + bu_y + cu = f(x,y)\ , \qquad u = u(x,y)\ ,\ a^2 + b^2 > 0\ , \tag{1}$$

where a, b and c are given constants and f(x,y) is a given continuous function. Our first main goal will be to find the general solution of (1). In the easy case, when $b = 0$, (1) is

$$au_x(x,y) + cu(x,y) = f(x,y)\ , \tag{2}$$

which (for each fixed y) is a first–order, linear ODE for u(x,y) regarded as a function of x. Following the procedure in the Summary of Section 1.1, we can solve (2), by first dividing by a ($a \neq 0$) and multiplying by the integrating factor $e^{cx/a}$. Thus,

$$e^{cx/a}\frac{\partial u}{\partial x}(x,y) + e^{cx/a}\frac{c}{a}\,u(x,y) \;=\; \frac{1}{a}\,f(x,y)e^{cx/a}$$

or

$$\frac{\partial}{\partial x}\Big[e^{cx/a}u(x,y)\Big] \;=\; \frac{1}{a}\,f(x,y)e^{cx/a}\ .$$

Integrating both sides with respect to x and multiplying by $e^{-cx/a}$, we obtain the general solution of (2), namely

$$u(x,y) = e^{-cx/a}\Big[\frac{1}{a}\int f(x,y)e^{cx/a}\,dx + C(y)\Big]\ , \tag{3}$$

where C(y) is an arbitrary C^1 function of y. The success of this method depends heavily on the fact that u_y is not present in (2). This is what enabled us to treat (2) as an ODE.

To handle the more general case when $b \neq 0$, we begin with the observation that $au_x + bu_y$ is the dot product of the vector **ai** + b**j** with the gradient $\nabla u = u_x\mathbf{i} + u_y\mathbf{j}$, and hence $au_x + bu_y$ is essentially the derivative of u in the direction of the vector **ai** + b**j**. If we introduce a new coordinate system for the xy–plane, so that one of the new axes is pointing in the direction **ai** + b**j**, then $au_x + bu_y$ will be proportional to the partial derivative of u with respect to the new variable labeling that axis, and we will have reduced (1) to the form of (2) in terms of new coordinates. To find an appropriate change of variables, first note that the family of lines of the form $bx - ay = d$ (where d is an arbitrary constant) all have slope b/a, and hence these lines are parallel to the direction $a\mathbf{i} + b\mathbf{j}$. We want to choose new coordinates, say (w,z), such that this family of lines becomes the family of new coordinate lines, say $w = d$. A simple change of variables (or coordinates) which has this effect is given by

$$w = bx - ay, \quad z = y. \tag{4}$$

The family of new coordinate lines $w = d$ then coincides with the family of lines $bx - ay = d$. The lines $z =$ const. are the same as the lines $y =$ const. which are parallel to the x-axis. We assume here that $b \neq 0$, so that the transformation (4) is invertible :

$$x = \frac{w+az}{b}, \quad y = z.$$

We define a new function v by

$$v(w,z) \equiv u(x,y) = u(\frac{w+az}{b}, z).$$

Note that $v(w,z)$ is just $u(x,y)$ expressed in terms of the new variables (w,z). By the above remarks, we expect that $au_x + bu_y$ will be proportional to v_z, since $au_x + bu_y$ is the derivative of u along the lines $w =$ const.. Indeed,

$$au_x + bu_y = a(v_w w_x + v_z z_x) + b(v_w w_y + v_z z_y) = (ab-ba)v_w + bv_z = bv_z.$$

Thus, equation (1) can be rewritten in terms of the variables (w,z) as

$$bv_z + cv = f(\frac{w+az}{b}, z). \tag{5}$$

This equation is of the simple form (2), namely an ODE depending on the parameter w. We know how to solve (5) for $v(w,z)$, and the solution of problem (1) will then be given by $u(x,y) = v(bx-ay,y)$, using $u(x,y) = v(w,z)$ and (4). We have converted problem (1) to the simpler form (5) by making a change of variables so that, when one of the new variables is held constant, we get a member of the family of lines $bx - ay = d$.

> The lines $bx - ay = d$, which are parallel to $ai + bj$ (i.e., have slope b/a), are called the **characteristic lines** of the PDE (1), $au_x + bu_y + cu = f(x,y)$.

Thus, a first–order linear PDE with constant coefficients becomes much simpler when expressed in terms of a new coordinate system with the set of characteristic lines as a set of coordinate lines.

Example 1. Find the general solution of the PDE

$$3u_x - 2u_y + u = x \; , \quad u = u(x,y) \; . \tag{6}$$

Solution. The characteristic lines have slope $-2/3$. They constitute the family of lines $2x + 3y = d$. Hence, we make the change of variables

$$\begin{cases} w = 2x + 3y \\ z = y \end{cases} ; \qquad \begin{cases} x = (w - 3z)/2 \\ y = z \; . \end{cases} \tag{7}$$

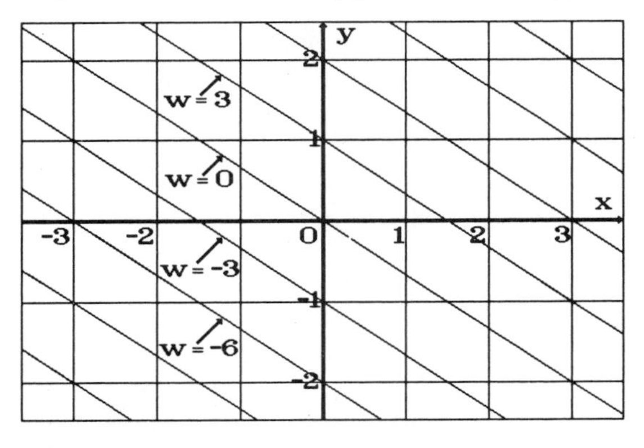

Figure 1

Setting $v(w,z) = u(x,y)$, we have $3u_x - 2u_y = 3(v_w w_x + v_z z_x) - 2(v_w w_y + v_z z_y) = 3(v_w \cdot 2) - 2(v_z + v_w \cdot 3) = -2v_z$. Thus, the PDE (6) becomes

$$-2v_z + v = (1/2)(w-3z) \; . \tag{8}$$

Dividing by -2 and multiplying by the integrating factor $e^{-z/2}$, we obtain

$$\frac{\partial}{\partial z}\left[e^{-z/2}v\right] = -\frac{1}{4}e^{-z/2}(w-3z) \; .$$

Integrating with respect to z, regarding w as fixed, we obtain

$$
\begin{aligned}
e^{-z/2}v(w,z) &= -\tfrac{1}{4}\,w\int e^{-z/2}\,dz + \tfrac{3}{4}\int ze^{-z/2}\,dz + C(w)\\
&= \tfrac{1}{2}\,we^{-z/2} + \tfrac{3}{4}\Big[ze^{-z/2}\cdot(-2) - \int e^{-z/2}(-2)\,dz\Big] + C(w)\\
&= e^{-z/2}[w/2 - 3z/2 - 3] + C(w)\,,
\end{aligned}
$$

where $C(w)$ is an arbitrary function of w, which we will assume is C^1 (i.e., $C'(w)$ is continuous). Thus, we have $v(w,z) = \tfrac{1}{2}[w - 3z - 6] + e^{z/2}C(w)$, and using (7) we get the general solution

$$
\begin{aligned}
u(x,y) &= \tfrac{1}{2}[2x + 3y - 3y - 6] + e^{y/2}C(2x+3y)\\
&= x - 3 + e^{y/2}C(2x+3y)\,,
\end{aligned}
\tag{9}
$$

where $C(2x+3y)$ is an arbitrary C^1 function of $(2x+3y)$, such as $(2x+3y)^2$, $\exp(2x+3y)$, $|\sin(2x+3y)|^{3/2}$, etc. . The C^1 assumption on C is needed so that $u(x,y)$ will be C^1 . □

Remark. In (9), if we choose C to be the zero function, then we obtain the particular solution $u_p(x,y) = x-3$. For any choice of the C^1 function C , we have a solution

$$u_h(x,y) \equiv e^{y/2}C(2x + 3y)$$

of the related homogeneous PDE $3u_x - 2u_y + u = 0$. Indeed,

$$
\begin{aligned}
&3\,\frac{\partial}{\partial x}\,[e^{y/2}C(2x+3y)] - 2\,\frac{\partial}{\partial y}\,[e^{y/2}C(2x+3y)] + e^{y/2}C(2x+3y)\\
&= 3e^{y/2}C'(2x+3y)\cdot 2 - 2e^{y/2}\cdot\tfrac{1}{2}\,C(2x+3y) - 2e^{y/2}C'(2x+3y)\cdot 3 + e^{y/2}C(2x+3y) \;=\; 0\,.
\end{aligned}
$$

The general solution (9) is the sum of the particular solution u_p and the general solution u_h of the related homogeneous equation. We can obtain other particular solutions by choosing specific functions for C. For example, setting $C(2x+3y) = 2x+3y$, we obtain the particular solution $x - 3 + e^{y/2}(2x+3y)$. The general solution of (6) can then be written as

$$u(x,y) \;=\; x - 3 + e^{y/2}(2x+3y) + e^{y/2}D(2x+3y)\,, \tag{10}$$

where $D(2x+3y)$ is an arbitrary C^1 function of $(2x+3y)$. The solutions (9) and (10) are both correct, and they are actually equivalent, in the sense that, as C and D range over all C^1 functions, both (9) and (10) generate the collection of all solutions. Thus, it can happen that two expressions for general solutions may look different and yet both are correct. *Keep this in mind when comparing your answers with the given answers for the exercises.* Note also that although we are essentially forced to take one of the new variables to be a function of $2x+3y$, the expression for the other new variable could be any linear combination of x and y which is not a multiple of $2x+3y$. For example, in place of the transformation (7), we could have used

$$\begin{cases} w = 2x + 3y \\ z = x \end{cases} \quad ; \quad \begin{cases} x = z \\ y = (w - 2z)/3 \ . \end{cases} \tag{11}$$

With this change of variables we obtain [in place of (8)] the equation $3v_z + v = z$, whose solution is $v(w,z) = z - 3 + e^{-z/3}G(w)$, where G is an arbitrary C^1 function. Thus,

$$u(x,y) \ = \ x - 3 + e^{-x/3}\, G(2x+3y) \ = \ x - 3 + e^{y/2}e^{-(2x+3y)/6}\, G(2x+3y) \ ,$$

which is equivalent to (9). □

In many cases, especially in applications, one is interested in finding a particular solution that satisfies a certain side condition. For equation (1), an appropriate side condition might be the requirement that $u(x,y)$ have specified values at points (x,y) that lie on a certain line. Such a condition has the form

$$u(x,mx + d) = g(x) \ , \tag{12}$$

where $g(x)$ is some given C^1 function, m is the slope of the line, and d is the y–intercept. In the case where the line is vertical (with infinite slope), condition (12) must be replaced by the condition $u(d,y) = g(y)$, where d is the x–intercept of the vertical line. In the following examples, we will see that such conditions usually suffice to completely determine the arbitrary function which is always present in the general solution. There is one important exception, however. If the line, on which the side condition is given, happens to be a characteristic line for the PDE, then the side condition does not uniquely determine a solution. Indeed, in this case, we will find that the function $g(x)$ in (12) must have a particular form, in order for a solution to exist. If $g(x)$ has this particular form, then we will find that there are infinitely many solutions of the PDE (1) with side condition (12).

Example 2. Solve the following PDE with the given side condition :

$$u_x - u_y + 2u = 1 \ , \qquad u(x,0) = x^2 \ . \tag{13}$$

Solution. Here the side condition specifies the values of u at points on the x–axis. First, we find the general solution of the PDE, and then we try to meet the side condition. The characteristic lines have slope −1 and are of the form $x + y = d$. Thus, we make the change of variables

$$\begin{cases} w = x + y \\ z = y \end{cases} \quad ; \quad \begin{cases} x = w - z \\ y = z \ . \end{cases}$$

and define $v(w,z) = u(x,y)$. The PDE in (13) becomes $-v_z + 2v = 1$, and we obtain $v(w,z) = \frac{1}{2} + e^{2z}C(w)$, where C is an arbitrary C^1 function. Thus we obtain the following general solution of the PDE in (13)

$$u(x,y) = \tfrac{1}{2} + e^{2y}C(x+y)\ . \tag{14}$$

Now, we must choose the function C, so that the side condition $u(x,0) = x^2$ is met. From (14), we have $u(x,0) = \frac{1}{2} + C(x)$. Thus, $\frac{1}{2} + C(x) = x^2$ or $C(x) = x^2 - \frac{1}{2}$. This completely determines the function C, in the sense that we know what C does to any real number [e.g., $C(3) = 3^2 - \frac{1}{2}$]. In particular, we know that for any values of x and y, $C(x+y) = (x+y)^2 - \frac{1}{2}$. Thus, the unique solution of the PDE, with the side condition, is

$$u(x,y) = \tfrac{1}{2} + e^{2y}[(x+y)^2 - 1/2]\ .$$

Note that $u(x,0) = x^2$, and one can check directly that $u(x,y)$ satisfies the PDE. It is a good idea to check your solutions directly, as this is the ultimate test of their validity. □

Remark. Some students run into difficulties with the function C. For example, do not make the mistake of thinking that just because $C(x) = x^2 - \frac{1}{2}$, we would also have $C(x+y) = x^2 - \frac{1}{2}$. Also, almost always, it is *not* true that $C(x+y) = C(x) + C(y)$. For example, $\log(x+y) \neq \log(x) + \log(y)$, $\sin(x+y) \neq \sin(x) + \sin(y)$, $(x+y)^2 \neq x^2 + y^2$, etc. . Some students find it objectionable to take the result $C(x) = x^2 - \frac{1}{2}$ and simply replace x by $x+y$ to get the correct result $C(x+y) = (x+y)^2 - \frac{1}{2}$. The objection is that $x+y$ is not the same as x, unless y happens to be zero (i.e., what gives us the right to replace x by something which is unequal to x ?). The objection may be circumvented, as follows. The formula $C(x) = x^2 - \frac{1}{2}$ tells us that C is the function that assigns to each number its square minus $\frac{1}{2}$. The formula $C(r) = r^2 - \frac{1}{2}$ describes the same function. In other words, the variable that is used to describe a function can be changed without changing the function. Thus, in place of $C(x) = x^2 - \frac{1}{2}$, use $C(r) = r^2 - \frac{1}{2}$, and then set $r = x+y$. Hence, the objection that $x+y \neq x$ is overcome. □

Example 3. Solve the PDE $u_x + 2u_y - 4u = e^{x+y}$ subject to the condition $u(x,4x + 2) = 0$.

Solution. Here the side condition requires that u vanish on the line $y = 4x+2$. The characteristic lines are of the form $2x-y = \text{const.}$, and we make the change of variables

$$\begin{cases} w = 2x - y \\ z = x + y \end{cases} \quad ; \quad \begin{cases} x = (w + z)/3 \\ y = (2z - w)/3\ . \end{cases}$$

Recall that the choice of z is essentially arbitrary. Our choice is motivated by the fact that e^{x+y} in the PDE will become simply e^z if we set $z = x+y$. For $v(w,z) = u(x,y)$, we have $u_x+2u_y = (v_w w_x+v_z z_x) + 2(v_w w_y+v_z z_y) = (2-2)v_w + (1+2)v_z = 3v_z$. Thus, the PDE becomes $3v_z - 4v = e^z$, from which we obtain

$$v(w,z) = -e^z + e^{4z/3}C(w) \quad \text{or} \quad u(x,y) = -e^{x+y} + e^{4(x+y)/3}C(2x-y)\,. \tag{15}$$

Note that the exponent is $4(x+y)/3 = -4(2x-y)/3 + 4x$. Thus, we could rewrite (15) as

$$u(x,y) = -e^{x+y} + e^{4x}[e^{-4(2x-y)/3}C(2x-y)] \quad \text{or} \quad u(x,y) = -e^{x+y} + e^{4x}D(2x-y)\,, \tag{16}$$

for an arbitrary C^1 function D. In order to meet the side condition, we need

$$0 = u(x,4x+2) = -e^{5x+2} + e^{4x}D(-2x-2)\,.$$

In other words, the function D must be chosen so that

$$D(-2x-2) = e^{5x+2}/e^{4x} = e^{x+2}.$$

To find the function D, we employ the following device. Set $r = -2x-2$, and note that $x = -(r+2)/2$. Then

$$D(r) = D(-2x-2) = e^{x+2} = e^{-(r+2)/2\,+2} = e^{(-r+2)/2}\,.$$

Thus, D is the function given by the formula $D(r) = e^{(-r+2)/2}$, and the solution of the PDE with side conditions is

$$u(x,y) = -e^{x+y} + e^{4x}e^{-(2x-y)/2\,+1} = -e^{x+y} + e^{3x+y/2\,+1}\,. \tag{17}$$

If we had used the form (15) for the general solution, then the side condition $u(x,4x+2) = 0$ would enable us to find the arbitrary function C. Although C would be different from D, the reader can check that the final result (17) would be the same. Thus, it is certainly not necessary to make any clever transformation of the exponent, although (16) looks tidier than (15). □

In Example 3, the side condition was given on the line $y = 4x+2$, which is not one of the characteristic lines of slope 2. Using the same PDE as in Example 3, we next illustrate (cf. Examples 4 and 5 below) what happens when the side condition is given on a characteristic line.

Example 4. Attempt to solve the PDE $u_x + 2u_y - 4u = e^{x+y}$ with side condition $u(x,2x-1) = 0$.

Solution. The general solution $u(x,y) = -e^{x+y} + e^{4x}D(2x-y)$ of the PDE was found in Example 3 (cf. (16)). The side condition is given on the characteristic line $y = 2x-1$, and it implies that

$$0 = u(x,2x-1) = -e^{3x-1} + e^{4x}D(1) \quad \text{or} \quad D(1) = e^{-x-1}\,. \tag{18}$$

Regardless of the choice of D, we have that D(1) is a constant, whereas e^{-x-1} is a nonconstant function of x. Hence, the side condition $u(x,2x-1) = 0$ can never be met, and the problem has no solution. We can see that the trouble here is that the side condition is given on a characteristic line, and $D(2x-y)$ is always constant on such a line. This makes it impossible to determine the function D, and typically (but not always; cf. Example 5) we arrive at a contradiction such as in (18), and the problem has no solution. When the side condition is on a line which is not a characteristic line, the argument (e.g., $2x-y$) of the arbitrary function is not constant, and the function can be found as in Example 3. □

Example 5. Solve the PDE $u_x+2u_y-4u = e^{x+y}$, subject to the condition $u(x,2x) = -e^{3x}+ e^{4x}$.

Solution. As in Example 3, the general solution of the PDE is $u(x,y) = -e^{x+y} + e^{4x}D(2x-y)$. The side condition (given on the characteristic line $y = 2x$) then tells us that

$$-e^{3x}+ e^{4x} = u(x,2x) = -e^{3x}+ e^{4x}D(0) .$$

In this case, the condition can be met, as long as the C^1 function D is chosen so that $D(0) = 1$. There are infinitely many C^1 functions D such that $D(0) = 1$, e.g.,

$$D(r) = r+1 \ , \quad D(r) = \cos(r) \ , \quad D(r) = e^r \ . \tag{19}$$

Corresponding to the choices (19), we have the respective solutions

$$u(x,y) = -e^{x+y} + e^{4x}(2x-y+1), \ u(x,y) = -e^{x+y} + e^{4x}\cos(2x-y), \ u(x,y) = -e^{x+y} + e^{4x}e^{2x-y}.$$

All of these functions satisfy both the PDE and the side condition. The fact that we did not get a contradiction as in Example 5 is due to the special choice of the function $e^{-3x} + e^{4x}$ in the side condition. Indeed, we would have to choose the function to be of the particular form $-e^{3x} + ke^{4x}$ for some constant k, in order that there be no contradiction. The C^1 function D would then be arbitrary, except for the requirement $D(0) = k$. Thus, we see that it is possible for a PDE, with a side condition given on a characteristic line, to have solutions, if the side condition has a particular form. In this case, there will be infinitely many solutions. However, if the function in the side condition does not have the correct particular form, then the problem has no solution, as in Example 4. □

There is a simple geometrical reason for the peculiarities that arise, when the side condition is given on a characteristic line. For if a coordinate system is chosen so that the coordinate lines (say $w = d$) are the characteristic lines, then the PDE becomes an ODE in the remaining variable z which acts as a position variable on each characteristic line. Since v (regarded as a function of z) satisfies a certain ODE on each characteristic line $w = d$, we know that v (or u) must have a particular form on such lines. Since solutions of first–order ODEs are typically determined by prescribing the value of the solution at a single point, the solution cannot be arbitrarily prescribed on the entire line, but only at a single point. The following example makes this clear in a special case when the change of variable is not necessary.

Example 6. Determine the form of the functions $g(x)$ for which the PDE $u_x - u = 0$, with side condition $u(x,1) = g(x)$, has a solution.

Solution. Here the characteristic lines are the horizontal lines $y = d$. The variable x serves as a position variable on each of these lines. Also, on the line $y = d$, the PDE becomes the ODE

$$\frac{\partial}{\partial x}[u(x,d)] - u(x,d) = 0 , \tag{20}$$

which has the solution $u(x,d) = C(d)e^x$, where $C(d)$ is an arbitrary constant that can vary with d. Thus, on each characteristic line, u must be a constant times e^x. In particular, when $d = 1$, $u(x,1) = C(1)e^x$, and so $g(x)$ must be of this form for a solution to exist. Along any characteristic line, u must have a particular form, because u is a solution of a particular ODE (20) on each of these lines. By piecing together the solutions $u(x,d) = C(d)e^x$ on each of the lines, we arrive at the general solution $u(x,y) = C(y)e^x$, where $C(y)$ is an arbitrary C^1 function of y. Note that if u is prescribed on a noncharacteristic line which intersects each line $y = d$ exactly once, then the constant $C(d)$ will be determined by the prescribed value of u at the intersection point. For this reason, a suitable side condition given on a noncharacteristic line will determine a unique solution, by piecing together the unique solutions on the characteristic lines. □

A side condition need not be given on a straight line.

Definition. If u is required to have specified values on some curve (e.g., a circle, a parabola, a line,etc.), then we call such a curve a **side condition curve**.

In order to ensure the existence of a unique solution of the PDE $au_x+bu_y+cu = f(x,y)$, which meets the side condition, various assumptions about the side condition curve are needed, and we will define a few terms. A **regular curve** is a curve with a unit tangent vector which turns continuously (if at all) with respect to arclength. A regular curve intersects a line **transversely**, if at each intersection point, the angle of intersection is nonzero. The following fact can be proved (cf. Problem 15).

Theorem 1 (Existence and Uniqueness). For the PDE $au_x + bu_y + cu = f(x,y)$, suppose that we are given a regular side condition curve which intersects each characteristic line of the PDE exactly once, and transversely. Assume also that the values of u are specified in a C^1 fashion along the side condition curve (i.e., the values define a C^1 function of arclength along the curve). Then there is a unique solution of the PDE which meets the given side condition.

Remark. Uniqueness follows easily from the fact that the values of a solution on a characteristic line is determined by its given value at the point of intersection with the side condition curve. The regularity of the side condition curve and the transversality of the intersections enter into the proof that the solutions on the individual characteristic lines can be pieced together to yield a C^1 solution of the PDE. From the previous examples, the interested reader should have little difficulty in proving Theorem 1 in the case where the side condition curve is a straight line. Note that the transversality condition guarantees that the side condition curve is not a characteristic line (Why ?). Examples 4 and 5 show that the transversality condition is necessary (cf. also Problem 9). In the next example, the side condition curve is $y = x^3$, instead of a line.

Example 7. Solve the PDE $u_x - u_y + u = 0$, subject to the condition $u(x,x^3) = e^{-x}(x + x^3)$.

Solution. Here u is specified on the curve $y = x^3$. This curve intersects each characteristic line $x+y = d$ exactly once, and transversely (since the slope of the curve x^3 is $3x^2$ which is never the same as the slope (-1) of the characteristic curves). With $v(w,z) \equiv u(x,y)$, where

$$\begin{cases} w = x + y \\ z = y \end{cases} \quad ; \quad \begin{cases} x = w - z \\ y = z\,. \end{cases}$$

the PDE becomes $-v_z + v = 0$. The general solution is $v(w,z) = C(w)e^z$, whence $u(x,y) = C(x+y)e^y = D(x+y)e^{-x}$. In order to meet the side condition, we need to choose D, so that $e^{-x}(x+x^3) = u(x,x^3) = D(x+x^3)e^{-x}$. The choice that works is the function $D(r) = r$, and the solution of the problem is then $u(x,y) = (x+y)e^{-x}$. □

An Application to Population or Inventory Analysis

Under certain natural assumptions, here we derive and solve a first–order PDE which governs the way in which the composition, with respect to age, of a population of individuals, changes with time. The individuals need not be biological organisms, but they could be manufactured items (e.g., light bulbs, transistors, food products) or more generally any collection of similar objects which become defective with age according to a statistical pattern. Thus, perhaps this first–order PDE has a greater variety of applications than the heat, wave and Laplace equations.

Suppose that at time t a certain population has approximately $P(y,t)\cdot\Delta y$ individuals between the ages of y and $y + \Delta y$. In other words, at a fixed time t, $P(y,t)$ is **the population density** with respect to the age variable y. At time t, the number of individuals between the ages of a and b is then $\int_a^b P(y,t)\,dy$. We suppose that the number of individuals of age between y and $y + \Delta y$, which die in the time interval from t to $t + \Delta t$ is approximately $D(y,t)\cdot P(y,t)\cdot\Delta y\cdot\Delta t$, for some function $D(y,t)$ which has been statistically determined, say by observation. One usually expects that the "death rate density" $D(y,t)$ increases as y increases (i.e., older individuals may be more likely to expire), and $D(y,t)$ could very well depend on t because of seasonal variations (e.g., air conditioners are more likely to die in the summer). If individuals never expire (i.e., $D(y,t) = 0$), then $P(y,t+\Delta t) = P(y-\Delta t,t)$ for any y and time interval Δt with $0 \le \Delta t \le y$, since the population density at time $t+\Delta t$ is just a translate, by the age difference Δt, of what it was at time t. However, if $D \neq 0$, then we must take into account that a number of individuals will die, as time advances from t to $t+\Delta t$. Indeed,

$$P(y,t+\Delta t) = P(y-\Delta t,t) - \int_0^{\Delta t} D(y-\Delta t+s,t+s)P(y-\Delta t+s,t+s)\,ds. \tag{21}$$

Differentiating both sides of (21) with respect to Δt, and then setting $\Delta t = 0$, we obtain

$$P_t(y,t) = -P_y(y,t) - D(y,t)P(y,t) \quad \text{or} \quad P_y + P_t + D(y,t)P = 0\,. \tag{22}$$

The coefficients for P_y and P_t are constant (both are 1), but the coefficient $D(y,t)$ of P is not necessarily constant. Nevertheless, all of the theory of this section still applies to PDEs of the form $au_x + bu_y + c(x,y)u = f(x,y)$ (i.e., only the constancy of the coefficients of u_x and u_y is needed to reduce this PDE to an ODE, by a linear change of variables). For equation (22), the family of characteristic lines is $y - t = d$. Hence, we make the change of variables

$$\begin{cases} w = y - t \\ z = y \end{cases} ; \qquad \begin{cases} t = z - w \\ y = z\,. \end{cases} \tag{23}$$

With $Q(w,z) \equiv P(y,t)$, we have $P_y + P_t = Q_w w_y + Q_z z_y + Q_w w_t + Q_z z_t = Q_z$, and (22) becomes $Q_z + D(z,z-w)Q = 0$. The integrating factor is $\exp[\int D(z,z-w)\,dz]$, and we obtain

$$Q(w,z) = C(w)\exp\left[-\int_w^z D(\zeta,\zeta-w)\,d\zeta\right], \tag{24}$$

where $C(w)$ is an arbitrary C^1 function, and the lower limit w in the integral is introduced for future convenience, but it can be replaced by any C^1 function of w (Why ?). Hence,

$$P(y,t) = C(y-t)\exp\left[-\int_{y-t}^{y} D(\zeta,\zeta-y+t)\,d\zeta\right].$$

If we set $t = 0$, then we obtain $P(y,0) = C(y)$. Thus, $C(y)$ is just the initial population density This is why we chose w for the lower limit in (24). We have (with $\tau \equiv \zeta-y+t$)

$$P(y,t) = P(y-t,0)\exp\left[-\int_{y-t}^{y} D(\zeta,\zeta-y+t)\,d\zeta\right] = P(y-t,0)\exp\left[-\int_0^t D(y-t+\tau,\tau)\,d\tau\right]. \tag{25}$$

Note that since $P(y,0)$ has not yet been defined for $y < 0$ (i.e. for negative ages), the solution (25) is undefined for $t > y$. For $y < 0$, it is convenient to define $P(y,0)$, so that $P(y,0)\cdot\Delta y$ is approximately the number of individuals to be produced between $|y|$ and $|y| + \Delta y$ time units into the future. In other words, for $y < 0$, $P(y,0)$ is the production rate at $-y$ time units into the future. Naturally, we take $D(y,t) = 0$ for $y < 0$. Then (25) defines $P(y,t)$ for all (y,t). In the case of a constant rate of production (say C) and when $D(y,t) = D(y)$ is time–independent (for $y > 0$), we can determine (using the middle expression in (25)) the

steady–state population density $P_\infty(y)$ which is obtained as $t \to \infty$:

$$P_\infty(y) \equiv \lim_{t\to\infty} P(y,t) = C \exp\left[-\int_0^y D(\zeta)\, d\zeta \right] . \tag{26}$$

Example 8. The number of avocados that a merchant acquires per day is a constant C. At any time, the probability that an avocado acquired y days ago is removed from the shelf (say due to spoilage or sale), during a small time interval of Δt days, is $(y/25)\Delta t$.

(a) What must C equal, so that there will be about 300 avocados on the shelf, in the long run?
(b) Assume that when the merchant took over the business at $t = 0$, the initial population density of avocados was $P(y,0) = 300e^{-y}$. Find $P(y,t)$, assuming that C is as in part (a)?

Solution. Here $D(y,t) \equiv y/25$, for $0 \le y$ (and $D(y) = 0$, for $y < 0$). For part (a), formula (26) implies that, in the long run, the population density is

$$P_\infty(y) = C \exp\left[-\int_0^y y/25\, dy \right] = C \exp\left(-\tfrac{1}{2} y^2/25\right) .$$

The number of avocados on hand is then $\int_0^\infty P_\infty(y)\, dy = C \int_0^\infty \exp(-\tfrac{1}{2}y^2/25)\, dy$ $= C\cdot 5\int_0^\infty \exp(-\tfrac{1}{2}x^2)\, dx = C\cdot 5\cdot\sqrt{\pi/2} \approx 6.27\cdot C$, using the fact that $\int_0^\infty e^{-\frac{1}{2}x^2}\, dx = \sqrt{\pi/2}$. Thus, $C \approx 300/6.27 \approx 48$. For part (b), we use (25) with $P(y,0) = C$ for $y < 0$, and $P(y,0) = 300e^{-y}$, for $y \ge 0$. Hence, for $y > t$, (25) yields

$$P(y,t) = 300e^{-(y-t)}\exp\left[-\int_{y-t}^y \zeta/25\, d\zeta \right] = 300 \exp\left[-(y-t) - t(2y-t)/50 \right] , \quad y > t.$$

Recall that $D(y) = 0$ when $y < 0$. Thus, when $y < t$, the interval of integration from $y-t$ to y in (25) can be replaced by the interval from 0 to y. Hence, (25) yields

$$P(y,t) = C \exp\left[-\int_0^y y/25\, d\tau \right] = C \exp\left(-\tfrac{1}{2}y^2/25\right) = P_\infty(y), \quad \text{for } 0 < y < t.$$

Thus, the steady–state density applies, as long as $0 < y < t$. □

Summary 2.1

1. Characteristic lines : The lines $bx - ay = \text{const.}$ which are parallel to $a\vec{i} + b\vec{j}$ (i.e., have slope b/a) are called the characteristic lines of the first–order PDE

$$au_x + bu_y + cu = f(x,y) , \qquad \text{(S1)}$$

with constant coefficients a, b and c, with $a^2+b^2 \neq 0$. The equation (S1) may be solved as a first–order linear ODE, if $a = 0$ or $b = 0$ (i.e., if the characteristic lines are $x = d$ or $y = d$). The geometrical significance of the characteristic lines is that $au_x + bu_y$ is essentially the directional derivative of u along these lines. Thus, on each characteristic line, (S1) is really an ODE for a function of a position variable along the line.

2. Change of variables : The PDE (S1) is converted to an ODE (with parameter w), when it is expressed in terms of new variables (z,w) for which the characteristic lines are the new coordinate lines $w = d$. Specifically, if $b \neq 0$, consider the transformation

$$\begin{cases} w = bx - ay \\ z = y \end{cases} ; \qquad \begin{cases} x = (w + az)/b \\ y = z . \end{cases} \qquad \text{(S2)}$$

(i.e., the characteristic lines are now given by $w = d$), and let $v(w,z) \equiv u(x,y) = u((w+az)/b,z)$ be the unknown function in terms of w and z. Then, by the chain rule (e.g., $u_x = v_w w_x + v_z z_x = bv_w$ and $u_y = v_w w_y + v_z z_y = -av_w + v_z$) , (S1) becomes

$$bv_z + cv = f((w+az)/b,z) , \qquad \text{(S3)}$$

in which there is no v_w. Then (S3) may be solved as a first–order ODE for v with w held constant, and the general solution of (S1) is then $u(x,y) \equiv v(bx-ay,y)$. Depending on the form of the function $f(x,y)$ in (S1), it may be more convenient to choose $z = x$ or some other linear combination of x and y (anything except a multiple of $bx-ay$). The resulting equation for v will still have no v_w term, although the coefficient of v_z may no longer be b as in (S3).

3. Side conditions on lines : The general solution of (S1) involves an arbitrary C^1 function. In order to single out a particular solution, an appropriate side–condition must be given. If we require that the solution have given values at points (x,y) on a line, say $y = mx + d$, then the side condition is

$$u(x,mx+d) = g(x) \qquad \text{(S4)}$$

where $g(x)$ is a given C^1 function. As long as $m \neq b/a$ (i.e., the line on which the side condition is given is *not* a characteristic line), the PDE (S1) will have a unique solution which meets the side condition (S4). In other words, if $m \neq b/a$, the side condition (S4) can be used to uniquely determine the arbitrary C^1 function in the general solution of (S1). However, if the side condition is given on a characteristic line (i.e., $m = b/a$), then there will be no solution of (S1) with (S4), unless $g(x)$ has a particular form. If $g(x)$ has this particular form, then there will be infinitely many solutions of (S1) satifying the side condition (S4).

4. Side conditions on curves : It is not necessary that a side condition be prescribed on a line. Indeed, a unique solution of (S1) is determined, if the values of u are given in a C^1 fashion on a regular curve which transversely intersects each characteristic curve exactly once. The essential idea is that the solution of (S1) is determined on a characteristic line by its value at a single point of the line, since (S1) is an ODE on the line. The regularity of the side condition curve and the transversality condition are needed to ensure that the solutions, on the individual characteristic lines, piece together to form a C^1 solution of (S1).

5. Application : An application to population and inventory analysis is given in the last subsection.

Exercises 2.1

1. Find the general solution of each of the following PDEs, where $u = u(x,y)$ in (a) – (d).

(a) $2u_x - 3u_y = x$, (b) $u_x + u_y - u = 0$

(c) $u_x + 2u_y - 4u = e^{x+y}$ (d) $3u_x - 4u_y = x + e^x$

(e) $v_z + 3v_w = 9w^2$, $v = v(w,z)$ (f) $g_t - cg_x = 0$, $g = g(x,t)$ (c constant).

2. Find the particular solution of $u_x + 2u_y - 4u = e^{x+y}$ satisfying the following side conditions.

(a) $u(x,0) = \sin(x^2)$ (b) $u(0,y) = y^2$ (c) $u(x,-x) = x$.

3. Show that the PDE $u_x + u_y - u = 0$ with side condition $u(x,x) = \tan(x)$ has no solution.

4. What form must $g(x)$ have in order that the following problem have a solution ?

$$u_x + 3u_y - u = 1 , \qquad u(x,3x) = g(x) .$$

If $g(x)$ has the required form, will there be more than one solution ?

5. Write down two different solutions of the PDE in Problem 4, when $g(x) = -1 + 2e^x$.

6. Solve the problem : $u_x - 2u_y = 0$, $u(x,e^x) = e^{2x}+4xe^x+4x^2$.

7. Let a, b and c be constants with $ab \neq 0$. Consider the homogeneous linear PDE $au_x + bu_y + cu = 0$. Bob says that the general solution is given by $u(x,y) = e^{-cx/a}f(bx-ay)$ for an arbitrary C^1 function f, while Jane says that it is $u(x,y) = e^{-cy/b}f(bx-ay)$. Who is correct ?

8. (a) Show that the PDE $u_x = 0$ has no solution which is C^1 everywhere and satisfies the side condition $u(x,x^2) = x$.

(b) Find a solution of the problem in (a) which is valid in the first quadrant $x > 0$, $y > 0$.

(c) Explain the results of (a) and (b) in terms of the intersections of the side condition curve and the characteristic lines.

9. (a) Show that the PDE $u_x = 0$ has no solution which is C^1 everywhere and satisfies the side condition $u(x,x^3) = x$, even though the side condition curve $y = x^3$ intersects each characteristic line $(y = d)$ only once.

(b) Part (a) demonstrates the necessity of the transversality condition on the intersections of the side condition curve with the characteristic lines. Explain why.
Hint. At what angle does the curve $y = x^3$ meet the x–axis ?

10. (a) Show that a solution of the homogeneous PDE $au_x + bu_y + cu = 0$ cannot be zero at one, and only one, point in the plane.

(b) If $c = 0$ in the PDE in (a), then show that the graph $z = u(x,y)$ of a solution u (defined everywhere) is a surface composed of horizontal parallel lines.

11. In Example 8 (b), how many of the original avocados (already present at $t = 0$) will remain after time t ? Your answer should be a function of t.

12. In Example 8, now assume that an avocado has a 10% chance of being removed from the shelf on any given day (i.e., more precisely $D(y) = .1$, for $y > 0$), regardless of its age.
(a) Show that in the long run, there will be about $Ce^{-y/10}$ y–day–old avocados on the shelf.

(b) According to part (a), what should the value of C be, if there are still to be about 300 avocados on the shelf in the long run. Does your answer agree with common sense ?

13. Air conditioners are produced at a constant rate of 100 per month beginning on New Years Day. The probability that an air conditioner will break down during a small time interval Δt months, assuming that t months have elapsed since New Years, is $(.2 - .1 \cdot \cos(\pi t/6))\Delta t$, regardless of the its age. Approximately how many y–month–old air conditioners will be operational at the end of the year, where $y \leq 12$? How could the total number of operational air

conditioners, at the end of the year, be determined ? If you have the resources for numerical integration, compute this number.
Hint. When using formula (25), remember that $D(y,t) = 0$ for $y < 0$.

14. A certain population has initial population density $C(y)$, for $y \geq 0$. The birth rate of the population at time t is proportional to its total size at time t, say $\alpha \cdot \int_0^\infty P(y,t)\,dy$ for some constant $\alpha > 0$. (We assume that $\int_0^\infty C(y)\,dy < \infty$.) The death rate density is constant, say $D(y,t) = k$ for $y \geq 0$ and some $k > 0$. Find the population density $P(y,t)$ for all $y, t > 0$.
Hint. First note that formula (25) applies, but $P(y-t,0)$ is not yet known for $t > y$. Let $f(t) = P(0,t)$ (i.e., $f(t)$ is the birth rate at time t). Once $f(t)$ is determined, then $P(y-t,0) = f(t-y)$ for $t > y$ (Why?), and the solution will then be explicitly given by (25). To find $f(t)$, note that

$$
\begin{aligned}
f(t) = \alpha \cdot \int_0^\infty P(u,t)\,du &= \alpha \cdot \int_0^t P(u,t)\,du + \alpha \cdot \int_t^\infty P(u,t)\,du \\
&= \alpha \cdot \int_0^t P(u-t,0) \cdot e^{-ku}\,du + \alpha \cdot \int_t^\infty C(u-t)\,e^{-kt}du \\
&= \alpha \cdot \int_0^t f(t-u) \cdot e^{-ku}\,du + \alpha \cdot e^{-kt} \int_0^\infty C(u)\,du .
\end{aligned}
$$

The last integral is the total initial population, say p_0. Change to the variable $v = t-u$ in the first integral, to obtain $e^{kt} f(t) = \alpha \cdot \int_0^t f(v) \cdot e^{kv}\,du + \alpha \cdot p_0$. Now differentiate.

15. By completing the following steps, prove Theorem 1. Let $(h(s),k(s))$ be a parametrization of the side condition curve in Theorem 1 by an arclength parameter s.

(a) For each point (x,y) in the plane, show that there are unique numbers $\sigma(x,y)$ and $\tau(x,y)$, such that $x = h(\sigma(x,y)) + a \cdot \tau(x,y)$ and $y = k(\sigma(x,y)) + b \cdot \tau(x,y)$. (Draw a picture.)

(b) Using the functions $\sigma(x,y)$ and $\tau(x,y)$ of part (a), show that with the change of variables

$$
\begin{cases} s = \sigma(x,y) \\ t = \tau(x,y) \end{cases} ; \qquad \begin{cases} x = h(s) + at \\ y = k(s) + bt , \end{cases}
$$

and with $v(s,t) \equiv u(x,y)$, the PDE $au_x + bu_y + cu = f(x,y)$ becomes $v_t + cv = F(s,t)$, where $F(s,t) \equiv f(h(s) + at, k(s) + bt)$. **Hint**. Note that $v_t = u_x x_t + u_y y_t$.

(c) Show that $v(s,t) = e^{-ct} \left[\int_0^t e^{cr} F(s,r)\,dr + U(s) \right]$, where $U(s) = v(s,0) = u(h(s),k(s))$ is the C^1 function which specifies the values of u on the side–condition curve. Thus, the unique solution of the problem in Theorem 1 is the C^1 function $u(x,y) = v(\sigma(x,y),\tau(x,y))$. (Note that the Jacobian $x_s y_t - y_s x_t = h'(s)b - k'(s)a \neq 0$ (Why?), so that σ and τ are C^1 by the Inverse Function Theorem which is covered in most advanced calculus books, e.g., [Taylor and Mann].)

2.2 Variable Coefficients

In many applications, we find first–order linear PDEs with variable coefficients

$$a(x,y)u_x + b(x,y)u_y + c(x,y)u = f(x,y)\ , \quad u = u(x,y)\ , \tag{1}$$

where a, b, c and f are given C^1 functions of x and y. Note that $a(x,y)u_x + b(x,y)u_y$ is the directional derivative of u at the point (x,y) in the direction of the vector

$$\mathbf{g}(x,y) \equiv a(x,y)\mathbf{i} + b(x,y)\mathbf{j}\ .$$

In Section 2.1, a and b were constants, and this vector had a fixed direction and magnitude, but now the vector can change as its base point (x,y) varies. Thus, $\mathbf{g}(x,y)$ is a vector field on the plane. It is helpful to think of $\mathbf{g}(x,y)$ as the velocity (at the point (x,y)) of a fluid flow in the xy–plane. When a and b are constants, the streamlines of the fluid are the straight lines with slope b/a (i.e., with tangent vectors parallel to $a\mathbf{i} + b\mathbf{j}$), and hence they are the characteristic lines. When a and b are not constant, the streamlines will be curved in general, and we refer to the streamlines as characteristic curves. More precisely, we make the following definition.

Definition 1. A curve in the xy–plane is called a **characteristic curve** for the PDE (1), if at each point (x_0,y_0) on the curve, the vector $\mathbf{g}(x_0,y_0) = a(x_0,y_0)\mathbf{i} + b(x_0,y_0)\mathbf{j}$ is tangent to the curve.

At each point on a characteristic curve, we have that $\mathbf{g}\cdot\nabla u$ (or $a(x,y)u_x + b(x,y)u_y$) is the directional derivative of u in the direction of the curve's tangent vector, and hence $\mathbf{g}\cdot\nabla u$ is proportional to the derivative of u , with respect to a position variable along the curve. Thus, as with the constant coefficient case, on each characteristic curve, the PDE (1) is actually an ODE for a function of a position variable along the curve. If the characteristic curves are graphs of functions y(x) (i.e., assuming that $a(x,y) \neq 0$), then Definition 1 implies that

$$\frac{dy}{dx} = \frac{b(x,y)}{a(x,y)}\ . \tag{2}$$

(i.e., the tangent line to the graph of y(x) at (x,y) is parallel to $\mathbf{g}(x,y) = a(x,y)\mathbf{i} + b(x,y)\mathbf{j}$).

The ODE (2) is known as the **characteristic equation** for the PDE (1). The solution curves of the the characteristic equation are the characteristic curves for (1).

In the case of constant coefficients a and b, the general solution of (2) is simply $y = \frac{b}{a}x + \text{const.}$ or $bx - ay = d$, where d is an arbitrary constant. In the general case of variable coefficients, (2) may be considerably more difficult to solve, but let us assume that (2) has been solved, and that the solution has been put in the implicit form $h(x,y) = d$, where d is an arbitrary constant. We can simplify the PDE (1), by making the change of variables

$$w = h(x,y) \quad \text{and} \quad z = y\,, \tag{3}$$

as we did when a and b were constant (e.g., $h(x,y) = bx-ay$, in that special case.). The rationale for this procedure is that w is constant on each characteristic curve and the PDE should become an ODE in the position variable z along these curves. As before, the choice $z = y$ is not necessary. Indeed, we can set z equal to any C^1 function of x and y, as long as the transformation can be inverted to give x and y in terms of z and w. Setting $v(w,z) = u(x,y)$, we can verify that (1) is transformed into an ODE in z, for w fixed. First compute,

$$\begin{aligned} au_x + bu_y &= a(v_w w_x + v_z z_x) + b(v_w w_y + v_z z_y) \\ &= (aw_x + bw_y)v_w + (az_x + bz_y)v_z\,. \end{aligned}$$

Thus, it suffices to show that $aw_x + bw_y = 0$, in order that v_w drop out of the transformed PDE for v. To this end, let (x_0,y_0) be a given point and let $y(x)$ be a solution of (2) such that $y(x_0) = y_0$. We know that $h(x,y(x)) = \text{const.}$, and hence using (2),

$$0 = \frac{d}{dx}h(x,y(x)) = h_x + h_y\frac{dy}{dx} = w_x + w_y\frac{b(x,y)}{a(x,y)}\,.$$

Thus, $aw_x + bw_y = 0$ at any point $(x,y(x))$. In particular, $aw_x + bw_y = 0$ at the arbitrary given point (x_0,y_0). Alternatively, recall that ∇h is normal to any level curve $h(x,y) = d$ and by construction $\mathbf{g} = a\mathbf{i} + b\mathbf{j}$ is tangent to this level curve. Thus, $aw_x + bw_y = \mathbf{g}\cdot\nabla h = 0$. Although we have shown that the method works, in many instances it turns out that (3) may be invertible only in a certain domain in the xy–plane. This signals the need for some care, as we will find in some of the examples below.

Example 1. Find the general solution of

$$-yu_x + xu_y = 0\,. \tag{4}$$

Solution. The characteristic equation is $dy/dx = -x/y$. This is a separable equation which is readily solved by separating the variables and integrating :

$$y\,dy = -x\,dx \;\Rightarrow\; \tfrac{1}{2}y^2 = -\tfrac{1}{2}x^2 + \tfrac{1}{2}d\,.$$

Thus, the characteristic curves form the family $x^2+y^2 = d$ of circles [$d > 0$] and the point $(0,0)$ [$d = 0$]. We make the change of variables

$$\begin{cases} w = x^2 + y^2 \\ z = y \end{cases} \quad ; \quad \begin{cases} x = \pm\,[w - z^2]^{\frac{1}{2}} \\ y = z\,. \end{cases} \tag{5}$$

In spite of the fact that the inverse transformation is double–valued and only defined for $w \geq z^2$, we will arrive at the correct solution anyway. Setting $v(w,z) = u(x,y)$, the PDE (4) becomes

$$0 = -yu_x + xu_y = -y(v_w w_x + v_z z_x) + x(v_w w_y + v_z z_y) = -(y\cdot 2x - x\cdot 2y)v_w + xv_z = xv_z\,,$$

(i.e., $xv_z = 0$). Thus, if v is a C^1 function of w, say $v = f(w)$, we suspect that

$$u(x,y) = f(x^2 + y^2) \tag{6}$$

will be a solution of the original PDE, in spite of the defects of the transformation (5). Indeed, we can check the solution (6) directly :

$$-yu_x + xu_y = -yf'(x^2+y^2)2x + xf'(x^2+y^2)2y = 0\,.$$

This shows that (6) is a solution. We will often resort to tentative procedures in deriving "hypothetical solutions", but until they are actually checked, we have no proof that they are actual solutions. We have still not conclusively demonstrated that (6) is the most general solution. A solution is of the form (6), if and only if it is a constant on each of the circles $x^2+y^2 = a^2$. We should check that *any* solution of (4) must be constant on these characteristic circles. To show this, we parameterize the circles via

$$x(t) = a\cos(t)\,, \quad y(t) = a\sin(t)\,, \quad a > 0\,. \tag{7}$$

As t varies, $(x(t),y(t))$ traces out the circle $x^2+y^2 = a^2$. The value of u at $(x(t),y(t))$ is a function of t that we denote by $U(t) \equiv u(x(t),y(t))$. We want to show that $U(t)$ is a constant, so that u will be constant on the arbitrary circle $x^2+y^2 = a^2$. For this, we compute :

$$\begin{aligned}\frac{dU}{dt} &= \frac{d}{dt}u(x(t),y(t)) = u_x x'(t) + u_y y'(t) = -u_x a\cdot\sin(t) + u_y a\cdot\cos(t) \\ &= -u_x(x(t),y(t))\cdot y(t) + u_y(x(t),y(t))\cdot x(t) = 0\,,\end{aligned}$$

by the PDE (4). Thus, the PDE tells us that the function u is constant on characteristic circles, and (6) is in fact the most general form for the solution. □

Remark. In Example 1, it is not necessary for the function $f(z)$ to be C^1 at $z = 0$. For example, if $f(z) = 3z^{2/3}$, then $u(x,y) = 3(x^2+y^2)^{2/3} = 3r^{4/3}$ $(r \equiv (x^2 + y^2)^{\frac{1}{2}})$, which is still C^1, even though $f'(z) = 2z^{-1/3}$ is undefined at $z = 0$. Technical difficulties such as this can occur at "critical points" (x_0,y_0), where $a(x_0,y_0) = 0$ and $b(x_0,y_0) = 0$. In Example 1, we can still say that all solutions are of the form (6), but the requirement that $f(z)$ be C^1 can be relaxed a bit at $z = 0$. The degree of concern over such matters will be left to the instructor. However, one should note that if $h(x,y)$ is C^1 and $\nabla h(x,y) \neq \mathbf{0}$ for all (x,y) in an open set S, then $f(h(x,y))$ will be C^1 on S, if and only if f is C^1 on $h(S)$. Thus, in order that the solutions of (1) be C^1, the arbitrary function of integration must certainly be C^1 near $h(x_0,y_0)$, if $\nabla h(x_0,y_0) \neq \mathbf{0}$. □

The preferred parametrization of a characteristic curve

In Example 1, we parametrized a characteristic curve via the functions $x(t)$ and $y(t)$, and we defined a function $U(t) = u(x(t),y(t))$ which is simply the value of u along the curve at "time" t. We found that $U(t)$ obeys the ODE $U'(t) = 0$, by virtue of the PDE (4). For the general first–order, linear PDE (1), the main goal of this subsection is to explicitly show that the function $U(t) \equiv u(x(t),y(t))$ must satisfy a certain ODE, if $(x(t),y(t))$ traces out a characteristic curve, as t varies.

There are many ways in which a particle can move along a characteristic curve (e.g., quickly, slowly, etc.). Perhaps the most natural way is to have the particle move in such a way that its velocity is $\mathbf{g}(x,y) = a(x,y)\mathbf{i} + b(x,y)\mathbf{j}$, when it is at the point (x,y) (i.e., as if it were carried along by the fluid flow with velocity $\mathbf{g}(x,y)$). For the particle to move in this way, the functions $x(t)$ and $y(t)$ which give the position $(x(t),y(t))$ of the particle at time t, must obey the system

$$\frac{dx}{dt} = a(x(t),y(t)) \, , \quad \frac{dy}{dt} = b(x(t),y(t)) \, . \tag{8}$$

Since the velocity vector $x'(t)\mathbf{i} + y'(t)\mathbf{j}$ of the particle is tangent to its path, we know that (8) ensures that the point $(x(t),y(t))$ traces out a characteristic curve for the PDE (1).

Definition 2. The system of equations (8) is called the **characteristic system** of the PDE (1). If $x(t)$ and $y(t)$ solve this system, then $(x(t),y(t))$ is said to be a **preferred parametrization** for the characteristic curve that is traced out as t varies.

In Section 1.1, we solved systems such as (8) in the simple case, where $a(x,y)$ and $b(x,y)$ were linear combinations of x and y. In general, we will only consider problems where the characteristic systems are easily solved. For the PDE considered in Example 1, the system (8) is $x'(t) = -y(t)$, $y'(t) = x(t)$. Differentiating the first equation and using the second, we have $x''(t) = -y'(t) = -x(t)$, or $x''(t) + x(t) = 0$. Thus,

$$x(t) = c_1\cos(t) + c_2\sin(t) \quad \text{and} \quad y(t) = -x'(t) = c_1\sin(t) - c_2\cos(t) \, . \tag{9}$$

For any choices of the constants c_1 and c_2, a characteristic curve (a circle or a point, in this case) is traced out. In (7) of Example 1, we chose $c_1 = a$ and $c_2 = 0$ for convenience.

Suppose that $(x(t),y(t))$ is a preferred parametrization for a characteristic curve of

$$a(x,y)u_x + b(x,y)u_y + c(x,y)u = f(x,y)\ . \tag{10}$$

Let $$U(t) \equiv u(x(t),y(t)),\ \ C(t) \equiv c(x(t),y(t)) \text{ and } F(t) \equiv f(x(t),y(t))\ . \tag{11}$$

In order to find the ODE which $U(t)$ must obey, we compute

$$\begin{aligned} U'(t) &= u_x(x(t),y(t))\, x'(t) + u_y(x(t),y(t))\, y'(t) && \text{(by the chain rule)} \\ &= u_x(x(t),y(t))\, a(x(t),y(t)) + u_y(x(t),y(t))\, b(x(t),y(t)) && \text{(by (8))} \\ &= -c(x(t),y(t))u(x(t),y(t)) + f(x(t),y(t)) && \text{(by the PDE (10))} \\ &= -C(t)\, U(t) + F(t). && \text{(by (11)).} \end{aligned}$$

Thus, we have shown that $U(t) \equiv u(x(t),y(t))$ must obey the ODE

$$U'(t) + C(t)U(t) = F(t)\ . \tag{12}$$

Letting $m(t) \equiv \exp\left[\int_0^t C(t)\,dt\right]$ be the integrating factor for (12), we obtain the solution

$$U(t) = \frac{1}{m(t)}\left[\int_0^t m(t)F(t)\,dt + U(0)\right]\ . \tag{13}$$

In this formula, $m(t)$ and $F(t)$ depend only on the values of $c(x,y)$ and $f(x,y)$ along the characteristic curve $x = x(t)$, $y = y(t)$. Thus, (13) shows that the values $U(t)$ of the solution u along the entire characteristic curve are completely determined, once the value $U(0) = u(x(0),y(0))$ is prescribed. If $c(x,y)$ and $f(x,y)$ are zero, as in Example 1, then (13) says that u will be constant on each characteristic curve. However, the constant can change, as one moves from one characteristic curve to another. For example, the particular solution $u(x,y) = (x^2+y^2)^3$ of Example 1 is 64 on the circle of radius 2, and 1 on the circle of radius 1. If $c(x,y)$ and $f(x,y)$ are not zero, then $U(t)$ need not be constant, although (13) shows that $U(t)$ must have a certain form which depends only on the choice of $U(0)$. Thus, in general we cannot specify the value of u at two distinct points of a characteristic curve, as the following example shows.

Example 2. Show that the problem $-yu_x + xu_y = 0$, $u(x,0) = 3x$ has no solution.

Solution. The side condition is given on the x–axis which intersects each of the characteristic circles $x^2+y^2 = a^2$ *twice*, at $(a,0)$ and $(-a,0)$ $[a \neq 0]$. We saw in Example 1 that $u(x,y)$ must be constant on such circles, and yet the side condition requires that $u(a,0) = 3a$ and $u(-a,0) = -3a$. Thus, this side condition can never hold for a solution of the PDE. The difficulty arises, because we cannot expect to prescribe u at more than one point on any characteristic curve. Given a value for u at one point, the values of u at the other points along the curve will be uniquely determined by the fact that u is a solution (13) of a certain ODE (12) along the curve. □

Remark. Sometimes, if the side condition is chosen carefully, a problem may still have a solution, even if the characteristic curves intersect the side condition curve more than once. For example, the problem, $-yu_x + xu_y = 0$ with $u(x,0) = 3x^2$, has the solution $u(x,y) = 3(x^2+y^2)$. The saving grace of this side condition is that $u(a,0) = 3a^2 = u(-a,0)$. □

The parametric form of solutions

We have seen that it is convenient to think of characteristic curves of the PDE

$$a(x,y)u_x + b(x,y)u_y + c(x,y)u = f(x,y) \tag{14}$$

as paths of particles moving with the flow of a fluid with velocity $\mathbf{g}(x,y) = a(x,y)\mathbf{i} + b(x,y)\mathbf{j}$. The position $(x(t),y(t))$ of a particle is completely determined by its starting position $(x(0),y(0))$ at time $t = 0$. If a side condition is given on some regular side condition curve which transversely intersects each characteristic curve exactly once, then it is convenient to take the starting position of the particle on a characteristic curve to be the point of intersection of the characteristic curve with the side condition curve. If we let s denote a position variable along the side condition curve, then we obtain a different characteristic curve for each value of s. For each fixed s, let $(X(s,t),Y(s,t))$ be the position, at time t, of the particle which begins at the point corresponding to s on the side condition curve, and flows with the fluid.

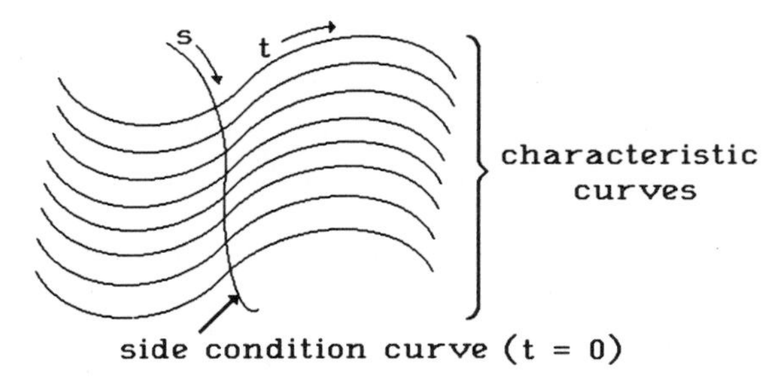

Figure 1

Note that the side condition curve itself is traced out by $(X(s,0),Y(s,0))$, as s varies and t is held fixed at 0. In other words, we have the following :
The functions $X(s,t)$ and $Y(s,t)$ are the solutions of the characteristic system (for each fixed s)

$$\frac{d}{dt}X(s,t) = a(X(s,t),Y(s,t)), \quad \frac{d}{dt}Y(s,t) = b(X(s,t),Y(s,t)), \tag{15}$$

with given initial values $X(s,0)$ and $Y(s,0)$.

Suppose that the values of u at points on the side condition curve are given by

$$u(X(s,0),Y(s,0)) = G(s)\ , \tag{16}$$

where $G(s)$ is a given C^1 function. We obtain $u(X(s,t),Y(s,t))$ as follows.

Let $U(s,t) \equiv u(X(s,t),Y(s,t)),\ \ C(s,t) \equiv c(X(s,t),Y(s,t)),\ \ F(s,t) \equiv f(X(s,t),Y(s,t))$

and

$$m(s,t) \equiv \exp\left[\int_0^t C(s,t)\, dt\right]. \tag{17}$$

Then we may apply the result (13) of the previous subsection, for each fixed s, to deduce that

$$U(s,t) = \frac{1}{m(s,t)}\left[\int_0^t m(s,t)\, F(s,t)\, dt + G(s)\right]. \tag{18}$$

From (17), we know that $U(s,t)$ is the value of u at the point $(X(s,t),Y(s,t))$. Thus, as s and t vary, the point (x,y,u), in xyu–space, given by

$$x = X(s,t)\ ,\quad y = Y(s,t)\ ,\quad u = U(s,t)\ , \tag{19}$$

traces out the surface of the graph of the solution u of the PDE (14) which meets the side condition (16).

The equations (19) constitute the **parametric form** of the solution of (14) with the condition (16).

Although (19) does not directly give us a formula for $u(x,y)$, it may be possible to solve the equations $x = X(s,t)$ and $y = Y(s,t)$ for s and t in terms of x and y, say $s = S(x,y)$, $t = T(x,y)$. Then $u(x,y) = U(S(x,y),T(x,y))$ will be the usual explicit form for the solution. It is often convenient to leave the solution in the form (18) for the purpose of generating three–dimensional computer plots of the graph of the solution in xyu–space (cf. Figure 2 below).

Example 3. Find the parametric form of the solution of the problem

$$-yu_x + xu_y = 0\ ,\quad u(s,s^2) = s^3 \quad (s > 0)\ . \tag{20}$$

Solution. By (15), the family of characteristic curves $(X(s,t),\ Y(s,t))$ are found by solving with initial conditions

$$\frac{d}{dt} X(s,t) = -Y(s,t), \quad \frac{d}{dt} Y(s,t) = X(s,t)$$
$$X(s,0) = s, \quad Y(s,0) = s^2 . \tag{21}$$

The general solution of this system is (cf. (9))

$$X(s,t) = c_1(s)\cos(t) + c_2(s)\sin(t) \quad \text{and} \quad Y(s,t) = c_1(s)\sin(t) - c_2(s)\cos(t) .$$

By the initial conditions, $c_1(s) = s$ while $c_2(s) = -s^2$. For the PDE in (20), we have $c(x,y) = 0$ and $f(x,y) = 0$ (cf. (14)). Thus, $m(s,t) \equiv 1$ and $F(s,t) \equiv 0$ in (17). According to (20) and (16), we have $G(s) = s^3$, and so $U(s,t) = s^3$ by (18). Hence, we have the parametric solution

$$X(s,t) = s\cos(t) - s^2\sin(t), \quad Y(s,t) = s\sin(t) + s^2\cos(t), \quad U(s,t) = s^3 . \tag{22}$$

When $t = 0$, and s $(s > 0)$ varies, we get the point (s,s^2,s^3) in xyu–space, which traces out the so–called twisted cubic. As t varies, the points on the curve move in circles about the u–axis. Thus, the graph of the solution is the surface obtained by revolving the twisted cubic about the u–axis, as we illustrate in Figure 2.

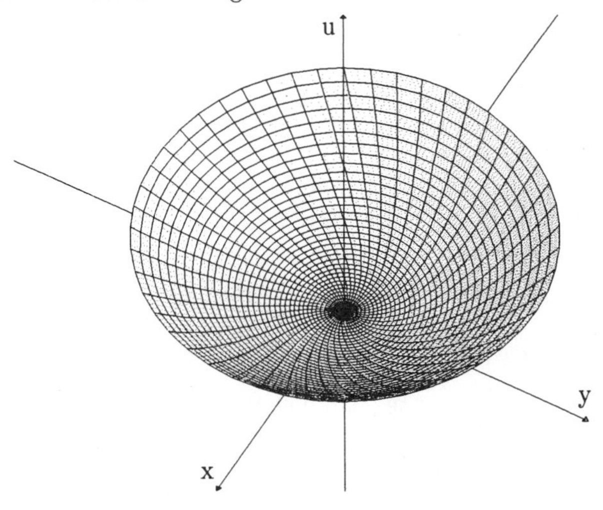

Figure 2

We can also obtain the solution in explicit form. We know from Example 1 that the general solution is of the form $u(x,y) = C(x^2+y^2)$. Thus, the side condition tells us that $C(s^2+s^4) = s^3$. Setting $r^2 = s^2 + s^4$, we have $s^2 = (-1 + \sqrt{1+4r^2})/2$. Thus, $C(r^2) = s^3$ yields

$$u(x,y) = \left[-1 + \sqrt{1+4(x^2+y^2)}\right]^{3/2} / \sqrt{8} .$$

If the side condition curve were replaced by (s,s^7), then the explicit form of the solution would be difficult (if not impossible) to obtain, due to the difficultly in solving for s^2 in terms of r^2. However, in the parametric form for the solution, we can simply replace s^2 by s^7 in (22). □

Example 4. Solve $(y+x)u_x + (y-x)u_y = u$, subject to $u(\cos(s),\sin(s)) = 1$ for $0 \le s \le 2\pi$.

Solution. The side condition states that u is to be 1 on the unit circle $x^2+y^2 = 1$. The characteristic equation $\frac{dy}{dx} = \frac{y-x}{y+x}$ (cf. (2)) is neither separable nor linear, although it becomes separable under the change of dependent variable $y(x) = x \cdot z(x)$. Instead, we opt for the parametric approach. The characteristic system (cf. (15)) is (for fixed s)

$$\frac{d}{dt}X(s,t) = X(s,t) + Y(s,t) \quad \text{and} \quad \frac{d}{dt}Y(s,t) = -X(s,t) + Y(s,t) ,$$

with initial conditions (23)

$$X(s,0) = \cos(s) , \; Y(s,0) = \sin(s) .$$

We solve the system using the method of Section 1.1. We differentiate the first equation in (23) with respect to t, and use the second equation to obtain $X'' = X' + Y' = X' + (-X + Y) = X' + (-X + X' - X)$, or $X'' - 2X' + 2X = 0$. The roots of the auxiliary equation $r^2 - 2r + 2 = 0$ are $1 \pm i$. Thus,

$$X(s,t) = c_1(s)e^t\cos(t) + c_2(s)e^t\sin(t) \quad \text{and} \quad Y(s,t) = -c_1(s)e^t\sin(t) + c_2(s)e^t\cos(t),$$

using $Y = X' - X$. For fixed s these curves $(X(s,t),Y(s,t))$ spiral away from the origin as t advances. The initial conditions yield $c_1(s) = \cos(s)$ and $c_2(s) = \sin(s)$. Thus,

$$X(s,t) = \cos(s)e^t\cos(t) + \sin(s)e^t\sin(t) = e^t\cos(s-t) \quad \text{and} \quad Y(s,t) = e^t\sin(s-t) .$$

Since $U(s,t) = u(X(s,t),Y(s,t))$ satisfies $\frac{d}{dt}U(s,t) = U(s,t)$ (cf. (12)) with $U(s,0) \equiv 1$, we have $U(s,t) = e^t$ (cf. (18), also). Thus, we have the parametric solution

$$x = e^t\cos(s-t), \quad y = e^t\sin(s-t), \quad u = e^t .$$

Since $x^2 + y^2 = e^{2t} = u^2$, we can write the solution in the explicit form $u(x,y) = \sqrt{x^2+y^2}$. Note that the graph of the solution is a cone and it is not C^1 at the origin. If the initial condition had not been so simple, then we might not have been able to get a simple formula for the solution in explicit form; but obtaining the parametric solution would be easy (e.g., consider $u(\cos(s),\sin(s)) = \cos(3s)$, in which case we would have $U(s,t) = e^t\cos(3s)$). □

Global Considerations

When the values of u are prescribed in a C^1 fashion along a regular side–condition curve which transversely intersects each characteristic curve once, then we have some way of piecing together solutions for u on the various characteristic curves. However, as we show in the next example, it can happen that there is no single side condition curve which transversely crosses each characteristic curve once. In such cases, constructing solutions which are defined throughout the xy–plane (i.e., globally) can lead to some interesting complications.

Example 5. Find the general solution of the PDE

$$xu_x - yu_y + yu = 0 . \tag{24}$$

Solution. The characteristic curves are found from

$$\frac{dy}{dx} = \frac{-y}{x} \quad \text{or} \quad \frac{dx}{x} + \frac{dy}{y} = 0 .$$

Integrating, we obtain $\log(|x|)+\log(|y|) = \log(|d|)$. Thus, the family of characteristic curves is the collection of curves $xy = d$ (hyperbolas, when $d \neq 0$). From the viewpoint of preferred parametrizations, the two branches of the hyperbola $xy = d$ should be regarded as distinct characteristic curves. Indeed, the system of equations for the preferred parametrization $(x(t),y(t))$ is $x'(t) = x(t)$, $y'(t) = -y(t)$, with solutions $x(t) = c_1e^t$, $y(t) = c_2e^{-t}$. For fixed nonzero c_1 and c_2, the point (c_1e^t,c_2e^{-t}) traces out only one branch of the hyperbola $xy = c_1c_2$. We proceed with the general solution process, as in Example 1, by making the change of variables

$$\begin{cases} w = xy \\ z = y \end{cases} ; \qquad \begin{cases} x = w/z \\ y = z . \end{cases} \tag{25}$$

The inverse transformation is not defined everywhere (only for $z \neq 0$). This will lead to some unexpected difficulties. Setting $v(w,z) = u(x,y)$ we have $xu_x - yu_y = x(v_ww_x + v_zz_x) - y(v_ww_y + v_zz_y) = (xy-yx)v_w - yv_z = -yv_z$. Thus, the PDE becomes $-zv_z + zv = 0$, which has the solution $v(w,z) = C(w)e^z$. Hence, we are led to the hypothetical solution

$$u(x,y) = C(xy)e^y , \tag{26}$$

where C is an arbitrary C^1 function. Although one can directly check that (26) is a solution, this time we demonstrate that (26) is *not* the most general solution. Observe that the solution (26) is the function $C(d)e^y$ on each of the branches of the hyperbola $xy = d$, and yet the branches are disconnected, so that there is no reason why the solution would have to be the same multiple of e^y on each branch. Note that if we restrict the domain of u to the upper half plane $y > 0$, then each hyperbola $xy = d$ has only one branch in this half plane, so that (11) will in fact be the most general solution in the half plane $y > 0$. Similar remarks apply, if we restrict ourselves to $y < 0$. Suppose that we try to glue together the two general solutions in each half plane by defining

$$u(x,y) = \begin{cases} C(xy)e^y & y \geq 0 \\ D(xy)e^y & y \leq 0 \end{cases}, \tag{27}$$

where C and D are C^1 functions. Certainly (27) is a C^1 solution of the PDE for $y > 0$ and for $y < 0$, but we must make sure that u is well defined and C^1 at points on the x–axis ($y = 0$). For u to be well–defined, we need $C(0) = D(0)$ [i.e., the formulas must agree when $y = 0$]. Also, for $y \geq 0$, we have $u_x(x,y) = C'(xy)ye^y$, and for $y \leq 0$, we have $u_x(x,y) = D'(xy)ye^y$. Hence, $u_x(x,0) = 0$ in both cases, and u_x is well–defined and continuous. Therefore,

$$u_y(x,y) = C'(xy)xe^y + C(xy)e^y, \text{ for } y \geq 0 \quad \text{and} \quad u_y(x,y) = D'(xy)xe^y + D(xy)e^y, \text{ for } y \leq 0.$$

In order that the formulas match at $y = 0$, we need $C'(0) = D'(0)$. Thus, (27) is a solution, provided we assume that $C(0) = D(0)$ and $C'(0) = D'(0)$. Indeed, (27) is the most general solution of the PDE, where C and D are arbitrary C^1 functions, subject only to the conditions $C(0) = D(0)$ and $C'(0) = D'(0)$. For example, taking $C(r) = r$ and $D(r) = \sin(r)$, we have the particular solution

$$u(x,y) = \begin{cases} xye^y & y \geq 0 \\ \sin(xy)e^y & y \leq 0 \end{cases}.$$

It is interesting to note that the functions C and D in (27) must *each* be defined for *all* real numbers in order that u(x,y) be defined for all x and y. Thus, we have an example of a first–order PDE that requires two arbitrary functions to express its general solution. This leads to an interesting question. Is there *one* side condition for the PDE (24) which uniquely determines a solution ? It is plausible that the answer is "no", because the side condition would determine *two* arbitrary functions, C and D. Indeed, the curve on which this side condition is given would have to cross each branch of every hyperbola $xy = d$ once and only once. Consider the four branches shown below in Figure 3. As is easily seen, there is no continuous curve that

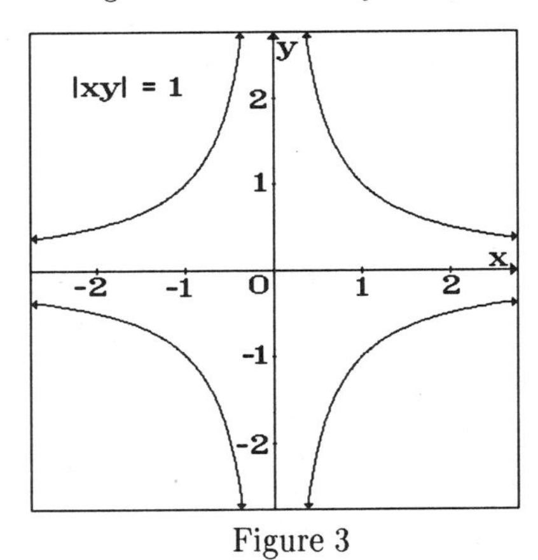

Figure 3

crosses each of these branches just once. This shows that there is no single side condition that will uniquely determine a solution of the PDE. One could get a unique solution by imposing two side conditions, say

$$u(x,1) = h(x) \quad \text{and} \quad u(x,-1) = k(x)\ , \quad \text{where } h(x) \text{ and } k(x) \text{ are } C^1 \text{ functions.}$$

Then we could determine the functions C and D in the general solution (27) via the relations

$$C(x)e = u(x,1) = h(x)$$
$$D(x)e^{-1} = u(x,-1) = k(x)\ .$$

Thus, $C(x) = h(x)/e$ and $D(x) = k(x)e$. Since we must have $C(0) = D(0)$ and $C'(0) = D'(0)$, we deduce that in order for a solution to exist the functions $h(x)$ and $k(x)$ must satisfy the rather weird conditions $h(0) = e^2k(0)$ and $h'(0) = e^2k'(0)$! □

Example 5 shows that, in constructing general solutions, complications can arise, when the family of characteristic curves has a "topologically nontrivial" configuration, in the sense that there is no regular side condition curve which transversely intersects each characteristic curve once. For this reason, it is difficult if not impossible (or at least very awkward) to formulate a specific procedure, whereby one can capture the completely general solution of a first–order linear PDE with variable coefficients. Essentially, one can form any solution by solving the ODE (12) on characteristic curves $(x(t),y(t))$ and piecing the solutions on the curves together. But in practice this is not as easy as it sounds. It is possible to give examples of first–order linear PDEs for which the general solution involves infinitely many arbitrary functions (cf. Problem 8).

An application to gas flow

Imagine a gas (or compressible medium) which flows parallel to a given line (say the x–axis). We denote the density (mass per unit volume) of the gas at the point (x,y,z) at time t by $\rho(x,t)$; we assume for simplicity that the density is independent of y and z. Let the velocity at the point (x,y,z) at time t be $v(x,t)\mathbf{i}$, where $\mathbf{i}$ is the unit vector in the positive x direction. We show that because of conservation of mass, the functions $\rho(x,t)$ and $v(x,t)$ must obey the so–called **continuity equation** :

$$\rho_t + (\rho v)_x = 0 \quad \text{or} \quad \rho_t + v\rho_x + v_x\rho = 0\ . \tag{28}$$

For the derivation, consider the space between x_0 and $x_0 + \Delta x$ (see Figure 4).

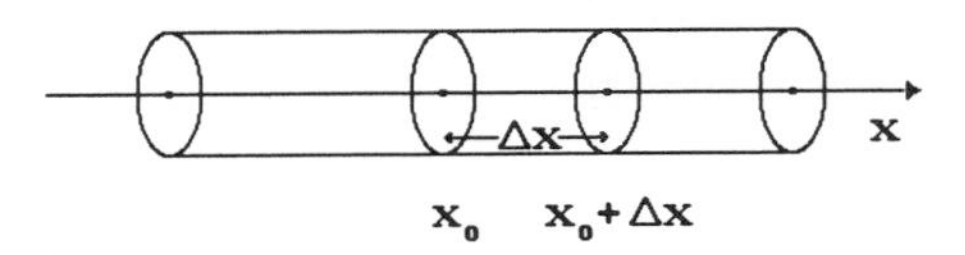

Figure 4

The amount of mass passing through a unit area of the plane $x = x_0$ into this space in the time interval Δt is $\rho(x_0,t)v(x_0,t)\Delta t$, and the amount that leaves through the face $x = x_0 + \Delta x$ (per unit area) in the time interval Δt is $\approx \rho(x_0+\Delta x,t)\cdot v(x_0+\Delta x,t)\Delta t$. Thus, the net change in the mass, per unit cross–sectional area, between the planes during Δt is given approximately by

$$\int_{x_0}^{x_0+\Delta x} [\rho(x,t+\Delta t) - \rho(x,t)]\,dx \approx -[\rho(x_0+\Delta x,t)v(x_0+\Delta x,t) - \rho(x_0,t)v(x_0,t)]\cdot\Delta t\,.$$

Dividing by $\Delta x\cdot\Delta t$, and taking the limits as Δx and Δt tend to zero, we obtain $\rho_t(x_0,t) = -(\rho v)_x(x_0,t)$, which gives us (28). By specifying $v(x,t)$ in advance, we can use (28) to figure out the "unknown" density $\rho(x,t)$, provided we know $\rho(x,0)$ [i.e., the density at the time $t = 0$]. In other words, finding $\rho(x,t)$ amounts to solving the problem

$$\rho_t + v(x,t)\rho_x + v_x(x,t)\rho = 0\,, \quad \rho(x,0) = \rho_0(x)\,, \tag{29}$$

where $\rho_0(x)$ is some given C^1 function (the initial density). We consider some special cases.

1. *Suppose that* $v(x,t) = v_0$, *a constant.* Then the PDE in (29) becomes $\rho_t + v_0\rho_x = 0$. The characteristic lines form the family of lines with slope $dx/dt = v_0$ (i.e., $x - v_0t = d$). Without going through the familiar change of variables, it is evident that the general solution of the PDE is $\rho(x,t) = C(x - v_0t)$, for an arbitrary C^1 function C. From the initial condition in (29), we get $\rho_0(x) = C(x - v_0\cdot 0) = C(x)$. Thus, the solution of the problem (29), in this case is

$$\rho(x,t) = \rho_0(x - v_0t)\,.$$

In other words, the density distribution is carried downwind with speed $|v_0|$.

2. *Suppose that* $v(x,t) = \alpha x$, *where* α *is some positive constant.* In this case, the velocity is in the direction $-\vec{i}$ for negative x and in the direction $+\vec{i}$ for positive x (i.e., the wind is blowing away from the point $x = 0$). Of course, we expect the density at $x = 0$ to decrease with time in this case. The PDE in (29) becomes $\rho_t + \alpha x\rho_x + \alpha\rho = 0$. The solutions of the characteristic equation $x'(t) = \alpha x$, form the family of characteristic curves $x = de^{\alpha t}$ or $xe^{-\alpha t} = d$ in the xt–plane. We make the change of variables

$$\begin{cases} w = xe^{-\alpha t} \\ z = t \end{cases}; \qquad \begin{cases} x = we^{\alpha z} \\ t = z\,, \end{cases}$$

and let $r(w,z) = \rho(x,t)$. The PDE then becomes

$$0 = \rho_t + \alpha x \rho_x + \alpha\rho = (r_w w_t + r_z z_t) + \alpha x(r_w w_x + r_z z_x) + \alpha r = r_z + \alpha r ,$$

or $r_z + \alpha r = 0$. The general solution is $r(w,z) = C(w)e^{-\alpha z}$, for an arbitrary C^1 function C , and this yields $\rho(x,t) = C(xe^{-\alpha t})e^{-\alpha t}$. Using the initial condition, we have $\rho_0(x) = \rho(x,0) = C(x)$, whence $C(x) = \rho_0(x)$. The unique solution of problem (14) in this case is then

$$\rho(x,t) = \rho_0(xe^{-\alpha t})e^{-\alpha t} . \tag{30}$$

The density at $x = 0$ decreases, as we expected. It is interesting to note that in the case where $\rho_0(x) = \rho_0 = \text{const.}$, we have that $\rho(x,t) = \rho_0 e^{-\alpha t}$ is independent of x, even though the wind velocity $v = \alpha x$ depends on x. Also, note that (30) shows that the graph of $\rho(x,t)$ is the graph of $\rho_0(x)$, after it has been stretched horizontally by a factor of $e^{\alpha t}$ and compressed vertically by a factor of $e^{-\alpha t}$. These operations (taken together) conserve the area under the graph, which means that the total mass (i.e., integral of the density with respect to x) is conserved.

A geometrical application

Here we will find all functions $u = u(x,y)$ such that the tangent plane to the graph $z = u(x,y)$ at *any* arbitrary point $(x_0,y_0,u(x_0,y_0))$ passes through the origin. Assume that u is C^1, and recall that the equation of the tangent plane to the graph at $(x_0,y_0,u(x_0,y_0))$ is

$$u_x(x_0,y_0)(x-x_0) + u_y(x_0,y_0)(y-y_0) - (z - u(x_0,y_0)) = 0 .$$

In order that (0,0,0) be on this tangent plane, we need $-u_x(x_0,y_0)x_0 - u_y(x_0,y_0)y_0 + u(x_0,y_0) = 0$. For this to hold for all (x_0,y_0) in the domain of u, the function u must satisfy the PDE

$$xu_x + yu_y - u = 0 . \tag{31}$$

The characteristic curves obey $dy/dx = y/x$, whose solution is $\log(y)-\log(x) = \log(d)$ or $y/x = d$, the family of rays from the origin in the xy–plane. To solve (31), we switch to a coordinate system such that one of the coordinates is constant on each such ray. Polar coordinates (r,θ) are perfectly suited, as $\theta = \text{constant}$ defines a ray. The transformation (for $(x,y) \neq (0,0)$) is

$$r = (x^2+y^2)^{\frac{1}{2}} \qquad\qquad x = r\cos\theta$$

with inverse

$$\theta = \begin{cases} \arccos[x/(x^2+y^2)^{\frac{1}{2}}] & y \geq 0 \\ \arccos[-x/(x^2+y^2)^{\frac{1}{2}}] + \pi & y < 0 \end{cases} \qquad\qquad y = r\sin\theta ,$$

The inverse transformation is simpler than the transformation itself (which is not defined at the origin). We know that (31) should become an ODE in r. Rather than computing $xu_x + yu_y$ using the transformation, we use the inverse transformation to compute v_r, where $v(r,\theta) = u(x,y)$:

$$v_r = u_x x_r + u_y y_r = u_x \cos(\theta) + u_y \sin(\theta) \qquad \text{or} \qquad rv_r = xu_x + yu_y \,.$$

Thus, (31) becomes $rv_r - v = 0$, whose solution is $v(r,\theta) = C(\theta)r$, where $C(\theta)$ is an arbitrary C^1 function of θ. The graph $z = v(r,\theta)$ consists of a family of rays in space issuing from the origin and forming a conical object, with possibly a vertex at the origin, as in Figure 5.

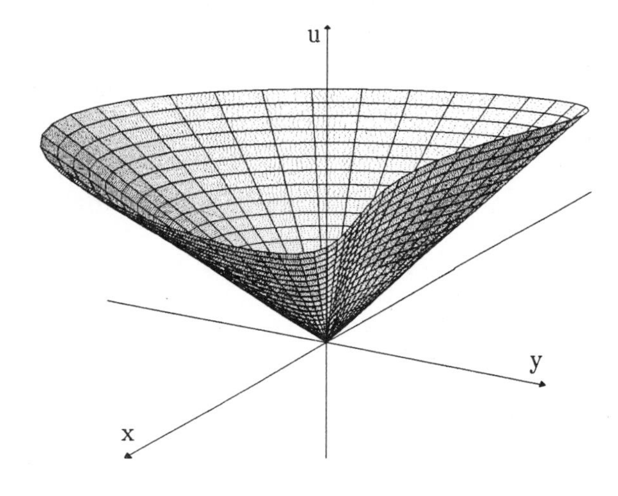

Figure 5

Note that the surface $\theta = \theta_0$ is a half–plane issuing from the z–axis and the graph intersects this half–plane in the line $z = C(\theta_0)r$. When there is a vertex at the origin, the graph will not have a well–defined tangent plane at (0,0,0), which implies that the function u(x,y), corresponding to $v(r,\theta) = C(\theta)r$, will not be C^1 at (0,0) even though C is C^1. This oddity arises, because the transformation $(x,y) \longrightarrow (r,\theta)$ is ill–defined at (0,0). When the cone degenerates to a plane through (0,0,0), we obtain a solution $u(x,y) = ax + by$ of the PDE (31) which is C^1 even at (0,0). We can easily demonstrate that any solution of (16) which is C^1 at *all* points (x,y) (even at (0,0)) *must* be of the form $u(x,y) = ax + by$ (i.e., with a planar graph). Indeed, if a solution is C^1 at (0,0), then it has a tangent plane, say P, at the origin. However, since all of the rays which form the surface are tangent to the surface at the origin, they must be tangent to P. A ray which is tangent to a plane must lie in the plane. Thus, these rays are contained in P, which is then the graph of u(x,y).

Summary 2.2

1. Characteristic curves : A curve in the xy–plane is called a characteristic curve for the PDE

$$a(x,y)u_x + b(x,y)u_y + c(x,y)u = f(x,y) , \quad u = u(x,y) , \tag{S1}$$

if at each point (x_0,y_0) on the curve, the vector $\mathbf{g}(x_0,y_0) = a(x_0,y_0)\mathbf{i} + b(x_0,y_0)\mathbf{j}$ is tangent to the curve. The family of characteristic curves can be found by solving the characteristic equation

$$\frac{dy}{dx} = \frac{b(x,y)}{a(x,y)} . \tag{S2}$$

The significance of the characteristic curves is that on each characteristic curve the PDE (S1) becomes an ODE for a function of a positional variable along the curve (cf. parts 2 and 3 below).

2. Change of variables : Let the family of solutions of (S2) (i.e. the family of characteristic curves) be written implicitly as $h(x,y) = d$, where d is an arbitrary constant. Then under the change of variables

$$w = h(x,y) , \quad z = y , \quad \text{and with} \quad v(w,z) \equiv u(x,y) , \tag{S3}$$

the PDE (S1) becomes a PDE for v which involves v_z but not v_w . In place of "$z = y$" in (S3), one can use $z = k(x,y)$, for any convenient C^1 function $k(x,y)$. But in general, the PDE for v will be equivalent to (S1) only for domains in the xy–plane where the transformation $w = h(x,y)$, $z = k(x,y)$ is uniquely invertible. In order to achieve the general solution, it is necessary to paste together solutions on such domains in such a way that the solution is C^1 across the borders of the domains (cf. Example 5).

3. The characteristic system : A characteristic curve may be thought of as a path that is traced out by a particle which flows with a fluid which has velocity $\mathbf{g}(x,y) = a(x,y)\mathbf{i} + b(x,y)\mathbf{j}$. The functions $x(t)$ and $y(t)$ which give the position $(x(t),y(t))$ of such a particle at time t, satisfy the characteristic system

$$\frac{dx}{dt} = a(x(t),y(t)) \quad \text{and} \quad \frac{dy}{dt} = b(x(t),y(t)) \tag{S4}$$

By definition, a solution $(x(t),y(t))$ of (S4) constitutes a preferred parametrization of the characteristic curve which it traces out. Let $u(x,y)$ be a solution of (S1), and let $U(t) \equiv u(x(t),y(t))$, $C(t) = c(x(t),y(t))$, and $F(t) \equiv f(x(t),y(t))$, then $U(t)$ satisfies the ODE

$$U'(t) + C(t)U(t) = F(t), \quad \text{with solution} \quad U(t) = \frac{1}{m(t)}\left[\int_0^t m(t)F(t)\,dt + U(0)\right] , \tag{S5}$$

where $m(t) \equiv \exp(\int_0^t C(t)\,dt)$. Thus, the value $U(t)$ of any solution u of the PDE (S1) at $(x(t),y(t))$ along a characteristic curve, is determined by its value $U(0) = u(x(0),y(0))$.

4. Parametric form of solutions : Let s be a position variable along a regular side–condition curve which transversely intersects each characteristic curve that it encounters exactly once. For each fixed s, let $(X(s,t),Y(s,t))$ be the position, at time t, of the particle which begins at the point corresponding to s on the side condition curve, and flows with the fluid. In other words, the functions $X(s,t)$ and $Y(s,t)$ are the solutions of the characteristic system (for each fixed s)

$$\frac{d}{dt} X(s,t) = a(X(s,t),Y(s,t)) \quad \text{and} \quad \frac{d}{dt} Y(s,t) = b(X(s,t),Y(s,t)) , \tag{S6}$$

with given initial values $X(s,0)$ and $Y(s,0)$, where $(X(s,0),Y(s,0))$ traces out the side condition curve as s varies. Given the PDE (S1) with side condition $u(X(s,0),Y(s,0)) = G(s)$, then by applying the formula (S5) for each fixed s, we have

$$u(X(s,t),Y(s,t)) = U(s,t) = \frac{1}{m(s,t)} \left[\int_0^t m(s,t)\, F(s,t)\, dt + G(s) \right] \tag{S7}$$

Consequently, we have the following solution of the PDE (S1) in parametric form

$$x = X(s,t) \ , \ y = Y(s,t) \ , \ u = U(s,t) \ . \tag{S9}$$

In the event that we can uniquely solve the first two equations in (S9) for s and t in terms of x and y , say $s = S(x,y)$ and $t = T(x,y)$, we obtain an explicit solution $u(x,y) = U(S(x,y),T(x,y))$. As s and t vary, the point $(X(s,t),Y(s,t),U(s,t))$ typically traces out a surface which contains the graph of an explicit solution $u(x,y)$, if such exists.

Exercises 2.2

1. Obtain the general solution of each of the following PDEs in the indicated domain.

(a) $xu_x + 2yu_y = 0$, for $x > 0, y > 0$ (b) $xu_x - 2yu_y + u = e^x$, for $x > 0$

(c) $xu_x - xyu_y - u = 0$, for all (x,y) (d) $yu_x - 4xu_y = 2xy$, for all (x,y).

2. Find the particular solution of the PDEs in Problem 1 satisfying the following respective side conditions.

(a) $u(x,1/x) = x \quad (x > 0)$ (b) $u(1,y) = y^2$

(c) $u(x,x) = x^2e^x$ (d) $u(x,0) = x^4$.

3. Find the parametric form of the solutions of the PDEs in Problem 1, which satisfy the following respective side conditions. In each case, note the futility of finding an explicit solution $u(x,y)$.

(a) $u(s,e^{-s}) = \sin(s)$, $s > 0$ (b) $u(s,\sinh(s)) = 0$, $s > 0$

(c) $u(s^2,s) = s^3$ (d) $u(s,s^3) = 1$.

4. Show that the PDE in Problem 1(d) has no solution satisfying the side condition $u(x,0) = x^3$. Explain this result in terms of characteristic curves.

5. Show that the only solutions of the PDE in Problem 1(a) that are C^1 and defined for all (x,y) are the constant functions (e.g., $u(x,y) \equiv 5$).
Hint. Note that the characteristic curves all issue from the origin.

6. Note that the general solution $u(x,y)$ of the PDE in Problem 1(c), namely $u(x,y) = xC(ye^x)$, has the property that $u(0,y) = 0$. Thus, $u(0,y)$ cannot be arbitrarily prescribed, even though the y–axis crosses each "characteristic curve" $y = de^{-x}$ only once. Explain this apparent discrepancy by giving the characteristic curves their preferred parametrizations $(x(t),y(t))$ with $x'(t) = x(t)$ and $y'(t) = -x(t)y(t)$. Note that each curve $y = de^{-x}$ is composed of three such characteristic curves, one of which is a point on the y–axis.

7. Construct a solution of the PDE $xu_x - 2yu_y = 0$, which is C^1 throughout the xy–plane, but which is not of the form $u(x,y) = C(yx^2)$ for a C^1 function C.

8. Consider the PDE $\sin(x)u_x - y\cos(x)u_y = 0$.

(a) Sketch the characteristic curves of this PDE .

(b) Show that any regular side condition curve, which transversly (i.e., at a nonzero angle) intersects, exactly once, any characteristic curve of this PDE that it encounters, must be contained in a vertical strip of width 2π.

(c) Deduce that infinitely many side condition curves are needed in order to uniquely determine a solution of this PDE, which is defined throughout the xy–plane.

(d) Show that, given an infinite family of C^1 functions, say $f_n(y)$, such that $f_n(0) = 0$ and $f_n'(0) = 0$ $(n = 0, \pm1, \pm2, \ldots)$, there is a solution $u(x,y)$ (C^1 for all (x,y)) of the PDE which satisfies each of the infinitely many side conditions $u([n+\frac{1}{2}]\pi,y) = f_n(y)$, for $n = 0, \pm1, \pm2, \ldots$.

9. In the continuity equation $\rho_t + v(x,t)\rho_x + v_x(x,t)\rho = 0$, suppose that $v(x,t) = \alpha x^n$ for an integer $n > 1$ and a constant $\alpha > 0$. Solve this equation, subject to the initial condition $\rho(x,0) = \rho_0(x)$. Show that contrary to expectations, the density $\rho(0,t)$ at the origin is independent of t.

Moreover, show that the solution exists provided $\alpha(n-1)tx^{n-1} > -1$, which is true for $t \geq 0$ and all x, if n is odd. What if n is even ? When n is even, discuss the nature and possible physical significance of the solution for negative t.

2.3 Higher Dimensions, Quasi–linearity, Applications

The method of characteristics also applies to the case of linear first–order PDEs in higher dimensions. For example, in dimension 3, the most general linear first–order PDE is

$$a(x,y,z)u_x + b(x,y,z)u_y + c(x,y,z)u_z + d(x,y,z,)u = f(x,y,z)\ , \quad u = u(x,y,z)\ , \tag{1}$$

for given C^1 functions a, b, c, d, and f . The characteristic curves $(x(t),y(t),z(t))$ [parametrized by the preferred parameter; cf. (8) of Section 2.2] are the solutions of the system

$$\frac{dx}{dt} = a(x(t),y(t),z(t))\ , \quad \frac{dy}{dt} = b(x(t),y(t),z(t))\ , \quad \frac{dz}{dt} = c(x(t),y(t),z(t)) \tag{2}$$

In practice, it is usually more convenient to treat x as the parameter instead of t, in which case the above system is reduced to the two equations (assuming that $a(x,y,z) \neq 0$)

$$\frac{dy}{dx} = \frac{b(x,y(x),z(x))}{a(x,y(x),z(x))} \quad \text{and} \quad \frac{dz}{dx} = \frac{c(x,y(x),z(x))}{a(x,y(x),z(x))}\ , \tag{3}$$

for the unknowns $y(x)$ and $z(x)$. The solutions of (3) typically depend on two arbitrary constants, say α and β. Writing the solutions as $y(x;\alpha,\beta)$ and $z(x;\alpha,\beta)$, the curve traced out by the point $(x,y(x;\alpha,\beta),z(x;\alpha,\beta))$, as x varies, is a characteristic curve for each fixed pair of values for α and β. Now, suppose that we can uniquely solve the two equations

$$y = y(x;\alpha,\beta) \quad \text{and} \quad z = z(x;\alpha,\beta) \tag{4}$$

simultaneously for α and β in terms of x, y and z. Say we find $\alpha = A(x,y,z)$ and $\beta = B(x,y,z)$ for some functions A and B. The characteristic curve corresponding to the pair of values (α,β) is the intersection of two surfaces $A(x,y,z) = \alpha$ and $B(x,y,z) = \beta$ (Why ?). On this characteristic curve, the functions A and B are constant (namely α and β, respectively). The PDE reduces to an ODE, if we change coordinates so that characteristic curves are obtained when two of the new coordinates (say $\bar{x}$ and $\bar{y}$) are fixed and the remaining coordinate $\bar{z}$ varies. Let

$$\bar{x} = A(x,y,z)\ , \quad \bar{y} = B(x,y,z) \quad \text{and} \quad \bar{z} = z\ . \tag{5}$$

Note that when $\bar{x}$ and $\bar{y}$ are fixed, we obtain a characteristic curve. Ideally, we hope that the transformation (5) is invertible, or else some difficulties can arise, as we have seen in dimension 2. Letting $\bar{u}(\bar{x},\bar{y},\bar{z}) = u(x,y,z)$, we have

$$au_x + bu_y + cu_z$$

$$= a\cdot(\bar{u}_{\bar{x}}\bar{x}_x + \bar{u}_{\bar{y}}\bar{y}_x + \bar{u}_{\bar{z}}\bar{z}_x) + b\cdot(\bar{u}_{\bar{x}}\bar{x}_y + \bar{u}_{\bar{y}}\bar{y}_y + \bar{u}_{\bar{z}}\bar{z}_y) + c\cdot(\bar{u}_{\bar{x}}\bar{x}_z + \bar{u}_{\bar{y}}\bar{y}_z + \bar{u}_{\bar{z}}\bar{z}_z)$$

$$= (aA_x + bA_y + cA_z)\bar{u}_{\bar{x}} + (aB_x + bB_y + cB_z)\bar{u}_{\bar{y}} + c\bar{u}_{\bar{z}} = c\bar{u}_{\bar{z}} \; , \qquad (6)$$

where the two terms in parentheses drop out for the following reason. Let (x_0,y_0,z_0) be any point and let $(x,y(x),z(x))$ be a characteristic curve passing through this point (i.e., $y(x_0) = y_0$, $z(x_0) = z_0$). Since A is constant on any characteristic curve, we have (for $a \neq 0$)

$$0 = \frac{d}{dx}A(x,y(x),z(x)) = A_x + A_y\frac{dy}{dx} + A_z\frac{dz}{dx} = A_x + A_y\frac{b}{a} + A_z\frac{c}{a} = \frac{1}{a}(aA_x + bA_y + cA_z) \, .$$

Thus, $aA_x + bA_y + cA_z = 0$ at the arbitrary point (x_0,y_0,z_0). Similarly, $aB_x + bB_y + cB_z = 0$. By (6), we see that the PDE (1) becomes an ODE in $\bar{z}$ (for fixed $\bar{x}$ and $\bar{y}$),

$$\bar{c}(\bar{x},\bar{y},\bar{z})\bar{u}_{\bar{z}} + \bar{d}(\bar{x},\bar{y},\bar{z})\bar{u} = \bar{f}(\bar{x},\bar{y},\bar{z}) \; , \qquad (7)$$

where $\bar{c}$, $\bar{d}$ and $\bar{f}$ are c, d and f, written in terms of $\bar{x}$, $\bar{y}$ and $\bar{z}$, using the inverse transformation (if such exists) of (5). Then (7) can be solved for $\bar{u}(\bar{x},\bar{y},\bar{z})$, and

$$u(x,y,z) = \bar{u}(A(x,y,z),B(x,y,z),z),$$

is a solution of the original PDE (1).

There are a number of technical obstacles to carrying out all of this. One must solve the system (3) which is not an easy matter in general, although sometimes the solution presents itself in the desired form $A(x,y,z) = \alpha$ and $B(x,y,z) = \beta$, making it unnecessary to solve (4) simultaneously for α and β. Also, the inverse transformation of (5) may be ill–defined or hard to obtain. We have seen difficulties that can arise in dimension 2 because of ill–defined inverse transformations. In dimension 3, the "global" situation can be further complicated because of the possibility that the characteristic curves can be knotted and linked. In the examples and exercises, we will keep things fairly simple.

Example 1. Find the general solution of the PDE

$$2u_x + 3u_y + 5u_z - u = 0 \, , \quad u = u(x,y,z) \, . \qquad (8)$$

Solution. The characteristic curves are found by solving the system

$$\frac{dy}{dx} = \frac{3}{2} \, , \qquad \frac{dz}{dx} = \frac{5}{2} \, .$$

We obtain $y = \frac{3}{2}x + \frac{\alpha}{2}$, $z = \frac{5}{2}x + \frac{\beta}{2}$. Alternatively, the characteristic curves, are the lines given by the intersection of the level surfaces (planes, here) $2y–3x = \alpha$ and $2z–5x = \beta$. Hence, we make the change of variables

$$\begin{cases} \bar{x} = 2y - 3x \\ \bar{y} = 2z - 5x \\ \bar{z} = z \end{cases} ; \qquad \begin{cases} x = -\bar{y}/5 + 2\bar{z}/5 \\ y = (5\bar{x} - 3\bar{y} + 6\bar{z})/10 \\ z = \bar{z}\,. \end{cases}$$

Here, the inverse transformation exists and was computed, although we will have no need for it. Setting $\bar{u}(\bar{x},\bar{y},\bar{z}) = u(x,y,z)$, we obtain

$$\begin{aligned} 2u_x + 3u_y + 5u_z &\\ &= 2(\bar{u}_{\bar{x}}\bar{x}_x + \bar{u}_{\bar{y}}\bar{y}_x + \bar{u}_{\bar{z}}\bar{z}_x) + 3(\bar{u}_{\bar{x}}\bar{x}_y + \bar{u}_{\bar{y}}\bar{y}_y + \bar{u}_{\bar{z}}\bar{z}_y) + 5(\bar{u}_{\bar{x}}\bar{x}_z + \bar{u}_{\bar{y}}\bar{y}_z + \bar{u}_{\bar{z}}\bar{z}_z) \\ &= (-6 + 6 + 0)\bar{u}_{\bar{x}} + (-10 + 0 + 10)\bar{u}_{\bar{y}} + 5\bar{u}_{\bar{z}}\,. \end{aligned}$$

Thus (8) becomes $5\bar{u}_{\bar{z}} - \bar{u} = 0$, whose solution is $\bar{u}(\bar{x},\bar{y},\bar{z}) = C(\bar{x},\bar{y})e^{\bar{z}/5}$, where $C(\bar{x},\bar{y})$ is an arbitrary C^1 function of $(\bar{x},\bar{y})$. In terms of x, y and z, the general solution of (8) is

$$u(x,y,z) = C(2y-3x,2z-5x)e^{z/5}\,. \ \square \tag{9}$$

Remark. The function C can be determined by imposing a side condition on a surface that cuts each characteristic line in a single point. Consider the following side condition on the xy–plane :

$$u(x,y,0) = x^2\sin(y)\,. \tag{10}$$

Equations (9) and (10) tell us that

$$C(2y-3x,-5x) = u(x,y,0) = x^2\sin(y)\,. \tag{11}$$

Now, set $r = 2y - 3x$, $s = -5x$. Then $x = -s/5$ and $y = (r - 3s/5)/2$, and (11) gives

$$C(r,s) = (-s/5)^2\sin(r/2 - 3s/10)\,.$$

The desired solution of the PDE (8), with side condition (10), is then

$$\begin{aligned} u(x,y,z) &= ((5x-2z)/5)^2\sin((2y-3x)/2 - 3(2z-5x)/10)\cdot e^{z/5} \\ &= (x-2z/5)^2\sin(y-3z/5)\cdot e^{z/5}\,. \ \square \end{aligned}$$

Example 2. Find the general solution of

$$u_x + zu_y + 6xu_z = 0\,, \quad u = u(x,y,z)\,. \tag{12}$$

Solution. The characteristic curves are found by solving the system

$$\frac{dy}{dx} = z\,, \qquad \frac{dz}{dx} = 6x\,,$$

for $y = y(x)$ and $z= z(x)$. Note that the first equation *cannot* be integrated to give $y = zx$, because z is an unknown function of x. The second equation can be integrated, yielding $z = 3x^2 + \alpha$. Then, the first equation gives $y = x^3 + \alpha x + \beta$. Thus, the characteristic curves are traced out by the point $(x,x^3 + \alpha x + \beta,3x^2 + \alpha)$, as x varies. We solve for α and β in terms of (x,y,z) to obtain $\alpha = z - 3x^2$, $\beta = y - x^3 - (z - 3x^2)x = y + 2x^3 - xz$. The characteristic curves are the intersections of the surfaces $\alpha = z - 3x^2$, $\beta = y + 2x^2 - xz$. We change variables :

$$\begin{cases} \bar{x} = z - 3x^2 \\ \bar{y} = y + 2x^3 - xz \\ \bar{z} = z \end{cases} \quad ; \quad \begin{cases} x = \pm[\tfrac{1}{3}(\bar{z} - \bar{x})]^{\frac{1}{2}} & (\bar{z} \geq \bar{x}) \\ y = \bar{y} - \pm 2[\tfrac{1}{3}(\bar{z} - \bar{x})]^{\frac{3}{2}} \pm [\tfrac{1}{3}(\bar{z} - \bar{x})]^{\frac{1}{2}}\,\bar{z} \\ z = \bar{z}\,. \end{cases}$$

Note that the inverse transformation is not well defined, unless we either restrict to the domain $x \geq 0$ (in which case we use "+" in the equation for x) or restrict to $x \leq 0$. Thus, we must be careful when claiming that any solution found via the transformation is the most general solution. Setting $\bar{u}(\bar{x},\bar{y},\bar{z}) = u(x,y,z)$, the PDE (12) becomes $6x\bar{u}_{\bar{z}} = 0$, and we arrive at the solution $\bar{u}(\bar{x},\bar{y},\bar{z}) = C(\bar{x},\bar{y})$ or

$$u(x,y,z) = C(z - 3x^2,y + 2x^3 - xz)\,, \tag{13}$$

where C is an arbitrary C^1 function. However, we can only assert with confidence that this is the general solution in the domain $x \geq 0$ (or in the domain $x \leq 0$), where the transformation is uniquely invertible. Actually, any solution must be of the form

$$u(x,y,z) = \begin{cases} C(z - 3x^2,\, y + 2x^3 - xz) & x \geq 0 \\ D(z - 3x^2,\, y + 2x^3 - xz) & x \leq 0\,. \end{cases} \qquad (C,\, D \in C^1)$$

However, in order for the solution to be well–defined at $x = 0$, we need $C(z,y) = D(z,y)$. Hence, the functions C and D must be the same, and (13) *is* in fact the most general solution which is defined for all (x,y,z). Since $u(0,y,z) = C(z,y)$, the function C would be immediately determined by a side condition specifying u on the plane $x = 0$. □

Quasi–linear First–Order PDEs and the Method of Lagrange

The general first–order PDE for $u = u(x,y)$ is of the form $F(x,y,u,u_x,u_y) = 0$, where F is a function of five variables (e.g., $F(x,y,u,u_x,u_y) = u^2u_x^3 - 3xu_xu_y + y^2$). While there is an extension of the method of characteristics which can be used to solve such equations (cf. Section

2.4), the explanation of the solution procedure is rather lengthy. However, there is one class of first–order PDEs which we can attack now.

A **first–order quasi–linear PDE** is an equation of the form

$$a(x,y,u)u_x + b(x,y,u)u_y - c(x,y,u) = 0 \quad (u = u(x,y)), \tag{14}$$

where a, b and c are given C^1 functions of three variables.

In the special case when a and b do not depend on u and $c(x,y,u) = -C(x,y)u + f(x,y)$, (14) becomes the first–order linear PDE $a(x,y)u_x + b(x,y)u_y + C(x,y)u = f(x,y)$, which we have already considered. However, when a and b depend on u, (14) is *nonlinear.* In 1779, Joseph Lagrange showed that solutions of (14) can be expressed implicitly as $\varphi(x,y,u) = 0$, where $\varphi(x,y,z)$ is a solution of the linear PDE (in dimension 3),

$$a(x,y,z)\varphi_x + b(x,y,z)\varphi_y + c(x,y,z)\varphi_z = 0 , \tag{15}$$

which is a special case of (1). First, suppose that $u(x,y)$ is a solution of (14). If we define $\varphi(x,y,z) = u(x,y) - z$, then at any point $(x,y,z) = (x,y,u(x,y))$, on the graph of u,

$$a(x,y,z)\varphi_x+b(x,y,z)\varphi_y+c(x,y,z)\varphi_z = a(x,y,u(x,y))u_x+b(x,y,u(x,y))u_y+c(x,y,u(x,y))\cdot(-1) = 0 ,$$

by virtue of equation (14). Conversely, suppose that φ is a solution of (15), such that the normal vector $\nabla\varphi = \varphi_x\mathbf{i} + \varphi_y\mathbf{j} + \varphi_z\mathbf{k}$ to the surface $\varphi(x,y,z) = 0$ at some point $p = (x_0,y_0,z_0)$ is not horizontal (i.e., $\varphi_z(p) \neq 0$). Then near p, the surface will be the graph of some function $u(x,y)$ [i.e., $\varphi(x,y,u(x,y)) = 0$]. We can show that $u(x,y)$ must be a solution of (14), as follows. Differentiating the equation $\varphi(x,y,u(x,y)) = 0$ with respect to x and y, we have

$$\varphi_x(x,y,u(x,y)) + \varphi_z(x,y,u(x,y))u_x(x,y) = 0 \text{ and } \varphi_y(x,y,u(x,y)) + \varphi_z(x,y,u(x,y))u_y(x,y) = 0 .$$

Thus, $u_x = -\varphi_x/\varphi_z$ and $u_y = -\varphi_y/\varphi_z$. Substituting these expressions for u_x and u_y into the left side of (14), we obtain

$$-[a(x,y,u(x,y))\varphi_x + b(x,y,u(x,y))\varphi_y + c(x,y,u(x,y))\varphi_z]/\varphi_z ,$$

which is 0, by the assumption that φ satisfies (15). Thus, $u(x,y)$ solves (14). In summary, the method of Lagrange yields the following fact.

Solutions $u = u(x,y)$ of the quasi–linear PDE (14) can be implicitly defined by $\varphi(x,y,u) = 0$, where $\varphi(x,y,z)$ solves the linear PDE (15), with $\varphi_z(p) \neq 0$ at some point p where $\varphi(p) = 0$.

Remark. There is a simple geometrical idea behind the the method of Lagrange. Let $\vec{v}(x,y,z) = a(x,y,z)\mathbf{i} + b(x,y,z)\mathbf{j} + c(x,y,z)\mathbf{k}$ be a given vector field in space. The PDE $a(x,y,u)u_x + b(x,y,z)u_y - c(x,y,u) = 0$ (i.e., $\mathbf{v}\cdot(u_x\mathbf{i} + u_y\mathbf{j} - \mathbf{k}) = 0$) says that $\mathbf{v}$ is tangent to the graph of u at all points $(x,y,u(x,y))$. Suppose that we think of the graph of u as a surface defined implicitly by $\varphi(x,y,z) = 0$. Since $\nabla\varphi$ is normal to this surface, the condition that $\mathbf{v}$ is tangent to the surface is implied by $\nabla\varphi\cdot\mathbf{v} = 0$, which is precisely the linear PDE (15). □

Example 3. Find a solution of the following quasi–linear PDE with the given side condition

$$u_x + u\cdot u_y = 6x\,, \quad u(0,y) = 3y\,. \tag{16}$$

Solution. The associated linear PDE in dimension 3, is

$$\varphi_x + z\cdot\varphi_y + 6x\varphi_z = 0\,. \tag{17}$$

This is the same PDE which was solved in Example 2. By (13), the general solution is $\varphi(x,y,z) = C(z - 3x^2, y + 2x^3 - xz)$, where C is an arbitrary C^1 function. Hence the solutions of the PDE in (16) are given implicitly by

$$C(u - 3x^2, y + 2x^3 - xu) = 0, \tag{18}$$

for various choices of the function C. For example, if $C(r,s) = r$, then we obtain the solution $u - 3x^2 = 0$, or explicitly $u(x,y) = 3x^2$, while if $C(r,s) = r - s$, then we get $u - 3x^2 - y - 2x^3 + xu = 0$, or explicitly $u(x,y) = (y + 3x^2 + 2x^3)/(1+x)$. The side condition in (16) can be used to determine the function C. Indeed, since this condition says that $u = 3y$ when $x = 0$, we substitute $3y$ for u, and 0 for x, in (18), arriving at the condition $C(3y,y) = 0$. There are many functions C which satisfy this condition, and a simple choice is $C(r,s) = r - 3s$. This yields the solution

$$u - 3x^2 - 3(y+2x^3-xu) = 0 \quad \text{or} \quad u(x,y) = 3(y+x^2+2x^3)/(1+3x)\,. \; □ \tag{19}$$

Remark. If we replace the side condition in (16) by the more general condition $u(0,y) = G(y)$ (for a given C^1 function G), then the function C must be chosen so that $C(G(y),y) = 0$. Again, the choice for C is not unique, but a simple possibility is $C(r,s) = r - G(s)$. Thus, a solution of the PDE in (16) might be obtained implicitly via the equation

$$u - 3x^2 - G(y+2x^3-xu) = 0\,. \tag{20}$$

In general, only rarely is it possible to solve (20) for u explicitly in terms of x and y. Nevertheless, at the point $(0,y,G(y))$ on the surface $\varphi(x,y,z) = z-3x^2 - G(y+2x^3-xz) = 0$, the $\mathbf{k}$ component of the normal vector $\nabla\varphi(0,y,G(y))$ is 1 (nonzero). Thus, the surface is not vertical at these points, and hence the surface is then the graph of some solution $u(x,y)$ (defined in a neighborhood of the y–axis) of the PDE in (16), even though the explicit formula for $u(x,y)$ may be elusive. Recall that in Section 2.2, we were able to determine parametric solutions $X(s,t)$, $Y(s,t)$, $U(s,t)$, even when finding explicit solutions $u(x,y)$ was futile. One can readily obtain parametric solutions of the quasi–linear equation (14) from the characteristic curves of the of the associated linear PDE (15) in xyz–space. We illustrate the method in the following example. □

Example 4. Find a parametric solution of the following quasi–linear PDE with side condition

$$u_x + u \cdot u_y = 6x\,, \quad u(0,y) = G(y)\,, \tag{21}$$

where $G(y)$ is an arbitrary C^1 function.

Solution. The characteristic system (cf. (2)) for the characteristic curves $(x(t),y(t),z(t))$ associated with the linear PDE $\varphi_x + z\varphi_y + 6x\varphi_z = 0$ is

$$\frac{dx}{dt} = 1\,, \quad \frac{dy}{dt} = z\,, \quad \frac{dz}{dt} = 6x\,. \tag{22}$$

The general solution is

$$x(t) = t + \alpha\,, \quad y(t) = t^3 + 3\alpha t^2 + \beta t + \gamma\,, \quad z(t) = 3t^2 + 6\alpha t + \beta\,, \tag{23}$$

where we solved first for x, then for z, and finally for y. The PDE $\varphi_x + z\varphi_y + 6x\varphi_z = 0$ implies that φ is constant on any characteristic curve, since

$$\frac{d}{dt}\varphi(x(t),y(t),z(t)) = \varphi_x \cdot x'(t) + \varphi_y \cdot y'(t) + \varphi_z \cdot z'(t) = \varphi_x + \varphi_y \cdot z(t) + \varphi_z \cdot 6x(t) = 0\,,$$

by the chain rule and (22). In other words, each characteristic curve lies on a surface of the form $\varphi(x,y,z) = \text{constant}$. The graph of the solution $u(x,y)$ is one of these surfaces, namely $\varphi(x,y,z) = 0$, where $\varphi(x,y,z) = z - 3x^2 - G(y+2x^3-xz)$. This suggests that the graph of $u(x,y)$ consists of a family of characteristic curves. Of course, we want each curve of this family to pass through a point of the form $(0,s,G(s))$, in order that that the side condition be met. Let $x = X(s,t)$, $y = Y(s,t)$, $z = Z(s,t)$ be the curve which passes through $(0,s,G(s))$ at "time" $t = 0$. By setting $x(0) = 0$, $y(0) = s$, and $z(0) = G(s)$ in (23), we find $\alpha = 0$, $\beta = G(s)$ and $\gamma = s$. Then

$$X(s,t) = t\,, \quad Y(s,t) = t^3 + G(s)t + s\,, \quad Z(s,t) = 3t^2 + G(s)\,.$$

As s and t vary, we get a surface which passes though the curve $(0,s,G(s))$, as required by the side condition. Since this surface is comprised of characteristic curves on which φ is constant (i.e. independent of t, we know that $\varphi(X(s,t),Y(s,t),Z(s,t)) = \varphi(X(s,0),Y(s,0),Z(s,0)) = \varphi(0,s,G(s)) = 0$. Thus, as s and t vary, the point $(X(s,t),Y(s,t),Z(s,t))$ traces out a set of points (x,y,u) such that $\varphi(x,y,u) = 0$ (i.e., the graph of u). In other words, we have the parametric solution

$$x = X(s,t) = t\,, \quad y = Y(s,t) = t^3 + G(s)t + s\,, \quad u = U(s,t) \equiv Z(s,t) = 3t^2 + G(s)\,. \quad \square$$

An Application to Traffic Flow

Let $\rho(x,t)$ be the density of cars at the point x at time t on a one–way road (i.e., $\int_a^b \rho(x,t)\,dx$ is the number of cars between $x = a$ and $x = b$). We make the simplifying assumption that $\rho(x,t)$ is C^1. Let M be the the legal speed limit plus the additional 5 mph which one can usually add with impunity. Let d be the density of bumper–to–bumper traffic. Then one might assume that the traffic velocity $v(x,t)$ at x at time t, is given by $v(x,t) = M\cdot(1 - \rho(x,t)/d)$. Note that $v = 0$ when $\rho = d$, and $v = M$ when $\rho = 0$. However, when $\rho = \frac{1}{2}d$ (i.e., there is about a car–length between cars), we have $v = \frac{1}{2}M$, which is rather unsafe if $M = 60$, but let us proceed. The equation of continuity $\rho_t + (v\rho)_x = 0$ (cf. (28) of Section 2.2) holds for traffic flow as well as for gas flow. Since $(v\rho)_x = [M(1 - \rho/d)\rho]_x = M(1 - 2\cdot\frac{\rho}{d})\rho_x$, we have the quasi–linear PDE for the traffic density

$$\rho_t + M(1 - 2\cdot\tfrac{\rho}{d})\rho_x = 0\,. \tag{24}$$

The associated linear PDE for $\varphi(x,t,z)$ is $\varphi_t + M(1 - 2\cdot\frac{z}{d})\varphi_x + 0\cdot\varphi_z = 0$. We directly obtain the parametric form of the solution, as follows. We use the parameter τ instead of t which is already used in the equation. The characteristic equations are $t'(\tau) = 1$, $x'(\tau) = M(1 - 2\cdot\frac{z}{d})$, and $z'(\tau) = 0$. Thus, for arbitrary constants α, β and γ, the characteristic curves are given by

$$t(\tau) = \tau + \alpha\,, \quad x(\tau) = M(1 - \tfrac{2}{d}\cdot\gamma)\tau + \beta\,, \quad z(\tau) = \gamma\,. \tag{25}$$

Suppose that we are given the initial density $\rho(x,0) = f(x)$, or $\rho(s,0) = f(s)$. For a fixed s, we want to choose a characteristic curve of the form (25) which runs through $(0,s,f(s))$ when $\tau = 0$. Thus, choose $\alpha = 0$, $\beta = s$ and $\gamma = f(s)$. The parametric solution is then

$$t = T(s,\tau) = \tau\,, \quad x = X(s,\tau) = M(1 - \tfrac{2}{d}\,f(s))\tau + s\,, \quad \rho = Z(s,\tau) = f(s)\,. \tag{26}$$

Let s have some fixed value, say x_0. Then (26) implies that the density $\rho(x,t)$ is a constant (namely $f(x_0)$) on the line

$$x = M(1 - \tfrac{2}{d}\cdot f(x_0))t + x_0 \tag{27}$$

in the xt–plane. If we change the value of x_0 to a new value x_1 and $f(x_0) \neq f(x_1)$, then the new line will intersect the old line at some point (x_2,t_2), since the slopes of the two lines will differ. Moreover, since $f(x_0) \neq f(x_1)$, the constant value of ρ on the new line will not equal the constant value of ρ on the old line. Thus, at the intersection point, we arrive at a contradiction, $\rho(x_2,t_2) = f(x_0)$ and $\rho(x_2,t_2) = f(x_1)$. Hence, the only solutions $\rho(x,t)$ of equation (24) which are defined for *all* (x,t) are the solutions where the initial density $\rho(x,0) = f(x)$ is constant, say $f(x) \equiv c$, in which case $\rho(x,t) \equiv c$ (i.e., the traffic moves with a uniform velocity $M(1 - \frac{c}{d})$, if $0 \leq c \leq d$). It may happen that a nonconstant solution will exist for all $t \geq 0$ (but not all $t < 0$). Indeed, if the initial density $f(x)$ of cars is chosen to be decreasing in the positive x direction (i.e., $f'(x) \leq 0$), then $f(x_0) > f(x_1)$ if $x_1 > x_0$, and the intersection point (x_2,t_2) will lie below the x–axis (i.e., $t_2 < 0$), since the slope of the line through $(x_1,0)$ is less than the slope of the line through $(x_0,0)$. However, if $f'(x)$ is positive at some point, say x_0, then for some $x_1 > x_0$, we have $f(x_1) > f(x_0)$ and the corresponding lines will intersect above the x–axis (i.e., the solution $\rho(x,t)$ will fail to exist at (x_2,t_2) where $t_2 \geq 0$). When $f'(x) > 0$ somewhere, we will now find the smallest time $t > 0$, for which the solution will fail to exist. At any fixed time t_0, the "graph" of the parametric solution (26) is the curve in the $x\rho$–plane given parametrically (as s varies) by

$$x(s) = M(1 - \tfrac{2}{d} f(s))t_0 + s\,, \quad \rho(s) = \rho(s,t_0) = f(s) \tag{28}$$

The tangent vector of this curve at $(x(s),\rho(s))$ is

$$x'(s)\mathbf{i} + \rho'(s)\mathbf{j} = \left[1 - \frac{2M}{d} f'(s)t_0\right] \mathbf{i} + f'(s)\mathbf{j}\,, \tag{29}$$

Thus, the tangent vector at $(x(s),\rho(s))$ will be vertical when $f'(s) \neq 0$ and $1 - \frac{2M}{d} f'(s)t_0 = 0$ (or $t_0 = d[2M \cdot f'(s)]^{-1}$). Let G be the largest value (or more precisely, the smallest upper bound) for $f'(x)$ (i.e., G is the maximal initial density gradient for the traffic). Assume that $G < \infty$. Then, as long as $t_0 < d/(2MG)$, there will be no vertical tangent to the density profile $\rho(x,t_0)$ and the solution will exist for all (x,t) with $t < d/(2MG)$. However, unless $f'(x) \equiv G$, the solution $\rho(x,t)$ fails to be C^1 for t "slightly" greater than $d/(2MG)$, since there will be vertical tangents at such times. Indeed, the density profile will typically double back on itself and cease to be the graph of a function, such as a wave that is breaking (cf. Figure 1 below). Note that $\rho(x,t)$ itself never exceeds d, if $f(x) \equiv \rho(x,0) < d$. However, the theory predicts that if $f'(x) > 0$ somewhere, the density will develop a sharp jump (i.e., a vertical tangent) or what is known as a **shock**. When a shock occurs, the density gradient is infinite, and this necessitates a rapid change in the velocity of cars approaching the shock point. Since car breaks can only act so fast, the theory suggests that accidents are likely to happen at the shock points. Moreover, as the next example illustrates, the theory can be used to predict where and when shocks are likely to occur.

Example 5. With the above notation, let the initial density of cars be $\rho(x,0) = a(1 + x^2)^{-1}$, for some positive constant $a < d$. Initially, the point of maximum density is at $x = 0$. Where is the point of maximum density at time t ? When and where does the first shock arise ?

Solution. For each x_0, the solution has the constant value $f(x_0) = a(1 + x_0^2)^{-1}$ on the line $x = M(1 - \frac{2}{d} \cdot f(x_0))t + x_0$ (cf. (28)). Setting $x_0 = 0$, we immediately see that the maximum density is at $x = M(1 - 2 \cdot \frac{a}{d})t$, at time t. Thus, the point of maximum density will move to the right if $a < d/2$, and to the left if $a > d/2$, and it stays at $x = 0$, if $a = d/2$. The maximum of $f'(x) = -2ax(1 + x^2)^{-2}$ occurs at a value of x where $0 = f''(x) = 2a(3x^2 - 1) \cdot (1 + x^2)^{-3}$, namely $x_1 = -1/\sqrt{3}$. The maximum value of $f'(x)$ is $G = f'(-1/\sqrt{3}) = 2a(4/3)^{-2}/\sqrt{3} = (9a/8)/\sqrt{3} = (3a/8)\sqrt{3}$. Thus, the first shock occurs at time $t_1 = d/(2MG) = d/(2M \cdot (3a/8)\sqrt{3}) = \frac{4d}{9aM}\sqrt{3}$, and at the position $x = M(1 - \frac{2}{d} \cdot f(x_1))t_1 + x_1 = M(1 - \frac{2}{d} \cdot 3a/4)\frac{4d}{9aM}\sqrt{3} - 1/\sqrt{3} = (1 - \frac{6a}{4d})\frac{4d}{9a}\sqrt{3} - \frac{1}{3}\sqrt{3} = (\frac{4d}{9a} - 1)\sqrt{3}$. As one might expect, this is less than the point of maximum density at time t_1, namely the point $M(1 - 2 \cdot \frac{a}{d}) \cdot \frac{4}{9}\sqrt{3}\, d/(aM) = (\frac{4d}{9a} - \frac{8}{9})\sqrt{3}$. □

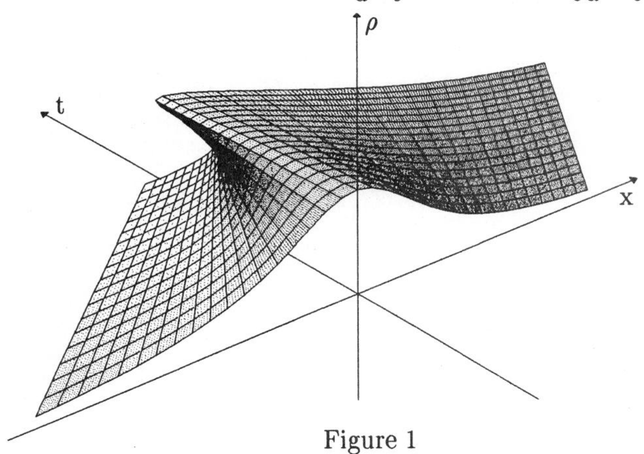

Figure 1

An application to continuum mechanics

Imagine a tube of gas or some possibly compressible medium (cf. the application to gas flow in Section 2.2) which has velocity $v(x,t)\mathbf{i}$. If $x(t)$ is the position of a small "gas element" at time t, then $x'(t) = v(x(t),t)$, and

$$x''(t) = \frac{d}{dt} v(x(t),t) = \frac{\partial v}{\partial x}\frac{dx}{dt} + \frac{\partial v}{\partial t}\frac{dt}{dt} = v_x(x(t),t) \cdot v(x(t),t) + v_t(x(t),t) .$$

Thus, the acceleration of of fluid particles at (x,t) is not simply $v_t(x,t)$, but rather $v \cdot v_x + v_t$.

A naive application of Newton's second law then yields the equation of motion

$$\rho(x,t)\cdot(v_t + v\cdot v_x) = f(x,t), \tag{30}$$

where $\rho(x,t)$ is the mass density and $f(x,t)$ is the force density (mass and force per unit length). While equation (30) is correct, it is not as obvious as we have led the reader to believe. Indeed, the more universal version of Newton's second law is that the total force on an object is the rate of change of the *momentum* (i.e., $f(t) = \frac{d}{dt}[m(t)v(t)] = m(t)v'(t) + m'(t)v(t)$) which is not simply mass times acceleration in the event that the mass changes with time. Since the density $\rho(x,t)$ does depend on time, equation (30) now *appears* somewhat doubtful. A careful guide through the correct derivation of (30) is supplied in Problem 13. In the absence of viscosity (which we assume is negligible), the force density $f(x,t)$ is the sum of the negative pressure gradient (i.e., $-p_x(x,t)$, where $p(x,t)$ is the pressure) and the external force density, say due to gravity. For simplicity, we assume that there are no external forces. Then **Euler's equations** are

$$\rho_t + (\rho v)_x = 0\ , \tag{31}$$

$$v_t + v\cdot v_x = -p_x/\rho\ , \tag{32}$$

$$p = f(\rho)\ . \tag{33}$$

Note that (31) is the equation of continuity (cf. (28) in Section 2.2), while (32) is (30). Equation (33) is known as the **equation of state** which gives us the pressure as a function of the density. The function f depends on the nature of the fluid or gas. For an ideal gas undergoing an adiabatic process (i.e., not giving up heat to the environment), we have $p = A\rho^\gamma$ (i.e., $f(\rho) = A\rho^\gamma$), where A and γ are positive constants which depend on the gas. For air, $\gamma \approx 1.4$, and usually $\gamma > 1$. By noting that $p_x = f'(\rho)\rho_x$, the equations (31) and (32) form the following system of PDEs :

$$\rho_t + v\cdot\rho_x = -v_x\rho \tag{34}$$

$$v_t + v\cdot v_x = -f'(\rho)\rho_x/\rho\ . \tag{35}$$

Finding the general solution of this system is a nontrivial undertaking. Through a process of linearization, the system can be approximately decoupled into two separate wave equations for v and ρ (cf. Problem 14). However, we can obtain some *exact* solutions and still remain within the context of first–order quasi–linear equations in a single unknown. Indeed, suppose that we attempt to find solutions for which $v(x,t) = V(\rho(x,t))$ for some function V (i.e., we search for solutions in which ρ and v are functionally related). Since $v_x = V'(\rho)\rho_x$ and $v_t = V'(\rho)\rho_t$,

$$\rho_t + V(\rho)\cdot\rho_x = -\rho V'(\rho)\rho_x\ , \tag{34'}$$

$$\rho_t + V(\rho)\cdot\rho_x = -f'(\rho)\rho_x/(\rho V'(\rho))\ . \tag{35'}$$

Since the left–hand sides of these equations are the same, we see that the function $V(\rho)$ must be chosen so that the right sides are equal, namely

$$\rho V'(\rho) = f'(\rho)/(\rho V'(\rho)) \quad \text{or} \quad V'(\rho) = \pm[f'(\rho)]^{\frac{1}{2}}/\rho \quad \text{or} \quad V(\rho) = \pm \int_{\rho_0}^{\rho} [f'(\rho)]^{\frac{1}{2}}/\rho \, d\rho \,. \tag{36}$$

where ρ_0 is the density when the velocity is $\mathbf{0}$. Thus, if we assume that v and ρ are functionally related, then the function which relates them is nearly determined by the equations (31), (32) and (33). However, there is no firm physical reason to assume a functional relation between v and ρ, as was done in the case of traffic flow. When $f(\rho) = A\rho^{\gamma}$, we obtain

$$V(\rho) = \pm \int_{\rho_0}^{\rho} [\gamma A \rho^{\gamma-1}]^{\frac{1}{2}}/\rho \, d\rho \;=\; \pm \frac{2\sqrt{\gamma A}}{\gamma-1} \cdot \left[\rho^{\frac{1}{2}(\gamma-1)} - \rho_0^{\frac{1}{2}(\gamma-1)}\right] . \tag{37}$$

Returning to the general case, since $V'(\rho) = \pm[f'(\rho)]^{\frac{1}{2}}/\rho$ is usually nonzero, we may reasonably assume that the function V has an inverse say R, so that $\rho = R(v)$. Since $R'(v) = V'(\rho)^{-1}$,

$$R'(v)/R(v) = 1/(\rho V'(\rho)) = \pm[f'(\rho)]^{-\frac{1}{2}} = \pm[f'(R(v)]^{-\frac{1}{2}} = \pm c(v)^{-1} ,$$

where

$$c(v) \equiv [f'(R(v)]^{\frac{1}{2}}. \tag{38}$$

Thus,

$$v_t + v \cdot v_x = -f'(\rho)\rho_x/\rho = -f'(R(v))R'(v)v_x/R(v) = - \pm c(v)v_x$$

or

$$v_t + (v \pm c(v))v_x = 0 \,. \tag{39}$$

This is a first–order quasi–linear PDE for v. The associated linear PDE in xtz–space is

$$\varphi_t + (z \pm c(z))\varphi_x - 0 \cdot \varphi_z = 0 \,.$$

The characteristic equations are $x'(t) = z \pm c(z)$ and $z'(t) = 0$. The solutions are $z = \alpha$ and $x - (z \pm c(z))t = \beta$. It follows that $\varphi(x,t,z) = C(z,x - (z \pm c(z))t)$, where C is an arbitrary C^1 function. The corresponding solution of (39) is then given implicitly by

$$\varphi(v,x - (v \pm c(v)) \cdot t) = 0 \,.$$

Suppose that the initial velocity is $v(x,0) = g(x)$ for some given function $g(x)$. Then we must choose φ such that $\varphi(g(x),x) = 0$. A simple choice for φ is then $\varphi(r,s) = r - g(s)$. As it may be difficult to solve $v - g(x - (v \pm c(v)) \cdot t) = 0$ for v in terms of x and t, the following parametric form (with τ as parameter, since t is already used in the equation) of the solution often proves more useful :

$$t = \tau , \quad x = [g(s) \pm c(g(s))]\tau + s , \quad v = g(s) . \tag{40}$$

As τ and s vary, the point (t,x,v) traces out the graph of the solution v(x,y) in txv–space. Observe that the value of v(x,t) on the line $x = [g(x_0) \pm c(g(x_0))]t + x_0$ is a constant, namely $g(x_0)$. In summary, we have shown the following :

The solution (where it exists) of the problem

$$\rho_t + (\rho v)_x = 0 , \tag{41}$$
$$v_t + v \cdot v_x = -p_x/\rho , \tag{42}$$
$$p = f(\rho) \tag{43}$$

with

$$v(x,0) = g(x) \quad \text{and} \quad \rho(x,0) = R(g(x)), \tag{44}$$

where R is the inverse function of V in (37), is given implicitly by

$$- g(x - (v \pm c(v))\cdot t) = 0 , \qquad \rho(x,t) = R(v(x,t)) , \tag{45}$$

or parametrically, by

$$t = \tau , \quad x = [g(s) \pm c(g(s))]\tau + s , \quad v = g(s) . \tag{46}$$

From the similarity of (46) with (26) for traffic flow, we expect to have the phenomenon of shocks. Note that (46) is the parametric solution for the velocity, while (26) is for the density. Since velocity and density are functionally related in our considerations for gas flow and traffic flow, it is easy to get the density and velocity solutions from one another in either situation.

Example 6. Assuming the equation of state $p = A\rho^\gamma$ $(\gamma > 1)$, find v(x,t) and $\rho(x,t)$ for the above problem (41) – (44) when $v(x,0) = \alpha x$ ($\alpha > 0$, constant), assuming $v = 0$ when $\rho = \rho_0$.

Solution. We take $g(x) = \alpha x$, and we need to determine R(v) and c(v). Since $f(\rho) = A\rho^\gamma$, (37) yields

$$V(\rho) = \pm \int_{\rho_0}^{\rho} [\gamma A \rho^{\gamma-1}]^{\frac{1}{2}}/\rho \, d\rho + V(\rho_0) = \pm \frac{2\sqrt{\gamma A}}{\gamma-1} \cdot \left[\rho^{\frac{1}{2}(\gamma-1)} - \rho_0^{\frac{1}{2}(\gamma-1)}\right] ,$$

$$\rho = R(v) = \left[\rho_0^{\frac{1}{2}(\gamma-1)} \pm \frac{\gamma-1}{2\sqrt{\gamma A}} \cdot v\right]^{2/(\gamma-1)}$$

and

$$c(v) \equiv [f'(R(v)]^{\frac{1}{2}} = \left[A\gamma R(v)^{\gamma-1}\right]^{\frac{1}{2}} = \sqrt{A\gamma}\, \rho_0^{\frac{1}{2}(\gamma-1)} \pm \tfrac{1}{2}(\gamma-1)v .$$

Let $c_0 = c(0) = [f'(\rho_0)]^{\frac{1}{2}} = \sqrt{A\gamma}\,\rho_0^{\frac{1}{2}(\gamma-1)}$. Then, we have the implicit solution

$$v - \alpha(x - (v \pm (c_0 \pm \tfrac{1}{2}(\gamma-1)v)\cdot t) = v - \alpha(x - (\pm c_0 + \tfrac{1}{2}(\gamma+1)v)\cdot t) = 0 .$$

Thus, $v\cdot(1 + \frac{1}{2}(\gamma+1)\alpha t) = \alpha(x - \pm c_0\cdot t)$, and explicitly

$$v(x,t) = \alpha(x - \pm c_0\cdot t)\cdot(1 + \tfrac{1}{2}(\gamma+1)\alpha t)^{-1} \tag{47}$$

and

$$\rho(x,t) = R(v(x,t)) = \left[\frac{1}{\sqrt{\gamma A}}[c_0 \pm \tfrac{1}{2}(\gamma-1)v(x,t)]\right]^{2/(\gamma-1)} \tag{48}$$

For definiteness, suppose that we have chosen the plus sign (cf. Problem 16, for the minus sign). Then for fixed t, the graph of $v(x,t)$ is a line which intersects the the x–axis at $c_0\cdot t$ and has slope $\alpha(1 + \frac{1}{2}(\gamma+1)\alpha t)^{-1}$. In Problem 14, we will see that c_0 is the velocity of sound in the gas when there is little wind (i.e., v is small). Thus we have shown that for our initial conditions, the intercept point $w(t)$ where $v(w(t),t) = 0$ moves with the velocity of sound, regardless of the value of α. The slope $\alpha(1 + \frac{1}{2}(\gamma+1)\alpha t)^{-1}$ tends to zero, as $t \to \infty$. Intuitively, this is because the wind is always blowing away from the from the zero velocity point, leaving slower moving wind behind. Note that the solution cannot be continued indefinitely backward in time, since the slope becomes infinite at $t_0 \equiv -[\frac{1}{2}(\gamma+1)\alpha]^{-1}$. This is a dramatic shock. As time is run backward the wind blows toward the zero velocity point and there is eventually a "big crunch" when all of the gas arrives at once. If we run time forward from time $t = t_0$, we get an explosion which issues from the point $x = c_0t_0$. Indeed, if $x(t)$ is the position of a gas element at time t, then by solving the ODE $x'(t) = v(x(t),t)$, we find (cf. Problem 15) that $x(t_0) = c_0t_0$, regardless of the choice of $x(0)$. In the case at hand, parametrically, the solution for v is given by (cf. (46))

$$t = \tau\,, \quad x = (c_0 + \tfrac{1}{2}(\gamma+1)\alpha s)\tau + s\,, \quad v = \alpha s\,. \tag{49}$$

In other words, v has the constant value αx_1 on the line $x = (c_0 + \frac{1}{2}(\gamma+1)\alpha x_1)t + x_1$. Observe that when $t = t_0 \equiv -[\frac{1}{2}(\gamma+1)\alpha]^{-1}$, we have $x = c_0t_0 - x_1 + x_1 = c_0t_0$. Thus, all of these lines pass through the shock point (c_0t_0,t_0), where v is then "grossly" undefined, as expected. We should also note that the solution is not really physically valid everywhere outside the shock point, because v must be at least $-2c_0/(\gamma-1)$ in order that the pressure be nonnegative, by (48). Substituting $-2c_0/(\gamma-1)$ for αs in (49), we see that the solution is only valid above the line which runs through the shock point (c_0t_0,t_0) with slope $(c_0 + \frac{1}{2}(\gamma+1)\cdot(-2c_0/(\gamma-1)))$ $= c_0(1 - \frac{\gamma+1}{\gamma-1}) = -2c_0/(\gamma-1)$. This then is the (constant!) velocity left–hand boundary of the expanding gas, where the pressure is zero. □

Summary 2.3

1. The method of characteristics in higher dimensions : The method of characteristic curves extends to the case of the first–order linear PDE

$$a(x,y,z)u_x + b(x,y,z)u_y + c(x,y,z)u_z + d(x,y,z)u = f(x,y,z) , \quad \text{with } u = u(x,y,z) , \qquad (S1)$$

for given C^1 functions a, b, c, d and f. With x as a positional variable along the characteristic curves (streamlines of the fluid flow with velocity vector field $a\mathbf{i} + b\mathbf{j} + c\mathbf{k}$) are traced out by $(x,y(x),z(x))$ as x varies, where $y(x)$ and $z(x)$ are solutions of the system of two equations (assuming that $a(x,y,z) \neq 0$)

$$\frac{dy}{dx} = \frac{b(x,y(x),z(x))}{a(x,y(x),z(x))} \quad \text{and} \quad \frac{dz}{dx} = \frac{c(x,y(x),z(x))}{a(x,y(x),z(x))} . \qquad (S2)$$

The solutions of (S2) typically depend on two arbitrary constants α and β, say $y = y(x;\alpha,\beta)$ and $z = z(x;\alpha,\beta)$. If it is possible to uniquely solve for α and β in terms of x, y and z, then the characteristic curves can be expressed as the intersection of the surfaces $A(x,y,z) = \alpha$ and $B(x,y,z) = \beta$. Then under the change of variables $\bar{x} = A(x,y,z)$, $\bar{y} = B(x,y,z)$, $\bar{z} = z$, the PDE (S1) is transformed (via the chain rule) to a PDE (cf. (7)), for $\bar{u}(\bar{x},\bar{y},\bar{z}) \equiv u(x,y,z)$, which does not involve $\bar{u}_{\bar{x}}$ and $\bar{u}_{\bar{y}}$ (i.e., (7) can be solved as an ODE for a function of $\bar{z}$). For first–order linear PDEs in any dimension, the basic idea is to introduce a change of variables in such a way that, when all but one of the new variables is held fixed, a characteristic curve results. Then the transformed PDE becomes an ODE for a function of the remaining new variable.

2. The method of Lagrange : Solutions $u = u(x,y)$ of the quasi–linear PDE

$$a(x,y,u)u_x + b(x,y,u)u_y - c(x,y,u) = 0, \quad u = u(x,y), \qquad (S3)$$

where a, b and c are given C^1 functions of three variables, can be implicitly defined by $\varphi(x,y,u) = 0$, where $\varphi(x,y,z)$ solves the linear PDE

$$a(x,y,z)\varphi_x + b(x,y,z)\varphi_y + c(x,y,z)\varphi_z = 0 , \qquad (S4)$$

with $\varphi_z(p) \neq 0$ at some point $p = (x_0,y_0,z_0)$ where $\varphi(p) = 0$. The solution for (S4) is typically of the form $\varphi(x,y,z) = C(A(x,y,z),B(x,y,z))$ for specific functions A and B, and an arbitrary C^1 function C. If one is given a side condition, say $u(x,0) = f(x)$, then even if it is possible to find a function C such that $C(A(x,0,f(x)),B(x,0,f(x)) = 0$, it may not be feasible to solve $\varphi(x,y,u) = 0$ for $u(x,y)$. Instead, one can aim for a parametric solution $x = X(s,t)$, $y = Y(s,t)$, $u = Z(s,t)$. The functions $X(s,t)$, $Y(s,t)$, $Z(s,t)$ are the solutions of the following characteristic system, with the initial conditions $X(s,0) = s$, $Y(s,0) = 0$, $Z(s,0) = f(s)$

$$\frac{dX}{dt} = a(X,Y,Z) , \quad \frac{dY}{dt} = b(X,Y,Z) , \quad \frac{dZ}{dt} = c(X,Y,Z) .$$

The graph of $u(x,y)$ is traced by the points $(X(s,t),Y(s,t),Z(s,t))$, as s and t vary. In applications to traffic flow theory and continuum mechanics, we found that the parametric solutions often do not define explicit solutions $u(x,y)$ for all (x,y), but rather the solutions tend to develop shocks where one or both partial derivatives become infinite. The parametric solutions can be used to locate the shock points.

Exercises 2.3

1. Solve the PDE $u_x + u_y + u_z = u$ for $u = u(x,y,z)$, subject to the side condition $u(x,y,0) = x^2 + y^2$.

2. Consider the PDE $u_x - u_y + u = z$ for $u(x,y,z)$.

(a) Solve this PDE subject to the side condition $u(0,y,z) = y^2 e^z$.

(b) Show that this PDE has no solution such that $u(x,y,x+y) = 0$.

(c) Find two (out of the infinitely many) solutions u such that $u(x,y,x+y) = x+y + e^y$.

(d) Explain the results of (b) and (c) in terms of the characteristic lines of the PDE in space and the plane $z = x + y$ on which the side condition is given.

3. (a) Find the general solution of the PDE $u_t = u_x + 2u_y - u_z$ for $u = u(x,y,z,t)$.

(b) What is the particular solution such that $u(x,y,z,0) = x^2 + y^2 + z^2$?

(c) For u as in (b), at a fixed time t, find the point (x,y,z) such that $u(x,y,z,t) = 0$?

4. (a) Find the general solution of $-yu_x + xu_y + u_z = 0$ for $u = u(x,y,z)$.
Hint. Use z as the parameter for the characteristic curves and note that $dx/dz = -y$ and $dy/dz = x$ imply that $d^2x/dz^2 = -x$. Hence, $x = \alpha\cos(z) + \beta\sin(z)$, etc. .

(b) Show that all but one of the characteristic curves are helixes which wind around the z–axis, and show that any solution of the PDE is constant on each one of these helixes.

5. Consider the PDE $xu_x + yu_y + zu_z = 0$.

(a) Solve the PDE subject to $u(x,y,1) = x^2 + y + 1$. Where is the solution defined ?

(b) By considering the characteristic curves, show that any solution of the PDE which is defined for all (x,y,z) must be constant.

(c) Find any nonconstant solution of the PDE which is C^1 for all (x,y,z) except $(0,0,0)$.

6. (a) Solve the quasi–linear PDE $2(u\cdot u_x + u\cdot u_y) = 1$ by expressing solutions implicitly in the form $C(A(x,y,u),B(x,y,u)) = 0$, for C^1 functions C.

(b) Find the solution of the PDE in part (a) that meets the side condition $u(x,2x) = 1$.

(c) Show that there is no solution of the PDE in part (a) such that $u(x,x) = 1$. **Hint.** Note that $\frac{d}{dx} u(x,x) = u_x(x,x) + u_y(x,x)$.

7. (a) Express solutions of the PDE $xu\cdot u_x - yu\cdot u_y = x^2$ in the form described in Problem 6(a).

(b) Find a solution $u(x,y)$ of the PDE in part (a) such that $u(1,y) = y^2 + 1$.

(c) Show that there are infinitely many solutions of the PDE in Part (a), such that $u(x,1/x) = x$, for $x > 0$. **Hint.** Take $C(r,s) = s-f(r)$, with $f(1) = 0$.

8. Find the parametric solution $x = X(s,t)$, $y = Y(s,t)$, $u = U(s,t)$ for the PDE in Problem 6(a) subject to the side condition $u(s,2s) = g(s)$, for a given C^1 function $g(s)$. Check that your answer is consistent with the answer to Problem 6(b), when $g(s) \equiv 1$.

9. Find the parametric solution $x = X(s,t)$, $y = Y(s,t)$, $u = U(s,t)$ for the PDE in Problem 7(a) subject to the side condition $u(1,s) = g(s)$, for a given function $g(s)$. For simplicity, assume $g(s) \geq 1$. Check that your answer is consistent with the answer to Problem 7(b), when $g(s) \equiv s^2 + 1$.

10. Solve the PDE $yuu_x + xuu_y = xy$, subject to the side condition $u(\cos(s),\sin(s)) = \sin(2s)$. Where is the solution valid ?

11. Using the notation and assumptions in the subsection on traffic flow, suppose that the initial density of cars is given by $\rho(x,0) = f(x)$. Suppose that x_0 is the only point where $f'(x)$ has an absolute maximum, say $f'(x_0) = G$. We have already seen that the first shock occurs at time $t_0 = d\cdot[2MG]^{-1}$. If initially there is a distance of about n car–lengths between cars around x_0 (i.e. $f(x_0) = \rho(x_0,0) = d/(1+n)$), then show that the first shock point is located at $x = x_0 + \frac{d(n-1)}{2G(n+1)}$, which is independent of the speed limit M.

12. In the subsection on traffic flow, we assumed that $v = M(1 - \frac{\rho}{d})$. More generally assume that $v = V(\rho)$ for some given function V.

(a) Show that the PDE for ρ becomes $\rho_t + (V(\rho) + \rho V'(\rho))\rho_x = 0$.

(b) Find the parametric solution of this equation, which meets the initial condition $\rho(x,0) = f(x)$.

(c) For what functions $V(\rho)$ (with $V(d) = 0$) does the PDE in (a) become linear ? For such V, find an explicit form $\rho(x,t)$ for the solution in (b). Do shocks develop in this case ?

13. In the following steps, we derive Euler's equation, $\rho(x,t)(v_t + vv_x) = f(x,t)$ (cf. (30)).

(a) Consider the portion of fluid between $x = a$ and $x = b$ at time $t = 0$. At time t, this portion of fluid will be between $x_1(t)$ and $x_2(t)$. Let $f(x,t)\mathbf{i}$ be the force per unit length acting on the fluid. Newton's equation states that the rate of change of the momentum of the fluid portion is equal to the total force on the fluid portion. Thus,

$$\frac{d}{dt}\int_{x_1(t)}^{x_2(t)} \rho(x,t)v(x,t)\,dx = \int_{x_1(t)}^{x_2(t)} f(x,t)\,dx\,.$$

Use Leibniz's rule (cf. Appendix A.3) and evaluate both sides at $t = 0$ to obtain,

$$\int_a^b \frac{\partial}{\partial t}(\rho v)\,dx + \rho(b,0)v(b,0)x_2'(0) - \rho(a,0)v(a,0)x_1'(0) = \int_a^b f(x,0)\,dx\,.$$

(b) Use $x_1'(0) = v(a,0)$, $x_2'(0) = v(b,0)$ and the fundamental theorem of calculus to deduce that

$$\int_a^b (\rho v)_t + (\rho v^2)_x \Big|_{t=0} dx = \int_a^b f(x,0)\,dx\,.$$

(c) Since a and b are arbitrary and the choice $t = 0$ is not necessary, we deduce that

$$(\rho v)_t + (\rho v^2)_x = f(x,t). \qquad (*)$$

Use the equation of continuity $\rho_t + (v\rho)_x = 0$ to convert the left–hand side of (*) to the desired form $\rho(v_t + v\cdot v_x)$.

14. In the notation of the subsection on continuum mechanics, show by completing the following steps that for small velocities $v(x,t)$ and densities $\rho(x,t)$ which deviate little from a constant density ρ_0, both $\rho(x,t)$ and $v(x,t)$ obey a wave equation.

(a) Let $v = \epsilon\bar{v}(x,t)$ and $\rho = \rho_0 + \epsilon\bar{\rho}(x,t)$, where ϵ is a small parameter. From the equation of state $p(x,t) = f(\rho(x,t)) = f(\rho_0 + \epsilon\bar{\rho}(x,t)) \approx f(\rho_0) + \epsilon f'(\rho_0)\bar{\rho}(x,t)$ (Why ?). By substituting these expressions into the equations $\rho_t + (\rho v)_x = 0$ and $\rho(v_t + vv_x) = -p_x$ and ignoring terms with factors of ϵ^2 (which are assumed small), obtain $\bar{\rho}_t\epsilon + \rho_0\bar{v}_x\epsilon = 0$ and $\rho_0\bar{v}_t\epsilon = -f'(\rho_0)\bar{\rho}_x(x,t)\epsilon$, or $\quad \bar{\rho}_t = -\rho_0 \bar{v}_x \quad$ and $\quad \rho_0 \bar{v}_t = -c_0^2 \bar{\rho}_x$, where $c_0^2 = f'(\rho_0)$. $\qquad$ (**)

(b) By differentiating the first equation in (**) with respect to t and the second equation with respect to x, obtain $\bar{\rho}_{tt} = c_0^2 \bar{\rho}_{xx}$, and similarly obtain $\bar{v}_{tt} = c_0^2 \bar{v}_{xx}$. For arbitrary C^2 functions f and g, $f(x + c_0 t) + g(x - c_0 t)$ is a generic solution of each of these equations. Thus, c_0 is interpreted as the speed at which disturbances are propagated in the medium under the above approximations (cf. Problem 12 of Section 1.3).

Remark. The above process of determining the equations which are satisfied by small deviations (with factors of ϵ) of known solutions (e.g., $\rho \equiv \rho_0$ and $v \equiv 0$) by ignoring terms with higher powers of ϵ is known as **linearization**, because the equations obtained in this way are linear. While this linearity makes the equations much easier to solve (e.g., because of the superposition principle), one should be aware that the linearized equations are only approximately correct, and certain important qualitative features of exact solutions may be lost in the process (e.g., the solutions of the linearized traffic flow equation do not have shocks). We observed this before in connection with the minimal surface equation (cf. Example 11 of Section 1.2) . □

15. In Example 6, the position $x(t)$ of a gas element at time t obeys the linear ODE $x'(t) = v(x(t),t) = \alpha(x(t) - c_0 t)\cdot(1 + \frac{1}{2}(\gamma+1)\alpha t)^{-1}$ (Why ?).

(a) Find the general solution of this ODE.

(b) Show that *every* solution $x(t)$ of this ODE approaches $-c_0/[\frac{1}{2}(\gamma+1)\alpha]$ as $t \to -[\frac{1}{2}(\gamma+1)\alpha]^{-1}$.

(c) Show that in spite of the fact that $v(x,t) \to \infty$ as $x \to \infty$, the velocity of each gas element eventually approaches $-2c_0/(\gamma-1)$, as $t \to \infty$.

16. Let $v_{\pm}(x,t) = \alpha(x - \pm c_0\cdot t)\cdot(1 + \frac{1}{2}(\gamma+1)\alpha t)^{-1}$. In Example 6, we elected to consider $v_+(x,t)$. Show that $v_-(x,t) = -v_+(-x,t)$. Why does this mean that $v_-(x,t)$ is the solution obtained by taking the mirror image of the physical setting for $v_+(x,t)$?

17. As we have done in the derivation of (34′) and (35′), assume that v and ρ are functionally related (i.e., $v = V(\rho)$).

(a) Assuming an equation of state of the form $p = f(\rho) > 0$, where $f'(\rho) > 0$, show that the equation $v_t + (v \pm c(v))v_x = 0$ is linear (not merely quasi–linear) if and *only* if $f(\rho) = C(1 - D/\rho)$, for some positive constants C and D. (Note that D represents a certain critical density, below which the pressure is presumably 0. As $\rho \to \infty$, the pressure approaches a maximum value of C.)

(b) Let $f(\rho) = C(1 - D/\rho)$. Assume that $v = 0$ when $\rho = D$, and that $v \geq 0$. How does v depend on ρ (i.e., what is the function $V(\rho)$?).

(c) Under the assumptions in (a) and (b), show that disturbances propagate with a constant speed of $[C/D]^{\frac{1}{2}}$ in the medium. In other words, $v(x,t) = g(x - [C/D]^{\frac{1}{2}}t)$, where $v(x,0) = g(x)$ and $0 < g(x) < [C/D]^{\frac{1}{2}}$. Why is it necessary to restrict the initial velocity $g(x)$?

2.4 Supplement on General Nonlinear First–Order PDEs (Optional)

The general first order PDE for $u = u(x,y)$ is of the form

$$F(x,y,u,u_x,u_y) = 0\ , \tag{1}$$

where F is some function [of five variables] which we assume is at least C^1. Since we do not usually think of u_x and u_y as variables, it is customary to denote u_x by p, and u_y by q, when referring to $F = F(x,y,u,p,q)$ as a function. For linear equations, F is of the form

$$F(x,y,u,p,q) = a(x,y)p + b(x,y)q + c(x,y)u - f(x,y)\ , \tag{2}$$

while for quasi–linear equations,

$$F(x,y,u,p,q) = a(x,y,u)p + b(x,y,u)q - c(x,y,u)\ . \tag{3}$$

We solved first–order linear PDEs by noting that they become an ODEs along the characteristic curves which may be regarded as the solutions of the system

$$X'(t) = a(X(t),Y(t)) \quad \text{and} \quad Y'(t) = b(X(t),Y(t)), \tag{4}$$

where the PDE was $F(x,y,u,u_x,u_y) = 0$ with F as in (2). Note that $F_p = a(x,y)$ and $F_q = b(x,y)$, in which case the system (4) may be written as $X' = F_p$ and $Y' = F_q$. To solve the general first–order PDE (1), with F an arbitrary given C^1 function, we might attempt to define characteristic curves as solutions of the system

$$\begin{aligned} X'(t) &= F_p(X(t),Y(t),U(t),P(t),Q(t)) \\ Y'(t) &= F_q(X(t),Y(t),U(t),P(t),Q(t))\ , \end{aligned} \tag{5}$$

where $U(t) = U(X(t),Y(t))$, $P(t) = u_x(X(t),Y(t))$, $Q(t) = u_y(X(t),Y(t))$. However, unlike the linear case, the right sides of (5) depend not only on $X(t)$ and $Y(t)$, but also on $U(t)$, $P(t)$ and $Q(t)$, which involve the unknown solution $u(x,y)$. But, we can think of (5) as being part of a larger system of 5 ODEs for the five unknown functions $X(t)$, $Y(t)$, $U(t)$, $P(t)$, $Q(t)$. We need to figure out what the remaining three equations should be. First, note that $U'(t) = d/dt[u(X(t),Y(t))] = u_xX'(t) + u_yY'(t) = P(t)X'(t) + Q(t)Y'(t) = P(t)F_p(\ldots) + Q(t)F_q(\ldots)$, where "..." denotes "X(t),Y(t),U(t),P(t),Q(t)". Thus, the equation for $U'(t)$ should be

$$U'(t) = P(t)F_p(\ldots) + Q(t)F_q(\ldots) .$$

The equation for $P'(t)$ is found by noting that $P'(t) = d/dt[u_x(X(t),Y(t)] = u_{xx}X'(t) + u_{xy}Y'(t) = u_{xx}F_p(\ldots) + u_{xy}F_q(\ldots)$. It seems as if we have reached an impasse, because the appearance of u_{xx} seems to require the introduction of yet another function $R(t) = u_{xx}(x(t),y(t))$, and this would lead to an even larger system. However, we have not yet used the fact that $u(x,y)$ should solve the PDE $F(x,y,u,u_x,u_y) = 0$. This fact tells us that

$$0 = \frac{d}{dx} F(x,y,u(x,y),u_x(x,y),u_y(x,y))$$
$$= F_x + F_u u_x + F_p u_{xx} + F_q u_{yx} .$$

Hence, $P'(t) = u_{xx}F_p(\ldots) + u_{xy}F_q(\ldots) = -[F_x(\ldots) + P(t)F_u(\ldots)]$, and similarly $Q'(t) = -[F_y(\ldots) + Q(t)F_u(\ldots)]$. Thus, we have finally arrived at

$$\begin{aligned} X'(t) &= F_p(\ldots) \\ Y'(t) &= F_q(\ldots) \\ U'(t) &= P(t)F_p(\ldots) + Q(t)F_q(\ldots) \\ P'(t) &= -[F_x(\ldots) + P(t)F_u(\ldots)] \\ Q'(t) &= -[F_y(\ldots) + Q(t)F_u(\ldots)] . \end{aligned} \tag{6}$$

Equations (6) constitute the **characteristic system** of the PDE $F(x,y,u,p,q) = 0$. A solution $(X(t),Y(t),U(t),P(t),Q(t))$ of (6) is called a **precharacteristic strip**; it defines a curve in xyupq–space, sometimes loosely referred to as "phase space" . The system can be regarded as the equations of motion for particles moving with a fluid flow in phase space. If

$$F(X(t),Y(t),U(t),P(t),Q(t)) = 0 \tag{7}$$

for all t, then the **precharacteristic strip** is known as a **characteristic strip**. If we omit $P(t)$ and $Q(t)$ in a characteristic strip, and just consider $(X(t),Y(t),U(t))$, then we obtain a curve [in xyu–space] which we call a **characteristic 3–curve**. (For linear equations the curve $(X(t),Y(t))$ was called a characteristic curve, but some books reserve this term for our notion of characteristic 3–curve.)

Of course, our ultimate goal here is to solve the PDE $F(x,y,u,u_x,u_y) = 0$ subject to an appropriate side condition (e.g., that u have prescribed values on some curve). The discussion about characteristic strips is irrelevant, unless it serves this goal. The basic idea is that we can obtain the graph of a solution $u(x,y)$ as the surface that is swept out in the xyu–space by a family of characteristic 3–curves which is constructed in such a way that the side condition will be met. We will describe the procedure and give some examples, but we omit the proof of the validity of the method (cf. Courant and Hilbert, Vol. II, p.75 ff.). The validity of a hypothetical solution obtained by this procedure could and should be directly checked anyway.

As s varies, let $(f(s),g(s))$ trace out a regular curve (in the xy–plane), which we regard as being a side condition curve. We seek a solution $u(x,y)$ of the problem.

$$F(x,y,u,u_x,u_y) = 0$$
$$u(f(s),g(s)) = G(s) , \tag{8}$$

where $G(s)$ is a given C^1 function. Such a problem may have no solution (e.g., consider the PDE $u_x^2 + u_y^2 + 1 = 0$). However, if a solution exists in some neighborhood of the side condition curve, then such a solution can often be found by completing the following steps.

1. If possible, find functions $h(s)$ and $k(s)$ such that

$$F(f(s),g(s),G(s),h(s),k(s)) = 0 , \tag{9}$$
$$G'(s) = h(s)f'(s) + k(s)g'(s) \quad \text{and} \tag{10}$$
$$F_p(f(s),g(s),...)g'(s) - F_q(f(s),g(s),...)f'(s) \neq 0 . \tag{11}$$

If $h(s)$ and $k(s)$ do not exist, then (8) has no solution. If there are several choices for $(h(s),k(s))$, then typically a solution of (8) exists for each such choice.

2. For each fixed value of s, solve the following characteristic system (cf. (6)) for $X(s,t)$, $Y(s,t)$, $U(s,t)$, $P(s,t)$, $Q(s,t)$, with the given initial conditions $P(s,0) = h(s)$, $Q(s,0) = k(s)$, where $h(s)$ and $k(s)$ are the functions found in step 1.

$$\begin{aligned}
\frac{d}{dt} X(s,t) &= F_p(X(s,t),Y(s,t),...) \\
\frac{d}{dt} Y(s,t) &= F_q(X(s,t),Y(s,t),...) \\
\frac{d}{dt} U(s,t) &= P(s,t)F_p(...) + Q(s,t)F_q(...) \\
\frac{d}{dt} P(s,t) &= -[F_x(...) + P(s,t)F_u(...)] \\
\frac{d}{dt} Q(s,t) &= -[F_y(...) + Q(s,t)F_u(...)] .
\end{aligned} \tag{12}$$

If it helps, assume that $F(X,Y,U,P,Q) \equiv 0$, for all (s,t), since by (9) this quantity is 0 at $t = 0$, and it is possible to prove that $F(X,Y,U,P,Q)$ is t–independent, if X, Y, U, P and Q solve the system (12) (i.e., solutions are always characteristic strips). Also, in view of the next step, it is unnecessary to solve for P and Q, if X, Y and U can be found without P and Q.

3. From the parametric viewpoint, as s and t vary, the point (x,y,u), defined by

$$x = X(s,t)\ , \quad y = Y(s,t)\ , \quad u = U(s,t)\ , \tag{13}$$

traces out the graph of a hypothetical solution u of (8) in the xyu–space, at least in a neighborhood of the curve traced out by $(f(s),g(s),G(s))$. In some cases, one can use the first two equations in (13) to solve for s and t in terms of x and y (say $s = S(x,y)$ and $t = T(x,y)$) to obtain a solution $u(x,y) = U(S(x,y),T(x,y))$, for (x,y) in a neighborhood of the curve $(f(s),g(s))$.

Remark. Some comment about the condition (11) is needed. The graph of the hypothetical solution is the surface of points in the xyu–space that we get from (13) by letting s and t vary. However, there is a potential problem here, because it can happen that a surface defined in this way will not project in a 1–1 fashion onto the xy–plane, in which case it will not be the graph of a function. In order to get a solution $u(x,y)$ at least in a neighborhood of the curve $(f(s),g(s))$, we need to know that the normal vector of the surface (13) at the point $(f(s),g(s),G(s))$ has nonzero **k** component for each value of s. This normal vector is the cross–product of two tangent vectors

$$\left[\frac{\partial X}{\partial s}\mathbf{i} + \frac{\partial Y}{\partial s}\mathbf{j} + \frac{\partial U}{\partial s}\mathbf{k}\right] \times \left[\frac{\partial X}{\partial t}\mathbf{i} + \frac{\partial Y}{\partial t}\mathbf{j} + \frac{\partial U}{\partial t}\mathbf{k}\right] \quad \text{at } (s,0).$$

The **k**–component is then $X_tY_s - X_sY_t$. Thus, (13) cannot be the graph of a solution in a neighborhood of the curve $(f(s),g(s))$, unless $X_t(s,0)k'(s) - Y_t(s,0)h'(s) \neq 0$. Since $X_t(s,0) = F_p(f(s),g(s),G(s),h(s),k(s))$ and $Y_t(s,0) = F_q(\ldots)$, this condition may be written as (11). In the case of *linear* PDEs, condition (11) is equivalent to the requirement that the side condition curve meet the characteristic curves transversely (Why ?). □

In Section 2.3, we solved the quasi–linear PDE $u_x + u \cdot u_y = 6x$, using the method of Lagrange in Example 3, and parametrically in Example 4. However, this PDE can be solved using the above steps for solving the general first–order PDE, and it is good to see how the method works for this familiar equation.

Example 1. Solve the PDE $u_x + u \cdot u_y = 6x$, subject to the side condition $u(0,s) = G(s)$.

Solution. Since the side condition curve is the y–axis traced out by $(0,s)$, we have $f(s) = 0$ and $g(s) = s$. Here $F(x,y,u,p,q) = p + uq - 6x$. We need to find functions $h(s)$ and $k(s)$ such that

$$h(s) + G(s)k(s) = 0\ , \qquad G'(s) = k(s) \quad \text{and} \quad 1 \neq 0,$$

which are (9), (10) and (11), respectively. Thus, we take $k(s) = G'(s)$ and $h(s) = -G(s)G'(s)$, and the condition (11) is automatic here. The characteristic system (12) is

$$\frac{dX}{dt} = 1\ , \quad \frac{dY}{dt} = U, \quad \frac{dU}{dt} = P + QU, \quad \frac{dP}{dt} = -(-6 + PQ)\ , \quad \frac{dQ}{dt} = -Q^2$$

with initial conditions

$$X(s,0) = 0,\ \ Y(s,0) = s,\ \ U(s,0) = G(s),\ \ P(s,0) = h(s) = -G(s)G'(s),\ \ Q(s,0) = k(s) = G'(s)\ .$$

Clearly, $X(s,t) = t$, and since we may assume that $F(X,Y,U,P,Q) = P + UQ - 6X = 0$, the equation for $\frac{dU}{dt}$ reduces to $\frac{dU}{dt} = 6X = 6t$, and so $U(s,t) = 3t^2 + G(s)$. Then, $\frac{dY}{dt} = U$ yields $Y(s,t) = t^3 + G(s)t + s$. Hence, we have obtained a parametric form of the solution

$$x = t, \qquad y = t^3 + G(s)t + s, \qquad u = 3t^2 + G(s)\ . \tag{14}$$

Recall that in the Remark following Example 3 in Section 2.3, we found that solutions of the problem may be expressed implicitly by

$$u-3x^2 - G(y+2x^3-xu) = 0. \tag{15}$$

If the expressions for x, y and u, in (14), are substituted into this implicit relation, we obtain the identity $G(s) - G(s) = 0$, meaning that the surface traced out, as s and t vary in (14), lies in the surface given by (15). For simple functions $G(s)$ (say, polynomials of degree 1 or 2), the first two equations in (14) can be easily solved for s and t in terms of x and y, in which case the third equation in (14) yields an explicit solution $u(x,y)$. In any case, the parametric solution (14) is usually superior to (15), when making a computer plot of the surface. □

Remark. For quasi–linear equations, where $F(x,y,u,p,q) \equiv a(x,y,u)p + b(x,y,u)q - c(x,y,u)$, observe that $pF_p + qF_q = a(x,y,u)p + b(x,y,u)q = c(x,y,u)$, when (x,y,u,p,q) satisfies $F(x,y,u,p,q) = 0$ (cf. (7)). It follows that the right sides of the first three equations in (12) can be expressed in terms which do not involve $P(t)$ and $Q(t)$, and thus these three equations suffice to determine X,Y and U (i.e., the last two equations in (12) may always be dropped if the PDE is quasi–linear). In the next example, all five equations in (12) are needed. □

Example 2. Solve the problem

$$u_x u_y - u = 0$$

$$u(s,-s) = G(s)\ ,$$

in the cases $G(s) = 1$, $G(s) = -1$, $G(s) = s$.

Solution. For each of the side conditions, we will need to solve the same system of characteristic equations (12), only the initial conditions are different in the three cases. Thus, first we find the general solution of the characteristic system (12), and then we determine the dependence on s from the side conditions in each case. We have $F(x,y,u,p,q,) = pq-u$, and the characteristic system (12) is

$$\frac{dX}{dt} = F_p = Q(t)\ , \quad \frac{dY}{dt} = F_q = P(t)\ , \quad \frac{dU}{dt} = QF_p + PF_q = 2P(t)Q(t)\ ,$$

$$\frac{dP}{dt} = -[F_x + P(t)F_u] = P(t)\ , \quad \frac{dQ}{dt} = -[F_y + Q(t)F_u] = Q(t)\ .$$

We easily find $P(t) = ae^t$ and $Q(t) = be^t$ for arbitrary constants a and b. Since we want a characteristic strip (i.e., $F(x,y,u,p,q) = 0$), we set $U(t) = P(t)Q(t) = abe^{2t}$ (and this does in fact solve the equation for $U(t)$). Finally, $X(t) = be^t+\alpha$ and $Y(t) = ae^t+\beta$ for constants α and β. The general characteristic strip is then given by

$$X(t) = be^t + \alpha\ , \quad Y(t) = ae^t + \beta\ , \quad U(t) = abe^{2t}\ , \quad P(t) = ae^t\ , \quad Q(t) = be^t\ .$$

Case $G(s) = 1$: Since the side condition is given on the line $y = -x$ traced out by $(s,-s)$, in (9), (10) and (11), we have $f(s) = s$ and $g(s) = -s$. We must find $h(s)$ and $k(s)$ such that

$$1 = G(s) = h(s)k(s) \qquad 0 = G'(s) = h(s) - k(s)\ ,$$

$$0 \neq F_p(...)(-1) - F_q(...)(1) = -k(s) - h(s)\ .$$

There are two choices, namely, $h(s) = 1$ and $k(s) = 1$, or $h(s) = -1$ and $k(s) = -1$. For the first choice we get

$$X(s,t) = e^t-1+s\ , \quad Y(s,t) = e^t-1-s\ , \quad U(s,t) = e^{2t}\ , \quad P(s,t) = e^t\ , \quad Q(s,t) = e^t\ .$$

We can solve for e^t in terms of x and y by adding the first two equations, obtaining $e^t = (x+y+2)/2$. Then the solution is $u(x,y) = e^{2t} = (x+y+2)^2/4$. If we choose $h(s) = -1$ and $k(s) = -1$, the reader may check that we get the solution $u(x,y) = (x+y-2)^2/4$.

Case $G(s) = -1$: In this case, there are no functions $h(s)$ and $k(s)$ that satisfy

$$-1 = h(s)k(s)\ , \quad 0 = G'(s) = h(s) - k(s),\ 0 \neq -k(s) - h(s).$$

Hence, in this case, there is no solution.

Case $G(s) = s$: Here, $h(s)$ and $k(s)$ must satisfy

$$s = h(s)k(s)\ , \quad 1 = G'(s) = h(s) - k(s), \quad 0 \neq -k(s) - h(s)\ .$$

From the first two equations, we obtain choices for the pair $(h(s),k(s))$

$$h(s) = [1 \pm (1+4s)^{\frac{1}{2}}]/2\ , \quad k(s) = [-1 \pm (1+4s)^{\frac{1}{2}}]/2\ ,$$

but these are only defined, and have a nonzero sum, for $s > -1/4$. Thus, there will be no solution, unless we only require the side condition $u(s,-s) = s$ to hold for $s > -1/4$. Assuming this,

$$X(s,t) = (1/2)[-1\pm(1+4s)^{\frac{1}{2}}][e^t-1] + s, \quad Y(s,t) = (1/2)[1\pm(1+4s)^{\frac{1}{2}}][e^t-1] - s, \quad U(s,t) = se^{2t}\ .$$

Now, while it is possible to solve the first two equations for s and e^t in terms of x and y (at least in a neighborhood of the side condition curve), this parametric form of the solution probably describes the graph of the explicit solution more conveniently than the explicit solution itself. □

Example 3. Suppose that each point (x,y) on a regular curve in the xy–plane, moves in a direction normal to the curve at a speed equal to $c(x,y)$, where $c(x,y)$ is a given positive C^1 function on the xy–plane. In geometric optics, a curve moving in this way represents a wave front curve moving in a two–dimensional medium, where the speed of light, $c(x,y)$, possibly depends on the position (x,y) (i.e., the index of refraction is variable). Each point on the moving curve traces out a (possibly curved) light ray in this medium. Show that, given a solution $u(x,y)$ of

$$u_x^{\ 2} + u_y^{\ 2} = c(x,y)^{-2} , \tag{16}$$

(known as the **eikonal equation** or the **Hamilton–Jacobi equation**, depending on the context) the curve $u(x,y) = \tau$ defines a wave front curve at time τ. For the PDE (16), show that the characteristic system (12) contains the equations of motion for a family of particles which have acceleration $\mathbf{A}(x,y) = -\nabla[-\tfrac{1}{2}c(x,y)^{-2}]$ (i.e., the particles are subject to force with potential $-\tfrac{1}{2}c(x,y)^{-2}$, if the particles have mass 1). Find the dual relation between the motion of the family of particles and the motion of the wave front curve, which foreshadows quantum mechanics.

Solution. Let $(x(\tau),y(\tau))$ be the position of a point on the curve $u(x,y) = \tau$ at time τ (i.e., $u(x(\tau),y(\tau)) = \tau$, and suppose that its velocity $\mathbf{v}(\tau) \equiv x'(\tau)\mathbf{i} + y'(\tau)\mathbf{j}$ is orthogonal to the curve $u(x,y) = \tau$ (i.e., $\mathbf{v}$ and ∇u are parallel; $\mathbf{v}\cdot\nabla u = \|\mathbf{v}\|\cdot\|\nabla u\|$). Assuming that u satisfies (16),

$$1 = \frac{d\tau}{d\tau} = \frac{d}{dt}\,u(x(\tau),y(\tau)) = u_x(x(\tau),y(\tau))x'(\tau) + u_y(x(\tau),y(\tau))y'(\tau) \qquad \text{(by the chain rule)}$$

$$= \mathbf{v}(\tau)\cdot\nabla u(x(\tau),y(\tau)) = \|\mathbf{v}(\tau)\|\cdot\|\nabla u(x(\tau),y(\tau))\| = \|\mathbf{v}(\tau)\|/c(x(\tau),y(\tau)) \qquad \text{(by the PDE (16))} .$$

Thus, $\|\mathbf{v}(\tau)\| = c(x(\tau),y(\tau))$, and the curve $u(x,y) = \tau$ does describe a wave front moving in the medium, as time τ varies.

If we take $F(x,y,u,p,q) = \tfrac{1}{2}p^2 + \tfrac{1}{2}q^2 - \tfrac{1}{2}c(x,y)^{-2}$, the characteristic system for (16) is

$$\frac{dX}{dt} = P(t), \quad \frac{dY}{dt} = Q(t), \quad \frac{dU}{dt} = P(t)^2 + Q(t)^2 = c(X(t),Y(t))^{-2},$$

$$\frac{dP}{dt} = A_1(X(t),Y(t)), \quad \frac{dQ}{dt} = A_2(X(t),Y(t)) ,$$

where $A_1(x,y)\mathbf{i} + A_2(x,y)\mathbf{j} = \mathbf{A}(x,y) \equiv -\nabla[-\tfrac{1}{2}c(x,y)^{-2}]$ (i.e., $A_1 = -c_x/c^3$ and $A_2 = -c_y/c^3$). Thus, we obtain

$$\frac{d^2X}{dt^2} = \frac{dP}{dt} = A_1(X(t),Y(t)) , \quad \frac{d^2Y}{dt^2} = \frac{dQ}{dt} = A_2(X(t),Y(t)) , \quad \frac{dU}{dt} = c(X(t),Y(t))^{-2} . \tag{17}$$

Hence, the characteristic 3–curves are of the form $(X(t),Y(t),U(t))$, where $(X(t),Y(t))$ is the position at time t of a particle of mass 1, which moves under the influence of the force $\mathbf{A}(x,y)$ with potential $-\tfrac{1}{2}c(x,y)^{-2}$. Suppose that we are given an initial curve $(f(s),g(s))$ of such particles, and a distribution of initial velocities $h(s)\mathbf{i} + k(s)\mathbf{j}$, such that $h(s)^2 + k(s)^2$

$- c(f(s),g(s))^{-2} = 0$ with $h(s)f'(s) + k(s)g'(s) = 0$ (i.e., the velocities are normal to the curve). Letting $X(s,t)$, $Y(s,t)$, $U(s,t)$ be the the solutions of (17) with initial conditions $X(s,0) = f(s)$, $Y(s,0) = g(s)$ and $U(s,0) \equiv 0$, we then have the parametric solution,

$$x = X(s,t)\ ,\quad y = Y(s,t)\ ,\quad u = U(s,t)\ , \tag{18}$$

of the PDE (16), which is 0 on the curve $(f(s),g(s))$. If the first two equations in (18) can be inverted, to give $s = S(x,y)$ and $t = T(x,y)$, then we have the explicit solution $u(x,y) = U(S(x,y),T(x,y))$. The initial wave front $u(x,y) = 0$ is traced out by the curve $(f(s),g(s))$ [i.e., $(X(s,0),Y(s,0))$], but in general, the wave front $u(x,y) = t$, at time $t \neq 0$, is *not* traced out by $(X(s,t),Y(s,t))$ as s varies. Indeed, if $u(X(s,t),Y(s,t)) \equiv t$, then

$$1 = \frac{dt}{dt} = \frac{d}{dt}\, u(X(s,t),Y(s,t)) = \frac{dU}{dt}(s,t) = c(X(s,t),Y(s,t))^{-2}$$

which is not true in general, unless $c(x,y) \equiv 1$. Note that while points $(x(\tau),y(\tau))$ on the wave fronts move with speed $c(x(\tau),y(\tau))$, the particles move with speed $[P(t)^2 + Q(t)^2]^{\frac{1}{2}} = c(X(t),Y(t))^{-1}$. Since $\frac{dX}{dt} = P(t) = u_x(X(t),Y(t))$ and $\frac{dY}{dt} = Q(t) = u_y(X(t),Y(t))$, the particles, as well the points on the wave fronts, move in the direction ∇u. Hence, the particles and light rays trace out the same paths, but at the (usually different) speeds c^{-1} and c, respectively. For fixed s, the particle which is at $(X(s,t),Y(s,t))$ at time t is on the wave front curve that exists at time $\tau = U(s,t)$. □

Remark 1. The particles in the family of unit mass particles constructed above are labeled by the parameter s. At time t, the particle with label s has kinetic energy $\frac{1}{2}[P(s,t)^2 + Q(s,t)^2]$ at time t and potential energy $-\frac{1}{2}c(X(s,t),Y(s,t))^{-2}$. The sum of these energies is zero (in particular, constant), because $F(X(s,t),Y(s,t),P(s,t),Q(s,t),U(s,t)) = 0$ (i.e., (X,Y,P,Q,U) is a characteristic strip), due to the fact that we chose $P(s,0)$, $Q(s,0)$ (i.e., $h(s)$ and $k(s)$) so that $F = 0$ when $t = 0$ (i.e., (9) holds). (The characteristic equations imply that F is constant on each precharacteristic strip.) In particle mechanics, the **instantaneous action** of a particle at any time is its kinetic energy minus its potenial energy. For the particles in our family, this difference is $c(X(s,t),Y(s,t))^{-2}$ (i.e., $\frac{1}{2}c^{-2} - (-\frac{1}{2}c^{-2}) = c^{-2}$). By (17) and the initial condition $U(s,0) = 0$, we have $U(s,t) = \int_0^t c(X(s,r),Y(s,r))^{-2}\,dr$. Thus, in the particle context, $U(s,t)$ is the (total) **action** of the particle s over the time interval $[0,t]$. In the wave context, $u(x,y) = U(s,t)$ is known as a **phase function** for the wave (i.e., the wave fronts can be thought of as curves of constant phase for the wave). Thus, the curves of constant phase for the wave (i.e., the wave fronts) are the curves of constant action for the associated family of particles. **Fermat's Principle** states that a light ray between two points takes that path between the points, such that the tip of the ray, moving from one point to the other with the speed, $c(x,y)$, of light along the path, will reach the other endpoint in the least amount of time. Since "time" (or phase) in the wave context corresponds to "action" in the particle context, we roughly see that Fermat's principle in the wave context corresponds to the **Principle of Least Action**, namely that if a particle is to move from one point to another in a given time interval, it takes the path of least (total) action. □

Remark 2. The intriguing wave/particle duality, which is built into the solution process for PDEs such as (16), was developed by the Irish mathematician Sir William Rowan Hamilton [1805–1865] and the German mathematician Karl Gustav Jacob Jacobi [1804–1851]. Perhaps the greatest significance of the Hamilton–Jacobi Theory was not realized until the advent of quantum mechanics in the early twentieth century, when it was discovered (in both theory and experiment) that the wave/particle duality is more than just a mathematical curiosity. Indeed, it forms a major foundation of modern physics in the atomic and subatomic domains. In Section 9.5, we discuss quantum mechanics and use Schrödinger's equation to determine the wave functions and energy levels of the electron in the hydrogen atom. □

We give a final specific example which illustrates the above generalities.

Example 4. Consider a medium in which the speed of light $c(x,y)$ is proportional to the distance from the x–axis, say $c(x,y) = y$, $y > 0$. Show that the wave front consisting of the positive y–axis moves in such a way that at later times it still forms a ray through the origin, and that the light rays (or associated particle trajectories) trace out semicircles which meet the x–axis orthogonally.

Solution. In this case the PDE (16) becomes ${u_x}^2 + {u_y}^2 = y^{-2}$. We take u to be 0 on the specified initial wave front; i.e., $f(s) = 0$, $g(s) = s$, $G(s) = 0$ (cf. (8)), where $s > 0$. To meet (9) and (10), namely $h(s)^2 + k(s)^2 - g(s)^{-2} = 0$ and $0 = G'(s) = h(s)f'(s) + k(s)g'(s) = k(s)$, we set $k(s) = 0$ and $h(s) = s^{-1}$. Taking $F(x,y,u,p,q) = \frac{1}{2}(p^2 + q^2 - y^{-2})$, the characteristic system (12) becomes,

$$\frac{dX}{dt} = P\,, \quad \frac{dY}{dt} = Q\,, \quad \frac{dU}{dt} = P^2 + Q^2\,, \quad \frac{dP}{dt} = 0\,, \quad \frac{dQ}{dt} = -Y^{-3}\,,$$

with initial conditions

$$X(s,0) = 0\,, \ Y(s,0) = s\,, \ U(s,0) = 0\,, \ P(s,0) = h(s) = s^{-1}, \ Q(s,0) = k(s) = 0\,.$$

Thus, $P(s,t) = s^{-1}$ and $X(s,t) = s^{-1}t$. From $\frac{dY}{dt} = Q$ and $\frac{dQ}{dt} = -Y^{-3}$, we get $-Y^{-3}\,dY = Q\,dQ$, and so $Y^{-2} = Q^2 + s^{-2}$. Then $\frac{dY}{dt} = Q = \pm(Y^{-2} - s^{-2})^{\frac{1}{2}}$ or $\pm Y\,(1 - s^{-2}Y^2)^{-\frac{1}{2}}\,dY = dt$. Integrating this, and using $Y(s,0) = s$, we obtain

$$\mp s^2\,(1 - s^{-2}Y^2)^{\frac{1}{2}} = t \quad \text{or} \quad Y(s,t) = [s^2 - t^2s^{-2}]^{\frac{1}{2}}\,.$$

Since $Y(s,t) = [s^2 - t^2s^{-2}]^{\frac{1}{2}} = [s^2 - X(s,t)^2]^{\frac{1}{2}}$, the trajectories of the light rays are semicircles which meet the x–axis orthogonally. From $\frac{dU}{dt} = P^2 + Q^2 = Y^{-2} = s^{-1}[1 - t^2s^{-4}]^{-1}$, we obtain $U(s,t) = \frac{1}{2}\log\left[\frac{1 + ts^{-2}}{1 - ts^{-2}}\right] = \tanh^{-1}(ts^{-2})$. Since $ts^{-2} = X(s,t)/[X(s,t)^2{+}Y(s,t)^2]^{\frac{1}{2}}$

$= \cos(\theta)$, in terms of the polar coordinate $\theta = \cos^{-1}[x/(x^2+y^2)^{\frac{1}{2}}]$, we have

$$u(x,y) = \tfrac{1}{2}\log\left[\frac{1 + ts^{-2}}{1 - ts^{-2}}\right] = \tfrac{1}{2}\log\left[\frac{1 + \cos(\theta)}{1 - \cos(\theta)}\right] = \log(\cot(\tfrac{1}{2}\theta)) = \tanh^{-1}(\cos(\theta)).$$

In particular, the wave front at time τ (i.e., $u(x,y) = \tau$) is the ray $\theta = \cos^{-1}(\tanh(\tau)) = 2\cot^{-1}(e^{\tau})$. As a byproduct, note that for each fixed s, the parametric curve $X(s,t) = s^{-1}t$, $Y(s,t) = [s^2 - t^2s^{-2}]^{\frac{1}{2}}$ is the position at time t of a particle of mass 1 subject to the force (or acceleration) $-\nabla(-\frac{1}{2}y^{-2}) = -y^{-3}\,\mathbf{j}$, with potential $-\frac{1}{2}y^{-2}$. Here the particle's initial position is (0,s) and its initial velocity is $s^{-1}\mathbf{i}$. Thus initially, the total energy (kinetic plus potential) of this particle is $\frac{1}{2}(s^{-1})^2 - \frac{1}{2}s^{-2} = 0$. At an arbitrary time t, the total energy is $\frac{1}{2}(P^2 + Q^2 - Y^{-2})$ which is easily computed to be identically 0, independent of t. This is an instance of the law of conservation of energy. Note that after time t, the ensemble of particles, beginning on the positive y–axis, is not on a ray. Indeed, the particle beginning at (0,s) hits the x–axis with infinite velocity at time $t = s^2$. The particles with $s^2 > t$ (i.e., $t^2/x^2 > t$ or $0 < x < \sqrt{t}$) are on the curve $y = [t^2x^{-2} - x^2]^{\frac{1}{2}}$ at time t. In Figure 1 below, these curves are plotted for $t = -3, -2.5, \ldots, 2.5, 3$. The wave fronts (rays issuing from the origin) have also been plotted for $\tau = -3, -2.5, \ldots, 2.5, 3$. Of course, the individual particles, as well as the light rays, travel on semicircles. However, when passing through a given point, the speed of the light ray is the reciprocal of the speed at which the corresponding particle passes though the point. □

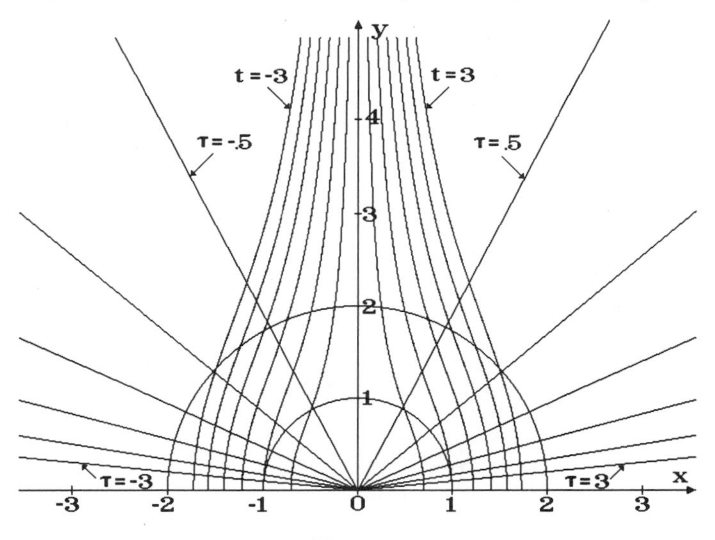

Figure 1

CHAPTER 3

THE HEAT EQUATION

Heat may be transferred by conduction, convection, and radiation. In conduction, the heat (molecular motion or vibration) is transferred locally by impacts of molecules with adjacent molecules. With convection, heat is carried from one region to another by a current flow, and heat radiation occurs via infrared electromagnetic waves. Here, we limit ourselves to heat conduction. In a homogeneous, solid, heat–conducting material, the temperature $u(x,y,z,t)$, at the point (x,y,z) at time t, very nearly obeys the heat equation $u_t = k(u_{xx} + u_{yy} + u_{zz})$, where k is a positive constant which measures the heat conductivity of the material. The function u can also have the interpretation of being the concentration of a chemical or dye in a liquid without currents, and hence the heat equation is often called the **diffusion equation**. In this chapter, we consider the case of one–dimensional heat flow, $u_t = ku_{xx}$, where $u = u(x,t)$. In the Classification Theorem (cf. Section 1.2), this PDE is an example of a parabolic PDE with constant coefficients, and it entails the main features of the general parabolic equation. In Section 3.1, we derive $u_t = ku_{xx}$ from physical laws, and solve the simplest initial/boundary–value problems. Uniqueness of solutions and the Maximum Principle are established in Section 3.2, where we also discuss variations of solutions with respect to variations in initial and boundary data. While a detailed proof of the Maximum Principle is provided, it may be omitted on first reading. Section 3.3 is devoted to the case where the boundary conditions do not change with time. In Section 3.4, we carefully motivate and use Duhamel's method to handle the inhomogeneous heat equation $(u_t - ku_{xx} = h(x, t))$ and time–dependent boundary conditions.

3.1 Derivation of the Heat Equation and Solutions of Standard Initial/Boundary–Value Problems

In what follows, we derive the heat equation from physically reasonable principles, namely conservation of energy and the fact that heat flows from hot regions to cold regions. It must be noted that we are treating the notion of temperature in an idealized sense when we speak of it as being a C^2 function of position and time. The temperature (i.e., absolute temperature measured on the Kelvin scale) of a small region of a substance is proportional to the average kinetic energy of the molecules in the region. Thus, temperature is a statistical notion. The concept of "temperature at a point" is a mathematical idealization that might be achieved by taking a limit as the regions become smaller and as the size of the molecules decreases (even more rapidly), while the number of molecules increases. The law that heat flows from hot to cold regions (i.e., essentially the second law of thermodynamics) is statistical in the sense that it can be violated, but only improbably ; in the above mathematically ideal limit, the probability of violation approaches zero. The success of the mathematical idealization apparently rests on the fact that molecules are very small relative to everyday objects, say the tip of a thermometer, that are used to measure temperatures. At any rate, with this awareness of the mathematical idealizations which we tacitly assume, we proceed with our derivation of the heat equation.

Consider a wire or rod which is made of some heat–conducting substance and which is insulated on the outside, except possibly over the ends at $x = 0$ and $x = L$.

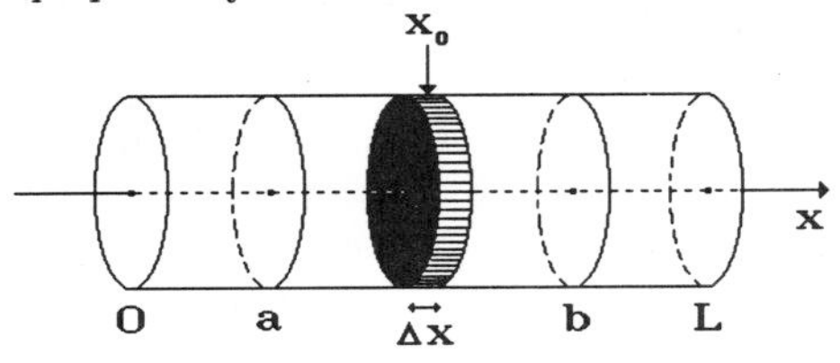

Figure 1

We assume that the temperature is constant on each cross section at each time, say $u(x,t)$ is the temperature at cross section x at time t. We keep the system of units arbitrary, so that $u(x,t)$ could be in Fahrenheit, Centigrade or Kelvin units. We introduce the following constants :

D = density of rod (i.e., the mass per unit volume),

C = specific heat of rod (i.e., the energy required to raise a unit of mass a unit in temperature),

L = length of the rod ,

A = area of cross section .

Consider a slab of the rod of small thickness Δx about some $x = x_0$ (cf. Figure 1). The mass of the slab is $DA(\Delta x)$. Hence the energy required to raise the temperature of the slab from 0 to $u(x_0,t)$ is $\approx u(x_0,t)CDA(\Delta x)$; if $u(x_0,t) < 0$, then this is negative, which means that energy must be extracted to lower the temperature. Letting $\Delta x \to 0$ and adding up the energies of all the slabs between $x = a$ and $x = b$, we arrive at the following expression for the **heat energy** of the portion of the rod between $x = a$ and $x = b$ at time t:

$$E(t) = \int_a^b CDA\, u(x,t)\, dx\ . \tag{1}$$

It is found experimentally that heat energy flows from hotter regions to colder regions, and that the rate of heat flow is proportional to the temperature difference, divided by the distance between the regions (i.e., the temperature gradient). In quantitative terms, the rate at which heat energy passes through the cross section at $x = a$ in the positive direction is

$$-KA\left[\frac{\partial u}{\partial x}\right]_{x=a} \quad \text{or} \quad -KA\, u_x(a,t)\ , \tag{2}$$

for some constant $K > 0$, called the **thermal conductivity** of the material in the rod. Note that if $u_x(a,t) < 0$, then the temperature to the left of $x = a$ is greater than the temperature to the right of $x = a$, whence (2) should be (and is) positive. The rate of heat flow is proportional to the temperature gradient $u_x(a,t)$.

Since we assumed that the rod is insulated on the outside curved surface, the only way that heat energy can enter the part of the rod between $x = a$ and $x = b$ is through the cross sections at $x = a$ and $x = b$. (We assume that there are no internal sources of heat such as chemical reactions or radioactivity.) Hence, the net rate at which heat energy enters this part of the rod is the rate at which it enters the end at $x = a$, minus the rate at which it leaves through the end at $x = b$. In other words, using (2), we get

$$\begin{aligned} E'(t) &= -KA\, u_x(a,t) - (-KA\, u_x(b,t)) \\ &= KA\, u_x(x,t)\Big|_a^b = \int_a^b KA \frac{\partial}{\partial x}[u_x(x,t)]\, dx\ , \end{aligned} \tag{3}$$

where the last equation comes from the fundamental theorem of calculus, $\int_a^b f'(x)\, dx = f(b) - f(a)$. Here we are assuming that u is C^2. On the other hand, we can also compute $E'(t)$ by differentiating (1) under the integral (cf. Appendix A.3) with respect to t:

$$E'(t) = \int_a^b CDA\, u_t(x,t)\, dx\ . \tag{4}$$

Using (3) and (4), we obtain

$$\int_a^b CDA\, u_t(x,t)\, dx = E'(t) = \int_a^b KA\, u_{xx}(x,t)\, dx\ .$$

Dividing by CDA and defining $k = K/CD$ (k is called the **diffusivity of heat** for the material in the rod), we get

$$\int_a^b [u_t(x,t) - ku_{xx}(x,t)]\, dx = 0 . \tag{5}$$

Since $[a,b]$ is an arbitrary subinterval of $[0,L]$, it follows that

$$u_t(x,t) - ku_{xx}(x,t) = 0 . \tag{6}$$

Indeed, if (6) were to fail at some point x_0, then (5) would fail if $[a,b]$ were chosen to be a small enough interval about x_0, so that the integrand is never zero on $[a,b]$.

> The PDE $u_t = ku_{xx}$ is the (one-dimensional) **heat equation**. We have shown (subject to the limitations of our mathematical idealization) that the temperature $u(x,t)$ obeys this PDE.

Example 1 (The fundamental source solution). A very important solution of $u_t = ku_{xx}$ is

$$u(x,t) = (4\pi kt)^{-\frac{1}{2}}\, e^{-x^2/(4kt)} , \quad t > 0 , \ -\infty < x < \infty . \tag{7}$$

At any fixed time $t > 0$, the graph (cf. Figure 2) of (7) in the xu-plane is a bell-shaped normal curve of the Gaussian distribution of probability theory. As t increases, the graph spreads out and decreases in height, always maintaining an area of 1 between it and the x-axis. Indeed, using

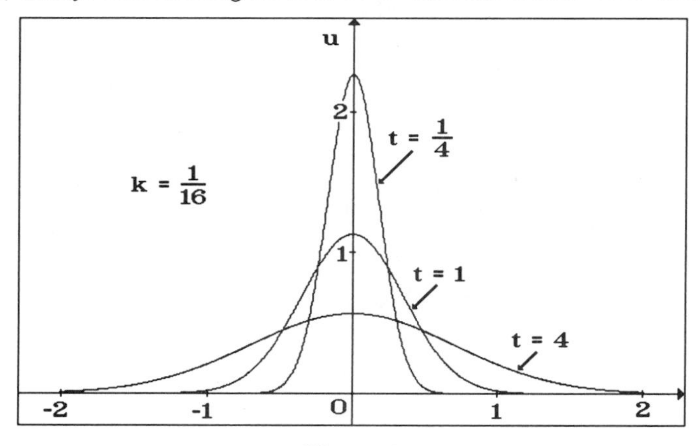

Figure 2

the standard result from statistics (also proved in Example 6 of Section 7.1) that $\int_{-\infty}^{\infty} e^{-ax^2} dx = \sqrt{\pi/a}$ $(a > 0)$, it follows that $\int_{-\infty}^{\infty} u(x,t)\, dx = 1$, so that the total heat energy in this infinite rod $(-\infty < x < \infty)$ is a constant, independent of t. As $t \longrightarrow 0^+$, observe that the graph of the temperature distribution $u(x,t)$, as a function of x, becomes more sharply peaked about $x = 0$. Indeed, the solution (7) represents the evolution of the temperature due to an initial heat source concentrated at $x = 0$ at $t = 0$. In Chapter 7, we will be led to this solution by Fourier transform methods (cf. also Problems 2 and 13). For now, we check that (7) is in fact a solution, using logarithmic differentiation. Note that

$$\log(u) = -\tfrac{1}{2}\log(4\pi k) - \tfrac{1}{2}\log(t) - \frac{x^2}{4kt}.$$

Thus, $\dfrac{u_x}{u} = -\dfrac{x}{2kt}$ or $u_x = -\dfrac{x}{2kt}u$.

Then, $u_{xx} = -\dfrac{1}{2kt}u - \dfrac{x}{2kt}u_x = -\left[\dfrac{1}{2kt} - \left[\dfrac{x}{2kt}\right]^2\right]u$,

while $\dfrac{u_t}{u} = -\dfrac{1}{2t} + \dfrac{x^2}{4kt^2} = -k\left[\dfrac{1}{2kt} - \left[\dfrac{x}{2kt}\right]^2\right]$. Thus, $u_t = ku_{xx}$. □

Remark (Probabilistic considerations). It is not unexpected that the temperature (7) due to an initial concentrated heat source has a standard normal Gaussian (bell–shaped) distribution, given the statistical foundations of heat flow. Indeed, a purely statistical derivation of (7), in terms of a random diffusion of a concentrated source at $x = 0$, is carried out in Problem 13. The idea is that if the particles at the source are allowed to move randomly to the left or to the right by a distance of Δx at regular time intervals of length Δt, then at time t the density of the particles will be a normal distribution of the form (7), as $\Delta t \to 0$, provided that $\tfrac{1}{2}\Delta x^2 = k\Delta t$. □

An initial/boundary–value problem

Physical intuition leads us to believe that if we specify the initial temperature distribution $u(x,0)$ in a rod $(0 \le x \le L)$ and if we specify the temperatures $u(0,t)$ and $u(L,t)$ at the ends, then at arbitrary (x,t) $[0 \le x \le L,\ t \ge 0]$, the temperature $u(x,t)$ should be determined. In other words, for "suitably nice" given functions $A(t)$, $B(t)$, $f(x)$, we expect that the following problem will have a unique solution :

D.E.	$u_t = ku_{xx}$	$0 \le x \le L,\ t \ge 0$	
B.C.	$u(0,t) = A(t)$	$u(L,t) = B(t)$	(8)
I.C.	$u(x,0) = f(x)$.		

Here, D.E. means "differential equation", while B.C. means "boundary conditions" (i.e., conditions at the ends $x = 0$ and $x = L$), and I.C. means "initial condition" (i.e., temperature distribution at $t = 0$). In Section 3.2, we prove that there is at most one solution of the "initial boundary–value problem" (8). In geometrical terms, we seek a function $u(x,t)$ defined on the semi–infinite strip $(0 \le x \le L,\ t \ge 0)$ in the xt–plane that satisfies the D.E. on the strip and has prescribed values on the border of the strip, as shown in Figure 3.

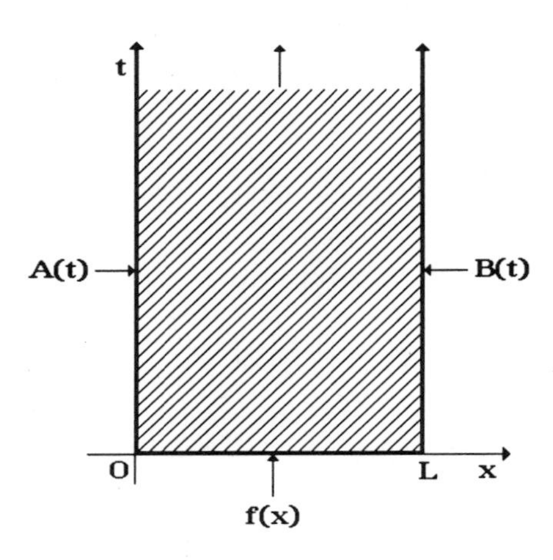

Figure 3

One can imagine the graph of $u(x,t)$ as a surface above this strip; the u–axis points out of the page. The temperature distribution at any time t_0 would then be obtained by taking a vertical slice of the graph of u with the plane $t = t_0$ in the xtu–space.

Separation of Variables

The solution process begins with the determination of all the product solutions $u(x,t) = X(x)T(t)$ of the D.E. $u_t = ku_{xx}$, where X is a function of x, and T is a function of t. For this, we use the method of "separation of variables" that was introduced in Section 1.3. We consider the B.C. and I.C. later. Now, substituting $u(x,t) = X(x)T(t)$ into $u_t = ku_{xx}$ and separating variables, we obtain

$$u_t = ku_{xx} \quad \Rightarrow \quad X(x)T'(t) = kX''(x)T(t) \quad \Rightarrow \quad \frac{T'(t)}{kT(t)} = \frac{X''(x)}{X(x)} = c\,,$$

for some constant c, since a function of t can equal a function of x only when both functions are constant (cf. Example 10 of Section 1.3.). We then have the two ODEs

$$T'(t) - kcT(t) = 0\,, \tag{9}$$

$$X''(x) - cX(x) = 0. \tag{10}$$

There are three cases ; $c < 0$, $c > 0$, $c = 0$. It is convenient to set $c = -\lambda^2$ when $c < 0$ and $c = \lambda^2$ when $c > 0$, for some constant $\lambda > 0$. (It would also be possible to assume that $\lambda < 0$, or simply $\lambda \neq 0$, but for future convenience we choose $\lambda > 0$.) The product solutions $u(x,t) = X(x)T(t)$ are obtained by solving the familiar ODEs (9) and (10) with constant coefficients (cf.

Section 1.1) in the three cases. The final results are:

Case 1 $(c = -\lambda^2 < 0)$:

$$u(x,t) = e^{-\lambda^2 kt}(c_1 \sin(\lambda x) + c_2 \cos(\lambda x)). \tag{11}$$

Case 2 $(c = \lambda^2 < 0)$:

$$u(x,t) = e^{\lambda^2 kt}(c_1 e^{\lambda x} + c_2 e^{-\lambda x}). \tag{12}$$

Case 3 $(c = 0)$:

$$u(x,t) = c_1 x + c_2 \ . \tag{13}$$

Remark. Every product solution of $u_t = ku_{xx}$ is one of the forms (11), (12) or (13). However, *not* every solution of $u_t = ku_{xx}$ is a product solution. Indeed, the solution (7) in Example 1 is not a product solution. Also, by the superposition principle, we can obtain other solutions of the homogeneous linear PDE $u_t - ku_{xx} = 0$, by forming linear combinations of the above product solutions for various values of λ, c_1 and c_2. Such linear combinations are solutions, but not always product solutions. For example, $e^{-kt}\sin(x) + e^{-4kt}\sin(2x)$ is a solution, but it cannot be expressed as a function of x times a function of t. Moreover, not every solution of $u_t = ku_{xx}$ is a linear combination of product solutions. Indeed, any such linear combination of solutions (11), (12) or (13) is defined at $t = 0$, whereas the solution (7) is not. Thus, solution (7) cannot be a linear combination of product solutions. However, as we will see, for most initial/boundary–value problems on a finite rod, solutions (if they exist) can be expressed as linear combinations (possibly infinite) of product solutions. □

Solving the simplest initial/boundary–value problem

The simplest initial/boundary–value problem for the heat equation is the standard problem (*where, for technical accuracy, we require that* $u(x,t)$ *have a* C^2 *extension to an "open" domain that strictly contains the strip* $0 \le x \le L$, $t \ge 0$) :

$$\begin{aligned} &\text{D.E.} \quad u_t = ku_{xx} \qquad 0 \le x \le L,\ t \ge 0 \\ &\text{B.C.} \quad u(0,t) = 0 \quad u(L,t) = 0 \\ &\text{I.C.} \quad u(x,0) = f(x)\ . \end{aligned} \tag{14}$$

Here, both ends of the rod are maintained at 0 (say by immersing the ends in ice water, if the units are measured on the Celsius scale). By shifting the temperature scale, we can make any temperature (above absolute zero) equal to zero on some new scale. The strategy now is to find all product solutions of the D.E. that also satisfy the B.C.. After this, we confront the I.C.. There are no nonzero Case 3 product solutions ($u(x,t) = c_1x + c_2$; cf. (13)) that satisfy the B.C.. Indeed, $0 = u(0,t) = c_1 \cdot 0 + c_2$ implies $c_2 = 0$, and $0 = u(L,t) = c_1L + c_2 = c_1L$ implies $c_1 = 0$. (At any time, the graph of a Case 3 product solution is a straight line in the xu–plane, and the only line which runs through (0,0) and (L,0) is the x–axis.) There are also no nonzero product solutions of the form (12) (Case 2). Indeed, using (12),

$$0 = u(0,t) = e^{\lambda^2 kt}(c_1e^0 + c_2e^0) = e^{\lambda^2 kt}(c_1 + c_2) \Rightarrow c_2 = -c_1$$

and

$$0 = u(L,t) = e^{\lambda^2 kt}(c_1e^{\lambda L} - c_1e^{-\lambda L}) = e^{\lambda^2 kt}c_1e^{-\lambda L}(e^{2\lambda L} - 1) .$$

Hence, $c_1 = 0$, since $e^{2\lambda L} - 1 > 0$, and $c_2 = -c_1 = 0$. Thus, our only hope for obtaining nonzero product solutions, satisfying the B.C., rests with the Case 1 solutions (11). We have

$$0 = u(0,t) = e^{-\lambda^2 kt}(c_1\sin(0) + c_2\cos(0)) = e^{-\lambda^2 kt}c_2 \Rightarrow c_2 = 0$$

and

$$0 = u(L,t) = e^{-\lambda^2 kt}c_1\sin(\lambda L) \ . \tag{15}$$

We are forced to take $c_2 = 0$, but we want to avoid setting $c_1 = 0$ in (15), since then we would just get another 0 solution. To avoid this, note that c_1 *can* be nonzero, but *only* if λ is chosen so that $\sin(\lambda L) = 0$. Since $\sin(z) = 0$, if and only if z is an integer multiple of π, the choices for (positive) λ are then given by $\lambda L = n\pi$ $(n = 1,2,3...)$ or $\lambda = n\pi/L$. Thus, the only product solutions of the D.E. which meet the B.C. of (14) are constant multiples of members of the following infinite family of product solutions

$$u_n(x,t) = e^{-(n\pi/L)^2 kt}\sin(n\pi x/L) , \quad n = 1, 2, 3,\dots . \tag{16}$$

Since the D.E. is linear and homogeneous, we may apply the superposition principle to deduce that the linear combination of any finite number, say N, of terms

$$u(x,t) = \sum_{n=1}^{N} b_n e^{-(n\pi/L)^2 kt}\sin(n\pi x/L) \tag{17}$$

is a solution of the D.E., for any choice of constants $b_1, b_2, \ldots, b_N$. Also observe that (17) meets the B.C. $u(0,t) = 0$ and $u(L,t) = 0$. Indeed, since these conditions are homogeneous and linear, the superposition principle also applies to the B.C. (More easily, one can check the B.C. directly.)

We now consider the I.C. of (14). If we set $t = 0$ in (17), we obtain

$$u(x,0) = \sum_{n=1}^{N} b_n \sin(n\pi x/L) . \tag{18}$$

If it happens that the f(x) in the I.C. of (14) is of the form of the right side of (18), then u(x,t) given by (17) will be a solution of problem (14). In other words, we have :

Proposition 1. Let $b_1, \ldots, b_N$ be given constants. A solution of the problem

$$\text{D.E.}\quad u_t = ku_{xx} \qquad 0 \le x \le L\ ,\ t \ge 0$$

$$\text{B.C.}\quad u(0,t) = 0\ , \quad u(L,t) = 0 \tag{19}$$

$$\text{I.C.}\quad u(x,0) = \sum_{n=1}^{N} b_n \sin(n\pi x/L)$$

is given by

$$u(x,t) = \sum_{n=1}^{N} b_n e^{-(n\pi/L)^2 kt} \sin(n\pi x/L) . \tag{20}$$

Note that all of the exponents $-(n\pi/L)^2 kt$ tend to $-\infty$, as $t \to \infty$, and hence all of the terms approach zero as $t \to \infty$. Thus, $u(x,t) \to 0$ as $t \to \infty$, and the temperature distribution of the rod eventually approaches zero, which is the temperature imposed at the ends. The rates at which the terms approach zero is different, because the magnitude of the exponents depend on n. The terms with higher values of n decrease more rapidly. This is physically reasonable, since the rate of heat flow is proportional to the temperature gradient, and the temperature gradients between hot and cold regions of the distribution $\sin(n\pi x/L)$ are proportional to n (i.e., the gradient of $\sin(n\pi x/L)$ is $n \cdot (\pi/L)\cos(n\pi x/L)$). This is illustrated in Figure 4.

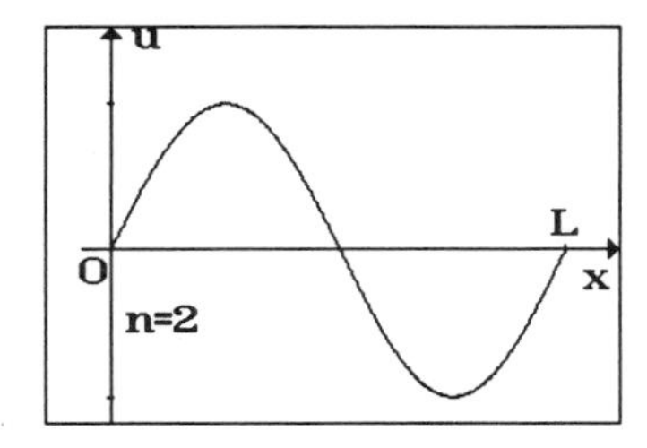

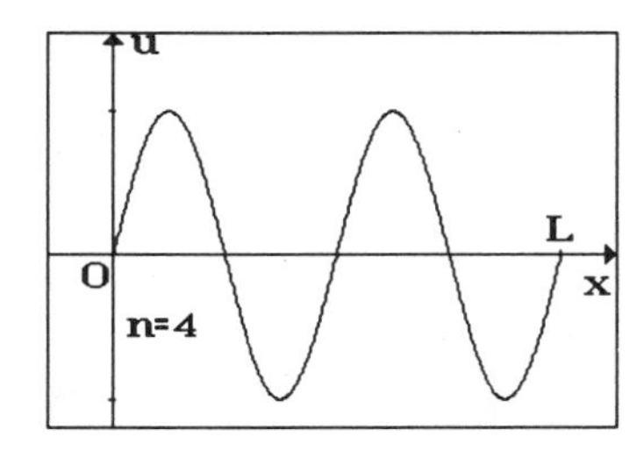

Figure 4

Example 2. Find a solution of the problem

$$\begin{aligned} &\text{D.E.} \quad u_t = 2u_{xx} \qquad 0 \le x \le \pi\,,\ t \ge 0 \\ &\text{B.C.} \quad u(0,t) = 0 \qquad u(\pi,t) = 0 \\ &\text{I.C.} \quad u(x,0) = 5\sin(2x) - 30\sin(3x)\,. \end{aligned} \tag{21}$$

Solution. Given Proposition 1 , we can immediately write down a solution. Indeed, here $L = \pi$, $k = 2$, $b_1 = 0$, $b_2 = 5$, $b_3 = -30$ and $N = 3$. Substituting these values into (20), we obtain

$$u(x,t) = 5e^{-8t}\sin(2x) - 30e^{-18t}\sin(3x)\,. \tag{22}$$

Observe that the ratio of the amplitude of the second term to that of the first is $30e^{-18t}/5e^{-8t} = 6e^{-10t}$. Thus, although the second term "dominates" the first at $t = 0$, eventually the second term is minuscule compared to the first (e.g., consider $t = 10$). Observe that in (22) and more generally in (20), $u(x,t)$ is obtained from $f(x)$ simply by inserting the appropriate exponentially decaying function of t. The constant which multiplies $-kt$ in the exponent is always the square of the coefficient of x in the sine factor, namely $(n\pi/L)^2$. Hence there is no need to figure out the value of n for each term. □

Remark. This example was easy to solve using Proposition 1, but suppose that you are taking a closed–book exam and cannot remember Proposition 1. You should still be able to solve the problem by carrying out the following steps, which pertain to a wide variety of problems with homogeneous and linear D.E. and B.C. :

1. Determine the product solutions of the D.E. via separation of variables.
2. Find those product solutions that meet the B.C. .
3. Form a linear combination of the product solutions in 2 that will meet the I.C. .

Most instructors will insist that you show that you can carry out these procedures in detail.

Example 3 (Heat flow in an insulated circular wire). Solve the following problem for a rod of length $2L$, from $x = L$ to $x = -L$.

$$\begin{aligned} &\text{D.E.} \quad u_t = ku_{xx} \qquad -L \le x \le L\,,\ t \ge 0 \\ &\text{B.C.} \quad u(-L,t) = u(L,t) \quad u_x(-L,t) = u_x(L,t) \\ &\text{I.C.} \quad u(x,0) = [\cos(\pi x/L)]^3\,. \end{aligned} \tag{23}$$

Solution. Note that the B.C. require that the temperatures of the ends $x = \pm L$ match and that the temperature gradients match as well. This is precisely what one would have, if the rod was bent into a circle and the ends were joined. Hence the problem is regarded as one for heat flow in an insulated circular wire. To solve the problem, note that the product solutions of the D.E. are found, via separation of variables, to be of the forms (11), (12) or (13), as before. We must determine which product solutions meet the B.C. . In Case 1 (cf. (11)), the first B.C. yields

$$e^{-\lambda^2 kt}(c_1\sin(-\lambda L) + c_2\cos(-\lambda L)) = e^{-\lambda^2 kt}(c_1\sin(\lambda L) + c_2\cos(\lambda L))$$

$$\Rightarrow\ 2e^{-\lambda^2 kt}c_1\sin(\lambda L) = 0 \ \Rightarrow\ c_1\sin(\lambda L) = 0\ .$$

For the second B.C., we first note that

$$u_x(x,t) = e^{-\lambda^2 kt}(c_1\lambda\cdot\cos(\lambda x) - c_2\lambda\cdot\sin(\lambda x))\ .$$

Thus, the second B.C. yields

$$e^{-\lambda^2 kt}(c_1\lambda\cos(-\lambda L) - c_2\lambda\sin(-\lambda L)) = e^{-\lambda^2 kt}(c_1\lambda\cos(\lambda L) - c_2\lambda\sin(\lambda L))$$

$$\Rightarrow\ 2e^{-\lambda^2 kt}c_2\sin(\lambda L) = 0 \ \Rightarrow\ c_2\sin(\lambda L) = 0\ .$$

Hence, we are forced to take c_1 and c_2 to be zero, unless we choose λ such that $\sin(\lambda L) = 0$, i.e., $\lambda = n\pi/L$, $n = 1, 2, 3, \ldots$. Thus, in Case 1, we have the following family of product solutions (where we have replaced c_1 and c_2 by b_n and a_n, since these constants depend on n) :

$$u_n(x,t) = e^{-(n\pi/L)^2 kt}[a_n\cos(n\pi x/L) + b_n\sin(n\pi x/L)]\ ,\quad n = 1, 2, 3, \ldots\ . \tag{24}$$

In Problem 5, we ask the reader to show that the only product solution in Case 2 (cf. (12)) that satisfies the B.C. is the zero solution. In Case 3 (cf. 13), where $u(x,t) = c_1 x + c_2$, we have

$$\begin{cases} u(-L,t) = u(L,t) \\ u_x(-L,t) = u_x(L,t) \end{cases} \Leftrightarrow \begin{cases} -c_1 L + c_2 = c_1 L + c_2 \\ c_1 = c_1 \end{cases} \Leftrightarrow \begin{cases} c_1 = 0 \\ c_2 \text{ arbitrary.} \end{cases}$$

Thus, $u(x,t) \equiv c_2$ is the only "surviving" product solution in Case 3. We can include these constant product solutions in the family (24) by letting $n = 0$; note $u_0(x,t) = e^0[a_0\cos(0) + b_0\sin(0)] = a_0$. By the superposition principle for the linear and homogeneous D.E. and B.C. , we have a rather general (not the *most* general) solution of the D.E. and B.C. , namely

$$u(x,t) = a_0 + \sum_{n=1}^{N} e^{-(n\pi/L)^2 kt}[a_n\cos(n\pi x/L) + b_n\sin(n\pi x/L)]\ . \tag{25}$$

For the I.C., we note that when $t = 0$, (25) becomes

$$u(x,0) = a_0 + \sum_{n=1}^{N} [a_n\cos(n\pi x/L) + b_n\sin(n\pi x/L)] . \tag{26}$$

In order to find the solution of (23), we must show that $[\cos(\pi x/L)]^3$ may be put in the form (26). This can be achieved through the use of trigonometric formulas such as

$$\begin{aligned} \cos(\alpha)\cos(\beta) &= \tfrac{1}{2}[\cos(\alpha+\beta) + \cos(\alpha-\beta)] \\ \cos(\alpha)\sin(\beta) &= \tfrac{1}{2}[\sin(\alpha+\beta) - \sin(\alpha-\beta)] \\ \sin(\alpha)\sin(\beta) &= \tfrac{1}{2}[\cos(\alpha-\beta) - \cos(\alpha+\beta)] . \end{aligned} \tag{27}$$

From these, we have $[\cos(\alpha)]^3 = \cos(\alpha)\cdot[\cos(\alpha)\cdot\cos(\alpha)] = \cos(\alpha)\cdot\frac{1}{2}[\cos(2\alpha) + \cos(0)] = \frac{1}{2}[\cos(\alpha)\cdot\cos(2\alpha)] + \frac{1}{2}\cos(\alpha) = \frac{1}{4}[\cos(3\alpha) + \cos(-\alpha)] + \frac{1}{2}\cos(\alpha) = \frac{1}{4}\cos(3\alpha) + \frac{3}{4}\cos(\alpha)$. Thus,

$$[\cos(\pi x/L)]^3 = \tfrac{3}{4}\cos(\pi x/L) + \tfrac{1}{4}\cos(3\pi x/L) ,$$

and referring to (26), we have $a_0 = 0$, $a_1 = \frac{3}{4}$, $a_2 = 0$, $a_3 = \frac{1}{4}$, $b_1 = b_2 = b_3 = 0$ and $N = 3$. From (25), we then obtain a solution of problem (23), namely

$$u(x,t) = \tfrac{3}{4}e^{-(\pi/L)^2kt}\cos(\pi x/L) + \tfrac{1}{4}e^{-(3\pi/L)^2kt}\cos(3\pi x/L) . \quad \square$$

Contained in Example 3 is the proof of the following result, which in the case of heat conduction in a circular wire, is the counterpart of Proposition 1.

Proposition 2. Let $a_0, a_1, \ldots, a_N$ and $b_1, \ldots, b_N$ be given constants. A solution of the problem

$$\begin{aligned} &\text{D.E.} \quad u_t = ku_{xx} \qquad -L \le x \le L ,\ t \ge 0 \\ &\text{B.C.} \quad u(-L,t) = u(L,t) \qquad u_x(-L,t) = u_x(L,t) \\ &\text{I.C.} \quad u(x,0) = a_0 + \sum_{n=1}^{N} [a_n\cos(n\pi x/L) + b_n\sin(n\pi x/L)] \end{aligned} \tag{28}$$

is given by

$$u(x,t) = a_0 + \sum_{n=1}^{N} e^{-(n\pi/L)^2kt}[a_n\cos(n\pi x/L) + b_n\sin(n\pi x/L)] . \tag{29}$$

Observe that (29) tends to the constant a_0 as $t \to \infty$. This constant is the average temperature in the circular wire. Indeed, integrating both sides of (29), we obtain

$$\frac{1}{2L}\int_{-L}^{L} u(x,t)\,dx = \frac{1}{2L}\int_{-L}^{L} a_0\,dx = a_0 ,$$

since all of the other terms have vanishing integrals, e.g.,

$$\frac{1}{2L}\int_{-L}^{L} \sin(n\pi x/L)\,dx = (-L/n\pi)\cos(n\pi x/L)\Big|_{-L}^{L} = 0 .$$

This average temperature is independent of t, which makes good physical sense, because no heat escapes from an insulated circular wire (i.e., heat passing through $x = L$ reenters at $x = -L$). Also it is physically reasonable that the temperature distribution will flatten itself out to its constant average as $t \to \infty$.

Some important questions arise from our treatment thus far. Foremost is the question of what to do in the event that the initial temperature distribution $u(x,0) = f(x)$ is not of the form given in Proposition 1 or 2. Certainly, not every continuous function is of this form, since a linear combination of sine and cosine functions is C^∞ (i.e., infinitely differentiable), while continuous functions may have corners where the derivatives will not exist. However, it does turn out to be true (cf. [Rudin (1976), p. 190])that any continuous function $f(x)$, defined on the closed interval $[-L,L]$, can be approximated to any degree of accuracy by a function of the form

$$a_0 + \sum_{n=1}^{N} [a_n\cos(n\pi x/L) + b_n\sin(n\pi x/L)] , \tag{30}$$

in the sense that the graph of (30) in the interval $[-L,L]$ can be made arbitrarily close to the graph of $f(x)$, by choosing N large enough and choosing appropriate constants a_n, b_n. Indeed, for "sufficiently nice" functions $f(x)$, we will develop formulas for the a_n and b_n , and we will provide estimates for the size of the number N which is needed to obtain the desired accuracy. This is done in Chapter 4 using Fourier series (cf. Theorem 2 of Section 4.1). In practice, the initial temperature distribution $f(x)$ is only known to within some experimental error. Hence, by approximating $f(x)$ to within this error by a sum of the form (30) we commit no observable error in replacing $f(x)$ by (30), and then Proposition 2 provides an adequate solution, as nearly as anyone can say. In Section 3.2, we will show that the solutions given in Propositions 1 and 2 are unique, as physical intuition suggests. Moreover, we show that if one approximates the initial temperature to within an experimental error by a sum, such as (30), then we commit no more than the experimental error in the solution.

Summary 3.1

1. The heat equation : The one–dimensional heat equation for the temperature $u(x,t)$ in a heat conducting medium, without currents or radiation, is the PDE $u_t = ku_{xx}$. The positive constant k is the diffusivity of heat, which measures the heat conductivity of the medium. This PDE is derived from two physical principles, namely the conservation of energy and the fact that the rate of heat energy flow is proportional to the temperature gradient.

2. The fundamental source solution: The fundamental source solution $u(x,t) = (4\pi kt)^{-\frac{1}{2}} e^{-x^2/4kt}$ $(t > 0 \,,\ x \text{ arbitrary})$ of the heat equation, $u_t = ku_{xx}$, describes the diffusion of the heat energy from an initial concentrated heat source at $x = 0$. It is not a sum of product solutions (cf. 4 below). For a fixed time t, the graph of u in the xu–plane is a normal probability curve which spreads as t increases, but $\int_{-\infty}^{\infty} u(x,t)\, dx = 1$ is independent of t.

3. An initial/boundary–value problem : Let $A(t)$, $B(t)$ and $f(x)$ be "well–behaved" functions. The problem

$$\begin{aligned} &\text{D.E.} \quad u_t = ku_{xx} \qquad 0 \le x \le L \,,\ t \ge 0 \\ &\text{B.C.} \quad u(0,t) = A(t) \qquad u(L,t) = B(t) \qquad\qquad (*) \\ &\text{I.C.} \quad u(x,0) = f(x) \,, \end{aligned}$$

is an example of an initial/boundary–value problem. Physical intuition suggests that this D.E. has a unique solution which meets the B.C. (boundary conditions) and the I.C. (initial condition). The mathematical proof of uniqueness is given in Section 3.2 .

4. Product solutions : The method of separation of variables is used to determine all the product solutions $u(x,t) = X(t)\, T(t)$ of the D.E. in (*). Every product solution of $u_t = ku_{xx}$ has one of the following forms, where c is the "separation constant" and $\lambda > 0$,

Case 1 $(\mathbf{c = -\lambda^2 < 0})$: $\quad u(x,t) = e^{-\lambda^2 kt}(c_1 \sin(\lambda x) + c_2 \cos(\lambda x)).$

Case 2 $(\mathbf{c = \lambda^2 < 0})$: $\quad u(x,t) = e^{\lambda^2 kt}(c_1 e^{\lambda x} + c_2 e^{-\lambda x}).$

Case 3 $(\mathbf{c = 0})$: $\quad u(x,t) = c_1 x + c_2 \,.$

5. Solutions of certain initial/boundary–value problems : The following Proposition 1 (for the rod with ends maintained at 0) and Proposition 2 (for the circular wire) are established by finding those product solutions (cf. 4 above) of the D.E. which meet the respective homogeneous, linear B.C., and then using the superposition principle to meet the I.C. . This procedure can be used to solve a wide variety of initial/boundary–value problems with homogeneous, linear B.C. .

Proposition 1. **Let** $b_1,...,b_N$ **be given constants. A solution of the problem**

$$\text{D.E.} \quad u_t = ku_{xx} \qquad 0 \le x \le L\ ,\ t \ge 0$$

$$\text{B.C.} \quad u(0,t) = 0 \quad u(L,t) = 0$$

$$\text{I.C.} \quad u(x,0) = \sum_{n=1}^{N} b_n \sin(n\pi x/L)$$

is given by

$$u(x,t) = \sum_{n=1}^{N} b_n e^{-(n\pi/L)^2kt} \sin(n\pi x/L)\ .$$

Proposition 2. **Let** $a_0,\ a_1,\ \dots\ ,\ a_N$ **and** $b_1,\ \dots\ ,\ b_N$ **be given constants. A solution of the problem**

$$\text{D.E.} \quad u_t = ku_{xx} \qquad -L \le x \le L\ ,\ t \ge 0$$

$$\text{B.C.} \quad u(-L,t) = u(L,t) \quad u_x(-L,t) = u_x(L,t)$$

$$\text{I.C.} \quad u(x,0) = a_0 + \sum_{n=1}^{N} [a_n \cos(n\pi x/L) + b_n \sin(n\pi x/L)]$$

is given by

$$u(x,t) = a_0 + \sum_{n=1}^{N} e^{-(n\pi/L)^2kt} [a_n \cos(n\pi x/L) + b_n \sin(n\pi x/L)]\ .$$

Exercises 3.1

1. Let $u(x,t)$ be a solution of $u_t = ku_{xx}$. Show that the following facts hold.

(a) For any constants a, x_0, t_0, the function $v(x,t) = u(ax - x_0, a^2t - t_0)$ satisfies $v_t = kv_{xx}$.
(b) For any constant k' , the function $v(x,t) = u(x,(k'/k)t)$ satisfies $v_t = k'v_{xx}$.
(c) The function $v(x,t) = t^{-\frac{1}{2}} \exp(-x^2/4kt) \cdot u(x/t,-1/t)$ satisfies $v_t = kv_{xx}$.

2. Solve the problem in Example 2 without using Proposition 1. That is, find the product solutions satisfying the D.E. and B.C. , and apply the superposition principle to meet the I.C. .

3. Use Proposition 1 to solve the problem

$$\text{D.E.}\quad u_t = 2u_{xx} \qquad 0 \le x \le 3\ ,\ t \ge 0$$

$$\text{B.C.}\quad u(0,t) = u(3,t) = 0$$

$$\text{I.C.}\quad u(x,0) = f(x)\ ,$$

in the following cases :

(a) $f(x) = 4\sin(2\pi x/3) - \sin(5\pi x/3)$

(b) $f(x) = 5\sin(4\pi x) + 2\sin(10\pi x)$

(c) $f(x) = \sin^3(\pi x/3)$

(d) $f(x) = -9\cos[(\pi/6)(2x+3)]$

(e) $f(x) = 3\cos[(8\pi x/3) - \frac{1}{2}\pi] - 3\cos[(8\pi x/3) + \frac{1}{2}\pi] + \sin(5\pi x)$.

4. (**A derivation of the fundamental source solution**) Let $u(x,t) = D\cdot b(t)\cdot f(b(t)^2x^2)$ for some positive constant D and some positive functions b and f, with $\int_{-\infty}^{\infty} D\cdot f(x^2)\,dx = 1$.

(a) Show that $\int_{-\infty}^{\infty} u(x,t)\,dx = 1$, for all t such that $b(t)$ is defined.

(b) Assuming that $u_t = ku_{xx}$ and that $b(t) \to \infty$ as $t \to 0^+$, show that $u(x,t) = (4\pi kt)^{-\frac{1}{2}}e^{-x^2/4kt}$. **Hint**. Substitute the form $b(t)\cdot f(b(t)^2x^2)$ for $u(x,t)$ into $u_t = ku_{xx}$ and set $x = 0$ to get an ODE for $b(t)$, from which $b(t) = (4k\alpha t)^{-\frac{1}{2}}$, where $\alpha = -f'(0)/f(0) > 0$. Show that if $y = b(t)^2x^2$, then $f'(y) + \alpha f(y) = -2y(f''(y) + \alpha f'(y))$. Thus, for $g(y) \equiv f'(y) + \alpha f(y)$, we have $g(y) = -2yg'(y)$. Solve for $g(y)$, and deduce that $f(y) = Ce^{-\alpha y}$.

5. Show that there are no nonzero Case 2 product solutions ($u(x,t) = e^{\lambda^2kt}(c_1e^{\lambda x} + c_2e^{-\lambda x})$) which satisfy the B.C. (23) for the circular wire.

6. Solve the problem

$$\text{D.E.}\quad u_t = u_{xx} \qquad -\pi \le x \le \pi\ ,\ t \ge 0$$

$$\text{B.C.}\quad u(-\pi,t) = u(\pi,t) \qquad u_x(-\pi,t) = u_x(\pi,t)$$

$$\text{I.C.}\quad u(x,0) = f(x)$$

in the following cases :

(a) $f(x) = 5\cos(x) + 3\sin(8x)$ (b) $f(x) = \frac{1}{2} + \cos(2x) - 6\sin(2x)$

(c) $f(x) = 4 + \cos^2(3x)$ (d) $f(x) = 6\sin(x) - 7\cos(3x) - 7\sin(3x)$

(e) $f(x) = [\sin(x) + 2\cos(x)]^2$.

7. (a) Show that for any particular solution of $u_t = ku_{xx}$ of the form $u(x,t) = e^{rx+st}$, where r and s are constants, we must have $s = kr^2$.

(b) Let $u(x,t) = u_1(x,t) + u_2(x,t)i$ be a complex solution of $u_t = ku_{xx}$. Then show that the real and imaginary parts, $u_1(x,t)$ and $u_2(x,t)$, are also solutions. (Note that $u_t = (u_1)_t + (u_2)_t i$ by definition, and if a, b, c and d are real, $a + bi = c + di$ if and only if $a = c$ and $b = d$.)

(c) Noting that r in (a) can be complex, use (b) and the formula $e^{\alpha+i\beta} = e^{\alpha}(\cos(\beta) + i\sin(\beta))$ to show that the following values of r yield the indicated solutions (where λ is real).

(i) $r = \pm i\lambda$: $e^{-k\lambda^2 t}\cos(\lambda x)$, $\pm e^{-k\lambda^2 t}\sin(\lambda x)$.

(ii) $r = \pm\lambda$: $e^{\pm\lambda x}e^{k\lambda^2 t}$.

(iii) $r = \lambda(1 \pm i)$: $e^{\lambda x}\cos(\lambda x + 2k\lambda^2 t)$, $\pm e^{\lambda x}\sin(\lambda x + 2k\lambda^2 t)$.

8. (a) Consider the problem

$$\begin{aligned} &\text{D.E.} \quad u_t = ku_{xx} \quad x \geq 0 , t \geq 0 \\ &\text{B.C.} \quad u(0,t) = \cos(\omega t) . \end{aligned} \qquad (*)$$

This is a heat conduction problem for a semi–infinite rod $(x \geq 0)$ whose end (at $x = 0$) is subjected to a periodic temperature variation $u(0,t) = \cos(\omega t)$. Use Problem 7c(iii) to find a solution of this problem which has both of the additional properties :

(P1) $u(x,t) \to 0$ as $x \to \infty$ and (P2) $u(x,t + 2\pi/\omega) = u(x,t)$ (i.e., $u(x,t)$ is periodic in time).

(b) Show that the solution of (*) is not unique, if either (P1) or (P2) is omitted.

(c) Assuming that $\omega = \pi/2$ and $k = \pi/4$, roughly sketch the graph of the temperature distribution in the xu–plane when t = 0, 1, 2, 3, 4, paying attention to where $u(x,t) = 0$.

(d) Show that at any fixed time t, the distance between consecutive local maxima, say x_1 and x_2 , of $u(x,t)$ is $2\pi(2k/\omega)^{\frac{1}{2}}$, and show that the ratio $u(x_2,t)/u(x_1,t)$ is $e^{-2\pi} \approx .00187$, regardless of the positive values for k and ω !

9. (a) Assume that the insulation around a rod is faulty in that the heat in each small slab leaks through the insulation at a rate proportional to the temperature of the slab. Show that the temperature $u(x,t)$ then obeys the equation $u_t = ku_{xx} - hu$ for some constant $h > 0$.

(b) Show that if $w(x,t)$ solves $w_t = kw_{xx}$, then $u(x,t) = e^{-ht}w(x,t)$ solves $u_t = ku_{xx} - hu$.

(c) Find a solution of the problem

$$\text{D.E.} \quad u_t = ku_{xx} - hu \quad 0 \le x \le L \ , t \ge 0$$

$$\text{B.C.} \quad u(0,t) = u(L,t) = 0$$

$$\text{I.C.} \quad u(x,0) = \sum_{n=1}^{N} b_n \sin(n\pi x/L) \ .$$

10. In the derivation of the heat equation, show that if K, D and C are well–behaved functions of x, then we obtain the PDE $C(x)D(x)u_t = \frac{\partial}{\partial x}\left[K(x)u_x\right]$.

11. (a) Assume that the temperature in a solid ball depends only on the distance r from the center of the solid (i.e., $u = u(r,t)$). By using the fact that the area of the sphere of radius r is $4\pi r^2$, show that $u_t = kr^{-2}(r^2u_r)_r = k(u_{rr} + 2r^{-1}u_r)$.

Hint. Use $E(t) = \int_a^b CD\cdot u(r,t)4\pi r^2\,dr$ and $E'(t) = K\left[4\pi b^2 u_r(b,t) - 4\pi a^2 u_r(a,t)\right]$.

(b) By setting $v(r,t) = r\cdot u(r,t)$, deduce that $u_t = kr^{-2}(r^2u_r)_r$ is equivalent to $v_t = kv_{rr}$, for $r > 0$. Thus, any solution $v(x,t)$ of $v_t = kv_{xx}$ gives rise to a radially symmetric symmetric heat flow $u(r,t) = v(r,t)/r$ in the ball, at least for $r > 0$. (cf. Section 9.4 for further information on heat flow for the sphere.)

12. Let u_0 and u_1 be constants and define (for $t > 0$)

$$u(x,t) = u_1 + (u_0 - u_1)\,\frac{2}{\sqrt{\pi}}\int_0^{\frac{1}{2}x/\sqrt{kt}} e^{-y^2}\,dy \ . \qquad (*)$$

(a) Verify that $u_t = ku_{xx}$ $(t > 0)$ by using the following result from the calculus :

$$\frac{d}{dz}\left[\int_0^{f(z)} g(y)\,dy\right] = g(f(z))\frac{df}{dz},$$

(b) Using the result $\int_0^\infty e^{-y^2}\,dy = \frac{1}{2}\sqrt{\pi}$, show that $\lim_{t\to 0^+} u(x,t) = u_0$, for any $x > 0$, while $u(0,t) \equiv u_1$. This suggests that (*) represents the temperature distribution of a semi–infinite rod $(0 \le x < \infty)$ when the temperature u_1 is imposed at the end $x = 0$ for $t \ge 0$, given that the initial temperature is the constant u_0. In other words, $u(x,t)$ solves

$$\text{D.E.} \quad u_t = ku_{xx} \qquad 0 \le x < \infty \ , t > 0$$

$$\text{B.C.} \quad u(0,t) = u_1 \qquad t > 0$$

$$\text{I.C.} \quad u(x,0^+) = u_0 \qquad x > 0 \ ,$$

where $u(x,0^+) \equiv \lim_{t\to 0} u(x,t)$ for $x > 0$. Note that the solution does not extend continuously to $(0,0)$, unless $u_0 = u_1$. (This solution can be used to estimate the temperature of the earth at a depth x below ground after an abrupt change $u_0 \longrightarrow u_1$ at ground level.)

13. Suppose that a particle, starting at the origin, has an equal chance of moving to the left or right by a distance Δx in a time interval of Δt.

(a) Let $n > 0$ be an integer, and let m be an integer, such that $-n \le m \le n$ and $n - m$ is even. By computing the number of ways that the particle can move a net distance of $m\Delta x$ in n time intervals Δt, show that the probability that it is at $x = m\Delta x$, after a time $t = n\Delta t$ is

$$\frac{n!(\frac{1}{2})^n}{(\frac{1}{2}(n+m))!\cdot(\frac{1}{2}(n-m))!} \ . \qquad (*)$$

(b) Use Stirling's formula $n! \approx \sqrt{2\pi}\, e^{-n}\, n^{n+\frac{1}{2}}$, for large n, to deduce that (*) is approximately

$$\frac{\sqrt{2\pi}\ e^{-n}\ n^{n+\frac{1}{2}}\ (\frac{1}{2})^n}{\sqrt{2\pi}\ e^{-\frac{1}{2}(n+m)}(\frac{1}{2}(n+m))^{\frac{1}{2}(n+m+1)}\ \sqrt{2\pi}\ e^{-\frac{1}{2}(n-m)}(\frac{1}{2}(n-m))^{\frac{1}{2}(n-m+1)}}$$

$$= \frac{2}{\sqrt{2\pi n}}\left[\frac{n}{n+m}\right]^{\frac{1}{2}(n+m+1)}\left[\frac{n}{n-m}\right]^{\frac{1}{2}(n-m+1)} = \frac{2}{\sqrt{2\pi n}}\left[1+\frac{m}{n}\right]^{-\frac{1}{2}(n+m+1)}\left[1-\frac{m}{n}\right]^{-\frac{1}{2}(n-m+1)}$$

$$= \frac{2}{\sqrt{2\pi n}}\left[1-\left[\frac{m}{n}\right]^2\right]^{-\frac{1}{2}(n+1)}\left[1+\frac{m}{n}\right]^{-\frac{1}{2}m}\left[1-\frac{m}{n}\right]^{\frac{1}{2}m}$$

$$= \frac{2}{\sqrt{2\pi t}}\left[1-\left[\frac{x}{t}\right]^2\cdot\left[\frac{\Delta t}{\Delta x}\right]^2\right]^{-\frac{1}{2}t/\Delta t-\frac{1}{2}}\left[1+\left[\frac{x}{t}\right]\cdot\left[\frac{\Delta t}{\Delta x}\right]\right]^{-\frac{1}{2}x/\Delta x}\left[1-\left[\frac{x}{t}\right]\cdot\left[\frac{\Delta t}{\Delta x}\right]\right]^{\frac{1}{2}x/\Delta x}(\Delta t)^{\frac{1}{2}}. \quad (**)$$

(c) Note that we get a well–defined density of order Δx, if Δt is proportional to Δx^2, say $\Delta x^2 = 2k\Delta t$. Then, dividing (**) by $2\Delta x$ and letting $\Delta t \to 0$, show that we obtain the fundamental source solution $(4\pi kt)^{-\frac{1}{2}}\exp(-x^2/(4kt))$. **Hint.** Recall that $\lim_{z\to\infty}(1 + az)^{1/z} = e^a$. Observe that we divide by $2\Delta x$ (instead of Δx), since the spacing between the possible values of $x = m\Delta x$ is $2\Delta x$, since $n - m$ must be even.

3.2 Uniqueness and the Maximum Principle

In Propositions 1 and 2 of the previous section, we found solutions of the standard initial/boundary–value problems for a rod with ends maintained at 0 and for the circular wire. Here we prove that the solutions found are in fact the only solutions. Without this uniqueness property, the theory loses its predictive power. We examine two methods for establishing uniqueness. The first method (cf. the proof of Theorem 1) is more straightforward than the second method which is based on the Maximum Principle. However, the second method gives us the finer result that, if we change the initial condition slightly, then any solution will change only slightly. This allows one to approximate initial conditions by more manageable functions, without disturbing the solution by a significant amount. Since initial conditions can never be exactly known in practice, this "stability" property is absolutely crucial in applications.

Theorem 1. (The Uniqueness Theorem) Let $u_1(x,t)$ **and** $u_2(x,t)$ **be** C^2 **solutions of the following problem, where** $a(t)$, $b(t)$ **and** $f(x)$ **are given** C^2 **functions :**

$$\begin{aligned} &\text{D.E.} \quad u_t = k u_{xx} \qquad 0 \le x \le L \ , t \ge 0 \\ &\text{B.C.} \quad u(0,t) = a(t) \qquad u(L,t) = b(t) \qquad\qquad (1) \\ &\text{I.C.} \quad u(x,0) = f(x) \ . \end{aligned}$$

Then $u_1(x,t) = u_2(x,t)$, **for all** $0 \le x \le L$ **and** $t \ge 0$.

Proof. Let $v(x,t) = u_1(x,t) - u_2(x,t)$. We need to show that $v(x,t) \equiv 0$. Note that $v(0,t) = u_1(0,t) - u_2(0,t) = a(t) - a(t) = 0$, and $v(L,t) = b(t) - b(t) = 0$, while $v(x,0) = u_1(x,0) - u_2(x,0) = f(x) - f(x) = 0$. Since the D.E. is linear and homogeneous, we know that $v = u_1 - u_2$ satisfies the D.E., by the superposition principle. Thus, $v(x,t)$ is a C^2 solution of the problem

$$\begin{aligned} &\text{D.E.} \quad v_t = k v_{xx} \qquad 0 \le x \le L \ , t \ge 0 \\ &\text{B.C.} \quad v(0,t) = 0 \qquad v(L,t) = 0 \qquad\qquad (2) \\ &\text{I.C.} \quad v(x,0) = 0 \ . \end{aligned}$$

We know that $v(x,t) \equiv 0$ is a solution of (2), but we must show that this is the *only* solution. We have reduced the demonstration of uniqueness of (1) to that of the simpler problem (2). Let $v(x,t)$ be a solution of (2) and define

$$F(t) = \int_0^L [v(x,t)]^2 \, dx \ , \quad t \ge 0 \ . \qquad (3)$$

Since $F(t)$ is the integral of a nonnegative (square) function, we have $F(t) \geq 0$. If $F(t) = 0$, then $v(x,t)$ must be equal zero for all x, $0 \leq x \leq L$. (The integral of a continuous, nonnegative function is positive unless the function is identiclly zero.) Thus, if we can show that $F(t) = 0$ for all $t \geq 0$, then $v(x,t) \equiv 0$, and we would be done. Observe that

$$F(0) = \int_0^L [v(x,0)]^2 \, dx = 0 , \tag{4}$$

from the I.C. of (2). We know that $F(t) \geq 0$. Thus, we need to show that $F(t)$ cannot increase (i.e., $F'(t) \leq 0$), and then $F(t) \equiv 0$ will follow. To show that $F'(t) \leq 0$, we begin by differentiating under the integral sign in (3). (Leibniz's rule of Appendix A.3 permits us to do this, because $v(x,t)$ is C^2.)

$$\begin{aligned} F'(t) &= \int_0^L \frac{\partial}{\partial t}\left[[v(x,t)]^2\right] dx = 2\int_0^L v(x,t)\cdot v_t(x,t)\, dx \\ &= 2\int_0^L v(x,t)\cdot k v_{xx}(x,t)\, dx , \quad \text{by the D.E.} . \end{aligned} \tag{5}$$

We apply integration by parts $\left[\int_a^b g(x)h'(x)\,dx = g(x)h(x)\Big|_a^b - \int_a^b g'(x)h(x)\,dx \right]$, where (for fixed t) $g(x) = v(x,t)$ and $h(x) = v_x(x,t)$. Then (5) becomes

$$F'(t) = 2\left[kv(x,t)v_x(x,t)\Big|_0^L - \int_0^L k[v_x(x,t)]^2\, dx \right]. \tag{6}$$

Since $v(L,t) = 0$ and $v(0,t) = 0$ by (2), the first term is 0. Now note that $-k[v_x(x,t)]^2 \leq 0$, whence $F'(t) \leq 0$. Thus, $F(t)$ can never increase. Since $F(0) = 0$ by (4) and $F(t) \geq 0$, the only possibility is that $F(t) \equiv 0$. Alternatively, $F(t) = F(t) - F(0) = \int_0^t F'(s)\, ds \leq 0$ and $F(t) \geq 0$ imply that $F(t) \equiv 0$. Thus (as mentioned above), $v(x,t) \equiv 0$, and $u_1(x,t) = u_2(x,t)$. □

Remarks. (1) The method used in the proof of Theorem 1 can also be used to prove uniqueness of solutions of the heat equation subject to other types of boundary conditions (and initial condition). We will consider these other B.C. and their physical importance in some detail in Section 3.3. Since we are already familiar with the B.C. for the circular wire (cf. Example 3 of Section 3.1), we use the method to prove uniqueness for this case, in Example 1 below.

(2) The proof of Theorem 1 also implies that $u(x,t)$ has a **continuous mean–square dependence** on the initial data. By this we mean the following :

If $u_1(x,t)$ and $u_2(x,t)$ satisfy the D.E. and the B.C. of (1), then

$$F(t) \equiv \int_0^L [u_1(x,t) - u_2(x,t)]^2\, dx \leq \int_0^L [u_1(x,0) - u_2(x,0)]^2\, dx = F(0) . \tag{7}$$

Indeed, as in the proof of Theorem 1 (cf. (6)) , $F'(t) \leq 0$ for $t \geq 0$, and so $F(t) \leq F(0)$. Inequality (7) says that if the *initial* temperature distributions are close, in the sense that the integral of the square of their difference is small (i.e., in the mean–square sense), then at any later time, the distributions are at least as close (again, in the mean–square sense). □

Example 1. State and prove a uniqueness theorem for the circular heat conduction problem (cf. Example 3 of Section 3.1).

Solution. The statement of the theorem is the following :

Let $u_1(x,t)$ and $u_2(x,t)$ be C^2 solutions of the problem

$$\text{D.E.}\quad u_t = ku_{xx} \qquad -L \leq x \leq L\ ,\quad t \geq 0$$

$$\text{B.C.}\quad u(L,t) = u(-L,t) \quad u_x(-L,t) = u_x(L,t) \tag{8}$$

$$\text{I.C.}\quad u(x,0) = f(x)\ .$$

Then $u_1(x,t) = u_2(x,t)$ for all $-L \leq x \leq L$, $t \geq 0$.

Proof. Set $v = u_1 - u_2$, and observe that v satisfies the D.E. and the B.C. of (*), and $v(x,0) = f(x) - f(x) \equiv 0$. Define $F(t) = \int_{-L}^{L} [v(x,t)]^2\, dx$, noting that $F(0) = 0$ and $F(t) \geq 0$. As in the proof of Theorem 1, it suffices to prove $F'(t) \leq 0$ (Why ?). The computation in (6) yields

$$F'(t) = 2\left[v(x,t)v_x(x,t)\Big|_{-L}^{L} - \int_{-L}^{L} k[v_x(x,t)]^2\, dx \right].$$

The endpoint terms cancel by the B.C., and $F'(t) \leq 0$. □

Remark. As a consequence of Theorem 1 and Example 1, we now know that the solutions of the initial/boundary–value problems in Proposition 1 and Proposition 2 of Section 3.1 are in fact the only solutions satisfying the given B.C. and I.C. . □

The Maximum Principle and its consequences

Of course, uniqueness is a crucial property that is necessary in a mathematical description of physical processes that have definite outcomes under prescribed conditions. However, there is a stronger property that successful mathematical models for such processes often have, namely "stability" or "continuity" with respect to small variations of prescribed conditions. Suppose that we change the initial temperature distribution and/or the boundary values slightly (say, by an experimentally undetectable amount). If it happens that the solution changes by a large (experimentally detectable) amount, then (in spite of uniqueness) our theory loses its predictive value. If this cannot happen, then we say (loosely) that our model has solutions that vary continuously (or is stable) with respect to variations in the prescribed conditions. Inequality (7) gives us stability in a certain weak sense. That is, when a deformation in the initial temperature distribution has a small mean–square, the deformation of the temperature distribution at any later time has a small mean–square. However, just because a function, say $g(x)$, has a small mean–square (i.e., $\int_0^L g(x)^2\,dx$ is small), it does *not* follow that $g(x)$ is small for *all* x in $[0,L]$. For example, let n be any positive integer such that $10^{-3n} < L$, and let $g(x) = 10^n$ for $0 \le x \le 10^{-3n}$ and set $g(x) = 0$ for $10^{-3n} < x \le L$. Then, in spite of the fact that $\int_0^L g(x)^2\,dx = 10^{-3n} \cdot 10^{2n} = 10^{-n}$, we have $g(x) = 10^n$ for $0 \le x \le 10^{-3n}$. More generally, take $g(x)$ to be zero, except for a spike in its graph. Regardless of how high the spike is, we can always take the spike to be narrow enough, so that $\int_0^L g(x)^2\,dx$ is as small as we wish. Thus, even if the mean–square of the difference of two temperature distributions is small, we cannot conclude that the difference is small at each point. To remedy this defect in the result (7), and to prove the type of stability which is uniformly valid for all points in the rod, we use the Maximum Principle which is stated in Theorem 2. However, we will defer the proof until the end of the section, by which time there will be ample motivation for it, in view of its many consequences and applications.

Theorem 2 (The Maximum Principle). Let $u(x,t)$ be a C^2 solution (there is at most one) of

$$\text{D.E.} \quad u_t = ku_{xx} \qquad 0 \le x \le L\ ,\ t \ge 0 \qquad (k > 0)$$

$$\text{B.C.} \quad u(0,t) = a(t) \quad u(L,t) = b(t) \tag{9}$$

$$\text{I.C.} \quad u(x,0) = f(x)\ ,$$

where a, b and f are given C^2 functions. Let $T > 0$ be any fixed future time. Define M to be the maximum value of the initial temperature, and let A and B be the maximum temperatures at the ends $x = 0$ and $x = L$ during the time interval from $t = 0$ to $t = T$, i.e.,

$$A = \max_{0 \le t \le T} \{a(t)\}\ , \quad B = \max_{0 \le t \le T} \{b(t)\} \quad \text{and} \quad M = \max_{0 \le x \le L} \{f(x)\}\ .$$

Let $\overline{M} = \max\{A,B,M\}$ (i.e., the largest). Then

$$u(x,t) \le \overline{M} \ \text{ for all } x \text{ and } t, \text{ with } 0 \le x \le L\ ,\ 0 \le t \le T\ . \tag{10}$$

Remark. In geometrical terms, the Maximum Principle states that if a solution of problem (9) is graphed in the xtu–space, as in Figure 1, then the surface $u = u(x,t)$ achieves its maximum height above one of the three sides $x = 0$, $x = L$, $t = 0$ of the rectangle $0 \le x \le L$, $0 \le t \le T$.

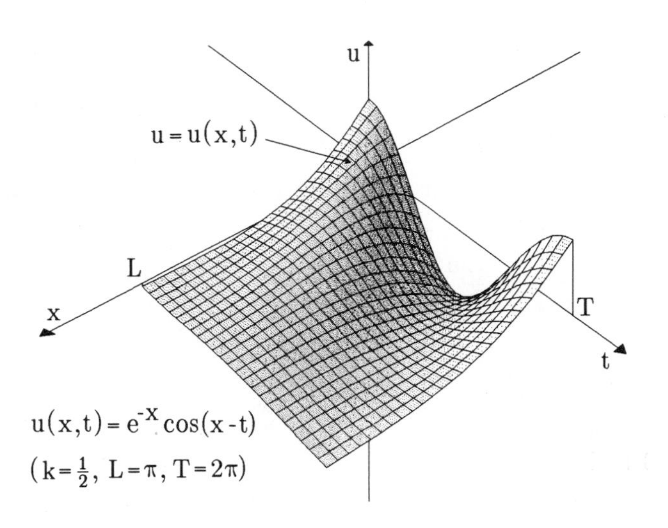

Figure 1

From a physical perspective, the Maximum Principle states that the temperature, at any point x inside the rod at any time t $(0 \le t \le T)$, is less than the maximum of the initial temperature distribution or the maximum of the temperatures prescribed at the ends during the time interval [0,T]. To those with good physical intuition, this is not too surprising, since one does not ordinarily expect heat to concentrate itself inside the rod, but rather it dissipates. Heat "avoids" itself, preferring to flow to colder regions. By the same token, cold "avoids" itself. Indeed, there is a Minimum Principle that follows easily from the Maximum Principle :

Corollary (The Minimum Principle). Let $u(x,t)$ **be a** C^2 **solution of** (9) **and let**

$$a = \min_{0 \le t \le T} \{a(t)\}, \quad b = \min_{0 \le t \le T} \{b(t)\} \quad \text{and} \quad m = \min_{0 \le x \le L} \{f(x)\}.$$

Defining $\underline{m} = \min\{a,b,m\}$ **, we have**

$$\underline{m} \le u(x,t) \quad \textbf{for all } x \textbf{ and } t\textbf{, where } 0 \le x \le L,\ 0 \le t \le T.$$

Proof. Let $v(x,t) = -u(x,t)$. Then $v(x,t)$ solves problem (9), if we replace $a(t)$, $b(t)$ and $f(x)$ by $-a(t)$, $-b(t)$ and $-f(x)$ respectively. Now, it is not hard to check that

$$\max_{0 \le t \le T} \{-a(t)\} = -\min_{0 \le t \le T} \{a(t)\} = -a, \quad \max_{0 \le t \le T} \{-b(t)\} = -b \quad \text{and} \quad \max_{0 \le x \le L} \{-f(x)\} = -m.$$

Thus, applying the Maximum Principle to $v(x,t)$, we have that $v(x,t) \le \max\{-a,-b,-m\} =$

$-\min\{a,b,m\} = -\underline{m}$. Then $v(x,t) \le -\underline{m}$, and multiplying through by -1 , we obtain the reverse inequality $u(x,t) = -v(x,t) \ge \underline{m}$, as desired. □

Example 2. Use the Maximum and Minimum Principles to deduce that the solution $u(x,t)$ of

$$\begin{aligned} &\text{D.E.} \quad u_t = 9u_{xx} \qquad 0 \le x \le 3 \ , t \ge 0 \\ &\text{B.C.} \quad u(0,t) = 0 \qquad u(3,t) = 0 \qquad\qquad (11) \\ &\text{I.C.} \quad u(x,0) = 6\sin(\pi x/3) + 2\sin(\pi x) \end{aligned}$$

satisfies the inequalities $0 \le u(x,t) \le 4\sqrt{2}$.

Solution. Here $k = 9$, $L = 3$ and the initial temperature distribution is of the form of Proposition 1 of Section 3.1. Thus, the (unique) solution of (11) is

$$u(x,t) = 6e^{-\pi^2 t}\sin(\pi x/3) + 2e^{-9\pi^2 t}\sin(\pi x) . \qquad (12)$$

From this it follows that $-8 \le u(x,t) < 8$ (Why ?), but we must do better than this. Ordinarily, one would use calculus (e.g., set $u_t(x,t) = 0$ and $u_x(x,t) = 0$ and solve simultaneously for x and t) to determine the local maxima, minima, or saddle points of this function strictly inside the strip $0 < x < 3, t > 0$. It is a nontrivial task to solve these equations, or prove that there is no solution. (Indeed, there is a critical point inside the strip (cf. Problem 14).) However, even if this were done, then one would still have to check the border of the strip for maxima and minima. The Maximum/Minimum Principles save us the work of checking for extrema strictly inside the strip, since they state that the maximum and minimum of a solution of the heat equation (up to any given time) is automatically achieved on the border. One might try to argue that the maximum for u must occur when $t = 0$, since the terms of (12) appear to be decreasing for all x. Actually, the temperature at $x = 3/2$ is *increasing* at $t = 0$:

$$u_t(3/2,0) = -6\pi^2\sin(\pi/2) - 18\pi^2\sin(3\pi/2) = 12\pi^2 > 0 . \qquad (13)$$

Thus, just because the amplitudes of the terms are decreasing, we cannot conclude that the values of $u(x,t)$ are decreasing. Hence, without the Maximum/Minimum Principles, we are essentially forced to look for local maxima and minima with $t > 0$. Instead, utilizing the Maximum Principle, we know that, in the strip, $u(x,t)$ cannot exceed the maximum on the edges $x = 0$, $x = 3$, $t = 0$ in the xt–plane. For $x = 0$ or $x = 3$, note that $u(0,t) = 0$ and $u(3,t) = 0$. To find the maximum and minimum of $u(x,0)$, we differentiate $f(x) = 6\sin(\pi x/3) + 2\sin(\pi x)$:

$$f'(x) = 2\pi[\cos(\pi x/3) + \cos(\pi x)] = 4\pi\cdot\cos(2\pi x/3)\cdot\cos(\pi x/3),$$

where we have used $\cos(\alpha) + \cos(\beta) = 2\cdot\cos(\frac{1}{2}(\alpha+\beta))\cdot\cos(\frac{1}{2}(\alpha-\beta))$. Thus, in the interval [0,3], $f'(x) = 0$, only when $x = 3/4, 3/2, 9/4$. The graph of $f(x)$ is shown in Figure 2 . Thus, the maximum values of f are $f(3/4) = f(9/4) = 4\sqrt{2}$. The point $x = 3/2$ is a local minimum with $f(3/2) = 4$. It is not surprising that the temperature at $x = 3/2$ will increase initially as the heat flows into this local minimum [cf. (13)]. The maximum of $u(x,t)$ on the borders $x = 0$, $x = 3$, $t = 0$ is then $4\sqrt{2}$, while the minimum on these borders is 0 . Hence, the Maximum Principle tells us that $u(x,t) \le 4\sqrt{2}$ for all $0 \le x \le 3$, $0 \le t$, while the Minimum Principle implies

$u(x,t) \geq 0$ for all $0 \leq x \leq 3$, $t \geq 0$, as desired. From (12), we see that $u(x,t) \to 0$ as $t \to \infty$, and $u(x,t)$ is very nearly $6e^{-\pi^2 t}\sin(\pi x/3)$, for large t. The graph of $u(x,t)$ in the xtu–space is shown in Figure 3. □

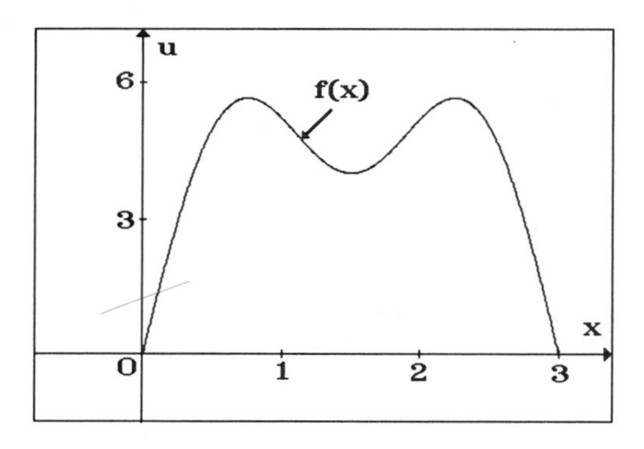

Figure 2

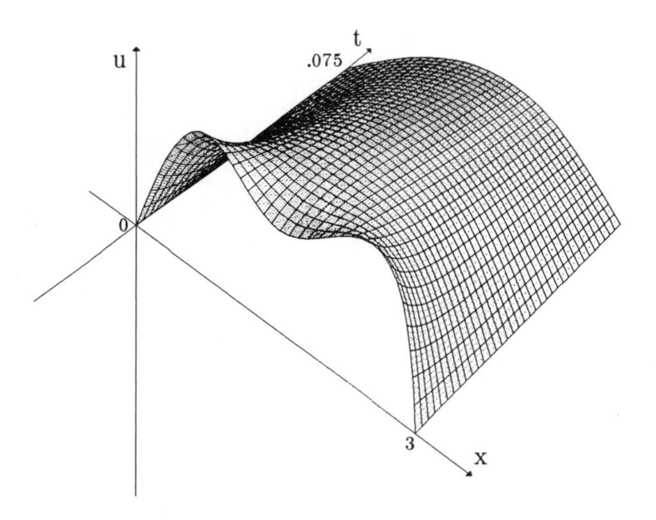

Figure 3

The following theorem states that if two initial/boundary–value problems for the heat equation have initial temperature distributions which are close at each point of the rod and if the prescribed boundary values are close up to a time T, then for each time t $(0 \leq t \leq T)$, the corresponding solutions (if they exist) must be at least as close at all points in the rod. This is the crucial property of continuous dependence of solutions on boundary and initial conditions which was discussed on earlier this section. We easily prove the result by using the Maximum Principle.

Theorem 3 (Continuous Dependence on the I.C. and the B.C.). Let $u_1(x,t)$ **and** $u_2(x,t)$ **be** C^2 **solutions of the respective problems** $(0 \le x \le L\,,\ t \ge 0)$

$$\begin{array}{llll} \text{D.E.} & u_t = ku_{xx} & \text{D.E.} & u_t = ku_{xx} \\ \text{B.C.} & u(0,t) = a_1(t) & \text{B.C.} & u(0,t) = a_2(t) \\ & u(L,t) = b_1(t) & & u(L,t) = b_2(t) \\ \text{I.C.} & u(x,0) = f_1(x) & \text{I.C.} & u(x,0) = f_2(x)\,. \end{array} \tag{14}$$

If, for some $\epsilon \ge 0$ **, we have**

$$|f_1(x) - f_2(x)| \le \epsilon \quad \textbf{for all } x,\ 0 \le x \le L$$

and $|a_1(t) - a_2(t)| \le \epsilon$ **and** $|b_1(t) - b_2(t)| \le \epsilon$ **for all** $t,\ 0 \le t \le T$

then $|u_1(x,t) - u_2(x,t)| \le \epsilon$ **, for all** x **and** t**, where** $0 \le x \le L\,,\ 0 \le t \le T$.

Proof. Let $v(x,t) = u_1(x,t) - u_2(x,t)$. Then $v_t = kv_{xx}$ and we have

$$|v(x,0)| = |f_1(x) - f_2(x)| \le \epsilon\,, \quad 0 \le x \le L\,,$$

$$\left.\begin{array}{l} |v(0,t)| = |a_1(t) - a_2(t)| \le \epsilon \\ |v(L,t)| = |b_1(t) - b_2(t)| \le \epsilon\,. \end{array}\right\}\,, \quad 0 \le t \le T$$

Thus, the maximum of v on the borders $t = 0$ $(0 \le x \le L)$ and $x = 0\,,\ x = L$ $(0 \le t \le T)$ is not greater then ϵ , while the minimum of v on these borders is not less than $-\epsilon$. Hence the Maximum/Minimum Principles yield the result

$$-\epsilon \le v(x,t) \le \epsilon \quad \text{or} \quad |u_1(x,t) - u_2(x,t)| = |v(x,t)| \le \epsilon\,. \quad \square$$

Remark. Observe that when $\epsilon = 0$, the problems in (14) are identical, and we have the conclusion $|u_1(x,t) - u_2(x,t)| \le 0$ (i.e. $u_1 = u_2$). Thus, the uniqueness result, Theorem 1, is the special case of Theorem 3 obtained by setting $\epsilon = 0$. The proofs of Theorem 3 and the Maximum Principle are entirely different from the proof of Theorem 1. To some readers, it may seem intuitively obvious that if a certain initial/boundary–value problem has unique solutions, then the solutions change only slightly, if the initial and/or boundary conditions are varied slightly. To show that this is *not* always the case, we offer the following counterexample.

Example 3. For any given constant $\alpha > 0$, consider the problem for $u(x,t)$

$$\text{D.E.} \quad (1-t)\cdot u_t = u \qquad 0 \le x \le 2,\ 0 \le t < 1$$

$$\text{B.C.} \quad u(0,t) = 0 \qquad u(2,t) = 0 \tag{15}$$

$$\text{I.C.} \quad u(x,0) = \alpha\cdot x(2-x)\ .$$

Show that the no matter how small the constant α is, the solution will become large as $t \to 1^-$.

Solution. For any fixed x, the PDE is a separable ODE, namely $du/u = (1-t)^{-1}dt$, and the general solution is $u(x,t) = f(x)/(1-t)$ for a C^1 function $f(x)$. Since, $f(x) = u(x,0) = \alpha\cdot x(2-x)$, the unique solution of problem (15) is $u(x,t) = \alpha x(2-x)/(1-t)$. In particular, $u(1,t) = \alpha/(1-t)$, which becomes arbitrarily large as $t \to 1^-$ if $\alpha > 0$, even if $\alpha = 10^{-9}$. Note that the maximum of the initial and boundary values for u is α, and thus there is no maximum principle for the solutions of the problem (15). □

Proof of the Maximum Principle

In the statement of the Maximum Principle, suppose that we replace the D.E. by the more general PDE $u_t = ku_{xx} - c$, where $c \ge 0$ is a fixed constant. We call the resulting theorem the **Generalized Maximum Principle**, or simply the **GMP**. Although we are primarily interested in the case where $c = 0$, we will first prove the GMP in the easier case where $c > 0$, which physically corresponds to a uniform heat loss along the rod. We then handle the case $c = 0$, by carefully considering a limit as $c \to 0^+$. Suppose that $u(x,t)$ solves the D.E. $u_t = ku_{xx} - c$, for $c > 0$. Let M_0 be the maximum of u on the closed rectangle R $(0 \le x \le L\ ,\ 0 \le t \le T)$ in Figure 4.

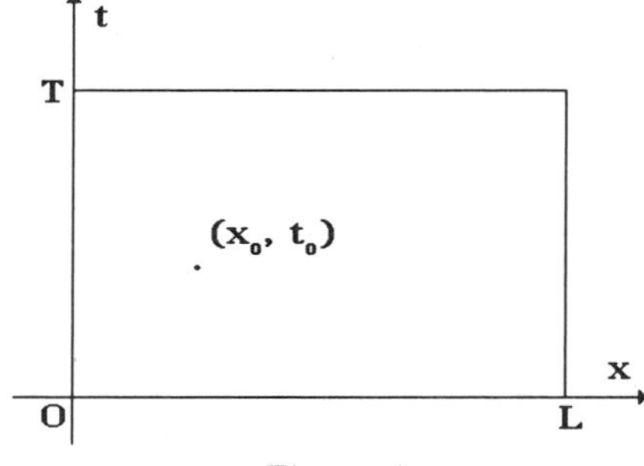

Figure 4

The existence of M_0 is guaranteed by the Maximum/Minimum Theorem in Appendix A.4. Certainly $M_0 \ge \overline{M}$, since $\overline{M}$ (by definition) is the maximum of u on the sides and the lower edge of R, but we need to prove that $M_0 = \overline{M}$. Let (x_0,t_0) be a point in R, such that $u(x_0,t_0) = M_0$. We assume that $0 < x_0 < L$ and $0 < t_0 \le T$, for otherwise $M_0 = \overline{M}$ (Why ?), and we would be done. However, we will show that this assumption leads to a contradiction, and hence the assumption must be false, and consequently $M_0 = \overline{M}$. Note that $u(x,t_0)$, as a function of x on the *open* interval (0,L), has a maximum at x_0, and so $u_x(x_0,t_0) = 0$. (We would not know this if x_0 were an endpoint.) Also, since the graph of $u(x,t_0)$ cannot be concave up at x_0 (Why not?), we must have $u_{xx}(x_0,t_0) \leq 0$. Using this fact and the equation $u_t = ku_{xx} - c$, we have

$$u_t(x_0,t_0) = ku_{xx}(x_0,t_0) - c \leq -c \ . \tag{16}$$

Thus, $u_t(x_0,t_0) < 0$. Hence, for $\hat{t}$ slightly less than t_0 (recall that $t_0 > 0$), we have $u(x_0,\hat{t}) > u(x_0,t_0) = M_0$, which is a contradiction (Why ?). Thus, the assumption that $0 < x_0 < L$ and $t_0 > 0$ must be false, and so the GMP has been proven in the case where $c > 0$. The proof in the case $c = 0$, is as follows. Let u be a solution of $u_t = ku_{xx}$ on R. Define a function v, of three variables, by setting $v(x,t,c) = u(x,t) - ct$. Note that $v_t = u_t - c = ku_{xx} - c = kv_{xx} - c$, which shows that for each fixed c, $v(x,t,c)$ satisfies the equation $v_t = kv_{xx} - c$. Thus, by the GMP, which we have proven for any fixed $c > 0$, we know that the maximum of v occurs at a point, say (x_c,t_c) , on the sides (i.e., $x_c = 0$ or L) or lower edge (i.e., $t_c = 0$) of R. Also, since $\overline{M}$ is (by definition) the maximum for u on the sides and lower edge of R, we have

$$v(x_c,t_c,c) = u(x_c,t_c) - c\cdot t_c \leq \overline{M} - c\cdot t_c \leq \overline{M} \ , \tag{17}$$

since $c\cdot t_c \geq 0$. If (x,t) is any point in R, then

$$u(x,t) = \lim_{c\to 0^+} [u(x,t) - ct] = \lim_{c\to 0^+} v(x,t,c) \ . \tag{18}$$

Since $v(x,t,c) \leq v(x_c,t_c,c)$ (Why ?) and $v(x_c,t_c,c) \leq \overline{M}$ by (17), we have $v(x,t,c) \leq \overline{M}$. Thus, by (18), u(x,t) is the limit of numbers $\leq \overline{M}$, and hence $u(x,t) \leq \overline{M}$, for any (x,t) in R. □

Remarks. There is also a Generalized Minimum Principle for solutions of $u_t = ku_{xx} - c$, but only for $c \leq 0$, instead of $c \geq 0$. Thus, only in the case $c = 0$, is there both a Minimum Principle and a Maximum Principle. It should be noted that although the Maximum Principle says that the maximum of a solution of the heat equation on R must occur on the sides or lower edge, it does not preclude the possibility that the maximum can occur strictly inside R as well. Indeed, any constant solution has this property. There is a stronger maximum principle, which implies that if the maximum does occur strictly inside R, say at (x_0,t_0), with $0 < x_0 < L$ and $t_0 > 0$, then the solution *must* be constant for $t \leq t_0$. For solutions which are finite sums of product solutions, this in turn implies that the solution must be constant everywhere (Why ?). In the supplement following the exercises, we sketch an "elementary" proof of the Strong Maximum Principle, which is intended for mature audiences only.

Summary 3.2

1. **The Uniqueness Theorem** (cf. Theorem 1) : **Let** $u_1(x,t)$ **and** $u_2(x,t)$ **be** C^2 **solutions of the following problem, where** $a(t)$, $b(t)$ **and** $f(x)$ **are given** C^2 **functions :**

$$\begin{aligned} &\text{D.E.} \quad u_t = ku_{xx} \qquad 0 \le x \le L\,,\ t \ge 0 \\ &\text{B.C.} \quad u(0,t) = a(t) \qquad u(L,t) = b(t) \qquad\qquad (S1) \\ &\text{I.C.} \quad u(x,0) = f(x)\,. \end{aligned}$$

Then $u_1(x,t) = u_2(x,t)$, **for all** x **and** t, **where** $0 \le x \le L$ **and** $t \ge 0$.

2. **The Maximum Principle** (cf. Theorem 2) : Let $u(x,t)$ be a C^2 solution of (S1). Let $T > 0$ and let $\overline{M}$ denote the maximum value that u achieves on the sides ($x = 0$ or $x = L$) or the lower edge of the rectangle $R : 0 \le x \le L\,,\ 0 \le t \le T$. Then $u(x,t) \le \overline{M}$, for all $0 \le x \le L$, $0 \le t \le T$. Similarly the Minimum Principle (cf. the Corollary of Theorem 2) asserts that if $\underline{m}$ denotes the minimum value that u achieves on the sides or the lower edge of the rectangle R, then $\underline{m} \le u(x,t)$ for all $0 \le x \le L\,,\ 0 \le t \le T$.

3. **Continuous Dependence on the I.C. and B.C.** : A consequence of the Maximum/Minimum Principle is that the solution, $u(x,t)$ of (S1), depends continuously on the initial and boundary conditions (cf. Theorem 3). The method of the proof of Theorem 1 provides a weaker version of this result, in that it shows that at each time $t > 0$, the mean–square of the difference of two temperature distributions (cf. (7)) is no greater than it was initially, provided that the two distributions have the same values at the ends prior to time t.

Exercises 3.2

1. (a) Let $v(x,t)$ be any C^2 solution of $v_t = kv_{xx}$ $(0 \le x \le L\,,\ t \ge 0)$, which satisfies the B.C. $v(0,t) = 0$ and $v(L,t) = 0$ (without initial condition). Show that for any t_1, t_2, with $t_2 \ge t_1 \ge 0$,

$$\int_0^L [v(x,t_2)]^2\,dx \le \int_0^L [v(x,t_1)]^2\,dx\,. \qquad (*)$$

Hint. Let $F(t) = \int_0^L [v(x,t)]^2\,dx$, and show that $F'(t) \le 0$, as in the proof of Theorem 1.

Then note that $F(t_2) - F(t_1) = \int_{t_1}^{t_2} F'(t)\,dt \le 0$ (Why ?).

(b) Explain why the conclusion (*) still holds when the B.C. are replaced by any of the following pairs of B.C. :

$$\begin{matrix} v_x(0,t) = 0 \\ v_x(L,t) = 0 \end{matrix} \;;\quad \begin{matrix} v_x(0,t) = 0 \\ v(L,t) = 0 \end{matrix} \;;\quad \begin{matrix} v_x(0,t) = h\cdot v(0,t) \;\; [\text{where } h > 0]. \\ v(L,t) = 0\;. \end{matrix}$$

2. State and prove a uniqueness theorem for the problem

$$\text{D.E.}\quad u_t = ku_{xx}$$

$$\text{B.C.}\quad u_x(0,t) = a(t) \qquad u_x(L,t) = b(t)$$

$$\text{I.C.}\quad u(x,0) = f(x)\;.$$

Hint. Use the method of proof of Theorem 1. Alternatively, use (*) for $v \equiv u_2 - u_1$ in Problem 1 with $t_1 = 0$ and $t_2 = t$.

3. Use the Maximum/Minimum Principles to deduce that the solution u of the problem

$$\text{D.E.}\quad u_t = ku_{xx} \qquad 0 \le x \le \pi\,,\; t \ge 0\;.$$

$$\text{B.C.}\quad u(0,t) = 0 \qquad u(\pi,t) = 0$$

$$\text{I.C.}\quad u(x,0) = \sin(x) + \tfrac{1}{2}\sin(2x)$$

satisfies $0 \le u(x,t) \le 3\sqrt{3}/4$ for all $0 \le x \le \pi$, $t \ge 0$.

4. Change the I.C. in Problem 3 to $u(x,0) = 5\sin(3x) - 3\sin(5x)$. Find the smallest constant D and the largest constant C, such that $C \le u(x,t) \le D$, for all $0 \le x \le \pi$, $t \ge 0$.
Hint. $\cos(\alpha) - \cos(\beta) = -2\cdot\sin(\tfrac{1}{2}(\alpha+\beta))\sin(\tfrac{1}{2}(\alpha-\beta))$.

5. Suppose that Joe adds the term $\frac{1}{10}\,x^3(\pi-x)^3$ to the initial temperature distribution in Problem 3. Assuming that he can find a solution of this new problem, show that this solution differs from the solution of the original problem by at most $\pi^6/640$ at *any* $0 \le x \le \pi$ and $t \ge 0$.
Hint. Apply Theorem 3, noting that $|f_1(x) - f_2(x)| = |\frac{1}{10}\,x^3(\pi-x)^3|$.

6. Consider the problem, where $\alpha > 0$,

$$\text{D.E.} \quad u_t = 2u_{xx} \qquad 0 \le x \le 2\,,\ t \ge 0$$

$$\text{B.C.} \quad u_x(0,t) = -\alpha \qquad u_x(2,t) = 0$$

$$\text{I.C.} \quad u(x,0) = -\alpha x(1 - \tfrac{1}{4}x)\,.$$

(a) By trying a function of the form $u(x,t) = Ax^2 + Bx + Ct$, deduce that $u(x,t) = \alpha x(\frac{1}{4}x - 1) + \alpha t$ is the required solution (unique by Problem 2).

(b) Take two values for α, say α_1 and α_2, and let $u_1(x,t)$ and $u_2(x,t)$ be the corresponding solutions, as in part (a). Show that $|u_1(x,0) - u_2(x,0)| \le |\alpha_1 - \alpha_2|$, and observe that $|(u_1)_x(0,t) - (u_2)_x(0,t)| \le |\alpha_1 - \alpha_2|$ and $|(u_1)_x(2,t) - (u_2)_x(2,t)| = 0 \le |\alpha_1 - \alpha_2|$.

(c) *If* the analog of Theorem 3 (for the B.C. here) were true, we would have $|u_1(x,t) - u_2(x,t)| \le |\alpha_1 - \alpha_2|$. Instead, check that in fact $|u_1(x,t) - u_2(x,t)| \to \infty$ as $t \to \infty$. Thus, α in this problem cannot be varied slightly without producing large variations in the solution for large t.

Remark. Physically, the B.C. $u_x(0,t) = -\alpha$ means that heat energy is added to the rod through the end at $x = 0$ at a constant rate proportional to α, since the heat flux through $x = x_0$ is $-KAu_x(x_0,t)$ (cf. (2) of Section 3.1). The end at $x = 2$ is insulated (i.e., there is no heat flux). Thus, the heat in the rod increases at a constant rate proportional to α as $t \to \infty$.

7. (Comparison Results). Let B denote the set of points which are on the lower edge ($t = 0$) or the sides ($x = 0$ and $x = L$) of the rectangle $R: 0 \le x \le L$, $0 \le t \le T$.

(a) Let C be a constant. From the Maximum Principle, deduce that if $u(x,t)$ is a C^2 solution of $u_t = ku_{xx}$ on R, and $u(\bar{x},\bar{t}) \le C$ for $(\bar{x},\bar{t})$ in B, then $u(x,t) \le C$ for all (x,t) in R.

(b) Let $u_1(x,t)$ and $u_2(x,t)$ be two C^2 solutions of the heat equation $u_t = u_{xx}$ on R. Show that if $u_1(\bar{x},\bar{t}) \le u_2(\bar{x},\bar{t})$ for all $(\bar{x},\bar{t})$) in B, then $u_1(x,t) \le u_2(x,t)$ for all (x,t) in R.
Hint. Use part (a) with $u = u_1 - u_2$ and $C = 0$.

8. Consider a rod of length π with ends maintained at zero. If the the initial temperature is given by $u(x,0) = [\sin(x)]^7$, then show that $u(x,t) \le e^{-kt}\sin(x)$ for all $0 \le x \le \pi$, $t \ge 0$.
Hint. Apply the Maximum Principle to $u(x,t) - e^{-kt}\sin(x)$ or use Problem 7(b), noting that $r^7 \le r$ for $0 \le r \le 1$. Do not bother to actually find the exact solution !

9. Explain what modifications of the proof (just before the Summary) of Theorem 2 (The Maximum Principle) are necessary in order to prove the following maximum principle for insulated ends (where the heat flux, proportional to u_x, is 0).

Let u be a C^2 solution of the problem

$$\text{D.E.}\quad u_t = ku_{xx} \qquad 0 \le x \le L,\ t \ge 0$$

$$\text{B.C.}\quad u_x(0,t) = 0 \qquad u_x(L,t) = 0$$

$$\text{I.C.}\quad u(x,0) = f(x)\,.$$

Then, $u(x,t) \le M \equiv \max_{0 \le x \le L} \{f(x)\}$.

Hint. In the proof of the generalized maximum principle for $u_t = ku_{xx} - c$ $(c > 0)$ with insulated ends, in order to reach a contradiction, we may still assume that the maximum occurs at (x_0,t_0) where $t_0 > 0$, but we may not assume that $0 < x_0 < L$, since we must now eliminate the possibility that $x_0 = 0$ or $x_0 = L$. However, the B.C. come to our rescue!

10. From the maximum principle in Problem 9, deduce the corresponding minimum principle.

11. Use Problem 9 in order to produce an alternate uniqueness proof for Problem 2.
Hint. First prove $u_1 - u_2 \le 0$, and then prove $u_2 - u_1 \le 0$.

12. Using Problems 9 and 10, prove the following result which is in the spirit of Theorem 3. Let $u_1(x,t)$ and $u_2(x,t)$ be solutions of the following respective problems (where $0 \le x \le L$, $t \ge 0$):

D.E. $u_t = ku_{xx}$	D.E. $u_t = ku_{xx}$
B.C. $u_x(0,t) = 0$	B.C. $u_x(0,t) = 0$
$u_x(L,t) = 0$	$u_x(L,t) = 0$
I.C. $u(x,0) = f_1(x)$	I.C. $u(x,0) = f_2(x)$,

where $|f_1(x) - f_2(x)| \le \epsilon$. Then $|u_1(x,t) - u_2(x,t)| \le \epsilon$, for all $0 \le x \le L$, $t \ge 0$.

13. State and prove a maximum principle for the circular wire (cf. (28) in Section 3.1).
Hint. In the proof of the generalized maximum principle for $u_t = ku_{xx} - c$ $(c > 0)$, note that if the maximum for u occurs at (x_0,t_0) with $t_0 > 0$ and $x_0 = \pm L$, then $u_x(\pm L,t_0) = 0$ (Why?).

14. (a) In Example 2, find the unique time $t_0 > 0$ for which $u(\frac{3}{2},t)$ is largest.

(b) For the time t_0 in part (a), show that $u(x,t_0)$ is largest only when $x = \frac{3}{2}$.

(c) Use the Maximum Principle to show, in spite of parts (a) and (b), that in any square region, $|x - \frac{3}{2}| \leq \epsilon$, $|t - t_0| \leq \epsilon$ $(\epsilon > 0)$ there are points (x,t), other than $(\frac{3}{2},t_0)$, where $u(x,t) \geq u(\frac{3}{2},t_0)$. (The Maximum Principle for the heat equation on a square holds regardless of the size or location of the square by the translation–invariance of the heat equation; cf. Problem 1 of Section 3.1).

(d) Here we show that $(\frac{3}{2},t_0)$ is not even a *local* maximum of $u(x,t)$. For $t < t_0$, let $x(t)$ be the left–most point in $[0,3]$ where $u_x(x(t),t) = 0$. Show that $\cos(\pi x(t)/3) = \frac{1}{2}(3 - e^{8\pi^2 t})^{\frac{1}{2}}$ for $t \leq t_0$. We have $\frac{d}{dt}\, u(x(t),t) = u_x(x(t),t)x'(t) + u_t(x(t),t) = ku_{xx}(x(t),t)$ (Why?). By an explicit computation show that $u_{xx}(x(t),t) < 0$ for $t < t_0$. Why does this show that $u(x(t),t) > u(\frac{3}{2},t_0)$ for $0 < t < t_0$? Why then is $(\frac{3}{2},t_0)$ not a local maximum of u?

Supplement on the Strong Maximum Principle for the heat equation

The key family of functions which enters the proof of the Strong Maximum Principle (cf. the Theorem below) is

$$w(x,t) = \exp[-B(x^2 + \alpha^2 t^2)] - \exp[-Br^2], \quad \text{for constants } \alpha, B, r > 0. \qquad (*)$$

Note that $-1 \leq w(x,t) \leq 1$ for all (x,t), and $w > 0$ only within the ellipse $x^2 + \alpha^2 t^2 = r^2$, which is inscribed in the rectangle $-r \leq x \leq r$, $-r/\alpha \leq t \leq r/\alpha$. Now,

$$w_x = \exp[-B(x^2+\alpha^2 t^2)](-2Bx), \qquad w_{xx} = \exp[-B(x^2+\alpha^2 t^2)]4B^2x^2 - \exp[-B(x^2+\alpha^2 t^2)]2B$$

and
$$w_t = \exp[-B(x^2+\alpha^2 t^2)](-2B\alpha^2 t).$$

Thus,
$$kw_{xx} - w_t = 2B\left[k(2Bx^2 - 1) + \alpha^2 t\right]\exp[-B(x^2+\alpha^2 t^2)].$$

Hence, $kw_{xx} - w_t > 0$ only for $t > -(k/\alpha^2)(2Bx^2 - 1)$. The curve defined by $kw_{xx} - w_t = 0$ is the parabola $t = -(k/\alpha^2)(2Bx^2 - 1)$, which opens downward and has vertex at $(0,k/\alpha^2)$, where the curvature is $4kB/\alpha^2$ (cf. Figure 5). Note that the vertex is strictly inside the ellipse if $r/\alpha > k/\alpha^2$. This will be the case if $\alpha > 2k/r$, and we will assume this henceforth. We can move the center of the ellipse to any point (x_0,t_0) by considering the function $w(x-x_0,t-t_0)$.

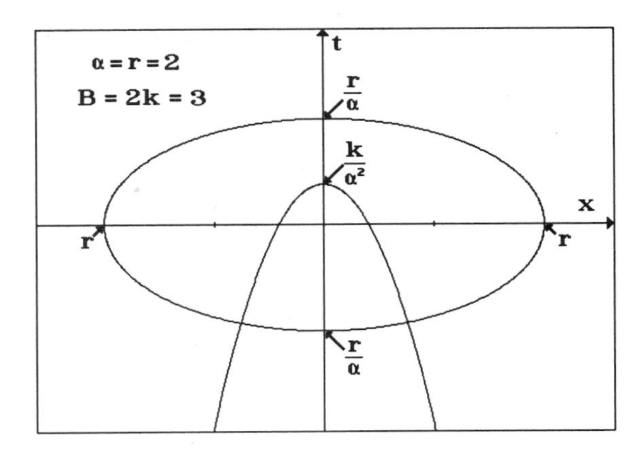

Figure 5

Lemma. Let u be a solution of the heat equation $u_t = ku_{xx}$ in a closed rectangle R that contains in its interior an elliptical region E of the form $(x-x_0)^2 + \alpha^2(t-t_0)^2 \leq r^2$, except possibly for the uppermost point $(x_0,t_0 + r/\alpha)$. We assume $\alpha > 2k/r$. If the maximum, say M, of u in R is achieved at some point P on the boundary of E, then it is also achieved at some point Q on the vertical segment running from the point $(x_0,t_0 + k/\alpha^2)$ [strictly inside the ellipse, since $r/\alpha > k/\alpha^2$] to the lowest point $(x_0,t_0 - r/\alpha)$.

Proof. We assume without loss of generality that $(x_0,t_0) = (0,0)$. For any constant $\epsilon > 0$, let $v = u + \epsilon w$, where w is defined by (*). For a point P′ which is in R but not in E, we have $v(P') = u(P') + \epsilon w(P') < M$, while $v(P) = M$, since $w(P) = 0$. Thus, the maximum of v is at some point Q_1 in E. At Q_1, we have $v_{xx} \leq 0$ and $v_t \geq 0$ (Why?). So at Q_1, $kw_{xx} - w_t = (kv_{xx} - v_t)/\epsilon \leq 0$. Hence Q_1 must be in the intersection, say J, of E with the parabolic region $t \leq -(k/\alpha^2)[2B\cdot x^2 - 1]$. Note that $v(Q_1) \geq v(P) = u(P) = M$. We have $u(Q_1) = v(Q_1) - \epsilon w(Q_1) \geq v(Q_1) - \epsilon \geq M - \epsilon$. Letting $\epsilon \to 0$, we deduce (cf. the Bolzano–Weierstrass Theorem of Appendix A.4) that there is a point Q_0 in J where $u(Q_0) = M$. By increasing the constant B, we can narrow the parabola indefinitely, thereby forcing the existence of a point Q on the segment from $(0,k/\alpha^2)$ to $(0,-r/\alpha)$, where $u(Q) = M$. □

Theorem (The Strong Maximum Principle). Let $u(x,t)$ **be a solution of the heat equation in the rectangle** R $(0 \leq x \leq L, 0 \leq t \leq T)$. **If** u **achieves its maximum at** $(\bar{x},T)$, **where** $0 < \bar{x} < L$, **then** u **must be constant in** R.

Proof. Let $M = u(\bar{x},T)$. First we prove that $u(\bar{x},t) = M$ for all t, $0 \le t \le T$. Suppose, on the contrary, that $u(\bar{x},t_1) < M$ for some t_1 with $0 \le t_1 < T$. Let t_2 be the smallest number bigger than t_1 such that $u(\bar{x},t_2) = M$ (i.e., $t_2 = \inf \{ t \mid t > t_1 , u(\bar{x},t) = M \}$; cf. Appendix A.4). Select an elliptical region E, as in the Lemma, with $(\bar{x},t_2)$ as the top point of E, and with (x_0,t_3) as the bottom point, where $t_1 < t_3 < t_2$. The Lemma implies that there is some point (x_0,t_4) with $t_3 \le t_4 < t_2$ such that $u(x_0,t_4) = M$. This contradicts the choice of t_2. Observe that the same argument shows that if the maximum M is achieved at some point P inside R and strictly between $x = 0$ and $x = L$, then u has the value M at all points in R directly below P. Now, select an ellipse E, as in the Lemma, which is tangent to the lines $t = T, x = \bar{x}$ and $x = \delta$, where $0 < \delta < \bar{x}$ (cf. Figure 6). Since the bottom point of the ellipse is at the level $t = T - \frac{1}{2}(\bar{x} - \delta)/\alpha$, we may confine the ellipse to an arbitrarily thin strip of the form $T-d \le t \le T$, $d > 0$, by choosing α large enough. By allowing δ to run between 0 and $\bar{x}$, we can use the Lemma and the above observation to guarantee that u has the value M at all points (x,t), where $\frac{1}{2}\bar{x} < x \le \bar{x}$ and $t \le T - d$. Then letting $d \to 0^+$ and using the continuity of u, we get that $u(x,t) = M$ for $\frac{1}{2}\bar{x} \le x \le \bar{x}$ and $0 \le t \le T$. Similarly, by considering ellipses to the right of $x = \bar{x}$, we can prove that $u(x,t) = M$ for $\bar{x} \le x \le \bar{x} + \frac{1}{2}(L-\bar{x})$ and $0 \le t \le T$. In the same way, by using ellipses between $x = 0$ and $x = \frac{1}{2}\bar{x}$, and also ellipses between $x = \bar{x} + \frac{1}{2}(L-\bar{x})$ and $x = L$, we can expand the domain where u has the value M to the rectangle $\frac{1}{4}\bar{x} \le x \le \bar{x} + \frac{3}{4}(L-\bar{x})$, $0 \le t \le T$. Repeating the process indefinitely, we get that u must be constant throughout R. (Note that the continuity of u allows us to deduce that u also has the value M on the sides $x = 0$ and $x = L$ of R). □

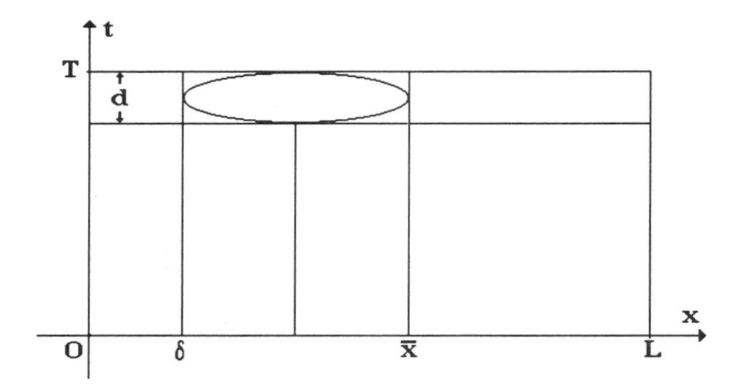

Figure 6

3.3 Time–Independent Boundary Conditions

We recall from equation (2) of Section 3.1, that the amount of heat energy (per unit time) that passes through the cross–section $x = a$ in the positive direction is $-KAu_x(a,t)$, where $K > 0$ is the heat conductivity, and A is the cross–sectional area. Hence, by prescribing u_x at $x = 0$ or $x = L$, we are imposing conditions on the rate at which heat energy passes through the ends of the rod. In particular, the condition $u_x(0,t) = 0$ means that no heat is flowing through the end $x = 0$, in other words, that this end is **insulated**. If we require that the ends be insulated or maintained at 0, then there are four sets of possible boundary conditions :

$$\text{(a)} \quad \begin{cases} u(0,t) = 0 \\ u(L,t) = 0 \end{cases} \qquad \text{(b)} \quad \begin{cases} u_x(0,t) = 0 \\ u(L,t) = 0 \end{cases}$$

$$\text{(b}'\text{)} \quad \begin{cases} u(0,t) = 0 \\ u_x(L,t) = 0 \end{cases} \qquad \text{(c)} \quad \begin{cases} u_x(0,t) = 0 \\ u_x(L,t) = 0\,. \end{cases} \tag{1}$$

We have considered (a) before in Section 3.1. In cases (b) and (b′), one end is insulated while the other is maintained at 0. We will treat (b) in some detail, but leave the consideration of (b′) as an exercise (alternatively, convert (b′) to (b) by turning the rod around, and thus reverse the ends). We have established the uniqueness of solutions of the heat equation with B.C. (a), (b) or (c) and some I.C. (cf. Problems 2 and 11 of Section 3.2). We begin with case (c) where the rod is completely insulated from its environment.

Example 1. Derive the solution of the problem

$$\begin{aligned} &\text{D.E.} \quad u_t = ku_{xx} \qquad 0 \le x \le L\,,\ t \ge 0 \\ &\text{B.C.} \quad u_x(0,t) = 0 \quad u_x(L,t) = 0 \\ &\text{I.C.} \quad u(x,0) = f(x)\,, \end{aligned} \tag{2}$$

for suitable initial distributions $f(x)$.

Solution. We use the same procedure as was used for (14) in Section 3.1. In other words, we find the product solutions of the D.E. that satisfy the B.C., and then obtain other solutions of the D.E. and the B.C. by forming linear combinations of these product solutions, using the superposition principle, in order to meet the I.C.. Regardless of the B.C. or I.C., the product solutions of the D.E. are found by the method of separation of variables. We have carried this out (once and for all) in Section 3.1, in the three cases. For Case 1 (cf. (11) of Section 3.1), we have

$$u(x,t) = e^{-\lambda^2 kt}(c_1\sin(\lambda x) + c_2\cos(\lambda x)) \tag{3}$$

$$u_x(x,t) = e^{-\lambda^2 kt}(c_1\lambda\cdot\cos(\lambda x) - c_2\lambda\cdot\sin(\lambda x)) . \tag{4}$$

Inserting $x = 0$ into (4), we see that $u_x(0,t) = \lambda e^{-\lambda^2 kt}c_1$. Since $\lambda > 0$, we are forced to take $c_1 = 0$ by the first B.C. of (2). Substituting $x = L$ into (4), we have

$$u_x(L,t) = -\lambda\cdot e^{-\lambda^2 kt}c_2\sin(\lambda L) .$$

This must be zero, by the second B.C. . The only way to avoid setting $c_2 = 0$ (producing only the trivial solution $u \equiv 0$) is to choose λ such that $\sin(\lambda L) = 0$. Hence, $\lambda L = n\pi$ or $\lambda = n\pi/L$, for $n= 1,2,3,\ldots$. Thus, we arrive at the family of [Case 1] product solutions of the D.E. and B.C. :

$$u_n(x,t) = a_n e^{-(n\pi/L)^2 kt}\cos(n\pi x/L) , \tag{5}$$

for $n = 1, 2, 3,\ldots$. In Problem 1, the reader is asked to check that there are no nonzero solutions of the D.E. and B.C. in Case 2 (cf. (12) of Section 3.1). There is a simple, yet important, Case 3 (cf. (13) of Section 3.1) solution of the D.E. and B.C. , namely the constant solution $u(x,t) = c_2$. This can be included in the family (5) by simply letting $n = 0$ be a possible value for n ; recall $\cos(0) = 1$. By the superposition principle, we have the more general solution of the (linear, homogeneous) D.E. and B.C. ,

$$u(x,t) = \sum_{n=0}^{N} a_n e^{-(n\pi/L)^2 kt}\cos(n\pi x/L) . \tag{6}$$

In order to meet the I.C. with such a solution, we need

$$f(x) = u(x,0) = \sum_{n=0}^{N} a_n\cos(n\pi x/L) . \tag{7}$$

Hence, if $f(x)$ is of this form, then the solution of problem (2) is given by (6). In Chapter 4 we will prove that any "reasonably nice" function $f(x)$ can be approximated to any degree of accuracy by a sum of the form (7). In practical terms, then we may simply assume that f(x) is of this form, because f(x) is only known within some experimental error. □

Remark. Note that the product solutions (5), with the exception of the case $n = 0$, all tend to zero as $t \to \infty$. The terms of (6) with larger values of n decrease more rapidly, because of the factor of n^2 in the exponents. This is to be expected, because the temperature gradients between hot and cold regions are greater for larger n, as shown in Figure 1 :

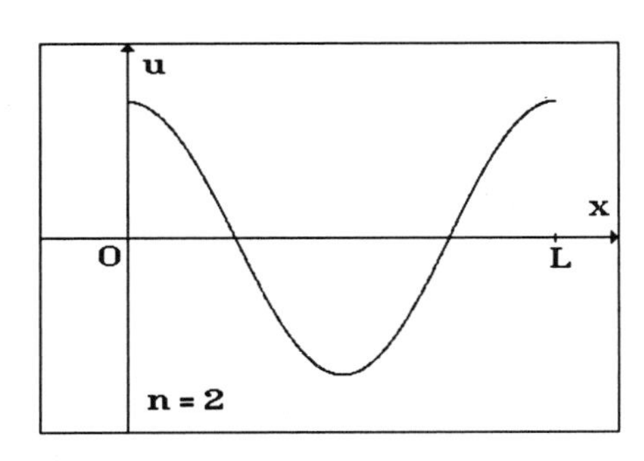

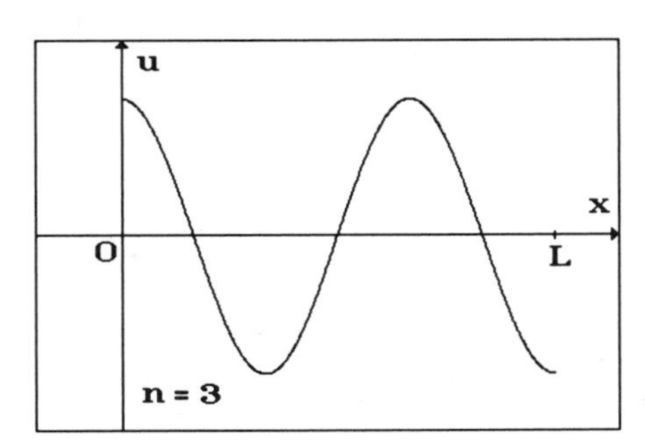

Figure 1

By (6) we have

$$\lim_{t\to\infty} u(x,t) = \frac{1}{L}\int_0^L f(x)\,dx = a_0 . \tag{8}$$

In other words, the temperature distribution for the insulated rod levels to a constant which is the average of the temperature initially (or at any later time, by Problem 2). □

Example 2. For suitable initial distributions f(x), derive the solution of the problem

$$\begin{aligned} &\text{D.E.} \quad u_t = ku_{xx} \qquad 0 \le x \le L\,,\ t \ge 0 \\ &\text{B.C.} \quad u_x(0,t) = 0 \qquad u(L,t) = 0 \\ &\text{I.C.} \quad u(x,0) = f(x)\,, \end{aligned} \tag{9}$$

where the end $x = 0$ is insulated and the end $x = L$ is maintained at 0.

Solution. Again, we find the product solutions of the D.E. that obey the B.C. . In Case 1 (cf. (11) of Section 3.1), we have (3) and (4) of Example 1, and the first B.C. yields $c_1 = 0$ as before. However, the second B.C. yields

$$0 = u(L,t) = e^{-\lambda^2 kt} c_2 \cos(\lambda L) . \tag{10}$$

To avoid setting $c_2 = 0$, we must choose λ so that $\cos(\lambda L) = 0$. Since $\cos(z) = 0$, if and only if only when z is an odd multiple of $\pi/2$, we find that (10) holds only when

$$L\lambda = (2n+1)\pi/2 \quad \text{or} \quad \lambda = (n + \tfrac{1}{2})\pi/L \qquad n = 0,1,2,\ldots .$$

(Recall that $\lambda > 0$.) One can check that there are no nonzero product solutions in Cases 2 and 3 (cf. (12) and (13) of Section 3.1) that satisfy the B.C. here. Hence, a complete family of product solutions is given by

$$u_n(x,t) = d_n \exp[-(n+\tfrac{1}{2})^2\pi^2 kt/L^2]\cdot\cos[(n+\tfrac{1}{2})\pi x/L]\ , \quad n = 0, 1, 2, \ldots$$

By the superposition principle, we have the more general solution

$$u(x,t) = \sum_{n=0}^{N} d_n \exp[-(n+\tfrac{1}{2})^2\pi^2 kt/L^2]\cdot\cos[(n+\tfrac{1}{2})\pi x/L] \tag{11}$$

which is the unique solution of problem (9), when

$$f(x) = u(x,0) = \sum_{n=0}^{N} d_n \cos[(n+\tfrac{1}{2})\pi x/L]\ . \quad \square \tag{12}$$

Remark. All of the terms of (11) tend to zero as $t \to \infty$. The physical reason is that heat is allowed to pass through the uninsulated end at $x = L$. The first term ($n = 0$) of (11) decays at a rate proportional to $\exp[-\frac{1}{4}\pi^2 kt/L^2]$ which is slower than the corresponding rate $\exp[-\pi^2 kt/L^2]$ for the case where both ends are maintained at 0 (cf. Section 3.1). In Chapter 4, we will prove that any "reasonably nice" function $f(x)$ can be approximated arbitrarily closely by a sum as in (12). Also, there is a maximum principle that governs solutions of (9), so that the solution is not unduly perturbed by approximating the initial temperature distribution (cf. Problem 11). Typical graphs of the terms of (11) are shown in Figure 2. □

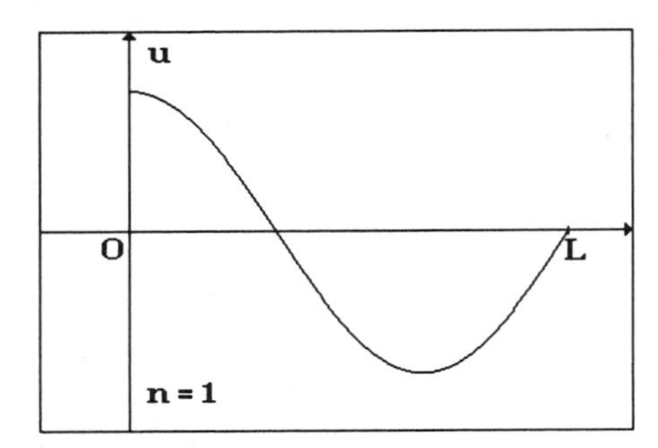

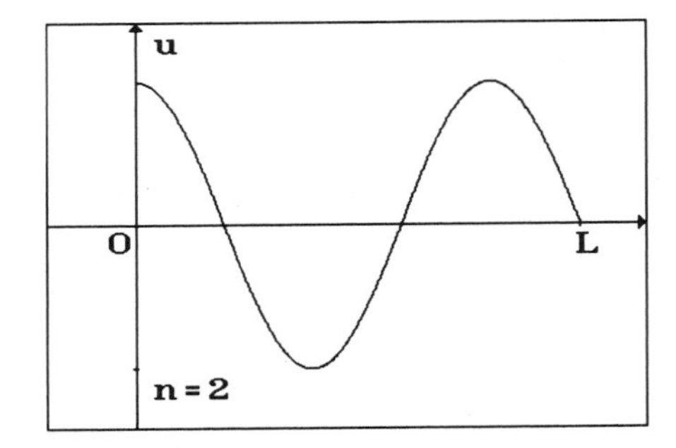

Figure 2

Inhomogeneous boundary conditions

The B.C. in Examples 1 and 2 [or (1)] are homogeneous, whence we were able to use the superposition principle in forming more general solutions of the D.E. and B.C. . We now turn to the situation where the B.C. are not both homogeneous, but are still time–independent.

> The method of solution is to first find a *particular solution* of the D.E. and B.C., and then add (to that particular solution) the solution of a related problem with homogeneous B.C. .

We have used this general idea before. The method is illustrated in the following example.

Example 3. For arbitrary real constants, a and b, and suitable g(x), solve the problem

$$\text{D.E.} \quad u_t = ku_{xx} \qquad 0 \le x \le L\,,\ t \ge 0$$

$$\text{B.C.} \quad u(0,t) = a \qquad u(L,t) = b \tag{13}$$

$$\text{I.C.} \quad u(x,0) = g(x)\,.$$

Solution. We first seek a particular solution $u_p(x,t)$ of the D.E. and B.C. . Since any particular solution will do, we may as well strive for simplicity. Indeed, a Case 3 product solution (cf. (13) of Section 3.1) $u_p(x,t) = cx+d$ will do, if c and d are chosen so that the B.C. are satisfied :

$$a = u_p(0,t) = c \cdot 0 + d = d\,,$$
$$b = u_p(L,t) = cL + d = cL + a\,.$$

Thus, $d = a$ and $c = (b-a)/L$ and

$$u_p(x,t) = (b-a)x/L + a \tag{14}$$

solves both the D.E. and B.C. . Consider now the **related homogeneous problem** (i.e., with homogeneous D.E. and B.C.)

$$\text{D.E.} \quad v_t = kv_{xx} \qquad 0 \le x \le L\,,\ t \ge 0$$

$$\text{B.C.} \quad v(0,t) = 0,\ \ v(L,t) = 0 \tag{15}$$

$$\text{I.C.} \quad v(x,0) = g(x) - u_p(x,0)\,.$$

If $g(x) - u_p(x,0)$ is of the form $\sum_{n=1}^{N} b_n \sin(n\pi x/L)$, then we can solve this problem, obtaining

$$v(x,t) = \sum_{n=1}^{N} b_n e^{-(n\pi/L)^2 kt} \sin(n\pi x/L) \tag{16}$$

(cf. Proposition 1, Section 3.1). Now, set $u(x,t) = u_p(x,t) + v(x,t)$. We easily check that $u(x,t)$ solves problem (13). Indeed, $u(x,t)$ solves the D.E. by the superposition principle, and we have

$$\text{B.C.} \quad u(0,t) = u_p(0,t) + v(0,t) = a + 0 = a$$

$$u(L,t) = u_p(L,t) + v(L,t) = b + 0 = b \tag{17}$$

$$\text{I.C.} \quad u(x,0) = u_p(x,0) + v(x,0) = u_p(x,0) + g(x) - u_p(x,0) = g(x)\,.$$

Observe that *it is necessary to subtract* $u_p(x,0)$ from $g(x)$, when forming the I.C. of the related homogeneous problem (15). Otherwise the cancellation of $u_p(x,0)$ in (17) does not occur, and $u(x,0) = g(x)$ will not hold. □

Example 4. Solve

$$\text{D.E.}\ \ u_t = ku_{xx} \qquad 0 \le x \le L\ ,\ t \ge 0$$
$$\text{B.C.}\ \ u(0,t) = 0 \qquad u(L,t) = L \tag{18}$$
$$\text{I.C.}\ \ u(x,0) = x + 3\sin(2\pi x/L)\ .$$

Solution. We can find a particular solution of the D.E. and B.C. of the form $u_p(x,t) = cx + d$. From the B.C., $0 = u(0,t) = c\cdot 0 + d = d$ and $L = u(L,t) = cL$. Thus, $d = 0$, $c = 1$ and $u_p(x,t) = x$. The related homogeneous problem is

$$\text{D.E.}\ \ v_t = kv_{xx} \qquad 0 \le x \le L\ ,\ t \ge 0$$
$$\text{B.C.}\ \ v(0,t) = 0 \quad v(L,t) = 0 \tag{19}$$
$$\text{I.C.}\ \ v(x,0) = x + 3\sin(2\pi x/L) - x = 3\sin(2\pi x/L)\ ,$$

whose solution is $v(x,t) = 3e^{-4\pi^2kt/L^2}\sin(2\pi x/L)$. The solution of the problem (19) is then

$$u(x,t) = u_p(x,t) + v(x,t) = x + 3e^{-4\pi^2kt/L^2}\sin(2\pi x/L)\ .\ \ \square$$

Remark. It is probably best not to commit to memory the formula (14). Instead the reader should consider a particular solution of the form $cx + d$, and find the constants, using the B.C. . The reason is that the formula only applies to the B.C. of (13). For other B.C., we obtain other particular solutions. For example,

$$\text{if}\ \begin{cases} u_x(0,t) = a \\ u(L,t) = b \end{cases},\ \text{then}\quad u_p(x,t) = a(x-L) + b\ .\ \ \square$$

Example 5. Solve

$$\text{D.E.}\ \ u_t = 2u_{xx} \qquad 0 \le x \le 1\ ,\ t \ge 0$$
$$\text{B.C.}\ \ u_x(0,t) = 1 \quad u(1,t) = -1$$
$$\text{I.C.}\ \ u(x,0) = x + \cos^2(3\pi x/4) - \tfrac{5}{2}\ .$$

Solution. We try a particular solution of the form $u_p(x,t) = cx + d$. The first B.C. yields $c = 1$, while $u_p(1,t) = 1 + d$ yields $d = -2$ by the second B.C. . Thus, $u_p(x,t) = x - 2$. The related homogeneous problem is

$$\text{D.E.}\ \ v_t = 2u_{xx} \qquad 0 \le x \le 1\ ,\ t \ge 0$$
$$\text{B.C.}\ \ v_x(0,t) = 0 \quad v(1,t) = 0$$
$$\text{I.C.}\ \ v(x,0) = [x + \cos^2(3\pi x/4) - \tfrac{5}{2}] - (x-2) = \tfrac{1}{2} + \tfrac{1}{2}\cos(3\pi x/2) - \tfrac{5}{2} + 2 = \tfrac{1}{2}\cos(3\pi x/2).$$

From Example 2, we know that $v(x,t) = \frac{1}{2}e^{-9\pi^2 t/2}\cos(3\pi x/2)$, and then

$$u(x,t) = x-2 + \tfrac{1}{2}e^{-9\pi^2 t/2}\cos(3\pi x/2) \ . \quad \square$$

Observe that in Examples 3,4 and 5 the particular solution was time–independent, or in the usual terminology, **steady–state** .

> Any steady–state solution of the heat equation $u_t = ku_{xx}$ is of the form $cx + d$ (Why ?).

In Examples 3,4 and 5 , the solutions $u(x,t) = u_p(x,t) + v(x,t)$ are sums of a steady–state particular solution of the D.E. and B.C. and the solution $v(x,t)$ (of the related homogeneous problem) which is **transient** in the sense that $v(x,t) \to 0$ as $t \to \infty$. Thus, in these examples,

$$u(x,t) = u_p(x,t) + v(x,t) \ \to \ u_p(x,t) \ , \quad \text{as } t \to \infty \ ,$$

(i.e., the solution u approaches the steady–state solution as $t \to \infty$). However, for some types of B.C., there are no steady–state particular solutions, as the next example shows.

Example 6. For given real constants, a and b, and suitable f(x), solve the problem

$$\begin{aligned} &\text{D.E.} \quad u_t = ku_{xx} \qquad 0 \le x \le L \ , \ t \ge 0 \\ &\text{B.C.} \quad u_x(0,t) = a \qquad u_x(L,t) = b \qquad\qquad (20) \\ &\text{I.C.} \quad u(x,0) = f(x) \ . \end{aligned}$$

Solution. First note that the B.C. state that heat is being drained out of the end $x = 0$ at a rate $u_x(0,t) = a$ (cf. (2) of Section 3.1 with $KA = 1$) and heat is flowing into the end $x = L$ at a rate $u_x(L,t) = b$. If $b > a$, then the heat energy is being added to the rod at a constant rate. (If $b < a$, the rod loses heat at constant rate). Thus, we cannot expect a steady–state solution of the D.E. and B.C. , unless it happens that $a = b$. Indeed, putting $u_p(x,t) = cx + d$, the B.C. tell us that $c = a$ and $c = b$, which is impossible unless $a = b$. The next simplest form for a particular solution, that reflects the fact that the heat energy is changing at a constant rate, is

$$u_p(x,t) = ct + h(x) \ , \qquad (21)$$

where c is a constant and $h(x)$ is a function of x . The constant c and the function $h(x)$ can be determined from the D.E. and B.C. . Indeed,

$$c = (u_p)_t = k(u_p)_{xx} = kh''(x) \Rightarrow h''(x) = \tfrac{c}{k} \Rightarrow h(x) = \tfrac{c}{2k}x^2 + dx + e ,$$

for constants d and e . The B.C. then yield

$$\left.\begin{array}{l} a = (u_p)_x(0,t) = h'(0) = d \\ b = (u_p)_x(L,t) = h'(L) = \tfrac{cL}{k} + d \end{array}\right\} \Rightarrow \left\{\begin{array}{l} d = a \\ c = \dfrac{(b-a)k}{L} \end{array}\right. .$$

Thus, we arrive at the following particular solution (where we have set $e = 0$, for simplicity) of the D.E. and B.C. of (20) :

$$u_p(x,t) = \tfrac{b-a}{L}\cdot kt + \tfrac{b-a}{2L}\cdot x^2 + ax = \tfrac{b-a}{L}\left[kt + \tfrac{1}{2}x^2\right] + ax . \qquad (22)$$

The related homogeneous problem is

$$\begin{array}{ll} \text{D.E.} \;\; v_t = kv_{xx} & 0 \le x \le L , \; t \ge 0 \\ \text{B.C.} \;\; v_x(0,t) = 0 & v_x(L,t) = 0 \\ \text{I.C.} \;\; v(x,0) = f(x) - u_p(x,0) = f(x) - \left[\tfrac{b-a}{2L}\cdot x^2 + ax\right] . & \end{array}$$

In the event that $f(x) - u_p(x,0)$ is of the form $\sum_{n=0}^{N} a_n\cos(n\pi x/L)$, we have the solution

$$\begin{aligned} u(x,t) &= u_p(x,t) + v(x,t) \\ &= u_p(x,t) + \sum_{n=0}^{N} a_n e^{-(n\pi/L)^2kt}\cos(n\pi x/L) , \end{aligned} \qquad (23)$$

where $u_p(x,t)$ is given by (22) . □

Boundary conditions of the third kind

There are many other types of boundary conditions that can be imposed. For example, $u(0,t) - u_x(L,t) = 1$, $u_x(0,t)-[u(0,t)]^2 = 0$ or, more generally, $F(u(0,t),u(L,t),u_x(0,t),u_x(L,t)) = 0$, where the function F of four variables can be chosen at will. Even by restricting ourselves to those B.C. that have some physical relevance, there are more B.C. than we have time, space or endurance to handle. However, we will consider one additional type of B.C. that has great physical relevance, namely

$$\text{B.C.} \quad \begin{array}{l} u_x(0,t) - c\cdot u(0,t) = 0 \\ \\ u_x(L,t) + c'\cdot u(L,t) = 0 , \end{array} \qquad (24)$$

where c and c′ are constants [usually positive in applications]. Note that as $c \to 0$, the first

B.C. becomes $u_x(0,t) = 0$, while as $c \to \infty$, we obtain $u(0,t) = 0$ (and similarly for the second B.C.). The B.C. (24) are known as **boundary conditions of the third kind**. The B.C. such as $u(0,t) = 0$ and $u_x(0,t) = 0$ are known as **B.C. of the first and second kinds** respectively ; thus, these are limiting cases of the B.C. of the third kind. Essentially, the B.C. (24) arise because perfect insulation and perfect thermal contact (e.g., maintaining an end exactly at 0 by means of an external medium) are difficult to achieve in practice. To understand this more exactly, consider the picture (cf. Figure 3) near the end $x = 0$:

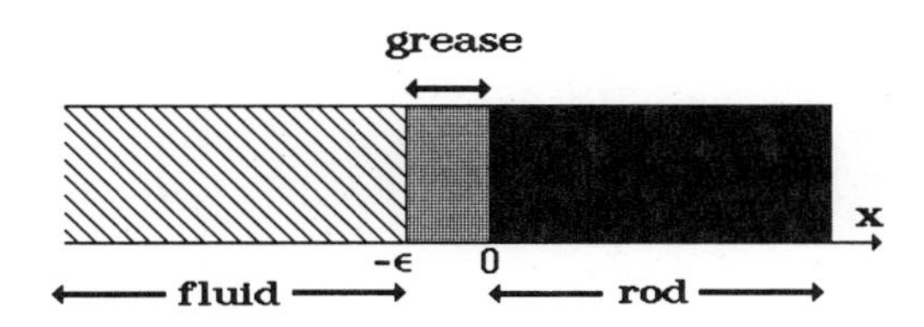

Figure 3

The region $x < -\epsilon$ is occupied by some fluid which is well circulated and maintained at temperature 0 , while the small region $-\epsilon \leq x \leq 0$ is occupied by some intermediate substance (say grease, oxide or imperfect insulation). Assume that this substance is in perfect thermal contact with the rod and the fluid so that $u(x,t)$ is continuous at the junctions $x = -\epsilon$ and $x = 0$. In order that the flux of heat energy across $x = 0$ be the same when measured on either side of $x = 0$, the temperature gradient must suffer a jump so that

$$k_1 u_x(0^-,t) = k u_x(0^+,t) \ , \tag{25}$$

where k_1 is the diffusivity constant for the intermediate substance (and k is that for the rod). Since the heat lost through $x = -\epsilon$ is assumed to be carried away immediately by circulating currents in the fluid, we have $u(-\epsilon,t) = 0$. If $u(0,t)$ is constant, then eventually a steady–state temperature distribution would result in $[-\epsilon,0]$. Indeed, the transient term decreases as $\exp[-\pi^2 k_1 t/\epsilon^2]$. This decrease is especially rapid if ϵ is small, so that $k_1\pi^2/\epsilon^2 >> 1$. If $u(0,t)$ is not constant, but varies rather slowly, then we still expect that the solution in the interval $[-\epsilon,0]$ will be rather close to a linear (steady–state) distribution, i.e.,

$$u(x,t) \approx \frac{x+\epsilon}{\epsilon}\, u(0,t) \qquad -\epsilon < x \leq 0 \ ,$$

when $k_1\pi^2/\epsilon^2 >> 1$. Then $u_x(0^-,t) \approx \frac{1}{\epsilon}\cdot u(0,t)$, and (25) yields (to good approximation)

$$u_x(0^+,t) - \tfrac{1}{\epsilon}(k_1/k)\cdot u(0,t) \ = \ 0 \ , \tag{26}$$

which is the first B.C. of (24), where $c = \frac{1}{\epsilon}(k_1/k)$. If $\frac{1}{\epsilon}(k_1/k)$ is small (say by virtue of k being large compared to k_1), then the end $x = 0$ of the rod behaves as if it were insulated. If $\frac{1}{\epsilon}(k_1/k)$

is large, then a close thermal contact exists between the rod and the fluid (i.e., $u(0,t) \approx 0$, where 0 is the temperature of the fluid). At the other end $x = L$, a similar analysis applies. Thus,

$$u_x(L^-,t) + \tfrac{1}{\epsilon}(k_1/k)\cdot u(L,t) = 0 , \tag{27}$$

since the derivatives change sign. Of course, it could happen that k_1 and ϵ may be different at the end $x = L$. Thus, c' need not equal c in (24), but it is true that c and c' are positive in this physical context. For the B.C. (24), it is not always easy to find the values λ for which there is a Case 1 product solution (cf. (11) of Section 3.1). In the exercises, the reader is asked to verify that λ (where $\lambda > 0$) must satisfy

$$(cc' - \lambda^2)\sin(\lambda L) + (c + c')\lambda\cos(\lambda L) = 0 . \tag{28}$$

Assume that c and c' are nonnegative. First note that $\lambda = \sqrt{cc'} > 0$ is a solution of (28), if and only if $\cos(\sqrt{cc'}\cdot L) = 0$ (i.e., $\sqrt{cc'} = (n+\frac{1}{2})\pi/L$, for some $n = 1, 2, 3, \ldots$). If a solution λ is not of this form, then $(cc' - \lambda^2)\cos(\lambda L) \neq 0$ and dividing both sides of (28) by this, we obtain

$$\tan(\lambda L) = \frac{(c + c')\lambda}{\lambda^2 - cc'} . \tag{29}$$

The (positive) solutions λ of this equation can be roughly determined from the points of intersections of the graphs of the functions of λ appearing on both sides of (29). If $\tan(\lambda L) = \infty$ (i.e., $\cos(\lambda L) = 0$) and $\lambda = (cc')^{\frac{1}{2}}$, then this exceptional solution must be included as well. Actually, this is simply the case where the graphs of the functions on either side of (29) approach a common vertical asymptote (i.e., they "intersect at ∞"). For example, if $c = c' = \pi/2$ and $L = 1$, then we have the graphs of $\tan(\lambda)$ and $\pi\lambda/(\lambda^2 - \frac{1}{4}\pi^2)$, shown in Figure 4.

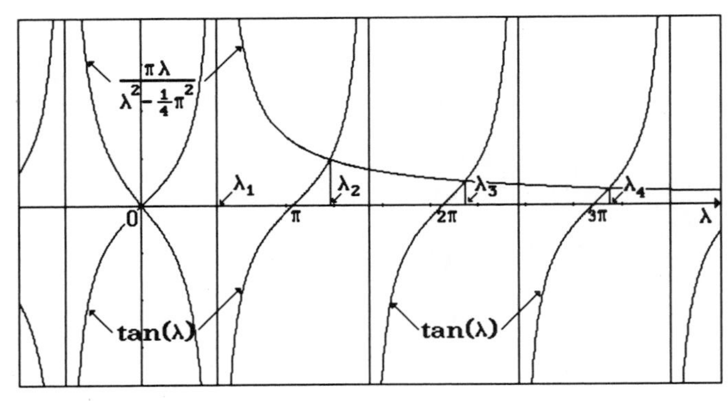

Figure 4

Here, $\lambda_1 = \pi/2$ corresponds to an intersection at infinity. The intersection at $\lambda = 0$ does not count, since we have assumed $\lambda > 0$. Numerical computation (cf. Problems 12 and 13 of Section 8.2) yields $\lambda_2 \approx 3.906$, $\lambda_3 \approx 6.741$, $\lambda_4 \approx 9.744$.

Remark. One can imagine physical situations where c and/or c′ are negative. For instance, consider a rod immersed in a medium in which a heat–producing chemical (or nuclear) reaction generates heat at a rate which is proportional to the local temperature. It is then possible that the portion of the medium which is close to a hot end of the rod may actually produce a heat influx into that end, at rate which is positively proportional to the temperature at the end. In such a situation, there are likely to be Case 2 product solutions which grow exponentially with time, possibly leading to a melt–down or explosion (cf. Problem 9, parts (c) and (d)). □

Summary 3.3

1. **Insulated ends :** The condition $u_x(0,t) = 0$ means that no heat is flowing through the end of the rod at $x = 0$. That is, this end is insulated. If each end of the rod is either insulated or maintained at 0, then there are four possible sets of (homogeneous) boundary conditions (cf. Example 1 and Example 2).

2. **Inhomogeneous B.C. :** For B.C. of the form $u(0,t) = a$ (or $u_x(0,t) = a$) and $u(L,t) = b$ (or $u_x(L,t) = b$) for constants a and b, the following method may be used. First, find a particular solution, $u_p(x,t)$, of the D.E. and the B.C. . Then the required solution is

$$u(x,t) = u_p(x,t) + v(x,t) ,$$

where $v(x,t)$ is the solution of the related problem with homogeneous B.C. and the modified I.C. given by $v(x,0) = u(x,0) - u_p(x,0)$. For illustrations, see Examples 3, 4, 5 and 6.

For B.C. of the above form, except in the case $u_x(0,t) = a$ and $u_x(L,t) = b$, there will be a particular solution of the D.E. $u_t = ku_{xx}$ and the B.C., which is of the form $u_p(x,t) = cx + d$, where c and d are constants . The solution $u(x,t)$ will then approach this steady–state (i.e., time–independent) solution as $t \to \infty$, since $v(x,t) \to 0$. When $u_x(0,t) = a$ and $u_x(L,t) = b$, with $a \neq b$, there are no steady–state particular solutions of the D.E. and B.C., but rather $u_p(x,t) = \frac{b-a}{L}\left[kt + \frac{1}{2}x^2\right] + ax$, which changes with time t at a constant rate. This is due to a nonzero net influx or outflux of heat through the ends (cf. Example 6).

3. **B.C. of the third kind :** Among the many different types of B.C., boundary conditions of the third kind, that is,

$$u_x(0,t) - cu(0,t) = 0 , \qquad u_x(L,t) + c'u(L,t) = 0 ,$$

where c and c' are positive constants, are important in physical applications. The determination of the values of λ for which there is a nontrivial product solution of the D.E. and B.C., is not always easy. Approximate values of λ may be obtained graphically or by numerical methods.

Exercises 3.3

1. Show that for the D.E. and B.C. of (2), there are no nonzero product solutions of the form $e^{k\lambda^2 t}(c_1e^{\lambda x} + c_2e^{-\lambda x})$ (i.e., the Case 2 product solution (12) in Section 3.1).

2. (a) Show that the solution (6) of Example 1 has the property that

$$\frac{1}{L}\int_0^L u(x,t)\,dx = a_0 \quad \text{for all } t \geq 0 , \tag{$*$}$$

by simply integrating (6). What does this say about the heat energy in the rod?

(b) *Without using the solution* (6), differentiate $\frac{1}{L}\int_0^L u(x,t)\,dx$ with respect to t under the integral, use the heat equation, the fundamental theorem of calculus, and the B.C., to prove that the integral in ($*$) of part (a) is a constant which is independent of t.

3. (a) Find all product solutions of the D.E. and B.C. for the problem

$$\begin{array}{lll} \text{D.E.} & u_t = ku_{xx} & 0 \leq x \leq L\,,\ t \geq 0 \\ \text{B.C.} & u(0,t) = 0 & u_x(L,t) = 0 \\ \text{I.C.} & u(x,0) = f(x)\,. & \end{array}$$

(b) Solve the problem, if $f(x) = \sum_{n=0}^{N} c_n \sin[(n+\frac{1}{2})\pi x/L]$, for some integer $N \geq 0$.

4. Use Problem 3 to solve

$$\begin{array}{lll} \text{D.E.} & u_t = 2u_{xx} & 0 \leq x \leq 1\,,\ t \geq 0 \\ \text{B.C.} & u(0,t) = -1 & u_x(1,t) = 1 \\ \text{I.C.} & u(x,0) = x + \sin(3\pi x/2) - 1\,. & \end{array}$$

5. Solve

$$\begin{array}{lll} \text{D.E.} & u_t = 5u_{xx} & 0 \leq x \leq 10\,,\ t \geq 0 \\ \text{B.C.} & u_x(0,t) = 2 & u_x(10,t) = 3 \\ \text{I.C.} & u(x,0) = \frac{1}{20}x^2 + 2x + \cos(\pi x)\,. & \end{array}$$

6. Solve

$$\begin{array}{lll} \text{D.E.} & u_t = ku_{xx} & 0 \leq x \leq L\,,\ t \geq 0 \\ \text{B.C.} & u(0,t) = a & u_x(L,t) = b \\ \text{I.C.} & u(x,0) = bx + a\,. & \end{array}$$

7. Solve

$$\begin{aligned} &\text{D.E.} \quad u_t = u_{xx} \qquad 0 \le x \le \pi\,,\ t \ge 0 \\ &\text{B.C.} \quad u_x(0,t) = 2 \qquad u(\pi,t) = 4 \\ &\text{I.C.} \quad u(x,0) = 4 - 2\pi + 2x + 7\cos(3x/2)\,. \end{aligned}$$

8. What condition on the constants a and b is necessary in order that the solution (if it exists) $u(x,t)$, of the initial/boundary–value problem (20) (with B.C. $u_x(0,t) = a$ and $u_x(L,t) = b$), will have the property that $\int_0^L u(x,t)\,dx$ is independent of t ? Why?

Hint. Show that $\frac{d}{dt}\int_0^L u(x,t)\,dx = k(b - a)$ (cf. Problem 2(b)).

9. Consider the problem

$$\begin{aligned} &\text{D.E.} \quad u_t = ku_{xx} \qquad 0 \le x \le L\,,\ t \ge 0 \\ &\text{B.C.} \quad u_x(0,t) - cu(0,t) = 0 \\ &\qquad\quad u_x(L,t) + c'u(L,t) = 0\,, \end{aligned}$$

where we assume that $c, c' \ge 0$, which is the physically relevant case.

(a) Show that a Case 1 (cf.(11) of Section 3.1) product solution of the D.E. and B.C. must be of the form

$$u_n(x,t) = a_n e^{-\lambda_n kt}\Big[\cos(\lambda_n x) + (c/\lambda_n)\sin(\lambda_n x)\Big]\,,$$

where λ_n is the n–th positive root of the equation

$$(cc' - \lambda^2)\sin(\lambda L) + (c+c')\lambda\cdot\cos(\lambda L) = 0 \qquad \text{(cf. (28))}.$$

(b) Show that there are no Case 2 (cf.(12), Section 3.1) solutions of the D.E. and B.C., and there is a Case 3 (cf.(13), Section 3.1) product solution only when $c = c' = 0$.

Hint. Show that λ must satisfy $\lambda(c+c')\cosh(\lambda L) + (\lambda^2 + c'c)\sinh(\lambda L) = 0$ in this case.

(c) Suppose now that $c < 0$ and $c' < 0$. Show that there is at least one nonzero Case 2 product solution of the D.E. and B.C. in this case. You need not actually find this solution explicitly.

(d) Suppose that $c < 0$ and $c' \ge 0$. Show that there is exactly one nonzero Case 2 product solution if $c + c' < 0$. If $c + c' > 0$, then show that there is a Case 2 product solution if and only if $L > -(c+c')/(c'c)$. Find all Case 2 product solutions when $c' = -c$.

10. For a fixed constant $c > 0$, the B.C. $u_x(0^+,t) - cu(0,t) = 0$ can be approximately achieved by adding to the end $(x = 0)$ a layer of substance of heat conductivity k_1 and thickness ϵ, which are chosen in such a way that $\frac{1}{\epsilon}(k_1/k) = c$ (cf. (26)). Explain why this approximation should become more and more accurate as $\epsilon \longrightarrow 0$ (and $k_1 = kc\epsilon \longrightarrow 0$).

Hint. Consider how rapidly the temperature distribution in $[-\epsilon,0]$ approaches a steady-state distribution for $u(-\epsilon,t) \equiv 0$ and a nearly constant $u(0,t)$. Note that $-k_1\pi^2/\epsilon^2 = -ck\pi^2/\epsilon$.

11. Prove the following maximum principle for the case where one end of the rod is insulated and the other has specified temperature.

Let u be a C^2 solution of the problem

$$\begin{aligned}&\text{D.E.}\quad u_t = ku_{xx} && 0 \le x \le L\,,\ t \ge 0\\ &\text{B.C.}\quad u_x(0,t) = 0 && u(L,t) = b(t)\\ &\text{I.C.}\quad u(x,0) = f(x)\ .\end{aligned}$$

Then for $M \equiv \max_{0 \le x \le L} \{f(x)\}$ and $B \equiv \max_{0 \le t \le T} \{b(t)\}$, we have

$$u(x,t) \le \max\{M,B\} \quad \text{for } 0 \le x \le L\,,\ 0 \le t \le T\ .$$

Hint. Modify the proof of the GMP at the end of Section 3.2. Use the fact that if $v(x,t)$ has a maximum at $(0,t_0)$, then $v_{xx}(0,t_0) \le 0$ follows from the B.C. $v_x(0,t_0) = 0$ (Why?).

Remark. This result leads to a minimum principle and a theorem on "continuity of solutions" with respect to variations of $f(x)$ and $b(t)$, as in Theorem 3 of Section 3.2 .

12. (a) Mimic the proof of Theorem 1 of Section 3.1, in order to prove uniqueness of solutions of the initial/boundary-value problem in Problem 9 (with B.C. of the third kind). Explain why the proof breaks down, if we do not have $c > 0$ and $c' > 0$. (Uniqueness can be established by other methods, however.)

(b) Show that solutions $u(x,t)$ of the initial/boundary-value problem in part (a), with $c > 0$ and $c' > 0$, achieve their maxima when $t = 0$, assuming that $u(x,0) \ge 0$ for some x in $[0,L]$. Give a plausible argument or an explicit example to show that the conclusion can be false if this assumption is dropped.

(c) When $c = 1$ and $c' = -1$, show (by an explicit example) that the maximum temperature need not occur when $t = 0$, even if $u(x,0) > 0$ for all x in $[0,L]$ (i.e., the maximum principle in part (b) fails in this case.). **Hint.** Find a Case 2 product solution.

3.4 Time–Dependent Boundary Conditions and Duhamel's Principle for Inhomogeneous Heat Equations

We have proved (cf. Theorem 1 of Section 3.2) that there is at most one C^2 solution of

$$\begin{aligned} &\text{D.E.}\quad u_t = ku_{xx} \qquad 0 \le x \le L\ ,\ t \ge 0 \\ &\text{B.C.}\quad u(0,t) = a(t) \qquad u(L,t) = b(t) \\ &\text{I.C.}\quad u(x,0) = f(x)\ . \end{aligned} \tag{1}$$

In Section 3.3 we found the solution of this problem in the case where $a(t)$ and $b(t)$ are constant functions (independent of t) and $f(x)$ has an appropriate form , i.e.

$$f(x) = u_p(x,0) + \sum_{n=1}^{N} b_n \sin(n\pi x/L)\ , \quad u_p(x,0) = \left[\frac{(b-a)}{L}\right]x + a\ .$$

As in the case where $a(t)$ and $b(t)$ are constant, we can find a function of the form $c(t)x + d(t)$ that satisfies the B.C. of (1). Indeed, from the B.C. we obtain $a(t) = c(t)0 + d(t) = d(t)$ and $b(t) = c(t)L + d(t)$, whence $d(t) = a(t)$ and $c(t) = [b(t)-a(t)]/L$. The function $w(x,t)$, defined by

$$w(x,t) = \left[\frac{b(t) - a(t)}{L}\right]x + a(t)\ , \tag{2}$$

then satisfies the B.C. of (1). However, $w(x,t)$ will *not* satisfy the D.E unless $a(t)$ and $b(t)$ are constant. Indeed,

$$w_t - kw_{xx} = \left[\frac{b'(t) - a'(t)}{L}\right]x + a'(t)\ . \tag{3}$$

For this reason, we write $w(x,t)$ instead of $u_p(x,t)$; the latter denotes a particular solution of the D.E. as well as the B.C. . We may still attempt to find a solution for problem (1) of the form $u(x,t) = w(x,t) + v(x,t)$. The function $v(x,t)$ must solve the following related problem with homogeneous B.C., but inhomogeneous D.E. :

$$\begin{aligned} &\text{D.E.}\quad v_t - kv_{xx} = -[b'(t)-a'(t)]x/L - a'(t) \qquad 0 \le x \le L\ ,\ t \ge 0 \\ &\text{B.C.}\quad v(0,t) = 0 \qquad v(L,t) = 0 \\ &\text{I.C.}\quad v(x,0) = u(x,0)-w(x,0) = f(x) - [a(0)-b(0)]x/L - a(0)\ . \end{aligned} \tag{4}$$

Indeed, $v = u-w$, and so $v_t - kv_{xx} = u_t - ku_{xx} - (w_t - kw_{xx}) = -(w_t - kw_{xx}) = -[b'(t)-a'(t)]x/L - a'(t)$ by (3). Also, $v(0,t) = u(0,t) - w(0,t) = a(t) - a(t) = 0$ and $v(L,t) = b(t) - b(t) = 0$. When $a(t)$ and $b(t)$ are constants, the D.E. is homogeneous and problem (4) is then familiar to us. However, in general, it appears that we have merely converted problem (1) into another problem (4), perhaps equally difficult. Observe that (4) is a special case of the general problem

$$\text{D.E.}\quad u_t - ku_{xx} = h(x,t) \qquad 0 \le x \le L,\ t \ge 0$$

$$\text{B.C.}\quad v(0,t) = 0 \quad v(L,t) = 0 \tag{5}$$

$$\text{I.C.}\quad v(x,0) = g(x)\ .$$

If we knew how to solve problem (5), then we could solve (4) and obtain the solution $u(x,t) = w(x,t) + v(x,t)$ of (1). Fortunately there is a method for solving (5) for suitable $h(x,t)$ and $f(x)$. This method was discovered by the French mathematician and physicist Jean–Marie–Constant Duhamel (1797–1872), who published his solution in 1833. There is one further simplification of (5) which we can easily make. Suppose u_1 and u_2 are solutions of the following respective problems :

(a) D.E. $(u_1)_t - k(u_1)_{xx} = 0$
B.C. $u_1(0,t) = 0$
$u_1(L,t) = 0$
I.C. $u_1(x,0) = g(x)$

(b) D.E. $(u_2)_t - k(u_2)_{xx} = h(x,t)$
B.C. $u_2(0,t) = 0$
$u_2(L,t) = 0$ (6)
I.C. $u_2(x,0) = 0$.

Then the reader may easily check that $v(x,t) = u_1(x,t) + u_2(x,t)$ solves (5). The problem for u_1 is familiar, and hence the only unfamiliar problem that remains is the problem for u_2, with homogeneous B.C. and I.C., but inhomogeneous D.E. . In summary of what has been accomplished so far, we state

Proposition 1. A solution of problem (1) is given by

$$u(x,t) = w(x,t) + u_1(x,t) + u_2(x,t)\ ,$$

where $w(x,t)$ **is the particular solution (2) of the B.C. and** $u_1(x,t)$ **solves (6a) with** $g(x) = f(x) - w(x,0)$ **and** $u_2(x,t)$ **solves (6b) with** $h(x,t) = -(w_t - kw_{xx}) = -[b'(t) - a'(t)]x/L - a'(t)$.

The reduction procedure embodied in Proposition 1 works for the other standard problems where the heat energy flux is prescribed at one or both ends, as the following example illustrates.

Example 1. Reduce the problem

$$\begin{aligned} &\text{D.E.} \quad u_t = ku_{xx} \qquad 0 \le x \le L\,,\; t \ge 0 \\ &\text{B.C.} \quad u(0,t) = a(t) \quad u_x(L,t) = b(t) \\ &\text{I.C.} \quad u(x,0) = f(x) \end{aligned} \tag{7}$$

to a problem of the form

$$\begin{aligned} &\text{D.E.} \quad (u_2)_t - k(u_2)_{xx} = h(x,t) \\ &\text{B.C.} \quad u_2(0,t) = 0 \qquad (u_2)_x(L,t) = 0 \\ &\text{I.C.} \quad u_2(x,0) = 0\,. \end{aligned} \tag{8}$$

Solution. Note that $w(x,t) = b(t)x + a(t)$ satisfies the B.C. of (7). Setting $v(x,t) = u(x,t) - w(x,t)$, we have $v_t - kv_{xx} = u_t - ku_{xx} - (w_t - kw_{xx}) = -b'(t)x - a'(t)$, and the related problem for v is then

$$\begin{aligned} &\text{D.E.} \quad v_t - kv_{xx} = -b'(t)x - a'(t) \qquad 0 \le x \le L\,,\; t \ge 0 \\ &\text{B.C.} \quad v(0,t) = 0 \quad v_x(L,t) = 0 \\ &\text{I.C.} \quad v(x,0) = f(x) - w(x,0) = f(x) - b(0)x - a(0)\,. \end{aligned} \tag{9}$$

Then, $v = u_1 + u_2$, where u_1 solves the familiar problem

$$\begin{aligned} &\text{D.E.} \quad (u_1)_t - k(u_1)_{xx} = 0 \\ &\text{B.C.} \quad u_1(0,t) = 0 \qquad (u_1)_x(L,t) = 0 \\ &\text{I.C.} \quad u_1(x,0) = f(x) - b(0)x - a(0)\,, \end{aligned}$$

and u_2 solves (8) with $h(x,t) = -b'(t)x - a'(t)$. □

Duhamel's method – the physical motivation

We now motivate Duhamel's method for the solution of a problem of the following form which arose [cf.(6b)] as the principal obstacle to solving (1) :

$$\begin{aligned} &\text{D.E.} \quad u_t - ku_{xx} = h(x,t) \qquad 0 \le x \le L\,,\; t \ge 0 \\ &\text{B.C.} \quad u(0,t) = 0 \qquad u(L,t) = 0 \\ &\text{I.C.} \quad u(x,0) = 0\;. \end{aligned} \tag{10}$$

Remark. The inhomogeneous heat equation in (10) has a simple physical interpretation. To this end, consider the derivation of the heat equation in Section 3.1 . Multiplying the D.E. by CDA (cf. Section 3.1 for definitions) and integrating over any subinterval $[a,b]$ of $[0,L]$, we obtain

$$\int_a^b CDA(u_t - ku_{xx})\,dx = \int_a^b CDA\cdot h(x,t)\,dx\ . \tag{11}$$

In the derivation of the heat equation, the left side of (11) was found to be the rate of heat energy gain in the portion $[a,b]$ of the rod, minus the net rate of heat influx through the cross sections $x = a$ and $x = b$. In the absence of any internal source of heat (e.g., chemical reactions, electrical currents, etc.), the left side of (11) is then 0. If, however, there is an internal source of heat such that $CDA\cdot h(x,t)\Delta x$ is the amount of heat produced in $[x,x+\Delta x]$ per unit time, then the right side of (11) is the rate of heat energy production in the portion $[a,b]$ due to this internal source, and equation (11) results. Since $[a,b]$ is an arbitrary subinterval of $[0,L]$, the integrands of (11) must be equal and we get the D.E. $u_t - ku_{xx} = h(x,t)$, where $h(x,t)$ is proportional to the internal heat source density distribution at time t . □

Now, suppose the rod is at temperature 0, when we turn on the heat source $h(x,0)$ for a very brief time interval from $t = -\Delta s$ to $t = 0$, and then switch it off. At $t = 0$ the temperature distribution in the rod will be very nearly equal to $h(s,0)\Delta s$ (i.e., if Δs is small, then only a small amount of heat will flow). The temperature at a later time $t \geq 0$ will then be $\approx v(x,t)\Delta s$, where $v(x,t)$ solves the problem

$$\text{D.E.}\quad v_t = kv_{xx} \qquad 0 \leq x \leq L\ ,\ t \geq 0$$

$$\text{B.C.}\quad v(0,t) = 0 \qquad v(L,t) = 0$$

$$\text{I.C.}\quad v(x,0) = h(x,0)\ .$$

(Note that $v(x,t)\Delta s$ solves this problem with initial temperature $h(x,0)\Delta s$, since Δs is a constant and the problem is linear.) More generally, if the source is turned on at the time $t = s - \Delta s$ and turned off at $t = s$, then the effect of this on the temperature at time $t \geq s$ is very nearly $v(x,t;s)\Delta s$, where s is a fixed parameter and $v(x,t;s)$ solves

$$\text{D.E.}\quad v_t = kv_{xx} \qquad 0 \leq x \leq L\ ,\ t \geq 0$$

$$\text{B.C.}\quad v(0,t;s) = 0 \qquad v(L,t;s) = 0 \tag{12}$$

$$\text{I.C.}\quad v(x,s;s) = h(x,s)\ .$$

The total effect of all these heat source contributions on the temperature at time t is then the integral (continuous superposition) of all the source contributions prior to t, namely

$$\int_0^t v(x,t;s)\,ds\ .$$

In other words, by this somewhat imprecise, physical reasoning, we are led to the conjecture that

$$u(x,t) = \int_0^t v(x,t;s)\, ds \tag{13}$$

will be the solution of (10) provided $v(x,t;s)$ is a solution of the problem (12) which has homogeneous D.E. and B.C. . In the sequel we will prove that (13) is in fact a solution of (10) under certain assumptions. But first we note that problem (12) is not entirely familiar because the initial condition is given at time $t = s$ instead of $t = 0$. This difficulty is easily handled by solving the following associated problem with I.C. at $t = 0$.

$$\begin{aligned} &\text{D.E.} \quad \tilde{v}_t = k\tilde{v}_{xx} \qquad 0 \le x \le L\,,\ t \ge 0 \\ &\text{B.C.} \quad \tilde{v}(0,t;s) = 0 \qquad \tilde{v}(L,t;s) = 0 \\ &\text{I.C.} \quad \tilde{v}(x,0;s) = h(x,s)\,. \end{aligned} \tag{14}$$

Then check that

$$v(x,t;s) = \tilde{v}(x,t-s;s)$$

solves (12). In other words, we just perform a translation in time to obtain an I.C. at $t = 0$, instead of $t = s$. Indeed, one can forget about (12), and write (13) in terms of $\tilde{v}$, namely

$$u(x,t) = \int_0^t \tilde{v}(x,t-s;s)\, ds \quad . \tag{15}$$

Thus, the hypothetical solution $u(x,t)$ of problem (10) may be obtained by solving the familiar problem (14) and performing the integration (15). Observe that the troublesome source term $h(x,t)$ in (10) has been transferred to a "harmless" initial condition in problem (12). This is the essence of Duhamel's method, and the general idea that sources can be converted to initial conditions of related problems is known as **Duhamel's principle**. Before formulating a precise mathematical statement and proof, we consider an example.

Example 2. Solve the following problem with heat source distribution $h(x,t) = t\cdot\sin(x)$.

$$\begin{aligned} &\text{D.E.} \quad u_t - ku_{xx} = t\cdot\sin(x) \qquad 0 \le x \le \pi\,,\ t \ge 0 \\ &\text{B.C.} \quad u(0,t) = 0 \qquad u(\pi,t) = 0 \\ &\text{I.C.} \quad u(x,0) = 0\,. \end{aligned} \tag{16}$$

Solution. We solve the related problem (14)

$$\begin{aligned} &\text{D.E.} \quad \tilde{v}_t = k\tilde{v}_{xx} \qquad 0 \le x \le \pi\,,\ t \ge 0 \\ &\text{B.C.} \quad \tilde{v}(0,t;s) = 0 \qquad \tilde{v}(\pi,t;s) = 0 \\ &\text{I.C.} \quad \tilde{v}(x,0;s) = h(x,s) = s\cdot\sin(x)\,. \end{aligned}$$

Here, s is treated like a constant, and so we easily obtain $\tilde{v}(x,t;s) = s \cdot e^{-kt}\sin(x)$. According to (15) the solution of (16) should be

$$u(x,t) = \int_0^t \tilde{v}(x,t-s;s)\,ds = \int_0^t s \cdot e^{-k(t-s)}\sin(x)\,ds$$
$$= e^{-kt}\sin(x)\int_0^t s \cdot e^{ks}\,ds = \left[k^{-1}t + k^{-2}(e^{-kt} - 1)\right]\sin(x)] .$$

This is the correct solution, as can be checked directly. □

Example 3. Solve the following problem, where the heat source distribution is $e^{-ct}\sin(x)$ for an arbitrary constant c. Does anything interesting happen when $c \approx 1$?

$$\text{D.E.}\quad u_t - u_{xx} = e^{-ct}\sin(x) \qquad 0 \le x \le \pi ,\ t \ge 0$$
$$\text{B.C.}\quad u(0,t) = 0 \qquad u(\pi,t) = 0$$
$$\text{I.C.}\quad u(x,0) = 0 .$$

Solution. We solve the related problem (14) with $h(x,s) = e^{-cs}\sin(x)$, obtaining $\tilde{v}(x,t;s) = e^{-cs}e^{-t}\sin(x)$. Then, by (15) the solution should be

$$u(x,t) = \int_0^t \tilde{v}(x,t-s;s)\,ds = \int_0^t e^{-t}e^{(1-c)s}\sin(x)\,ds = \begin{cases} \frac{1}{1-c}\,(e^{-ct} - e^{-t})\sin(x) & c \ne 1 \\ t \cdot e^{-t}\sin(x) & c = 1 \end{cases} .$$

As $t \to \infty$, we have the approximate behavior

$$u(x,t) \approx \begin{cases} [1-c]^{-1}e^{-ct}\sin(x) & c < 1 \\ t \cdot e^{-t}\sin(x) & c = 1 \\ [c-1]^{-1}e^{-t}\sin(x) & c > 1 \end{cases} .$$

There are qualitative changes in the behavior of $u(x,t)$ as c passes through 1. □

Proving Duhamel's principle

In the proof of our rigorous statement of Duhamel's principle (see Theorem 1 below), we will need the next result concerning the differentiation of integrals with respect to a variable occurring as a limit of integration as well as inside the integral (cf. the Appendix A.3).

Lemma 1. Suppose $g(t,s)$ **and** $g_t(t,s)$ **are continuous functions. Then**

$$\frac{d}{dt}\left[\int_0^t g(t,s)\,ds\right] = g(t,t) + \int_0^t g_t(t,s)\,ds \qquad (17)$$

Proof. Let $H(t,y)$ be defined by

$$H(t,y) = \int_0^y g(t,s)\, ds\ .$$

We compute $\frac{d}{dt} H(t,t)$, since this is the left–hand side of (17). Let $y(t) = t$, and note that

$$\begin{aligned}\frac{d}{dt} H(t,t) &= \frac{d}{dt} H(t,y(t)) = H_t(t,t)\frac{dt}{dt} + H_y(t,y(t))\frac{dy}{dt} \\ &= H_t(t,t) + H_y(t,t)\ . \qquad (18)\end{aligned}$$

We have $H_t(t,y) = \int_0^y g_t(t,s)\, ds$ by Leibniz's rule (cf. Appendix A.3). Also, $H_y(t,y) = g(t,y)$, since differentiating with respect to an upper limit yields the integrand evaluated at the upper limit, [i.e., $\frac{d}{dx}\int_0^x f(s)\, ds = f(x)$]. Thus, $H_y(t,t) = g(t,t)$, $H_t(t,t) = \int_0^t g_t(t,s)\, ds$, and (17) then follows from (18). □

Theorem 1 (Duhamel's principle). Suppose that $h(x,t)$ **is a given** C^2 **function for** $0 \le x \le L$, $t \ge 0$. **Assume that for each** $s \ge 0$ **the problem**

$$\begin{aligned}&\text{D.E.} \quad v_t = kv_{xx} \qquad 0 \le x \le L\ ,\ t \ge s \\ &\text{B.C.} \quad v(0,t;s) = 0 \qquad v(L,t;s) = 0 \qquad (19) \\ &\text{I.C.} \quad v(x,s;s) = h(x,s)\end{aligned}$$

has a C^2 **solution** $v(x,t;s)$, **where** $v(x,t;s)$, $v_t(x,t;s)$ **and** $v_{xx}(x,t;s)$ **are continuous [jointly with** (x,t)**] in** s, **as well. Then the unique solution of the problem**

$$\begin{aligned}&\text{D.E.} \quad u_t - ku_{xx} = h(x,t) \qquad 0 \le x \le L\ ,\ t \ge 0 \\ &\text{B.C.} \quad u(0,t) = 0 \qquad u(L,t) = 0 \qquad (20) \\ &\text{I.C.} \quad u(x,0) = 0\end{aligned}$$

is given by

$$u(x,t) = \int_0^t v(x,t;s)\, ds\ , \qquad (21)$$

or equivalently, by (15), **where** $\tilde{v}$ $[\ v(x,t;s) = \tilde{v}(x,t-s;s)\ ]$ **solves** (14).

Proof. The function $u(x,t)$ defined by (21) satisfies the I.C. $u(x,0) = 0$. It also satisfies the B.C. of (20), since $v(x,t;s)$ satisfies the B.C. of (19). Now use Lemma 1, with $g(t,s) = v(x,t;s)$, where x fixed. Then by (17)

$$
\begin{aligned}
u_t(x,t) &= v(x,t;t) + \int_0^t v_t(x,t;s)\, ds \\
&= h(x,t) + \int_0^t k v_{xx}(x,t;s)\, ds \ ,
\end{aligned}
\tag{22}
$$

where we have used the I.C. of (19) with $s = t$ and the D.E. of (19). If we apply Leibniz's rule to the final integral of (22), then we obtain $u_t(x,t) = h(x,t) + ku_{xx}(x,t)$ by (21). By our hypotheses on $v(x,t;s)$, we know that $u(x,t)$ in (21) is C^2 (by Leibniz's rule again). Concerning uniqueness, see Problem 10. □

Remarks. At present, we only know how to solve problem (20) [or (14)] in the case where the function $h(x,t)$ is of the form

$$
h(x,t) = \sum_{n=0}^{N} b_n(t) \sin(n\pi x/L) \ . \tag{23}
$$

On the other hand, the function $h(x,t)$ in Proposition 1 , given by

$$
h(x,t) = -[b'(t)-a'(t)]x/L - a'(t) \ , \tag{24}
$$

is not of the form (23). Indeed, (23) vanishes at $x = 0$ or $x = L$ only when $a'(t) = b'(t) = 0$ (i.e., only when $a(t)$ and $b(t)$ are constant functions). Thus, it seems that we have fallen short of solving problem (1), except in the case where $a(t)$ and $b(t)$ are constants, which was already considered in Section 3.3 . Even if one attempts to represent $h(x,t)$ in (24) by its "Fourier sine series" (covered in Chapter 4), the difficulty with the endpoints remains. However, the series representation can be shown to yield the physically correct solution for $0 < x < L$, $t > 0$, and it has the correct limiting values as $x \to 0^+$, $x \to L^-$ and $t \to 0^+$, for suitably nice functions $a(t)$, $b(t)$ and $f(x)$. The following examples are contrived in order to avoid any difficulties with infinite series solutions. Indeed, we have added sources that cancel with (24) after transforming to the related problems with homogeneous B.C. . In Chapter 4, we will consider the formal Fourier series solutions of problems which are not contrived. □

Example 4. Solve the problem

$$
\begin{aligned}
&\text{D.E.} \quad u_t - ku_{xx} = t[\sin(2\pi x) + 2x] \qquad 0 \le x \le 1 \ , \ t \ge 0 \\
&\text{B.C.} \quad u(0,t) = 1 \qquad u(1,t) = t^2 \\
&\text{I.C.} \quad u(x,0) = 1 + \sin(3\pi x) - x \ .
\end{aligned}
\tag{25}
$$

Solution. A simple function that satisfies the B.C. is $w(x,t) = (t^2-1)x + 1$. Then $u(x,t) = w(x,t) + v(x,t)$, where $v(x,t)$ solves the related problem

$$\text{D.E.}\quad v_t - kv_{xx} = u_t - ku_{xx} - (w_t - kw_{xx}) = t\cdot\sin(2\pi x)$$

$$\text{B.C.}\quad v(0,t) = u(0,t) - w(0,t) = 0$$

$$v(1,t) = u(1,t) - w(1,t) = 0$$

$$\text{I.C.}\quad v(x,0) = u(x,0) - w(x,0) = \sin(3\pi x)\ .$$

Now, $v = u_1 + u_2$, where u_1 and u_2 solve respectively

$$\begin{aligned} &\text{D.E.}\quad (u_1)_t - k(u_1)_{xx} = 0 \\ &\text{B.C.}\quad u_1(0,t) = 0 \\ &\qquad\quad u_1(1,t) = 0 \\ &\text{I.C.}\quad u_1(x,0) = \sin(3\pi x) \end{aligned}
\qquad\qquad
\begin{aligned} &\text{D.E.}\quad (u_2)_t - k(u_2)_{xx} = t\cdot\sin(2\pi x) \\ &\text{B.C.}\quad u_2(0,t) = 0 \\ &\qquad\quad u_2(1,t) = 0 \\ &\text{I.C.}\quad u_2(x,0) = 0\ . \end{aligned}$$

We know that $u_1(x,t) = e^{-9\pi^2kt}\sin(3\pi x)$ (cf. Section 3.1). The function u_2 is found via Duhamel's principle. Indeed, $u_2(x,t) = \int_0^t \tilde{v}(x,t-s;s)\,ds$, where $\tilde{v}$ solves the problem

$$\text{D.E.}\quad \tilde{v}_t - k\tilde{v}_{xx} = 0$$

$$\text{B.C.}\quad \tilde{v}(0,t;s) = 0 \qquad \tilde{v}(1,t;s) = 0$$

$$\text{I.C.}\quad \tilde{v}(x,0;s) = s\cdot\sin(2\pi x)\ .$$

We know that $\tilde{v}(x,t;s) = s\cdot e^{-4\pi^2kt}\sin(2\pi x)$, and so

$$u_2(x,t) = \int_0^t s\cdot e^{-4\pi^2k(t-s)}\sin(2\pi x)\,ds = e^{-4\pi^2kt}\sin(2\pi x)\int_0^t s\cdot e^{4\pi^2ks}\,ds$$

$$= (4\pi^2k)^{-2}[4\pi^2kt + e^{-4\pi^2kt} - 1]\cdot\sin(2\pi x)\ .$$

The solution of (25) is then given by $u(x,t) = w(x,t) + u_1(x,t) + u_2(x,t)$. □

Duhamel's principle for other boundary conditions

Duhamel's principle also applies to problems with D.E. $u_t - ku_{xx} = h(x,t)$ and homogeneous B.C. of the three forms

$$\begin{cases} u_x(0,t) = 0 \\ u(L,t) = 0 \end{cases};\qquad \begin{cases} u(0,t) = 0 \\ u_x(L,t) = 0 \end{cases};\qquad \begin{cases} u_x(0,t) = 0 \\ u_x(L,t) = 0\ . \end{cases}$$

The only difference is that the function $\tilde{v}$ (or v) satisfies the same corresponding B.C. . Then the proof of Duhamel's principle proceeds just as in the standard case. We illustrate Duhamel's principle for these other B.C. in the next examples.

Example 5. Solve the problem

$$\text{D.E.}\quad u_t - 8u_{xx} = \cos(t) + e^t \sin(x/2) \qquad 0 \le x \le \pi\,,\ t \ge 0$$

$$\text{B.C.}\quad u(0,t) = \sin(t) \qquad u_x(\pi,t) = 0 \tag{26}$$

$$\text{I.C.}\quad u(x,0) = 0\ .$$

Solution. Note that $w(x,t) = \sin(t)$ satisfies the B.C. . Letting $v(x,t) = u(x,t) - w(x,t)$, we obtain the related problem

$$\text{D.E.}\quad v_t - 8v_{xx} = u_t - 8u_{xx} - (w_t - 8w_{xx}) = e^t \sin(x/2)$$

$$\text{B.C.}\quad v(0,t) = 0 \qquad v_x(\pi,t) = 0 \tag{27}$$

$$\text{I.C.}\quad v(x,0) = 0\ .$$

We can solve this via Duhamel's principle. Of course, one should not confuse the v here with the v that was used in the statement of Theorem 1. The solution of (27) is $\int_0^t \tilde{v}(x,t-s;s)\, ds$, where $\tilde{v}$ solves the problem

$$\text{D.E.}\quad \tilde{v}_t - 8\tilde{v}_{xx} = 0 \qquad 0 \le x \le \pi\,,\ t \ge 0$$

$$\text{B.C.}\quad \tilde{v}(0,t;s) = 0 \qquad \tilde{v}_x(\pi,t;s) = 0$$

$$\text{I.C.}\quad \tilde{v}(x,0;s) = e^s \sin(x/2)\ .$$

From Problem 3 of Exercises 3.3 , we know that $\tilde{v}(x,t;s) = e^s e^{-2t} \sin(x/2)$. Then

$$u(x,t) = w(x,t) + \int_0^t \tilde{v}(x,t-s;s)\, ds \ = \ \sin(t) + e^{-2t} \sin(x/2) \int_0^t e^{3s}\, ds$$

$$= \sin(t) + \tfrac{1}{3}[e^t - e^{-2t}] \sin(x/2). \quad \square$$

Example 6. Solve the following inhomogeneous problem with insulated ends.

$$\text{D.E.}\quad u_t - u_{xx} = (2t+1)\cos(3x) \qquad 0 \le x \le \pi\,,\ t \ge 0$$

$$\text{B.C.}\quad u_x(0,t) = 0 \qquad u_x(\pi,t) = 0 \tag{28}$$

$$\text{I.C.}\quad u(x,0) = 0\ .$$

Solution. We apply Duhamel's principle when the ends are insulated, to obtain $u(x,t) = \int_0^t \tilde{v}(x,t-s;s)\,ds$, where $\tilde{v}(x,t;s)$ solves the problem

$$\begin{aligned} &\text{D.E.} \quad \tilde{v}_t - \tilde{v}_{xx} = 0 \qquad 0 \le x \le \pi,\ t \ge 0 \\ &\text{B.C.} \quad \tilde{v}_x(0,t;s) = 0 \qquad \tilde{v}_x(\pi,t;s) = 0 \\ &\text{I.C.} \quad \tilde{v}(x,0;s) = (2s+1)\cos(3x)\,. \end{aligned}$$

We can easily check that $\tilde{v}(x,t;s) = (2s+1)e^{-9t}\cos(3x)$, whence

$$\begin{aligned} u(x,t) &= \int_0^t (2s+1)e^{-9(t-s)}\cos(3x)\,ds = e^{-9t}\cos(3x)\int_0^t (2s+1)e^{9s}ds \\ &= \frac{1}{81}\left[18t + 7 - 7e^{-9t}\right]\cos(3x)\,. \quad \square \end{aligned}$$

Summary 3.4

1. **Reduction of time–dependent B.C. (Proposition 1) : A solution of the problem**

$$\text{D.E.} \quad u_t = ku_{xx} \qquad 0 \le x \le L\,,\ t \ge 0$$

$$\text{B.C.} \quad u(0,t) = a(t) \qquad u(L,t) = b(t)$$

$$\text{I.C.} \quad u(x,0) = f(x)\ .$$

is given by $u(x,t) = w(x,t) + u_1(x,t) + u_2(x,t)$, **where** $w(x,t) = [b(t) - a(t)]x/L + a(t)$, **and where** $u_1(x,t)$ **and** $u_2(x,t)$ **are solutions of the following respective problems :**

(a) D.E. $(u_1)_t - k(u_1)_{xx} = 0$

B.C. $u_1(0,t) = 0$

$u_1(L,t) = 0$

I.C. $u_1(x,0) = g(x)$

(b) D.E. $(u_2)_t - k(u_2)_{xx} = h(x,t)$

B.C. $u_2(0,t) = 0$

$u_2(L,t) = 0$

I.C. $u_2(x,0) = 0$,

where $g(x) = f(x) - w(x,0)$ **and** $h(x,t) = -[b'(t) - a'(t)]x/L - a'(t)$.

2. **Theorem 1 (Duhamel's Principle) : Suppose that** $h(x,t)$ **is a given** C^2 **function for** $0 \le x \le L$, $t \ge 0$. **Assume that for each** $s \ge 0$ **the problem**

$$\text{D.E.} \quad v_t = kv_{xx} \qquad 0 \le x \le L\,,\ t \ge s$$

$$\text{B.C.} \quad v(0,t;s) = 0 \qquad v(L,t;s) = 0$$

$$\text{I.C.} \quad v(x,s;s) = h(x,s)$$

has a C^2 **solution** $v(x,t;s)$, **where** $v(x,t;s)$, $v_t(x,t;s)$ **and** $v_{xx}(x,t;s)$ **are continuous [jointly with** (x,t)**] in** s, **as well. Then the unique solution of the problem**

$$\text{D.E.} \quad u_t - ku_{xx} = h(x,t) \qquad 0 \le x \le L\,,\ t \ge 0$$

$$\text{B.C.} \quad u(0,t) = 0 \qquad u(L,t) = 0$$

$$\text{I.C.} \quad u(x,0) = 0$$

is given by $u(x,t) = \int_0^t v(x,t;s)\, ds$. (Other types of homogeneous B.C. are possible.)

Exercises 3.4

1. (a) Find a particular solution of the form $u_p(x,t) = w(x)$ [i.e., u_p is steady–state] of the D.E. and the B.C. of the following problem

$$\text{D.E.}\quad u_t - ku_{xx} = h(x) \qquad 0 \le x \le L\,,\ t \ge 0$$

$$\text{B.C.}\quad u(0,t) = a \qquad u(L,t) = b$$

$$\text{I.C.}\quad u(x,0) = f(x)\,.$$

(b) Use the result of (a) to solve the following problem *without using Duhamel's principle.*

$$\text{D.E.}\quad u_t - 2u_{xx} = \sin(3x) \qquad 0 \le x \le \pi\,,\ t \ge 0$$

$$\text{B.C.}\quad u(0,t) = 0 \qquad u(\pi,t) = 0$$

$$\text{I.C.}\quad u(x,0) = 0\,.$$

c) Obtain the solution of part (b), but now use Duhamel's principle.

2. Verify that if $\tilde{v}(x,t;s)$ solves (14), then $v(x,t;s) = \tilde{v}(x,t-s;s)$ solves (12).

3. Solve

$$\text{D.E.}\quad u_t - u_{xx} = e^{-4t}\cos(t)\sin(2x) \qquad 0 \le x \le \pi\,,\ t \ge 0$$

$$\text{B.C.}\quad u(0,t) = 0 \qquad u(\pi,t) = 0$$

$$\text{I.C.}\quad u(x,0) = \sin(3x)\,.$$

4. Solve

$$\text{D.E.}\quad u_t - u_{xx} = t\cdot\cos(x) \qquad 0 \le x \le \pi\,,\ t \ge 0$$

$$\text{B.C.}\quad u_x(0,t) = 0 \qquad u_x(\pi,t) = 0$$

$$\text{I.C.}\quad u(x,0) = 0\,.$$

5. (a) Find a particular solution $w(x,t)$ of the D.E. and B.C. of the problem (where $h(t)$ is C^1)

$$\text{D.E.}\quad u_t - ku_{xx} = h(t) \qquad 0 \le x \le L\,,\ t \ge 0$$

$$\text{B.C.}\quad u_x(0,t) = a \qquad u_x(L,t) = b \qquad (*)$$

$$\text{I.C.}\quad u(x,0) = f(x)\,.$$

Hint. Try the form $w(x,t) = c_1x^2 + c_2x + g(t)$.

(b) Show that the solution of (*) is given by $u = w + v$, where w is given in (a) and v is the solution (if it exists) of

$$\text{D.E.}\quad v_t - kv_{xx} = 0 \qquad 0 \le x \le L\,,\ t \ge 0$$

$$\text{B.C.}\quad v_x(0,t) = 0 \qquad v_x(L,t) = 0$$

$$\text{I.C.}\quad v(x,0) = f(x) - w(x,0)\,.$$

Thus, (*) can be solved without Duhamel's principle.

6. Solve

$$\text{D.E.}\quad u_t - ku_{xx} = \cos(3t) \qquad 0 \le x \le 1\,,\ t \ge 0$$

$$\text{B.C.}\quad u_x(0,t) = -1 \qquad u_x(1,t) = 0$$

$$\text{I.C.}\quad u(x,0) = \cos(\pi x) + \tfrac{1}{2}x^2 - x\,.$$

Hint. Use Problem 5 .

7. Solve

$$\text{D.E.}\quad u_t - u_{xx} = xe^t/\pi + t[2 - 2x/\pi + \sin(x)] \qquad 0 \le x \le \pi\,,\ t \ge 0$$

$$\text{B.C.}\quad u(0,t) = t^2 \qquad u(\pi,t) = e^t$$

$$\text{I.C.}\quad u(x,0) = x/\pi + \sin(2x)\,.$$

8. Solve

$$\text{D.E.}\quad u_t - 4u_{xx} = e^t\sin(x/2) - \sin(t) \qquad 0 \le x \le \pi\,,\ t \ge 0$$

$$\text{B.C.}\quad u(0,t) = \cos(t) \qquad u_x(\pi,t) = 0$$

$$\text{I.C.}\quad u(x,0) = 1\,.$$

9. Solve

$$\text{D.E.}\quad u_t - u_{xx} = x - x^2 + 2t + e^{-4\pi^2 t}\cos(2\pi x) \qquad 0 \le x \le 1\,,\ t \ge 0$$

$$\text{B.C.}\quad u_x(0,t) = t \qquad u_x(1,t) = -t$$

$$\text{I.C.}\quad u(x,0) = 0\,.$$

10. Show that the uniqueness theorem (cf. Theorem 1 of Section 3.2) remains true if the D.E. is replaced by $u_t - ku_{xx} = h(x,t)$. Is it true for other types of B.C. (e.g., $u_x(0,t) = a(t)$) ?

11. Suppose that $h(x,t)$ in the D.E. of (10) is changed to $q(x,t)$. By applying the Maximum/Minimum Principle to (14) (and using (15)), show that the change in the solution at time t will be at most $t \cdot \max|h(x,s) - q(x,s)|$, where the maximum is taken over the rectangle: $0 \le x \le L$, $0 \le s \le t$.

12. Consider

$$\text{D.E.} \quad u_t = ku_{xx} \qquad 0 \le x \le L,\ t \ge 0$$

$$\text{B.C.} \quad u(0,t) = a(t) \quad u(L,t) = b(t) \qquad (*)$$

$$\text{I.C.} \quad u(x,0) = f(x),$$

where $a(t)$ and $b(t)$ are polynomials, say $a(t) = \sum_{i=0}^{n} A_i t^i$ and $b(t) = \sum_{i=0}^{n} B_i t^i$. Here we show that there is a unique particular solution of the D.E. and B.C. which is of the form $u_p(x,t) = \sum_{i=0}^{n} F_i(x)t^i$, where each $F_i(x)$ is a polynomial of degree at most $2(n-i) + 1$. This shows that problem $(*)$ can be converted into a problem for $v(x,t) = u(x,t) - u_p(x,t)$ with homogeneous D.E. and B.C., so that Duhamel's principle need not be used in this case.

(a) By substituting $u_p(x,t) = \sum_{i=0}^{n} F_i(x)t^i$ into $u_t = ku_{xx}$ and equating coefficients, deduce that $F_n''(x) = 0$, $kF_{n-1}''(x) = nF_n(x)$, ..., $kF_{n-i}''(x) = (n-i+1)F_{n-i+1}(x)$, ..., $kF_0''(x) = F_1(x)$.

(b) Show that the B.C. imply that $F_n(x) = (b_n - a_n)x/L + a_n$. More generally, show that $F_i(x)$ is uniquely determined by $F_{i+1}(x)$ and the B.C. . Thus, by induction, $u_p(x,t) = \sum_{i=0}^{n} F_i(x)t^i$ is uniquely determined. (We thank Kenneth Rogers for this problem.)

CHAPTER 4

FOURIER SERIES AND STURM–LIOUVILLE THEORY

In Chapter 3 we discovered that it was easy to obtain solutions of initial/boundary–value problems for the heat equation, provided that the initial temperature can be expressed (within experimental error) by a sum of sine or cosine functions which meet the boundary conditions. The problems in Chapter 3 were specially designed, so that $f(x)$ could be put in this form by means of trigonometric identities. In this chapter, we learn how to find an approximation, of the appropriate form, for any "reasonably nice" function on a finite interval. In Chapters 5 and 6, we find that the theory of Fourier series also applies to problems for the wave equation and Laplace's equation.

In Section 4.1, we define the Fourier series of a function, and we bring out the analogy between this series and the decomposition of an ordinary vector into its components relative to an orthogonal basis. Section 4.2 contains statements and proofs of various convergence results for Fourier series. In view of practical applications, we pay attention to the error introduced by truncating a Fourier series at a finite number of terms. In Section 4.3, different types of Fourier series (e.g., cosine series and sine series) are considered. These series are used to solve problems for the heat equation with various boundary conditions, when the initial conditions are not in the special forms found in Chapter 3. In Section 4.4, we introduce a generalization of the theory of Fourier series, namely the Sturm–Liouville theory. This theory can be used when the heat conducting material in a rod has a possibly variable thermal conductivity, density, specific heat, or source term which is proportional to the temperature. It applies also to wave problems, and more generally, it is used to analyze the solutions of the ODEs with homogeneous boundary conditions, which result when separation of variables is used.

4.1 Orthogonality and the Definition of Fourier Series

We found in Chapter 3 that it is important to represent initial temperature distributions $f(x)$, $0 \le x \le L$, accurately (say within experimental error) by functions of the following forms (depending on the type of B.C.):

$$f(x) \approx \begin{cases} \sum_{n=1}^{N} b_n \sin(n\pi x/L) & (1) \\ \sum_{n=0}^{N} a_n \cos(n\pi x/L) & (2) \\ \sum_{n=0}^{N} c_n \sin[(n+\frac{1}{2})\pi x/L] & (3) \\ \sum_{n=0}^{N} d_n \cos[(n+\frac{1}{2})\pi x/L] . & (4) \end{cases}$$

Form (1) is appropriate when both ends are maintained at 0 ; (2) when both ends are insulated ; (3) and (4) when one end is insulated, while the other end is maintained at 0. We will concentrate on (1) and (2), which are most conveniently treated simultaneously by working with functions on the larger interval $[-L,L]$. Indeed, it is this larger interval which was used in the case of heat conduction problems for the circular wire (cf. Proposition 2 in Section 3.1). In that case, we needed to approximate initial temperatures by linear combinations of both functions $\sin(n\pi x/L)$ and $\cos(n\pi x/L)$. Our first objective is to show that the functions $\sin(n\pi x/L)$ and $\cos(n\pi x/L)$, $n = 1, 2, 3, \ldots$, can be regarded as orthogonal vectors in an infinite–dimensional "space" of functions defined on $[-L,L]$. The sums in (1) and (2) are linear combinations of these "vectors". Of course, in three–dimensional space, any vector is a linear combination of three orthogonal vectors. Analogous results in function space require much more work and preparation. Thus, we will study the problem of how to represent functions $f(x)$ as such linear combinations.

At first we will be concerned with **trigonometric polynomials**, which are finite sums of the form

$$\tfrac{1}{2}a_0 + \sum_{n=1}^{N} a_n \cos(n\pi x/L) + b_n \sin(n\pi x/L) , \qquad x \text{ real} ,$$

where $a_0, a_1, \ldots, a_N$ and $b_1, \ldots, b_N$ are real constants. Subsequently, we will study infinite series of the form

$$\tfrac{1}{2}a_0 + \sum_{n=1}^{\infty} a_n\cos(n\pi x/L) + b_n\sin(n\pi x/L) \ , \qquad (*)$$

where

$$a_n = \frac{1}{L}\int_{-L}^{L} f(x)\cos(n\pi x/L)\,dx \ , \quad n = 0, 1, 2, \dots \ ,$$

and

$$b_n = \frac{1}{L}\int_{-L}^{L} f(x)\sin(n\pi x/L)\,dx \ , \quad n = 1, 2, 3, \dots \ .$$

The series $(*)$, called the **Fourier series of** $f(x)$, is named after the outstanding French mathematical physicist Joseph Fourier (1768–1830). In the early part of the nineteenth century, Fourier worked on the theory of heat conduction and in 1822 he published his *magnum opus*, *La Théorie Analytique de la Chaleur*, in which he made extensive use of the series that now bears his name. Actually, Fourier worked only with trigonometric polynomials, and the consensus among historians seems to be that Fourier contributed nothing whatever to the *mathematical* theory of Fourier series. Indeed, these series were well–known much earlier to Leonhard Euler (1707–1783), Daniel Bernoulli (1700–1782), Joseph Lagrange (1736–1813) and others .

Orthogonality in function space

Let $f(x)$ and $g(x)$ be two real–valued continuous functions defined for $-L \le x \le L$. We define the **inner product** of f and g to be the real number $\langle f,g\rangle$ given by

$$\langle f,g\rangle = \int_{-L}^{L} f(x)g(x)\,dx \quad . \qquad (5)$$

This is similar to the "dot product" of vectors in 3–dimensional space :

$$\mathbf{a}\cdot\mathbf{b} = a_1b_1 + a_2b_2 + a_3b_3 = \sum_{n=1}^{3} a_nb_n \ .$$

Since the integral of the sum of two functions is the sum of the integrals, it follows that

$$\langle f,g+h\rangle = \langle f,g\rangle + \langle f,h\rangle \quad . \qquad (6)$$

Also, we have $\langle cf,g\rangle = c\langle f,g\rangle$, for any real constant c, and the obvious symmetry $\langle f,g\rangle = \langle g,f\rangle$. The **norm** (or length) of the function f is

$$\|f\| = \sqrt{\langle f,f\rangle} = \left[\int_{-L}^{L} [f(x)]^2\,dx \right]^{\frac{1}{2}} . \qquad (7)$$

This is analogous to the length $\|\mathbf{a}\| = [\mathbf{a}\cdot\mathbf{a}]^{\frac{1}{2}}$ of a vector **a** in the xyz–space. Two functions f and g are said to be **orthogonal** on $[-L,L]$ if $<f,g> = 0$.

A family of continuous functions, $f_1, f_2, f_3, \ldots$, is called an **orthogonal family of norm–square** L **on** $[-L,L]$, if for any members f_m and f_n, we have

$$<f_n,f_m> = \begin{cases} 0 & \text{if } m \neq n \\ L & \text{if } m = n\,. \end{cases} \tag{8}$$

In Proposition 1 below, we prove that the family $\sin(n\pi x/L)$, $n = 1, 2, 3, \ldots$, is an orthogonal family of norm–square L, by using the corollary of the following integral formula of Green. Green's formula will often be employed in subsequent computations.

Green's Formula. Let $f(x)$ **and** $g(x)$ **be** C^2 **functions defined on** $[a,b]$. **Then**

$$\int_a^b f''(x)g(x)\,dx - \int_a^b f(x)g''(x)\,dx = [f'(x)g(x) - f(x)g'(x)]\Big|_a^b\,. \tag{9}$$

Proof. From the product rule, we have

$$\frac{d}{dx}[f'(x)g(x) - f(x)g'(x)] = f''(x)g(x) + f'(x)g'(x) - f'(x)g'(x) - f(x)g''(x)\,,$$

and integrating both sides from a to b , we obtain (9). □

Corollary. Let $f(x)$ **and** $g(x)$ **be** C^2 **functions defined on** $[-L,L]$, **and suppose that** $f(-L) = f(L)$, $f'(-L) = f'(L)$, $g(-L) = g(L)$, **and** $g'(-L) = g'(L)$. **Then,**

$$<f'',g> - <f,g''> = 0 \quad \text{or} \quad <f'',g> = <f,g''>\,. \tag{10}$$

Proof. Recall the definition (5), and apply Green's formula (9), noting that the right side of (9) vanishes by the assumptions on f and g at $x = \pm L$. □

Proposition 1. The family of functions $s_n(x) \equiv \sin(n\pi x/L)$ $(n = 1,2,3\ldots)$ **is orthogonal of norm–square** L **on the interval** $[-L,L]$ (cf. (8)).

Proof. We apply (10) with $f(x) = s_n(x)$ and $g(x) = s_m(x)$, noting that $s_n'' = -(n\pi/L)^2 s_n$ and $s_m'' = -(m\pi/L)^2 s_m$:

$$0 = <s_n'',s_m> - <s_n,s_m''> = <-(n\pi/L)^2 s_n,s_m> - <s_n,-(m\pi/L)^2 s_m> = (\pi/L)^2(m^2-n^2)<s_n,s_m>. \quad (11)$$

Thus, if $m \neq n$, we may divide by m^2-n^2 to obtain $<s_n,s_m> = 0$. For $m = n$, we compute

$$\begin{aligned} <s_n,s_n> &= \int_{-L}^{L} \sin^2(n\pi x/L)\, dx = \int_{-L}^{L} \tfrac{1}{2}[1 - \cos(2n\pi x/L)]\, dx \\ &= \left[\tfrac{1}{2} x - (L/4n\pi)\cdot\sin(2n\pi x/L) \right]\Big|_{-L}^{L} = L\ . \quad \square \end{aligned}$$

Note that the same proof may be used to show that the family of functions $c_n(x) = \cos(n\pi x/L)$ $(n = 1, 2, 3,\ldots)$ is orthogonal of norm–square L on the interval $[-L,L]$. Moreover, the same computation as (11), with s_m replaced by c_m (and s_n left as it is), yields $<s_n,c_m> = 0$ for $n \neq m$. Also, we have $<s_n,c_n> = 0$, because

$$\begin{aligned} <c_n,s_n> &= \int_{-L}^{L} \cos(n\pi x/L)\cdot\sin(n\pi x/L)\, dx = \int_{-L}^{L} \tfrac{1}{2}\sin(2n\pi x/L)\, dx \\ &= (-L/4n\pi)\cos(2n\pi x/L)\Big|_{-L}^{L} = 0\ . \end{aligned}$$

Thus, we can shuffle the families $s_1, s_2, \ldots$ and $c_1, c_2, \ldots$ to obtain the larger orthogonal family $s_1, c_1, s_2, c_2, \ldots$ on $[-L,L]$. The foregoing integrals of products of sines and/or cosines could also have been computed without Green's formula, by using the trigonometric identities (27) in Section 3.1. However, the proof using Green's formula easily generalizes to certain multiple integrals, when trigonometric identities are not available.

Example 1. Show that the constant function $c_0(x) = \cos(0\pi x/L) \equiv 1$ is orthogonal to each member of the family $s_1, c_1, s_2, c_2, \ldots$. However, note that c_0 does not have norm–square L. Remedy this defect by multiplying c_0 by an appropriate constant.

Solution. Note that $<c_0,s_n> = \int_{-L}^{L} 1\cdot\sin(n\pi x/L)\, dx = 0$, and similarly $<c_0,c_n> = 0$, $n = 1, 2, 3, \ldots$. Thus, c_0 is orthogonal to each of the functions $s_1, c_1, s_2, c_2 \ldots$ on $[-L,L]$. Also, for any constant b, we have $<bc_0,bc_0> = b^2<c_0,c_0> = b^2\int_{-L}^{L} 1^2\,dx = b^2 2L$. Thus, by choosing $b = \sqrt{1/2}$ so that $b^2 = 1/2$, we have $<\sqrt{1/2}\,c_0,\sqrt{1/2}\,c_0> = L$. Note that $\sqrt{1/2}\,c_0$ is still orthogonal to $s_1, c_1, s_2, c_2, \ldots$ on $[-L,L]$, since $<\sqrt{1/2}\,c_0,s_n> = \sqrt{1/2}<c_0,s_n> = 0$, etc.. □

In summary of the preceding analysis, we have

Proposition 2. The following functions form an orthogonal family of norm square L **on** $[-L,L]$:

$$\sqrt{1/2}\,c_0(x) = \sqrt{1/2}\,,\quad s_n(x) = \sin(n\pi x/L)\,,\quad c_n(x) = \cos(n\pi x/L)\,,\quad n = 1, 2, 3, \ldots\,. \tag{12}$$

Remark. We will eventually prove that there is no C^1 function on $[-L,L]$ (other than the zero function) which is orthogonal to all the members in (12). Hence, the family (12) cannot be enlarged. This suggests that the family (12) might serve as a basis of orthogonal vectors in an infinite–dimensional "space" of functions defined on $[-L,L]$. In the xyz–space, there is a familiar basis of orthogonal vectors of norm–square L, namely $\mathbf{e}_1 = \sqrt{L}\,\mathbf{i}$, $\mathbf{e}_2 = \sqrt{L}\,\mathbf{j}$, $\mathbf{e}_3 = \sqrt{L}\,\mathbf{k}$. Any vector $\mathbf{v}$ in the xyz–space can be written as a linear combination of $\mathbf{e}_1$, $\mathbf{e}_2$ and $\mathbf{e}_3$, say $\mathbf{v} = v_1\mathbf{e}_1 + v_2\mathbf{e}_2 + v_3\mathbf{e}_3$, for some scalars v_1, v_2 and v_3. We can express these scalars in terms of dot products of $\mathbf{v}$ with the basis vectors. For example, to find the formula for v_1, we compute

$$\mathbf{v}\cdot\mathbf{e}_1 = (v_1\mathbf{e}_1 + v_2\mathbf{e}_2 + v_3\mathbf{e}_3)\cdot\mathbf{e}_1 = v_1\mathbf{e}_1\cdot\mathbf{e}_1 = v_1\|\mathbf{e}_1\|^2 = v_1 L\,,\ \text{ so that }\ v_1 = L^{-1}\,\mathbf{v}\cdot\mathbf{e}_1.$$

In general, $v_n = L^{-1}\mathbf{v}\cdot\mathbf{e}_n$ $(n = 1,2,3)$, and we see that v_n is uniquely determined by this formula. The next result and its proof are strictly analogous. □

Theorem 1. Suppose that $f(x)$ **is of the form**

$$f(x) = \tfrac{1}{2}a_0 + \sum_{n=1}^{N} a_n\cos(n\pi x/L) + b_n\sin(n\pi x/L)\,. \tag{13}$$

Then, the coefficients a_n **and** b_n **are uniquely determined by the formulas**

$$a_n = \frac{1}{L}\int_{-L}^{L} f(x)\cos(n\pi x/L)\,dx = L^{-1}<f,c_n>\,,\qquad n = 0, 1, \ldots\,, \tag{14}$$

$$b_n = \frac{1}{L}\int_{-L}^{L} f(x)\sin(n\pi x/L)\,dx = L^{-1}<f,s_n>\,,\qquad n = 1, 2, \ldots\,, \tag{15}$$

where $s_n(x) = \sin(n\pi x/L)$ **and** $c_n(x) = \cos(n\pi x/L)$.

Proof. In terms of s_n and c_n, we can write (13) in the form

$$f = \tfrac{1}{2}a_0 + \sum_{n=1}^{N} a_n c_n + b_n s_n . \tag{13'}$$

Taking the inner product of both sides of (13′) with c_m (m fixed , $1 \leq m \leq N$) we obtain

$$<f,c_m> = <\tfrac{1}{2}a_0,c_m> + \sum_{n=1}^{N} a_n<c_n,c_m> + b_n<s_n,c_m> . \tag{*}$$

By orthogonality (cf. Proposition 2), all of the terms on the right side of (*) vanish, except for $a_m<c_m,c_m>$. In particular, note that $<\tfrac{1}{2}a_0,c_m> = 0$, since c_m is orthogonal to any constant function on [–L,L]. Thus, (*) reduces to $<f,c_m> = a_m<c_m,c_m> = a_mL$, and so $a_m = L^{-1}<f,c_m>$, $m = 1, 2, 3,\ldots$. By using s_m instead of c_m , we arrive at $b_m = L^{-1}<f,s_m>$, $m = 1,2,3,\ldots$. Thus, we have formula (15) and formula (14), except when $n = 0$. To handle $n = 0$, we take the inner product of (13′) with the constant function $c_0(x) \equiv 1$, obtaining $<f,c_0> = <\tfrac{1}{2}a_0,c_0> = a_0L$ or $a_0 = L^{-1}<f,c_0> = \frac{1}{L}\int_{-L}^{L} f(x)\,dx$. □

Fourier series — the definition and examples

Not every function f(x) can be written in the form (13). The right side of (13) is smooth (i.e., C^∞) , but many functions have graphs with jumps or corners. We will encounter functions f(x) for which the integrals (14) and (15) are not zero for infinitely many values of n. In such cases, f(x) cannot be represented as a finite sum as in (13). Also, even if N approaches ∞ , the sum (13) might not converge to f(x), unless some additional assumptions are made. In the case where $N = \infty$, the sum is the Fourier series of f(x) on [–L,L] , which is defined as follows.

Definition (Fourier Series). Let f(x) be a function defined on [–L,L] , such that the integrals

$$a_n \equiv \frac{1}{L}\int_{-L}^{L} f(x)\cdot\cos(n\pi x/L)\,dx , \quad n = 0, 1, 2, \ldots, \tag{16}$$

and

$$b_n \equiv \frac{1}{L}\int_{-L}^{L} f(x)\cdot\sin(n\pi x/L)\,dx , \quad n = 1, 2, 3, \ldots, \tag{17}$$

exist and are finite. Then the **Fourier series** of f on [–L,L] is the expression

$$\text{FS } f(x) = \tfrac{1}{2}a_0 + \sum_{n=1}^{\infty} a_n\cos(n\pi x/L) + b_n\sin(n\pi x/L) . \tag{18}$$

The coefficients a_0, a_n, b_n ($n = 1, 2, 3,\ldots$) are known as the **Fourier coefficients** of f.

Remark 1. Note that no claim is made that the sum in (18) actually converges, when values for x are inserted. Indeed, there are functions $f(x)$, such that all the integrals in (16) and (17) exist, but FS $f(x)$ diverges for every value of x. Moreover, there are functions $f(x)$ which are continuous on $[-L,L]$, such that FS $f(x)$ diverges for all rational numbers x. For a survey of such results, see [Coppel]. Of course, one can always change the value of $f(x)$ at any finite number of points to produce many different new functions without changing the integrals in (16) and (17), and hence without changing the Fourier series. However, clearly FS $f(x)$ can converge, at every point in $[-L,L]$, to at most one (if any) function. Thus, we see that a function is not uniquely determined by its Fourier series, without some further assumptions (e.g., continuity) about the function. □

Remark 2. The notation FS $f(x)$ for the Fourier series of $f(x)$ is unique to this book. Some books fail to make clear the distinction between a function and its Fourier series. Surely, one hopes that, for "sufficiently nice" functions, FS $f(x) = f(x)$, but in view of Remark 1, this is not the case for all functions. The most common notation used is

$$f(x) \sim \tfrac{1}{2}a_0 + \sum_{n=1}^{\infty} a_n\cos(n\pi x/L) + b_n\sin(n\pi x/L),$$

where the symbol " $\sim$ " means "has the Fourier series". We do not insist that this notation be avoided. However, the symbol " $\sim$ " has other connotations, i.e., it most often stands for "approximately (or asymptotically) equals". While our notation is not standard, it is at least unambiguous, and it is convenient, since it frees one from rewriting the infinite sum in (18). □

Example 2. Find the Fourier series of the function $f(x) = x$ for $-L \le x \le L$.

Solution. We compute the Fourier coefficients a_n first for $n \ge 1$,

$$a_n = \frac{1}{L}\int_{-L}^{L} x\cdot\cos(n\pi x/L)\,dx = \frac{x}{n\pi}\sin(n\pi x/L)\Big|_{-L}^{L} - \frac{1}{n\pi}\int_{-L}^{L}\sin(n\pi x/L)\,dx$$

$$= 0 + \frac{L}{(n\pi)^2}\cos(n\pi x/L)\Big|_{-L}^{L} = 0\ ,\quad n = 1, 2, 3, \ldots .$$

For $n = 0$, we get $a_0 = \frac{1}{L}\int_{-L}^{L} x\,dx = L^{-1}x^2/2\Big|_{-L}^{L} = 0$. Thus, all of the a_n $(n = 0, 1, 2, \ldots)$ vanish. We could have arrived at this result by noting that the integrands $x\cdot\cos(n\pi x/L)$ are transformed to their negatives when x is replaced by $-x$; i.e., they are "odd" functions. For such functions the integral from $-L$ to 0 cancels with the integral from 0 to L. As for the b_n,

$$b_n = \frac{1}{L}\int_{-L}^{L} x\cdot\sin(n\pi x/L)\,dx = \frac{-x}{n\pi}\cos(n\pi x/L)\Big|_{-L}^{L} + \frac{1}{n\pi}\int_{-L}^{L}\cos(n\pi x/L)\,dx$$

$$= \frac{-2L}{n\pi}\cos(n\pi) + \frac{L}{(n\pi)^2}\sin(n\pi x/L)\Big|_{-L}^{L} = \frac{2L}{n\pi}(-1)^{n+1}\ ,\quad n = 1, 2, 3, \ldots ,$$

since $\cos(n\pi) = (-1)^n$ (check this for $n = 0, 1, 2 \ldots$). Thus, we obtain

$$\text{FS } f(x) = \sum_{n=1}^{\infty} (-1)^{n+1} \frac{2L}{n\pi} \sin(n\pi x/L) = \frac{2L}{\pi} \sum_{n=1}^{\infty} (-1)^{n+1} \frac{1}{n} \sin(n\pi x/L)$$

$$= \frac{2L}{\pi} \left[\sin(\pi x/L) - \tfrac{1}{2}\sin(2\pi x/L) + \tfrac{1}{3}\sin(3\pi x/L) - \ldots\right] . \tag{19}$$

Note that FS $f(0) = 0 = f(0)$. However, FS $f(\pm L) = 0 \neq \pm L = f(\pm L)$, and so FS $f(x)$ is not $f(x)$ at $x = \pm L$. We will see from Theorem 3 in Section 4.2, that FS $f(x) = f(x) = x$, for $-L < x < L$. Thus, for $x = L/2$, we have $\frac{2L}{\pi}[1 - \frac{1}{3} + \frac{1}{5} - \ldots] = L/2$, or equivalently, $1 - \frac{1}{3} + \frac{1}{5} - \ldots = \pi/4$, a sure (but tedious) way of computing π. □

It is interesting to observe the way that the Fourier series (19) approaches the function $f(x) = x$ in $(-L,L)$, as more terms of the series are added. Let $S_N(x)$ denote the sum of the first N terms of FS $f(x)$. In Figure 1 below, we have plotted $S_1(x)$, $S_2(x)$ and $S_3(x)$ in relation to $f(x) = x$. The approximation becomes better with increasing N (see, for example, $S_{10}(x)$ in Figure 2). However, the approximation is poor near $x = \pm L$, since $S_N(\pm L) = 0$. Also, note that outside the interval $[-L,L]$, $S_N(x)$ bears little resemblance to $f(x) = x$, since $S_N(x)$ is periodic (i.e., $S_N(x+2L) = S_N(x)$ for all x), but $f(x) = x$ is not periodic. The results of Section 4.2 imply that the graph of FS $f(x)$ [or $S_\infty(x)$] is given by Figure 3 below.

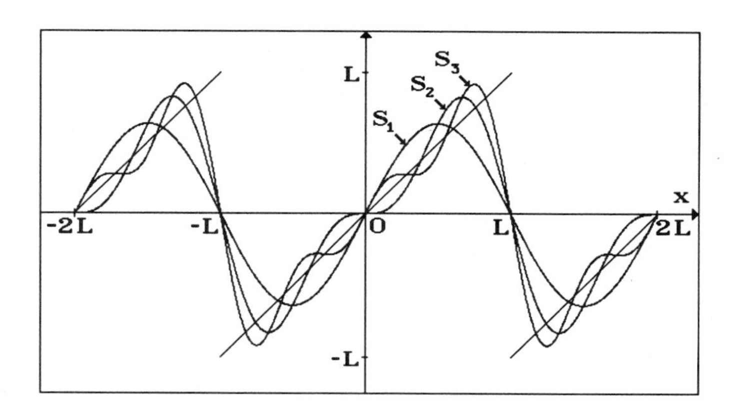

Figure 1

Graphs of the first three partial sums of FS $f(x)$ for $f(x) = x$, $-L \leq x \leq L$.

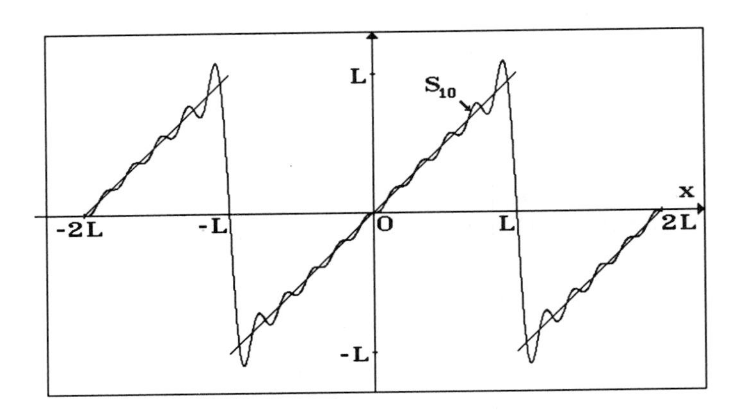

Figure 2

The graph of the partial sum $S_{10}(x)$.

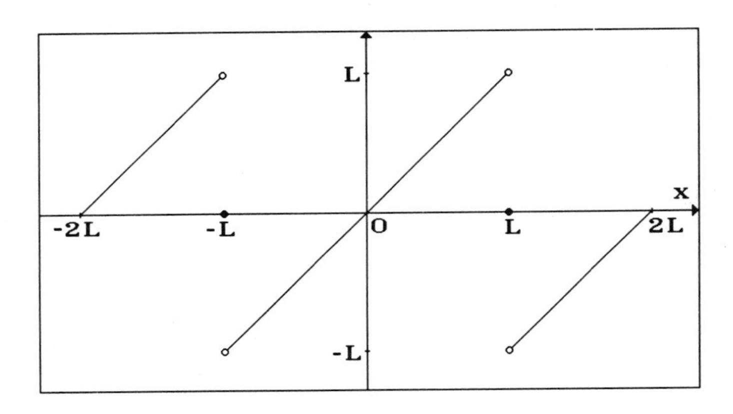

Figure 3

The graph of $S_\infty(x)$ or FS f(x) .

If f(x) is defined on [0,2L], then the Fourier series of f(x) is defined by (18), provided the interval of integration [–L,L] in (16) and (17) is replaced by [0,2L]. There is nothing to prevent one from defining the Fourier series of a function defined on any closed interval of positive length.

Example 3. Compute the Fourier series of $f(x) = x$ defined on the interval $[0,2L]$.

Solution. The integrals for the Fourier coefficients are now from 0 to 2L. Thus, $a_0 = \frac{1}{L}\int_0^{2L} x\,dx = 2L$, while for $n = 1, 2, 3, \ldots$, we have

$$a_n = \frac{1}{L}\int_0^{2L} x\cdot\cos(n\pi x/L)\,dx = \frac{x}{n\pi}\sin(n\pi x/L)\Big|_0^{2L} - \frac{1}{n\pi}\int_0^{2L}\sin(n\pi x/L)\,dx\ ,$$

whence $a_n = 0$ for $n \geq 1$. For b_n, we compute

$$b_n = \frac{1}{L}\int_0^{2L} x\cdot\sin(n\pi x/L)\,dx = -\frac{x}{n\pi}\cos(n\pi x/L)\Big|_0^{2L} + \frac{1}{n\pi}\int_0^{2L}\cos(n\pi x/L)\,dx = \frac{-2L}{n\pi}.$$

Thus,

$$\text{FS } f(x) = L - \frac{2L}{\pi}\sum_{n=1}^{\infty}\frac{1}{n}\sin(n\pi x/L)\ . \quad \square \tag{20}$$

Remark. Examples 2 and 3 show that the Fourier series of a function depends not only on the length of the interval on which the function is defined, but also on the position of the interval. It can be shown that the graphs of the partial sums of (20) are obtained from the graphs of the partial sums of (19) [e.g., from Figures 1,2 and 3], by shifting those graphs up by L units and over to the right by L units. □

Example 4. Let $f(x) = \begin{cases} x & 0 \leq x \leq \pi \\ 0 & -\pi \leq x < 0 \end{cases}$. The graph of $f(x)$ is shown in Figure 4.

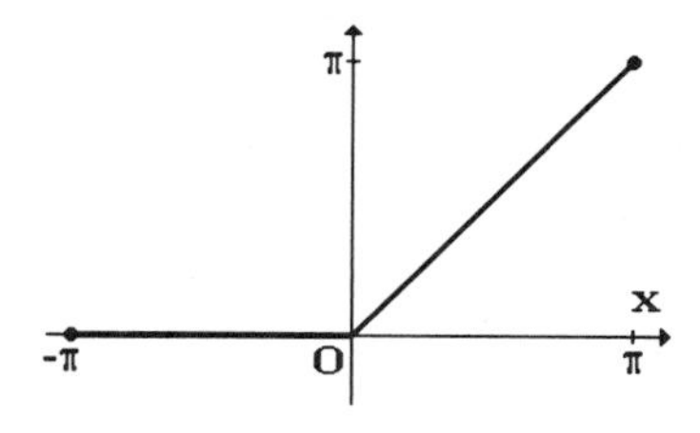

Figure 4

Compute FS $f(x)$, noting that $L = \pi$.

Solution. Since $f(x) = 0$ for $-\pi \leq x \leq 0$, we have (for $n > 0$)

$$a_n = \frac{1}{\pi}\int_{-\pi}^{\pi} f(x)\cos(nx)\,dx \ = \frac{1}{\pi}\int_0^{\pi} x\cdot\cos(nx)\,dx$$

$$= \frac{1}{n\pi} x \cdot \sin(nx)\Big|_0^\pi + \frac{1}{\pi n^2}\cos(nx)\Big|_0^\pi = \frac{1}{\pi n^2}[(-1)^n - 1]$$

and

$$b_n = \frac{1}{\pi}\int_0^\pi x \cdot \sin(nx)\, dx = \frac{-1}{n\pi} x \cdot \cos(nx)\Big|_0^\pi + \frac{1}{\pi n^2}\sin(nx)\Big|_0^\pi = (-1)^{n+1} \cdot \frac{1}{n}.$$

We must compute a_0 separately : $a_0 = \frac{1}{\pi}\int_0^\pi x\, dx = \pi/2$. Thus, $\frac{1}{2}a_0 = \pi/4$, and

$$\text{FS } f(x) = \frac{\pi}{4} + \sum_{n=1}^{\infty}\left[\frac{1}{\pi n^2}[(-1)^n - 1]\cos(nx) + (-1)^{n+1}\frac{1}{n}\sin(nx)\right]$$

$$= \frac{\pi}{4} + \left[-\frac{2}{\pi}\cos(x) + \sin(x) + 0 \cdot \cos(2x) - \frac{1}{2}\sin(2x) - \frac{2}{9\pi}\cos(3x) + \frac{1}{3}\sin(3x) + \dots\right].$$

Results from Section 4.2 imply that FS $f(x) = f(x)$ for $-\pi < x < \pi$. Thus, at $x = 0$, we obtain

$$0 = \frac{\pi}{4} - \frac{2}{\pi}\left[1 + \frac{1}{3^2} + \frac{1}{5^2} + \dots\right] \quad \text{or} \quad 1^{-2} + 3^{-2} + 5^{-2} + \dots = \pi^2/8 . \quad \square$$

Remark. In Example 4, the sine terms of FS $f(x)$ all vanish when we set $x = \pm\pi$, and we obtain FS $f(\pm\pi) = \pi/4 + (2/\pi)[1^{-2} + 3^{-2} + 5^{-2} + \dots] = \pi/2$, but $f(-\pi) = 0$ and $f(\pi) = \pi$. Thus, as in Example 2, we find that FS $f(x)$ is not always equal to $f(x)$ for all x in $[-L,L]$. However, in these examples, the failure of equality occurs only at the endpoints $\pm L$. Observe that $\cos(n\pi x/L)$ and $\sin(n\pi x/L)$ are unchanged when x is changed from $-L$ to L. Thus, for any function f, FS $f(-L)$ = FS $f(L)$, and so unless $f(-L) = f(L)$, we cannot possibly have FS $f(x) = f(x)$ at both endpoints. Indeed, under fairly general circumstances, we show in Section 4.2 that $\text{FS}(\pm L) = \frac{1}{2}[f(L) + f(-L)]$. This is the case in Examples 2 and 4. $\square$

The convergence of Fourier series

Under various assumptions (e.g., we have just seen that $f(-L) = f(L)$ is a necessary condition) on a function $f(x)$, one can prove that FS $f(x)$ does converge to $f(x)$ for all x in $[-L,L]$. For example, the following result is proved in Section 4.2 .

Theorem 2 (same as Theorem 2 of Section 4.2). **Let** $f(x)$ **be a** C^2 **function on the interval** $[-L,L]$, **such that** $f(-L) = f(L)$ **and** $f'(-L) = f'(L)$. **Let** a_n **and** b_n **be the Fourier coefficients of** $f(x)$ (cf. (16) **and** (17)), **and let** $M = \max_{-L \le x \le L} |f''(x)|$. **Then for any** $N \ge 1$,

$$\left| f(x) - \left[\tfrac{1}{2}a_0 + \sum_{n=1}^{N} a_n\cos(n\pi x/L) + b_n\sin(n\pi x/L)\right]\right| \le \frac{4L^2M}{\pi^2 N}, \tag{21}$$

for all x **in** $[-L,L]$.

Remarks. (1) As N gets larger, the right side of (21) approaches 0 , which implies that the sum in brackets on the left side of (21) converges to f(x) as N approaches ∞ , i.e.,

$$f(x) = \text{FS } f(x) = \tfrac{1}{2}a_0 + \sum_{n=1}^{\infty} a_n\cos(n\pi x/L) + b_n\sin(n\pi x/L). \qquad (22)$$

(2) Actually (21) tells us more than (22). Indeed, (21) tells us how many terms of FS f(x) will suffice, in order to approximate f(x) to within a certain error. However, by itself, (22) does not contain this information. Note also that the right side of (21) is independent of x, which means that we can guarantee that the error can be made small simultaneously for *all* x in [–L,L] , by taking N sufficiently large.

(3) The conditions $f(-L) = f(L)$ and $f'(-L) = f'(L)$ remind one of the boundary conditions for heat conduction in a circular wire (cf. Example 3 in Section 3.1). These conditions insure that if the interval [–L,L] is bent into a circle by joining $x = -L$ to $x = L$, then the graph of f(x) above this circle is continuous and has a well defined tangent line (or derivative) at the juncture where –L and L are identified. The reason why such conditions arise is seen roughly as follows. The terms of FS f(x) are all periodic of period 2L, in the sense that

$$\cos[n\pi(x+2L)/L] = \cos(n\pi x/L + 2n\pi) = \cos(n\pi x/L) \quad \text{and} \quad \sin[n\pi(x+2L)/L] = \dots = \sin(n\pi x/L)\ .$$

Thus, FS f(x+2L) = FS f(x) , and so FS f(x) is a periodic function of x of period 2L , assuming that FS f(x) converges. In particular, FS f(–L) = FS f(–L + 2L) = FS f(L), as we have observed before. Thus, again, in order that FS f(x) = f(x) at $x = \pm L$, it is necessary that $f(-L) = f(L)$. Moreover, if FS f(x) is differentiable, then $f'(-L) = f'(L)$ is also a necessary condition. Observe that even though f(x) is specified only on the interval [–L,L], FS f(x) is defined for all x, provided it converges. Sometimes the formula which defines f(x) (e.g., f(x) = x) still makes sense for values of x outside of [–L,L], but there is no chance that FS f(x) = f(x) outside of [–L,L] , unless f(x) is a periodic function of period 2L. Note that periodic functions of period 2L correspond to functions defined on a circle of circumference 2L. Thus, Fourier series are most naturally used for representing periodic functions or functions defined on circles. As we will see, roughly speaking, the smoother the function f(x) $(-L \le x \le L)$ appears to be, when it is transferred to a circle (or periodically extended), the more rapidly the Fourier series converges to f(x). □

Example 5. Take $L = 1$ and $f(x) = x^3 - x$, $-1 \le x \le 1$. Apply Theorem 2 to get an estimate on the number of terms of FS f(x) needed to approximate f(x) within an error of .01 .

Solution. First, we check that f(x) satisfies the hypotheses of Theorem 2. Note that $f(-1) = f(1) = 0$ and $f'(-1) = f'(1) = 2$, since $f'(x) = 3x^2-1$; also, $f''(x) = 6x$ is continuous (i.e., f is C^2). Now $M = \max_{-1 \le x \le 1} |f''(x)| = 6$. According to Theorem 2, the truncation of FS f(x) at the N–th term will be within .01 of f(x) , provided

$$4(1)^2 \cdot 6/(\pi^2 N) < .01 \quad \text{or} \quad N > 2400/\pi^2 \approx 243.17\ .$$

Thus taking N = 244 , we will certainly achieve the desired accuracy. □

Error estimates by means of integral approximations – more examples

The estimate for N in Example 5 is grossly conservative, since we will now directly show that 5 terms suffice. The Fourier coefficients a_n vanish since $f(x) = x^3 - x$ is odd, (i.e., $f(-x) = -f(x)$), while in Example 7 below, we compute $b_n = (-1)^n 12/(n\pi)^3$. Thus, for $-1 \le x \le 1$,

$$\text{FS } f(x) = \frac{12}{\pi^3}\sum_{n=1}^{\infty}(-1)^n \frac{1}{n^3}\sin(n\pi x) = x^3 - x, \tag{23}$$

where the last equality is due to (22). Using (23) and the fact $|(-1)^n \sin(n\pi x)| \le 1$, we have

$$\left| x^3 - x - \frac{12}{\pi^3}\sum_{n=1}^{N}(-1)^n \frac{1}{n^3}\sin(n\pi x)\right|$$
$$= \left|\frac{12}{\pi^3}\sum_{n=N+1}^{\infty}(-1)^n \frac{1}{n^3}\sin(n\pi x)\right| \le \frac{12}{\pi^3}\sum_{n=N+1}^{\infty}\frac{1}{n^3}. \tag{24}$$

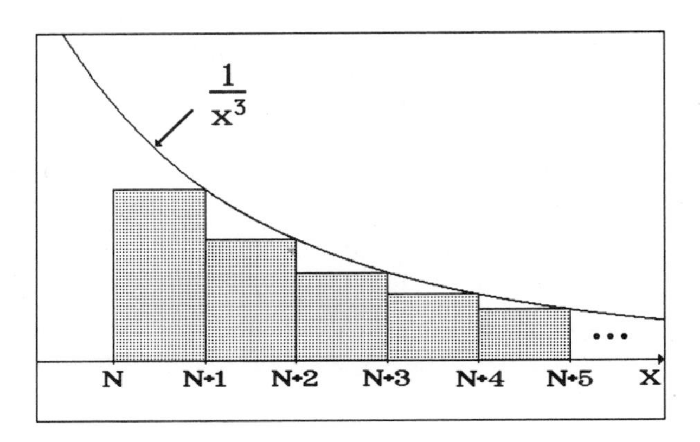

Figure 5

The final sum in (24) is the same as the sum of the areas of the shaded blocks in Figure 5, which is less than the area $\int_N^{\infty}\frac{1}{x^3}\,dx$ under the curve. Hence, (24) yields

$$\left| x^3 - x - \frac{12}{\pi^3}\sum_{n=1}^{N}(-1)^n \frac{1}{n^3}\sin(n\pi x)\right| \le \frac{12}{\pi^3}\int_N^{\infty}\frac{1}{x^3}\,dx = \frac{6}{\pi^3 N^2}.$$

Setting $6/(\pi^3 N^2) \le .01$, we have $N \ge [600/\pi^2]^{\frac{1}{2}} \approx 4.4$. Thus, $N = 5$ suffices. Actually, explicit numerical tabulation suggests that $N = 4$ suffices, while $N = 3$ does not suffice. For the function $f(x) = x^3 - x$, we see that the value of N provided by Theorem 2 is much larger than it needs to be. Indeed, in most cases, the explicit estimates based on integral comparisons, as above, using the computed Fourier coefficients, are likely to yield better results than (21) in Theorem 2. However, (21) is still valuable, since it yields an estimate, even when the Fourier coefficients cannot be computed (say, because the integration is too difficult). Also, there are infinitely many C^2 functions that satisfy the hypotheses of Theorem 2, and it would be impossible to treat them one at a time. Theorem 2 handles all of them at once. □

Example 6. Compute the Fourier series of $f(x) = x^3$, $-L \le x \le L$.

Solution. We have $a_n = L^{-1}\langle c_n,f\rangle = \frac{1}{L}\int_{-L}^{L} x^3 \cos(n\pi x/L)\,dx = 0$, because the integrand is odd (i.e., it is changed to its negative under replacing x by $-x$). To compute $b_n = L^{-1}\langle s_n,f\rangle$, we could integrate by parts three times, a rather unpleasant ordeal. Instead, we will use the following version of Green's formula (cf. (9))

$$\langle g'',f\rangle = [g'(x)f(x) - g(x)f'(x)]\Big|_{-L}^{L} + \langle g,f''\rangle \quad . \tag{25}$$

Now, $b_n = L^{-1}\langle s_n,f\rangle$, and in order to apply (25), we write s_n in terms of s_n'' (i.e., $s_n = -(L/n\pi)^2 s_n''$). Thus, $b_n = -L^{-1}(L/n\pi)^2\langle s_n'',x^3\rangle$. Using (25) with $g(x) = s_n(x)$ and $f(x) = x^3$,

$$\langle s_n'',x^3\rangle = [s_n'(x)\cdot x^3 - s_n(x)\cdot 3x^2]\Big|_{-L}^{L} + \langle s_n,6x\rangle$$

$$= \frac{n\pi}{L} c_n(x)\cdot x^3\Big|_{-L}^{L} + \langle s_n,6x\rangle = (-1)^n\, 2L^2 n\pi - (L/n\pi)^2 \langle s_n'',6x\rangle \,.$$

Applying (25) again to $\langle s_n'',6x\rangle$, we get

$$\langle s_n'',6x\rangle = [s_n'(x)\cdot 6x - s_n(x)\cdot 6]\Big|_{-L}^{L} + \langle s_n,0\rangle = (n\pi/L)c_n(x)\cdot 6x\Big|_{-L}^{L} = (-1)^n 12n\pi \,.$$

Thus, $b_n = -L^{-1}(L/n\pi)^2\langle s_n'',x^3\rangle = \frac{-L}{n^2\pi^2}\langle s_n'',x^3\rangle = \frac{-L}{n^2\pi^2}\left[(-1)^n 2L^2 n\pi - (L/n\pi)^2(-1)^n 12n\pi\right]$

$$= (-1)^n\left[\frac{-2L^3}{n\pi} + \frac{12L^3}{n^3\pi^3}\right] = 2\frac{L^3}{\pi^3}(-1)^n\left[\frac{6}{n^3} - \frac{\pi^2}{n}\right],$$

and

$$\mathrm{FS}\, x^3 = 2\frac{L^3}{\pi^3}\sum_{n=1}^{\infty}(-1)^n\left[\frac{6}{n^3} - \frac{\pi^2}{n}\right]\sin(n\pi x/L) \,. \quad □$$

Example 7. Show that for $f(x) = x^3 - x$, defined on $[-1,1]$, we have

$$\text{FS } f(x) = \frac{12}{\pi^3}\sum_{n=1}^{\infty}(-1)^n \frac{1}{n^3}\sin(n\pi x) .$$

Solution. We computed FS x in Example 2, and FS x^3 in Example 6. Taking $L = 1$ in these examples, we have $\text{FS } f(x) = \text{FS }(x^3 - x) = \text{FS } x^3 - \text{FS } x$, where the last equation follows from $\langle x^3 - x, s_n\rangle = \langle x^3, s_n\rangle - \langle x, s_n\rangle$. In other words, the Fourier series of $x^3 - x$ can be computed by subtracting FS x from FS x^3:

$$\begin{aligned}\text{FS }(x^3 - x) &= \frac{2}{\pi^3}\sum_{n=1}^{\infty}(-1)^n\left[\frac{6}{n^3} - \frac{\pi^2}{n}\right]\cdot\sin(n\pi x/L) - \frac{2}{\pi}\sum_{n=1}^{\infty}(-1)^{n+1}\frac{1}{n}\sin(n\pi x)\\ &= \frac{12}{\pi^3}\sum_{n=1}^{\infty}(-1)^n \frac{1}{n^3}\sin(n\pi x) .\end{aligned}$$

Alternatively, we can compute FS $(x^3 - x)$ directly, using method of Example 6 based on Green's formula (25). This is easier then computing FS x^3 and FS x separately ! Of course, we have that $a_n = L^{-1}\langle c_n, x^3 - x\rangle = 0$, since $x^3 - x$ is odd. Note also that $b_n = L^{-1}\langle s_n, x^3 - x\rangle = -(1/n\pi)^2\langle s_n'', x^3 - x\rangle$. $(L = 1)$, and (25) yields

$$\begin{aligned}\langle s_n'', x^3 - x\rangle &= [s_n'(x)(x^3 - x) - s_n(x)(3x^2 - 1)]\Big|_{-1}^{1} + \langle s_n, 6x\rangle = 0 - (1/n\pi)^2\langle s_n'', 6x\rangle\\ &= -(1/n\pi)^2\left[[s_n'(x)\cdot 6x - s_n(x)\cdot 6]\Big|_{-1}^{1} - \langle s_n, 0\rangle\right] = \frac{-1}{n\pi}c_n(x)\cdot 6x\Big|_{-1}^{1} = -\frac{12}{n\pi}(-1)^n .\end{aligned}$$

Thus, $b_n = -(1/n\pi)^2\left[-\frac{12}{n\pi}(-1)^n\right] = \frac{12}{\pi^3}(-1)^n\frac{1}{n^3}$, as required. □

Summary 4.1

1. **Orthogonality** : The inner product of two functions f and g defined on $[-L,L]$ is defined by $\langle f,g\rangle = \int_{-L}^{L} f(x)g(x)\,dx$, and the norm of f is $\|f\| = \sqrt{\langle f,f\rangle}$. A family of functions f_1, f_2, f_3, ... is an **orthogonal family of norm–square** L on $[-L,L]$, if for any members f_m and f_n,

$$\langle f_n,f_m\rangle = \begin{cases} 0 & \text{if } m \neq n \\ L & \text{if } m = n \ . \end{cases}$$

If $c_n(x) \equiv \cos(n\pi x/L)$ and $s_n(x) \equiv \sin(n\pi x/L)$, then the functions $\sqrt{1/2}\cdot c_0$, c_1, s_1, c_2, s_2, ... form an orthogonal family of norm–square L on $[-L,L]$ (cf. Proposition 2).

2. **Green's formula** : A key result which is useful in establishing orthogonality is Green's formula,

$$\int_a^b f''(x)g(x)\,dx - \int_a^b f(x)g''(x)\,dx = [f'(x)g(x) - f(x)g'(x)]\Big|_a^b \ .$$

In particular, $\langle g'',f\rangle = [g'(x)f(x) - g(x)f'(x)]\Big|_{-L}^{L} + \langle g,f''\rangle$, which is the most useful form for the computation of Fourier coefficients (cf. Examples 6 and 7).

3. **Fourier Series** : Let f(x) be a function defined on $[-L,L]$, such that the integrals

$$\begin{aligned} a_n &\equiv \tfrac{1}{L}\int_{-L}^{L} f(x)\cdot\cos(n\pi x/L)\,dx = L^{-1}\langle f,c_n\rangle\ , \quad n = 0,1,2,\ldots \\ b_n &\equiv \tfrac{1}{L}\int_{-L}^{L} f(x)\cdot\sin(n\pi x/L)\,dx = L^{-1}\langle f,s_n\rangle\ , \quad n = 1,2,3,\ldots \end{aligned} \tag{S1}$$

exist and are finite. Then the **Fourier series** of f on $[-L,L]$ is the expression

$$\text{FS } f(x) = \tfrac{1}{2}a_0 + \sum_{n=1}^{\infty} a_n\cos(n\pi x/L) + b_n\sin(n\pi x/L) \ . \tag{S2}$$

The coefficients a_0, a_n, b_n $(n = 1, 2, 3,\ldots)$ are known as the **Fourier coefficients** of f .

4. Representing functions by Fourier series : If $f(x)$ is expressible as a **trigonometric polynomial**, say (for some *finite* integer $N \geq 1$)

$$f(x) = \tfrac{1}{2}a_0 + \sum_{n=1}^{N} a_n\cos(n\pi x/L) + b_n\sin(n\pi x/L) , \tag{S3}$$

then Theorem 1 states that the a_n and b_n *must* be the Fourier coefficients (S1) of $f(x)$. Thus, such functions are equal to their Fourier series (i.e., FS $f(x) = f(x)$, for trigonometric polynomials f). Examples 2 and 4 show that there are functions such that FS $f(x) \neq f(x)$ for some x in $[-L,L]$. Also, there are functions $f(x)$ such that FS $f(x)$ converges, but FS $f(x) \neq f(x)$ for *all* x in $[-L,L]$. However, there are results, established in Section 4.2, that guarantee that FS $f(x) = f(x)$, under certain assumptions. One of these results is

Theorem 2 (same as Theorem 2 of Section 4.2). **Let** $f(x)$ **be a** C^2 **function on the interval** $[-L,L]$, **such that** $f(-L) = f(L)$ **and** $f'(-L) = f'(L)$. **Let** a_n **and** b_n **be the Fourier coefficients of** $f(x)$ (cf. (S1)), **and let** $M = \max_{-L \leq x \leq L} |f''(x)|$. **Then for any** $N \geq 1$,

$$\left| f(x) - \left[\tfrac{1}{2}a_0 + \sum_{n=1}^{N} a_n\cos(n\pi x/L)+b_n\sin(n\pi x/L)\right] \right| \leq \frac{4L^2M}{\pi^2 N} , \tag{S4}$$

for all x **in** $[-L,L]$.

If the Fourier coefficients of a function $f(x)$ can be computed and FS $f(x) = f(x)$, then estimates for the number of terms of FS $f(x)$, which suffice to approximate $f(x)$ to within a given error, can be obtained by applying an integral comparison (cf. Figure 5). Moreover, these estimates are typically much sharper than (S4).

Exercises 4.1

1. Recall that through the use of trigonometric identities (cf. (27) in Section 3.1), we obtained the result $\cos^3(x) = \frac{3}{4}\cos(x) + \frac{1}{4}\cos(3x)$. Deduce that

$$\int_{-\pi}^{\pi} \cos^3(x)\cos(x)\,dx = 3\pi/4 \quad \text{and} \quad \int_{-\pi}^{\pi} \cos^3(x)\cos(3x)\,dx = \pi/4 .$$

Hint. Use Proposition 2 or Theorem 1 .

2. Find the Fourier series of the following functions without computing any integrals.

(a) $f(x) = \cos^2(\pi x)\cdot\sin^2(\pi x)$, $-1 \le x \le 1$
(b) $f(x) = \sin(x)[\sin(x) + \cos(x)]^2$ $-\pi \le x \le \pi$.

Hint. Use trigonometric identities and Theorem 1 which says FS $f(x) = f(x)$ for these functions.

3. Suppose $f(x) = \sum_{n=1}^{N} a_n\cos(n\pi x/L)$ and $g(x) = \sum_{n=1}^{N} \alpha_n\cos(n\pi x/L)$ for constants a_n and α_n, $n = 1, 2,\ldots, N$. Show that $\int_{-L}^{L} f(x)g(x)\,dx = \langle f,g\rangle = L\sum_{n=1}^{N} a_n\alpha_n$.

4. Compute the Fourier series of the following functions defined on $[-L,L]$.

(a) $f(x) = \begin{cases} 1 & 0 \le x \le L \\ 0 & -L \le x < 0 \end{cases}$ (b) $f(x) = |x| = \begin{cases} x & 0 \le x \le L \\ -x & -L \le x \le 0 \end{cases}$.

5. Let $f(x) = x^2$ for $-L \le x \le L$.

(a) Compute the Fourier series of f(x) via integration by parts.

(b) Recompute the series using Green's formula, as was done in Example 6.

6. Assuming that FS $f(x) = f(x) = x^2$ (in Problem 5) for $-L \le x \le L$, obtain the results

$$1 - 1/2^2 + 1/3^2 - 1/4^2 + \ldots = \pi^2/12$$
$$1 + 1/2^2 + 1/3^2 + 1/4^2 + \ldots = \pi^2/6 . \qquad (*)$$

From these results, obtain

$$1 + 1/3^2 + 1/5^2 + 1/7^2 + \dots = \pi^2/8$$
$$1/2^2 + 1/4^2 + 1/6^2 + 1/8^2 + \dots = \pi^2/24 \,. \qquad (**)$$

Hint. For (*), consider $x = 0$ and $x = L$. For (**), add and subtract the results in (*).

7. Compute the Fourier series for

$$\text{(a)}\ f(x) = e^x \quad -\pi \le x \le \pi \qquad \text{(b)}\ f(x) = e^x \quad 0 \le x \le 2\pi \,.$$

Hint. For (a), note that $(1+n^2)<f,c_n> = <f'',c_n> - <f,c_n''> = [f'(x)c_n(x) - f(x)c_n'(x)]\,|_{-\pi}^{\pi} = \dots$, and similarly for $<f,s_n>$, where $s_n(x) = \sin(nx)$ and $c_n(x) = \cos(nx)$. For (b) use Green's formula as in (a), but on $[0,2\pi]$.

8. Compute FS f(x) for the function $f(x) = \begin{cases} 0 & -\pi \le x < 0 \\ \sin(x) & 0 \le x \le \pi \end{cases}$.

9. Let $f(x) = (x^2-1)^2$ for $-1 \le x \le 1$.

(a) Use Green's formula (as in Example 7) to compute FS f(x) with relative ease.

(b) Verify that f(x) satisfies the hypotheses of Theorem 2. How many terms of FS f(x) suffice to approximate f(x) to within an error of .001, according to Theorem 2 ?

(c) Use the method of the Remark after Example 5, to dramatically improve the estimate in (b).

(d) Assuming that FS f(x) = f(x) (which follows from Theorem 2), show that

$$1 + 1/2^4 + 1/3^4 + 1/4^4 + \dots = \pi^4/90 \,.$$

4.2 Convergence Theorems for Fourier Series

Recall that in order to solve initial/boundary–value problems for the heat equation, one must approximate initial temperature functions by linear combinations of sine and/or cosine functions of an appropriate form. If the Fourier series of a function converges to the function, then we may achieve the desired approximation by considering a sufficient number of terms of the Fourier series. We stated a convergence result (Theorem 2 of Section 4.1) that is valid for C^2 functions $f(x)$ $(-L \leq x \leq L)$ such that $f(-L) = f(L)$ and $f'(-L) = f'(L)$. In this section, we prove this result, as well as other convergence theorems that apply to a more general class of functions. Roughly speaking, this broader class consists of functions which are C^1 on $[-L,L]$, except for a finite number of points where their graphs have jumps or corners (i.e., piecewise C^1). Although almost everyone who plans to use Fourier series should clearly understand the statements of these convergence theorems, their proofs (as well as proofs of preliminary results) are not necessarily of great utility in applications. Nevertheless, there are enough details supplied in these proofs so that they can be understood by the interested reader.

Periodic functions

Before proving the convergence theorems, we will need to establish some preliminary results. We begin with some facts concerning periodic functions and periodic extensions of functions.

Definition. A function $g(x)$, defined for all real x, is said to be **periodic** (of **period** 2L), if $g(x+2L) = g(x)$ for all x.

This means that if the graph of $g(x)$ is moved to the right or left by a distance of 2L (or any multiple of 2L), then the graph falls exactly on the top of itself. The functions $\sin(n\pi x/L)$ and $\cos(n\pi x/L)$ are examples of periodic functions of period 2L. Indeed, $\sin(n\pi(x+2L)/L) = \sin(n\pi x/L + 2\pi) = \sin(n\pi x/L)$, and similarly for $\cos(n\pi x/L)$. Since finite linear combinations of periodic functions of period 2L are also periodic, we know that functions of the form

$$S_N(x) = \tfrac{1}{2}a_0 + \sum_{n=1}^{N} a_n\cos(n\pi x/L) + b_n\sin(n\pi x/L)$$

(e.g., partial sums of Fourier series) are also periodic of period 2L. A function need not be continuous to be periodic. For example, the function, whose graph is shown in Figure 1,

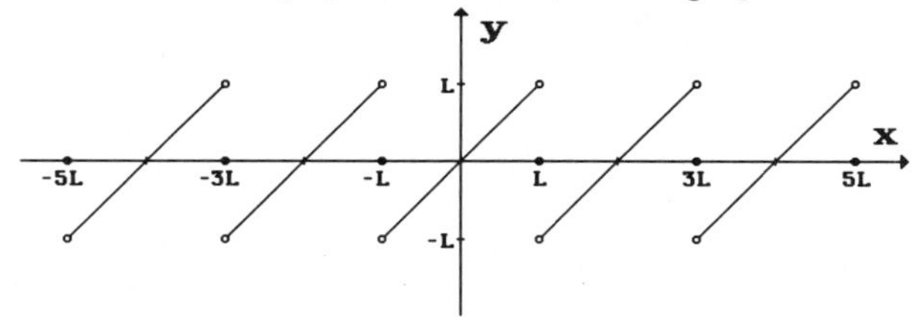

Figure 1

is periodic of period 2L. (The open circles mean that the points they "enclose" are omitted ; the solid dots on the x–axis are included in the graph.) This is actually the graph of FS f(x), where $f(x) = x \quad (-L \le x \le L)$, as we stated in Section 4.1 and will eventually prove (cf. Theorem 3). The partial sums $S_N(x)$ (with N = 1, 2, 3 and 10) of FS f(x) were graphed in Figures 1 and 2 of Section 4.1 . Note that these partial sums are smooth (i.e., C^∞) , since they are *finite* sums of smooth functions. The infinite sum, FS f(x), is not even continuous, although it is still periodic.

Definition. Let f(x) be a function defined on the closed interval [–L,L], such that f(–L) = f(L) . Then the **periodic extension** of f(x) is the unique periodic function $\tilde{f}(x)$ of period 2L, such that $\tilde{f}(x) = f(x)$, for $-L \le x \le L$.

Remark. In order for the periodic extension to exist, we must have f(–L) = f(L) . Thus, note that the function f(x) = x , defined for $-L \le x \le L$, does not have a periodic extension. To remedy this situation, one can redefine f(x) at the endpoints $x = \pm L$ to be the average of the limits (assuming that they exist)

$$f(L^-) \equiv \lim_{x \uparrow L} f(x) \qquad \text{and} \qquad f(-L^+) \equiv \lim_{x \downarrow -L} f(x) ,$$

obtained by taking the one–sided limits of f(x), as x approaches the endpoints from within [–L,L] . The reason for doing this is that, if a "piecewise C^1" function f(x) on [–L,L] is adjusted in this way at $x = \pm L$ and also averaged at jumps in (–L,L) , then FS f(x) converges to the periodic extension of this adjusted function. See Theorem 3, for a precise statement of the result. □

Proposition 1. If g(x) is a periodic function of period 2L, then the integral (if it exists) of g(x) over an interval of length 2L is the same as the integral of g(x) over any other interval of length 2L. In other words, for any real number c,

$$\int_{-L+c}^{L+c} g(x)\, dx = \int_{-L}^{L} g(x)\, dx . \tag{1}$$

Proof. Since the interval [L,L+c] is obtained from [–L,–L+c] by shifting to the right by 2L and g(x) is periodic of period 2L, we have

$$\int_{-L}^{-L+c} g(x)\, dx = \int_{L}^{L+c} g(x)\, dx . \tag{2}$$

Now,
$$\int_{-L+c}^{L+c} g(x)\, dx = \int_{-L+c}^{-L} g(x)\, dx + \int_{-L}^{L} g(x)\, dx + \int_{L}^{L+c} g(x)\, dx ,$$

but the first and third integrals on the right side cancel by (2). □

Example 1. Compute $\int_0^{2\pi} \sin^5(x)\cdot\cos^{100}(x)\,dx$.

Solution. The integrand is periodic of period 2π. Thus, according to Proposition 1, the integral is

$$\int_{-\pi}^{\pi} \sin^5(x)\cdot\cos^{100}(x)\,dx .$$

The integrand is odd (i.e., it is changed to its negative, by replacing x by $-x$). Thus, the integral from $-\pi$ to 0 cancels the integral from 0 to π, yielding the value 0. □

Bessel's inequality and the Riemann–Lebesgue Lemma

The square of the length of a vector in space equals the sum of the squares of its components. For any function $f(x)$, defined on $[-L,L]$ with finite norm square $\|f\|^2 \equiv \int_{-L}^{L} f(x)^2\,dx$, we expect a similar result, but for now, we will prove only an inequality. We will also use it to show that the Fourier coefficients a_n and b_n of a function, with finite norm, approach zero as $n \to \infty$. This fact will be needed in the proofs of the convergence results for Fourier series.

Bessel's Inequality. Let $f(x)$ be defined on $[-L,L]$, and suppose that $\int_{-L}^{L} [f(x)]^2\,dx$ exists and is finite. Assume that the Fourier coefficients

$$a_n = \frac{1}{L}\int_{-L}^{L} f(x)\cos(n\pi x/L)\,dx \quad\textbf{and}\quad b_n = \frac{1}{L}\int_{-L}^{L} f(x)\sin(n\pi x/L)\,dx \tag{3}$$

exist so that $\text{FS } f(x) = \frac{1}{2}a_0 + \sum_{n=1}^{\infty} a_n\cos(n\pi x/L) + b_n\sin(n\pi x/L)$ **is defined formally (i.e., it might not converge). Then, we have Bessel's inequality,**

$$\tfrac{1}{2}a_0^2 + \sum_{n=1}^{\infty}(a_n^2 + b_n^2) \le \frac{1}{L}\int_{-L}^{L} [f(x)]^2\,dx . \tag{4}$$

Proof. Let $S_N(x) \equiv \frac{1}{2}a_0 + \sum_{n=1}^{N} a_n\cos(n\pi x/L) + b_n\sin(n\pi x/L)$ denote the N–th partial sum of FS $f(x)$. Then

$$0 \le \int_{-L}^{L} [f(x) - S_N(x)]^2\,dx = \int_{-L}^{L} [f(x)]^2\,dx - 2\int_{-L}^{L} f(x)S_N(x)\,dx + \int_{-L}^{L} [S_N(x)]^2\,dx . \tag{5}$$

By computing the last two integrals on the right in (5), we show that (5) yields (4).

$$\int_{-L}^{L} f(x)S_N(x)\,dx = \int_{-L}^{L} f(x)\left[\tfrac{1}{2}a_0 + \sum_{n=1}^{N} a_n\cos(n\pi x/L) + b_n\sin(n\pi x/L)\right] dx$$

$$= \tfrac{1}{2}a_0\cdot\int_{-L}^{L} f(x)\,dx + \sum_{n=1}^{N}\left[a_n\int_{-L}^{L} f(x)\cos(n\pi x/L)\,dx + b_n\int_{-L}^{L} f(x)\sin(n\pi x/L)\,dx\right]$$

$$= \tfrac{1}{2}a_0\cdot La_0 + \sum_{n=1}^{N}(a_n\cdot L\cdot a_n + b_n\cdot L\cdot b_n) = L\cdot\left[\tfrac{1}{2}a_0^2 + \sum_{n=1}^{N}(a_n^2 + b_n^2)\right].$$

Next, we replace $f(x)$ in this computation by $S_N(x)$. Then the first N Fourier coefficients of the trigonometric polynomial $S_N(x)$ are the same as those of $f(x)$ (cf. Theorem 1 of Section 4.1). Thus, we obtain the same result as above, namely

$$\int_{-L}^{L} [S_N(x)]^2\,dx = L\cdot\left[\tfrac{1}{2}a_0^2 + \sum_{n=1}^{N}(a_n^2 + b_n^2)\right].$$

Inequality (5) then becomes

$$0 \le \int_{-L}^{L} [f(x)]^2\,dx - L\cdot\left[\tfrac{1}{2}a_0^2 + \sum_{n=1}^{N}(a_n^2 + b_n^2)\right] \quad\text{or}\quad \tfrac{1}{2}a_0^2 + \sum_{n=1}^{N}(a_n^2 + b_n^2) \le \frac{1}{L}\int_{-L}^{L} [f(x)]^2\,dx\,.$$

Since N can be chosen arbitrarily large, we then have (4) (Why ?). □

Remark. We can rewrite Bessel's inequality (4) in the form

$$\frac{1}{L^2}\left[\tfrac{1}{2}\langle f,c_0\rangle^2 + \sum_{n=1}^{N}\langle f,c_n\rangle^2 + \langle f,s_n\rangle^2\right] \le \frac{1}{L}\langle f,f\rangle\,, \qquad (6)$$

where $c_n(x) = \cos(n\pi x/L)$, $s_n(x) = \sin(n\pi x/L)$, and we have used the inner product notation of Section 4.1 . Multiplying by L and rewriting the first term, (6) becomes

$$\frac{1}{L}\left[\langle f,\sqrt{1/2}\cdot c_0\rangle^2 + \sum_{n=1}^{N}\langle f,c_n\rangle^2 + \langle f,s_n\rangle^2\right] \le \langle f,f\rangle = \|f\|^2\,. \qquad (7)$$

In other words, we have a relation between the square of the length of f and the sum of the squares of the components of f relative to the orthogonal family $\sqrt{1/2}\cdot c_0$, c_n, s_n $(n = 1, 2, 3,\ldots)$ of Proposition 2 in Section 4.1. In fact, it is possible (but more difficult) to prove that "$\le$" in (4) and (7) may be replaced by "$=$". Then we obtain the so–called **Parseval's equality** , which is essentially an infinite dimensional version of the Pythagorean Theorem in "function space" . □

Example 2. Use Bessel's inequality for $f(x) = x$ $(-L \le x \le L)$ to prove that

$$\sum_{n=1}^{\infty}\frac{1}{n^2} \le \frac{\pi^2}{6}\,. \qquad (8)$$

Solution. We computed FS f(x) in Example 2 of Section 4.1, and found that

$$\text{FS } f(x) = \frac{2L}{\pi}\sum_{n=1}^{\infty}(-1)^{n+1}\frac{1}{n}\sin(n\pi x/L) \ ,$$

(i.e., $a_n = 0$, $n \geq 0$ and $b_n = (-1)^{n+1}\, 2L/(n\pi)$, $n \geq 1$). Bessel's inequality then yields

$$\sum_{n=1}^{\infty}(2L/n\pi)^2 \leq \frac{1}{L}\int_{-L}^{L} x^2\, dx = 2L^2/3 \ ,$$

and multiplying by $\pi^2/(4L^2)$, we obtain (8). Parseval's equality (which we have not proved yet) implies that actually $\sum_{n=1}^{\infty} 1/n^2 = \pi^2/6$. □

Bessel's inequality (4) tells us that the sum $\frac{1}{2}a_0^2 + \sum_{n=1}^{\infty}(a_n^2 + b_n^2)$ converges to some finite value, if $\|f\|$ is finite. This implies that both a_n and b_n must approach zero as $n \to \infty$, for otherwise infinitely many terms of the series would be greater then some positive number and the sum would be infinite. Thus, the following result, which will be crucial in proving convergence theorems for Fourier series, is an immediate consequence of Bessel's inequality :

The Riemann–Lebesgue Lemma. Let f(x) **be defined on** [–L,L]. **Suppose that** $\int_{-L}^{L}[f(x)]^2\, dx < \infty$ **and the Fourier coefficients of** f(x) **exist. Then, these coefficients approach zero as** $n \to \infty$, **i.e.**

$$\lim_{n\to\infty} \frac{1}{L}\int_{-L}^{L} f(x)\cos(n\pi x/L)\, dx = 0$$

and (9)

$$\lim_{n\to\infty} \frac{1}{L}\int_{-L}^{L} f(x)\sin(n\pi x/L)\, dx = 0 \ .$$

Remark. Although the following considerations are no substitute for the proof of the Riemann–Lebesgue Lemma, based on Bessel's inequality, they may help the reader to gain a better understanding of why one would expect the limits (9) to hold. If we multiply f(x) by $\sin(n\pi x/L)$ [or $\cos(n\pi x/L)$], then we obtain a new function whose graph oscillates between f(x) and –f(x), as shown in Figure 2 (where n = 6).

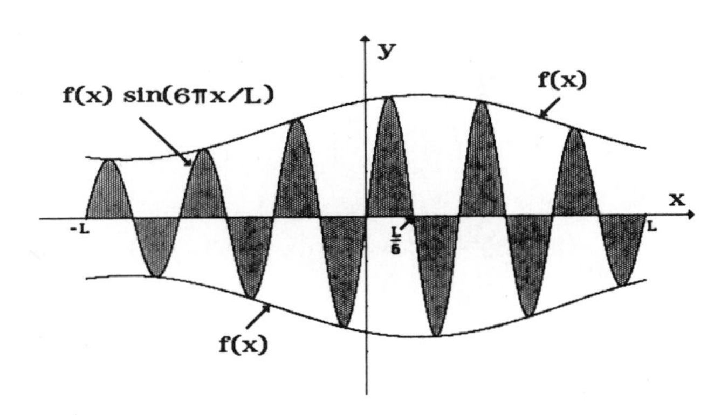

Figure 2

As $n \to \infty$, one expects the shaded areas above the x–axis to very nearly cancel the shaded area below the x–axis. In other words, we intuitively expect the limits in (9) will be zero. □

It is often useful to estimate the rate at which the Fourier coefficients of a well–behaved function tend to zero. The following proposition provides a sample result in this direction. It is proved directly using Green's formula.

Proposition 2. Let $f(x)$ **be a** C^2 **function on** $[-L,L]$ **such that** $f(-L) = f(L)$ **and** $f'(-L) = f'(L)$. **Let** M **be the maximum of** $|f''(x)|$, **for** $-L \le x \le L$. **Then**

$$|a_n| = \left| \frac{1}{L}\int_{-L}^{L} f(x)\cos(n\pi x/L)\,dx \right| \le \frac{2L^2M}{\pi^2 n^2}$$

and (10)

$$|b_n| = \left| \frac{1}{L}\int_{-L}^{L} f(x)\sin(n\pi x/L)\,dx \right| \le \frac{2L^2M}{\pi^2 n^2}, \quad \textbf{for all } n \ge 1 .$$

Proof. We use Green's formula (cf. Example 6 in Section 4.1) to get, for $n \ge 1$,

$$a_n = L^{-1}\langle f,c_n\rangle = -L^{-1}(L/n\pi)^2\langle f,c_n''\rangle = \frac{-L}{n^2\pi^2}\left[[f'(x)c_n(x) - f(x)c_n'(x)]\Big|_{-L}^{L} + \langle f'',c_n\rangle \right]$$

$$= \frac{-L}{n^2\pi^2}\int_{-L}^{L} f''(x)\cos(n\pi x/L)\,dx , \tag{11}$$

where the endpoint evaluations cancel by the assumptions $f(-L) = f(L)$ and $f'(-L) = f'(L)$. Since $|f''(x)\cos(n\pi x/L)| \le M$, we obtain

$$|a_n| \leq \frac{L}{n^2\pi^2}\int_{-L}^{L} M\,dx = \frac{2L^2M}{\pi^2 n^2},$$

as desired. For $|b_n|$, merely substitute s_n for c_n in the above argument. □

Remarks. Note that the assumptions on $f(x)$ tell us that if the interval $[-L,L]$ were bent into a circle with $-L$ and L identified, then $f(x)$ would define a C^2 function on this circle, except possibly at the point $\pm L$, where the function would still be at least C^1, by the assumptions $f(-L) = f(L)$ and $f'(-L) = f'(L)$. Generally speaking, the smoother $f(x)$ is on the circle, the more rapidly the Fourier coefficients tend to zero as $n \to \infty$. For example, if $f(x)$ is C^4 and $f(-L) = f(L)$, $f'(-L) = f'(L)$, $f''(-L) = f''(L)$ and $f^{(3)}(-L) = f^{(3)}(L)$, then Green's formula could be applied twice to yield

$$a_n = -(L/n^2\pi^2)\langle f'',c_n\rangle = -(L^3/n^4\pi^4)\langle f^{(4)},c_n\rangle,$$

and we would have $|a_n| \leq (2L^4/n^4\pi^4) \cdot \max_{-L\leq x\leq L}|f^{(4)}(x)|$, with the same result for $|b_n|$. More generally, if $f(x)$ is C^{2k} with f and its first $2k-1$ derivatives matching at $x = \pm L$, then the Fourier coefficients of f will decrease at least as fast as $(\text{const.})\cdot n^{-2k}$. Note also that the estimates (10) can be strengthened by an application of the Riemann–Lebesgue Lemma, as follows. We write the first estimate in (10) as $n^2|a_n| \leq 2L^2M/\pi^2$, which says that $n^2|a_n|$ is bounded above by a constant. In fact, $n^2 a_n \to 0$ as $n \to \infty$, because

$$n^2 a_n = \frac{-L}{\pi^2}\int_{-L}^{L} f''(x)\cos(n\pi x/L)\,dx \to 0,$$

by (11) and (9) applied to $f''(x)$. (Similarly, $n^2 b_n \to 0$.) Indeed, Bessel's inequality applied to $f''(x)$, yields the still stronger result that $\sum_{n=1}^{\infty} n^4(a_n^2 + b_n^2) < \infty$. □

Some technical preliminaries for the convergence proof

We will need the following formula for the proof of convergence theorems.

Proposition 3. For any real θ such that $\sin(\theta/2) \neq 0$,

$$\tfrac{1}{2} + \cos(\theta) + \cos(2\theta) + \ldots + \cos(n\theta) = \frac{\sin([n+\frac{1}{2}]\theta)}{2\sin(\theta/2)}. \tag{12}$$

Proof. Multiplying the left side of (12) by $2\sin(\theta/2)$, we obtain

$$\sin(\theta/2) + 2\sin(\theta/2)\cos(\theta) + 2\sin(\theta/2)\cos(2\theta) + \ldots + 2\sin(\theta/2)\cos(n\theta)\ . \tag{13}$$

We want to prove that this is $\sin([n+\frac{1}{2}]\theta)$. Using the identity $2\sin(\alpha)\cos(\beta) = \sin(\beta+\alpha) - \sin(\beta-\alpha)$, we have $2\sin(\theta/2)\cos(k\theta) = \sin([k+\frac{1}{2}]\theta) - \sin([k-\frac{1}{2}]\theta)$. Applying this to each term of (13), except the first , we obtain

$$\sin(\theta/2) + [\sin(3\theta/2) - \sin(\theta/2)] + [\sin(5\theta/2) - \sin(3\theta/2)] + \ldots$$
$$+ [\sin([n+\tfrac{1}{2}]\theta) - \sin([n-\tfrac{1}{2}]\theta) = \sin([n+\tfrac{1}{2}]\theta)\ ,$$

by noting that all terms, except for $\sin([n+\frac{1}{2}]\theta)$, cancel. □

Remark. Although we have proved (12), it is not clear how one might have arrived at formula (12) in the first place. A derivation can be based on the observation that the left side of (12) is the real part of the complex geometric series $\frac{1}{2} + \sum_{k=1}^{n} e^{ik\theta}$ which can be summed and simplified, before extracting its real part that turns out to be the right side of (12). Perhaps it is simpler to note that we produce a "collapsing" or "telescoping" sum when we multiply the left side of (12) by $\sin(\theta/2)$. □

Definition (the **n–th Dirichlet kernel**). For any integer $n \geq 0$, we define

$$D_n(x) = \tfrac{1}{2} + \cos(\pi x/L) + \cos(2\pi x/L) + \ldots + \cos(n\pi x/L). \tag{14}$$

The function $D_n(x)$ is known as the **n–th Dirichlet kernel.**

As a consequence of Proposition 3 and the fact that $D_n(2kL) = n + \frac{1}{2}$ by (14), we have

$$D_n(x) = \begin{cases} \dfrac{\sin([n+\frac{1}{2}]\,\pi x/L)}{2\sin(\,\pi x/2L)}\,, & \sin(\pi x/2L) \neq 0 \\[2ex] n + \frac{1}{2}\,, & \sin(\pi x/2L) = 0\ . \end{cases} \tag{14'}$$

From (14) we see that $D_n(x)$ is periodic of period $2L$, since it is a sum of periodic functions of period $2L$. Thus, if we graph $D_n(x)$ for $-L \leq x \leq L$, then we know the graph for all x. For $n = 3$ and $L = 3.5$, the graph of $D_n(x)$ $(-L \leq x \leq L)$ is shown in Figure 3A. In general, $D_n(0) = n + \frac{1}{2}$, and as n increases, the peak at $x = 0$ becomes more pronounced than the smaller flanking wavy portions, as Figure 3B illustrates in the case $n = 10$.

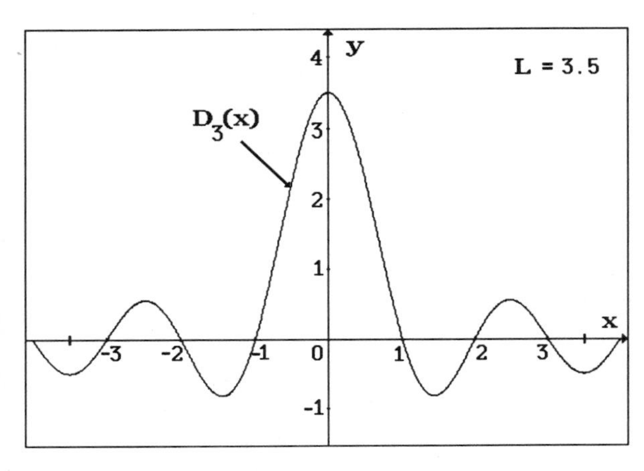

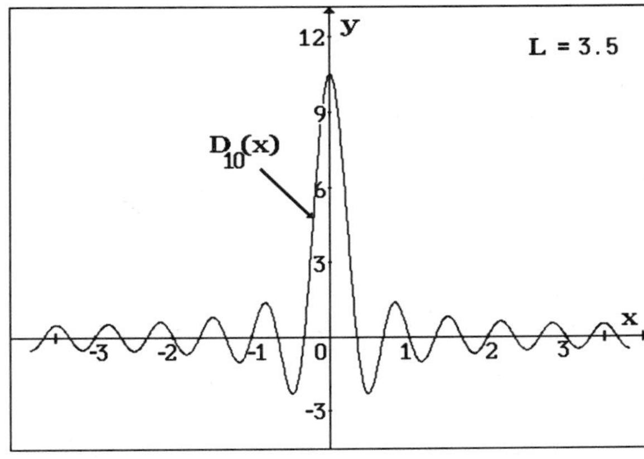

Figure 3A

Figure 3B

If we multiply a reasonably nice function $h(x)$ by $D_n(x)$, then for large n, the graph of $D_n(x)h(x)$ is relatively pronounced near $x = 0$, compared to the original graph of $h(x)$. Thus, if we integrate $D_n(x)h(x)$ from $-L$ to L (as in (15) below), then we expect that the limit of the integral, as $n \to \infty$, will only depend on the values of $h(x)$ near $x = 0$. The following result makes this rigorous. We will see that this result is the primary reason why the Fourier series of a "nice" function converges to the function. Some readers may appreciate the fact that equation (15) essentially says that $L^{-1}D_n(x)$ approaches the Dirac delta function $\delta(x)$ as $n \to \infty$ (cf. Section 7.5 for a discussion of $\delta(x)$).

Proposition 4. Let $h(x)$ be a C^1 function, for $-L \le x \le L$. Then

$$\lim_{n\to\infty} \frac{1}{L}\int_{-L}^{L} D_n(x)h(x)\,dx = h(0)\,, \tag{15}$$

where $D_n(x)$ is the n–th Dirichlet kernel (14).

Proof. We need to prove that the difference

$$\frac{1}{L}\int_{-L}^{L} D_n(x)h(x)\,dx - h(0) \tag{16}$$

approaches 0, as $n \to \infty$. We will use the Riemann–Lebesgue lemma to do this, but first we need to combine the terms of (16) into a single integral. Note that

$$\frac{1}{L}\int_{-L}^{L} D_n(x)\,dx = \frac{1}{L}\int_{-L}^{L} [\,\tfrac{1}{2} + \cos(\pi x/L) + \ldots + \cos(n\pi x/L)\,]\,dx = 1\,,$$

since all of the cosines integrate to zero. Multiplying by the constant $h(0)$, we have

$$\frac{1}{L}\int_{-L}^{L} h(0)D_n(x)\,dx = h(0)\,. \tag{17}$$

Using (17), we can rewrite the difference (16) as follows :

$$\begin{aligned}\frac{1}{L}\int_{-L}^{L} D_n(x)h(x)\,dx - h(0) &= \frac{1}{L}\int_{-L}^{L} D_n(x)h(x)\,dx - \frac{1}{L}\int_{-L}^{L} D_n(x)h(0)\,dx \\ &= \frac{1}{L}\int_{-L}^{L} D_n(x)[h(x)-h(0)]\,dx\,.\end{aligned} \tag{18}$$

In order to apply the Riemann–Lebesgue lemma to this last integral, we write $D_n(x)$ in terms of $\cos(n\pi x/L)$ and $\sin(n\pi x/L)$:

$$\begin{aligned}D_n(x) &= \frac{\sin([n+\frac{1}{2}]\,\pi x/L)}{2\sin(\pi x/2L)} = \frac{\sin(\pi x/2L)\cos(n\pi x/L) + \cos(\pi x/2L)\sin(n\pi x/L)}{2\sin(\pi x/2L)} \\ &= \tfrac{1}{2}\cos(n\pi x/L) + \tfrac{1}{2}\cot(\pi x/2L)\sin(n\pi x/L)\,.\end{aligned}$$

Thus, the final integral in (18) can be written as a sum of integrals

$$\frac{1}{L}\int_{-L}^{L} \tfrac{1}{2}[h(x)-h(0)]\cos(n\pi x/L)\,dx + \frac{1}{L}\int_{-L}^{L}\Big\{\tfrac{1}{2}\cot(\pi x/2L)\,[h(x)-h(0)]\Big\}\sin(n\pi x/L)\,dx\,. \tag{19}$$

Now, $\frac{1}{2}[h(x)-h(0)]$ is continuous (indeed, C^1) in $[-L,L]$, and hence it satisfies the hypotheses of the Riemann–Lebesgue lemma. Thus, the first integral in (19) tends to zero as $n \to \infty$. The second integral in (19) is somewhat of a problem, because $\cot(\pi x/2L)$ is infinite at $x = 0$ and consequently the function in braces must be analyzed more closely near $x = 0$. We apply L'Hospital's rule to compute the limit of this function as $x \to 0$:

$$\lim_{x\to 0} \frac{h(x)-h(0)}{2\tan(\pi x/2L)} = \lim_{x\to 0} \frac{h'(x)}{\sec^2(\pi x/2L)\cdot \pi/L} = \frac{L}{\pi}h'(0)\,,$$

where we have used the fact that $h(x)$ is C^1. Thus, the function in braces in (19) is well–behaved near $x = 0$, and the Riemann–Lebesgue lemma also applies to the second integral in (19). Hence, (16) (which is the same as (19)) tends to 0 as $n \to \infty$. □

The first convergence theorem

We are now ready to prove our first convergence theorem. As will be clear from the proof, the crux of the matter is really Proposition 4, which has just been proved. In other words, the convergence of Fourier series is primarily due to the fact that when a C^1 function is integrated against the n–th Dirichlet kernel, the result approaches the value of the function at 0.

Theorem 1 (Pointwise Convergence of Fourier Series). Let $f(x)$ be C^1 on $[-L,L]$, and assume that $f(-L) = f(L)$ and $f'(-L) = f'(L)$ [so that $f(x)$ may be regarded as a C^1 function on the circle of length 2L]. Then FS $f(x) = f(x)$ for all x in $[-L,L]$. In other words, if the N–th partial sum of FS $f(x)$ is denoted by

$$S_N(x) = \tfrac{1}{2}a_0 + \sum_{n=1}^{N}[a_n\cos(n\pi x/L) + b_n\sin(n\pi x/L)] , \tag{20}$$

where

$$a_n \equiv \frac{1}{L}\int_{-L}^{L} f(y)\cos(n\pi y/L)\,dy \quad \textbf{and} \quad b_n \equiv \frac{1}{L}\int_{-L}^{L} f(y)\sin(n\pi y/L)\,dy, \tag{21}$$

then (for any fixed x in $[-L,L]$)

$$\text{FS } f(x) \equiv \tfrac{1}{2}a_0 + \sum_{n=1}^{\infty}[a_n\cos(n\pi x/L) + b_n\sin(n\pi x/L)] \equiv \lim_{N\to\infty} S_N(x) = f(x) . \tag{22}$$

Proof. We demonstrate that the limit (22) holds by virtue of (15) in Proposition 4. We write $S_N(x)$ in terms of an integral involving D_N. Using the definition (20) of $S_N(x)$ and keeping x fixed, so that $\cos(n\pi x/L)$ and $\sin(n\pi x/L)$ are constants, we compute :

$$S_N(x) = \frac{1}{2}\left\{\frac{1}{L}\int_{-L}^{L} f(y)\,dy\right\} + \sum_{n=1}^{N}\left[\left\{\frac{1}{L}\int_{-L}^{L} f(y)\cos(n\pi y/L)\,dy\right\}\cos(n\pi x/L)\right.$$
$$\left. + \left\{\frac{1}{L}\int_{-L}^{L} f(y)\sin(n\pi y/L)\,dy\right\}\sin(n\pi x/L)\right]$$

$$= \frac{1}{L}\int_{-L}^{L} \frac{1}{2}f(y)\,dy + \sum_{n=1}^{N}\frac{1}{L}\int_{-L}^{L}\left\{\cos(n\pi x/L)\cos(n\pi y/L) + \sin(n\pi x/L)\sin(n\pi y/L)\right\}f(y)\,dy$$

$$= \frac{1}{L}\int_{-L}^{L} \frac{1}{2}f(y)\,dy + \sum_{n=1}^{N}\frac{1}{L}\int_{-L}^{L}\cos(n\pi(y-x)/L)f(y)\,dy = \frac{1}{L}\int_{-L}^{L}\left[\frac{1}{2} + \sum_{n=1}^{N}\cos(n\pi(y-x)/L)\right]f(y)\,dy$$

$$= \frac{1}{L}\int_{-L}^{L} D_N(y-x)f(y)\,dy .$$

Thus, $S_N(x)$ is compactly expressed as an integral involving the N–th Dirichlet kernel D_N :

$$S_N(x) = \frac{1}{L}\int_{-L}^{L} D_N(y-x)f(y)\,dy\,.$$

Hence, to establish (22) it remains to prove that

$$\lim_{N\to\infty} \frac{1}{L}\int_{-L}^{L} D_N(y-x)f(y)\,dy = f(x)\,. \tag{23}$$

This will be proved using Proposition 4 , but we need to rewrite the integral in (23) so that Proposition 4 may be applied directly. Note that $D_N(y-x)$ and $\tilde{f}(y)$ (the periodic extension of $f(y)$) are both periodic functions of y of period $2L$. Using the fact that $f(y) = \tilde{f}(y)$ for y in $[-L,L]$ and also Proposition 1 with c being the fixed value x, we obtain

$$\int_{-L}^{L} D_N(y-x)f(y)\,dy = \int_{-L}^{L} D_N(y-x)\tilde{f}(y)\,dy = \int_{-L+x}^{L+x} D_N(y-x)\tilde{f}(y)\,dy\,. \tag{24}$$

In the last integral, we change the variable of integration from y to $z = y - x$ to get

$$\int_{-L+x}^{L+x} D_N(y-x)\tilde{f}(y)\,dy = \int_{-L}^{L} D_N(z)\tilde{f}(z+x)\,dz\,. \tag{25}$$

Now, let $h(z) \equiv \tilde{f}(z+x)$. Using (24) and (25), we obtain

$$\lim_{N\to\infty} \frac{1}{L}\int_{-L}^{L} D_N(y-x)f(y)\,dy \;=\; \lim_{N\to\infty} \frac{1}{L}\int_{-L}^{L} D_N(z)h(z)\,dz. \tag{26}$$

To apply Proposition 4 to this final limit, we need to know that $h(z)$ is a C^1 function of z. Since f is C^1 and $f(-L) = f(L)$ and $f'(-L) = f'(L)$, it follows easily that $\tilde{f}$ is C^1. Then, $h(z) = \tilde{f}(z+x)$ is C^1, as required. Thus, Proposition 4 says that the final limit in (26) is $h(0) = \tilde{f}(0+x) = \tilde{f}(x) = f(x)$, since x is in $[-L,L]$. Hence, (23) and (22) are proved. □

Uniform convergence

There are several directions in which Theorem 1 could stand some improvement. Although it says that $\mathrm{FS}\, f(x) = f(x)$ for appropriate functions f, there is no indication of how many terms are needed in order that $S_N(x)$ be within a certain error ϵ of $f(x)$. Indeed, there are functions such that $\mathrm{FS}\, f(x) = f(x)$, and yet there is no N which will guarantee that $|S_N(x) - f(x)| \le \epsilon$ *simultaneously* for all x in $[-L,L]$ (cf. Example 6, and replace f by $\tilde{f}$). In other words, N might be arbitrarily large, depending on the value of x. In order to clarify this, we state the following.

Definitions. We say that the sequence of functions $S_1(x), S_2(x), \ldots$ **converges uniformly** to $f(x)$ on the interval $[-L,L]$, if

$$\lim_{N\to\infty}\left[\max_{-L\leq x\leq L}|f(x)-S_N(x)|\right]=0\,, \tag{$*$}$$

i.e., the maximum vertical separation between the graphs of $S_N(x)$ and $f(x)$ approaches 0 as $N\to\infty$. The sequence of functions $S_1(x), S_2(x), \ldots$ **converges pointwise** to $f(x)$ if, for each fixed x in $[-L,L]$, we have

$$\lim_{N\to\infty}|f(x)-S_N(x)|=0 \quad \text{or equivalently} \quad \lim_{N\to\infty}S_N(x)=f(x).$$

(For readers who know the distinction, the "max" in $(*)$ should be interpreted as "sup" or "least upper bound", but understandably, we do not want to belabor this point here.)

Remark. Although we are primarily concerned with the sequence of partial sums $S_1(x), S_2(x), \ldots$ of the Fourier series of $f(x)$, the above definitions still make sense for arbitrary sequences of functions defined on $[-L,L]$. Also, it is clear that a sequence of functions which converges uniformly to $f(x)$ will also converge pointwise to $f(x)$. However, the next example shows that pointwise convergence need not imply uniform convergence. Eventually (cf. Theorem 4), we prove that if $f(x)$ is continuous and "piecewise C^1" with $f(-L)=f(L)$, then the sequence $S_1(x), S_2(x), \ldots$ will converge to $f(x)$ uniformly. However, Theorem 1 only tells us that this sequence converges to $f(x)$ pointwise. □

Example 3. Let $f_n(x)$ and $g_n(x)$ be the functions defined on $[-2,2]$ with graphs as in Figure 4.

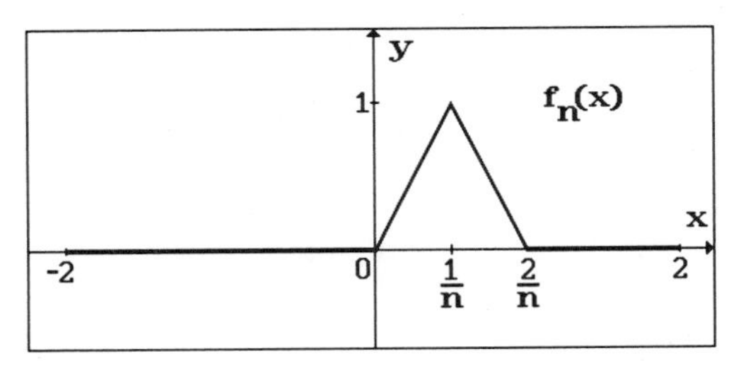

$f_n(x)$

$g_n(x)$,

Figure 4

where $n = 1, 2, 3, \ldots$. Show that the sequences $f_1(x), f_2(x), f_3(x), \ldots$ and $g_1(x), g_2(x), g_3(x), \ldots$

both converge pointwise to the zero function $h(x) \equiv 0$. However, show that the sequence $g_1(x)$, $g_2(x)$, $g_3(x)$, ... converges uniformly to $h(x)$, whereas the sequence $f_1(x)$, $f_2(x)$, $f_3(x)$, ... does not converge uniformly to $h(x)$.

Solution. We have $g_n(x) = (x + 2)/(4n)$, and so $\lim_{n\to\infty} g_n(x) = \lim_{n\to\infty} [(x + 2)/(4n)] = 0 = h(x)$ for each x in $[-2,2]$, which shows that $g_1(x)$, $g_2(x)$, $g_3(x)$, ... converges pointwise to $h(x)$. The graph of $g_n(x)$ shows that the maximum of $|h(x) - g_n(x)|$ for $-2 \le x \le 2$ is $1/n$, and so

$$\lim_{n\to\infty} \left[\max_{-2\le x\le 2} |h(x) - g_n(x)| \right] = \lim_{n\to\infty} \frac{1}{n} = 0 .$$

However, from the graph of $f_n(x)$, we see that, for each n, the maximum of $|h(x) - f_n(x)|$ in $[-2,2]$ is 1, and it occurs at $x = 1/n$ (directly below the tip of the spike). Thus,

$$\lim_{n\to\infty} \left[\max_{-2\le x\le 2} |h(x) - f_n(x)| \right] = \lim_{n\to\infty} 1 = 1 \neq 0.$$

Hence, $f_1(x)$, $f_2(x)$, $f_3(x)$,... does not converge uniformly to $h(x)$. Nevertheless, surprisingly, $f_1(x)$, $f_2(x)$, $f_3(x)$,..., does converge to $h(x)$ pointwise. Indeed, let x_0 be any *fixed* value for x in $[-2,2]$. First suppose that $-2 \le x_0 \le 0$, then $f_n(x_0) = 0$, and so

$$\lim_{n\to\infty} f_n(x_0) = 0 = h(x_0) . \qquad (*)$$

Now assume that $0 < x_0 \le 2$. Then we can find an integer N (e.g., select any $N > 2/x_0$) such that x_0 is not in the interval $(0,2/n)$ for all $n \ge N$. Then $f_n(x_0) = 0$ for all $n \ge N$ (Why ?) . In other words, (*) also holds for $0 < x_0 \le 2$! Since x_0 is fixed, we are not allowed to take $x_0 = 1/n$ (i.e., x_0 cannot depend on n). □

Remark. In applications it is usually *uniform* convergence of the partial sums $S_N(x)$ to $f(x)$ that is desired. For example, in order to use the Maximum Principle to deduce that the solution $u_N(x,t)$ of a heat problem for an approximation $S_N(x)$ of the true initial temperature $f(x)$, is within ϵ of the true solution, it is necessary to know that the *maximum* of $|f(x) - S_N(x)|$ on the rod is less than ϵ (cf. Theorem 3 of Section 3.2). This is not guaranteed by pointwise convergence, but rather by uniform convergence if N is chosen large enough. □

As we will find (cf. Theorem 4), the assumptions in Theorem 1 are actually strong enough to yield uniform convergence. However, an easier proof is available, if we strengthen these assumptions somewhat, as in the following result.

Theorem 2. **Let $f(x)$ be defined and C^2 on the interval $[-L,L]$, with $f(-L) = f(L)$ and $f'(-L) = f'(L)$. Then FS $f(x)$ converges uniformly to $f(x)$ [i.e., the sequence of partial sums $S_1(x), S_2(x), \ldots$ of FS $f(x)$ converge uniformly to $f(x)$]. Indeed, with the notation of (20),**

$$|f(x) - S_N(x)| \le \frac{4L^2M}{\pi^2 N}, \quad \text{for } \textbf{\textit{all}} \ x \text{ in } [-L,L], \text{ where } M \equiv \max_{-L \le x \le L} |f''(x)| \tag{27}$$

Proof. We know from Theorem 1 that FS $f(x) = f(x)$. Thus we have

$$f(x) - S_N(x) = \text{FS } f(x) - S_N(x) = \sum_{n=N+1}^{\infty} [a_n\cos(n\pi x/L) + b_n\sin(n\pi x/L)]$$

and

$$|f(x) - S_N(x)| \le \sum_{n=N+1}^{\infty} \Big[|a_n| + |b_n|\Big] \le \frac{4L^2M}{\pi^2} \sum_{n=N+1}^{\infty} \frac{1}{n^2},$$

where we have used (10) of Proposition 2. It remains to show that $\sum_{n=N+1}^{\infty} \frac{1}{n^2} \le \frac{1}{N}$. For this, we consider the graph of the function $\frac{1}{x^2}$ (cf. Figure 5).

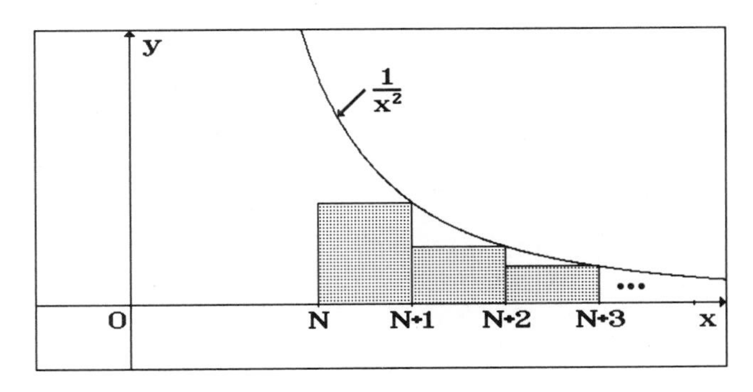

Figure 5

The sum $\sum_{n=N+1}^{\infty} \frac{1}{n^2}$ is the sum of the areas of the shaded blocks, which is less than the area $\int_N^{\infty} \frac{1}{x^2}\, dx = \frac{1}{N}$ under the curve $y = \frac{1}{x^2}$, as required. □

Remark. We have already used the estimate (27) in Example 5 and Problem 9 of Section 4.1, for the functions $x^3 - x$ and $(x^2-1)^2$ $(-1 \le x \le 1)$, respectively. In both cases, we found that much better estimates could be obtained by using an integral comparison with the explicitly computed Fourier coefficients. If the Fourier coefficients of a function can be computed, it is likely that the error estimate will be better if computed in this fashion, rather then using (27). The underlying weakness in (27) can be traced back to the weakness of (10) which was already noted in the Remark following Proposition 2. Nevertheless, Theorem 2 establishes the uniform convergence of FS $f(x)$ to $f(x)$ for *all* functions $f(x)$ satisfying the hypotheses, a fact that could not be established separately for each of the infinitely many such functions. □

Convergence of Fourier series for piecewise C^1 functions

Often, one is interested in the Fourier series of a function $f(x)$ which is not C^1 or for which $f(-L) \neq f(L)$ or $f'(-L) \neq f'(L)$. For example, if $f(x) = x$, then $f(-L) \neq f(L)$. We computed the Fourier series of this function in Example 2 of Section 4.1, and from the graphs of the partial sums (cf. Figures 1 and 2 in Section 4.1), it appears that FS $f(x)$ does converge to $f(x)$, except when $x = \pm L$, but this does not follow from Theorem 1. Indeed, none of the functions in Problems 4 through 8 of Exercises 4.1 satisfy the hypotheses of Theorem 1. However, by graphing partial sums, it would appear that the Fourier series for these functions converge to the functions except at "breaks" in the graphs where the series converge to the average value. [If $f(-L^+) \neq f(L^-)$ then we consider this to be a "break" at $x = \pm L$.] These functions are all examples of "piecewise C^1" functions which we define precisely below. We will establish a convergence theorem that applies to such functions (cf. Theorem 3).

Recall that if a function f is defined on some open interval containing a point x_0, then we say that " $f(x_0^-)$ is the left–hand limit of $f(x)$ as x approaches x_0 from the left " if $f(x)$ can be made arbitrarily close to $f(x_0^-)$ by requiring that x is sufficiently close to, but less than, x_0. Similarly, one defines the right–hand limit $f(x_0^+)$. We write

$$f(x_0^-) = \lim_{x \uparrow x_0} f(x) \quad \text{and} \quad f(x_0^+) = \lim_{x \downarrow x_0} f(x) .$$

These limits, if they exist, need not depend on the value of f at x_0. Indeed, $f(x_0^-)$ and $f(x_0^+)$, may exist, even if $f(x_0)$ is undefined. In the event that $f(x_0)$ is defined and $f(x_0^-) = f(x_0) = f(x_0^+)$, then f is **continuous** at x_0. If x_0 is the left–hand endpoint of an interval on which $f(x)$ is defined, then we can still define $f(x_0^+)$. Similarly $f(x_0^-)$ can be defined, if x_0 is the right–hand endpoint of an interval on which $f(x)$ is defined.

Definition. A function $f(x)$ is **piecewise continuous** on $[a,b]$, if all of the following hold :

(A) $f(x)$ is defined and continuous at all but a finite number of points in $[a,b]$.
(B) For all x_0 in (a,b), the limits $f(x_0^+)$ and $f(x_0^-)$ exist.
(C) $f(a^+)$ and $f(b^-)$ exist.

Definition. $f(x)$ is **piecewise C^1** on $[a,b]$, if $f(x)$ and $f'(x)$ are piecewise continuous on $[a,b]$.

Remark. If $f'(x)$ is piecewise continuous on $[a,b]$, then $f(x)$ is automatically piecewise continuous on $[a,b]$, but we will not bother to prove this. □

The function with the graph shown in Figure 6 exhibits many of the pathologies of a piecewise C^1 function.

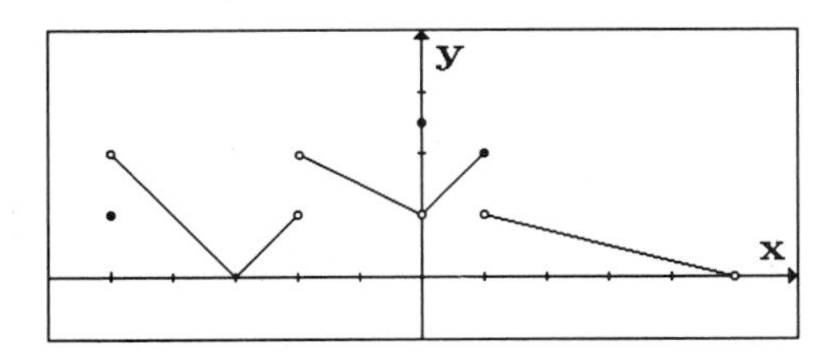

Figure 6

There are plenty of examples of functions which are not piecewise continuous. Unbounded functions such as $1/x$ $(-1 \leq x \leq 1,\ x \neq 0)$ cannot be piecewise continuous because left (and/or right)–hand limits cannot exist (i.e., are infinite) where the function "blows up" (e.g., $1/x$ blows up at $x = 0$). It is also possible to have bounded functions which are not piecewise continuous. For example, $f(x) = \sin(1/x)$ $(-1 \leq x \leq 1,\ x \neq 0)$ is bounded (since $|f(x)| \leq 1$), but $f(0^+)$ and $f(0^-)$ do not exist, since $\sin(1/x)$ oscillates infinitely often between the values ± 1, as $x \to 0$. □

Example 4. Find two functions $f(x)$ and $g(x)$ whose graphs are congruent, such that $f(x)$ is C^∞ and $g(x)$ is not even piecewise C^1.

Solution. Let $f(x) = x^3$ and let $g(x) = \sqrt[3]{x}$. The graphs are congruent, since the the graphs of inverse functions are reflections of each other in the line $y = x$. Clearly, $f(x)$ is C^∞, but note that $g'(x) = \frac{1}{3}x^{-2/3}$, which has no limit as x approaches zero from either side. □

We introduce one last definition, which is convenient in the formulation of the convergence theorem for Fourier series of piecewise C^1 functions.

Definition. Let $f(x)$ be a piecewise C^1 function on the interval $[-L,L]$. By changing the values of $f(x)$ at a finite number of points, we arrive at the **adjusted function** $\bar{f}(x)$ defined as follows :

$$\bar{f}(x) = \begin{cases} \frac{1}{2}[f(x^+) + f(x^-)] & -L < x < L \\ \frac{1}{2}[f(-L^+) + f(L^-)] & x = \pm L \quad . \end{cases} \tag{28}$$

In other words, $\bar{f}(x)$ coincides with $f(x)$ at all points in $(-L,L)$ where $f(x)$ is continuous, but $\bar{f}(x)$ is the average of the left–hand and right–hand limits of $f(x)$ at points of discontinuity in $(-L,L)$. The value of $\bar{f}(x)$ at $x = \pm L$ can also be thought of as an average of left–hand and right–hand limits, if we bend the interval $[-L,L]$ into a circle.

Example 5. Let $f(x)$ be the function whose graph is shown on the left–hand side of Figure 7.

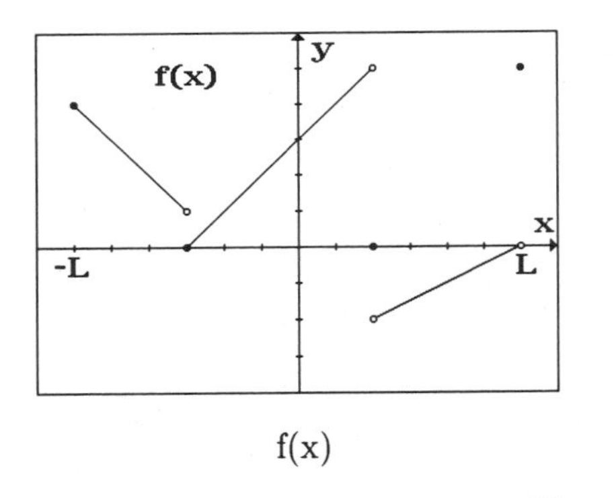

f(x)

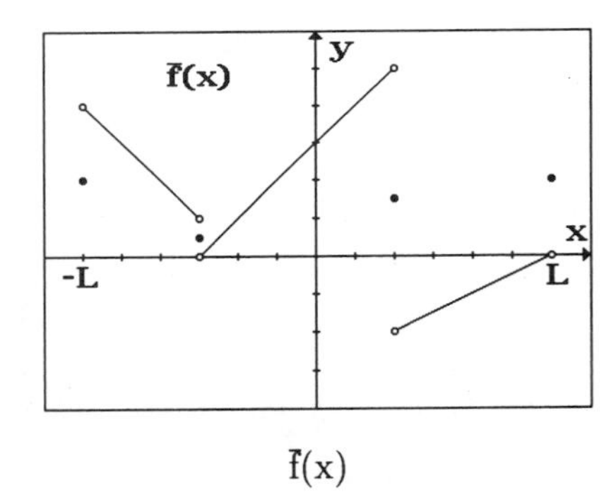

f̄(x)

Figure 7

The adjusted function $\bar{f}(x)$ is graphed on the right–hand side of Figure 7.

Theorem 3. Let $f(x)$ be a piecewise C^1 function on $[-L,L]$ and let $\bar{f}(x)$ be the adjusted function in (28). Then FS $f(x) = \bar{f}(x)$, for all x in $[-L,L]$. Indeed, we have

$$\text{FS } f(x) = \tilde{\bar{f}}(x) \qquad -\infty < x < \infty , \tag{29}$$

where $\tilde{\bar{f}}(x)$ is the periodic extension of the adjusted function $\bar{f}(x)$.

Proof. The proof proceeds in the same way as the proof of Theorem 1, until we reach (23) which is now replaced by the following limit that must be established for each x in [–L,L] :

$$\lim_{N\to\infty} \frac{1}{L}\int_{-L}^{L} D_N(y-x)f(x)\,dx = \bar{f}(x)\,. \tag{30}$$

Letting $h(z) \equiv \tilde{\bar{f}}(z+x)$, the limit in (30) is equivalent to the limit

$$\lim_{N\to\infty} \frac{1}{L}\int_{-L}^{L} D_N(z)h(z)\,dz = h(0)\,. \tag{31}$$

Indeed, (30) is transformed to (31) in the same way that (23) was reduced to Proposition 4. However, we cannot simply apply Proposition 4, because now h is not necessarily C^1. Nevertheless, we know that (since $\bar{f}$ is piecewise C^1 and h is a translate of $\tilde{\bar{f}}$) the limits $h(0^+)$, $h(0^-)$, $h'(0^+)$ and $h'(0^-)$ all exist, and $h(0) = \frac{1}{2}[h(0^-) + h(0^+)]$. Thus, in order to prove (31), it only remains to prove the analog (cf. Proposition 5, below) of Proposition 4. As in the proof of Theorem 1, this is equivalent to the desired result (29). □

Proposition 5. Let $h(x)$ be a piecewise C^1 function on $[-L,L]$. If D_n denotes the n–th Dirichlet kernel (14), then

$$\lim_{n\to\infty} \frac{1}{L}\int_{-L}^{L} D_n(x)h(x)\,dx = \tfrac{1}{2}[h(0^-) + h(0^+)]\,. \tag{32}$$

Proof. The result (32) follows from adding the two results :

$$\lim_{n\to\infty} \frac{1}{L}\int_{-L}^{0} D_n(x)h(x)\,dx = \tfrac{1}{2}h(0^-) \quad\text{and}\quad \lim_{n\to\infty} \frac{1}{L}\int_{0}^{L} D_n(x)h(x)\,dx = \tfrac{1}{2}h(0^+)\,.$$

We prove the second of these; the proof of the first is similar. Observe that

$$\frac{1}{L}\int_0^L D_n(x)\,dx = \frac{1}{L}\int_0^L \left\{ \tfrac{1}{2} + \cos(\pi x/L) + \ldots + \cos(n\pi x/L)\right\} dx = \frac{1}{2}\,,$$

and multiplying by the constant $h(0^+)$, we have $\frac{1}{L}\int_0^L D_n(x)h(0^+)\,dx = \frac{1}{2}h(0^+)$. Thus,

$$\frac{1}{L}\int_0^L D_n(x)h(x)\,dx - \tfrac{1}{2}h(0^+) = \frac{1}{L}\int_0^L D_n(x)[h(x) - h(0^+)]\,dx\,, \tag{33}$$

and it suffices to prove that the integral on the right approaches 0 as $n \to \infty$. To this end, we write $D_n(x) = \frac{1}{2}\cos(n\pi x/L) + \frac{1}{2}\cot(\pi x/2L)\sin(n\pi x/L)$, as in the proof of Proposition 4. Then we apply the Riemann–Lebesgue lemma to each part, after we split the integral (33) into two integrals as was done before [cf. (19)]. This time, we use a one–sided version of L'Hospital's rule to obtain

$$\lim_{x \downarrow 0} \frac{h(x)-h(0^+)}{2\tan(\pi x/2L)} = \frac{L}{\pi} h'(0^+) .$$

Thus, the Riemann–Lebesgue lemma will apply to both integrals, as before. Thus, the right–hand side of (33) tends to zero as $n \to \infty$. □

Example 6. Using Theorem 3, graph the Fourier series FS f(x) (for $-4L \le x \le 4L$) of the function

$$f(x) = \begin{cases} 1 & 0 \le x \le L \\ 0 & -L \le x < 0 \ . \end{cases} \tag{34}$$

Verify directly that FS f(x) converges to the adjusted function at the discontinuities of f(x), but show that the convergence of FS f(x) to $\bar{f}(x)$ on [–L,L] is not uniform.

Solution. It is not necessary to compute the Fourier coefficients to draw the graph of FS f(x). According to Theorem 3, FS f(x) is the periodic extension $\tilde{\bar{f}}(x)$ which is graphed in Figure 8.

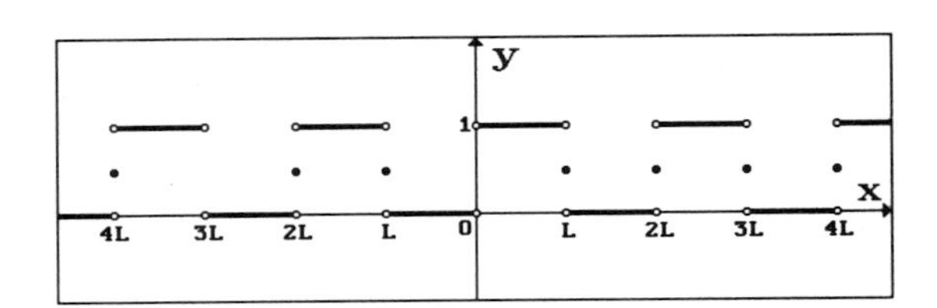

Figure 8

The Fourier series of f(x) was computed in Problem 4(a) of Exercises 4.1 to be

$$\text{FS } f(x) = \frac{1}{2} + \frac{1}{\pi}\sum_{n=1}^{\infty} \frac{1}{n}\,[1+(-1)^{n+1}]\sin(n\pi x/L). \tag{35}$$

We have FS $f(0) = \frac{1}{2}$, and this is the value $\frac{1}{2}[f(0^-) + f(0^+)]$ of the adjusted function $\bar{f}(x)$ at $x = 0$. Note also that FS $f(\pm L) = \frac{1}{2} = \bar{f}(\pm L)$. We can show that the convergence here is not uniform as follows. Let $S_N(x)$ be the N–th partial sum of (35). Since $S_N(0) = \frac{1}{2}$ and $S_N(x)$ is continuous, we know that $S_N(\epsilon)$ must be close to $\frac{1}{2}$, if $|\epsilon|$ is small enough. In particular, we can choose ϵ $(0 < \epsilon < L)$ small enough, so that $S_N(\epsilon) < \frac{3}{4}$. However, $\tilde{f}(\epsilon) = f(\epsilon) = 1$, and so $|S_N(\epsilon) - \tilde{f}(\epsilon)| > \frac{1}{4}$ (i.e., the maximum vertical separation between the graphs of S_N and $\tilde{f}$ is always greater than $\frac{1}{4}$). For uniform convergence, we would need to have the maximum vertical separation tending to 0, as $N \to \infty$. Geometrically, it is impossible for the graph of the continuous function $S_N(x)$ to be close to $y = 0$ just to the left of $x = 0$, and close to $y = 1$ just to the right of $x = 0$. □

Remark 1. More generally, it is possible to prove that a sequence of continuous functions [e.g., $S_N(x)$] cannot converge uniformly to a discontinuous function [e.g., $\tilde{f}(x)$ in Example 6]. Thus, the Fourier series of a function cannot converge uniformly to $\tilde{f}$, if $\tilde{f}$ is discontinuous.

Remark 2 (The Gibbs phenomenon). In Example 6, there is also a more subtle nonuniformity in the convergence of $S_N(x)$ to $\tilde{f}(x)$. In Figure 9 below, observe that the graph of $S_7(x)$ (where we have taken $L = 2$) overshoots the value 1, as it heads upward from the origin.

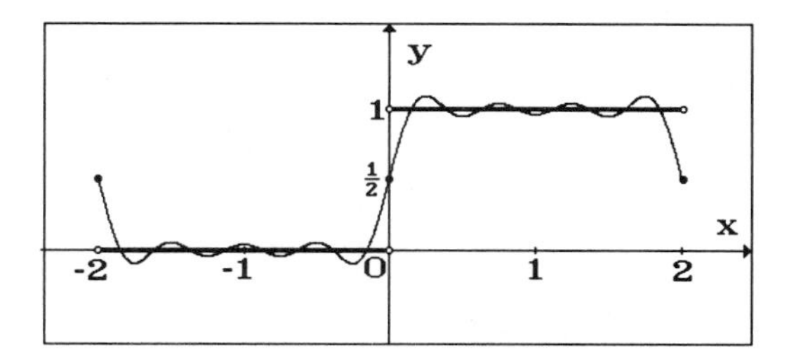

Figure 9

The maximum value of $S_7(x)$ occurs at $x = .25$, and it is about 1.0921 . One might expect that the overshoot of $S_N(x)$ will decrease to zero, as $N \to \infty$. However, this is not the case. Indeed, the "overshoot maximum" for $S_{2n-1}(x)$ occurs at $x = 1/n$ and

$$\lim_{n\to\infty} S_{2n-1}(1/n) = \frac{1}{2} + \frac{1}{\pi}\int_0^{\pi} \frac{\sin x}{x}\, dx \approx 1.0895 .$$

We lead the reader to this conclusion in Problem 9. The general fact that the overshoot discrepancy does not tend to zero at a jump in the graph of $f(x)$ is known as the **Gibbs phenomenon**. The Gibbs phenomenon is not in violation of Theorem 3. For a piecewise C^1 function $f(x)$, the overshoot maximum of $S_N(x)$ does not occur at a fixed value of x which is independent of N. At any fixed value of x, $S_N(x)$ converges to $\tilde{f}(x)$. This is quite similar to the case of the pointwise (but nonuniform) convergence of the functions $f_n(x)$ in Example 3. □

Uniform convergence of FS f(x) for *continuous* piecewise C^1 functions f(x) with f(L) = f(–L)

Theorem 3 gives us the pointwise convergence of FS f(x) to $\tilde{\bar{f}}(x)$ under weaker hypotheses than Theorem 2. However, the convergence in Theorem 2 is uniform. The above Example 6 and the remarks, show that the Fourier series of a piecewise C^1 function f(x) on [–L,L] cannot converge uniformly to $\tilde{f}(x)$, if f(x) has discontinuities (i.e., jumps) or if $f(-L^+) \neq f(L^-)$. If f(x) is piecewise C^1 and has no discontinuities in [–L,L] and f(–L) = f(L) , then there is still hope for uniform convergence of FS f(x) to f(x) in [–L,L]. The following Theorem 4 provides the uniform convergence of FS f(x) to f(x) for such functions. The main advantage of Theorem 4 over Theorem 2 is that f(x) need not be C^2 (or even C^1).

Theorem 4. Let f(x) **be a *continuous* piecewise** C^1 **function on** [–L,L], **such that** f(–L) = f(L). **Then** FS f(x) **converges uniformly to** f(x) **on** [–L,L]. **In other words,**

$$\max_{-L \leq x \leq L} |f(x) - S_N(x)| \to 0 \quad \text{as} \quad N \to \infty , \tag{36}$$

where $S_N(x)$ **is the N–th partial sum of** FS f(x).

Remark. Note that it is not assumed that f′(x) is continuous everywhere, so that the graph of f(x) may have corners. However, the continuity assumption ensures that the graph of f(x) will have no gaps. Moreover, the assumption f(–L) = f(L) ensures that the periodic extension $\tilde{f}(x)$ has no gaps (i.e., f(x) yields a continuous function on the circle of circumference 2L). □

Proof. From Theorem 3, we already know that FS f(x) = f(x) for x in [–L,L] , since the adjusted function $\tilde{f}(x)$ is the same as f(x) by the continuity assumptions on f(x). Thus,

$$f(x) - S_N(x) = \text{FS } f(x) - S_N(x) = \sum_{n=N+1}^{\infty} a_n\cos(n\pi x/L) + b_n\sin(n\pi x/L) . \tag{37}$$

If we can prove that

$$\sum_{n=1}^{\infty} \Big[|a_n| + |b_n| \Big] < \infty , \tag{38}$$

then the "tail" $\sum_{n=N+1}^{\infty} \Big[|a_n| + |b_n| \Big]$ of the series (38) would tend to 0, as $N \to \infty$, and by (37),

$$\lim_{N\to\infty} \Big[\max_{-L \leq x \leq L} |f(x) - S_N(x)| \Big] \leq \lim_{N\to\infty} \Big[\sum_{n=N+1}^{\infty} |a_n| + |b_n| \Big] = 0 , \tag{39}$$

as required by (36). Thus, it suffices to prove (38). To this end, we need to examine the Fourier coefficients a_n and b_n. For b_n, we have

$$b_n = \frac{1}{L}\int_{-L}^{L} f(x)\sin(n\pi x/L)\,dx = -\frac{1}{L} f(x)\cdot\frac{L}{n\pi}\cos(n\pi x/L)\Big|_{-L}^{L} + \frac{1}{L}\int_{-L}^{L} f'(x)\cdot\frac{L}{n\pi}\cos(n\pi x/L)\,dx$$

$$= \frac{L}{n\pi}\,\frac{1}{L}\int_{-L}^{L} f'(x)\cos(n\pi x/L)\,dx = \frac{L}{n\pi}A_n\,, \tag{40}$$

where A_n is the Fourier cosine coefficient of $f'(x)$, and we have used the fact that "integration by parts" is justified for a *continuous* piecewise C^1 function. Similarly, we have

$$a_n = -\frac{L}{n\pi}B_n = -\frac{L}{n\pi}\frac{1}{L}\int_{-L}^{L} f'(x)\sin(n\pi x/L)\,dx\ . \tag{41}$$

Since $f'(x)$ is piecewise continuous, it satisfies the hypotheses of Bessel's inequality, and so

$$\sum_{n=1}^{\infty}\left[A_n^2 + B_n^2\right] \le \frac{1}{L}\int_{-L}^{L}[f'(x)]^2\,dx - \tfrac{1}{2}A_0 < \infty\,. \tag{42}$$

Given any sequences $\alpha_1,\dots,\alpha_k$ and $\beta_1,\dots,\beta_k$ of real numbers, **the Cauchy–Schwarz inequality** (cf. Problem 9) says that

$$|\alpha_1\beta_1 + \dots + \alpha_k\beta_k| \le (\alpha_1^2 + \dots + \alpha_k^2)^{\frac{1}{2}}\cdot(\beta_1^2 + \dots + \beta_k^2)^{\frac{1}{2}}\,. \tag{43}$$

For $k = 3$, (43) is just the familiar result $|\mathbf{a}\cdot\mathbf{b}| \le \|\mathbf{a}\|\cdot\|\mathbf{b}\|$, but the result holds for any k. In particular, we consider the sequences

$$|A_1|,\ |B_1|,\ |A_2|,\ |B_2|,\ |A_3|,\ |B_3|,\ \dots \quad\text{and}\quad \tfrac{1}{1},\tfrac{1}{1},\tfrac{1}{2},\tfrac{1}{2},\tfrac{1}{3},\tfrac{1}{3},\ \dots\,.$$

By (43), applied to this pair of sequences, we have for any $k = 1, 2, 3, \dots$,

$$|A_1| + |B_1| + |A_2|\cdot\tfrac{1}{2} + |B_2|\cdot\tfrac{1}{2} + \dots + |A_k|\cdot\tfrac{1}{k} + |B_k|\cdot\tfrac{1}{k}$$

$$\le \left[|A_1|^2 + |B_1|^2 + |A_2|^2 + |B_2|^2 + \dots + |A_k|^2 + |B_k|^2\right]^{\frac{1}{2}}\cdot\left[2\cdot[(\tfrac{1}{1})^2 + (\tfrac{1}{2})^2 + \dots + (\tfrac{1}{k})^2]\right]^{\frac{1}{2}}.$$

Letting $k \to \infty$, we obtain (using the result $\sum_{n=1}^{\infty}\frac{1}{n^2} \le \frac{\pi^2}{6}$ of Example 2)

$$\sum_{n=1}^{\infty}\frac{1}{n}\left[|A_n| + |B_n|\right] \le \left[\sum_{n=1}^{\infty}(A_n^2 + B_n^2)\right]^{\frac{1}{2}}\cdot\left[2\cdot\frac{\pi^2}{6}\right]^{\frac{1}{2}} < \infty\,, \tag{44}$$

by virtue of (42). Thus, using (40), (41) and (44),

$$\sum_{n=1}^{\infty}\Big[|a_n| + |b_n|\Big] = \frac{L}{\pi}\sum_{n=1}^{\infty}\frac{1}{n}\Big[|A_n| + |B_n|\Big] < \infty ,$$

which gives us the result (38), and thus (39) holds. □

Remark (Fejér's theorem). There are continuous (but not piecewise C^1) functions $f(x)$ defined on $[-L,L]$, with $f(-L) = f(L)$, whose Fourier series diverge at an infinite number of points in $[-L,L]$ (cf. [Coppel]). Nevertheless, it is possible to uniformly approximate such a function to within any given positive error by a trigonometric polynomial. In other words, given $\epsilon > 0$, there is a trigonometric polynomial, say $P_N(x) = c_0 + \sum_{n=1}^{N} c_n\cos(n\pi x/L) + d_n\sin(n\pi x/L)$, such that $|f(x) - P_N(x)| \le \epsilon$, for all $-L \le x \le L$. If FS $f(x)$ does not converge uniformly to $f(x)$, then we might not be able to find a partial sum $S_N(x)$ of FS $f(x)$, such that $|f(x) - S_N(x)| \le \epsilon$. However, the following remarkable result is proved in [Bari, vol. I, p. 135].

Theorem 5 (Fejér's theorem). Let $f(x)$ be a continuous (but not necessarily piecewise C^1) function defined on $[-L,L]$, such that $f(-L) = f(L)$. Let $S_N(x)$ denote the partial sum (from $n = 0$ to N) of the Fourier series of $f(x)$. Let $A_N(x) = \frac{1}{N+1}\Big[S_0(x) + S_1(x) + \dots + S_N(x)\Big]$ (i.e., the *average* of the first $N+1$ partial sums of FS $f(x)$). Then the sequence of trigonometric polynomials $A_0(x)$, $A_1(x)$, $A_2(x)$, ... converges uniformly to $f(x)$ on the interval $[-L,L]$, although the sequence $S_0(x)$, $S_1(x)$, $S_2(x)$, ... may not even converge pointwise to $f(x)$.

It can be easily shown (cf. Problem 13) that among all trigonometric polynomials $P_N(x)$ of "degree" N , $S_N(x)$ makes the mean–square error $\int_{-L}^{L} [f(x) - P_N(x)]^2\, dx$ the smallest. However, Fejér's theorem suggests that $S_N(x)$ may not make the "uniform error" $\max_{-L\le x\le L} |f(x) - S_N(x)|$ the smallest among among all trigonometric polynomials of degree N, as a concrete counterexample in Problem 14 shows. □

Summary 4.2

1. **Bessel's inequality : Let $f(x)$ be defined on $[-L,L]$, and suppose that $\int_{-L}^{L} [f(x)]^2 dx$ exists and is finite. Assume that the Fourier coefficients**

$$a_n = \frac{1}{L}\int_{-L}^{L} f(x)\cos(n\pi x/L)\, dx \quad \text{and} \quad b_n = \frac{1}{L}\int_{-L}^{L} f(x)\sin(n\pi x/L)\, dx \tag{S1}$$

exist, so that $\text{FS } f(x) = \frac{1}{2}a_0 + \sum_{n=1}^{\infty} a_n\cos(n\pi x/L) + b_n\sin(n\pi x/L)$ **may be defined formally (possibly not convergent). Then we have Bessel's Inequality**

$$\tfrac{1}{2}a_0^2 + \sum_{n=1}^{\infty}(a_n^2 + b_n^2) \le \frac{1}{L}\int_{-L}^{L} [f(x)]^2\, dx\ . \tag{S2}$$

If this inequality is replaced by equality, it is called Parseval's equality. For continuous piecewise C^1 functions $f(x)$, Parseval's equality is proved in Problems 5–7.

2. **The Riemann–Lebesgue Lemma : Let $f(x)$ be defined on $[-L,L]$. If $\int_{-L}^{L} [f(x)]^2\, dx < \infty$, and the Fourier coefficients of $f(x)$ exist, then these coefficients tend to zero, as $n \to \infty$, i.e.,**

$$\lim_{n\to\infty} \frac{1}{L}\int_{-L}^{L} f(x)\cos(n\pi x/L)\, dx = 0 \quad \textbf{and} \quad \lim_{n\to\infty} \frac{1}{L}\int_{-L}^{L} f(x)\sin(n\pi x/L)\, dx = 0\ . \tag{S3}$$

The smoother $f(x)$ is, when viewed as a function on a circle of circumference $2L$, the more rapidly the Fourier coefficients of $f(x)$ will decay as $n \to \infty$ (cf. Proposition 2 and remarks after it).

3. **Pointwise convergence of FS f(x) (Theorem 1) : Let $f(x)$ be C^1 on $[-L,L]$, and assume that $f(-L) = f(L)$ and $f'(-L) = f'(L)$ [so that $f(x)$ may be regarded as a C^1 function on the circle of circumference $2L$]. Then $\text{FS } f(x) = f(x)$ for all x in $[-L,L]$. In other words, writing**

$$S_N(x) = \tfrac{1}{2}a_0 + \sum_{n=1}^{N}[a_n\cos(n\pi x/L) + b_n\sin(n\pi x/L)]\ , \tag{S4}$$

where a_n and b_n are given by (S1), we have (for any fixed x in $[-L,L]$)

$$\text{FS } f(x) \equiv \tfrac{1}{2}a_0 + \sum_{n=1}^{\infty}[a_n\cos(n\pi x/L) + b_n\sin(n\pi x/L)] \equiv \lim_{N\to\infty} S_N(x) = f(x)\ . \tag{S5}$$

4. Uniform convergence of FS f(x) **(Theorem 2) : Let** f(x) **be defined and** C^2 **on the interval** [–L,L] , **with** f(–L) = f(L) **and** f′(–L) = f′(L). **Then** FS f(x) **converges uniformly to** f(x) **on** [–L,L]. **Indeed, with** $S_N(x)$ **given by** (S4),

$$|f(x) - S_N(x)| \le \frac{4L^2M}{\pi^2 N}, \quad \text{for } \textbf{\textit{all}} \ x \text{ in } [-L,L], \text{ where } M \equiv \max_{-L \le x \le L} |f''(x)| . \qquad \text{(S6)}$$

5. Pointwise convergence of FS f(x) **(Theorem 3) : Let** f(x) **be a piecewise** C^1 **function on** [–L,L] **and let** $\bar{f}(x)$ **be the adjusted function in** (28). **Then** FS f(x) = $\bar{f}(x)$, **for all** x **in** [–L,L]. **Indeed,**

$$\text{FS } f(x) = \tilde{\bar{f}}(x) , \quad \text{for } -\infty < x < \infty, \qquad \text{(S7)}$$

where $\tilde{\bar{f}}(x)$ **is the periodic extension of the adjusted function** $\bar{f}(x)$.

6. Uniform convergence of FS f(x) **(Theorem 4) : Let** f(x) **be a *continuous* piecewise** C^1 **function on** [–L,L], **such that** f(–L) = f(L). **Then** FS f(x) **converges uniformly to** f(x) **on** [–L,L]. **In other words,**

$$\max_{-L \le x \le L} |f(x) - S_N(x)| \to 0 \quad \text{as} \quad N \to \infty , \qquad \text{(S8)}$$

where $S_N(x)$ **is the** N–th **partial sum of** FS f(x) **given by** (S4).

Exercises 4.2

1. Let f(x) = |x| for –1 ≤ x ≤ 1 .

(a) Sketch the periodic extension $\tilde{f}(x)$ for –4 ≤ x ≤ 4 .

(b) What theorem ensures that the partial sums $S_N(x)$ of FS f(x) converge to $\tilde{f}(x)$ uniformly ?

(c) Why can we not use Theorem 2 to establish uniform convergence of FS f(x) to f(x) ?

(d) Show directly that FS f(x) converges uniformly to f(x) by means of an integral comparison, as in the proof of Theorem 2 . You may assume that FS f(x) = f(x) by Theorem 3.

2. Find a nonconstant polynomial of lowest degree which satisfies the hypotheses of Theorems 1 and 2 for the interval [–L,L]. **Hint.** The degree must be at least 3. Why?

3. Let
$$f(x) = \begin{cases} 2(x+1) & \text{for } -1 \le x \le 0 \\ x & \text{for } \ \ 0 < x \le 1 \end{cases} .$$

Sketch the graph of FS f(x) in the interval [–4,4]. What theorem did you use ?

4. (a) In Problem 6 of Section 4.1, we showed that $\sum_{n=1}^{\infty} \frac{1}{n^2} = \frac{\pi^2}{6}$, asuming that FS f(x) = f(x), for $-L \leq x \leq L$, where $f(x) = x^2$. What theorem(s) imply that this assumption is correct ?

(b) Use part (a) to show that Parseval's equality is valid for the function $f(x) = x$ (cf. Example 2), even though we have not proved Parseval's equality in general.

(c) Verify directly that Parseval's equality holds for the function of Example 6 :

$$f(x) = \begin{cases} 1 & 0 \leq x \leq L \\ 0 & -L \leq x < 0 \end{cases} .$$

Hint. Refer to Problem 6 of Section 4.1 , again.

5. (a) Verify the following result (cf. the assumptions and the proof of Bessel's inequality).

$$\int_{-L}^{L} [f(x) - S_N(x)]^2 \, dx = \int_{-L}^{L} f(x)^2 \, dx - L\left[\tfrac{1}{2}a_0^2 + \sum_{n=1}^{N} a_n^2 + b_n^2\right] .$$

(b) Use the result in part (a) to conclude that a function $f(x)$, $-L < x < L$, with $\int_{-L}^{L} f(x)^2 \, dx < \infty$ and whose Fourier coefficients exist, will satisfy Parseval's equality if and only if

$$\lim_{N \to \infty} \int_{-L}^{L} [f(x) - S_N(x)]^2 \, dx = 0 .$$

6. Find a sequence of functions $f_1(x)$, $f_2(x)$, ... defined on $[-L,L]$, such that for each x in $[-L,L]$, $\lim_{n \to \infty} f_n(x) = 0$, and yet

$$\lim_{n \to \infty} \int_{-L}^{L} f_n(x) \, dx \neq \int_{-L}^{L} \left[\lim_{n \to \infty} f_n(x)\right] dx .$$

Hint. Modify the functions $f_n(x)$ in Example 3 so that $f_n(1/n) = n$ (i.e., increase the height of the vertex to n). Observe that then $\int_{-L}^{L} f_n(x) \, dx = 1$.

7. (a) In view of Theorem 2 and Problem 5 , show that a C^2 function $f(x)$ on $[-L,L]$, such that $f(-L) = f(L)$ and $f'(-L) = f'(L)$ will satisfy Parseval's equality.

(b) In view of Problem 6, why can we *not* use Theorem 3 and Problem 5 to conclude at once that Parseval's equality holds for any piecewise C^1 function $f(x)$ on $[-L,L]$?

(c) Use Problem 5 and Theorem 4 to prove that Parseval's equality holds for any continuous piecewise C^1 function $f(x)$ on $[-L,L]$ such that $f(-L) = f(L)$.

8. Use Problems 5 and 4(c) to conclude that it is sufficient, but *not* necessary, that $S_N(x) \to f(x)$ uniformly, in order to conclude that Parseval's equality holds for $f(x)$ as in Problem 5.

Remark. In fact, Parseval's equality holds for any piecewise continuous function $f(x)$ defined on $[-L,L]$. A proof is in [Rudin, Principles of Mathematical Analysis, 3rd ed., p. 191].

9. (**The Cauchy–Schwarz inequality**; cf. (43)). Let $\mathbf{a} = (\alpha_1,\alpha_2,\ldots,\alpha_k)$ and $\mathbf{b} = (\beta_1,\beta_2,\ldots,\beta_k)$, where the α_i and β_i are real numbers. Define $\mathbf{a}\cdot\mathbf{b} = \alpha_1\beta_1 + \alpha_2\beta_2 + \ldots + \alpha_k\beta_k$, and let $\|\mathbf{a}\| = (\mathbf{a}\cdot\mathbf{a})^{\frac{1}{2}} = (\alpha_1^2 + \alpha_2^2 + \ldots + \alpha_k^2)^{\frac{1}{2}}$ be the length or norm of $\mathbf{a}$. For any real number r, let $\mathbf{a} + r\mathbf{b} = (\alpha_1,\alpha_2,\ldots,\alpha_k) + r(\beta_1,\beta_2,\ldots,\beta_k) = (\alpha_1 + r\beta_1,\ \alpha_2 + r\beta_2,\ \ldots,\alpha_k + r\beta_k)$.

(a) Let $h(r) = \|(\mathbf{a} + r\mathbf{b})\|^2 = (\mathbf{a} + r\mathbf{b})\cdot(\mathbf{a} + r\mathbf{b})$. Show that $h(r) = (\mathbf{b}\cdot\mathbf{b})r^2 + 2(\mathbf{a}\cdot\mathbf{b})r + \mathbf{a}\cdot\mathbf{a}$ and note that $h(r) \geq 0$, for all r (Why?).

(b) If $\mathbf{b}\cdot\mathbf{b} \neq 0$, explain why the graph of $h(r)$ cannot intersect the r–axis more than once (if at all), and hence why the quadratic equation $h(r) = 0$ cannot have two distinct real roots.

(c) Use parts (a) and (b) and the quadratic formula to conclude that $2[(\mathbf{a}\cdot\mathbf{b})]^2 - 4(\mathbf{b}\cdot\mathbf{b})(\mathbf{a}\cdot\mathbf{a}) \leq 0$ or $|\mathbf{a}\cdot\mathbf{b}| \leq \|\mathbf{a}\|\ \|\mathbf{b}\|$, which is the Cauchy–Schwarz inequality.

10. Use the same idea as in Problem 9, to show that for any two piecewise continuous functions $f(x)$ and $g(x)$ defined on $[-L,L]$, we have $|\langle f,g\rangle| \leq \|f\|\ \|g\|$. In other words,

$$\left|\int_{-L}^{L} f(x)g(x)\,dx\right| \leq \left[\int_{-L}^{L} f(x)^2\,dx\right]^{\frac{1}{2}} \left[\int_{-L}^{L} g(x)^2\,dx\right]^{\frac{1}{2}}.$$

The piecewise continuity of f(x) and g(x) ensures that the integrals exist, which you may assume.
Hint. Consider $h(r) = \|f + rg\|^2$, for real r.

11. Show that there is no piecewise C^1 function f defined on $[-L,L]$, such that $\|f\| \neq 0$ and f is orthogonal to all of the functions $c_0, c_1, s_1, c_2, s_2, \ldots$. (Recall $c_n(x) = \cos(n\pi x/L)$, etc. .)
Hint. Consider the Fourier series of such a function and use Theorem 3.

12. Complete the following steps to establish the claims made in Remark 2, following Example 6 concerning the Gibbs phenomenon (overshoot at a discontinuity).

(a) Let $S_{2n-1}(x)$ be the partial sum (n = 1,2,3,,...)

$$S_{2n-1}(x) = \frac{1}{2} + \frac{2}{\pi}\left[\sin(\pi x/L) + \frac{1}{3}\sin(3\pi x/L) + \dots + \frac{1}{2n-1}\sin([2n-1]\pi x/L)\right]$$

of FS f(x) in (35). Compute $S'_{2n-1}(x)$ and multiply by $\sin(\pi x/L)$ to produce a collapsing sum, as in the proof of Proposition 3, thus obtaining $\sin(\pi x/L)\cdot S'_{2n-1}(x) = \frac{1}{L}\sin(2n\pi x/L)$.

(b) Conclude that the first positive value of x for which $S'_{2n-1}(x) = 0$ is $x = L/(2n)$, and hence the overshoot maximum value is

$$S_{2n-1}(\tfrac{L}{2n}) = \frac{1}{2} + \frac{1}{\pi}\left[2\sin(\pi/2n) + \frac{2}{3}\sin(3\pi/n) + \dots + \frac{2}{2n-1}\sin([2n-1]\pi/2n)\right]. \qquad (*)$$

(c) Show that the sum in brackets in (*) is the sum of areas of the shaded blocks shown below.

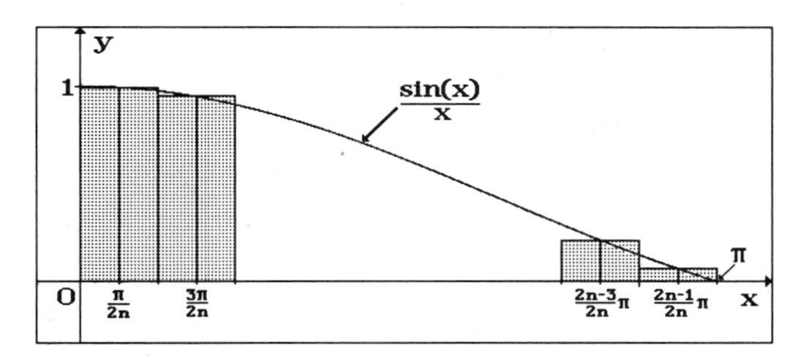

Figure 10

(d) Conclude that the limit of the overshoot maximum values, as $n \to \infty$, is given by

$$\lim_{n\to\infty} S_{2n-1}(L/2n) = \frac{1}{2} + \frac{1}{\pi}\int_0^{\pi} \frac{\sin(x)}{x}\,dx \approx 1.0895 .$$

Hint. Use the trapezoidal rule to estimate the integral in (d).

13. Let $f(x)$ be a piecewise continuous function on $[-L,L]$, and for $N \geq 0$, let $S_N(x) = \frac{1}{2}a_0 + \sum_{n=1}^{N} a_n\cos(n\pi x/L) + b_n\sin(n\pi x/L)$ be the N–th partial sum of FS f(x). Let $P_N(x) = \frac{1}{2}c_0 + \sum_{n=1}^{N} c_n\cos(n\pi x/L) + d_n\sin(n\pi x/L)$ be an arbitrary trigonometric polynomial of degree N.

(a) Show that $\int_{-L}^{L} [f(x) - P_N(x)]^2 \, dx = \int_{-L}^{L} [f(x)]^2 \, dx - 2\int_{-L}^{L} P_N(x)S_N(x) \, dx + \int_{-L}^{L} P_N(x)^2 \, dx$

$= \int_{-L}^{L} [f(x)]^2 \, dx - \int_{-L}^{L} S_N(x)^2 \, dx + \int_{-L}^{L} [P_N(x) - S_N(x)]^2 \, dx$.

(b) Why does part (a) show that the mean–square error $\int_{-L}^{L} [f(x) - P_N(x)]^2 \, dx$ is least when $P_N(x)$ is chosen to be $S_N(x)$?

14. Let $f(x) = x^2$ for $-L \le x \le L$.

(a) Show that the constant c, such that the mean–square error $\int_{-L}^{L} [f(x) - c]^2 \, dx$ is smallest, is $\frac{1}{2}a_0 = L^2/3$.

(b) Show that the constant d, such that the "uniform error" $\max_{-L \le x \le L} |f(x) - d|$ is smallest, is $L^2/2$. Compare $\max_{-L \le x \le L} |f(x) - L^2/2|$ with $\max_{-L \le x \le L} |f(x) - L^2/3|$.

(c) Why do parts (a) and (b) imply that the best uniform approximation of a function by a trigonometric polynomial of given degree is not necessarily a partial sum of its Fourier series ?

4.3 Sine and Cosine Series and Applications

One of our goals in this chapter is to approximate functions on an interval by linear combinations of sine and/or cosine functions of a form (depending on the B.C.) that is appropriate for the solution of certain heat conduction problems. The Fourier series treated in Section 4.2 dealt with functions defined on $[-L,L]$, and this series is ideally suited for solving the following problem for the circular wire :

$$\begin{aligned} &\text{D.E.} \quad u_t = ku_{xx} \qquad -L \le x \le L\,,\ t \ge 0 \\ &\text{B.C.} \quad u(-L,t) = u(L,t) \qquad u_x(-L,t) = u_x(L,t) \\ &\text{I.C.} \quad u(x,0) = f(x)\ . \end{aligned} \tag{1}$$

If the function $f(x)$ is continuous and piecewise C^1 with $f(-L) = f(L)$, then by Theorem 4 of Section 4.2, we can find a partial sum of FS $f(x)$, say

$$S_N(x) = \tfrac{1}{2}a_0 + \sum_{n=1}^{N} \left\{a_n\cos(n\pi x/L) + b_n\sin(n\pi x/L)\right\},$$

such that $|f(x)-S_N(x)| < \epsilon$, for any preassigned "experimental error" ϵ and all x in $[-L,L]$. Replacing $f(x)$ by $S_N(x)$ in (1), we then obtain a solution

$$u_N(x,t) = \tfrac{1}{2}a_0 + \sum_{n=1}^{N} e^{-(n\pi/L)^2kt}\left\{a_n\cos(n\pi x/L) + b_n\sin(n\pi x/L)\right\} \tag{2}$$

via Proposition 2 of Section 3.1. By using a maximum principle we could establish that $u_N(x,t)$ given by (2) is within ϵ of the exact C^2 solution [if such exists] of problem (1) with the original function $f(x)$.

If we replace N in (2) by ∞, then one is tempted to assert that (2) gives us the exact solution of the original problem (1). There are many difficulties in proving this assertion, or even in formulating the sense in which such a statement is correct. For example, if $f(x)$ is not C^2, then there can be no solution of (1) which is C^2 at $t = 0$ (e.g., the D.E. cannot be satisfied at $t = 0$). Moreover, even if $f(x)$ is C^2 [with $f(-L) = f(L)$ and $f'(-L) = f'(L)$], there are substantial problems in verifying that, with $N = \infty$, the sum (2) actually converges to a C^2 function that satisfies the D.E. . In particular, recall that it is quite common for an infinite sum of smooth (i.e., C^∞) functions to converge to a discontinuous function (cf. Example 6 of Section 4.2). From the viewpoint of applications, an extensive effort to resolve these difficulties is unwarranted, because $f(x)$ is only known to within some experimental error, and there is nothing to be gained by obtaining an exact solution of (1). In mathematics, it is often interesting and challenging to find exact answers to precisely formulated problems, but in applications one rarely finds precisely formulated problems, and it makes no sense to seek exact solutions. For the sake of the mathematically inclined, later in this section we will briefly address the pitfalls that may result when $N = \infty$ in (2). However, such difficulties are best handled by considering an alternate manner of expressing solutions in terms of integral formulas. This will be covered in Chapter 7.

A motivation for Fourier sine and cosine series

In the heat conduction problem for a rod of length L $(0 \le x \le L)$, where both ends are maintained at 0, recall that we wish to approximate the initial temperature by a linear combination $\sum_{n=1}^{N} b_n \sin(n\pi x/L)$. Roughly speaking, we will do this by extending the initial temperature distribution $f(x)$ defined on $[0,L]$, with $f(0) = 0$, to a new function (the odd extension) $f_o(x)$ defined on $[-L,L]$, in such a way that $f_o(x) = f(x)$ for $0 \le x \le L$, and $f_o(x) = -f(-x)$ for $-L \le x \le 0$. Since $f_o(x)$ is defined on $[-L,L]$, it makes sense to speak of its Fourier series FS $f_o(x)$. Since $f_o(x)$ is odd [i.e., $f_o(-x) = -f_o(x)$], all of the cosine terms will be absent in FS $f_o(x)$. Under suitable assumptions on $f_o(x)$, the partial sums of FS $f_o(x)$ will approximate $f_o(x)$ on $[-L,L]$, and hence also $f(x)$ on $[0,L]$, as desired. The series FS $f_o(x)$ is called the "Fourier sine series" of $f(x)$. By considering the "even extension" of $f(x)$, we analogously obtain the "Fourier cosine series" of $f(x)$. This would be of use in problems where both ends are insulated. Through the use of more elaborate extensions, we can obtain series approximations involving $\cos([n + \frac{1}{2}]\pi x/L)$ or $\sin([n + \frac{1}{2}]\pi x/L)$ for problems where one end is insulated and the other is maintained at 0. These new types of series are all modifications of the "standard" Fourier series of Section 4.2. Consequently, nearly all of the convergence properties established so far, carry over to these new series without much difficulty. In what follows, we supply the details of the constructions motivated by the above discussion.

Properties of even and odd functions

Definition. Let $f(x)$ be a function defined for $-L \le x \le L$. Then $f(x)$ is called **even**, if $f(-x) = f(x)$, and $f(x)$ is called **odd**, if $f(-x) = -f(x)$, for all x in $[-L,L]$.

Note that if $(x,f(x))$ is on the graph of an even function $f(x)$, then $(-x,f(x))$ will also be on the graph (i.e., the graph is invariant under reflection in the y–axis), as in Figure 1.

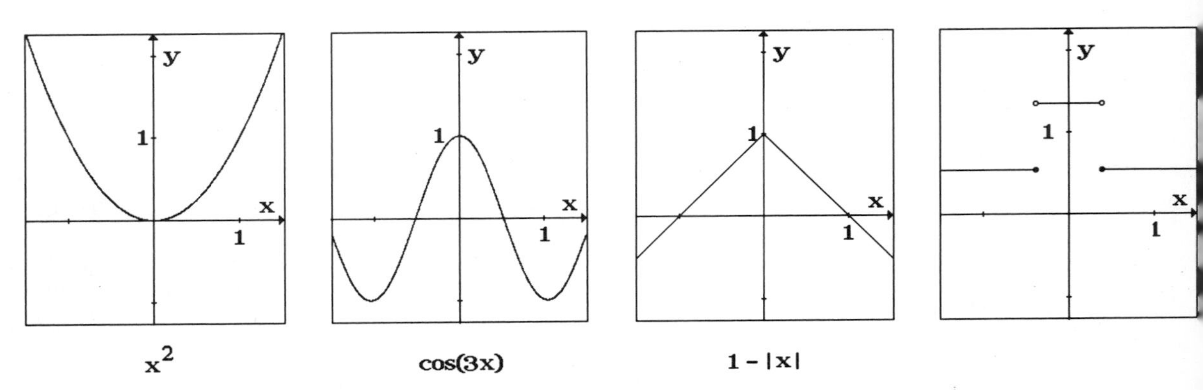

Figure 1 (Examples of even functions)

If $f(x)$ is odd, then $(x,f(x))$ is on the graph if and only if $(-x,-f(x))$ is on the graph (cf. Figure 2) ; i.e., the graph is invariant under reflection through the origin, $(x,y) \longleftrightarrow (-x,-y)$.

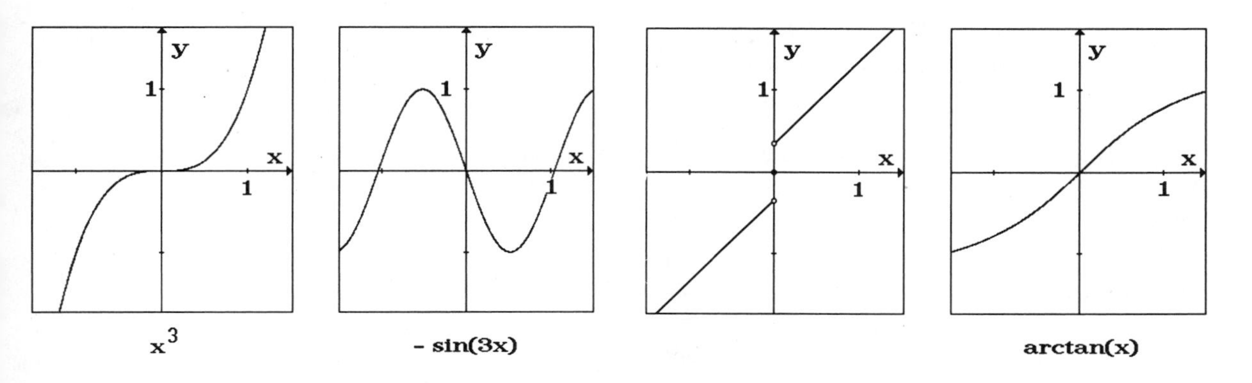

Figure 2 (Examples of odd functions)

Definition. Let $f(x)$ be a function defined for $0 \le x \le L$. The **even extension** of $f(x)$ is the unique even function $f_e(x)$ defined for x in $[-L,L]$ with $f_e(x) = f(x)$ for x in $[0,L]$, i.e.,

$$f_e(x) = \begin{cases} f(x) & \text{if } \quad 0 \le x \le L \\ f(-x) & \text{if } -L \le x \le 0 . \end{cases}$$

If $f(0) = 0$, we can also define the **odd extension** $f_o(x)$. It is the unique odd function defined for x in $[-L,L]$, such that $f_o(x) = f(x)$ for x in $[0,L]$, i.e.,

$$f_o(x) = \begin{cases} f(x) & \text{if } \quad 0 \le x \le L \\ -f(-x) & \text{if } -L \le x \le 0 . \end{cases}$$

Note that "$f(0) = 0$" is needed for consistency.

For example, suppose that $f(x) = \sqrt{x}$, $0 \le x \le L$. Then we have the graphs in Figure 3 ($L = 1.5$).

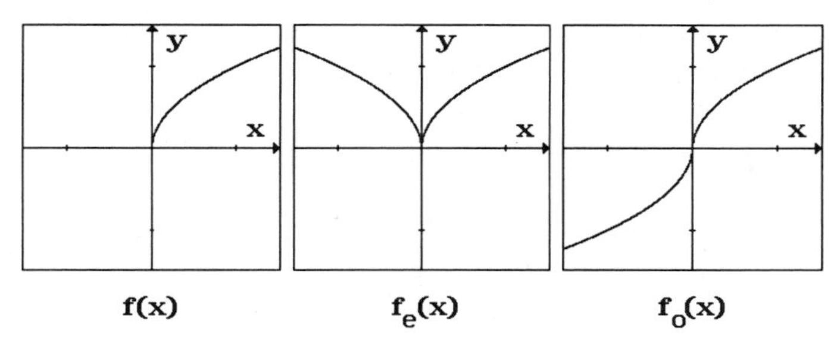

Figure 3

We collect some obvious facts concerning even and odd functions :

(A) The product of two even functions is even.
(B) The product of two odd functions is even.
(C) The product of an odd function and an even function is odd.
(D) If $f(x)$ is odd $(-L \le x \le L)$, then $\int_{-L}^{L} f(x)\,dx = 0$, if the integral exists.
(E) If $f(x)$ is even $(-L \le x \le L)$, then $\int_{-L}^{L} f(x)\,dx = 2\int_{0}^{L} f(x)\,dx$, if the integrals exist.

Proposition 1. Let $f(x)$ **be a function, defined for** $-L \le x \le L$**, with Fourier coefficients**

$$a_n = \frac{1}{L}\int_{-L}^{L} f(x)\cos(n\pi x/L)\,dx \quad \textbf{and} \quad b_n = \frac{1}{L}\int_{-L}^{L} f(x)\sin(n\pi x/L)\,dx\,.$$

If $f(x)$ **is even, then** $b_n = 0$ $(n = 1, 2, 3,...)$ **and**

$$a_n = \frac{2}{L}\int_{0}^{L} f(x)\cos(n\pi x/L)\,dx\,, \qquad (n = 0, 1, 2,...). \tag{3}$$

If $f(x)$ **is odd, then** $a_n = 0$ $(n = 0, 1, 2,...)$ **and**

$$b_n = \frac{2}{L}\int_{0}^{L} f(x)\sin(n\pi x/L)\,dx\,, \qquad (n = 1, 2, 3,...). \tag{4}$$

Proof. If $f(x)$ is even, then $b_n = 0$ by facts (C) and (D), since $\sin(n\pi x/L)$ is odd. Formula (3) follows from (A) and (E). The case when $f(x)$ is odd is handled similarly. □

Fourier sine and cosine series

Definition. Let $f(x)$ be a function defined on $[0,L]$, such that the integrals (3) and (4) exist. Then the **Fourier sine series** of $f(x)$ is the expression

$$\text{FSS } f(x) = \sum_{n=1}^{\infty} b_n\sin(n\pi x/L)\,, \quad \text{where} \quad b_n = \frac{2}{L}\int_{0}^{L} f(x)\sin(n\pi x/L)\,dx\,. \tag{5}$$

The **Fourier cosine series** of $f(x)$ is the expression

$$\text{FCS } f(x) = \tfrac{1}{2}a_0 + \sum_{n=1}^{\infty} a_n\cos(n\pi x/L)\,, \quad \text{where} \quad a_n = \frac{2}{L}\int_{0}^{L} f(x)\cos(n\pi x/L)\,dx\,. \tag{6}$$

Proposition 2. **Let $f(x)$ be defined for $0 \le x \le L$, and suppose that the integrals in (5) and (6) exist. Then (redefining $f(0)$ to be 0) the Fourier sine series of $f(x)$ is the Fourier series of the odd extension $f_o(x)$ defined on $[-L,L]$. The Fourier cosine series of $f(x)$ is the Fourier series of the even extension $f_e(x)$ defined on $[-L,L]$, i.e.,**

$$\text{FSS } f(x) = \text{FS } f_o(x) \quad \textbf{and} \quad \text{FCS } f(x) = \text{FS } f_e(x) .$$

Proof. We simply check that the Fourier coefficients of $f_o(x)$ are given by $a_n = 0$ and b_n as in (5). Indeed, $a_n = 0$ (for all $n = 0, 1, 2, \ldots$) by Proposition 1, and

$$b_n = \frac{1}{L}\int_{-L}^{L} f_o(x)\sin(n\pi x/L)\,dx = \frac{2}{L}\int_0^L f_o(x)\sin(n\pi x/L)\,dx \quad \text{(by property (E))},$$

which is as in (5), since $f_o(x) = f(x)$ for $0 \le x \le L$. Similarly, $\text{FCS } f(x) = \text{FS } f_e(x)$. □

Since $\text{FSS } f(x) = \text{FS } f_o(x)$ and $\text{FCS } f(x) = \text{FS } f_e(x)$, we can obtain convergence results for $\text{FSS } f(x)$ and $\text{FCS } f(x)$ by applying the theorems of Section 4.2 to the extensions $f_o(x)$ and $f_e(x)$. The following theorems suffice for most of our applications of Fourier sine and cosine series. We recall the notation for left–hand and right–hand limits :

$$f(x_0^+) = \lim_{x \downarrow x_0} f(x) \quad \text{and} \quad f(x_0^-) = \lim_{x \uparrow x_0} f(x) .$$

Theorem 1. **Let $f(x)$ be a piecewise C^1 function defined on $[0,L]$. Then**

$$\text{FSS } f(x) = \begin{cases} \frac{1}{2}[f(x^-) + f(x^+)] & 0 < x < L \\ 0 & x = 0 \ \textbf{ or } \ x = L \end{cases} . \tag{8}$$

If $f(x)$ is also continuous on $[0,L]$ with $f(0) = 0$ and $f(L) = 0$, then the partial sums $S_N(x)$ of $\text{FSS } f(x)$ converge uniformly to $f(x)$ on $[0,L]$, i.e.,

$$\max_{0 \le x \le L} \left\{ |f(x) - S_N(x)| \right\} \to 0 \quad \textbf{as} \quad N \to \infty . \tag{9}$$

Proof. If necessary, redefine $f(0)$ to be 0, and let $f_o(x)$ be the odd extension of $f(x)$. Then

$f_o(x)$ is a piecewise C^1 function defined on $[-L,L]$. By Theorem 3 of Section 4.2, we know that FS $f_o(x)$ converges to the adjusted function $\bar{f}_o(x)$. Note that $\bar{f}_o(0) = \frac{1}{2}[f_o(0^+) + f_o(0^-)] = \frac{1}{2}[f(0^+) - f(0^+)] = 0$, and $\bar{f}_o(L) = \frac{1}{2}[f_o(-L^+) + f_o(L^-)] = \frac{1}{2}[-f(L^-) + f(L^-)] = 0$, and for $0 < x < L$, $\bar{f}_o(x) = \frac{1}{2}[f(x^+) + f(x^-)]$. Thus, by Proposition 2 above and Theorem 3 of Section 4.2, we have FSS $f(x) =$ FS $f_o(x) = \bar{f}_o(x)$, from which (8) follows. If $f(x)$ is also continuous, with $f(0) = 0$ and $f(L) = 0$, then $f_o(x)$ is continuous with $f_o(L) = f_o(-L) = 0$. Thus, Theorem 4 of Section 4.2 applies to $f_o(x)$, and we have (9), since $f_o(x) = f(x)$ on $[0,L]$. □

Theorem 2. Let $f(x)$ be a piecewise C^1 function on $[0,L]$. Then

$$\text{FCS } f(x) = \begin{cases} \frac{1}{2}[f(x^-) + f(x^+)] & 0 < x < L \\ f(0^+) & x = 0 \\ f(L^-) & x = L\,. \end{cases} \tag{10}$$

If $f(x)$ is also continuous on $[0,L]$, then the partial sums $S_N(x)$ of FCS $f(x)$ converge uniformly to $f(x)$ in the sense of (9).

Proof. We apply Theorem 3 of Section 4.2 to the (piecewise C^1) even extension $f_e(x)$ in order to obtain FCS $f(x) = \bar{f}_e(x)$. Note that $\bar{f}_e(0) = \frac{1}{2}[f_e(0^-) + f_e(0^+)] = \frac{1}{2}[f(0^+) + f(0^+)] = f(0^+)$, and $\bar{f}_e(L) = \frac{1}{2}[f_e(L^-) + f_e(-L^+)] = \frac{1}{2}[f(L^-) + f(L^-)] = f(L^-)$, and for $0 < x < L$, $\bar{f}_e(x) = \frac{1}{2}[f(x^-) + f(x^+)]$. Hence (10) follows from FCS $f(x) =$ FS $f_e(x) = \bar{f}_e(x)$. If $f(x)$ is continuous, then $f_e(x)$ is continuous, and Theorem 4 of Section 4.2 yields the uniform convergence. □

Example 1. Find the Fourier sine and cosine series for the function $f(x) = L - x$ $(0 \le x \le L)$, and sketch the graphs of FSS $f(x)$ and FCS $f(x)$ in the interval $[-3L,3L]$.

Solution. We compute FSS $f(x)$, using Green's formula (cf. (9) of Section 4.1). Recall that $s_n(x) = \sin(n\pi x/L)$. Using the inner product notation $\langle g,h\rangle = \int_0^L g(x)h(x)\,dx$ (now on $[0,L]$!),

$$b_n = \frac{2}{L}\int_0^L f(x)\sin(n\pi x/L)\,dx = \frac{2}{L}<f,s_n> = -\frac{2}{L}(L/n\pi)^2<f,s_n''>$$
$$= -\frac{2L}{\pi^2 n^2}\left[[f(x)s_n'(x) - f'(x)s_n(x)]\Big|_0^L + <f'',s_n>\right]$$
$$= -\frac{2L}{\pi^2 n^2}(-L\cdot n\pi/L + 0) = \frac{2L}{n\pi}\,.$$

Thus,
$$\text{FSS } f(x) = \frac{2L}{\pi}\sum_{n=1}^{\infty}\frac{1}{n}\sin(n\pi x/L)\,. \tag{11}$$

We know from Proposition 2 that FSS $f(x)$ = FS $f_o(x)$ which is $\tilde{\bar{f}}_o(x)$ by Theorem 3 of Section 4.2. Thus, the graph of FSS $f(x)$ is the same as the graph of $\tilde{\bar{f}}_o(x)$ (cf. Figure 4).

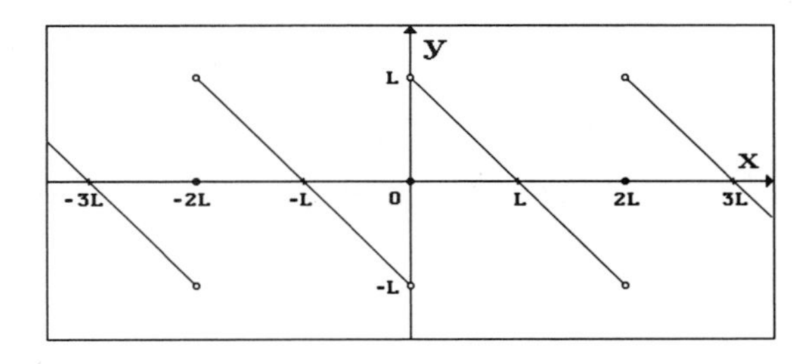

Figure 4

The restriction of the graph to the interval $[-L,L]$ is the graph of $\bar{f}_o(x)$ which is properly "adjusted" at $x = 0$. The series (11) does not converge uniformly, because $\tilde{\bar{f}}_o$ is not continuous. Note that $f(x)$ is continuous, but $f(0) \neq 0$, as is needed for uniform convergence in Theorem 1.

For FCS $f(x)$, we compute $a_0 = \frac{2}{L}\int_0^L (L-x)\,dx = L$, and for $n \geq 1$,

$$a_n = \frac{2}{L}\int_0^L f(x)\cos(n\pi x/L)\,dx = \frac{2}{L}<f,c_n> = -\frac{2}{L}(L/n\pi)^2<f,c_n''>$$
$$= -\frac{2L}{\pi^2 n^2}\left[[f(x)c_n'(x) - f'(x)c_n(x)]\Big|_0^L + <f'',c_n>\right] = \frac{2L}{\pi^2 n^2}\left[-(-1)^n + 1\right].$$

Thus,
$$\text{FCS } f(x) = \frac{L}{2} + \frac{2L}{\pi^2}\sum_{n=1}^{\infty}\frac{1}{n^2}[(1-(-1)^n]\cos(n\pi x/L) \tag{12}$$
$$= \frac{L}{2} + \frac{4L}{\pi^2}\sum_{k=0}^{\infty}\frac{1}{(2k+1)^2}\cos[(2k+1)\pi x/L]\,.$$

We know from Proposition 2 above and Theorem 3 of Section 4.2, that FCS $f_e(x) = \tilde{\bar{f}}_e(x)$. Thus, the graph of FCS f(x) is the same as the graph of $\tilde{\bar{f}}_e(x)$ (cf. Figure 5).

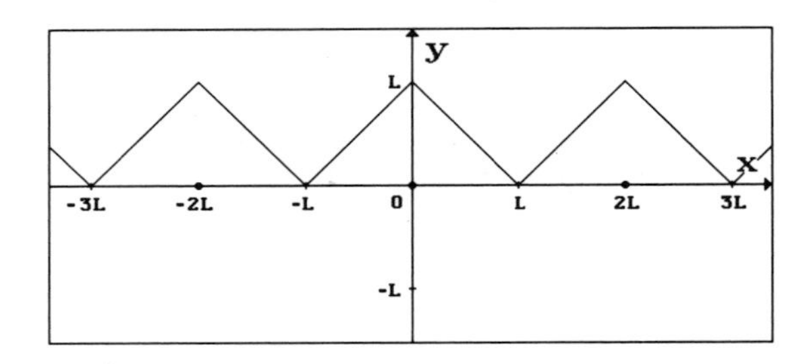

Figure 5

In this case, the series (12) converges uniformly, as can be inferred directly from the coefficients that decrease as n^{-2}. Alternatively, uniform convergence is guaranteed by Theorem 2 (which does not require $f(0) = 0$ as in Theorem 1), or we could use Theorem 4 of Section 4.2, since $f_e(x)$ is a continuous piecewise C^1 function and $f_e(-L) = f_e(L)$. □

Example 2. Find the Fourier cosine series of $f(x) = \sin(x)$ $[0 \le x \le \pi]$, and sketch the graph of FCS f(x), for $-2\pi \le x \le 2\pi$.

Solution. We use Green's formula to compute a_n , for $n = 2, 3, 4, \ldots$:

$$
\begin{aligned}
a_n &= \frac{2}{\pi}\int_0^\pi f(x)\cos(nx)\,dx = \frac{2}{\pi}\langle f,c_n\rangle = -\frac{2}{\pi n^2}\langle f,c_n''\rangle \\
&= -\frac{2}{\pi n^2}\Big[\big[f(x)c_n'(x) - f'(x)c_n(x)\big]\Big|_0^\pi + \langle f'',c_n\rangle\Big] \\
&= -\frac{2}{\pi n^2}\Big[[(-1)^n + 1] - \langle f,c_n\rangle\Big] = -\frac{2}{\pi n^2}\Big[(-1)^n + 1 - \frac{\pi}{2}a_n\Big] .
\end{aligned}
$$

Thus, $n^2 a_n = -\frac{2}{\pi}[(-1)^n + 1] + a_n$, and solving for a_n yields

$$
a_n = \frac{-2[(-1)^n+1]}{\pi(n^2-1)}, \quad \text{for } n = 2, 3, 4, \ldots .
$$

Separate calculations reveal that $a_0 = \frac{4}{\pi}$ and $a_1 = 0$. Thus,

$$\text{FCS } f(x) = \frac{2}{\pi} - \frac{2}{\pi}\sum_{n=2}^{\infty} \frac{[(-1)^n+1]}{(n^2-1)} \cos(nx) = \frac{2}{\pi} - \frac{4}{\pi}\sum_{k=1}^{\infty} \frac{1}{4k^2-1} \cos(2kx) . \tag{13}$$

The graphs of FCS f(x) and $\tilde{f}_e(x)$ coincide (cf. Figure 6).

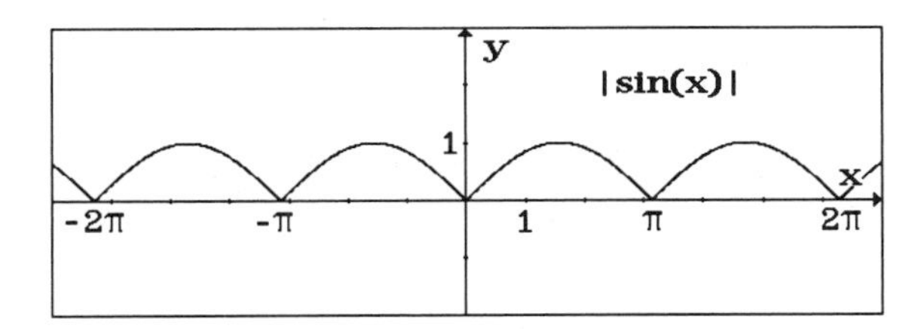

Figure 6

Of course, the Fourier sine series of f(x) is just sin(x). One reason why the cosine series might be desired, instead of the much simpler sine series, is that f(x) might be the initial temperature distribution in a rod with *insulated* ends, in which case we need to approximate f(x) by a linear combination of the functions cos(nx) , n = 0,1,2, ... □

Recall (cf. Problem 3 of Section 4.2) that for the problem

$$\begin{aligned} &\text{D.E.} \quad u_t = ku_{xx} \qquad 0 \le x \le L ,\ t \ge 0 \\ &\text{B.C.} \quad u(0,t) = 0 \quad u_x(L,t) = 0 \\ &\text{I.C.} \quad u(x,0) = f(x) , \end{aligned} \tag{14}$$

where the end at x = 0 is held at zero and the end at x = L is insulated, we found that if $f(x) = \sum_{n=1}^{N} c_n \sin[(n + \frac{1}{2})\pi x/L]$, then $u(x,t) = \sum_{n=1}^{N} c_n e^{-\lambda_n^2 kt} \sin(\lambda_n x)$, $\lambda_n = (n + \frac{1}{2})\pi/L$. Thus, for the problem (14), it is desirable to approximate f(x) by a linear combination of the functions $\sin[(n + \frac{1}{2})\pi x/L]$, instead of $\sin(n\pi x/L)$. Fortunately, it is not necessary to develop from scratch a whole new theory of Fourier series in terms of $\sin[(n + \frac{1}{2})\pi x/L]$. Indeed, we can obtain the new theory from the old one, as follows.

Let $f^e(x)$ denote the function defined on [0,2L] by

$$f^e(x) = \begin{cases} f(x) & 0 \le x \le L \\ f(2L-x) & L \le x \le 2L \end{cases} .$$

Pictorially, the graph of $f^e(x)$ is obtained by reflecting the graph of f(x) in the vertical line through x = L , as in Figure 7.

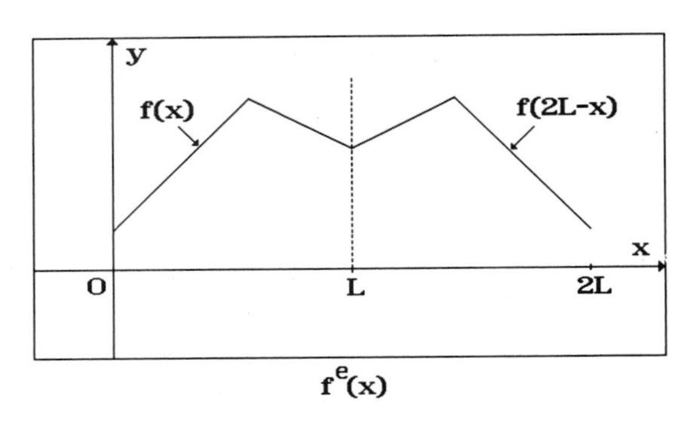

Figure 7

For this reason, we call $f^e(x)$ the **even extension of f(x) about $x = L$** (as opposed to $f_e(x)$ which is the even extension (about 0) considered earlier). We now show that the desired series for $f(x)$ in terms of $\sin[(n+\frac{1}{2})\pi x/L]$ is obtained by taking the usual Fourier sine series of $f^e(x)$ on the interval $[0,2L]$. In Problem 7, the reader is asked to state and prove an analogous theorem for series representations in terms of $\cos[(n+\frac{1}{2})\pi/L]$.

Theorem 3. The Fourier sine series of $f^e(x)$ on $[0,2L]$ is given by

$$\text{FSS } f^e(x) = \sum_{n=0}^{\infty} c_n \sin[(n+\tfrac{1}{2})\pi x/L] \ , \tag{15}$$

where

$$c_n = \frac{2}{L}\int_0^L f(x)\sin[(n+\tfrac{1}{2})\pi x/L]\, dx \ , \quad n = 0, 1, 2, \dots \ . \tag{16}$$

If $f(x)$ [and hence $f^e(x)$] is piecewise C^1, then

$$\text{FSS } f^e(x) = \begin{cases} \frac{1}{2}[f^e(x^-) + f^e(x^+)] & 0 < x < 2L \\ 0 & x = 0 \quad \text{or} \quad x = 2L \ . \end{cases} \tag{17}$$

In particular, for $0 \le x \le L$, we have

$$\text{FSS} f^e(x) = \begin{cases} \frac{1}{2}[f(x^-) + f(x^+)] & 0 < x < L \\ 0 & x = 0 \\ f(L^-) & x = L \ . \end{cases}$$

Moreover, if $f(x)$ is piecewise C^1 and *continuous* with $f(0) = 0$, then the partial sums $S_N(x)$ of FSS $f^e(x)$ converge uniformly to $f(x)$ on $[0,L]$.

Proof. By definition, the Fourier sine series of $f^e(x)$ defined on $[0,2L]$ is given by

$$\text{FSS } f^e(x) = \sum_{k=1}^{\infty} b_k \sin(k\pi x/2L) , \quad \text{where} \quad b_k = \frac{2}{2L}\int_0^{2L} f^e(x)\sin(k\pi x/2L)\, dx .$$

In order to prove (15) and (16), we must show that

$$b_k = \begin{cases} c_n & k = 2n+1 \quad n = 0, 1, 2, \ldots \\ 0 & k = 2n \quad n = 0, 1, 2, \ldots \end{cases} .$$

[Note that when $k = 2n+1$, $\sin(k\pi x/2L) = \sin[(n + \frac{1}{2})\pi x/L]$.] First, we show that $b_k = 0$ for k even, say $k = 2n$. Indeed,

$$b_{2n} = \frac{1}{L}\int_0^{2L} f^e(x)\sin(n\pi x/L)\, dx = 0 ,$$

because $\sin(n\pi x/L)$ is odd about $x = L$ and $f^e(x)$ is even about $x = L$ (i.e., consequently, the product is odd about $x = L$, and the integral vanishes). On the other hand, for $k = 2n+1$, $\sin[(n + \frac{1}{2})\pi x/L]$ is even about $x = L$, whence

$$b_{2n+1} = \frac{1}{L}\int_0^{2L} f^e(x)\sin[(n + \tfrac{1}{2})\pi x/L]\, dx = \frac{2}{L}\int_0^{L} f(x)\sin[(n + \tfrac{1}{2})\pi x/L]\, dx = c_n .$$

Thus, (15) and (16) follow. Now, (17) follows directly from Theorem 1, while (18) follows directly from (17), upon noting that $\frac{1}{2}[f^e(L^-) + f^e(L^+)] = f(L^-)$. Finally, the uniform convergence statement is immediate from Theorem 1, with $f^e(x)$ replaced by $f(x)$. □

Example 3. Let $f(x) = x(2L-x)$, for $0 \le x \le L$. Find a series representation for $f(x)$ of the form

$$f(x) = \sum_{n=0}^{\infty} c_n \sin[(n + \tfrac{1}{2})\pi x/L] . \tag{19}$$

Is the series uniformly convergent ? Estimate the error if this series is truncated.

Solution. According to Theorem 3, the series (19) with c_n defined by (16) will converge uniformly to $f(x)$ on $[0,L]$, since $f(x)$ is piecewise C^1, continuous and $f(0) = 0$. We compute the coefficients c_n using Green's formula, as follows.

$$c_n = \frac{2}{L}\int_0^L f(x)\sin[(n+\tfrac{1}{2})\pi x/L]\,dx = \frac{2}{L}<f,s_{n+\frac{1}{2}}> = -\frac{2}{L}\frac{L^2}{\pi^2(n+\frac{1}{2})^2}<f,s''_{n+\frac{1}{2}}>$$

$$= -\frac{2L}{\pi^2(n+\frac{1}{2})^2}\left[[s'_{n+\frac{1}{2}}(x)f(x) - s_{n+\frac{1}{2}}(x)f'(x)]\Big|_0^L + <f'',s_{n+\frac{1}{2}}>\right]$$

$$= -\frac{2L}{\pi^2(n+\frac{1}{2})^2}\left[0 - \int_0^L 2\cdot\sin[(n+\tfrac{1}{2})\pi x/L]\,dx\right]$$

$$= -\frac{4L^2}{\pi^3(n+\frac{1}{2})^3}\cos[(n+\tfrac{1}{2})\pi x/L]\Big|_0^L = \frac{4L^2}{\pi^3(n+\frac{1}{2})^3}.$$

Letting $S_N(x) = \sum_{n=0}^{N} c_n\sin[(n+\frac{1}{2})\pi x/L]$, we have

$$|f(x) - S_N(x)| \le \sum_{n=N+1}^{\infty}|c_n| \le \frac{4L^2}{\pi^3}\int_N^{\infty}(x+\tfrac{1}{2})^{-3}\,dx = \frac{2L^2}{\pi^3}(N+\tfrac{1}{2})^{-2},$$

which provides an error estimate for the truncation $S_N(x)$. □

Formal solutions versus exact solutions

In the following examples, we illustrate the use (and potential abuse) of various Fourier series representations of solutions of heat conduction problems, in the case where the initial temperature distribution is not a *finite* sum of sine and/or cosine functions, as was assumed in Chapter 3.

Example 4. Attempt to find an exact solution of the problem

$$\begin{array}{lll} \text{D.E.} & u_t = u_{xx} & 0 \le x \le L\,,\ t \ge 0 \\ \text{B.C.} & u(0,t) = 0 & u(L,t) = 0 \\ \text{I.C.} & u(x,0) = x(L-x)\ . & \end{array} \tag{20}$$

Effort. Although $f(x) = x(L-x)$ $[0 \le x \le L]$ is not a *finite* linear combination of the functions $\sin(n\pi x/L)$, we know from Theorem 1 that FSS $f(x) = f(x)$ on $[0,L]$. The Fourier sine coefficients of $f(x)$ can be computed using Green's formula , or two integrations by parts. For x in $[0,L]$, we have

$$f(x) = \text{FSS } f(x) = \frac{8L^2}{\pi^3}\sum_{k=0}^{\infty}\frac{1}{(2k+1)^3}\sin[(2k+1)\pi x/L]\ . \tag{21}$$

At this point, we are inclined to assert that

$$u(x,t) = \frac{8L^2}{\pi^3}\sum_{k=0}^{\infty}\frac{e^{-[2k+1]^2\pi^2t/L^2}}{(2k+1)^3}\sin[(2k+1)\pi x/L] \tag{22}$$

is the solution of problem (20). While (22) clearly satisfies the B.C. and I.C., it is *not* the case that (22) is a C^2 function that satisfies the D.E., for $0 \le x \le L$ and $t \ge 0$. Indeed, the D.E. is not satisfied at $x = 0$ when $t = 0$. Note that $u_{xx}(0,0) = f''(0) = -2$, while $u_t(0,0) = 0$, since $u(0,t) \equiv 0$. Thus, $u_t \ne u_{xx}$ at $(x,t) = (0,0)$. Actually, problem (20) has no C^2 solution defined on the strip $0 \le x \le L$, $t \ge 0$, because we have just seen that the I.C. and the B.C. contradict the D.E. at $(x,t) = (0,0)$. Here, we also have an example where *the superposition principle fails for infinite sums.* Indeed each term of (22) satisfies the D.E., but the entire sum does not satisfy the the D.E. ! The sum (22) is also not C^2 at $(x,t) = (0,0)$, because $u_{xx}(0^+,0) = -2$, while $u_{xx}(0^-,0) = f_o''(0^-) = 2$, since $f_o(x) = x(x+L)$ for $-L \le x \le 0$. □

Remark. In the above example, we attempted to obtain a solution of a heat problem, by expanding the initial temperature in terms of an appropriate type of Fourier series (e.g., as in (21)) and inserting the correct time–dependent exponential factors (e.g., as was done to form (22)). We have seen that the resulting infinite sum $u(x,t)$ need not be a solution of the problem Indeed, there might not be any C^2 solution. Nevertheless, we refer to such infinite sums as formal solutions of the problem. More precisely, we state the following definition.

> **Definition.** Given an initial/boundary–value problem for the heat equation, the expression $u(x,t)$, obtained from the appropriate type of Fourier series (depending on the B.C.) of the initial temperature by inserting the correct time–dependent exponential factors, is known as the **formal solution** of the problem. (More generally, the term "formal" describes any plausible result or procedure which may be unjustified or unjustifiable.)

In many cases it is true, but rather difficult to show, that the formal solution does satisfy the D.E. when $t > 0$, even if it fails to satisfy the D.E. at $t = 0$, say due to an initial temperature which is not C^2. We will demonstrate this in Chapter 7. Frequently, if one truncates a formal solution after a sufficiently large *finite* number of terms, then one obtains an exact solution of the D.E. and B.C. which satisfies the I.C. to within any prescribed error, which is all that is necessary in applications. In other words, determining whether the formal solution is a solution in the strict sense (e.g., whether it is a C^2 solution of the D.E.) may be only of mathematical interest. □

Example 5. Find the formal solution of the problem

$$\begin{aligned} &\text{D.E.} \quad u_t = ku_{xx} \qquad 0 \le x \le \pi,\ t \ge 0 \\ &\text{B.C.} \quad u_x(0,t) = 0 \quad u_x(\pi,t) = 0 \\ &\text{I.C.} \quad u(x,0) = \sin(x)\ . \end{aligned} \tag{23}$$

By truncating the formal solution, find a solution of the D.E. and B.C. that meets the I.C. to within an experimental error of .01 . Is there any exact solution of this problem which is C^2 for all (x,t) with $0 \le x \le \pi$, $t \ge 0$?

Solution. From the B.C., we know (cf. Example 1 of Section 3.3) that we should expand $\sin(x)$ $(0 \le x \le \pi)$ into a cosine series. In Example 2, we found

$$\sin(x) = \frac{2}{\pi} - \frac{4}{\pi}\sum_{n=1}^{\infty} \frac{\cos(2nx)}{4n^2 - 1}, \quad 0 \le x \le \pi. \tag{24}$$

The *formal* solution of (23) is then

$$u(x,t) = \frac{2}{\pi} - \frac{4}{\pi}\sum_{n=1}^{\infty} e^{-4n^2kt} \cdot \frac{\cos(2nx)}{4n^2 - 1}. \tag{25}$$

There is no C^2 solution of problem (23), because $u_x(0^+,0) = \cos(0) = 1$ by the I.C. , and yet $u_x(0,0^+) = 0$ by the B.C. (i.e., the B.C. contradicts the I.C. at (0,0)). In particular, (25) is not a solution in the strict sense. The N–th partial sum $u_N(x,t) = \frac{2}{\pi} - \frac{4}{\pi}\sum_{n=1}^{N} e^{-4n^2kt} \cdot \frac{\cos(2nx)}{4n^2 - 1}$ of (25) is an exact (indeed, C^∞) solution of the D.E. and B.C., and

$$|u_N(x,0) - \sin(x)| \le \frac{4}{\pi}\sum_{n=N+1}^{\infty} \frac{1}{4n^2-1} \le \frac{4}{\pi}\int_N^\infty \frac{dx}{4x^2-1} = \frac{1}{\pi}\log_e[(2N+1)/(2N-1)],$$

which is less then .01 for $N \ge 32$. □

Example 6. Find a formal solution of the following problem :

D.E. $u_t = u_{xx} \qquad 0 \le x \le \pi, \ t \ge 0$

B.C. $u(0,t) = \sin(t) \qquad u(\pi,t) = 0$

I.C. $u(x,0) = 0$.

Solution. We apply the methods of Section 3.4 involving Duhamel's principle, and we treat the infinite sum which arises in a formal manner. A particular solution of the B.C. is easily found to be $w(x,t) = (1 - \frac{x}{\pi})\sin(t)$. The related problem for the function $v(x,t) = u(x,t) - w(x,t)$ is

$$\begin{aligned} &\text{D.E. } v_t - v_{xx} = u_t - u_{xx} - w_t + w_{xx} = (\tfrac{x}{\pi} - 1)\cos(t) \\ &\text{B.C. } v(0,t) = 0 \qquad v(\pi,t) = 0 \\ &\text{I.C. } v(x,0) = u(x,0) - w(x,0) = 0 . \end{aligned} \tag{26}$$

A formal use of Duhamel's principle [cf. Theorem 1 of Section 3.4] tells us that

$$v(x,t) = \int_0^t \tilde{v}(x,t-s;s)\, ds, \tag{27}$$

where $\tilde{v}(x,t;s)$ is the formal solution of the associated problem

$$\begin{aligned} &\text{D.E.} \quad \tilde{v}_t = \tilde{v}_{xx} \qquad 0 \le x \le \pi \, , \; t > 0 \\ &\text{B.C.} \quad \tilde{v}(0,t;s) = 0 \qquad \tilde{v}(\pi,t;s) = 0 \\ &\text{I.C.} \quad \tilde{v}(x,0;s) = (\tfrac{x}{\pi} - 1)\cos(s) \, . \end{aligned} \tag{28}$$

The formal solution of (28) is found by computing the Fourier sine series of $(\frac{x}{\pi} - 1)\cos(s)$ and then inserting the time dependent exponential factors. The result is

$$\tilde{v}(x,t;s) = -\frac{2}{\pi}\cos(s)\sum_{n=1}^{\infty} \frac{1}{n}\, e^{-n^2 t}\sin(nx) \, .$$

Using this [with t replaced by $t-s$] in (27) and formally interchanging the sum and the integral, we obtain the formal solution of (26) :

$$\begin{aligned} v(x,t) &= -\frac{2}{\pi}\sum_{n=1}^{\infty} \frac{1}{n}\left[\int_0^t \cos(s)\, e^{-n^2(t-s)} ds\right]\sin(nx) \\ &= -\frac{2}{\pi}\sum_{n=1}^{\infty} \frac{1}{n}\left[\frac{n^2\cos(t)+\sin(t)}{n^4+1} - \frac{n^2 e^{-n^2 t}}{n^4+1}\right]\sin(nx) \, . \end{aligned}$$

The formal solution of the original problem is then

$$u(x,t) = (1 - \tfrac{x}{\pi})\sin(t) + v(x,t) \; .$$

If we truncate the series for $v(x,t)$ at some integer $N > 0$, we obtain a C^∞ function $v_N(x,t)$ and the associated $u_N(x,t) = (1 - \frac{x}{\pi})\sin(t) + v_N(x,t)$. It is easy to check that $u_N(x,t)$ satisfies the B.C. and I.C. of the original problem exactly, regardless of the value of N. However, a computation shows

$$(u_N)_t - (u_N)_{xx} = \left[(1 - \tfrac{x}{\pi}) - \frac{2}{\pi}\sum_{n=1}^{N} \frac{1}{n}\sin(nx)\right]\cos(t) \, . \tag{29}$$

For any $0 < x \le \pi$, the expression in brackets approaches 0 as $N \to \infty$, since FSS $(1 - \frac{x}{\pi})$ $= \frac{2}{\pi}\sum_{n=1}^{\infty} \frac{1}{n}\sin(nx)$, and Theorem 1 applies. At $x = 0$, (29) reduces to $\cos(t)$, regardless of N, and so at $x = 0$, $u_N(x,t)$ will not nearly solve the D.E. as $N \to \infty$. The given problem has no exact solution. (Why ?). □

Remark. The form of a formal solution to a problem with inhomogeneous B.C. is not unique. In Example 6, if we had chosen a different particular solution of the B.C., say $\overline{w}(x,t) = (1 - \frac{x}{\pi})^3\sin(t)$, then we would obtain a formal solution of the form $u(x,t) = (1 - \frac{x}{\pi})^3\sin(t) + \overline{v}(x,t)$, which would have a different form than the original formal solution. □

Summary 4.3

1. Even and odd extensions : If $f(x)$ is defined on $[0,L]$, then the even extension $f_e(x)$ of $f(x)$ is the unique even function on $[-L,L]$ which is equal to $f(x)$ on $[0,L]$. If $f(0) = 0$, then the odd extension $f_o(x)$ of $f(x)$ is the unique odd function on $[-L,L]$ which is equal to $f(x)$ on $[0,L]$.

2. Fourier sine and cosine series : If $f(x)$ is defined on $[0,L]$, then the Fourier sine series of $f(x)$ is the expression

$$\text{FSS } f(x) = \sum_{n=1}^{\infty} b_n \sin(n\pi x/L) , \quad \text{where} \quad b_n = \frac{2}{L}\int_0^L f(x)\sin(n\pi x/L)\, dx . \qquad \text{(S1)}$$

The Fourier cosine series of $f(x)$ is the expression

$$\text{FCS } f(x) = \tfrac{1}{2}a_0 + \sum_{n=1}^{\infty} a_n \cos(n\pi x/L) , \quad \text{where} \quad a_n = \frac{2}{L}\int_0^L f(x)\cos(n\pi x/L)\, dx . \qquad \text{(S2)}$$

By Proposition 2, we have FSS $f(x)$ = FS $f_o(x)$ and FCS $f(x)$ = FS $f_e(x)$. Consequently, the convergence theorems for Fourier series in Section 4.2 can be used to prove convergence theorems for Fourier sine and cosine series (cf. Theorems 1 and 2).

3. Other Fourier series and extensions : If a function $f(x)$ defined on $[0,L]$ is extended to a function $f^e(x)$ on $[0,2L]$ in such a way that $f^e(x)$ is even about L, then (cf. Theorem 3)

$$\text{FSS } f^e(x) = \sum_{n=0}^{\infty} c_n \sin[(n + \tfrac{1}{2})\pi x/L] , \quad \text{where} \quad c_n = \frac{2}{L}\int_0^L f(x)\sin[(n + 1/2)\pi x/L]\, dx . \qquad \text{(S3)}$$

This modified series can be used to represent initial temperature functions $f(x)$ for a rod with B.C. $u(0,t) = 0$ and $u_x(L,t) = 0$ (cf. also Problem 7).

4. Formal solutions : In an initial/boundary–value problem for the heat equation with standard homogeneous B.C., if one writes down the appropriate type of Fourier series for the initial temperature $f(x)$, and inserts the correct time–dependent exponential factors, then the resulting expression $u(x,t)$ is called the formal solution. The formal solution might not be a solution in the strict sense, because an infinite sum of product solutions of the D.E. might not be a C^2 function which satisfies the D.E. (cf. Example 4). In other words, the superposition principle does not always hold for infinite sums. By truncating the formal solution after a finite number of terms, one has an exact solution of the D.E. and B.C. . This finite sum will satisfy the I.C. to within experimental error, provided the appropriate Fourier series of the initial temperature function $f(x)$ converges uniformly to $f(x)$, and sufficiently many terms are considered. In this sense, the question of the validity of formal solutions is not a serious issue in applications, although the question is resolved in Chapter 7. For problems with inhomogeneous B.C. the form of a formal solution depends on the choice of a particular solution of the B.C. .

Exercises 4.3

1. Verify the basic facts (A)–(E) preceding Proposition 1 concerning even and odd functions.

2. Show that any function $f(x)$ defined on $[-L,L]$ can be written as the sum of an even function and an odd function. **Hint.** Suppose $f(x) = g(x) + h(x)$, where $g(x)$ is even and $h(x)$ is odd. Then solve the equations, $f(x) = g(x) + h(x)$ and $f(-x) = g(x) - h(x)$, simultaneously for $g(x)$ and $h(x)$ in terms of $f(x)$ and $f(-x)$. Then check to see that $g(x)$ is even and $h(x)$ is odd.

3. Assuming that $f(x)$ and $f'(x)$ are defined on $[-L,L]$, show that $f'(x)$ is even if $f(x)$ is odd, and vice versa.

4. Let $f(x) = x$, $0 \le x \le L$.

(a) Compute the Fourier cosine series FCS $f(x)$.

(b) Compute the Fourier sine series FSS $f(x)$.

(c) Graph FCS $f(x)$ and FSS $f(x)$ for $-3L \le x \le 3L$.

Hint [For (c)]. FCS $f(x) = \tilde{\bar{f}}_e(x)$ and FSS $f(x) = \tilde{\bar{f}}_o(x)$ for all x, as observed in Example 1.

5. Repeat Problem 4 in the cases :

(a) $f(x) = 1$, $0 \le x \le L$. (b) $f(x) = \cos(x)$, $0 \le x \le \pi$.

6. In Problem 4, where $f(x) = x$ $(0 \le x \le L)$, explain why FCS $f(x)$ converges uniformly to $f(x)$ in $[0,L]$, whereas FSS $f(x)$ does not. **Hint.** See the proofs of Theorems 1 and 2.

7. Let $f(x)$ be defined on $[0,L]$, and define $f^o(x)$ on $[0,2L]$ by

$$f^o(x) = \begin{cases} f(x) & 0 \le x < L \\ 0 & x = L \\ -f(2L-x) & L < x \le 2L. \end{cases}$$

State and prove the analog of Theorem 3, for FCS $f^o(x)$. In heat problems, what sort of B.C. will dictate the use of FCS $f^o(x)$?

8. (a) Find the *formal* solution of the problem

$$\begin{array}{lll} \text{D.E.} & u_t = ku_{xx} & -\pi \le x \le \pi,\ t \ge 0 \\ \text{B.C.} & u(-\pi,t) = u(\pi,t) & u_x(-\pi,t) = u_x(\pi,t) \\ \text{I.C.} & u(x,0) = x^2. & \end{array} \qquad (*)$$

(b) Show that there can be no solution of problem (*) which is C^2 for $-\pi \le x \le \pi$ and $t \ge 0$. However, show that there is a C^∞ solution of the D.E. and B.C. that will satisfy the I.C. to within any specified error $\epsilon > 0$. **Hint.** Consider $u_x(\pm\pi,0)$.

9. (a) Find the formal solution of the problem

$$\begin{aligned} &\text{D.E.} \quad u_t = u_{xx} \qquad 0 \le x \le 1 \\ &\text{B.C.} \quad u(0,t) = 0 \qquad u(1,t) = 0 \\ &\text{I.C.} \quad u(x,0) = f(x) = \begin{cases} x & 0 \le x \le 1/2 \\ 1-x & 1/2 \le x \le 1 . \end{cases} \end{aligned}$$

(b) If $u_t(x,t)$ is formally computed by differentiating each term of the formal solution with respect to t, then show that $u_t(1/2,0) = -\infty$ results. Provide a physical explanation of this result by considering the flux of heat through the ends of a small interval centered at $x = 1/2$.

10. (a) Find a formal solution of the problem

$$\begin{aligned} &\text{D.E.} \quad u_t = ku_{xx} \qquad 0 \le x \le \pi \\ &\text{B.C.} \quad u(0,t) = 0 \qquad u_x(\pi,t) = 0 \\ &\text{I.C.} \quad u(x,0) = f(x) = \begin{cases} 0 & 0 \le x < \pi/2 \\ 1 & \pi/2 \le x \le \pi . \end{cases} \end{aligned}$$

(b) Does this formal solution satisfy the I.C. at $x = \pi/2$? Why not ? **Hint.** See Theorem 3.

(c) If we redefine $f(\pi/2) = 1/2$, and truncate the formal solution to N terms, is it possible to ensure that the truncated function $u_N(x,t)$ satisfies $|u_N(x,0)-f(x)| < .1$, for large N ? Why not ?

11. Find a formal solution of the problem

$$\begin{aligned} &\text{D.E.} \quad u_t = ku_{xx} \qquad 0 \le x \le 1 ,\ t \ge 0 \\ &\text{B.C.} \quad u(0,t) = -1 \qquad u_x(1,t) = 1 \\ &\text{I.C.} \quad u(x,0) = 0 . \end{aligned}$$

12. Find a formal solution of the problem

$$\begin{aligned} &\text{D.E.} \quad u_t = ku_{xx} \qquad 0 \le x \le 10 ,\ t \ge 0 \\ &\text{B.C.} \quad u_x(0,t) = 2 \qquad u_x(10,t) = 3 \\ &\text{I.C.} \quad u(x,0) = 0. \end{aligned}$$

13. Find a formal solution of the problem

$$\text{D.E.}\quad u_t = ku_{xx} \qquad 0 \le x \le \pi\,,\ t \ge 0$$
$$\text{B.C.}\quad u_x(0,t) = 2 \qquad u(\pi,t) = 1 - e^{-t}$$
$$\text{I.C.}\quad u(x,0) = 7\cdot\cos(3x/2).$$

Hint. Use Duhamel's principle as in Example 6.

14. (a) Recall from (24) that $\sin(x) = \frac{2}{\pi} - \frac{4}{\pi}\sum_{n=1}^{\infty}\frac{1}{4n^2-1}\cos(2nx)$, $0 \le x \le \pi$. Differentiate both sides (term–by–term for the right–hand side) with respect to x, and set $x = 0$. Deduce that the derivative of an infinite sum cannot always be computed by differentiating termwise.

(b) Show that the Fourier sine series for $\cos(x)$, $0 \le x \le \pi$, can be obtained by differentiating the terms in Fourier sine series (in part (a)) for $\sin(x)$. (You may use your answer to 5(b).) Does FSS $\cos(x) = \cos(x)$ at $x = 0$ and $x = \pi$? What if $0 < x < \pi$?

(c) Show that the Fourier cosine series for $\sin(x)$ *cannot* be computed by differentiating the terms of Fourier sine series for $\cos(x)$, $0 \le x \le \pi$.

Remark. The different results of parts (b) and (c) are explained in Problem 15.

15. (**Differentiation of Fourier series**)

(a) Show that if $f(x)$ is continuous and piecewise C^1 on $[-L,L]$ with $f(-L) = f(L)$, then the Fourier series of $f'(x)$ can be computed by differentiating term–by–term FS $f(x)$.

Hint. Integration by parts implies $\frac{1}{L}\int_{-L}^{L} f'(x)\cos(n\pi x/L)\,dx = (n\pi/L)\frac{1}{L}\int_{-L}^{L} f(x)\sin(n\pi x/L)dx$, and $\frac{1}{L}\int_{-L}^{L} f'(x)\sin(n\pi x/L)\,dx = -(n\pi/L)\frac{1}{L}\int_{-L}^{L} f(x)\cos(n\pi x/L)dx$. Note that integration by parts will not be valid in general if $f(x)$ is piecewise C^1 but *not* continuous (Why?).

(b) In view of part (a), explain the results of parts (b) and (c) in Problem 14, by considering the continuity of the even extension of $\sin(x)$ and the odd extension of $\cos(x)$ $(0 \le x \le L)$.

16. (**Integration of Fourier series**)

(a) Show that if $f(x)$ is piecewise continuous on $[-L,L]$ and $\int_{-L}^{L} f(x)\,dx = 0$, then its antiderivative $G(x) \equiv \int_0^x f(x)\,dx$ is continuous and piecewise C^1 with $G(-L) = G(L)$. Conclude directly from part (a) of Problem 15 that FS $f(x)$ can be obtained via term–by–term differentiation of FS $G(x)$, and hence that (aside from the constant term) FS $G(x)$ can be obtained from FS $f(x)$ via term–by–term integration of FS $f(x)$. The constant term $\frac{1}{2L}\int_{-L}^{L} G(x)\,dx$ of FS $G(x)$ must be computed separately.

(b) Find FS x^2 from FS x $(-L \le x \le L)$ using the method of part (a).

(c) Suppose that we omit the constant term in FS x^2 $(-L \le x \le L)$ and integrate all of the other terms of FS $f(x)$. What function has the resulting series as its Fourier series ?

17. (**Wirtinger's inequality**). Let $f(x)$ be a continuous, piecewise C^1 function defined on $[-L,L]$, such that $\int_{-L}^{L} f(x)\,dx = 0$ and $f(-L) = f(L)$. Use Problem 14(a) and Parseval's equality to show

$$\int_{-L}^{L} f(x)^2\,dx \le \frac{L^2}{\pi^2}\int_{-L}^{L} f'(x)^2\,dx \qquad \text{(Wirtinger's inequality)},$$

with equality holding only when $f(x) = a_1\cos(\pi x/L) + b_1\sin(\pi x/L)$ (i.e., $a_n = b_n = 0,\ n \ge 2$).

(b) If $f(x)$ is defined and piecewise C^1 on $[0,L]$ with $f(0) = f(L) = 0$, then show that Wirtinger's inequality holds with the lower limits 0 instead of $-L$ in the integrals.
(For related inequalities, see [Beckenbach and Bellman, Chapter 5] and the references therein.)

18. (**The Isoperimetric Inequality**) Here we use Wirtinger's inequality of Problem 13 to show that a "sufficiently nice" closed, curve of length 2π encloses an area of at most π (the Isoperimetric Inequality), where equality holds only in the case of the circle. Assume that the curve is parametrized by the periodic C^1 functions $x(t)$, $y(t)$ of period 2π, with unit speed (i.e., $[x'(t)^2 + y'(t)^2]^{\frac{1}{2}} \equiv 1$). Moreover, assume that the center of gravity of the curve is at the origin (i.e., $\int_{-\pi}^{\pi} x(t)\,dt = \int_{-\pi}^{\pi} y(t)\,dt = 0$). Let $\mathbf{r}(t) = x(t)\mathbf{i} + y(t)\mathbf{j}$, and let $|\mathbf{r}(t)| = [x(t)^2 + y(t)^2]^{\frac{1}{2}}$. Suppose that $\mathbf{r}(t) \ne \mathbf{0}$ and that the angle $\alpha(t)$ from $\mathbf{r}(t)$ to $\mathbf{r}'(t)$ satisfies $0 \le \alpha(t) \le \pi$.

(a) Show that the area enclosed by the curve is $A = \int_{-\pi}^{\pi} \frac{1}{2}|\mathbf{r}(t)|\sin(\alpha(t))\,dt$, where $\alpha(t)$ is the angle from $\mathbf{r}(t)$ to $\mathbf{r}'(t)$, which we assume is between 0 and π. (Draw a picure of a "triangular" element of the area, with vertices at the ends of position vectors $\mathbf{r}(t)$, $\mathbf{r}(t+\Delta t)$ and $\mathbf{0}$.)

(b) Use the Cauchy–Schwarz inequality $\langle f,g\rangle^2 \le \|f\|^2\|g\|^2$ (cf. Problem 9 of Section 4.2) to show that $A^2 \le \frac{1}{2}\pi\int_{-\pi}^{\pi} |\mathbf{r}(t)|^2\,dt$ (Take $f(t) = \frac{1}{2}\sin(\alpha(t)) \le \frac{1}{2}$ and $g(t) = |\mathbf{r}(t)|$).

(c) Use part (b) and Wirtinger's inequality (cf. Problem 17) to deduce that

$$A^2 \le \tfrac{1}{2}\pi\int_{-\pi}^{\pi} |\mathbf{r}(t)|^2\,dt = \tfrac{1}{2}\pi\int_{-\pi}^{\pi} x(t)^2 + y(t)^2\,dt \le \tfrac{1}{2}\pi\int_{-\pi}^{\pi} |\mathbf{r}'(t)|^2\,dt = \pi^2 .$$

(d) Show that $A = \pi$ only in the case of a circle. (Note that one of the inequalities above will be strict if $\sin(\alpha(t)) \not\equiv 1$.)

Remarks. (1) For more on isoperimetric inequalities, see [Pòlya and Szegö, 1951] and [Osserman, 1978] and the references therein.
(2) In [W. Blaschke, 1916] it is proved that if L is the length of a polygon with n equal sides (but not necessarily equal interior angles) which encloses an area A, then $L^2 \geq 4n \tan(\pi/n) \cdot A$. As $n \to \infty$, we obtain $L^2 \geq 4\pi A$ for curves whose lengths and enclosed areas can be approximated arbitrarily closely by such polygons.

4.4 Sturm–Liouville Theory

In Chapter 3, and in the chapters to follow, the method of separation of variables is used to obtain product solutions which satisfy the D.E. and B.C. of initial/boundary–value problems. This method leads to new boundary–value problems involving ODEs and the purpose of this section is to study these problems. For example, in the case of the initial/boundary–value problem

$$\begin{aligned} &\text{D.E.}\quad u_t = ku_{xx} \qquad 0 \le x \le L\,,\ t \ge 0 \\ &\text{B.C.}\quad u(0,t) = 0 \quad u(L,t) = 0 \\ &\text{I.C.}\quad u(x,0) = f(x)\,, \end{aligned} \tag{1}$$

the process of separation of variables, where the product solutions $u(x,t) = X(x)T(t)$ satisfy the D.E. and B.C. of (1), led to the problem

$$\begin{aligned} &\text{D.E.}\quad X''(x) + \lambda X(x) = 0 \qquad 0 \le x \le L \\ &\text{B.C.}\quad X(0) = 0\,,\quad X(L) = 0\,, \end{aligned} \tag{2}$$

where λ denotes the separation constant. We obtain a more general boundary–value problem, if we consider the derivation of the heat equation (cf. Section 3.1) in the case when the material of the rod varies from cross section to cross section. Thus, if the linear density, D, the specific heat, C, and the thermal conductivity, K, of the rod are functions of x, then we obtain (cf. Problem 8 of Section 3.1) the D.E.

$$C(x)D(x)u_t = \frac{\partial}{\partial x}(K(x)u_x)\,. \tag{3}$$

In addition, if we allow a heat source of the form $Q(x)u(x,t)$, then in place of (3) we obtain

$$\text{D.E.}\quad C(x)D(x)u_t = \frac{\partial}{\partial x}(K(x)u_x) + Q(x)u\,. \tag{4}$$

Let the ends of the rod be at $x = a$ and $x = b$, and consider the homogeneous B.C. given by

$$\text{B.C.}\quad \begin{aligned} &c_1u(a,t) + c_2u_x(a,t) = 0 \quad (c_1^2 + c_2^2 \ne 0) \\ &c_3u(b,t) + c_4u_x(b,t) = 0 \quad (c_3^2 + c_4^2 \ne 0)\,, \end{aligned} \tag{5}$$

where c_1, c_2, c_3 and c_4 are real constants. For a physical interpretation of such B.C., see Example 6 of Section 3.3). Substituting $u(x,t) = X(x)T(t)$ into (4), we get

$$\frac{T'}{T} = \frac{\frac{d}{dx}(KX')}{CDX} + \frac{Q}{CD} = -\lambda\,, \tag{6}$$

by separation of variables, where λ is a constant. In particular, we obtain the D.E.

$$\frac{d}{dx}(KX') + (Q + \lambda CD)X = 0\,. \tag{7}$$

Setting $g(x) = C(x)D(x)$ and $q(x) = Q(x)$ and using the B.C. (5), we get

$$\begin{aligned} &\text{D.E.} \quad \frac{d}{dx}[K(x)X'(x)] + (q(x) + \lambda g(x))X(x) = 0\,, \quad a \le x \le b \\ &\text{B.C.} \quad c_1X(a) + c_2X'(a) = 0\,, \quad c_3X(b) + c_4X'(b) = 0\,, \end{aligned} \tag{8}$$

where we assume that $K(x)$, $g(x) > 0$ on $[a,b]$, and that $q(x)$, $g(x)$, $K'(x)$ and $X''(x)$ are continuous on $[a,b]$. The boundary value problem (8) is called a **Sturm–Liouville problem**. [The French mathematician, Joseph Liouville (1809–1882), was first to solve a boundary value problem by solving an equivalent integral equation. Jacques Charles François Sturm (1803–1855) was a Swiss mathematician, who in collaboration with Liouville, studied boundary value problems.]

Definition. The D.E. in (8) is called a **Sturm–Liouville equation.** A value of the parameter λ for which a *nontrivial solution* (i.e. $X \not\equiv 0$) exists is called an **eigenvalue** of the problem and a corresponding nontrivial solution $X(x)$ of (8) is called an **eigenfunction** which is associated with that eigenvalue. Problem (8) is also called an **eigenvalue problem**.

Remark. Since both the D.E. and the B.C. of (8) are homogeneous, the trivial function $X(x) \equiv 0$ satisfies (8), regardless of the value of λ. However, as we will show, for most values of λ, the only solution is the trivial solution. For example, in Section 3.1 we found that (2) has a nontrivial solution only when λ is of the form $\lambda_n = (n\pi/L)^2$, $n = 1, 2, 3, \ldots$. (Note that the symbol "λ_n" was used to denote $n\pi/L$ in Section 3.1 .) The eigenfunction corresponding to the eigenvalue λ_n is of the form $X_n(x) = A_n \sin(n\pi x/L)$, $n = 1, 2, 3, \ldots$, where A_n is any nonzero constant (cf. Example 1 below). The eigenvalue problem (2) is easy to solve, because we can explicitly solve not only the D.E. but also the equation $\sin(L\sqrt{\lambda}) = 0$. However, in general, one cannot find explicit formulas for the eigenfunctions or the eigenvalues of Sturm–Liouville problems. Although explicit solutions are rare, in this section we will establish a number of qualitative properties of the solutions of (8). □

Properties of eigenvalues and eigenfunctions

It is customary to use the dependent variable y instead of X in (8). Thus, (8) becomes

$$\text{D.E.}\quad \frac{d}{dx}\left(K(x)\frac{dy}{dx}\right) + (q(x) + \lambda g(x))y(x) = 0\,,\quad a \le x \le b$$

$$\text{B.C.}\quad \begin{cases} c_1 y(a) + c_2 y'(a) = 0 & (c_1^2 + c_2^2 \neq 0) \\ c_3 y(b) + c_4 y'(b) = 0 & (c_3^2 + c_4^2 \neq 0)\,, \end{cases} \tag{9}$$

where, we assume that $K'(x)$, $g(x)$, $q(x)$ and $y''(x)$ are continuous on $[a,b]$, and that $K(x)$ and $q(x)$ are positive on $[a,b]$.

Remark. Problem (9) is different from the initial–value problems considered in Chapter 1. The solution, $y(x)$, of an initial value problem for a linear second–order ODE, is required to satisfy *two* conditions at a *single* value (e.g, $y(x_0) = y_0$ and $y'(x_0) = y_1$). For the reader's convenience, we state the following existence and uniqueness theorem proved in [Simmons, Chapter 11]. □

Theorem 1 (Existence and Uniqueness for Initial–Value Problems). Let $P(x)$, $Q(x)$ and $R(x)$ be continuous on the interval $a \le x \le b$. If x_0 is a point in this interval and y_0 and y_1 are arbitrary real numbers, then the initial–value problem

$$\text{D.E.}\quad \frac{d^2y}{dx^2} + P(x)\frac{dy}{dx} + Q(x)y = R(x)$$

$$\text{I.C.}\quad y(x_0) = y_0\,,\quad y'(x_0) = y_1$$

has a unique solution on the interval $a \le x \le b$.

In case of a boundary–value problem, the solution $y(x)$ must satisfy conditions at *two* distinct values of x. For example, for the D.E. $y'' + y = 0$, the boundary conditions may be given by $y(x) = 0$ and $y(L) = 1$. But such a problem need *not* have a solution. In fact, this problem has no solution, if $L = 2\pi$ (cf. Problems 4 and 5). Proofs of the following existence theorem for Sturm–Liouville problems are in advanced texts (e.g., [Zalman Rubinstein, p. 173]). However, we will sketch part of the argument in the proof of Theorem 9 and in the Remark following it.

Theorem 2 (Existence of Eigenvalues of the Sturm–Liouville Problem (9)). The Sturm–Liouville problem (9) has an infinite number of eigenvalues, which can be written in increasing order as $\lambda_1 < \lambda_2 < \ldots < \lambda_n < \ldots$, such that $\lim_{n\to\infty} \lambda_n = \infty$. The eigenfunction $y_n(x)$ corresponding to λ_n has exactly $n-1$ zeros in (a,b).

To motivate the subsequent results of this section, we first consider some examples.

Example 1. Determine the eigenvalues and eigenfunctions of the Sturm–Liouville problem

$$\begin{aligned} &\text{D.E.} \quad y'' + \lambda y = 0, \qquad 0 \le x \le L \\ &\text{B.C.} \quad y(0) = 0, \qquad y(L) = 0. \end{aligned} \tag{10}$$

Solution. We first note that (9) reduces to (10), if $K(x) \equiv 1$, $q(x) \equiv 0$, $g(x) \equiv 1$, $a = 0$, $b = L$, $c_1 = 1$, $c_2 = 0$, $c_3 = 1$, $c_4 = 0$. We consider the three cases $\lambda > 0$, $\lambda = 0$ and $\lambda < 0$. But if $\lambda = 0$ or $\lambda < 0$, then the only solution of (10) is $y(x) \equiv 0$ (cf. Section 3.1), which is not an eigenfunction. If $\lambda > 0$, then the general solution of the D.E. is

$$y(x) = A\sin(x\sqrt{\lambda}) + B\cos(x\sqrt{\lambda}). \tag{11}$$

By the B.C., $y(0) = B = 0$ and $y(L) = A\sin(L\sqrt{\lambda}) = 0$. Since we want nontrivial solutions, $A \neq 0$, and we set $\sin(L\sqrt{\lambda}) = 0$, obtaining $L\sqrt{\lambda} = n\pi$. Thus, the eigenvalues are $\lambda = \lambda_n = (n\pi/L)^2$, with corresponding eigenfunctions $y_n(x) = A_n\sin(n\pi x/L)$, $A_n \neq 0$, $n = 1, 2, 3, \ldots$. □

Remark. For the product solutions of the heat equation in Section 3.1, the symbol "λ" denotes the square root of the eigenvalue here. To avoid confusion, one may wish to mentally replace "λ" by "β" everywhere "λ" appears in Chapter 3. Since ultimately mathematics is invariant under a change of notation, one should not become too attached to a particular use of some symbol. Note that in Example 1 (as well as in the next example), all of the real eigenvalues are positive, and $\lambda_n \to \infty$ as $n \to \infty$ (cf. Theorem 2). Also, $y_n(x)$ has $n-1$ real zeros in $(0,L)$, and by Green's formula on $[0,L]$, the eigenfunctions $y_1(x)$, $y_2(x)$, ... are orthogonal on $[0,L]$. Orthogonality results for the general problem (9) are established in Theorem 5 below. Moreover, although we do not consider the possibility of complex eigenvalues in Examples 1 and 2, the fact that there are none is a consequence of Theorem 7 below. □

Example 2. For the eigenvalue problem

$$\begin{aligned} &\text{D.E.} \quad y'' + \lambda y = 0, \qquad 0 \le x \le L, \; L < \pi/2 \\ &\text{B.C.} \quad y(0) - y'(0) = 0, \quad y(L) + y'(L) = 0 \end{aligned} \tag{12}$$

determine the equation whose zeros (roots) are the eigenvalues.

Solution. If $\lambda = 0$ or $\lambda < 0$, then it is easy to check that the only solution of (12) is the trivial solution. If $\lambda > 0$, then the general solution of the D.E. is

$$y(x) = A\cos(x\sqrt{\lambda}) + B\sin(x\sqrt{\lambda}). \tag{13}$$

Since $y'(x) = \sqrt{\lambda}\,(-A\sin(x\sqrt{\lambda}) + B\cos(x\sqrt{\lambda}))$, the first B.C. implies that $A = B\sqrt{\lambda}$. This, in conjunction with the second B.C. yields

$$0 = B\,[2\sqrt{\lambda}\cos(L\sqrt{\lambda}) + (1-\lambda)\sin(L\sqrt{\lambda})]. \tag{14}$$

Thus, if $B \neq 0$ and $\lambda \neq 1$ (note that $L < \pi/2$ in (12)), then (14) becomes

$$\tan(L\sqrt{\lambda}) = \frac{2\sqrt{\lambda}}{\lambda - 1} \quad (\lambda > 0) , \tag{15}$$

which is the required equation. It can be approximately solved numerically or graphically (cf. Figure 4 in Section 3.3, but $\lambda^2 \leftrightarrow \lambda$ by the above Remark). Note that if we had allowed $L = \pi/2$ (or any positive odd integral multiple of $\pi/2$), then $\lambda = 1$ would have been an eigenvalue (Why ?). □

In order to a study the general linear, homogeneous second–order D.E. when the coefficients are not all constants, we introduce the second–order, linear, differential operator

$$L \equiv p_2(x)\frac{d^2}{dx^2} + p_1(x)\frac{d}{dx} + p_0(x) . \tag{16}$$

If y is a C^2 function on some interval, then $L[y]$ is defined by

$$L[y] \equiv p_2(x)\frac{d^2y}{dx^2} + p_1(x)\frac{dy}{dx} + p_0(x)y , \tag{17}$$

and the corresponding linear, homogeneous second–order ODE can be written as $L[y] = 0$. When trying to solve the ODE $L[y] = 0$ with coefficients p_0, p_1 and p_2 which are not all constant, one might try to introduce the notion of an "integrating factor" that was used in case of linear, first–order ODEs. In Section 1.1, we found that multiplication by the integrating factor $m(x) = \exp\left[\int p(x)\,dx\right]$ reduced the ODE $\frac{dy}{dx} + p(x)y = 0$ to the simple form

$$0 = m(x)\frac{dy}{dx} + p(x)m(x)y = \frac{d}{dx}[m(x)y] . \tag{18}$$

In the next example, for the second–order ODE $L[y] = 0$, we similarly try to find a function, say $z = z(x)$, such that $zL[y]$ is the derivative of a certain combination of y and y' .

Integrating factors, Lagrange's identity and self–adjoint operators

Example 3. By means of formal calculations, attempt to find an "integrating factor" , $z = z(x)$, such that $zL[y]$ is the derivative of a combination of y and y' , where $L[y]$ is given by (17).

Solution. In order to find $z = z(x)$, we consider $\int z(x)L[y]\,dx$ and integrate by parts :

$$\begin{aligned}\int z(x)L[y]\,dx &= \int (zp_2y'' + z\,p_1y' + zp_0y)\,dx \\ &= (zp_2)y' - (zp_2)'y + \int (zp_2)''\,y\,dx + (zp_1)y - \int (zp_1)'y\,dx + \int zp_0y\,dx .\end{aligned} \tag{19}$$

Combining the terms in (19) yields

$$\int z(x)L[y]\,dx = (zp_2)y' - (zp_2)'y + (zp_1)y + \int yL^*[z]\,dx\,, \tag{20}$$

where

$$L^*[z] \equiv \frac{d^2}{dx^2}(zp_2(x)) - \frac{d}{dx}(zp_1(x)) + zp_0(x)\,. \tag{21}$$

Thus, if we choose $z = z(x)$ such that

$$L^*[z] = 0\,, \tag{22}$$

then, upon differentiating (20) with respect to x, we obtain

$$zL[y] = \frac{d}{dx}[\,(zp_2)y' - (zp_2)'y + zp_1y\,]\,. \tag{23}$$

While (23) is the desired result, the problem is that the determination of $z(x)$ requires that we solve the second–order ODE given in (22), which may be just as hard as solving $L[y] = 0$. □

Although the attempt to find an integrating factor has led to an equally difficult problem, we will find that the interplay between the operators L and L^* is quite helpful in obtaining properties of solutions of Sturm–Liouville problems. In particular, the following identity will be used to establish orthogonality results for eigenfunctions associated with different eigenvalues.

Example 4. Consider the differential operators L and L^* defined by (17) and (21) respectively. If p_2, p_1, p_0, y and z are C^2 functions, verify the following identity

$$zL[y] - yL^*[z] = \frac{d}{dx}[p_2\cdot(y'z - yz') + (p_1 - p_2')zy]\,, \tag{24}$$

which is known as **Lagrange's identity.**

Solution. By (20) we have

$$\int (zL[y] - yL^*[z])\,dx = (zp_2)y' - (zp_2)'y + (zp_1)y\,. \tag{25}$$

Thus, if we differentiate both sides of (25) with respect to x and rearrange the terms on the right–hand side of (25), then we obtain Lagrange's identity (24). □

Lagrange's identity (24) provides a relation between L and L^*. In the next example, we consider the case where $L[y] = \frac{d}{dx}\left[K(x)\frac{dy}{dx}\right] + q(x)y$. This operator appears in the D.E. of the Strum–Liouville problem (9), which can be written in the form $L[y] + \lambda g(x)y = 0$. Examples 5 and 6 will show that such operators are the only ones having the property that $L = L^*$.

Example 5. Let L denote the linear, second–order differential operator defined by

$$L[y] = \frac{d}{dx}[K(x)\frac{dy}{dx}] + q(x)y , \tag{26}$$

where $K(x)$ and $q(x)$ satisfy the requirements in (9) and y is a C^2 function. Find L^* for this operator. Show that $L = L^*$, in this case.

Solution. By comparing the expression

$$L[y] = K(x)\frac{d^2y}{dx^2} + K'(x)\frac{dy}{dx} + q(x)y \tag{27}$$

with (17), we see that $p_2(x) = K(x)$, $p_1(x) = K'(x)$ and $p_0(x) = q(x)$. Hence, by (21)

$$L^*[y] = \frac{d^2}{dx^2}(yK(x)) - \frac{d}{dx}(yK'(x)) + q(x)y . \tag{28}$$

Carrying out the differentiations, we find that

$$L[y] = L^*[y] , \tag{29}$$

for every C^2 function y. This says that the operators L and L^* are the same. □

Definition. Let L and L^* denote the linear, second–order differential operators defined by (17) and (21), respectively. Then L^* is called the **adjoint** of L and the differential equation $L^*[y] = 0$ is called the **adjoint equation**. The operator L is said to be **self–adjoint**, if L and L^* are the same, that is, $L = L^*$. A homogeneous, linear, second–order ODE is said to be in **self–adjoint form** if the ODE has the form

$$\frac{d}{dx}[P(x)\frac{dy}{dx}] + Q(x)y = 0 . \tag{30}$$

Remark. Note that the Sturm–Liouville equation (9) is in self–adjoint form and that, by Example 5, the operator L in (26) is self–adjoint. It turns out that some of the most common ODEs of mathematical physics are in self–adjoint form (cf. Bessel's equation and Legendre's differential equation which we study in Chapter 9). □

Example 6. Show that the linear, second–order differential operator

$$L[y] = p_2(x)y'' + p_1(x)y' + p_0(x)y \tag{31}$$

is self–adjoint (i.e., $L = L^*$) if and only if $p_2'(x) = p_1(x)$, i.e.,

$$L[y] = \frac{d}{dx}[p_2(x)\frac{dy}{dx}] + p_0(x)y . \tag{32}$$

Solution. The adjoint L^* of L is given by (cf. (21))

$$L^*[y] = \frac{d^2}{dx^2}(yp_2) - \frac{d}{dx}(yp_1) + yp_0 = p_2y'' + (2p_2' - p_1)y' + (p_2'' - p_1' + p_0)y . \tag{33}$$

Thus, $L = L^*$ if and only if $2p_2' - p_1 = p_1$ or $p_2' = p_1$. If $p_2' = p_1$, then (32) holds (Why ?). □

Remark. In Problem 6 we show that every linear, homogeneous second–order ODE

$$p_2(x)y'' + p_1(x)y' + p_0(x)y = 0 , \tag{34}$$

where $p_2(x) > 0$ and p_0, p_1 and p_2' are continuous, can be transformed into an ODE in self–adjoint form, by multiplying both sides of (34) by $\exp\left[\int (p_1 - p_2')/p_2 \, dx\right]$. □

The Sturm–Liouville differential operator

In the sequel, we will focus on the **Sturm–Liouville differential operator** defined by

$$L[y] \equiv \frac{d}{dx}(K(x)\frac{dy}{dx}) + q(x)y , \tag{35}$$

where $K(x) > 0$, $g(x) > 0$, and $K'(x)$, $q(x)$ and $g(x)$ are continuous on $[a,b]$. (By modifying the proofs given below, we can relax the condition that $K(x)$ be strictly positive, by requiring that $K(x) \geq 0$ and that $K(x)$ vanish at most at a finite number of points in $[a,b]$). We now state and prove several fundamental properties of the solution of the following Sturm–Liouville problem, with L defined by (35).

$$\begin{aligned} &\text{D.E.} \quad L[y] + \lambda g(x)y = 0 && a \leq x \leq b \\ &\text{B.C.} \quad c_1y(a) + c_2y'(a) = 0 && (c_1^2 + c_2^2 \neq 0) \\ &\qquad\quad c_3y(b) + c_4y'(b) = 0 && (c_3^2 + c_4^2 \neq 0) , \end{aligned} \tag{36}$$

where c_1, c_2, c_3 and c_4 are real constants.

> **Theorem 3 (A Uniqueness Theorem). Consider the Sturm–Liouville problem (36). If $y(x)$ and $Y(x)$ are two eigenfunctions corresponding to the *same* eigenvalue λ, then $y(x) = \alpha Y(x)$, $a \leq x \leq b$, for some nonzero constant α (i.e., $y(x)$ and $Y(x)$ are linearly dependent).**

Proof. We consider the function

$$\omega(x) = Y'(a)y(x) - y'(a)Y(x) , \tag{37}$$

and suppose that

$$[Y'(a)]^2 + [y'(a)]^2 \neq 0 . \tag{38}$$

(The case when $[Y'(a)]^2 + [y'(a)]^2 = 0$ is considered in Problem 9). Then, it is straightforward to check that $\omega(x)$ satisfies the following initial–value problem

$$\begin{aligned} &\text{D.E.} \quad L[\omega] + \lambda g(x)\omega = 0 \ , \quad a \le x \le b \\ &\text{I.C.} \quad \omega(a) = \omega'(a) = 0 \ , \end{aligned} \tag{39}$$

where L is the Sturm–Liouville operator defined by (35). But then, by the uniqueness theorem for initial–value problems (cf. Theorem 1 and Problem 3), $\omega(x) \equiv 0$. Therefore,

$$Y'(a)y(x) - y'(a)Y(x) \equiv 0 \ , \quad a \le x \le b \ . \tag{40}$$

Since $y(x)$ and $Y(x)$ are eigenfunctions, $y(x) \not\equiv 0$ and $Y(x) \not\equiv 0$. Hence, (38) and (40) imply that $y'(a)Y'(a) \ne 0$. Thus, by (40), $y(x) = \alpha Y(x)$, where $\alpha = y'(a)/Y'(a)$. □

Remark. In Theorem 3, we showed that, for the Sturm–Liouville problem (36), there is only one linearly independent eigenfunction associated with each eigenvalue λ. For this reason, λ is said to be **simple** or to have **multiplicity one.** □

Theorem 4 (Green's Formula for L). Let L be the Sturm–Liouville differential operator defined by (35). **If** $y(x)$ **and** $z(x)$ **are** C^2 **functions on** [a,b] **, then**

$$\int_a^b (zL[y] - yL[z])\, dx = \Big[K(x)\,(y'z - yz')\Big]_{x=a}^{x=b} \ . \tag{41}$$

Proof. Since L is self–adjoint (cf. Example 5), Lagrange's identity (cf. (24)) becomes

$$zL[y] - yL[z] = \frac{d}{dx}[K(x)\,(y'z - yz')] \ . \tag{42}$$

Thus, upon integrating both sides of (42) with respect to x from a to b , we obtain (41). □

Remark. We note that with $K(x) \equiv 1$ and $q(x) \equiv 0$, formula (41) yields Green's formula (for the operator d^2/dx^2) in Section 4.1, as a special case. □

The following definition extends the notion of orthogonality, which was introduced in Section 4.1 .

Definition. A positive, continuous function $g(x)$ defined on [a,b] is called a **weight function.** Two continuous functions $f(x)$ and $h(x)$ defined on [a,b] are said to be **orthogonal on** [a,b] **with respect to the weight function** $g(x)$, if

$$\int_a^b f(x)h(x)\, g(x)\, dx = 0 \ . \tag{43}$$

In particular, if $g(x) \equiv 1$, then (43) reduces to the definition in Section 4.1 .

Theorem 5 (Orthogonality of Eigenfunctions). Let λ_m and λ_n be two distinct eigenvalues of the Sturm–Liouville problem (36). Then the corresponding eigenfunctions $y_m(x)$ and $y_n(x)$ are orthogonal on $[a,b]$ with respect to the weight function $g(x)$.

Proof. Since $L[y_m] = -\lambda_m g y_m$ and $L[y_n] = -\lambda_n g y_n$, we have

$$y_n L[y_m] - y_m L[y_n] = (\lambda_n - \lambda_m) y_m y_n g \ . \tag{44}$$

Integrating both sides of (44) from a to b and using Green's formula (41),

$$\Big[K(y_m' y_n - y_m y_n')\Big]_{x=a}^{x=b} = (\lambda_n - \lambda_m)\int_a^b y_m(x) y_n(x)\, g(x) dx \ . \tag{45}$$

The B.C. in (36) ensure that the left side of (45) vanishes (e.g., if $c_2 \neq 0$, then $y_m'(a)y_n(a) - y_m(a)y_n'(a) = [c_2 y_m'(a)y_n(a) - y_m(a)c_2 y_n'(a)]/c_2 = [-c_1 y_m(a)y_n(a) + y_m(a)c_1 y_n(a)]/c_2] = 0$). Thus,

$$0 = (\lambda_m - \lambda_n) \int_a^b y_m(x) y_n(x) g(x)\, dx \ . \tag{46}$$

Since $\lambda_m \neq \lambda_n$, (46) says that $y_m(x)$ and $y_n(x)$ are orthogonal on $[a,b]$, with respect to the weight function $g(x)$. □

Remark. The previous notions and results (e.g., Lagrange's identity, Green's formula, etc.) make sense and are valid, when $y(x)$ is allowed to be complex–valued and the eigenvalues are not assumed to be real. In the next theorem, we prove that the eigenvalues must in fact be real. Note that the constant α in Theorem 3 can be complex, but each of the eigenfunctions associated with λ is a product of a complex constant and a real–valued eigenfunction (Why ?). □

Theorem 6. All of the eigenvalues of the Sturm–Liouville problem (36) are real.

Proof. Let $\lambda = \alpha + i\beta$ (α,β real) be an arbitrary eigenvalue of the Sturm–Liouville problem (36), and let $y(x)$ be a complex–valued eigenfunction corresponding to λ. Since $K(x)$, $q(x)$ and $g(x)$ (in the equation $L[u] + \lambda g(x) y = 0$) are real, complex conjugation yields

$$\frac{d}{dx}\left(K(x)\frac{d\bar{y}}{dx}\right) + (q(x) + \bar{\lambda} g(x))\bar{y} = 0 \ . \tag{47}$$

Thus, $\overline{\lambda} = \alpha - i\beta$ is also an eigenvalue of the problem (36) with corresponding eigenfunction $\overline{y}(x)$. (Recall that the constants c_1, c_2, c_3 and c_4 appearing in the B.C. of (36) are all real). Now suppose that $\lambda \neq \overline{\lambda}$. Then using (46), and the fact that $y\overline{y} = |y|^2$, we have

$$(\lambda - \overline{\lambda})\int_a^b g(x)\,|y(x)|^2\,dx = 0 \ . \qquad (48)$$

Since $g(x) > 0$, $|y(x)|^2 \geq 0$ and $y(x) \not\equiv 0$, (48) yields $\lambda = \overline{\lambda}$, and thus λ is real. □

The next theorem shows that under some additional hypotheses, the eigenvalues of the Sturm–Liouville problem are not only real, but nonnegative.

Theorem 7. In the Sturm–Liouville problem (36), suppose that $q(x) \leq 0$ for $a \leq x \leq b$ and that the real constants c_j $(j = 1,\ldots,4)$ satisfy the inequalities

$$c_1 \cdot c_2 \leq 0 \quad \textbf{and} \quad c_3 \cdot c_4 \geq 0 \ , \qquad (49)$$

then all the eigenvalues of (36) are nonnegative. Moreover, if 0 is an eigenvalue, then $q(x) \equiv 0$, $c_1 = c_3 = 0$, and any eigenfunction with eigenvalue 0 must be constant.

Proof. Suppose that $L[y] + \lambda g y = 0$, where $y(x) \not\equiv 0$. Then

$$0 = \int_a^b y \cdot (L[y] + \lambda g y)\,dx = \int_a^b y(x)\,\frac{d}{dx}\left(K\,\frac{dy}{dx}\right)dx + \int_a^b q(x)y(x)^2\,dx + \lambda\int_a^b g(x)y(x)^2\,dx \ . \qquad (50)$$

Using integration by parts for the first integral on the right–hand side, we obtain

$$\int_a^b y\,\frac{d}{dx}\left(K\,\frac{dy}{dx}\right)dx = \Big[y(x)y'(x)K(x)\Big]_{x=a}^{x=b} - \int_a^b y'(x)^2\,K(x)\,dx \ . \qquad (51)$$

Substituting this expression for the first integral in (50) and rearranging, we obtain

$$\lambda\int_a^b g(x)y(x)^2\,dx = \int_a^b -q(x)y(x)^2\,dx + \int_a^b y'(x)^2\,K(x)\,dx - \Big[y(x)y'(x)K(x)\Big]_{x=a}^{x=b} . \qquad (52)$$

Now it is straightforward to verify that (49) and the following B.C.

$$c_1 y(a) + c_2 y'(a) = 0 \quad \text{and} \quad c_3 y(b) + c_4 y'(b) = 0 \qquad (53)$$

imply that $y(a)y'(a) \geq 0$ and $y(b)y'(b) \leq 0$. Consequently, since $K(x) > 0$, we have

$$-\Big[y(x)y'(x)K(x)\Big]_{x=a}^{x=b} = y(a)y'(a)K(a) - y(b)y'(b)K(b) \geq 0 \ . \qquad (54)$$

Since $-q(x) \geq 0$ and $g(x) > 0$, we conclude from (52) and (54) that $\lambda\int_a^b g(x)y(x)^2\,dx \geq 0$, and so $\lambda \geq 0$. Note that λ can be 0, only when all of the terms on the right–hand side of (52) vanish. However, this implies $q(x) \equiv 0$ and $y'(x) \equiv 0$ (i.e., y is a nonzero constant function). Moreover, a nonzero constant function cannot meet the B.C. (53), unless $c_1 = c_3 = 0$. □

The Sturm Comparison Theorem

In general, one cannot obtain explicit solutions to an eigenvalue problem. However, the Sturm–Liouville theory provides a wealth of information concerning the eigenvalues and the zeros and oscillation properties of the corresponding eigenfunctions. One of the most useful and remarkable results in this theory is the Sturm Comparison Theorem. To motivate this theorem, we first consider an example.

Example 7. Suppose that we have constants $\lambda_2 > \lambda_1 > 0$. Let $y_1(x)$ be any nonzero solution of $y'' + \lambda_1 y = 0$ and let $y_2(x)$ be any nonzero solution of $y'' + \lambda_2 y = 0$. Show that between any two consecutive zeros of $y_1(x)$, there is a zero of $y_2(x)$.

Solution. The general solution of $y'' + \lambda_1 y = 0$ can be written in the form $A\sin(x\sqrt{\lambda_1} + \delta)$, where A and δ are arbitrary constants. Thus, any interval, say J, joining consecutive zeros of $y_1(x)$ has length $\pi/\sqrt{\lambda_1}$, while the distance between consecutive zero of $y_2(x)$ is $\pi/\sqrt{\lambda_2}$. Since $\pi/\sqrt{\lambda_2} < \pi/\sqrt{\lambda_1}$, J cannot be entirely between two consecutive zeros of $y_2(x)$ (i.e., J must contain a zero of $y_2(x)$). □

As in Example 7, the intuitive idea in the general setting (cf. Theorem 8 below) is that the size of λg in $L[y] + \lambda g y = 0$ governs the frequency of oscillation of solutions. However, in the general case, the distance between one pair of consecutive zeros need not be the same as the distance between another pair. Note that each interior zero of a nontrivial solution has a smallest successor (and largest predecessor) for otherwise the derivative would vanish at the zero and the solution would be trivial by uniqueness (cf. Theorem 3 and Problem 11).

Theorem 8 (The Sturm Comparison Theorem). Let $L[y] \equiv \frac{d}{dx}\left[K(x)\frac{dy}{dx}\right] + q(x)y$, **where** $K(x) > 0$ **on** $[a,b]$, **and** $K'(x)$ **and** $q(x)$ **are continuous on** $[a,b]$. **Suppose that** $y_1(x)$ **and** $y_2(x)$ **are** C^2 **solutions of the respective Sturm–Liouville equations**

$$L[y_1] + \lambda_1 g_1(x) y_1 = 0 \quad \textbf{and} \quad L[y_2] + \lambda_2 g_2(x) y_2 = 0\,, \tag{55}$$

where $\lambda_2 g_2(x) \geq \lambda_1 g_1(x)$ **on** $[a,b]$, **and** $g_1(x)$ **and** $g_2(x)$ **are continuous on** $[a,b]$. **Then between any two consecutive zeros** α **and** β **(where** $a \leq \alpha < \beta \leq b$**) of** $y_1(x)$ **there is at least one zero of** $y_2(x)$, **provided that** $\lambda_1 g_1(x) \not\equiv \lambda_2 g_2(x)$ **on** $[\alpha,\beta]$.

Proof. By (55) and Green's formula (41) for the operator L on $[\alpha,\beta]$, we obtain

$$\int_\alpha^\beta [\lambda_2 g_2 - \lambda_1 g_1] y_1 y_2 \, dx = \int_\alpha^\beta y_2 L[y_1] - y_1 L[y_2] \, dx = \Big[K\cdot(y_1' y_2 - y_1 y_2')\Big]_{x=\alpha}^{x=\beta}$$

or

$$\int_\alpha^\beta [\lambda_2 g_2 - \lambda_1 g_1] y_1 y_2 \, dx = K(\beta) y_1'(\beta) y_2(\beta) - K(\alpha) y_1'(\alpha) y_2(\alpha) \; , \tag{56}$$

where we have used $y_1(\alpha) = y_1(\beta) = 0$. Since α and β are *consecutive* zeros of $y_1(x)$, we may assume that $y_1(x) > 0$ on $\alpha < x < \beta$ (otherwise, replace y_1 by $-y_1$ which is also a solution of $L[y] + \lambda_1 g_1(x)y = 0$ with the same zeros as y_1). Then $y_1'(\alpha) > 0$ and $y_2'(\beta) < 0$ (Why ?). If $y_2(x)$ has no zero in (α,β), then $y_2(x) > 0$ on (α,β), or $y_2(x) < 0$ on (α,β). If $y_2(x) > 0$, then the integral on the left–hand side of (56) is strictly positive (Why ?), while the right–hand side of (56) is nonpositive (Why ?). Similarly, if $y_2(x) < 0$ on (α,β), then the left–hand side of (56) is negative, while the right–hand side of (56) is positive. In either case, we arrive at a contradiction, and so $y_2(x)$ must have a zero in (α,β). □

Existence of eigenvalues

One can use the Sturm Comparison Theorem to prove Theorem 2 on the existence of an infinite number of eigenvalues of the Sturm–Liouville problem (36). We illustrate the idea in the following special case of Theorem 2.

Theorem 9. Consider the Sturm–Liouville problem

$$\begin{aligned} &\text{D.E.} \quad y'' + q(x)y + \lambda g(x)y = 0 \qquad a \le x \le b \\ &\text{B.C.} \quad y(a) = 0 \; , \quad y(b) = 0 \; , \end{aligned} \tag{57}$$

where $q(x)$ **and** $g(x)$ **are continuous, and** $g(x) > 0$ **on** $[a,b]$. **Assume that the solution** $Y(x,\lambda)$ **(which depends on the parameter** λ**) of the initial value problem,**

$$\begin{aligned} &\text{D.E.} \quad Y'' + q(x)Y + \lambda g(x)Y = 0 \\ &\text{I.C.} \quad Y(a) = 0 \; , \; Y'(a) = 1 \end{aligned} \tag{58}$$

is a continuous function of (x,λ). **Then problem** (57) **has an infinite number of eigenvalues, say** $\lambda_1 < \lambda_2 < \ldots$, **with** $\lim_{n\to\infty} \lambda_n = \infty$.

Proof (sketch). The eigenvalues of problem (57) are precisely those values of λ for which $Y(b,\lambda) = 0$ (Why ?). We must show that there is a sequence of such values of λ which tend to ∞. For $c > 0$, let $\lambda(c)$ be a value of λ such that $q(x) + \lambda(c)g(x) \ge c^2$, for all x in $[a,b]$.

This is possible, since $g(x)$ has a positive minimum on $[a,b]$, and $q(x)$ is bounded (cf. Appendix A.4). The Sturm Comparison Theorem implies that between any two zeros of a nontrivial solution of $y'' + c^2y = 0$, say $\sin[c(x-a)]$, there is a zero of $Y(x,\lambda(c))$. In particular, if $c(b-a) > n\pi$ or $c > n\pi/(b-a)$, there are at least n zeros of $Y(x,\lambda(c))$ in (a,b). The Sturm Comparison Theorem implies that as λ increases, each zero of $Y(x,\lambda)$ in (a,b) will move to the left (Why ?). Moreover, as λ increases, say at a steady rate, no zero can suddenly appear strictly inside the interval, say by means of the graph dipping down between two consecutive zeros. Indeed, if this were to happen, then the graph would contact the x–axis tangentially (Here we have used the assumption that $Y(x,\lambda)$ is continuous, so that the graph does not jump suddenly.) However, the graph always meets the x–axis at a non–zero angle (Why ?). Thus, each of the infinitely many new zeros which appear as $\lambda \to \infty$, must first appear at the right endpoint b and move to the left. The values of λ at the times of appearance are the eigenvalues of the problem (57) (Why ?). Moreover, at no finite time can there ever be an infinite number of zeros of $Y(x,\lambda)$ in $[a,b]$ (cf. Problem 11). Thus, the eigenvalues form a sequence, tending to ∞. □

Remark. There are some details left out of the above sketch. We did not wish to deprive the interested reader of the challenge of filling them in, nor did we want to obsure the main idea of the the proof with the full details. The assumption, that $Y(x,\lambda)$ is continuous, can in fact be proved and hence it is superfluous (cf. Theorem 3). The argument can be generalized to handle the case where the B.C. are of the form $c_1y(a) + c_2y'(a) = 0$ and $c_3y(b) + c_4y'(b) = 0$, with $c_2 \neq 0$ or $c_4 \neq 0$. Indeed, we can always change the I.C. in (58), so that the first of these B.C. is met, and the second B.C. is met at some time between the appearance of one zero at the right endpoint and the appearance of the next zero. (Simply note that $Y'(b,\lambda)/Y(b,\lambda)$ ranges from $-\infty$ to ∞ between the times of appearances of zeros.) Moreover, although we have implicitly assumed that $K(x) = 1$ in Theorem 9, one can change the independent variable, say $t = \int_a^x \frac{ds}{K(s)}$, in order to convert $\frac{d}{dx}\left[K(x)\frac{dy}{dx}\right] + q(x)y + \lambda g(x)y = 0$, into the form $z'' + Kqz + \lambda Kgz = 0$, where $z(t) = y(x(t))$, and K, q and g are expressed in terms of t. Also converting the B.C., the eigenvalues of the new problem are the same as those of the old problem. Thus, we "nearly" have a proof of Theorem 2, except for the proof of the continuity of $Y(x,\lambda)$. The proof of the fact that the eigenfunction $y_n(x)$, corresponding to the n–th eigenvalue λ_n, has exactly $n-1$ zeros in (a,b) follows easily from the case $n = 1$ (cf. Problem 16). □

In the study of higher dimensional PDEs in Chapter 9, we will encounter more eigenvalue problems. In particular in Section 9.5, we consider **Bessel's equation of order m**,

$$x^2y'' + xy' + (x^2 - m^2)y = 0 . \qquad (59)$$

which arises in problems for a vibrating circular drum and for heat flow in a disk. For each nonnegetive integer m the (normalized) C^2 solution of (59) is denoted by $J_m(x)$ and is called the **Bessel function of the first kind of order m** . (For an explicit formula see Section 9.5 or Appendix A.5). Using the Sturm Comparison Theorem, in the next example we will show, without the benefit of an explicit formula for $J_m(x)$, that $J_m(x)$ has an infinite number of positive zeros !

Example 8. Using the fact that $J_m(x)$ satisfies the ODE (59), verify that $y_m(x) = x^{\frac{1}{2}}J_m(x)$ $(x > 0)$, satisfies the ODE

$$y'' + \left[1 + \frac{\frac{1}{4} - m^2}{x^2}\right] y = 0 \quad , \quad x > 0 \ . \tag{60}$$

Use this to show that $J_m(x)$ has an infinite number of positive zeros.

Solution. First compute y_m'' in terms of $J_m(x)$. Then for $x > 0$, we have by (59)

$$x^{\frac{3}{2}}\left[y_m'' + \left[1 + \frac{\frac{1}{4} - m^2}{x^2}\right] y_m\right] = x^2J_m'' + xJ_m' + (x^2 - m^2)J_m = 0 \ .$$

Next, fix a nonnegative integer m, and let

$$g_m(x) = 1 + \frac{\frac{1}{4} - m^2}{x^2} \qquad (x > 0).$$

For each fixed $m \geq 0$ and $0 < \epsilon < 1$, we have

$$g_m(x) > 1 - \epsilon \ , \quad \text{for all } x > m/\sqrt{\epsilon} > 0 \ .$$

We now apply the Sturm Comparison Theorem to the ODE (60) and the ODE $y'' + (1-\epsilon)\, y = 0$, with $g(x) \equiv 1-\epsilon$. The zeros of the solution $\sin((1-\epsilon)^{\frac{1}{2}}x)$ of $y'' + (1-\epsilon)y = 0$, are of the form $n\pi(1-\epsilon)^{-\frac{1}{2}}$, $n = 0, \pm1, \pm2, \ldots$. Between any two such zeros in the range $x > m/\sqrt{\epsilon}$, there is a zero of $J_m(x)$. In particular, $J_m(x)$ has infinitely many zeros, and we have some idea of where they can be found. For more detailed information, see Appendix 5. □

Example 9. For a continuous function $g(x)$ on $[a,b]$, consider the Sturm–Liouville problem

$$\begin{aligned} &\text{D.E.} \quad y'' + \lambda g(x)y = 0 \ , \quad a \leq x \leq b \\ &\text{B.C.} \quad y(a) = 0 \ , \quad y(b) = 0 \ , \end{aligned} \tag{61}$$

where

$$0 < m < g(x) < M \ . \tag{62}$$

Deduce that the following estimates holds for the eigenvalues λ_n :

$$\frac{1}{M}\left[\frac{n\pi}{b-a}\right]^2 < \lambda_n < \frac{1}{m}\left[\frac{n\pi}{b-a}\right]^2 \ , \quad n = 1, 2, \ldots \ . \tag{63}$$

In particular, $\lambda_n \to \infty$ as $n \to \infty$.

Solution. By Theorem 2 (or Theorem 9), we know that (61) has an infinite number of eigenvalues $\lambda_1 < \lambda_2 < \ldots < \lambda_n < \ldots$. Moreover, by Theorem 2 and the B.C. in (61), the eigenfunction $y_n(x)$ associated to λ_n has $n+1$ zeros in $[a,b]$, say $a = x_1 < x_2 < \ldots < x_{n+1} = b$. For $n \geq 1$

$$\pi/\sqrt{\lambda_n M} < x_{j+1} - x_j < \pi/\sqrt{\lambda_n m}, \quad j = 1, \ldots, n, \tag{64}$$

by the Sturm Comparison Theorem applied to the ODEs $y'' + \lambda_n my = 0$, $y'' + \lambda_n g(x)y = 0$ and $y'' + M\lambda_n y = 0$ (cf. Problem 16). From (64) and the fact that the sum of the lengths of the n subintervals $[x_j,x_{j+1}]$, $j = 1,\ldots,n$, is the total length $b-a$, it follows that

$$n\pi/\sqrt{\lambda_n M} < \sum_{j=1}^{n}(x_{j+1} - x_j) = (b-a) < n\pi/\sqrt{\lambda_n m}\,. \tag{65}$$

Thus, by (65) we have the desired estimate (63). □

Heat problems for the inhomogeneous rod and eigenfunction expansions

This section was motivated by the initial/boundary–value problem

$$\begin{array}{lll}
\text{D.E.} & g(x)u_t = \dfrac{\partial}{\partial x}\left[K(x)u_x\right] + q(x)u & a \leq x \leq b, \ t \geq 0 \\
 & c_1u(a,t) + c_2u_x(a,t) = 0 \quad (c_1^2 + c_2^2 \neq 0) & \\
\text{B.C.} & & \\
 & c_3u(b,t) + c_4u_x(b,t) = 0 \quad (c_3^2 + c_4^2 \neq 0)\,, & \\
\text{I.C.} & u(x,0) = f(x) &
\end{array} \tag{66}$$

for the inhomogeneous rod, where $g(x) \equiv C(x)D(x)$ (specific heat, times linear density), $K(x)$ is the thermal conductivity, and $q(x)u$ is a temperature–dependent source or sink. We saw that separation of variables, with $u(x,t) = X(x)T(t)$, led to the Sturm–Liouville problem (8). Now let $\lambda_1 < \lambda_2 < \ldots < \lambda_n < \ldots$ denote the eigenvalues of (8) (cf. Theorem 2), with corresponding eigenfunctions $y_n(x) \equiv X_n(x)$. Then for each n, the function

$$u_n(x,t) = d_n e^{-\lambda_n t} y_n(x)$$

satisfies the D.E. and B.C. of (66). *Formally*, we consider the "infinite superposition"

$$u(x,t) = \sum_{n=1}^{\infty} d_n e^{-\lambda_n t} y_n(x) , \tag{67}$$

where we strive to attain

$$u(x,0) = f(x) = \sum_{n=1}^{\infty} d_n y_n(x) . \tag{68}$$

In order to find the constants d_n , we use the orthogonality of the eigenfunctions $y_n(x)$ on $[a,b]$, with respect to the weight function $g(x)$ (cf. Theorem 5). Formally using term–by–term integration, we have

$$\int_a^b f(x)y_m(x)g(x)\,dx = \int_a^b \sum_{n=0}^{\infty} d_n y_n(x)y_m(x)g(x)\,dx$$

$$= \sum_{n=0}^{\infty} d_n \int_a^b y_n(x)y_m(x)g(x)\,dx = d_m \int_a^b [y_m(x)]^2 g(x)\,dx ,$$

and hence

$$d_m = \frac{\int_a^b f(x)y_m(x)g(x)\,dx}{\int_a^b [y_m(x)]^2 g(x)\,dx} , \quad m = 0,1,2,\dots . \tag{69}$$

Consequently, the formal solution of the initial/boundary–value problem (66) is given by (67), where d_m is determined by (69). When the initial temperature is (say within experimental error) a *finite* linear combination of the eigenfunctions, then all but finitely many of the d_n are zero, and the formal solution (67) is then an exact C^2 solution of (66).

The foregoing formal considerations suggest a generalization of Fourier series. To see this, let $f(x)$ be defined on $[a,b]$, and suppose that the integrals in (69) are finite. Then the formal series

$$E\,f(x) = \sum_{m=0}^{\infty} d_m y_m(x) , \tag{70}$$

where d_m is defined by (69), is called the **eigenfunction expansion** of f on $[a,b]$, with respect to the eigenfunctions $\{y_n(x)\}_{n=0}^{\infty}$. For example, in the special case when $a = 0$, $b = L$, $g(x) \equiv 1$ and $y_n(x) = \sin(n\pi x/L)$, (70) reduces to the Fourier sine series of f. All of the various types of Fourier series which we have considered so far are special cases of (70). In Chapter 9, we will encounter other types of eigenfunction expansions, as for example, Laplace series and Fourier–Bessel series, in connection with our study of PDEs in higher dimensions. These series are used to represent functions of several variables, say on a sphere, a disk, or a solid ball. This section is an introduction to the one dimensional case. At this point, pure mathematicians are likely to pursue some convergence theorems for eigenfunction expansions (cf. [Titchmarsh]), while individuals in applied fields are probably wondering what numerical techniques are available for computing approximations for eigenfunctions and eigenvalues (cf. [Keller]).

Summary 4.4

1. Sturm–Liouville problems : A heat conduction problem for a nonuniform rod with a heat source led to the following Sturm–Liouville problem :

$$\text{D.E.}\quad \frac{d}{dx}\left(K(x)\frac{dy}{dx}\right) + (q(x) + \lambda g(x))y = 0 \qquad a \le x \le b$$

$$\text{B.C.}\quad \begin{aligned} &c_1 y(a) + c_2 y'(a) = 0 \qquad (c_1^2 + c_2^2 \neq 0) \\ &c_3 y(b) + c_4 y'(b) = 0 \qquad (c_3^2 + c_4^2 \neq 0) \end{aligned} \qquad (*)$$

where $K(x) > 0$, $g(x) > 0$, and $g(x)$, $q(x)$, $K'(x)$ and $y''(x)$ are continuous, on $[a,b]$.

2. Definitions. The D.E. in (*) is called a **Sturm–Liouville equation.** The values of the parameter λ for which a *nontrivial solution* (i.e. $y \neq 0$) of (*) exists is called an **eigenvalue** of the problem and a corresponding solution $y(x)$ of (*) is called an **eigenfunction** (associated with that eigenvalue). Problem (*) is also called an **eigenvalue problem.**

3. Properties of eigenvalues and eigenfunctions : According to Theorem 2, the eigenvalues for the Sturm–Liouville problem (*) form an infinite sequence $\lambda_1 < \lambda_2 < \lambda_3 \ldots$ with $\lambda_n \to \infty$, as $n \to \infty$. Moreover, any eigenfunction $y_n(x)$ associated with λ_n has $n-1$ zeros in (a,b).
(cf. also items 5, 7, 8 and 9 below).

4. Self–adjoint form : A homogeneous, linear second–order ODE is in self–adjoint form if the ODE has the form

$$\frac{d}{dx}\left(P(x)\frac{dy}{dx}\right) + Q(x)y = 0\ .$$

In particular, the Sturm–Liouville equation in (*) is in self–adjoint form.

5. Uniquenes : The eigenfunctions of the Sturm–Liouville problem (*) are uniquely determined up to a nonzero multiplicative constant (cf. Theorem 3).

6. Sturm–Liouville differential operator, Lagrange's identity and Green's formula : The operator

$$L[y] = \frac{d}{dx}\left(K(x)\frac{dy}{dx}\right) + (q(x) + \lambda g(x))y$$

is called the Sturm–Liouville differential operator. Lagrange's identity for L is (cf.(42))

$$zL[y] - yL[z] = \frac{d}{dx}\Big[K(x)\,(y'z - yz')\Big]\ ,$$

which, upon integration, yields Green's formula for L (cf. Theorem 4) :

$$\int_a^b (zL[y] - yL[z])\,dx = \Big[K(x)\,(y'z - yz')\Big]_{x=a}^{x=b} ,$$

where $y(x)$ and $z(x)$ are C^2 functions on $[a,b]$.

7. Orthogonality of Eigenfunctions : If λ_m and λ_n are two distinct eigenvalues of the Sturm–Liouville problem (∗), with corresponding eigenfunctions $y_m(x)$ and $y_n(x)$, then $y_m(x)$ and $y_n(x)$ are orthogonal on [a,b] with respect to the (positive) weight function $g(x)$ (cf. Theorem 5), i.e.,

$$\int_a^b g(x)y_m(x)y_n(x)\,dx = 0 , \text{ if } m \neq n .$$

8. Reality of eigenvalues : All of the eigenvalues of the Sturm–Liouville problem (∗) are real (cf. Theorem 6).

9. Sturm Comparison Theorem (cf. Theorem 8) : One of the most useful and remarkable results in this theory is the Sturm Comparison Theorem (cf. Theorem 8). It can be used (cf. Theorem 9 and the remarks following it) in establishing the properties of eigenvalues and eigenfunctions in Theorem 2. In Example 9, we use the Sturm Comparison Theorem to show that for the problem

$$\text{D.E.} \quad y'' + \lambda g(x)y = 0 \qquad a \leq x \leq b$$

$$\text{B.C.} \quad y(a) = 0 , \quad y(b) = 0 ,$$

where $0 < m < g(x) < M$ and $g(x)$ is continuous on [a,b], the eigenvalues λ_n , $n \geq 1$, satisfy the estimates :

$$\frac{1}{M}\left[\frac{n\pi}{b-a}\right]^2 < \lambda_n < \frac{1}{m}\left[\frac{n\pi}{b-a}\right]^2 .$$

In Example 8, the Sturm Comparison Theorem is used to prove that Bessel functions have an infinite number of zeros.

Exercises 4.4

1. Find the eigenvalues λ_n and the eigenfunctions $y_n(x)$ of the Sturm–Liouville problem $y'' + \lambda y = 0$ $(0 \le x \le L)$ with the given B.C.

   (a) $y(0) = 0$, $y'(L) = 0$ (b) $y'(0) = 0$, $y(L) = 0$ (c) $y'(0) = 0$, $y'(L) = 0$.

2. Find the eigenvalues λ_n and corresponding eigenfunctions $y_n(x)$ of the eigenvalue problem $y'' + \lambda y = 0$ with the given B.C.

   (a) $y(0) - y'(0) = 0$, $y(\pi) - y'(\pi) = 0$ (b) $y(0) - y(2\pi) = 0$, $y'(0) - y'(2\pi) = 0$

   (c) $y(0) + y(1) = 0$, $y'(0) + y'(1) = 0$.

3. (a) Prove that the unique solution of the initial–value problem in Theorem 1 , with $R(x) \equiv 0$, $y(x_0) = y'(x_0) = 0$, is the trivial solution.

   (b) Let $y(x)$ be an eigenfunction of the Sturm–Liouville problem (9). Show that all the zeros of $y(x)$ in $[a,b]$ are simple, i.e., if $y(x_0) = 0$, $a \le x_0 \le b$, then $y'(x_0) \ne 0$.

4. (a) Show that every real value λ is an eigenvalue for the problem

   D.E. $y'' + \lambda y = 0 \qquad 0 \le x \le 1$

   B.C. $y(0) - y(1) = 0 \qquad y'(0) + y'(1) = 0$.

   (b) Show that the problem

   D.E. $y'' + \lambda y = 0 \qquad 0 \le x \le \pi$

   B.C. $\pi y(0) - y(\pi) = 0 \qquad \pi y'(0) + y'(\pi) = 0$

   has no real eigenvalues.

Remark. Note that the B.C. in Problem 4 (or in Problem 2 (b), (c)) are *not* of the prescribed form given by the B.C. of the Strum–Liouville problem (9).

5. (a) Show that the following boundary–value problem has no solution :

   D.E. $y'' + y = 0$

   B.C. $y(0) = 0 \quad y(2\pi) = 1$.

   (b) If in part (a), the B.C. are changed to $y(0) = 0$ and $y(L) = 1$, then determine all values of $L > 0$ such that the boundary–value problem will have a solution.

6. Show that every linear, homogeneous second–order ODE,

$$p_2(x)y''(x) + p_1(x)y'(x) + p_0(x)y(x) = 0 \,, \qquad (*)$$

where $p_2(x) > 0$ and $p_0(x)$, $p_1(x)$ and $p_2'(x)$ are continuous can be transformed into an equation which is in self–adjoint form.

Hint. Multiply (*) by $\exp\left[\int \left[\frac{p_1 - p_2'}{p_2}\right] dx\right]$ and check that the resulting ODE is of the form $q_2y'' + q_1y' + q_0y = 0$, where $q_2' = q_1$.

7. Let m be a nonnegative integer. Use Problem 6 to transform each of the following equations in self–adjoint form :

(a) $x^2y'' + xy' + (x^2-m^2)y = 0 \quad , \; x > 0$

(b) $(1-x^2)y'' - 2xy' + m(m+1)y = 0 \quad , \; -1 < x < 1$

(c) $(1-x^2)y'' - xy' + m^2y = 0 \quad , \quad -1 < x < 1$

(d) $y'' - 2xy' + 2my = 0 \quad , \; -\infty < x < \infty$

(e) $xy'' + (1-x)y' + my = 0 \quad , \quad 0 \le x < \infty$.

Remark. Equations (a)–(e) are called **Bessel's equation, Legendre's equation, Chebyshev's equation, Hermite's equation,** and **Laguerre's equation,** respectively.

8. Consider the linear, homogeneous second–order ODE $y'' + P(x)y' + Q(x)y = 0$. Use the substitution $y(x) = z(x)\exp[-\frac{1}{2}\int P(x)\, dx]$ to reduce this ODE to the form $z'' + G(x)z = 0$, where $G = -\frac{1}{4}P^2 - \frac{1}{2}P' + Q$.

9. Prove Theorem 3 in the case where $[Y'(a)]^2 + [y'(a)]^2 = 0$.
Hint. In the proof of Theorem 3, let $w(x) = Y(a)y(x) - Y(x)y(a)$.

10. For each nonnegative integer n, let $y_n(x)$ denote a (polynomial) solution of the ODE $[(1-x^2)y'(x)]' + n(n+1)y = 0$. Show that if $n \neq m$, then $y_n(x)$ and $y_m(x)$ are orthogonal on $[-1,1]$ with respect to the weight function $g(x) \equiv 1$.

11. Let $y(x)$ be a nontrivial solution of the Sturm–Liouville equation (cf. the D.E. of (36)) for $a \le x \le b$. Show that $y(x)$ can have at most a finite number of zeros in $a \le x \le b$.

Hint. If $y(x)$ has an infinite number of zeros in $[a,b]$ then by the Bolzano–Weierstrass theorem (cf. Appendix A4), there exists a point x_0 in $[a,b]$ and a sequence $\{x_n\}$ of zeros of $y(x)$ such that $x_n \to x_0$ as $n \to \infty$. Next form the Newton quotient of $y(x)$ to conclude that $y'(x_0) = 0$. This is impossible by Problem 3.

12. (a) If the inequalities $c_1 \cdot c_2 \le 0$ and $c_3 \cdot c_4 \ge 0$ in Theorem 7 are replaced by $c_1 \cdot c_2 > 0$ and $c_3 \cdot c_4 < 0$, explain why these new inequalities are unrealistic for a heat problem.

(b) Give a specific example which shows that some of the eigenvalues can be negative, if $c_1 \cdot c_2 > 0$ and $c_3 \cdot c_4 < 0$. Can all of the eigenvalues be negative ?

(c) By considering (67), describe the set of initial temperatures $f(x)$ for which such a rod will still have a bounded temperature as $t \to \infty$.

13. Without solving (explicitly) the eigenvalue problem

$$\text{D.E.} \quad y'' + \lambda y = 0 \qquad 0 \le x \le 1$$

$$\text{B.C.} \quad y(0) = 0\,, \quad y(1) + y'(1) = 0\,,$$

show that all the eigenvalues are positive.

14. (a) Example 3 of Section 3.1 involves the heat flow in an insulated circular wire. Show that the D.E. and B.C. in this example (cf. (23) of Section 3.1) leads to the eigenvalue problem

$$\text{D.E.} \quad y'' + \lambda y = 0 \qquad -L \le x \le L$$

$$\text{B.C.} \quad y(-L) - y(L) = 0\,, \quad y'(-L) - y'(L) = 0\,.$$

(b) Verify that the eigenvalues are $\lambda_n = (n\pi/L)^2$, $n \ge 0$. For $n \ge 1$, find two linearly independent eigenfunctions $y_n(x)$ and $z_n(x)$. Why does this not violate Theorem 3 ?

15. Let $g(x)$ be continuous on $[a,\infty)$, and suppose that $0 < m < g(x)$ on $[a,\infty)$. Prove that any solution of the ODE $y'' + g(x)y = 0$ has infinitely many zeros.

16. (a) Use the Sturm Comparison Theorem to prove inequality (64).

(b) Following the considerations in the proof of Theorem 9, show that the eigenfunction $y_1(x)$ (for (57)), associated with the smallest eigenvalue λ_1, has no zero in (a,b).

CHAPTER 5

THE WAVE EQUATION

Physical scientists, engineers and applied mathematicians regard the wave equation

$$u_{tt} = a^2(u_{xx} + u_{yy} + u_{zz}) , \quad u = u(x,y,z) , \ a > 0 \qquad (*)$$

in dimension 3, as one of the most important PDEs, because this equation describes the vibrations of continuous mechanical systems, and the propagation of electromagnetic and sound waves. We have discussed some applications of the wave equation in Section 1.2 of Chapter 1. In this chapter, we will study a special case of (*), the one–dimensional wave equation

$$u_{tt} = a^2 u_{xx} , \qquad u = u(x,t) , \qquad (**)$$

when u does not depend on y and z (cf. Chapter 9 for a study of (*)). For definiteness, we interpret $u(x,t)$ as the transverse displacement (in a direction perpendicular to the x–axis) of a vibrating string at position x, at time t. With this model in mind, in Section 5.1, we use Newton's second law to derive (**), and solve this equation when the initial profile and velocity of the string is specified by finite Fourier sine series. We also establish the uniqueness of solutions by proving that the energy is conserved. In Section 5.2, we derive D'Alembert's formula for the solution of initial value problems for the infinite string. The method of images is used with D'Alembert's formula to solve several problems for the semi–infinite and finite strings, and to prove certain maximum principles which are needed to analyze the error of a solution due to an error in the initial conditions. We begin Section 5.3 with a discussion of the various standard types of B.C. . The same techniques which were used in Chapter 3 to solve heat problems with inhomogeneous B.C. are shown to also work for wave problems. Moreover, a version of Duhamel's method is motivated and used to solve problems for the inhomogeneous wave equation which results when external forces act on the string.

Historical Remarks. The subject of partial differential equations, and in particular the study of the wave equation, had its beginning in the eighteenth century. Among many other topics, the mathematicians and physicists of this era were interested in boundary value problems involving vibrations of strings stretched between fixed points and vibrations of columns of air in organ pipes, with regard to certain mathematical theories of music. The earliest contributors to such theories include Brook Taylor (1685–1731), Daniel Bernoulli (1700–1782), Leonhard Euler (1707–1783) and Jean D'Alembert (1717–1783). In the nineteenth century the wave equation was applied in the burgeoning field of elasticity, and subsequently in the study of the propagation of sound and light waves. During this period, important contributions were made by Simon D. Poisson (1781–1840), Georg F.B. Riemann (1826–1866), Hermann von Helmholtz (1821–1894), Gustav R. Kirchhoff (1824–1887) and John W.S. Rayleigh (1842–1919). In the twentieth century, the wave equation and many associated equations (e.g., the Dirac equation, the Klein–Gordon equation, Maxwell's equations, etc.) arise not only in the modern equivalents of previous applications, but also in the classical and quantum description of every known elementary particle, as well as in general relativity. More recent applications of the wave equation appear in the theories of superconductivity, superfluidity, and now, completing the circle, "superstrings" which is one of the latest unified field theories that has been proposed.

5.1 The Wave Equation – Derivation and Uniqueness

For a positive constant a, the one–dimensional wave equation

$$u_{tt} = a^2 u_{xx}\ , \quad u = u(x,t)\ , \tag{**}$$

bears a resemblance to the heat equation $u_t = ku_{xx}$. The essential difference is due to the presence of the second time derivative. Note that if $u(x,t)$ is a solution of (**), then the time–reversed function $v(x,t) = u(x,-t)$ is also a solution. This property does not generally hold for solutions of the heat equation, except for the time–independent solutions $u(x,t) = cx + d$. We will show that under certain ideal conditions, a solution of the wave equation can be interpreted as the time–dependent profile (or amplitude) of a vibrating string. When a real string is plucked, eventually air resistance and the conversion of the energy of motion into internal heat will reduce the amplitudes of vibrations. However, in an ideal situation where such dissipative forces are absent, we could not tell whether a movie of the string was being run forward or backward. This is not the case for heat distributions, where the characteristic dampening and leveling tells us that time is advancing. In spite of these differences, we will find that the basic method for solving the heat equation carries over to the wave equation.

The derivation of the wave equation

Consider a homogeneous string which stretches and offers negligible resistance to bending. At rest, the string is stretched between $x = 0$ and $x = L$. The string can vibrate in many ways. For example, points on the string might simply move back and forth on the x–axis, so that the profile of the string remains flat. Such vibrations are known as **longitudinal vibrations**. Suppose that the position (at time t) of a certain point, say at $(x,0,0)$ when the string is at rest, is given by $(r_1(x,t),r_2(x,t),r_3(x,t))$, for some functions r_1, r_2 and r_3. For longitudinal vibrations, we have $r_2 \equiv r_3 \equiv 0$. Generally, the functions r_1, r_2 and r_3 can be shown to obey a system of three PDEs each of which is typically nonlinear (cf. [Antman] and Problem 7). In order to simplify matters, we assume that the string vibrates in a plane, say the xy–plane (i.e., $r_3(x,t) \equiv 0$). Also, we will assume that the string under consideration is vibrating in such a way that $r_1(x,t) \equiv x$, so that the point corresponding to x is displaced by $r_2(x,t)$, only in the y–direction at time t. Such vibrations are said to be **transverse vibrations**. To simplify the notation, we denote $r_2(x,t)$ by $u(x,t)$. In the process of showing that $u(x,t)$ must obey the wave equation, we will prove that only a certain class of strings will admit nontrivial transverse vibrations. We refer to such strings as being "linearly elastic" within a given range of stretching. (The precise definition is given below, but roughly, this means that the tension at any point in the string is proportional to the the local stretching factor). Although there are no real physical strings which are linearly elastic for an arbitrarily large range of stretching, we can at least say that solutions of the resulting wave equation are valid as long as the degree of stretching is not too extreme. In the following derivation, at no time will we make the usual (but actually unnecessary) assumption that terms such as $u_x(x,t)^2$ can be neglected, in order to arrive at a linear wave equation.

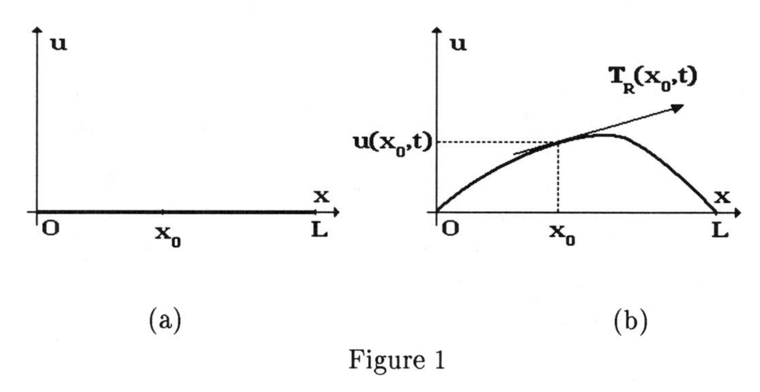

(a) (b)

Figure 1

In Figure 1(b), $u(x_0,t)$ is the vertical displacement, at time t, of the point on the string that would be at x_0 on the x–axis, if the string were in the rest position of Figure 1(a). By assumption of transverse vibrations, there is no horizontal displacement. The vector $\mathbf{T}_R(x_0,t)$ (cf. Figure 2(a)) is the force that the portion of string to the right of x_0 exerts on the portion to the left of x_0. We call $\mathbf{T}_R$ the **right–tension** at x_0, at time t. Similarly, we define the **left–tension** $\mathbf{T}_L(x_0,t)$. We assume that $\mathbf{T}_R(x_0,t)$ is tangent at x_0 to the graph of $u(x,t)$, as a function of x, and we assume that $\mathbf{T}_L(x_0,t) = -\mathbf{T}_R(x_0,t)$. This last assumption might seem to imply the points of the string will never move, since the net force on each point is $\mathbf{T}_L + \mathbf{T}_R = \mathbf{0}$. However, since the mass of a point is 0, the acceleration of a point is not determined by Newton's equation $\mathbf{F} = m\mathbf{a}$. To obtain the acceleration of a point x_0, we need to take the limit of the ratio of the net force on the portion of the string above the interval $[x_0-h,\ x_0+h]$ to the mass of this portion, as $h \to 0$ (cf. Figure 2(a)).

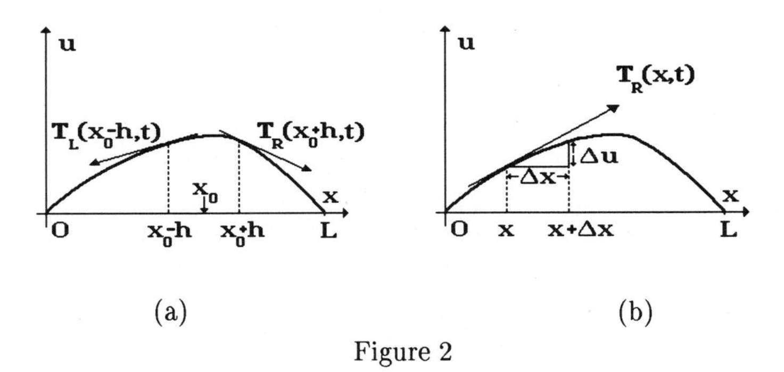

(a) (b)

Figure 2

We compute this limit as follows. Let the *magnitude* of the left(or right)–tension at x_0 at time t, be denoted by $T(x_0,t)$. This is called the **tension** at x_0 at time t. We assume that the tension is the positive constant T_0, when the string is straight (as in Figure 1(a)). If the string is stretched by a factor of s, relative to its length at rest, then the new tension will be of the form

$g(s) \cdot T_o$, for some function $g(s)$, which depends on the nature of the string. That is, we are assuming that the function g depends only on s, and not on other quantities such as the rate of stretching or temperature variations due to bending. Of course, $g(1) = 1$, because when the string has not been stretched (i.e., $s = 1$), the tension is T_o . We assume that $g(s)$ is C^1 for stretches s which are encountered during vibrations. Using these assumptions together with the restriction that the vibrations are transverse, we will show that Newton's second law impies that $g(s) = s$ (i.e., the tension in the string is proportional to the stretching factor s). Then we will see that the PDE for $u(x,t)$ which results (without any of the usual dubious statements about u_x^2 being negligible) is necessarily a linear PDE, namely the wave equation.

Definition. A string is **linearly elastic** for $a \le s \le b$, provided that, when the string, at tension T_o, is stretched by a factor of s in $[a,b]$, the tension becomes $T_o s$ (i.e., $g(s) = s$).

More precisely, we will show that, in conjunction with Newton's law, the assumption that the string admits nontrivial transverse vibrations actually implies that the string is linearly elastic for s in any range which is encountered during such a vibration. In Figure 2(b) above, the string in the interval $[x, x + \Delta x]$ has been stretched by about a factor of $(\Delta x^2 + \Delta u^2)^{\frac{1}{2}}/\Delta x$ which tends to $s \equiv (1 + u_x^2)^{\frac{1}{2}}$, as $\Delta x \to 0$. The direction of the right tension at x is approximately the same as that of the unit vector $(\Delta x \mathbf{i} + \Delta u \mathbf{j}) / (\Delta x^2 + \Delta u^2)^{\frac{1}{2}}$, which approaches the exact unit tangent vector $(\mathbf{i} + u_x \mathbf{j})/(1 + u_x^2)^{\frac{1}{2}} = \frac{1}{s}(\mathbf{i} + u_x \mathbf{j})$, as $\Delta x \to 0$. Thus,

$$\mathbf{T}_R(x,t) = g(s) \cdot T_o \cdot \frac{1}{s} [\mathbf{i} + u_x(x,t)\mathbf{j}] , \quad \text{where} \quad s = (1 + u_x^2)^{\frac{1}{2}} .$$

Consequently, the net force on the string between $x - h$ and $x + h$, as in Figure 2(a), is

$$\mathbf{F}(x,t) = \mathbf{T}_L(x-h,t) + \mathbf{T}_R(x+h,t) = \mathbf{T}_R(x+h,t) - \mathbf{T}_R(x-h,t) .$$

Since we are considering only transverse vibrations, the mass of the portion of the string, from $x-h$ to $x+h$, remains the same as it would be if the portion were in the straight rest position, namely $2hD$, where D is the mass per unit length of the string when straight (i.e., $D \equiv$ the linear density of the straight string in Figure 1(a)). By Newton's law and the assumption of purely transverse vibrations, the acceleration of the string at x is

$$u_{tt}(x,t)\mathbf{j} = \lim_{h \to 0} \frac{\mathbf{F}}{2hD} = \frac{1}{D} \lim_{h \to 0} \frac{1}{2h} [\mathbf{T}_R(x+h,t) - \mathbf{T}_R(x-h,t)]$$

$$= \frac{1}{D} \frac{\partial}{\partial x} \left[\mathbf{T}_R \right] = \frac{1}{D} \frac{\partial}{\partial x} \left[T_o \frac{g(s)}{s} \right] \mathbf{i} + \frac{1}{D} \frac{\partial}{\partial x} \left[T_o \frac{g(s)}{s} u_x \right] \mathbf{j}$$

Equating the $\mathbf{i}$ and $\mathbf{j}$ components of both sides, we have

$$0 = (T_o/D)\,\frac{\partial}{\partial x}\left[\frac{g(s)}{s}\right] \tag{I}$$

and

$$u_{tt} = (T_o/D)\cdot\frac{g(s)}{s}\,u_{xx} + (T_o/D)\cdot\frac{\partial}{\partial x}\left[\frac{g(s)}{s}\right]u_x\,, \tag{J}$$

where $s = (1 + u_x^2)^{\frac{1}{2}}$. Thus, (I) implies that $G(s) \equiv g(s)/s$ is independent of x. Now

$$0 = \frac{\partial}{\partial x}\left[G((1 + u_x^2(x,t))^{\frac{1}{2}})\right] \quad \text{or} \quad G(s(x_1,t)) = G(s(x_2,t))\,, \tag{I$'$}$$

for all x_1 and x_2 in $[0,L]$ (Why ?). For any fixed time t, we may choose x_1 to be a point such that $u_x(x_1,t)$ is 0. (Here we assume that the ends of the string are fixed and apply Rolle's theorem.) Thus $s(x_1,t) = 1$, and since $G(1) = g(1)/1 = 1$, (I$'$) yields $1 = G(s(x_2,t))$ for any x_2 and any t during the vibration. In other words, $g(s) = s$ for any stretch s which is encountered during a transverse vibration. Thus, the PDE (J) reduces to the wave equation

$$u_{tt} = a^2 u_{xx}\,, \quad \text{where} \quad a^2 = T_o/D\,. \tag{**}$$

Remark. In many derivations, the PDE (J) is linearized (cf. Section 1.2) by the outright assumption that u_x^2 is negligible, so that one can (all too conveniently !) set $s = 1$ in the PDE (J), which then becomes the linear wave equation (**). However, the assumption that u_x^2 is small presupposes some knowledge about the solution of the typically nonlinear PDE (J) (cf. Example 11 of Section 1.2, for the pitfalls of linearization). Rather than introducing this questionable assumption, we have shown that it is unnecessary to do so, by demonstrating that $g(s) = s$ follows from the assumption of transverse vibrations and Newton's law. The property $g(s) = s$ means that the string behaves as a spring or a rubber band which is stretched, but not by too much, so that Hooke's law holds (cf. Example 6 of Section 1.1). However, not all strings satisfy $g(s) = s$, except in a very short interval about $s = 1$. For example, the tension in a piece of twine can increase enormously even for small stretches, whereas for a string of taffy the tension may even eventually decrease upon stretching. In essence, we have shown that such strings do not admit purely transverse vibrations $u(x,t)$ for which the stretching factor, $(1 + u_x(x,t)^2)^{\frac{1}{2}}$, lies outside the range of linear elasticity for some (x,t). □

Solving the standard initial value problem for a string with fixed ends

As with the heat equation, we first find all of the product solutions of $u_{tt} = a^2u_{xx}$, using the method of separation of variables. Substituting $u(x,t) = X(x)T(t)$ into $u_{tt} = a^2u_{xx}$, we get

$$X(x)T''(t) = a^2X''(x)T(t) \quad \text{or} \quad \frac{T''(t)}{a^2T(t)} = \frac{X''(x)}{X(x)} = c = \pm\lambda^2\,,$$

where λ is some nonnegative constant. The ODE $X'' = \pm \lambda^2 X$ for $X(x)$ is exactly the same as for the heat equation in Section 3.1, but the solutions of the ODE $T'' = \pm\lambda^2 a^2 T$ for $T(t)$ are different, because of the second time derivative. The possible product solutions fall into the following three cases, where c_1, c_2, d_1 and d_2 are arbitrary constants :

Case 1 $(c = -\lambda^2 < 0)$:

$$u(x,t) = [d_1\sin(\lambda at) + d_2\cos(\lambda at)][c_1\sin(\lambda x) + c_2\cos(\lambda x)]. \tag{1}$$

Case 2 $(c = \lambda^2 > 0)$:

$$u(x,t) = (d_1e^{\lambda at} + d_2e^{-\lambda at})(c_1e^{\lambda x} + c_2e^{-\lambda x}). \tag{2}$$

Case 3 $(c = \lambda^2 = 0)$:

$$u(x,t) = (d_1t + d_2)(c_1x + c_2). \tag{3}$$

We will now formulate one of the simplest standard initial/boundary–value problems for the wave equation. Recall that in solving Newton's equation $mx''(t) = F(t)$, it is necessary to specify $x(t_0)$ and $x'(t_0)$ in order to obtain a unique solution for the position $x(t)$ of a particle. For the wave equation (whose derivation was based on Newton's equation), it is also necessary to specify not only the initial profile $u(x,0)$ of the string, but also the initial velocity $u_t(x,0)$. Otherwise, we will not obtain a unique solution $u(x,t)$. First we assume that the ends of the string are fixed (i.e., $u(0,t) = 0$, $u(L,t) = 0$), although in Section 5.3, we consider the cases where one or both of the ends are allowed to slide vertically. We expect that under reasonable circumstances the following standard problem will have a unique solution:

$$\begin{array}{ll} \text{D.E.} & u_{tt} = a^2u_{xx}, \quad 0 \le x \le L\,,\ -\infty < t < +\infty\,, \\ \text{B.C.} & u(0,t) = 0,\ u(L,t) = 0, \\ \text{I.C.} & \begin{cases} u(x,0) = f(x), & \text{(initial position)} \\ u_t(x,0) = g(x), & \text{(initial velocity).} \end{cases} \end{array} \tag{4}$$

We require that $u(x,t)$ have a C^2 extension to an *open* domain that contains the strip $0 \le x \le L$, $-\infty < t < \infty$. As with the heat equation, the only product solutions which satisfy the B.C., are of the form (1), where $c_2 = 0$ and $\lambda = n\pi/L$, $n = 1, 2, 3, \ldots$. By taking linear combinations of such solutions of the D.E. and B.C., we obtain a solution of the form

$$u(x,t) = \sum_{n=1}^{N} [A_n \sin(\tfrac{n\pi at}{L}) + B_n \cos(\tfrac{n\pi at}{L})] \sin(\tfrac{n\pi x}{L}). \tag{5}$$

Note that

$$u_t(x,t) = \sum_{n=1}^{N} \frac{n\pi a}{L} [A_n \cos(\tfrac{n\pi at}{L}) - B_n \sin(\tfrac{n\pi at}{L})] \sin(\tfrac{n\pi x}{L}). \tag{6}$$

Substituting $t = 0$ in (5) and (6) yields

$$u(x,0) = \sum_{n=1}^{N} B_n \sin(\tfrac{n\pi x}{L}), \quad u_t(x,0) = \sum_{n=1}^{N} A_n \frac{n\pi a}{L} \sin(\tfrac{n\pi x}{L}). \tag{7}$$

In summary, we have shown the following result.

Proposition 1. A solution of the problem

$$\begin{aligned}
&\text{D.E.} \quad u_{tt} = a^2 u_{xx}, \qquad 0 \le x \le L, \ -\infty < t < +\infty, \\
&\text{B.C.} \quad u(0,t) = 0, \ u(L,t) = 0, \\
&\text{I.C.} \quad \begin{cases} u(x,0) = f(x) = \sum_{n=1}^{N} B_n \sin(n\pi x/L), \\ u_t(x,0) = g(x) = \sum_{n=1}^{N} \overline{A}_n \sin(n\pi x/L) \end{cases}
\end{aligned} \tag{8}$$

is

$$u(x,t) = \sum_{n=1}^{N} \left[\frac{L}{n\pi a} \overline{A}_n \sin(\tfrac{n\pi at}{L}) + B_n \cos(\tfrac{n\pi at}{L})\right] \sin(\tfrac{n\pi x}{L}). \tag{9}$$

Remark. Here we have expressed $u(x,t)$ in terms of the Fourier sine coefficients B_n and $\overline{A}_n$ of $f(x)$ and $g(x)$ respectively. Note that A_n in (5) is $(L/n\pi a)\overline{A}_n$. □

Of course, one can pose problems where $f(x)$ and $g(x)$ are not finite sine series as above. But if $f(x)$ and $g(x)$ are continuous and piecewise C^1 on $[0,L]$, with $f(0) = f(L) = 0$ and $g(0) = g(L) = 0$, then we know that $f(x)$ and $g(x)$ may be uniformly approximated to within any small positive error by partial sums of their Fourier sine series (cf. Theorem 1, Section 4.3). Thus, such $f(x)$ and $g(x)$ can be replaced by finite sine series within experimental error. Later we will show that two solutions must be close, if the initial profiles and velocity distributions of the two solutions are close (cf. the Corollary to Theorem 5 in Section 5.2).

Example 1. Solve the initial/boundary–value problem

$$\begin{aligned} &\text{D.E.} \quad u_{tt} = a^2 u_{xx}, \qquad 0 \le x \le L, \quad -\infty < t < +\infty, \\ &\text{B.C.} \quad u(0,t) = 0, \quad u(L,t) = 0, \\ &\text{I.C.} \quad \begin{cases} u(x,0) = f(x) = 2\sin(\frac{3\pi x}{L}), \\ u_t(x,0) = g(x) = \sin(\frac{\pi x}{L}) - 3\sin(\frac{5\pi x}{L}). \end{cases} \end{aligned}$$

Solution. We simply apply Proposition 1 with $B_3 = 2$, $\overline{A}_1 = 1$, $\overline{A}_5 = -3$, and with all of the other $\overline{A}_n$ and B_n equal to zero. Then

$$u(x,t) = \frac{L}{\pi a}\sin(\frac{\pi at}{L})\sin(\frac{\pi x}{L}) + 2\cos(\frac{3\pi at}{L})\sin(\frac{3\pi x}{L}) - \frac{3L}{5\pi a}\sin(\frac{5\pi at}{L})\sin(\frac{5\pi x}{L}).$$

Note that to get the solution quickly, one can simply insert a factor of $\cos(n\pi at/L)$ in the terms involving $\sin(n\pi x/L)$ in $f(x)$ and insert a factor of $(L/n\pi a)\sin(n\pi at/L)$ in the terms involving $\sin(n\pi x/L)$ in $g(x)$, and then add the results to get $u(x,t)$. □

Harmonics

The individual terms of the series (5), namely (for $n = 1, 2, 3, \ldots$)

$$u_n(x,t) = [A_n\sin(\frac{n\pi at}{L}) + B_n\cos(\frac{n\pi at}{L})]\sin(\frac{n\pi x}{L}) \tag{10}$$

are called the **harmonics** or **overtones** of the string with fixed ends at $x = 0$ and $x = L$.

These constitute a complete family of product solutions of the D.E. $u_{tt} = a^2 u_{xx}$ with B.C. $u(0,t) = u(L,t) = 0$, as n runs through the values $1, 2, \ldots$. If A_n and B_n are not both zero, we can rewrite $u_n(x,t)$ as follows. Let $R_n \equiv (A_n^2 + B_n^2)^{\frac{1}{2}}$. Then there is a θ_n, such that $\cos(\theta_n) = A_n/R_n$ and $\sin(\theta_n) = B_n/R_n$, since $(A_n/R_n)^2 + (B_n/R_n)^2 = 1$. Thus,

$$u_n(x,t) = R_n[\cos(\theta_n)\sin(\frac{n\pi at}{L}) + \sin(\theta_n)\cos(\frac{n\pi at}{L})]\sin(\frac{n\pi x}{L}) = R_n\sin(\frac{n\pi at}{L} + \theta_n)\sin(\frac{n\pi x}{L}).$$

We see that $u_n(x,t)$ oscillates between $\pm R_n\sin(n\pi x/L)$ as t varies. R_n is called the **amplitude** of $u_n(x,t)$ and θ_n is the **phase** of $u_n(x,t)$ (cf. Figure 3, where $R_n = 1$, $L = 6$).

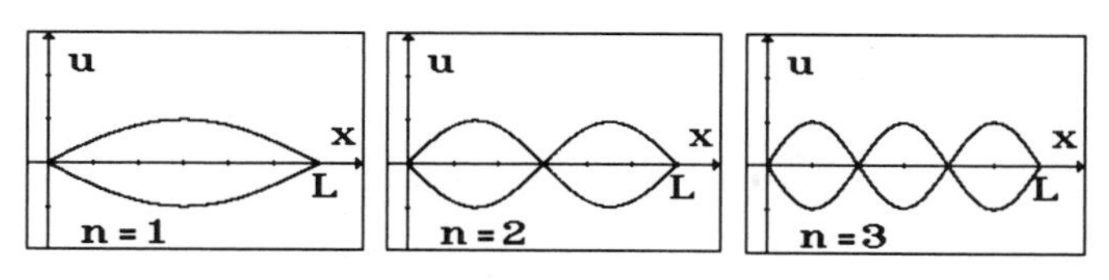

Figure 3

The time that it takes a harmonic to complete one oscillation is called its **period**. The period is inversely proportional to n. To find the period of the n–th harmonic, we set $n\pi at/L = 2\pi$ and solve for t, obtaining $t = 2L/na$. The **frequency** of a harmonic is the number of oscillations per unit time, and it is just the reciprocal of the period. Hence, the period and frequency of the n–th harmonic of the string (with fixed ends) are respectively given by

$$P_n = \frac{2L}{na} \quad \text{and} \quad \nu_n = \frac{1}{P_n} = \frac{na}{2L}\,. \tag{11}$$

Uniqueness and the energy integral

As with the heat equation, there is the uniqueness question. Can there be two different solutions to problem (4) for a given f(x) and g(x) ?

Theorem 1 (Uniqueness). Let $u_1(x,t)$ **and** $u_2(x,t)$ **be** C^2 **solutions of the following problem**

$$\text{D.E.} \quad u_{tt} = a^2 u_{xx}, \qquad 0 \le x \le L, \quad -\infty < t < +\infty,$$

$$\text{B.C.} \quad u(0,t) = A(t)\ , \quad u(L,t) = B(t)\ ,$$

$$\text{I.C.} \quad u(x,0) = f(x)\ , \quad u_t(x,0) = g(x)\ .$$

Then $u_1(x,t) = u_2(x,t)$ **for all** $0 \le x \le L,\ -\infty < t < +\infty.$

Proof. Let $v(x,t) = u_1(x,t) - u_2(x,t)$. Note that v satisfies the related problem with homogeneous B.C. and I.C. . In particular, $v(x,0) = 0$ and $v_t(x,0) = 0$. We need to show that $v(x,t) = 0$ for all t. In the proof of the corresponding Theorem 1, in Section 3.2, we accomplished this by proving that $\int_0^L [v(x,t)]^2\, dx = 0$ via differentiation with respect to t. Here, we will prove that the function

$$H(t) = \int_0^L \left[a^2[v_x(x,t)]^2 + [v_t(x,t)]^2\right] dx \tag{12}$$

is zero. Once this is done, then $v_t(x,t) = 0$ for all x in $[0,L]$ and all real t, and (as required)

$$v(x,t) = v(x,t) - v(x,0) = \int_0^t v_t(x,t)\, dt = 0.$$

We compute $H'(t)$, using $v_{tt} = a^2 v_{xx}$ and differentiating under the integral (cf. Appendix A.3) :

$$H'(t) = \int_0^L [a^2 2v_x v_{xt} + 2v_t v_{tt}]\, dx = 2a^2 \int_0^L [v_x v_{xt} + v_t v_{xx}]\, dx$$

$$= 2a^2 \int_0^L \frac{\partial}{\partial x}(v_x v_t)\, dx = 2a^2 [v_x(x,t) v_t(x,t)]\Big|_0^L = 0 ,$$

since (by the B.C. $v(0,t) = 0$ and $v(L,t) = 0$) $v_t(0,t) = \frac{d}{dt} v(0,t) = 0$, and similarly $v_t(L,t) = 0$. Since $H'(t) = 0$, we know that $H(t)$ is constant, but this constant is 0, since $H(0) = 0$ according to the initial conditions for v (note that $v_x(x,0) = \frac{d}{dx} v(x,0) = 0$). □

Remark. The function $H(t)$ (cf. (12)) in the above proof is actually proportional to the total energy of the string given by $v(x,t)$. Indeed, the kinetic energy of the segment of the string from x to $x + \Delta x$ is approximately $\frac{1}{2}(D\Delta x)[u_t(x,t)]^2$, where D is the mass per unit length. The work done (energy expended or potential energy) in stretching the segment from Δx to $(\Delta x^2 + \Delta u^2)^{\frac{1}{2}} \approx [1 + u_x(x,t)^2]^{\frac{1}{2}} \Delta x = s\Delta x$, is the following integral of the force (tension) with respect to the increase r in the length of the segment during the stretching process.

$$\int_0^{(s-1)\Delta x} T_0 \frac{\Delta x + r}{\Delta x}\, dr = T_0 \cdot \left(r + \frac{r^2}{2\Delta x}\right)\Big|_0^{(s-1)\Delta x}$$

$$= T_0 \Delta x (s - 1 + \tfrac{1}{2}(s-1)^2) = \tfrac{1}{2} T_0 (s^2 - 1)\Delta x = \tfrac{1}{2} T_0 [u_x(x,t)]^2\, \Delta x ,$$

where we have assumed, as in the derivation of the wave equation, that the string is linearly elastic, with rest tension T_0. Hence, the potential energy of the segment of string from x to $x + \Delta x$ is $\approx \frac{1}{2} T_0 [u_x(x,t)]^2\, \Delta x$. When this is added to the kinetic energy and the result is integrated, we get the **energy integral** (or simply, **energy**) of the string at time t

$$E(t) = \frac{1}{2} \int_0^L \left[T_0 [u_x(x,t)]^2 + D[u_t(x,t)]^2\right] dx, \tag{13}$$

which is $\frac{1}{2}D$ times $H(t)$ in (12), using v for u. For an unforced string, we expect $E(t)$ to be constant. Thus, it is not surprising that $H'(t) = 0$ in the preceding proof. Of course, the physical intuition is no substitute for the explicit computation of $H'(t)$ above, but it does lead us to consider the function $H(t)$ in the first place. In other words, the physical basis for the proof of the uniqueness (Theorem 1) is the law of conservation of energy. □

Example 2. Calculate the energy of the n–th harmonic

$$u(x,t) = [A_n \sin(\tfrac{n\pi at}{L}) + B_n \cos(\tfrac{n\pi at}{L})]\sin(\tfrac{n\pi x}{L}) .$$

Solution. The energy $E(t)$ is defined by (13), and we know from the proof of Theorem 1 that $E(t)$ is constant. Thus, $E(t) = E(0)$, and so we only need to compute $E(0)$:

$$\begin{aligned} E(0) &= \frac{1}{2}\int_0^L \Big[T_o[u_x(x,0)]^2 + D[u_t(x,0)]^2\Big]\,dx \\ &= \frac{1}{2}\int_0^L \Big[T_o[\tfrac{n\pi}{L} B_n\cos(\tfrac{n\pi x}{L})]^2 + D[\tfrac{n\pi a}{L} A_n \sin(\tfrac{n\pi x}{L})]^2\Big]\,dx \\ &= \frac{1}{2}(\tfrac{n\pi}{L})^2 \Big[T_o B_n^2 \int_0^L \cos^2(\tfrac{n\pi x}{L})\,dx + Da^2A_n^2\int_0^L \sin^2(\tfrac{n\pi x}{L})\,dx\Big] \\ &= \frac{n^2\pi^2}{4L}(T_oB_n^2 + Da^2A_n^2) = \frac{\pi^2}{4L}T_o(B_n^2 + A_n^2)\,n^2 = \frac{\pi^2}{4L}T_oR_n^2\,n^2 , \end{aligned}$$

where we recall that $a \equiv (T_o/D)^{\frac{1}{2}}$ and $R_n^2 = A_n^2 + B_n^2$ is the square of the amplitude of the harmonic. It is possible, but rather lengthy, to compute $E(t)$ directly by considering $[u_x(x,t)]^2$ and $[u_t(x,t)]^2$. Of course, we know that the end result will be $E(0)$ anyway. □

Remark. Note that the energy of the n–th harmonic is proportional to n^2 for a given amplitude, and it is proportional to the tension and the square of the amplitude. However, unlike the frequency $\nu_n = \frac{1}{2}nL\sqrt{T_o/D}$, the energy is independent of the linear density D. □

Example 3. (The motion of the plucked string) Find the formal solution of the problem for the motion of the plucked string:

$$\begin{aligned} &\text{D.E.}\quad u_{tt} = a^2u_{xx}, \qquad 0 \le x \le L,\ -\infty < t < \infty, \\ &\text{B.C.}\quad u(0,t) = 0,\ u(L,t) = 0, \qquad\qquad (14) \\ &\text{I.C.}\quad u(x,0) = f(x),\ u_t(x,0) = 0, \end{aligned}$$

where $f(x)$ is the function with graph shown in Figure 4.

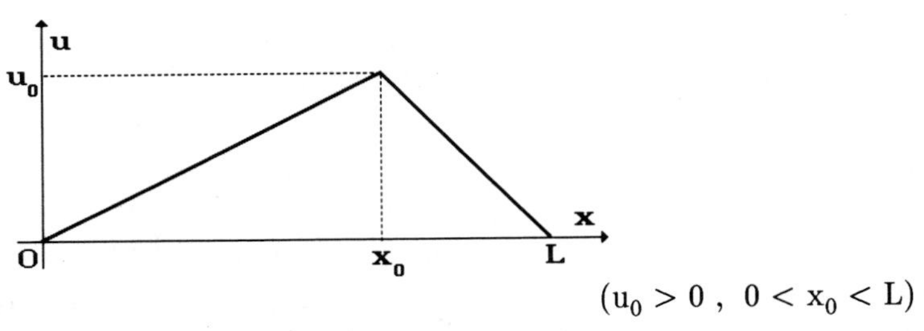

$(u_0 > 0\ ,\ 0 < x_0 < L)$

Figure 4

(i.e., the string is "plucked" at some fixed x_0 in $(0,L)$, lifted to the displacement u_0 and released with zero initial velocity, i.e., $u_t(x,0) = 0$).

Solution. We have

$$f(x) = \begin{cases} (u_0/x_0)x, & 0 \le x \le x_0, \\ u_0(x-L)/(x_0-L), & x_0 \le x \le L. \end{cases}$$

The formal solution of problem (14) is obtained by computing the Fourier sine coefficients B_n of $f(x)$ (cf. Proposition 1) and using formula (9) with $N = \infty$. (The notion of a formal solution was introduced in Section 4.3.) Note that f is continuous and piecewise C^1. Integration by parts is valid for such functions. Thus,

$$B_n = \frac{2}{L}\int_0^L f(x)\sin(\tfrac{n\pi x}{L})\,dx = \frac{2}{L}\, f(x)(\tfrac{-L}{n\pi})\cos(\tfrac{n\pi x}{L})\Big|_0^L + \frac{2}{L}(\tfrac{L}{n\pi})\int_0^L f'(x)\cos(\tfrac{n\pi x}{L})\,dx.$$

The endpoint evaluation is zero, since $f(0) = f(L) = 0$, and from the fact that

$$f'(x) = \begin{cases} u_0/x_0, & 0 \le x < x_0 \\ u_0/(x_0-L), & x_0 < x \le L, \end{cases}$$

we obtain

$$\begin{aligned} B_n &= \frac{2}{L}\frac{L}{n\pi}\Big[\int_0^{x_0} \frac{u_0}{x_0}\cos(\tfrac{n\pi x}{L})\,dx + \int_{x_0}^L \frac{u_0}{x_0-L}\cos(\tfrac{n\pi x}{L})\,dx\Big] \\ &= \frac{2}{L}\Big[\frac{L}{n\pi}\Big]^2 \Big[\frac{u_0}{x_0}\sin(\tfrac{n\pi x_0}{L}) - \frac{u_0}{x_0-L}\sin(\tfrac{n\pi x_0}{L})\Big] = \frac{2Lu_0}{n^2\pi^2}\Big[\frac{1}{x_0} + \frac{1}{L-x_0}\Big]\sin(\tfrac{n\pi x_0}{L}) \\ &= \frac{2L^2u_0}{\pi^2 x_0(L-x_0)}\,\frac{1}{n^2}\,\sin(\tfrac{n\pi x_0}{L}). \end{aligned}$$

The formal solution is

$$u(x,t) = \frac{2L^2 u_0}{\pi^2 x_0 (L-x_0)} \sum_{n=1}^{\infty} \frac{\sin(n\pi x_0/L)}{n^2} \cos(\frac{n\pi at}{L}) \sin(\frac{n\pi x}{L}). \tag{15}$$

Let $u_N(x,t)$ denote the sum of the first N terms of (15). Then $u_N(x,t)$ is a C^∞ solution of the D.E. with the given B.C.. Moreover, $u_N(x,0)$ is the N–th partial sum $S_N(x)$ of FSS f(x), and thus we know that for N sufficiently large $u_N(x,0)$ will approximate $f(x)$ to within any preassigned error (cf. Theorem 1 of Section 4.3). Hence, for practical purposes, the problem of the plucked string is solved. While it is possible to prove that the sum (15) converges (i.e., $u(x,t)$ is defined for all (x,t)), $u(x,t)$ is not C^2 and hence is not a strict solution of the D.E. . Using the techniques of Section 5.3, one can show that for most times t, the graph of $u(x,t)$ in (15) consists of three line segments whose slopes do not match at the two interior corners where pairs of them join. Indeed, the two corners move in opposite directions (and with horizontal speed a) around the parallelogram formed by the original profile and its reflection through the point $\frac{1}{2}L$ on the x–axis (cf. Figure 5 below). Observe that $u(x,t + 2L/a) = u(x,t)$. In particular, the string returns to the original plucked position at $t = 2L/a$ and repeats its motion. The values L/n, $2L/n$, $3L/n$, ..., $(n-1)L/n$, where $\sin(n\pi x/L)$ vanishes are called the **nodes** of the harmonic $\cos(n\pi at/L)\sin(n\pi x/L)$. Formula (15) shows that the harmonics, with x_0 as a node, drop out of the sum, because of the factor $\sin(n\pi x_0/L)$. □

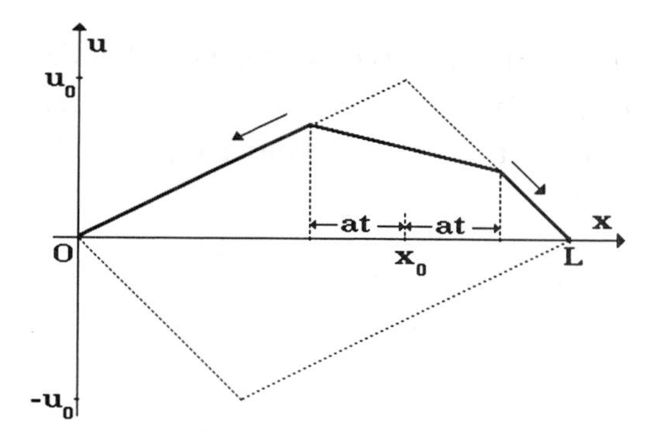

Figure 5

Summary 5.1

1. Derivation of the wave equation : Under the assumption that a string of linear density D, stretched with tension T_o between two points, is executing only transverse vibrations, we demonstrate (via Newton's equation $\mathbf{F} = m\mathbf{a}$) that the amplitude $u(x,t)$ of the vibration must obey the wave equation $u_{tt} = a^2 u_{xx}$, where $a^2 \equiv T_o/D$. In the course of the derivation, we prove that a string which undergoes transverse vibrations must be linearly elastic (i.e., the tension at any point is of the form sT_o, where $s \equiv (1 + u_x^2)^{\frac{1}{2}}$ is the local stretching factor). Consequently, it is *not* necessary to assume that u_x^2 is negligible in order to achieve a linear wave equation for $u(x,t)$, for transversely vibrating strings.

2. The standard problem for fixed ends (Proposition 1) : A solution of the problem

$$
\begin{aligned}
&\text{D.E.} \quad u_{tt} = a^2 u_{xx}, \qquad 0 \le x \le L, \ -\infty < t < +\infty, \\
&\text{B.C.} \quad u(0,t) = 0, \ u(L,t) = 0, \\
&\text{I.C.} \quad \begin{cases} u(x,0) = f(x) = \sum_{n=1}^{N} B_n \sin(\frac{n\pi x}{L}), \\ u_t(x,0) = g(x) = \sum_{n=1}^{N} \bar{A}_n \sin(\frac{n\pi x}{L}) \end{cases}
\end{aligned}
\tag{S1}
$$

is

$$
u(x,t) = \sum_{n=1}^{N} \left[\frac{L}{n\pi a} \bar{A}_n \sin(\frac{n\pi at}{L}) + B_n \cos(\frac{n\pi at}{L}) \right] \sin(\frac{n\pi x}{L}). \tag{S2}
$$

3. Harmonics : The product solutions of the D.E. and B.C. of (S1) are called **harmonics** and they are of the form (where $n = 1, 2, 3, \ldots$)

$$
u_n(x,t) = [A_n \sin(\frac{n\pi at}{L}) + B_n \cos(\frac{n\pi at}{L})] \sin(\frac{n\pi x}{L}) = R_n \sin(\frac{n\pi at}{L} + \theta_n) \sin(\frac{n\pi x}{L}), \tag{S3}
$$

where $R_n = (A_n^2 + B_n^2)^{\frac{1}{2}}$ is the **amplitude** of the harmonic, and $A_n/R_n = \cos\theta_n$, $B_n/R_n = \sin\theta_n$ define the **phase** θ_n.

4. Energy : The energy at time t of any solution u(x,t) of (S1) is given by

$$E(t) = \frac{1}{2}\int_0^L \left[T_o[u_x(x,t)]^2 + D[u_t(x,t)]^2\right] dx, \qquad (S4)$$

which was shown to be constant in the proof of the uniqueness theorem (Theorem 1). The energy of the harmonic $u_n(x,t)$ in (S3) was computed to be $\frac{\pi^2}{4L} T_o R_n^2 n^2$ in Example 2.

5. Uniqueness : Theorem 1 implies that (S2) is the unique solution of the problem (S1). More generally, even if the B.C. in (S1) are replaced by $u(0,t) = A(t)$ and $u(x,t) = B(t)$ and if f(x) and g(x) are not necessarily finite Fourier sine series, Theorem 1 states that there is at most one (possibly no solutions) C^2 solution of the resulting problem. The proof proceeds by showing that the energy of the difference of two solutions is time–independent and initially zero.

6. The plucked string : The formal solution for the problem of the plucked string is found in Example 3. By truncating the formal solution, one can produce C^∞ solutions of the D.E. and B.C. which meet the I.C. to within any preassigned experimental error. However, it can be shown that the full sum of the formal solution converges to a function (which is not even C^1) with corners which move in opposite directions around a parallelogram, as in Figure 5.

Exercises 5.1

1. Solve the problem

$$\begin{aligned} &\text{D.E.}\quad u_{tt} = a^2 u_{xx}, \qquad 0 \le x \le L, \quad -\infty < t < \infty, \\ &\text{B.C.}\quad u(0,t) = 0, \quad u(L,t) = 0, \\ &\text{I.C.}\quad u(x,0) = f(x), \quad u_t(x,0) = g(x). \end{aligned}$$

in the following cases:

(a) $f(x) = 3\sin(\frac{\pi x}{L}) - \sin(\frac{4\pi x}{L})$, $g(x) = \frac{1}{2}\sin(\frac{2\pi x}{L})$, (b) $f(x) = [\sin(\frac{\pi x}{L})]^3$, $g(x) = 0$,
(c) $f(x) = 0$, $g(x) = \sin(\frac{\pi x}{L})\cos^2(\frac{\pi x}{L})$, (d) $f(x) = [\sin(\frac{\pi x}{L})]^3$, $g(x) = \sin(\frac{\pi x}{L})\cos^2(\frac{\pi x}{L})$.

Hint. For (b) and (c), use trigonometric identities. For (d), use the superposition principle.

2. Suppose u(x,t) solves $u_{tt} = a^2 u_{xx}$, $(a \neq 0)$.
(a) Let α, β, x_0 and t_0 be constants, with $\alpha \neq 0$. Show that the function v(x,t), given by $v(x,t) = u(\alpha x + x_0, \beta t + t_0)$, satisfies $v_{tt} = (\beta a/\alpha)^2 v_{xx}$.

(b) For any constant ω, let $\bar{x} = \cosh(\omega)x + a\cdot\sinh(\omega)t$ and $\bar{t} = a^{-1}\cdot\sinh(\omega)x + \cosh(\omega)t$. Recalling that $\cosh^2(\omega) - \sinh^2(\omega) = 1$, show that $x = \cosh(\omega)\bar{x} - a\cdot\sinh(\omega)\bar{t}$ and $t = -a^{-1}\cdot\sinh(\omega)\bar{x} + \cosh(\omega)\bar{t}$ (i.e., $(x,t) \to (\bar{x},\bar{t})$ is an invertible change of variables.)

(c) Define $\bar{u}(\bar{x},\bar{t}) = u(x,t)$. Show that $u_{tt} - a^2u_{xx} = \bar{u}_{\bar{t}\bar{t}} - a^2\bar{u}_{\bar{x}\bar{x}}$. **Hint**. By the chain rule, $u_x = \bar{u}_{\bar{x}}\bar{x}_x + \bar{u}_{\bar{t}}\bar{t}_x = \bar{u}_{\bar{x}}\cosh(\omega) + \bar{u}_{\bar{t}}a^{-1}\sinh(\omega)$, and compute u_{xx} and u_{tt} similarly.

Remark. The transformations $(x,t) \to (\bar{x},\bar{t})$, known as **Lorentz transformations**, mix space and time. Part (c) shows that the wave equation $u_{tt} - a^2u_{xx} = 0$ retains its form under Lorentz transformations. In this way, Albert Einstein (1879–1955) was led to his famous unification of space and time (i.e., relativity). In most physics books, $\cosh(\omega)$ is written as $(1-v^2/a^2)^{-\frac{1}{2}}$, and $\sinh(\omega)$ is then $\pm \frac{v}{a}\cdot(1-v^2/a^2)^{-\frac{1}{2}}$ (Why ?), where vi is the relative velocity of two observers, and $a = c \equiv$ the speed of light $\approx 2.99 \times 10^8\ m\cdot s^{-1}$.

3. In the derivation of the wave equation (Section 5.1) we did not consider the effect of gravity which exerts an additional force of $-2hDg\,\mathbf{j}$ on the segment of string between $x_0 - h$ and $x_0 + h$, where $g = 32\ ft/sec^2$ is the acceleration due to gravity.

(a) Deduce that $u(x,t)$ obeys $u_{tt} = a^2u_{xx} - g$, if the effect of gravity is considered.

(b) Find a solution of $u_{tt} = a^2u_{xx} - g$ which satisfies the B.C. $u(0,t) = 0$ and $u(L,t) = 0$ and which is time–independent [i.e., $u(x,t) = U(x)$]. What does this solution represent?

4. In Section 5.3, we show that the B.C. $u_x(0,t) = 0$ means that the end $x = 0$ of the string is free to slide vertically (and similarly, for the end $x = L$). Show that the proof of Theorem 1 yields uniqueness in the cases when one or both ends are free to slide.

5. Let $v(x,t)$ and $w(x,t)$ be two C^2 solutions of the problem

$$\text{D.E.}\quad u_{tt} = a^2u_{xx}, \qquad 0 \le x \le L,\ -\infty < t < \infty,$$

$$\text{B.C.}\quad u(0,t) = 0,\ u(L,t) = 0.$$

(a) Use the technique in the proof of Theorem 1 to show that

$$\frac{d}{dt}\int_0^L \left[a^2v_x(x,t)w_x(x,t) + v_t(x,t)w_t(x,t)\right] dx = 0.$$

Hint. $v_{xt}w_x + v_xw_{xt} + v_{xx}w_t + v_tw_{xx} = (v_tw_x + v_xw_t)_x$.

(b) Let $B(v,w) = \int_0^L [\frac{1}{2}T_0 v_x w_x + \frac{1}{2}D v_t w_t]\, dx$.

In part (a), we proved that $B(v,w)$ is a constant, independent of t. Note that the energy E_u of $u(x,t)$ is $B(u,u)$. Establish the result

$$B(v+w,v+w) = B(v,v) + B(w,w) + 2B(v,w). \qquad (*)$$

(i.e., $E_{v+w} \neq E_v + E_w$, unless $B(v,w) = 0$).

Hint. Rather than writing out $B(v+w,v+w)$ in terms of an integral, note that B clearly has the properties $B(u_1,u_2) = B(u_2,u_1)$ and $B(u_1+u_2,u) = B(u_1,u) + B(u_2,u)$, which are all that is needed to get $(*)$.

(c) Show that for any two harmonics, say $u_n(x,t)$ (formula (10)) and $u_m(x,t)$ where $m \neq n$, we have $B(u_m,u_n) = 0$. Conclude from (b) that the energy of $u_m + u_n$ is the sum of the energies of u_n and u_m. Is this still true if $n = m$? Why not ?

(d) Show that the energy of $u(x,t) = \sum_{n=1}^{N} \left[\frac{L}{n\pi a} \bar{A}_n \sin(\frac{n\pi at}{L}) + B_n \cos(\frac{n\pi at}{L})\right] \sin(\frac{n\pi x}{L})$ is

$$E_u = \frac{\pi^2 T_0}{4L} \sum_{n=1}^{N} n^2 [B_n^2 + (L\bar{A}_n/n\pi a)^2].$$

Hint. $E_u = B(u,u) = B(\sum_{n=1}^{N} u_n, \sum_{n=1}^{N} u_n) = \sum_{n=1}^{N} B(u_n,u_n) + \sum_{m<n\le N} 2B(u_m,u_n) = \sum_{n=1}^{N} B(u_n,u_n)$.

6. Consider the problem

$$\begin{aligned} &\text{D.E.} \quad u_{tt} = u_{xx}, \qquad 0 \le x \le \pi, \quad -\infty < t < \infty, \\ &\text{B.C.} \quad u(0,t) = 0, \quad u(\pi,t) = 0, \\ &\text{I.C.} \quad u(x,0) = x(\pi - x), \quad u_t(x,0) = 0. \end{aligned}$$

(a) Find a solution of the D.E. and B.C. that satisfies the I.C. to within an error of .001.

(b) By computing u_{tt} and u_{xx} at $(x,t) = (0,0)$, show that there is no C^2 solution of the problem.

7. Consider a string which vibrates in the xy–plane, but not necessarily transversely. When the string is at rest the points on the string have coordinates of the form $(x,0)$ $(0 \le x \le L)$. Suppose that at time t, the point which was in the rest position $(x,0)$, has xy–coordinates $(r(x,t),u(x,t))$ [e.g., for transverse vibrations $r(x,t) = x$ and for longitudinal vibrations $u(x,t) = 0$].

(a) Show that the local stretching factor s at the point corresponding to x is $(r_x^2 + u_x^2)^{\frac{1}{2}}$.

(b) Suppose that the tension at the point corresponding to x is still of the form $g(s)T_0$. Show that by virtue of Newton's equation ($\mathbf{F} = m\mathbf{a}$) $r(x,t)$ and $u(x,t)$ satisfy the system

$$r_{tt} = \frac{1}{D}\frac{\partial}{\partial x}\left[T_0 \frac{g(s)}{s} r_x\right] \quad \text{and} \quad u_{tt} = \frac{1}{D}\frac{\partial}{\partial x}\left[T_0 \frac{g(s)}{s} u_x\right],$$

where $s = (r_x^2 + u_x^2)^{\frac{1}{2}}$, with B.C. $r(0,t) = u(0,t) = u(L,t) = 0$ and $r(L,t) = L$.

(c) When will these equations decouple, in the sense that they can be solved separately ?

(d) The equations are still valid if D and T_0 are allowed to be positive C^1 functions of x, but then $T_0(x)$ cannot be brought outside of the parentheses. Assuming that $g(s) = s$, find the unique time–independent (steady–state) solution $(R(x),U(x))$ of the system with the given B.C., assuming that $\int_0^L [T_0(x)]^{-1}\, dx < \infty$. When is $R(x) \equiv x$? (i.e., when the string is in the standard configuration $(x,0)$, $0 \le x \le L$, under what circumstances is the string really at rest ?)

8. For a string that vibrates transversally in a medium, say air, one must take air resistance into account. Assuming that the force due to the air resistance is proportional (but oppositely directed) to the velocity $u_t(x,t)\,\mathbf{j}$, show that $u(x,t)$ obeys the equation $u_{tt} = a^2 u_{xx} - ku_t$, for some real $k > 0$.

9. Use separation of variables to find *all* product solutions of the problem (with $k > 0$)

$$\text{D.E.}\quad u_{tt} = a^2 u_{xx} - ku_t, \qquad 0 \le x \le L,\ -\infty < t < \infty,$$

$$\text{B.C.}\quad u(0,t) = 0,\ u(L,t) = 0,$$

for the string with air resistance and fixed ends.

5.2 D'Alembert's Solution for Wave Problems

We have seen that one can represent the solution of the problem (where $N < \infty$)

$$\begin{aligned}
&\text{D.E.} \quad u_{tt} = a^2 u_{xx}\,, \qquad 0 \le x \le L,\ -\infty < t < \infty,\\
&\text{B.C.} \quad u(0,t) = 0 \quad u(L,t) = 0\,,\\
&\text{I.C.} \quad u(x,0) = f(x) = \sum_{n=1}^{N} B_n \sin(\tfrac{n\pi x}{L})\,, \quad u_t(x,0) = g(x) = \sum_{n=1}^{N} \bar{A}_n \sin(\tfrac{n\pi x}{L})
\end{aligned} \tag{1}$$

by the series

$$u(x,t) = \sum_{n=1}^{N} \left[\frac{L}{n\pi a}\bar{A}_n \sin(\tfrac{n\pi at}{L}) + B_n \cos(\tfrac{n\pi at}{L})\right] \sin(\tfrac{n\pi x}{L}), \tag{2}$$

If $f(x)$ and $g(x)$ are not finite Fourier sine series, but are continuous and piecewise C^1 and vanish at $x = 0$ and $x = L$, we may approximate $f(x)$ and $g(x)$ to within any (positive) experimental error by truncations of their Fourier sine series (cf. Theorem 1 of Section 4.3). Thus, problem (1) has been solved for all practical purposes, for such $f(x)$ and $g(x)$, as nearly as anyone can say. However, in order that the theory have predictive value, we need to know that small changes in $f(x)$ and $g(x)$ induce small changes in the solution. Otherwise, two different approximations, both within experimental error, may lead to significantly different solutions. This property of "continuity of solutions with respect to variations in the I.C." was established for the heat equation by the use of the Maximum Principle (cf. Theorem 2 of Section 3.2). However, a direct translation of the Maximum Principle to the case of the wave problem (1) is false, as the following example shows.

Example 1. In problem (1) take $f(x) = 0$ and $g(x) = \sin(\pi x/L)$. Show that the maximum of the solution $u(x,t)$ does not occur when $t = 0$, $x = 0$ or $x = L$.

Solution. The solution is $u(x,t) = \frac{L}{\pi a}\sin(\pi at/L)\sin(\pi x/L)$. Note that $u(x,t)$ vanishes at the ends and also initially (i.e., $u(0,t) = 0$, $u(L,t) = 0$, $u(x,0) = 0$). However, $u(x,t) \not\equiv 0$ as a direct translation of the Maximum Principle for the heat equation would imply. Indeed, the maximum of $u(x,t)$ occurs at $x = \frac{1}{2}L$, $t = \frac{1}{2}L/a$ (as well as $t = (2n + \frac{1}{2})L/a$, $n = 0, \pm 1, \ldots$). □

The reason for the failure of the direct translation of the Maximum Principle is that it does not take into account *both* of the I.C. in (1). Eventually (cf. Theorem 5), we prove that the solution $u(x,t)$ of (1) obeys the following type of maximum principle :

$$\max_{\substack{0 \le x \le L\\ -\infty < t < \infty}} |u(x,t)| \le \max_{0 \le x \le L} |f(x)| + \frac{L}{2a} \max_{0 \le x \le L} |g(x)|. \tag{3}$$

Note that this involves both $f(x)$ and $g(x)$, and the absolute values cannot be removed. Using (3), we will establish (cf. Corollary of Theorem 5) the desired result that small changes in the I.C. produce small changes in solutions. This was done for the heat equation in Theorem 3 of Section 3.2. The key to obtaining (3) is **D'Alembert's formula** for the solution of the wave equation on an infinite string $(-\infty < x < \infty)$ with I.C. $u(x,0) = f(x)$ and $u_t(x,0) = g(x)$, namely,

$$u(x,t) = \frac{1}{2}[f(x-at) + f(x+at)] + \frac{1}{2a}\int_{x-at}^{x+at} g(r)\,dr . \tag{4}$$

This formula is of great interest in itself, and it avoids the problem of convergence of infinite series in the Fourier series approach. Our first goal is to derive D'Alembert's formula (4).

Derivation of D'Alembert's formula

We can write the wave equation in the form

$$\left[\frac{\partial^2}{\partial t^2} - a^2\frac{\partial^2}{\partial x^2}\right]u(x,t) = 0 , \tag{5}$$

where the expression in parentheses is a differential operator which operates on the function u to yield $u_{tt} - a^2u_{xx}$. This operator can be factored into two first–order operators:

$$\left[\frac{\partial^2}{\partial t^2} - a^2\frac{\partial^2}{\partial x^2}\right]u = \left[\frac{\partial}{\partial t} - a\frac{\partial}{\partial x}\right]\left[\frac{\partial}{\partial t} + a\frac{\partial}{\partial x}\right]u . \tag{6}$$

We can use this factorization to find the general solution of the wave equation. Suppose that $u(x,t)$ is any C^2 solution of (5). Note that

$$\left[\frac{\partial}{\partial t} - a\frac{\partial}{\partial x}\right][u_t + au_x] = u_{tt} - a^2u_{xx} = 0 .$$

In other words, the function $y(x,t) \equiv u_t + au_x$ solves the PDE $y_t - ay_x = 0$. Since the characteristic lines (cf. Section 2.1) of this PDE are of the form $x+at = \text{const.}$, we know that the solution y must be of the form $y(x,t) = h(x+at)$ for some C^1 function h. Thus,

$$u_t + au_x = y(x,t) = h(x+at) .$$

The characteristic lines for this first–order PDE for u are $x-at = \text{const.}$. In view of the form of the right–hand side $h(x+at)$, we make the change of variables

$$w = x - at \quad \text{and} \quad z = x + at\ .$$

Letting $v(w,z) = u(x,t)$, we obtain

$$u_t + au_x = v_w w_t + v_z z_t + a(v_w w_x + v_z z_x) = 2av_z = h(z)\ .$$

Thus, $v(w,z) = \int \frac{1}{2a} h(z)\, dz + G(w) = F(z) + G(w)$, or

$$u(x,t) = F(x+at) + G(x-at)\ . \tag{7}$$

Hence, we have shown that an arbitrary solution of the wave equation can be written in the form (7), where F and G are arbitrary C^2 functions. Conversely, one can easily check that any function of the form (7) is a solution of the wave equation (i.e., (7) is the general solution). If we graph the function $F(x+at)$ as a function of x at a fixed time t, we obtain the graph of $F(x)$ translated to the left by a distance of "at", as the reader may verify. Thus, $F(x+at)$ describes a wave with initial profile $F(x)$ moving to the left with speed a. Similarly, $G(x-at)$ yields a wave traveling to the right with speed a. We have shown that the general solution of $u_{tt} = a^2 u_{xx}$ is a superposition of two waves traveling in opposite directions with speed a.

Example 2. We know that $\cos(\lambda at)\sin(\lambda x)$ is a (product) solution of $u_{tt} = a^2 u_{xx}$. Hence, it must be possible to rewrite $\cos(\lambda at)\sin(\lambda x)$ in the form $F(x+at) + G(x-at)$. Do it.

Solution. Using the identity $\cos(\beta)\sin(\alpha) = \frac{1}{2}[\sin(\alpha+\beta) + \sin(\alpha-\beta)]$, with $\alpha = \lambda x$ and $\beta = \lambda at$,

$$\cos(\lambda at)\sin(\lambda x) = \frac{1}{2}\left[\sin(\lambda(x+at)) + \sin(\lambda(x-at))\right]\ .$$

This exhibits the "standing wave" product solution, as a superposition of waves traveling to the right and left with speed a. □

Theorem 1 (D'Alembert's Formula). Let $f(x)$ **be** C^2 **and let** $g(x)$ **be** C^1 $(-\infty < x < \infty)$. **Then the unique solution of the problem**

$$\text{D.E.}\quad u_{tt} = a^2 u_{xx}, \qquad -\infty < x < \infty,\ -\infty < t < \infty,$$

$$\text{I.C.}\quad u(x,0) = f(x),\ \ u_t(x,0) = g(x), \tag{8}$$

is given by

$$u(x,t) = \frac{1}{2}\left[f(x+at) + f(x-at)\right] + \frac{1}{2a}\int_{x-at}^{x+at} g(r)\, dr\ . \tag{9}$$

Proof. We know that *if* a solution of (8) exists, then it must be of the form $u(x,t) = F(x+at) + G(x-at)$, where F and G are C^2 functions. The I.C. will be satisfied precisely when

$$f(x) = u(x,0) = F(x + 0\cdot a) + G(x - 0\cdot a) = F(x) + G(x), \tag{10}$$

$$g(x) = u_t(x,0) = F'(x)a - G'(x)a. \tag{11}$$

Integrating the second equation, we obtain the following pair of equations for the unknown functions $F(x)$ and $G(x)$:

$$F(x) + G(x) = f(x) \quad \text{and} \quad F(x) - G(x) = \frac{1}{a}\int_0^x g(r)\,dr + C,$$

where C is an arbitrary constant. Adding and subtracting yields

$$F(x) = \frac{1}{2}\left[f(x) + \frac{1}{a}\int_0^x g(r)dr + C\right], \tag{12}$$

and

$$G(x) = \frac{1}{2}\left[f(x) - \frac{1}{a}\int_0^x g(r)dr - C\right] = \frac{1}{2}\left[f(x) + \frac{1}{a}\int_x^0 g(r)dr - C\right]. \tag{13}$$

These equations are identities in the sense that they hold for all values of x, as in $\sin^2(x) = 1 - \cos^2(x)$. Because of this, we may substitute $x + at$ for x in (12) and $x - at$ for x in (13), and obtain valid results [e.g., $\sin^2(x+at) = 1 - \cos^2(x+at)$], namely,

$$F(x+at) = \frac{1}{2}\left[f(x+at) + \frac{1}{a}\int_0^{x+at} g(r)dr + C\right], \quad G(x-at) = \frac{1}{2}\left[f(x-at) + \frac{1}{a}\int_{x-at}^0 g(r)dr - C\right]. \tag{15}$$

Adding these expressions, we obtain (9). However, the above argument was based on the assumption that a solution of the problem (8) exists. We have just shown that *if* a solution exists, then it must be given by the D'Alembert's formula (9). We must finally show that $u(x,t)$, given by (9), actually solves problem (8). Note that $F(x)$ and $G(x)$ defined by (12) and (13) are C^2, since $f(x)$ is C^2 and $g(x)$ is C^1 ($\int_0^x g(r)\,dr$ is C^2, since its derivative $g(x)$ is C^1). The right–hand side of (9) is $F(x+at) + G(x-at)$ for the C^2 functions F and G, and hence we know that $u(x,t)$ defined by (9) solves the D.E.. The initial conditions of (8) are met, since $F(x)$ and $G(x)$ given by (12) and (13) satisfy (10) and (11), by construction. □

Remark. There are no boundary conditions in problem (8), because the string has no ends. For a finite string, say $0 \le x \le L$, the functions $f(x)$ and $g(x)$ would only be defined on $[0,L]$, and $f(x\pm at)$ would be undefined for t large, making the solution (9) undefined. We overcome this problem later by extending (in various ways that depend on the B.C.) $f(x)$ and $g(x)$ to functions which are defined for all x. □

Example 3. Solve

$$\text{D.E.}\quad u_{tt} = a^2 u_{xx}, \qquad -\infty < x,\ t < \infty,$$
$$\text{I.C.}\quad u(x,0) = \frac{1}{1+x^2}, \quad u_t(x,0) = 0\ .$$

Solution. This is problem (8) with $f(x) = (1+x^2)^{-1}$ and $g(x) = 0$. By (9), the solution is

$$u(x,t) = \frac{1}{2}\left[\frac{1}{1+(x+at)^2} + \frac{1}{1+(x-at)^2}\right]. \tag{16}$$

The graph of the initial profile $u(x,0) = (1+x^2)^{-1}$ is shown in Figure 1(a) below.

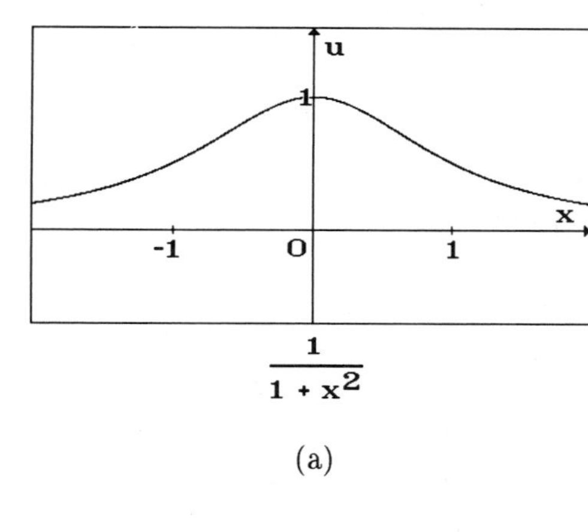

$$\frac{1}{1+x^2}$$

(a)

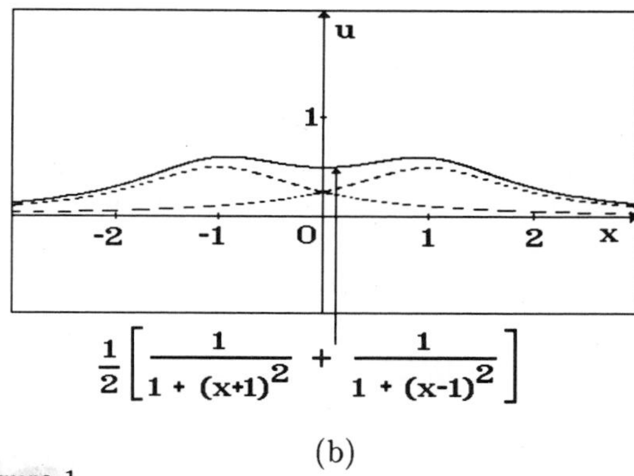

$$\frac{1}{2}\left[\frac{1}{1+(x+1)^2} + \frac{1}{1+(x-1)^2}\right]$$

(b)

Figure 1

At $t = 1/a$ the graph of $u(x,t)$ is the solid curve in Figure 1(b), obtained by graphically adding the two dashed profiles corresponding to the two terms of solution (16) at $t = 1/a$. As t increases, the graphs of the two terms move apart. Eventually, the solution looks like two waves shaped as in (a) but with half the amplitude, one moving with speed "a" to the left and the other moving with speed "a" to the right. You may wish to verify that this behavior actually takes place by experimenting with a long rope. □

Example 4. Solve

$$\text{D.E.}\quad u_{tt} = a^2 u_{xx}, \qquad -\infty < x,\ t < \infty,$$
$$\text{I.C.}\quad u(x,0) = 0, \quad u_t(x,0) = \frac{2}{1+x^2}\ .$$

What is the limit of the amplitude $u(x,t)$ at any fixed x, as $t \to \infty$?

Solution. Here the string is initially straight $(u(x,0) = 0)$, but has a variable upward velocity at $t = 0$. The upward velocity at x is $2(1+x^2)^{-1}$. By (9),

$$u(x,t) = \frac{1}{2a}\int_{x-at}^{x+at} 2(1+r^2)^{-1}\,dr = \frac{1}{a}[\arctan(x+at) - \arctan(x-at)].$$

At $t = 1/a$ the profile is the sum (solid) of the two dashed curves shown in Figure 2(a) below.

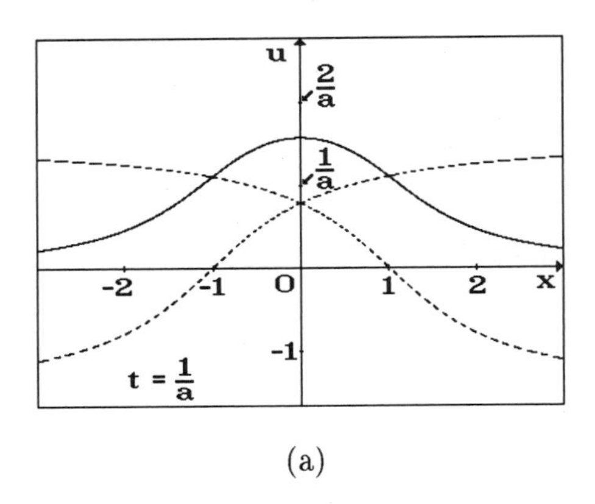

(a)

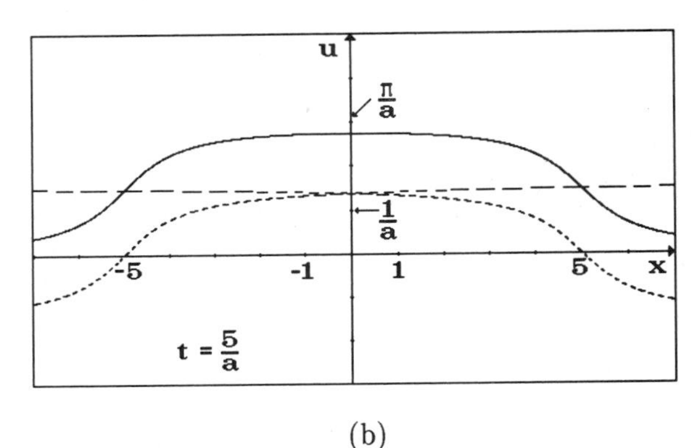

(b)

Figure 2

As time advances, the dashed curves move to the right and left and the graph of the profile appears as in Figure 2(b). Note that for each *fixed* x we have

$$\lim_{t\to\infty} u(x,t) = \frac{1}{a}[\arctan(\infty) - \arctan(-\infty)] = \frac{\pi}{a}. \quad \square$$

Theorem 2 **(A maximum magnitude principle). Let** $f(x)$ **be** C^2 **and** $g(x)$ **be** C^1 $(-\infty < x < \infty)$. **Suppose that** $M_f = \max_{-\infty<x<\infty} |f(x)| < \infty$, **and that** I_g **is the maximum of all of the absolute values** $|\int_c^d g(x)\,dx|$ **as** c **and** d **vary over all possible values. Then the solution** $u(x,t)$ **of**

$$\text{D.E.} \quad u_{tt} = a^2 u_{xx}, \qquad -\infty < x,\ t < \infty,$$

$$\text{I.C.} \quad u(x,0) = f(x), \quad u_t(x,0) = g(x),$$

satisfies

$$|u(x,t)| \le M_f + \frac{1}{2a} I_g, \qquad (-\infty < x,\ t < \infty).$$

Proof. Using D'Alembert's formula (9), we obtain

$$|u(x,t)| \le \frac{1}{2}(|f(x+at)| + |f(x-at)|) + \frac{1}{2a}\left|\int_{x-at}^{x+at} g(r)\,dr\right| \le \frac{1}{2}\,2M_f + \frac{1}{2a} I_g. \quad \square$$

Remark. When $g(x) \equiv 0$, we have $|u(x,t)| \le M_f$, i.e., the magnitude of $u(x,t)$ is never greater than the largest magnitude of u at $t = 0$. Indeed, when $g(x) \equiv 0$, we have the following maximum/minimum principle in the usual sense (cf. Problem 8).

$$\min_{-\infty<x<\infty} f(x) = \min_{-\infty<x<\infty} u(x,0) \le u(x,t) \le \max_{-\infty<x<\infty} u(x,0) = \max_{-\infty<x<\infty} f(x) \quad (\text{for } g(x) \equiv 0) .$$

When $g(x) \ge 0$ for all x, note that $I_g = \int_{-\infty}^{\infty} g(x)\,dx$. In Example 4, we have $I_g = \int_{-\infty}^{\infty} 2(1+x^2)^{-1}\,dx = 2\pi$, and hence $|u(x,t)| \le \pi/a$. It is important to realize that $\int_{-\infty}^{\infty} |g(x)|\,dx$ can be infinite, yet I_g (defined as in Theorem 2) might be finite. Indeed, this is the case for $g(x) = \sin(x)$, where $I_g = \int_0^{\pi} \sin(x)dx = 2$, but $\int_{-\infty}^{\infty} |\sin(x)|\,dx = \infty$. □

In addition to the maximum magnitude principle (Theorem 2), D'Alembert's formula

$$u(x,t) = \frac{1}{2}[f(x+at) + f(x-at)] + \frac{1}{2a}\int_{x-at}^{x+at} g(r)\,dr,$$

yields a number of properties of solutions of the wave problem for the infinite string.

Property 1. Disturbances propagate with speed a.

The value $u(x_0,t_0)$ depends only on the values of g in the interval $[x_0 - at_0, x_0 + at_0]$ and on the values of f at the endpoints of this interval. Geometrically, this is the interval cut out by the characteristic lines that pass through the point (x_0,t_0), as shown in Figure 3.

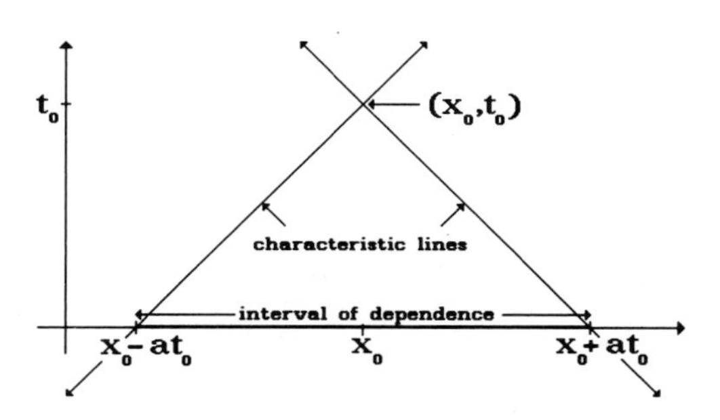

Figure 3

Definition. The interval $[x_0 - at_0, x_0 + at_0]$ is called the **interval of dependence** for the point (x_0,t_0), [since $u(x_0,t_0)$ depends only on the values $u(x,0)$ and $u_t(x,0)$ for x in this interval].

The physical interpretation is that "a" represents the speed at which a disturbance moves along the string. Such a disturbance cannot reach the point x_0 on the string within time t_0, unless it is no further than at_0 units away when $t = 0$, (i.e., unless it is within the interval of dependence of (x_0,t_0)). In Chapter 7, we will find that the interval dependence of the point (x_0,t_0) for the heat equation for the infinite rod consists of the entire rod if $t_0 > 0$ (i.e., according to the heat equation [but not in "reality", since the speed limit is c], heat travels with infinite speed). □

Property 2. Odd/even initial data yield odd/even solutions.

If f(x) and g(x) are odd, then u(x,t) is odd in the x–variable, since

$$\begin{aligned}
u(-x,t) &= \tfrac{1}{2}[f(-x + at) + f(-x - at)] + \frac{1}{2a}\int_{-x-at}^{-x+at} g(r)\,dr \\
&= \tfrac{1}{2}[-f(x - at) - f(x + at)] - \frac{1}{2a}\int_{x+at}^{x-at} g(-s)\,ds \\
&= -\tfrac{1}{2}[f(x - at) + f(x + at)] + \frac{1}{2a}\int_{x+at}^{x-at} g(s)\,ds \\
&= -\tfrac{1}{2}[f(x + at) + f(x - at)] - \frac{1}{2a}\int_{x-at}^{x+at} g(s)\,ds \\
&= -\,u(x,t).
\end{aligned}$$

Actually, we can get $u(-x,t) = -u(x,t)$ by a uniqueness argument. Simply note that the function $v(x,t) = -u(-x,t)$ satisfies the D.E. $v_{tt} = a^2 v_{xx}$ with $v(x,0) = -u(-x,0) = -f(-x) = f(x)$, and $v_t(x,0) = -u_t(-x,0) = -g(-x) = g(x)$, i.e., v(x,t) and u(x,t) solve the same problem (8) of Theorem 1. Uniqueness and the definition of v imply that $u(x,t) = v(x,t) = -u(-x,t)$. Hence $u(-x,t) = -u(x,t)$. With a similar argument one can show that if f(x) and g(x) are even so is u(x,t) [i.e., $u(-x,t) = u(x,t)$]. □

Property 3. Periodic initial data yield periodic solutions.

If f(x) and g(x) are periodic functions of period 2L, then u(x,t) is also periodic of period 2L in x. This follows easily from D'Alembert's formula, but there is also a uniqueness argument [$v(x,t) \equiv u(x+2L,t)$ solves the same problem]. This fact is useful in dealing with finite strings. It can be shown (cf. Problem 11) that if f(x) and g(x) are periodic of period 2L *and* $\int_{-L}^{L} g(x)\,dx = 0$, then u(x,t) is not only periodic in x of period 2L, but also periodic in t of period 2L/a. □

Example 5. Let $f(x)$ and $g(x)$ be C^2 functions which are 0 for $|x| \geq 10$. Suppose that $u(x,t)$ is the solution of the wave equation $u_{tt} = 4u_{xx}$ with $u(x,0) = f(x)$ and $u_t(x,0) = g(x)$. Verify that $u(40,t) = 0$ for $t \leq 15$.

Solution. Here $a = 2$ and intuitively we do not expect the initial disturbances $f(x)$ and $g(x)$ to spread faster than with speed 2. Since these disturbances are confined to the interval $(-10,10)$ for $t = 0$ (cf. Figure 4), we expect $u(x,t)$ to vanish outside the interval $(-10-2t,10+2t)$. The point $x = 40$ lies in this interval only if $t > 15$, and therefore $u(40,t) = 0$ for $t \leq 15$. Indeed, using the D'Alembert's formula, we have

$$u(40,t) = \frac{1}{2}[f(40+2t) + f(40-2t)] + \frac{1}{4}\int_{40-2t}^{40+2t} g(r)\,dr,$$

which vanishes for $t \leq 15$, since $f(x)$ and $g(x)$ vanish on $[40 - 2t, 40 + 2t]$ for $t \leq 15$. □

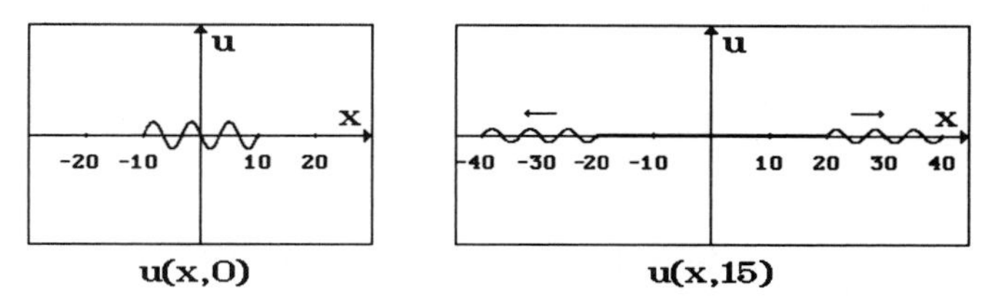

Figure 4

Example 6 (The semi–infinite string). Let $f(x)$ be a C^2 function defined for $x \geq 0$ such that $f(0) = 0$ and $f''(0) = 0$. Solve the following problem for the semi–infinite string $(0 \leq x < \infty)$, with fixed end at $x = 0$.

$$\begin{aligned} &\text{D.E.} \quad u_{tt} = a^2 u_{xx} \qquad 0 \leq x < \infty,\ -\infty < t < \infty, \\ &\text{B.C.} \quad u(0,t) = 0 \\ &\text{I.C.} \quad u(x,0) = f(x),\quad u_t(x,0) = 0. \end{aligned} \tag{17}$$

Solution. We exploit Property 2. Note that $f(x)$ is defined for $x \geq 0$, but we can consider (cf. Figure 5) the odd extension $f_o(x)$, $-\infty < x < \infty$ (i.e., $f_o(x) = f(x)$ for $x \geq 0$, and $f_o(x) = -f(-x)$ for $x \leq 0$).

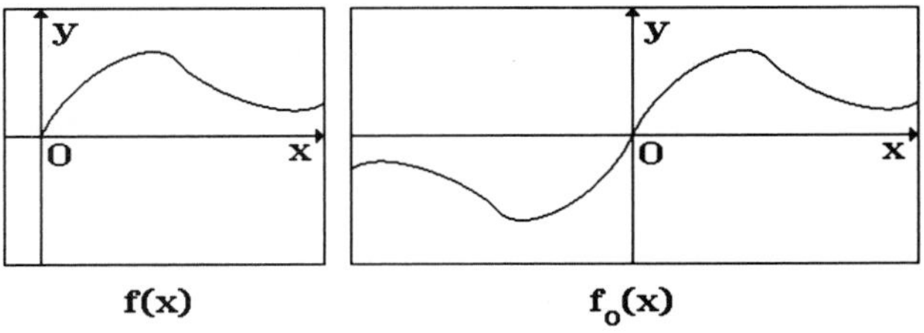

Figure 5

The "related extended problem" is

$$\text{D.E. } u_{tt} = a^2 u_{xx}, \qquad -\infty < x, t < \infty,$$
$$\text{I.C. } u(x,0) = f_o(x), \ u_t(x,0) = 0, \tag{18}$$

According to D'Alembert's formula, the solution of this problem is

$$u(x,t) = \tfrac{1}{2}[f_o(x+at) + f_o(x-at)] . \tag{19}$$

By Property 2, we know that $u(x,t)$ in (19) is odd in x, since $f_o(x)$ is odd. Thus, $u(0,t) = 0$ (as can be seen directly from (19)), and so $u(x,t)$ in (19) satisfies the B.C. in (17). The assumption that $f''(0) = 0$ is necessary in order that $f_o(x)$ be C^2 at $x = 0$ (otherwise $f_o''(0^+) \neq f''(0^-)$). Hence, $u(x,t)$ is a C^2 solution of $u_{tt} = a^2 u_{xx}$. Moreover,

$$u(x,0) = \tfrac{1}{2}[f_o(x + a\cdot 0) + f_o(x - a\cdot 0)] = f_o(x),$$

which is the same as $f(x)$ when $x \geq 0$ (i.e., the I.C. of (17) is met). □

Remark. The technique used in Example 6 is known as **the method of images.** We began with problem (17) on the interval $0 \leq x < \infty$ with a boundary condition at $x = 0$. By considering the inverted mirror image (odd extension) of the initial data, we were able to convert the problem (17) to the familiar problem (18) on the infinite string. The odd extension forces (18) to vanish at $x = 0$. If the B.C. of (17) were changed to $u_x(0,t) = 0$ (physically, this means that the end $x = 0$ is free to slide vertically; cf. Section 5.3), then we would select the even extension f_e of f (assuming $f'(0) = 0$) in the related extended problem. The solution of (18), with $f_e(x)$ instead of $f_o(x)$, would be $u(x,t) = \frac{1}{2}[f_e(x+at) + f_e(x-at)]$, and $u_t(0,t) = a(f_e)'(at) + (-a)(f_e)'(-at) = 0$. If $g(x)$ is not zero, one uses the same sort of extension for $g(x)$ as is used for $f(x)$. This is illustrated in the next example. □

Example 7. On a semi-infinite string with fixed end, a wave has the profile $f(x)$ (cf. Figure 6) at $t = 0$. We assume that for small t, $u(x,t) = f(x+at)$, i.e., the wave is moving to the left with speed a. At $t = 2$, the wave makes contact with the end. What happens after $t = 2$?

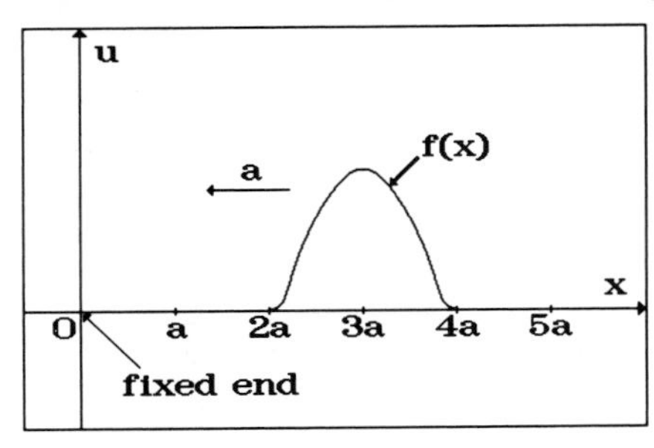

Figure 6

Solution. We have $u(x,0) = f(x)$ and $u_t(x,t) = af'(x+at)$, whence $u_t(x,0) = af'(x)$. Thus, we seek a solution of the problem

$$\text{D.E.} \quad u_{tt} = a^2 u_{xx}, \qquad 0 \le x \le \infty, \ -\infty < t < \infty,$$

$$\text{B.C.} \quad u(0,t) = 0,$$

$$\text{I.C.} \quad u(x,0) = f(x), \ u_t(x,0) = af'(x).$$

Because of the B.C., we take the odd extensions of $f(x)$ and $g(x) = af'(x)$. The odd extension of $g(x)$ is *not* $a(f_o)'(x)$. Indeed, $(f_o)'(x)$ is even (cf. Problem 3 in Section 4.3, where one shows that the derivative of an odd function is even and vice–versa). Instead, the odd extension of $f'(x)$ is $(f_e)'(x)$, since $(f_e)'(x)$ is odd and agrees with $f'(x)$ when $x \ge 0$. Thus, the related extended problem for the infinite string is

$$\text{D.E.} \quad u_{tt} = a^2 u_{xx}, \qquad -\infty < x, t < \infty,$$

$$\text{I.C.} \quad u(x,0) = f_o(x), \ u_t(x,0) = a(f_e)'(x).$$

The solution of this extended problem is given by D'Alembert's formula

$$\begin{aligned} u(x,t) &= \tfrac{1}{2}[f_o(x+at) + f_o(x-at)] + \frac{1}{2a}\int_{x-at}^{x+at} a(f_e)'(r)\,dr \\ &= \tfrac{1}{2}[f_o(x+at) + f_e(x+at)] + \tfrac{1}{2}[f_o(x-at) - f_e(x-at)] \\ &= \tfrac{1}{2}(f_o+f_e)(x+at) + \tfrac{1}{2}(f_o-f_e)(x-at)\,. \end{aligned} \tag{20}$$

In Figure 7, we have graphed $f_o(x)$, $f_e(x)$, $\frac{1}{2}(f_o+f_e)(x)$ and $\frac{1}{2}(f_o-f_e)(x)$.

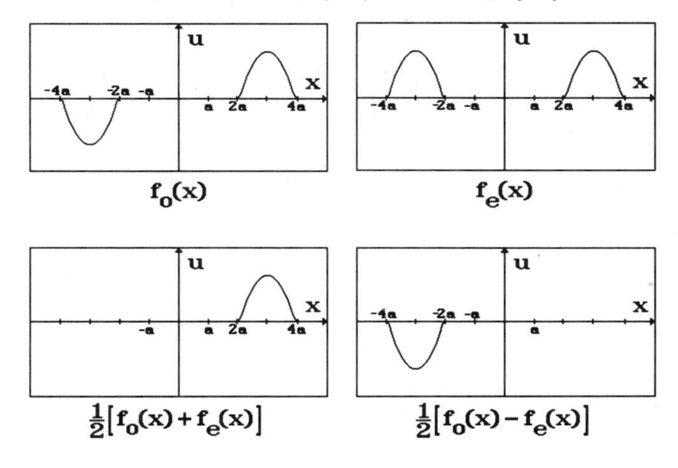

Figure 7

We now see that the solution (20) is the superposition of a wave with initial profile $\frac{1}{2}[f_o(x)+f_e(x)]$ moving to the left, and a wave with initial profile $\frac{1}{2}[f_o(x)-f_e(x)]$ moving to the right. These waves contact each other at $t = 2$ and partially cancel each other for $2 < t < 4$. When $t > 4$, the wave moving to the left has completely passed into the "imaginary" domain $x < 0$ and the originally fictitious wave, moving to the right, lies entirely in the domain $x > 0$. Therefore, the wave heading toward the end $x = 0$ in our original picture appears to bounce back, but it will be *inverted* upon its return. □

Solving finite string problems by the method of images and D'Alembert's formula

The method of images can also be used to solve the problem for the *finite* string:

$$\begin{aligned}
&\text{D.E.} \quad u_{tt} = a^2 u_{xx}\,, \qquad 0 \le x \le L,\ -\infty < t < \infty,\\
&\text{B.C.} \quad u(0,t) = 0,\ \ u(L,t) = 0, \qquad\qquad (21)\\
&\text{I.C.} \quad u(x,0) = f(x),\ \ u_t(x,0) = g(x),
\end{aligned}$$

which we have previously solved (within experimental error) using Fourier sine series, assuming that $f(x)$ and $g(x)$ are continuous and piecewise C^1 and vanish at $x = 0$ and at $x = L$. Here we solve the problem in a different way. In the case of the semi–infinite string, we extend $f(x)$ and $g(x)$, and formulate a related extended problem on the infinite string. The function $f(x)$, $0 \le x \le L$, is extended in two steps:

(a) Let $f_o(x)$ be the odd extension of $f(x)$ to $[-L,L]$.

(b) Let $\tilde{f}_o(x)$, $-\infty < x < \infty$, be the periodic extension of $f_o(x)$ of period $2L$.

The steps are illustrated in Figure 8.

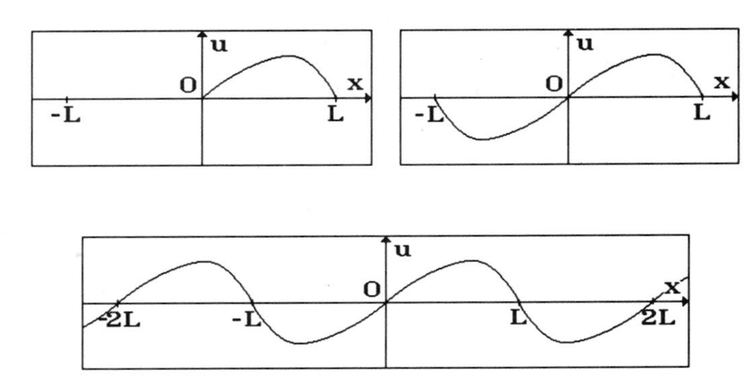

Figure 8

Thus, we have imaginatively extended the initial profile $f(x)$ of the finite string to the profile $\tilde{f}_o(x)$ of an infinite string. We can do exactly the same for the initial velocity $g(x)$, obtaining $\tilde{g}_o(x)$. The next theorem states that the finite string $(0 \leq x \leq L)$ behaves as if it were part of the infinite string with I.C. given by $\tilde{f}_o(x)$ and $\tilde{g}_o(x)$.

Theorem 3. Let $f(x)$ **and** $g(x)$ **be functions defined for** $0 \leq x \leq L$ **and let** $\tilde{f}_o(x)$ **and** $\tilde{g}_o(x)$ **be the periodic extensions of the odd extensions** of $f(x)$ **and** $g(x)$. **Assume that** $\tilde{f}_o(x)$ **is** C^2 **and** $\tilde{g}_o(x)$ **is** C^1. **(This holds if** $f(0) = f(L) = 0$, $f''(0) = f''(L) = 0$ **and** $g(0) = g(L) = 0$ **with** $f(x)$ C^2 **and** $g(x)$ C^1 **for** $0 \leq x \leq L$). **Then the (unique) solution of the problem (21) is given by**

$$u(x,t) = \frac{1}{2}[\tilde{f}_o(x+at) + \tilde{f}_o(x-at)] + \frac{1}{2a}\int_{x-at}^{x+at} \tilde{g}_o(r)\, dr, \tag{22}$$

which is the solution for the infinite string with initial conditions $u(x,0) = \tilde{f}_o(x)$ **and** $u_t(x,0) = \tilde{g}_o(x)$, $-\infty < x < \infty$.

Proof. By Theorem 1, with $f(x)$ and $g(x)$ replaced by $\tilde{f}_o(x)$ and $\tilde{g}_o(x)$, we know that (22) solves the D.E. with I.C. $u(x,0) = \tilde{f}_o(x)$ and $u_t(x,0) = \tilde{g}_o(x)$. Now, $\tilde{f}_o(x) = f(x)$ and $\tilde{g}_o(x) = g(x)$ for $0 \leq x \leq L$, whence (22) satisfies the I.C. of (21), as well as the D.E.. Since Theorem 1 of Section 5.1 implies uniqueness, we only need to check that the B.C. are satisfied by (22). Indeed,

$$u(0,t) = \frac{1}{2}[\tilde{f}_o(at) + \tilde{f}_o(-at)] + \frac{1}{2a}\int_{-at}^{at} \tilde{g}_o(r)\, dr = 0,$$

since $\tilde{f}_o$ and $\tilde{g}_o$ are odd. To check that the B.C. $u(L,t) = 0$ is satisfied, we observe that $\tilde{f}_o$ and $\tilde{g}_o$ are "odd about $x = L$" in the sense that $\tilde{f}_o(L-x) = -\tilde{f}_o(L+x)$. This is easily seen from Figure 8, but one can also compute $\tilde{f}_o(L-x) = -\tilde{f}_o(-L+x) = -\tilde{f}_o(2L-L+x) = -\tilde{f}_o(L+x)$. Thus,

$$u(L,t) = \frac{1}{2}[\tilde{f}_o(L+at) + \tilde{f}_o(L-at)] + \frac{1}{2a}\int_{L-at}^{L+at} \tilde{g}_o(r)\, dr = 0, \text{ as required. } \square$$

The next theorem shows that (22) is the same as the solution obtained using Fourier sine series.

Theorem 4. **If $\tilde{f}_o(x)$ is C^2 and $\tilde{g}_o(x)$ is C^1, then the solution of problem (21) is given by the following formula (which is equivalent to (22) by Theorem 1, Section 5.1)**

$$u(x,t) = \sum_{n=1}^{\infty} [A_n \sin(\tfrac{n\pi at}{L}) + B_n \cos(\tfrac{n\pi at}{L})] \sin(\tfrac{n\pi x}{L}), \tag{23}$$

where

$$B_n = \frac{2}{L}\int_0^L f(x)\sin(\tfrac{n\pi x}{L})\,dx \quad \textbf{and} \quad A_n = \frac{2}{n\pi a}\int_0^L g(x)\sin(\tfrac{n\pi x}{L})\,dx\,, \quad n = 1, 2, 3, \ldots\ . \tag{24}$$

Proof. By Property 3, $u(x,t)$, given by (22), is a C^2 periodic function in x for each fixed t, since $\tilde{f}_o(x)$ is C^2 and $\tilde{g}_o(x)$ is C^1. Also, $u(0,t) = u(L,t) = 0$. Thus, Theorem 1 of Section 4.3 implies that $u(x,t)$ is equal to its Fourier sine series for each fixed t, i.e.,

$$u(x,t) = \sum_{n=1}^{\infty} b_n(t)\sin(\tfrac{n\pi x}{L}), \tag{25}$$

where

$$b_n(t) = \frac{2}{L}\int_0^L u(x,t)\sin(\tfrac{n\pi x}{L})\,dx. \tag{26}$$

It remains to show that

$$b_n(t) = A_n\sin(\tfrac{n\pi at}{L}) + B_n\cos(\tfrac{n\pi at}{L}). \tag{27}$$

Since $u(x,t)$ is C^2, we may differentiate under the integral in (26) (cf. Appendix A.3) to obtain

$$\begin{aligned} b_n''(t) &= \frac{2}{L}\int_0^L u_{tt}(x,t)\sin(\tfrac{n\pi x}{L})\,dx = \frac{2a^2}{L}\int_0^L u_{xx}\sin(\tfrac{n\pi x}{L})\,dx \\ &= \frac{2a^2}{L}\left\{u_x(x,t)s_n(x) - u(x,t)s_n'(x)\Big|_0^L + \int_0^L u(x,t)s_n''(x)\,dx\right\} \\ &= -\frac{2a^2}{L}(\tfrac{n\pi}{L})^2\int_0^L u(x,t)\sin(\tfrac{n\pi x}{L})\,dx = -(\tfrac{n\pi a}{L})^2\, b_n(t), \end{aligned}$$

where we have used Green's formula (cf. (9) of Section 4.1). Thus, $b_n(t)$ is a solution of the ODE $b_n''(t) + (n\pi a/L)^2 b_n(t) = 0$, and so $b_n(t)$ must be of the form

$$b_n(t) = c_n \sin(\tfrac{n\pi at}{L}) + d_n \cos(\tfrac{n\pi at}{L}). \tag{28}$$

Using (24), (26) and (28), we have

$$d_n = b_n(0) = \frac{2}{L}\int_0^L u(x,0)\sin(\tfrac{n\pi x}{L})dx = \frac{2}{L}\int_0^L f(x)\sin(\tfrac{n\pi x}{L})dx = B_n$$

and

$$\frac{n\pi a}{L} c_n = b_n'(0) = \frac{2}{L}\int_0^L u_t(x,0)\sin(\tfrac{n\pi x}{L})dx = \frac{2}{L}\int_0^L g(x)\sin(\tfrac{n\pi x}{L})dx = \frac{n\pi a}{L} A_n.$$

Thus, $d_n = B_n$ and $c_n = A_n$, and (27) is established. □

Remark. Theorem 4 is the first instance where we have proven that a formal infinite series is an actual solution of a PDE under certain assumptions ($\tilde{f}_o(x)$ is C^2 and $\tilde{g}_o(x)$ is C^1). Note that we did *not* prove this by verifying that each term of (23) satisfies the D.E.. In general, there is no infinite superposition principle. Instead, we were able to establish the *existence* of a C^2 solution using the method of images and D'Alembert's formula, and then (knowing the existence of the solution) we determined what its Fourier sine series had to be. We have not yet done such a thing for the heat equation, because we have not yet established anything like D'Alembert's formula for the heat equation. We eventually do this in Chapter 7, thereby justifying infinite series solutions, under certain assumptions, for the heat equation. □

We can also use D'Alembert's formula to prove the following maximum magnitude principle for the wave problem for the *finite* string with fixed ends.

Theorem 5 (A maximum magnitude principle). If $\tilde{f}_o(x)$ is C^2 and $\tilde{g}_o(x)$ is C^1, then the solution $u(x,t)$ of the problem

D.E. $u_{tt} = a^2 u_{xx}, \quad 0 \le x \le L, \ -\infty < t < \infty,$
B.C. $u(0,t) = 0, \ u(L,t) = 0,$

I.C. $u(x,0) = f(x), \ u_t(x,0) = g(x),$

satisfies

$$|u(x,t)| \le M_f + \frac{L}{2a} M_g \ , \quad \textbf{where} \quad M_f = \max_{0\le x\le L} |f(x)| \ \textbf{and} \ M_g = \max_{0\le x\le L} |g(x)|.$$

Proof. We know from (22) in Theorem 3, that

$$u(x,t) = \tfrac{1}{2}[\tilde{f}_o(x+at) + \tilde{f}_o(x-at)] + \frac{1}{2a}\int_{x-at}^{x+at} \tilde{g}_o(r)\, dr. \tag{29}$$

Clearly, $|\tilde{f}_o(x)| \le M_f$, whence $\frac{1}{2}|\tilde{f}_o(x+at) + \tilde{f}_o(x-at)| \le M_f$. It remains to prove that

$$\left|\int_{x-at}^{x+at} \tilde{g}_o(r)\, dr\right| \le LM_g. \tag{30}$$

Since $\int_{-L}^{L} \tilde{g}_o(r)\, dr = 0$ and $\tilde{g}_o$ is periodic, we know that the integral of $\tilde{g}_o$ over any interval of length $2L$ is 0 (Proposition 1, Section 4.2). By deleting a subinterval of length $2kL$ ($k \ge 0$, an integer) from the interval $[x - at, x + at]$, starting from the left endpoint $x - at$, we can conclude that $\int_{x-at}^{x+at} \tilde{g}_o(r)dr$ reduces to an integral of $\tilde{g}_o(r)$ over an interval of length $< 2L$. If this remaining interval has length $\le L$, then the absolute value of the integral of $\tilde{g}_o(r)$ over it, is no greater than LM_g. If the interval has length between L and $2L$, then it must contain at least one of the points $0, \pm L, \pm 2L, \pm 3L, \ldots$. However, recall that $\tilde{g}_o$ is odd about these points. Thus, cancellations of the integral of $\tilde{g}_o$, over subintervals placed symmetrically about these points, will reduce the interval of integration to a length $\le L$. Hence, we have shown (30). □

Corollary (Continuous dependence on initial data). Let $u_1(x,t)$ and $u_2(x,t)$ be solutions of the following respective problems $(0 \le x \le L,\ -\infty < t < \infty)$:

D.E. $u_{tt} = a^2 u_{xx}$, D.E. $u_{tt} = a^2 u_{xx}$,

B.C. $\begin{cases} u(0,t) = A(t), \\ u(L,t) = B(t), \end{cases}$ B.C. $\begin{cases} u(0,t) = A(t), \\ u(L,t) = B(t), \end{cases}$

I.C. $\begin{cases} u(x,0) = f_1(x), \\ u_t(x,0) = g_1(x), \end{cases}$ I.C. $\begin{cases} u(x,0) = f_2(x), \\ u_t(x,0) = g_2(x). \end{cases}$

If $|f_1(x) - f_2(x)| \le \epsilon$ **and** $|g_1(x) - g_2(x)| \le \delta$, **for some** $\epsilon, \delta > 0$, **then**

$$|u_1(x,t) - u_2(x,t)| \le \epsilon + \frac{L\delta}{2a}. \tag{31}$$

Proof. Apply Theorem 5 to the case when $u = u_1 - u_2$, $f = f_1 - f_2$, and $g = g_1 - g_2$. Note that $M_f \le \epsilon$ and $M_g \le \delta$. □

Remark. This corollary shows that a small changes in the initial position and initial velocity will lead to a small change in the solution. Since u, f and ϵ have units of length, while u_t, g and δ

have units of velocity, the "conversion factor" $\frac{1}{2}L/a$ is necessary, so that $\frac{1}{2}L\delta/a$ has units of length. Actually, we can choose our time scale so that $\frac{1}{2}L/a = 1$. With such units, the time that it takes for a disturbance to travel from one end to the other is $L/a = 2$. Using this time scale, (31) becomes $|u_1 - u_2| \leq \epsilon + \delta$. □

Example 8. In Example 3 of Section 5.1, suppose that the maximum initial height u_0 of the plucked string is only known to within an error of Δu_0 and the initial velocity of the plucked point is only known to be zero within an error of Δv_0 (say, due to an unsteady hand). Estimate the error $\Delta u(x,t)$ in the "solution" due to these uncertainties. (We ignore the fact that the problem has no C^2 solution, since $f(x)$ is not C^1. One could overcome this by replacing $f(x)$ by a truncation of FSS $f(x)$ which is valid to within an error of much less than Δu_0 or Δv_0).

Solution. The error in the initial profile is greatest at x_0, where it is Δu_0, because the error tapers off to 0 linearly as one approaches the ends. Similarly, the maximum error for the initial velocity is Δv_0. Inequality (31) yields $|\Delta u(x,t)| \leq \Delta u_0 + \frac{L}{2a}\Delta v_0$. □

Remark. An analysis of the motion of the plucked string, by means of D'Alembert's formula, is the subject of Problem 10. □

Summary 5.2

1. The general solution of the wave equation : By taking advantage of the factorization $\partial_t^2 - a^2\partial_x^2 = (\partial_t - a\partial_x)(\partial_t + a\partial_x)$ of the wave operator, one can successively solve two first–order PDEs to obtain the general solution

$$u(x,t) = F(x+at) + G(x-at)$$

of the wave equation $u_{tt} = a^2u_{xx}$ on the entire xt–plane.

2. D'Alembert's formula (Theorem 1) : Let $f(x)$ **be** C^2 **and let** $g(x)$ **be** C^1 $(-\infty < x < \infty)$. **Then the unique solution of the problem** (for the infinite string)

$$\text{D.E.}\quad u_{tt} = a^2u_{xx}, \qquad -\infty < x, t < \infty,$$

$$\text{I.C.}\quad u(x,0) = f(x),\ \ u_t(x,0) = g(x), \tag{S1}$$

is given by

$$u(x,t) = \frac{1}{2}[f(x+at) + f(x-at)] + \frac{1}{2a}\int_{x-at}^{x+at} g(r)\,dr. \tag{S2}$$

3. A maximum magnitude principle for infinite strings (Theorem 2) : Let $M_f = \max_{-\infty<x<\infty} |f(x)|$, **where** $f(x)$ **is** C^2 **and suppose that** I_g **is the maximum of all of the absolute values** $|\int_c^d g(x)\,dx|$ **as** c **and** d **vary over all real numbers. Then the solution** $u(x,t)$ **of the problem**

$$\text{D.E.}\quad u_{tt} = a^2u_{xx}, \qquad -\infty < x, t < \infty,$$

$$\text{I.C.}\quad u(x,0) = f(x),\ \ u_t(x,0) = g(x),$$

satisfies

$$|u(x,t)| \le M_f + \frac{1}{2a}I_g, \qquad (-\infty < x, t < \infty).$$

4. Properties of solutions of the infinite string problem (S1) :

Property 1. The value $u(x_0,t_0)$ of the solution only depends on the values of $f(x)$ and $g(x)$ in the **interval of dependence** $[x_0-at_0, x_0+at_0]$ (i.e., disturbances propagate with speed a).

Property 2. If $f(x)$ and $g(x)$ are both odd (or both even), then the solution $u(x,t)$ is odd (or even) in x. In particular, $u(0,t) = 0$ in the odd case and $u_x(0,t) = 0$ in the even case.

Property 3. If $f(x)$ and $g(x)$ are periodic of period $2L$, then the solution $u(x,t)$ is periodic in x of period $2L$. If $\int_{-L}^{L} g(x)\,dx = 0$ also, then $u(x,t)$ is periodic in t of period $2L/a$.

5. **The method of images** : For a semi–infinite string problem $(x \geq 0)$ with a fixed end (i.e., with B.C. $u(0,t) = 0$), one extends the initial data $f(x)$ and $g(x)$ (originally defined for $x \geq 0$) oddly to form a related problem for the infinite string with initial data $f_o(x)$ and $g_o(x)$. By Property 2 in part 4 above, the solution of this related problem will be odd in x, and hence will meet the B.C. . For the B.C. $u_x(0,t) = 0$ (a free end), the even extensions $f_e(x)$ and $g_e(x)$ are used for the initial data in the related problem for the infinite string. The problem of the finite string with fixed ends at $x = 0$ and $x = L$, is solved by using the periodic extensions $\tilde{f}_o(x)$ and $\tilde{g}_o(x)$ in the related infinite string problem (cf. Theorems 3 and 4).

6. **A maximum magnitude principle for the finite string with fixed ends (Theorem 5) : If $\tilde{f}_o(x)$ is C^2 and $\tilde{g}_o(x)$ is C^1, then the solution $u(x,t)$ of the problem**

$$\begin{aligned} &\text{D.E. } u_{tt} = a^2 u_{xx}, \qquad 0 \leq x \leq L,\ -\infty < t < \infty,\\ &\text{B.C. } u(0,t) = 0,\ u(L,t) = 0,\\ &\text{I.C. } u(x,0) = f(x),\ u_t(x,0) = g(x), \end{aligned}$$

satisfies

$$|u(x,t)| \leq M_f + \frac{L}{2a} M_g, \quad \textbf{where} \quad M_f = \max_{0\leq x\leq L} |f(x)| \quad \textbf{and} \quad M_g = \max_{0\leq x\leq L} |g(x)|.$$

Exercises 5.2

1. Find the solution of

$$\begin{aligned} &\text{D.E. } u_{tt} = a^2 u_{xx}, \qquad -\infty < x, t < \infty,\\ &\text{I.C. } u(x,0) = f(x),\ u_t(x,0) = g(x), \end{aligned}$$

in the following cases:

(a) $f(x) = x^2$, $g(x) = x$ (b) $f(x) = e^{-x^2}$, $g(x) = 2axe^{-x^2}$ (c) $f(x) = 0$, $g(x) = 1$

(d) $f(x) = 1$, $g(x) = 0$ (e) $f(x) = \sin(x)$, $g(x) = a\cos(x)$ (f) $f(x) = 0$, $g(x) = \sin^2(x)$.

2. Suppose $u(x,t)$ solves the problem in Problem 1, where $f(x)$ and $g(x)$ vanish for $|x| \leq 10$ and $a = 2$. Show that $u(0,4) = 0$, and explain this in terms of the interval of dependence for the point $(0,4)$ (cf. Property 1).

3. Let $u(x,t)$ be the solution of the problem in Problem 1, where f and g are even.

(a) Use D'Alembert's formula to show that $u(x,t)$ is even in x (i.e., $u(-x,t) = u(x,t)$).

(b) Let $v(x,t) = u(-x,t)$. Show directly (without D'Alembert's formula) that $v(x,t)$ satisfies the D.E. and I.C. .

(c) Why can we conclude from (b) that $v(x,t) = u(x,t)$, thereby obtaining the result in (a)?

Hint. See the arguments in the discussion of Property 2.

4. Solve

$$\text{D.E.} \quad u_{tt} = a^2 u_{xx}, \qquad 0 \le x \le \infty, \ -\infty < t > \infty,$$

$$\text{B.C.} \quad u_x(0,t) = 0,$$

$$\text{I.C.} \quad u(x,0) = x^3, \ u_t(x,0) = 0.$$

Hint. See the Remark following Example 6. Note that $f_e(x) = |x|^3$.

5. Redo Example 7 in the case when the end $x = 0$ is free to slide vertically (i.e., with the B.C. $u_x(0,t) = 0$). In particular, show that the wave "bounces off", without being inverted.

6. Suppose that in Problem 1, $f(x)$ is C^2, $g(x)$ is C^1, and both $f(x)$ and $g(x)$ vanish outside some finite interval, say $[-b,b]$.

(a) Prove that for any fixed x, we have

$$\lim_{t\to\infty} u(x,t) = \frac{1}{2a}\int_{-\infty}^{\infty} g(r)\,dr = \frac{1}{2a}\int_{-b}^{b} g(r)\,dr.$$

(b) Show that regardless of how large some fixed value t_0 is, there is some value x_0 for which $u(x_0,t_0) = 0$.

(c) For fixed t, prove that $\lim_{x\to\pm\infty} u(x,t) = 0$.

7. Consider the problem

$$\text{D.E.} \quad u_{tt} = a^2 u_{xx}, \qquad 0 \le x \le L, \ -\infty < t < \infty,$$

$$\text{B.C.} \quad u(0,t) = 0, \ u(L,t) = 0,$$

$$\text{I.C.} \quad u(x,0) = 3\sin(\pi x/L) - \sin(4\pi x/L), \ u_t(x,0) = \tfrac{1}{2}\sin(2\pi x/L).$$

Verify explicitly that D'Alembert's formula (22) and the Fourier series solution (23) are indeed equal (as is guaranteed by uniqueness). (Both solutions will reduce to the answer of Exercise 1(a), Section 5.1).

8. (a) For a solution $u(x,t)$ of

$$\text{D.E.} \quad u_{tt} = a^2 u_{xx}, \qquad -\infty < x,\ t < \infty,$$
$$\text{I.C.} \quad u(x,0) = f(x), \quad u_t(x,0) = 0,$$

show that we have the maximum/minimum principle

$$\min_{-\infty<x<\infty} f(x) \le u(x,t) \le \max_{-\infty<x<\infty} f(x), \qquad -\infty < x,\ t < \infty.$$

(b) Show that the analogous maximum/minimum principle for the problem of a *finite* string with fixed ends and the same I.C. is *false*, but we still have $|u(x,t)| \le \max_{0\le x\le L} |f(x)|$.

Hint. Consider $\cos(\pi at/L)\sin(\pi x/L)$.

9. (a) Show that, in view of the B.C. and I.C., the following problem has no solution by considering the D.E. at $x = 0$, $t = 0$:

$$\text{D.E.} \quad u_{tt} = a^2 u_{xx}, \qquad 0 \le x \le L,\ -\infty < t < \infty,$$
$$\text{B.C.} \quad u(0,t) = 0, \quad u(L,t) = 0,$$
$$\text{I.C.} \quad u(x,0) = f(x) = x(L - x), \quad u_t(x,0) = g(x) = 0.$$

(b) Suppose we reverse $f(x)$ and $g(x)$ in the above problem in part (a). Prove that there *is* a solution in this case (cf. Theorem 3).

10. By using the method of images and D'Alembert's formula, show that the plucked string of Example 3 in Section 5.1 behaves as in Figure 5 in that example. **Hint.** Consider $\tilde{f}_0(x)$, and first examine the case when the string is plucked at $x_0 = \frac{1}{2}L$.

11. Show that if $f(x)$ [a C^2 function] and $g(x)$ [a C^1 function] are periodic of period 2L and $\int_{-L}^{L} g(x)\,dx = 0$, then $u(x,t)$, given by D'Alembert's formula, is periodic in t of period $2L/a$.

5.3 Other Boundary Conditions and Inhomogeneous Wave Equations

We have mentioned before that the boundary condition $u_x(0,t) = 0$ means that the end $x = 0$ is free to slide vertically. We now derive this result, by showing that if $F(t)\mathbf{j}$ is a vertical force applied to the end $x = 0$ and T_0 is the rest tension, then

$$u_x(0,t) = -\frac{F(t)}{T_0}. \tag{1}$$

The vertical component of the force acting on the portion of the string above the interval $[0,\Delta x]$ is $F_{\Delta x}(t) \equiv T_0 u_x(\Delta x,t) + F(t)$, which is the sum of the vertical component of the right tension at $x = \Delta x$ (cf. the derivation of the wave equation in Section 5.1) and the vertical component of the applied force at the end $x = 0$. The average vertical acceleration of the portion $0 \le x \le \Delta x$ is then $F_{\Delta x}(t)/(D\Delta x)$ by Newton's second law (where D = linear density of the string). Then, the vertical acceleration at the end $x = 0$ is given by the limit

$$\lim_{\Delta x \to 0} \frac{F_{\Delta x}(t)}{D\Delta x} = \lim_{\Delta x \to 0} \left[\frac{T_0[u_x(\Delta x,t) - u_x(0,t)]}{D\Delta x} + \frac{T_0 u_x(0,t) + F(t)}{D\Delta x} \right]$$

$$= \frac{T_0}{D} u_{xx}(0,t) + \lim_{\Delta x \to 0} \frac{T_0 u_x(0,t) + F(t)}{D\Delta x}.$$

This limit can only exist (i.e., the end $x = 0$ can only have a well–defined acceleration) if the numerator $T_0 u_x(0,t) + F(t)$ vanishes. In other words, $u_{tt}(0,t)$ will not exist unless (1) holds. In particular, $u_x(0,t) = 0$ if the end is free, namely when $F = 0$. In a similar way it can be shown that if $G(t)\mathbf{j}$ is the external vertical force applied to the end $x = L$, then

$$u_x(L,t) = \frac{G(t)}{T_0}. \tag{2}$$

As with the heat equation, the following standard possibilities for homogeneous boundary conditions arise :

(a) both ends are fixed (i.e., $u(0,t) = u(L,t) = 0$),

(b) one end is free and the other end is fixed (e.g., $u_x(0,t) = 0$ and $u(L,t) = 0$),

(c) both ends are free (i.e., $u_x(0,t) = 0$ and $u_x(L,t) = 0$).

With some appropriate modifications, nearly all of the results and techniques that we have covered in case (a) carry over to cases (b) and (c). For example, the reader may check that the proof of Theorem 1 (uniqueness) in Section 5.2, still applies in the cases when, in the B.C., $u(0,t)$ and/or $u(L,t)$ are replaced by $u_x(0,t)$ and/or $u_x(L,t)$ respectively (cf. Exercise 4 of Section 5.1). In the examples that follow, we illustrate the use of the Fourier series approach and the method of images in the solution of problems with one or both ends free. We then move on to the case of inhomogeneous B.C., and finally cover a version of Duhamel's principle which is used to solve the inhomogeneous wave equation that arises when a string is subject to external driving forces between the ends.

Example 1. By using separation of variables and Fourier cosine series, solve the following problem for a finite string with free ends for appropriate initial data $f(x)$ and $g(x)$:

$$\begin{aligned} &\text{D.E.} \quad u_{tt} = a^2 u_{xx}, \qquad 0 \le x \le L,\ -\infty < t < \infty, \\ &\text{B.C.} \quad u_x(0,t) = 0,\ \ u_x(L,t) = 0, \qquad\qquad (3) \\ &\text{I.C.} \quad u(x,0) = f(x),\ \ u_t(x,0) = g(x). \end{aligned}$$

Solution. Separation of variables leads to the three cases of product solutions of the D.E. (cf. (1), (2) and (3) of Section 5.1). For the Case 1 product solution

$$u_x(x,t) = [d_1\sin(\lambda at) + d_2\cos(\lambda at)]\,[c_1\lambda\cos(\lambda x) - c_2\lambda\sin(\lambda x)].$$

To avoid the useless trivial solution with $d_1 = d_2 = 0$, the B.C. $u_x(0,t) = 0$ forces $c_1 = 0$ (recall $\lambda > 0$). The B.C. $u_x(L,t) = 0$ yields the condition $\sin(\lambda L) = 0$ or $\lambda = n\pi/L$, $n = 1, 2, \ldots$. This gives an infinite family of product solutions of the D.E. and B.C., namely the harmonics

$$u_n(x,t) = [A_n\sin(\tfrac{n\pi at}{L}) + B_n\cos(\tfrac{n\pi at}{L})]\,\cos(\tfrac{n\pi x}{L}). \tag{4}$$

The reader may check that no Case 2 product solution meets the B.C., except $u(x,t) = 0$. However, there is a Case 3 product solution meeting the B.C.,

$$u_0(x,t) = A_0 t + B_0\ . \tag{5}$$

This solution represents a straight string drifting vertically with velocity A_0. We do not encounter this solution when one or both ends are fixed. By applying the superposition principle, we obtain the more general solution of the D.E. and B.C.

$$u(x,t) = \tfrac{1}{2}(A_0 t + B_0) + \sum_{n=1}^{N} [A_n\sin(\tfrac{n\pi at}{L}) + B_n\cos(\tfrac{n\pi at}{L})]\,\cos(\tfrac{n\pi x}{L}). \tag{6}$$

We have

$$u(x,0) = \tfrac{1}{2}B_0 + \sum_{n=1}^{N} B_n\cos(\tfrac{n\pi x}{L}), \tag{7}$$

$$u_t(x,0) = \tfrac{1}{2}A_0 + \sum_{n=1}^{N} \tfrac{n\pi a}{L} A_n\cos(\tfrac{n\pi x}{L}). \tag{8}$$

Thus, in the event that f(x) and g(x) are finite cosine series of the form (7) and (8), then u(x,t) is given by (6). If f(x) and g(x) are continuous and piecewise C^1, then Theorem 2 of Section 4.3 ensures that the Fourier cosine series of f(x) and g(x) converge uniformly to f(x) and g(x). Thus, within any positive experimental error, such functions f(x) and g(x) can be represented in the form (7) and (8) by truncating their Fourier cosine series at a sufficient number of terms. □

Remark. If we let N tend to ∞, then the expression (6) might not converge to a C^2 solution of the problem (i.e., (6) might only be a formal solution), and even if it does, a direct proof would be difficult without appealing to theorems which justify differentiating an infinite sum term–by–term. Instead, as we have found in the case of fixed ends (cf. Theorem 3 of Section 5.2), the method of images and D'Alembert's formula can be used (cf. the next example) to express solutions in a different way, without infinite sums and the difficulties associated with them. □

Example 2. Solve the problem in Example 1, by the method of images and D'Alembert's formula.

Solution. The method of images for problem (3) with free ends differs only in one respect from the corresponding treatment for the case of fixed ends (cf. (21) in Section 5.2). Indeed, all we do is replace the *odd* periodic extensions $\tilde{f}_o(x)$ and $\tilde{g}_o(x)$ by the *even* periodic extensions $\tilde{f}_e(x)$ and $\tilde{g}_e(x)$. The extension $\tilde{f}_e(x)$ of a function f(x) on [0,L] is illustrated in Figure 1.

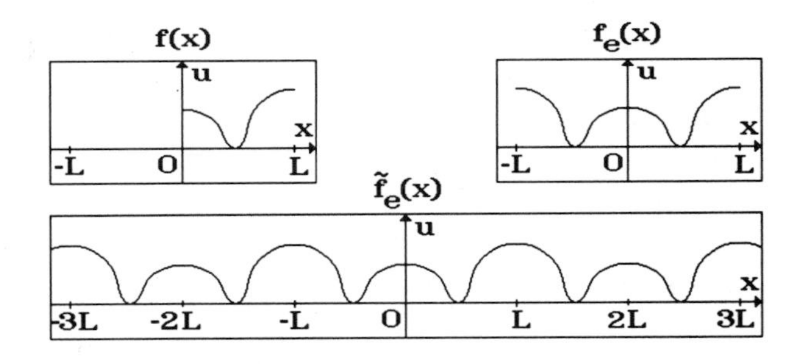

Figure 1

If the extension $\tilde{f}_e(x)$ is C^2 and $\tilde{g}_e(x)$ is C^1, then one can show (see the proof of Theorem 3 in Section 5.2) that the C^2 solution of problem (3) is given by

$$u(x,t) = \frac{1}{2}[\tilde{f}_e(x+at) + \tilde{f}_e(x-at)] + \frac{1}{2a}\int_{x-at}^{x+at} \tilde{g}_e(r)\, dr. \tag{9}$$

The initial data f(x) and g(x) have been evenly extended about $x = 0$ and $x = L$, and thus the function u(x,t) will be even about $x = 0$ and $x = L$. Hence, $u_x(0,t) = 0$ and $u_x(L,t) = 0$ (Why ?). Thus, (9) gives us a solution of (3) for $0 \le x \le L$, as long as $\tilde{f}_e(x)$ is C^2 and $\tilde{g}_e(x)$ is C^1, which will be the case if f(x) is C^2 with $f'(0) = f'(L) = 0$ and g(x) is C^1 with $g'(0) = g'(L) = 0$ (Why ?). □

Remark. Having established the existence of the C^2 solution $u(x,t)$, given by (9), when $\tilde{f}_e$ is C^2 and $\tilde{g}_e$ is C^1, the analog of Theorem 4 in Section 5.2 (with the analogous proof) yields

$$u(x,t) = \tfrac{1}{2}(A_0 t + B_0) + \sum_{n=1}^{\infty} [A_n \sin(\tfrac{n\pi at}{L}) + B_n \cos(\tfrac{n\pi at}{L})] \cos(\tfrac{n\pi x}{L}) , \tag{10}$$

where

$$A_0 = \frac{2}{L}\int_0^L g(x)\,dx\ ,\quad A_n = \frac{2}{n\pi a}\int_0^L g(x)\cos(\tfrac{n\pi x}{L})\,dx\ \ (n \geq 1)\ ,\quad B_n = \frac{2}{L}\int_0^L f(x)\cos(\tfrac{n\pi x}{L})\,dx\ \ (n \geq 0)\ .$$

In this way, we establish that the Fourier series approach in Example 1 is valid even when $N = \infty$ (i.e., (10) does indeed converge to a C^2 function), provided that $\tilde{f}_e$ is C^2 and $\tilde{g}_e$ is C^1. □

Example 3. Let $u(x,t)$ solve the problem

$$\begin{aligned}
&\text{D.E.}\quad u_{tt} = u_{xx}, && 0 \leq x \leq 2,\ -\infty < t < \infty,\\
&\text{B.C.}\quad u(0,t) = 0,\ u_x(2,t) = 0, && \\
&\text{I.C.}\quad u(x,0) = f(x) = \tfrac{1}{2}x^3(2-x)^3\ ,\quad u_t(x,0) = g(x) = 0. &&
\end{aligned} \tag{11}$$

Thus, the end at $x = 0$ is fixed, and the end at $x = 2$ is free. Use the method of images to determine the profile $u(x,t)$ at $t = 0, 2, 4, 6, 8$.

Solution. The graph of $u(x,0) = \tfrac{1}{2}x^3(2-x)^3$ is shown in Figure 2.

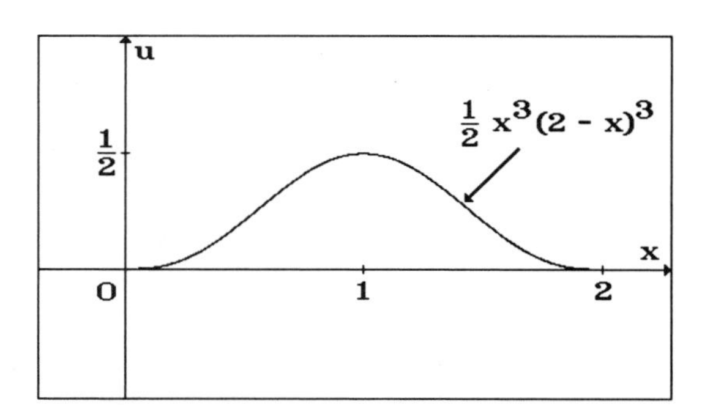

Figure 2

The method of images dictates that we extend this initial profile to all x, $-\infty < x < \infty$, in such a way that the extension is odd about $x = 0$ (to insure $u(0,t) = 0$) and even about $x = 2$ (to insure $u_x(2,t) = 0$). The only extension with these properties is graphed below if Figure 3.

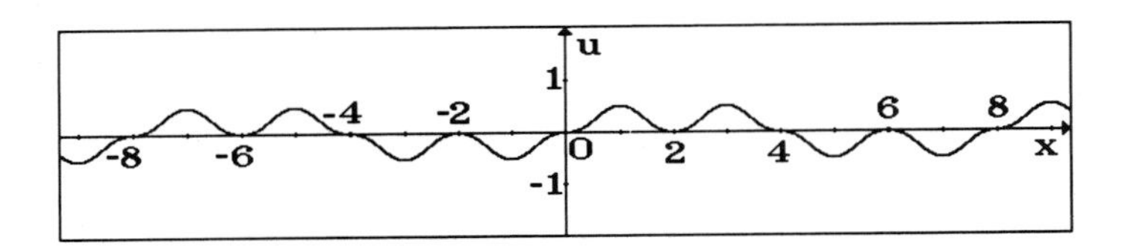

Figure 3

Let us denote the function having this graph by F(x). Note that F(x) is periodic of period 8, odd about 0, and even about 2. It is awkward to write down some explicit formulas for F(x) in all of the various intervals. Instead, it is easier to simply find the value of F(x) from its graph and f(x) (e.g., $F(-9) = -1/2$, $F(4.25) = -(.25)^3(1.75)^3/2$, etc.). Of course, the appropriate extension for $g(x) = 0$ $(0 \le x \le 2)$ is $G(x) = 0$ $(-\infty < x < \infty)$. The solution of the associated infinite string problem

$$\text{D.E.} \quad u_{tt} = u_{xx}, \qquad -\infty < x, t < \infty,$$

$$\text{I.C.} \quad u(x,0) = F(x), \ u_t(x,0) = G(x) = 0,$$

is given, via D'Alembert's formula, by

$$u(x,t) = \tfrac{1}{2}[F(x+t) + F(x-t)] . \tag{12}$$

This yields the solution of (11), if we restrict u(x,t) to the interval $0 \le x \le 2$. (Note that F is C^2 and therefore u(x,t) in (12) is C^2. If we had used x^2 instead of x^3 in the definition of f(x), there would be a problem because $F''(0^+) \ne F''(0^-)$ in this case, and F would not be C^2). Now (12) says that each of the "bumps" in the initial profile splits into two waves of half the original amplitude moving in opposite directions with unit speed. At $t = 2$ the left–moving bump from the interval [2,4] will meet the right–moving (inverted) bump from the interval [–2,0] in the "real" interval [0,2]. These bumps will exactly cancel each other at $t = 2$, so that $u(x,2) = 0$ for $0 \le x \le 2$. Similarly, at $t = 4$ the (inverted) left–moving bump from [4,6] will meet the (inverted) right–moving bump from [–4,–2] to produce $u(x,4) = -f(x)$ for $0 \le x \le 2$. The reader can check that $u(x,6) = 0$ for $0 \le x \le 2$. And at $t = 8$ the string returns to its original position. Indeed, we know this by (12) and the fact that F(x) is periodic of period 8. □

Remark. It would be possible, although much more tedious, to solve this problem using the Fourier series approach as was done in Example 1. The product solutions of the D.E. and B.C. of (11), with 2 replaced by an arbitrary length L (and D.E. $u_{tt} = a^2u_{xx}$) are found to be

$$u_n(x,t) = \left[A_n \sin[(n+\tfrac{1}{2})\tfrac{\pi at}{L}] + B_n \cos[(n+\tfrac{1}{2})\tfrac{\pi at}{L}]\right] \sin[(n+\tfrac{1}{2})\tfrac{\pi x}{L}], \tag{13}$$

for $n = 0, 1, 2, \dots$. The formal solution of

$$
\begin{aligned}
&\text{D.E.}\quad u_{tt} = a^2 u_{xx}, \qquad 0 \le x \le L,\ -\infty < t < \infty,\\
&\text{B.C.}\quad u(0,t) = 0,\ u_x(L,t) = 0, \\
&\text{I.C.}\quad u(x,0) = f(x),\ u_t(x,0) = g(x),
\end{aligned}
\tag{14}
$$

is given by

$$
u(x,t) = \sum_{n=0}^{\infty} u_n(x,t), \tag{15}
$$

where $u_n(x,t)$ is given by (13), and for $n = 0, 1, 2, \dots,$

$$
B_n = \frac{2}{L}\int_0^L f(x)\sin[(n+\tfrac{1}{2})\tfrac{\pi x}{L}]\,dx,\quad A_n = \frac{2}{(n+\frac{1}{2})\pi a}\int_0^L g(x)\sin[(n+\tfrac{1}{2})\tfrac{\pi x}{L}]\,dx. \tag{16}
$$

For the specific $f(x)$ and $g(x)$, given in (11) with $L = 2$ and $a = 1$, we could (with some effort) compute B_n. Of course $A_n = 0$. Even without computing B_n, we can answer the questions posed in Example 3. Indeed, $u(x,2) = 0$, because $\cos[(n+\frac{1}{2})\pi t/2] = 0$ when $t = 2$. Also, $u(x,4) = -f(x)$, because $\cos[(n+\frac{1}{2})4\pi/2] = -1$. Similarly, $u(x,6) = 0$, and $u(x,8) = f(x)$. Note that $\cos[(n+\frac{1}{2})\pi t/2]$ is periodic of period 8. For arbitrary times t, the alternative formula (12) is vastly superior to (15), since (15) not only involves an infinite sum, but also the B_n must be computed in order to determine the terms for arbitrary values of t. □

The methods of Section 3.3 for the standard time–independent B.C. carry over without difficulty to the case of the wave equation. We illustrate this in the next examples.

Example 4. Solve the problem

$$
\begin{aligned}
&\text{D.E.}\quad u_{tt} = a^2 u_{xx}, \qquad 0 \le x \le \pi,\ -\infty < t < \infty,\\
&\text{B.C.}\quad u(0,t) = -1,\ u_x(\pi,t) = 2, \\
&\text{I.C.}\quad u(x,0) = \sin(x/2) + 2x - 1,\ u_t(x,0) = -2\sin(3x/2).
\end{aligned}
\tag{17}
$$

Solution. As in Section 3.3, we choose a particular solution of the D.E. and B.C.. The simplest choice is the steady–state (time–independent) function $u_p(x,t) = 2x - 1$. The solution of (17) is then $u(x,t) = u_p(x,t) + v(x,t)$, where $v(x,t)$ is the solution of the related homogeneous problem

$$
\begin{aligned}
&\text{D.E.}\quad v_{tt} = a^2 v_{xx}, \qquad 0 \le x \le \pi,\ -\infty < t < \infty,\\
&\text{B.C.}\quad v(0,t) = 0,\ v_x(\pi,t) = 0, \\
&\text{I.C.}\quad \begin{cases} v(x,0) = u(x,0) - u_p(x,0) = \sin(x/2), \\ v_t(x,0) = u_t(x,0) - (u_p)_t(x,0) = -2\sin(3x/2). \end{cases}
\end{aligned}
\tag{17'}
$$

By applying the superposition principle to the product solutions (13) (cf., the D.E. and B.C. in (11) and (17′)), we obtain

$$v(x,t) = \cos(\tfrac{at}{2})\sin(\tfrac{x}{2}) - \tfrac{4}{3a}\sin(\tfrac{3at}{2})\sin(\tfrac{3x}{2}),$$

Thus, $u(x,t) = 2x - 1 + v(x,t)$ is the solution of (17). If the I.C. of (17′) had not been a finite $\sin[(n+\frac{1}{2})x]$–series, then we would have a formal solution, which hopefully could be truncated, still meeting the I.C. within an error of experimental size. We could also attempt to solve the related homogeneous problem for v by the method of images, as in Example 3. □

Example 5. Solve

$$\text{D.E.}\quad u_{tt} = a^2u_{xx}, \qquad 0 \le x \le L,\ -\infty < t < \infty,$$

$$\text{B.C.}\quad u_x(0,t) = c,\ u_x(L,t) = d, \tag{18}$$

$$\text{I.C.}\quad u(x,0) = f(x),\ u_t(x,0) = g(x).$$

Solution. Note that the B.C. mean that the end $x = 0$ is subject to a downward force c/T_0 and the end $x = L$ is subject to an upward force d/T_0 (cf. equations (1) and (2)). Thus, we expect the string to drift vertically if $c \ne d$. Indeed, there is no steady–state particular solution of the D.E. and B.C., unless $c = d$. Instead, we try a particular solution of the form $u_p(x,t) = kt + H(x)$. This will not work, since the D.E. implies $H'' = 0$ and therefore $H(x) = c_1x + c_2$, but then the B.C. are not met unless $c = c_1 = d$. The next guess is $u_p(x,t) = kt^2 + H(x)$. Then the D.E. yields $2k = a^2H''(x)$, whence $H(x) = kx^2/a^2 + c_1x + c_2$. The constants k and c_1 can be found in terms of c and d using the B.C., and we simply set $c_2 = 0$. Thus,

$$u_p(x,t) = \frac{a^2(d-c)}{2L}t^2 + \frac{d-c}{2L}x^2 + cx. \tag{19}$$

The solution of (18), if it exists, is then $u(x,t) = u_p(x,t) + v(x,t)$, where $v(x,t)$ solves the following familiar related homogeneous problem (cf. Example 1) :

$$\text{D.E.}\quad v_{tt} = a^2v_{xx}, \qquad 0 \le x \le L,\ -\infty < t < \infty,$$

$$\text{B.C.}\quad v_x(0,t) = 0,\ v_x(L,t) = 0,$$

$$\text{I.C.}\quad v(x,0) = f(x) - u_p(x,0),\ v_t(x,0) = g(x) - (u_p)_t(x,0)\ .$$ □

The inhomogeneous wave equation and related problems

As with heat problems (cf. Section 3.4), the introduction of time–*dependent* B.C., such as $u(0,t) = c(t)$, in wave problems leads to an inhomogeneous D.E. . For example, consider

$$\begin{array}{ll} \text{D.E.} & u_{tt} = a^2 u_{xx}, \qquad 0 \le x \le L,\ -\infty < t < \infty, \\ \text{B.C.} & u(0,t) = c(t),\ u(L,t) = d(t), \\ \text{I.C.} & u(x,0) = f(x),\ u_t(x,0) = g(x)\,. \end{array}$$

The function $w(x,t) = \frac{x}{L}(d(t) - c(t)) + c(t)$ satisfies the B.C., but not the D.E. . Hence, $v(x,t) = u(x,t) - w(x,t)$ will not satisfy the wave equation, but rather the D.E. in the related problem

$$\begin{array}{ll} \text{D.E.} & v_{tt} - a^2 v_{xx} = -(w_{tt} - a^2 w_{xx}) = -\frac{x}{L}(d''(t) - c''(t)) - c''(t), \\ \text{B.C.} & v(0,t) = u(0,t) - w(0,t) = 0, \quad v(L,t) = u(L,t) - w(L,t) = 0 \\ \text{I.C.} & \begin{cases} v(x,0) = u(x,0) - w(x,0) = f(x) - \frac{x}{L}(d(0) - c(0)) - c(0), \\ v_t(x,0) = u_t(x,0) - w_t(x,0) = g(x) - \frac{x}{L}(d'(0) - c'(0)) - c'(0). \end{cases} \end{array} \tag{20}$$

Thus, we are led to consider the inhomogeneous wave equation

$$u_{tt} - a^2 u_{xx} = h(x,t)\,, \tag{21}$$

where $h(x,t)$ is a given function which is proportional to the vertical component of an external force density applied to the string. Indeed, we may rewrite (21) in the form

$$D\,\Delta x\, u_{tt} = T_0\,\Delta x\, u_{xx} + D\,\Delta x\, h(x,t)\,, \tag{22}$$

where D is the linear density, T_0 is the tension of the string at rest and Δx is the length of a small portion of string centered at x. Equation (22) is then Newton's equation for this portion. The first term on the right–hand side of (22) is nearly the force due to the tensions at the ends of this portion (cf. the derivation in Section 5.1, where $\Delta x = 2h$). The second term must then be the external force applied to this portion. Thus, $D\,h(x,t)$ is the **linear force density** (i.e., force per unit length) imposed on the string. In the case of gravity, we would have $D\,h(x,t)\,\Delta x = -D\,\Delta x\, g$ or $h(x,t) = -g = -32\ \text{ft/sec}^2$. Note that $h(x,t)$ may be referred to as the **applied acceleration at** (x,t).

Heat problems with an inhomogeneous D.E., due to internal heat sources, were solved in Section 3.4 using Duhamel's principle, whereby solving the inhomogeneous D.E. was accomplished by solving a family of related problems in which the source appears in the initial conditions instead of the D.E.. The same idea works for wave problems with inhomogeneous D.E. (21). For simplicity, we first consider the infinite string problem

$$\text{D.E.} \quad u_{tt} - a^2 u_{xx} = h(x,t) , \qquad -\infty < x,\ t < \infty,$$

$$\text{I.C.} \quad u(x,0) = 0, \ \ u_t(x,0) = 0 . \tag{23}$$

The motivation for the method of Duhamel is as follows. Suppose the acceleration $h(x,s)$ is applied to the string at $t = s - \Delta s$ and that the acceleration is promptly turned off at $t = s$. The string will acquire a velocity of $h(x,s)\Delta s$, and its position change is $h(x,s)(\Delta s)^2/2$. Assuming that Δs is small, the change in position is "negligible". The effect of the imposed acceleration is then $v(x,t;s)\Delta s$, where $v(x,t;s)$ is the solution of

$$\text{D.E.} \quad v_{tt} = a^2 v_{xx} , \qquad -\infty < x < \infty,\ \ t \geq s,$$

$$\text{I.C.} \quad v(x,s;s) = 0, \ \ v_t(x,s;s) = h(x,s). \tag{24}$$

This problem has initial conditions given at the arbitrary time $t = s$, instead of $t = 0$. We can write $v(x,t;s) = \tilde{v}(x,t-s;s)$, where $\tilde{v}(x,t;s)$ solves the familiar problem with I.C. given at $t = 0$.

$$\text{D.E.} \quad \tilde{v}_{tt} = a^2 \tilde{v}_{xx} , \qquad -\infty < x < \infty,\ \ t \geq 0,$$

$$\text{I.C.} \quad \tilde{v}(x,0;s) = 0, \ \ \tilde{v}_t(x,0;s) = h(x,s). \tag{25}$$

By D'Alembert's formula, we know that the solution of (25) is given by

$$\tilde{v}(x,t;s) = \frac{1}{2a} \int_{x-at}^{x+at} h(r,s)\ ds , \tag{26}$$

and therefore the solution of (24) is

$$v(x,t;s) = \tilde{v}(x,t-s;s) = \frac{1}{2a} \int_{x-a(t-s)}^{x+a(t-s)} h(r,s)\ dr.$$

We expect that the solution of (23) will be the integral from $s = 0$ to $s = t$ or superposition of all the effects $v(x,t;s)\Delta s$ with respect to s. In other words, let us hypothesize that the solution of (23) is given by

$$u(x,t) = \int_0^t v(x,t;s)\ ds = \int_0^t \tilde{v}(x,t-s;s)\ ds = \frac{1}{2a} \int_0^t \int_{x-a(t-s)}^{x+a(t-s)} h(r,s)\ dr\ ds . \tag{27}$$

The above heuristic argument is not a proof. A rigorous statement and proof are as follows.

Theorem 1 (Duhamel's principle for the wave equation). Let $h(x,t)$ **be a** C^1 **function,** $-\infty < x,\ t < \infty$. **Then (27) is the unique solution of the problem**

$$\text{D.E.}\quad u_{tt} - a^2u_{xx} = h(x,t) \qquad -\infty < x\ ,\ t < \infty$$

$$\text{I.C.}\quad u(x,0) = 0,\ \ u_t(x,0) = 0\ . \tag{28}$$

Proof. We know from (26) that $\tilde{v}(x,t;s)$ is C^2, since $h(x,t)$ is assumed to be C^1. We can then twice apply Lemma 1 of Section 3.4, once with $g(t,s) = v(x,t;s) = \tilde{v}(x,t-s;s)$, and then with $g(t,s) = \tilde{v}_t(x,t;s)$ to obtain

$$\begin{aligned} u_t(x,t) &= \tilde{v}(x,0;s) + \int_0^t \tilde{v}_t(x,t-s;s)\,ds = \int_0^t \tilde{v}_t(x,t-s;s)\,ds, \\ u_{tt}(x,t) &= \tilde{v}_t(x,0;t) + \int_0^t \tilde{v}_{tt}(x,t-s;s)\,ds = h(x,t) + \int_0^t a^2\tilde{v}_{xx}(x,t-s;s)\,ds \\ &= h(x,t) + a^2u_{xx}(x,t), \end{aligned} \tag{29}$$

where we have used the D.E. of (25) and Leibniz's rule (cf. Appendix A.3) in the final equation. This shows that $u(x,t)$ in (27) is a C^2 solution of the D.E. in (28). By (27), $u(x,0) = 0$, while (29) yields $u_t(x,0) = 0$. Uniqueness is evident from the fact that if u_1 and u_2 are two solutions of (28), then $v = u_1 - u_2$ satisfies $v_{tt} = a^2v_{xx}$ with I.C. $v(x,0) = 0$ and $v_t(x,0) = 0$. Hence, $v \equiv 0$ by the previous uniqueness result (Theorem 1 in Section 5.2). □

Remarks. The double integral (27) with respect to r and s admits a nice geometrical interpretation. Indeed, it is the integral of the function h over the **characteristic triangle** (or **region of influence**) **of the point** (x,t) shown in Figure 4 :

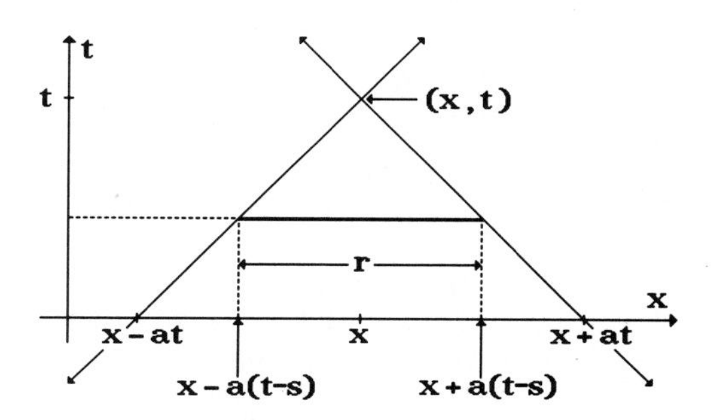

Figure 4

The integral with respect to r "sums" the values of h along the horizontal segments of the triangle and the integral with respect to s "sums" the results for these segments. Equation (27) says that $u(x,t)$ is the integral of h over the characteristic triangle of the point (x,t). □

When the I.C. in (28) are no longer trivial, the problem can be treated by splitting the problem into two pieces, as we have done previously in Section 3.4 in the context of heat problems. We illustrate this in the following example.

Example 6. Solve

$$\text{D.E.} \quad u_{tt} - u_{xx} = x - t, \qquad -\infty < x, t < \infty,$$
$$\text{I.C.} \quad u(x,0) = x^2, \quad u_t(x,0) = \sin(x). \tag{30}$$

Solution. We split the problem up into two familiar problems for functions $u_1(x,t)$ and $u_2(x,t)$:

$$\text{D.E.} \quad (u_1)_{tt} - (u_1)_{xx} = 0, \qquad\qquad \text{D.E.} \quad (u_2)_{tt} - (u_2)_{xx} = x - t,$$

$$\text{I.C.} \begin{cases} u_1(x,0) = x^2, \\ (u_1)_t(x,0) = \sin(x), \end{cases} \qquad\qquad \text{I.C.} \begin{cases} u_2(x,0) = 0, \\ (u_2)_t(x,0) = 0\,. \end{cases}$$

The solution of (30) is then $u(x,t) = u_1(x,t) + u_2(x,t)$ (Why ?). From D'Alembert's formula,

$$u_1(x,t) = \tfrac{1}{2}[(x+t)^2 + (x-t)^2] - \tfrac{1}{2}[\cos(x+t) - \cos(x-t)]\ .$$

We compute $u_2(x,t)$ using Theorem 1 and (27):

$$u_2(x,t) = \frac{1}{2}\int_0^t \int_{x-(t-s)}^{x+(t-s)} (r-s)\, dr\, ds = \frac{1}{2}\int_0^t \left[\frac{r^2}{2} - sr\right]_{x-t+s}^{x+t-s} ds$$

$$= \frac{1}{2}\int_0^t \left[\frac{(x+t-s)^2}{2} - \frac{(x+s-t)^2}{2} - s(x+t-s) + s(x+s-t)\right] ds$$

$$= \frac{1}{2}\int_0^t \left[2s^2 - 2s(x+t) + \frac{(x+t)^2}{2} - \frac{(x-t)^2}{2}\right] ds$$

$$= \frac{t^3}{3} - \frac{t^2(x+t)}{2} + t^2 x = -\frac{t^3}{6} + \frac{t^2 x}{2}.$$

The solution $u(x,t) = u_1(x,t) + u_2(x,t)$ may be checked directly. □

Duhamel's principle also applies in the case of a finite string, as the following examples illustrate. In Example 7, there is an applied force density, but the B.C. and I.C. are homogeneous. Example 8 is an instance of the type problem with time–dependent B.C. (but homogeneous D.E.) which was the original motivation for this subsection (cf. (20)). As in Example 6, one can handle the case where both the D.E. and B.C. are inhomogeneous, by splitting the problem into two parts, and adding the solutions of the two parts.

Example 7. A string of length π with fixed ends and initially at rest is driven by a harmonic vertical force density proportional to $\sin(\omega t)\sin(x)$, where ω is a positive constant. Find the displacement $u(x,t)$ by solving the problem

$$\begin{aligned} &\text{D.E.} \quad u_{tt} = a^2 u_{xx} + A\sin(\omega t)\sin(x)\,, \qquad 0 \le x \le \pi,\ t \ge 0, \\ &\text{B.C.} \quad u(0,t) = 0,\ u(\pi,t) = 0, \\ &\text{I.C.} \quad u(x,0) = 0,\ u_t(x,0) = 0. \end{aligned} \tag{31}$$

Solution. Duhamel's principle works in the presence of homogeneous linear boundary conditions such as we have here. In other words, we expect that the solution of (31) is given by $u(x,t) = \int_0^t \tilde{v}(x,t-s,s)\,ds$, where $\tilde{v}(x,t;s)$ is the solution of

$$\begin{aligned} &\text{D.E.} \quad \tilde{v}_{tt} = a^2 \tilde{v}_{xx}\,, \qquad 0 \le x \le \pi,\ t \ge 0, \\ &\text{B.C.} \quad \tilde{v}(0,t;s) = 0,\quad \tilde{v}(\pi,t;s) = 0, \\ &\text{I.C.} \quad \tilde{v}(x,0;s) = 0\,,\quad \tilde{v}_t(x,0;s) = A\sin(\omega s)\sin(x). \end{aligned} \tag{32}$$

Since s is just a constant, we easily obtain

$$\tilde{v}(x,t;s) = \frac{A}{a}\sin(\omega s)\sin(at)\sin(x)\,.$$

Then,

$$\begin{aligned} u(x,t) &= \int_0^t \tilde{v}(x,t-s;s)\,ds = \int_0^t \frac{A}{a}\sin(\omega s)\sin[a(t-s)]\sin(x)\,ds \\ &= \frac{A}{a}\sin(x)\int_0^t \sin(\omega s)\sin[a(t-s)]\,ds. \end{aligned}$$

Let the last integral be denoted by $I(t)$. By using Green's formula,

$$\int_0^t [f''(s)g(s) - f(s)g''(s)]\,ds = [f'(s)g(s) - f(s)g'(s)]\Big|_0^t\,,$$

$$(a^2-\omega^2)\,I(t) = \Big[\omega\cos(\omega s)\sin[a(t-s)] + a\sin(\omega s)\cos[a(t-s)]\Big]_0^t = a\sin(\omega t) - \omega\sin(at).$$

For $\omega \ne a$, we then obtain

$$u(x,t) = \frac{A}{a(a^2-\omega^2)}\,[a\sin(\omega t) - \omega\sin(at)]\sin(x). \tag{33}$$

To find the solution for $\omega = a$, we take the limit of (33) as $\omega \to a$, using L'Hospital's rule :

$$u(x,t) = \frac{A}{a}\sin(x) \lim_{\omega \to a} \frac{at\cos(\omega t) - \sin(at)}{-2\omega} = -\frac{A}{2a}\left[t\cos(at) - \frac{\sin(at)}{a}\right]\sin(x). \quad (34)$$

Any doubts concerning these methods may be dispelled by directly checking that (33) or (34) solve problem (31). Note that (33) remains bounded as a function of t, and its amplitude becomes larger as $\omega \to a$. In other words, we have the phenomenon of resonance as the driving term $A\sin(\omega t)\sin(x)$ approaches the natural harmonic $A\sin(at)\sin(x)$ of the undriven string. Indeed, for $\omega = a$ we obtain the solution (34) whose amplitude grows without bound as $t \to \infty$. Incidentally, bridges (which, in a certain sense, might be thought of as strings) have been known to collapse under the influence of periodic winds, because of the phenomenon of resonance. □

Remark. For finite string problems, as an alternative to using Duhamel's principle, one can carry out the following steps.

Step 1. Try to approximate the forcing term $h(x,t)$ in the D.E. $u_{tt} = a^2u_{xx} + h(x,t)$ by a Fourier series (of the type appropriate to the B.C.) whose Fourier coefficients are functions of t, say $h(x,t) = \sum_{n=1}^{N} h_n(t)\sin(n\pi x/L)$, when the ends are fixed [$h(x,t)$ is of this form in (31)].

Step 2. Assume that $u(x,t) = \sum_{n=1}^{N} u_n(t)\sin(n\pi x/L)$ (say when the ends are fixed) and substitute this form into the D.E. $u_{tt} - a^2u_{xx} = h(x,t)$. Equate coefficients of the two sides, obtaining the ODEs $u_n''(t) + (an\pi/L)^2u_n(t) = h_n(t)$, $n = 1, 2, \ldots N$.

Step 3. From the initial conditions $u(x,0) = f(x) = \sum_{n=1}^{N} B_n\sin(n\pi x/L)$ and $u_t(x,0) = \sum_{n=1}^{N} A_n\sin(n\pi x/L)$ (finite sums, say within experimental error), we know that $u_n(0) = B_n$ and $u_n'(0) = A_n$. Solve the ODEs in Step 3 subject to these initial conditions.

Step 4. With $u_n(t)$ found in Step 3, the solution is $u(x,t) = \sum_{n=1}^{N} u_n(t)\sin(n\pi x/L)$.

In Example 7, we obtain $u_1''(t) + a^2u_1(t) = A\sin(\omega t)$, with initial conditions $u_1(0) = 0$ and $u_1'(0) = 0$, whose solution is the coefficient of $\sin(x)$ in (33) [or (34), if $\omega = a$]. For $n > 0$, $u_n(t) \equiv 0$. If it happens that $h(x,t)$ is not a *finite* Fourier series in x, then the infinite series obtained by this procedure must be regarded as a *formal* solution (cf. the definition in Section 4.3), until proven otherwise. One advantage of using Duhamel's principle (in conjunction with the method of images) is that infinite sums can be avoided, if desired.

Example 8. Find a formal solution of the problem

$$\begin{aligned}
&\text{D.E.} \quad u_{tt} = a^2 u_{xx}, \qquad 0 \le x \le L,\ -\infty < t < \infty, \\
&\text{B.C.} \quad u(0,t) = 0\,,\ u(L,t) = A\sin(\omega t), \qquad (35)\\
&\text{I.C.} \quad u(x,0) = 0\,,\ u_t(x,0) = A\omega x/L\,.
\end{aligned}$$

Solution. The function $w(x,t) \equiv A(x/L)\sin(\omega t)$ satisfies the B.C. of (35). Thus, by writing $u(x,t) = w(x,t) + v(x,t)$, we obtain the following related problem for $v(x,t)$ with homogeneous B.C. .

$$\begin{aligned}
&\text{D.E.} \quad v_{tt} - a^2 v_{xx} = A\tfrac{x}{L}\,\omega^2\sin(\omega t), \quad 0 \le x \le L\,, -\infty < t < \infty\,, \\
&\text{B.C.} \quad v(0,t) = 0\,,\ v(L,t) = 0, \qquad (36)\\
&\text{I.C.} \quad v(x,0) = 0\,,\ v_t(x,0) = A\omega x/L - A\omega x/L = 0\,.
\end{aligned}$$

Note that the choice of I.C. $u_t(x,0) = A\omega x/L$ is compatible with the B.C. $u(L,t) = A\sin(\omega t)$ at $x = L$. The cancellation, producing $v_t(x,0) = 0$, is fortuitous. There is some freedom in choosing $w(x,t)$ satisfying the B.C. (i.e., $w(x,t)$ does not have to be linear in x). Perhaps the choice of $w(x,t)$ should be motivated to achieve simplicity in the related problem for $v(x,t)$. We formally solve problem (36), using Duhamel's principle, but we could achieve the same result by using the procedure in the preceding remark. According to Duhamel's principle, $v(x,t) = \int_0^t \tilde{v}(x,t-s;s)\,ds$, where $\tilde{v}(x,t;s)$ solves

$$\begin{aligned}
&\text{D.E.} \quad \tilde{v}_{tt} = a^2\tilde{v}_{xx}, \qquad 0 \le x \le L,\ -\infty < t < \infty, \\
&\text{B.C.} \quad \tilde{v}(0,t;s) = 0,\ \tilde{v}(L,t;s) = 0, \qquad (37)\\
&\text{I.C.} \quad \tilde{v}(x,0;s) = 0,\ \tilde{v}_t(x,0;s) = A\tfrac{x}{L}\,\omega^2\sin(\omega s).
\end{aligned}$$

This problem has no exact solution, since the second B.C. contradicts the second I.C. at $x = L$. One could construct a formal solution by taking the odd, periodic extension (of period 2L) of the initial velocity for $\tilde{v}$ and using D'Alembert's formula, but the extension is not continuous, let alone C^1. Instead, we construct a formal solution of (37) from the Fourier sine series of x in [0,L]. Indeed,

$$\text{FSS } x = \frac{2L}{\pi}\sum_{n=1}^{\infty}\frac{(-1)^{n+1}}{n}\sin\left(\frac{n\pi x}{L}\right)$$

yields

$$\tilde{v}(x,t;s) = \frac{2AL\omega^2}{\pi^2 a}\sum_{n=1}^{\infty}\frac{(-1)^{n+1}}{n^2}\sin(\omega s)\sin\left(\frac{n\pi at}{L}\right)\sin\left(\frac{n\pi x}{L}\right).$$

Using the result (assuming $\omega \ne n\pi a/L$)

$$\int_0^t \sin(\omega s)\sin[\tfrac{n\pi a}{L}(t-s)]\,ds = \frac{(n\pi a/L)\sin(\omega t) - \omega\sin(n\pi at/L)}{(n\pi a/L)^2 - \omega^2},$$

we obtain the formal solution

$$u(x,t) = \frac{A}{L}x\sin(\omega t) - \frac{2AL\omega^2}{\pi^2 a}\sum_{n=1}^{\infty}(-1)^n \frac{\frac{n\pi a}{L}\sin(\omega t) - \omega\sin(\frac{n\pi at}{L})}{n^2[(\frac{n\pi a}{L})^2 - \omega^2]}\sin(\tfrac{n\pi x}{L}).$$

The reader may check that the infinite sum is the same formal solution for $v(x,t)$ that would be obtained, if one were to use the procedure in the preceding remark. Also, note that we again (cf. Example 7) have the phenomenon of resonance, as $\omega \to n\pi a/L$ for some n. If $\omega = n\pi a/L$, then the n–th term in the sum should be replaced by its limiting value as $\omega \to n\pi a/L$. This term is unbounded as $t \to \infty$. If the sum is truncated at $n = N$, then the resulting function $u_N(x,t)$ satisfies the B.C. and I.C. of (35), but it does not quite meet the D.E. :

$$(u_N)_{tt} - a^2(u_N)_{xx} = \frac{A\omega^2}{L}\left[x - \frac{2L}{\pi}\sum_{n=1}^{N}\frac{(-1)^{n+1}}{n}\sin(\tfrac{n\pi x}{L})\right]\sin(\omega t).$$

The quantity in brackets approaches 0 as $N \to \infty$, provided that $0 \le x < L$. If $x = L$, the right side is $A\omega^2\sin(\omega t)$, regardless of N, suggesting that there is no exact solution of the original problem. This is probably not a serious difficulty. Indeed, if we restrict x to the interval $[0,L-\delta]$ for any small $\delta > 0$, then for N large enough, $u_N(x,t)$ satisfies the wave equation to within any given experimental error. Choosing δ to be much less than the radius of an atom, it seems unlikely that anyone will consider this to be a significant defect in applications. □

Summary 5.3

1. Boundary conditions : If the end $x = 0$ of a string is allowed to side vertically (i.e., transversely) and if a vertical external force $F(t)\mathbf{j}$ is applied, then $u_x(0,t) = -F(t)/T_0$. Thus, if no vertical force is applied to the end (i.e., the end is free), then the boundary condition $u_x(0,t) = 0$ holds, and similarly $u_x(L,t) = 0$ for a free end at $x = L$.

2. Solving wave problems with fixed or free ends : Wave problems, where each end is free or fixed, can be solved by two different methods :

A. The Fourier series method : Find the product solutions of the D.E. which meet the B.C., and write (or approximate) each of the functions $f(x)$ and $g(x)$ in the I.C. $u(x,0) = f(x)$, $u_t(x,t) = g(x)$ as Fourier series of the type which is appropriate to the B.C. (e.g., a Fourier cosine series, if both ends are free). Then form a superposition of the product solutions in order to meet the I.C. (cf. Example 1).

B. The method of images : Extend the initial displacement $u(x,0) = f(x)$ $(0 \leq x \leq L)$ to a function defined for all x, which is even about each free end and odd about each fixed end. (Such an extension is unique.) Do the same for the initial velocity $u_t(x,0) = g(x)$. Then the solution for the finite string problem is given by D'Alembert's formula for the infinite string with initial data being the above extensions defined for all x (cf. Examples 2 and 3).

Solutions obtained, using either method A or B, give the same values for $u(x,t)$, even though the respective formulas for $u(x,t)$ may appear different. If the extension of $f(x)$ is C^2 and the extension of $g(x)$ is C^1, then a unique (C^2) solution for the wave problem will exist.

3. Inhomogeneous time–independent B.C. : For wave problems with B.C. of the form $u(0,t) = a$ or $u_x(0,t) = a$, and $u(L,t) = b$ or $u_x(L,t) = 0$, first find a function $u_p(x,t)$ which solves the D.E. and B.C. . Then form the related problem for $v(x,t) = u(x,t) - u_p(x,t)$, with homogeneous B.C. , which can be solved using either of the techniques in 2 above.

4. The inhomogeneous wave equation : In problems where there are time–dependent B.C. (e.g., $u(0,t) = A(t)$), one typically encounters the inhomogeneous wave equation of the form

$$u_{tt} = a^2 u_{xx} + h(x,t) , \tag{S1}$$

when considering the related problem with homogeneous B.C. . The function $h(x,t)$ can be interpreted as being proportional to a time–dependent externally applied force density. For the infinite string, with zero initial amplitude and velocity, the solution of the inhomogeneous wave equation is

$$u(x,t) = \frac{1}{2a}\int_0^t \int_{x-a(t-s)}^{x+a(t-s)} h(r,s)\,dr\,ds . \tag{S2}$$

which is the integral of h over the characteristic triangle of the point (x,t). If the initial data is not trivial, then we simply add D'Alembert's solution to (S2). The solution (S2) was obtained by applying Duhamel's principle in the context of the wave equation. Duhamel's principle also can be used to solve wave problems for finite strings with an external forcing term (cf. Examples 7 and 8). Solutions can also be obtained (cf. Remark preceding Example 8) by first writing the source $h(x,t)$ as a Fourier series (of the type appropriate to the B.C.) with time-dependent coefficients $h_n(t)$, and then assuming a solution $u(x,t)$ of the same form, where the coefficients $u_n(t)$ are determined by solving the second-order ODEs $u_n''(t) + (n\pi a/L)^2 u_n(t) = h_n(t)$, with initial values $u_n(0)$ and $u_n'(0)$ determined by the I.C. .

Exercises 5.3

1. (a) Show that $u_x(L,t) = G(t)/T_0$ (cf. (2)), where $G(t)\mathbf{j}$ is the force applied to the end $x = L$.
(b) Explain how a boundary condition of the form $u_x(0,t) = b \cdot u(0,t)$ $(b > 0)$ would arise.

2. Solve

$$\text{D.E. } u_{tt} = a^2 u_{xx}, \quad 0 \le x \le \pi, \ -\infty < t < \infty,$$
$$\text{B.C. } u_x(0,t) = 0, \ u_x(\pi,t) = 0,$$
$$\text{I.C. } u(x,0) = \cos^2(x), \ u_t(x,0) = \sin^2(x).$$

(a) using the Fourier series approach,

(b) using the method of images.

3. Give a simple argument to demonstrate that the function $F(x)$ of Example 3 is the *only* extension of $f(x)$ which is even about 2 and odd about 0.

4. Redo Example 3 in the cases where the B.C. are replaced by

(a) $u(0,t) = 0$, $u(2,t) = 0$, (b) $u_x(0,t) = 0$, $u_x(2,t) = 0$.

Determine the profile $u(x,t)$ in these cases for $t = 1, 2, 3, 4$ and $0 \le x \le 2$.

5. (a) Sketch a derivation of the fact that the harmonics $u_n(x,t)$ in (13) form a complete family of product solutions for the problem

$$\text{D.E. } u_{tt} = a^2 u_{xx},$$
$$\text{B.C. } u(0,t) = 0, \ u_x(0,t) = 0.$$

(b) What is the lowest frequency of the harmonics in (13) ? How does this compare with the lowest frequency when both ends are fixed ?

6\. Solve

D.E. $u_{tt} = 4u_{xx}$, $0 \le x \le \pi$, $-\infty < t < \infty$,

B.C. $u_x(0,t) = -1$, $u_x(\pi,t) = 1$,

I.C. $u(x,0) = \frac{x^2}{\pi} - x + 2\cos(3x)$, $u_t(x,0) = \cos(x)$.

7\. Solve

D.E. $u_{tt} = a^2 u_{xx} + e^{-t}\cos(x)$, $-\infty < x, t < \infty$,

I.C. $u(x,0) = 0$, $u_t(x,0) = 0$.

8\. Solve

D.E. $u_{tt} = a^2 u_{xx} + e^{-t}\cos(x)$ $-\infty < x, t < \infty$,

I.C. $u(x,0) = f(x)$, $u_t(x,0) = g(x)$.

9\. Solve

D.E. $u_{tt} = a^2 u_{xx} + \cos(\omega t)\sin(3x)$, $0 \le x \le \frac{1}{2}\pi$, $\omega > 0$, $-\infty < t < \infty$,

B.C. $u(0,t) = 0$ (fixed end), $u_x(\pi/2,t) = 0$ (free end),

I.C. $u(x,0) = 0$, $u_t(x,0) = 0$.

(a) Using Duhamel's principle.

(b) By assuming a solution of the form $\sum_{n=0}^{N} u_n(t)\sin((2n+1)x)$.

(c) For what value of the constant ω is resonance obtained ?

10\. Find a formal solution of the problem

D.E. $u_{tt} = a^2 u_{xx}$, $0 \le x \le L$, $-\infty < t < \infty$,

B.C. $u_x(0,t) = 0$, $u_x(L,t) = \sin(\omega t)$,

I.C. $u(x,0) = 0$, $u_t(x,0) = \frac{\omega}{2L} x^2$.

For what values of ω do we have resonance ?

11. Find a formal solution of the problem

$$\text{D.E.}\quad u_{tt} = a^2 u_{xx}, \qquad 0 \le x \le 1,\ -\infty < t < \infty,$$

$$\text{B.C.}\quad u(0,t) = \sin(\omega t),\ u_x(1,t) = 0,$$

$$\text{I.C.}\quad u(x,0) = 0,\ u_t(x,0) = \omega\cos(\pi x).$$

For what values of ω do we have resonance ?

CHAPTER 6

LAPLACE'S EQUATION

Laplace's equation plays an important role in a vast array of applications in gravitation theory, electrostatics, steady–state temperature problems, fluid mechanics, etc. . Some of these applications were already discussed, in broad terms, in Section 1.2 . Here, we concentrate on Laplace's equation in dimension 2, namely $u_{xx} + u_{yy} = 0$. In Section 6.1, we outline the applications in this two–dimensional setting, discuss the invariance of Laplace's equation under translations and rotations of coordinates, and introduce the two basic boundary–value problems for Laplace's equation, namely the Dirichlet and Neumann problems. In Section 6.2, we solve boundary–value problems for rectangular regions. We express Laplace's equation in terms of polar coordinates in Section 6.3, in order to obtain the mean–value theorem for harmonic functions and solve the Dirichlet problem for the annular region between concentric circles. Also in Section 6.3, the Poisson integral formula is established for the solution of the Dirichlet problem for the disk, with given continuous data on the circular boundary. While the Maximum/Minimum Principle for Laplace's equation is used in various places in Sections 6.2 and 6.3, the proof is deferred to Section 6.4, by which time there is ample motivation for results of a more theoretical nature. In Section 6.5, we utilize the close relationship between Laplace's equation and complex variable theory to solve problems of two–dimensional ideal fluid flow, steady–state temperatures and electrostatics. The applications of conformal mappings are kept at a very concrete level, and we do not assume any prior knowledge of complex–variable theory.

Historical Remarks on Laplace, Dirichlet, Poisson and Neumann

Here we include a few brief biographical sketches of the key individuals who contributed to the early development of potential theory (i.e., the theory of boundary–value problems for Laplace's and Poisson's equation).

The French theoretical physicist and mathematician **Pierre Simon de Laplace** (1749–1827) was so famous in his own time that he was known as the Newton of France. With the support of D'Alembert, Laplace became a professor at the École Militaire of Paris. Laplace's primary interests were celestial mechanics, probability theory, and his personal advancement (but not necessarily in this order). Indeed, it seems that his political views shifted with the volatile social climate in France during his lifetime, so as to maximize his acquisition of titles and wealth. However, his flexibility probably also saved him from imprisonment or execution during the French Revolution. In his *magnum opus*, the *Mécanique Céleste*, published in five volumes (1799–1825), Laplace developed potential theory. While Laplace did not always acknowledge contributions of other mathematical physicists (a notable omission in this work was Lagrange's name), he expanded the frontiers of potential theory so extensively, that the key equation $\Delta u = 0$ bears his name. We note, however, that this equation had been found earlier (in 1752) by L. Euler in his studies of hydrodynamics. In Laplace's other masterpiece, *Théore Analytique de Probabilités* (1812), he included many of his discoveries in probability theory. While Laplace pursued honors, titles and wealth, he also gave generous assistance to many younger scientists. Among his many protégés were Cauchy and Poisson.

The eminent German mathematician **Gustav Peter Dirichlet** (1805–1859) was a pupil of Georg Simon Ohm (1787–1854), the German physicist who is best known for his discovery of Ohm's law ($E = RI$). While Dirichlet's main interest was in number theory, he made significant contributions to algebra, Fourier series, and theoretical mechanics. At the beginning of his career, Dirichlet was inspired by the works of Laplace and Poisson whom he met in Paris. In 1850, Dirichlet published an important paper that deals with the boundary–value problem which bears his name. In 1855, when the great mathematical genius Karl Friedrich Gauss (1777–1855) died, Dirichlet became Gauss' successor at the University of Göttingen. After Dirichlet's death, his notes on applied mathematics, as well as his celebrated *Vorlesungen über Zahlentheorie* were edited and published by his pupil and friend Richard Dedekind.

Another brilliant French mathematical physicist of this period was **Siméon–Denis Poisson** (1781–1840). Poisson was a student and protégé of Laplace and owed to his teacher his first appointment at the newly founded École Polytechnique. Throughout his life, Poisson pursued many weighty administrative and pedagogical responsibilities. Nevertheless, his list of nearly 300 original publications is very impressive. Most of Poisson's books were published during his last ten years, and according to historians these books exhibit an uncommon gift for clear exposition of the state of mathematical physics at that time. Poisson worked closely with Laplace. In 1833, Poisson pointed out that the gravitational potential u, within a region of density ρ, obeys the inhomogeneous version of Laplace's equation, namely $\Delta u = 4\pi G\rho$ (cf. (1) of Section 1.2), which is now known as Poisson's equation. As an indication of the breadth of his studies, there is the Poisson distribution in probability theory, the Poisson bracket in theoretical mechanics, the Poisson ratio in elasticity theory, the Poisson Integral Formula (cf. Section 6.3), and the Poisson Summation Formula of Fourier transform theory (cf. Section 7.5, Problem 10), and more.

The Neumann problem (cf. (11) of Section 6.1) was named after the German physicist **Carl Gottfried Neumann** (1832–1925). Neumann, who led a quiet life, was a productive researcher and successful professor at the University of Leipzig from 1868 until his retirement in 1911. He was especially prominent in the field of potential theory. Neumann is also remembered for his service in founding and editing the prestigious German mathematical periodical, the *Mathematische Annalen*.

6.1 General Orientation

Laplace's equation in dimension two for a function $u = u(x,y)$ is

$$u_{xx} + u_{yy} = 0 \quad \text{or} \quad \Delta u = 0\,, \tag{1}$$

where $\Delta \equiv \frac{\partial^2}{\partial x^2} + \frac{\partial^2}{\partial y^2}$ is the **Laplace operator** or the **Laplacian**. A function u, which solves (1) at all points (x,y) in some open region D, is said to be **harmonic** on D. (The same term applies to solutions in higher–dimensions, as well.) Recall that solutions of second–order equations are required to have continuous second partial derivatives (e.g., harmonic functions on D are C^2 on D.) As we will demonstrate in Section 6.3, harmonic functions actually turn out to be infinitely differentiable (i.e., C^∞).

Although we live in three dimensions, equation (1) arises in many applications, and the associated boundary–value problems for (1) are usually easier to solve than the three–dimensional analogs. As the following example illustrates, it often happens that the desired solution $u(x,y,z)$ of the three–dimensional Laplace equation $u_{xx} + u_{yy} + u_{zz} = 0$ is already known to be independent of z, in which case we need only to solve (1) for $u(x,y)$.

Example 1. Suppose a uniform electrical charge density is applied to the z–axis. Find the most general form for the resulting harmonic electrostatic potential $u(x,y,z)$.

Solution. Since the physical situation is unchanged by translations in the z–direction, we deduce that $u(x,y,z)$ does not depend on z, say $u(x,y,z) = u(x,y)$. Thus, we seek appropriate solutions of (1). Note also that the physical situation is unchanged by rotations about the z–axis, in which case we deduce that $u(x,y) = f(r)$, where f is a C^2 function and $r = (x^2 + y^2)^{1/2}$ is the distance from (x,y,z) to the z–axis, i.e., the distance from (x,y) to $(0,0)$. By the same computation as in Example 1 of Section 1.2, we find that in terms of $f(r)$, (1) becomes

$$f''(r)(x^2+y^2)\,r^{-2} + f'(r)[2r^{-1} - (x^2+y^2)\,r^{-3}] = f''(r) + \tfrac{1}{r}f'(r) = 0\,. \tag{2}$$

Letting $g(r) = f'(r)$, (2) becomes a first–order linear (or separable) equation for g. We obtain $g(r) = Cr^{-1}$, and so $f(r) = C\log(r) + K$, for arbitrary constants C and K. Thus,

$$u(x,y) = \tfrac{1}{2}C\log(x^2 + y^2) + K \tag{3}$$

is the general form of the potential. For $C > 0$, note that, unlike the potential $-C(x^2 + y^2 + z^2)^{-\frac{1}{2}} + K$ of Example 1 of Section 1.2, the potential (3) increases without bound as $x^2 + y^2 \to \infty$. This means that no finite amount of energy will suffice to transport an oppositely charged particle arbitrarily far from the z–axis. In the gravitational context, escape velocities are infinite in dimension two. □

As we have already observed in Example 2 of Section 1.2, the heat equation $u_t = k(u_{xx} + u_{yy} + u_{zz})$ reduces to Laplace's equation, when u is a steady–state temperature distribution, since $u_t = 0$ if u is time–independent. Moreover, if u does not depend on z, say as in a flat plate which is insulated on its upper and lower surfaces, then a steady–state temperature distribution satisfies (1). The reader may still wonder why the higher–dimensional heat equations are of the form $u_t = k\Delta u$, involving the Laplacian Δu, instead of some other combination of the spatial derivatives, say $u_{xx} + u_{xy} + u_{yy}$. The next example shows how symmetry considerations dictate that the heat equation for two–dimensional heat flow is given by $u_t = k(u_{xx} + u_{yy})$. This argument is more elementary than the usual derivation which is based on the nontrivial higher–dimensional analog of Green's formula.

Example 2. Show that, if the temperature $u(x,y,t)$ in a flat, homogeneous heat conducting plate (without heat sources) obeys a second–order linear PDE, then this PDE must be of the form

$$u_t = k(u_{xx} + u_{yy}), \tag{4}$$

for some constant $k > 0$.

Solution. The general second–order linear PDE for $u = u(x,y,t)$ is

$$q_1 u_{xx} + q_2 u_{yy} + r u_{xy} + r_1 u_{xt} + r_2 u_{yt} + s u_{tt} + a_1 u_x + a_2 u_y + b u_t + cu = f. \tag{5}$$

All of the coefficients and f must be constants, because the plate is homogeneous and the physical circumstances (e.g., heat conductivity constant) do not depend on time. Since (5) must reduce to the one–dimensional heat equation $(ku_{xx} - u_t = 0)$ when u does not depend on y, we must have $r_1 = 0$, $s = 0$, $a_1 = 0$, $c = 0$, and $f = 0$, and we may take $q_1 = k$ and $b = -1$ (Why?). Similarly, when u does not depend on x, (5) must reduce to $-ku_{yy} + u_t = 0$, in which case $r_2 = 0$, $a_2 = 0$ and $q_2 = k$. So far, we have deduced that (5) must be of the form

$$k(u_{xx} + u_{yy}) + r u_{xy} - u_t = 0. \tag{6}$$

Thus, we need only to show that $r = 0$, in the hypothetical heat equation (6). Note that the function $u(x,y,t) = rt + xy$ satisfies (6). If we rotate the xy–coordinate axes, in the clockwise direction, by 90°, the point which had coordinates (x,y) will have coordinates $(y,-x)$. Since the laws of heat conduction are insensitive to the angular position that we choose for the axes, we know that u should still be a solution of (6) under the replacement of x by y and y by $-x$. However, under this replacement, the function u becomes $rt - xy$ which is not a solution of (6) unless $r = 0$. Thus, (4) is the only second–order linear PDE that can govern heat flow in the given circumstances. □

Remark. In the same way, we can deduce that the only possible second–order linear PDE for the amplitude $u(x,y,t)$ of a homogeneous transversely vibrating membrane is

$$u_{tt} = a^2(u_{xx} + u_{yy}), \tag{7}$$

the two–dimensional wave equation with disturbance speed a. A similar argument yields the three–dimensional heat and wave equations. In certain situations, a more accurate heat or wave equation might be nonlinear. For example, the amplitude $u(x,y)$ of a steady membrane, say a soap film spanning a nonplanar loop, is known to obey the minimal surface equation (24) in Example 11 of Section 1.2, instead of Laplace's equation, as (7) suggests. However, (7) is the only possible linear approximation to the "true" PDE (if any) for the amplitude, and (1) is the best linear approximation of the minimal surface equation (cf. Example 11 of Section 1.2). □

Rotational Invariance of the Laplacian Δ

We now demonstrate that at any point $p = (x_0,y_0)$, we have that $u_{xx} + u_{yy}$ is twice the average of the second directional derivatives of u at p along all of the lines in the xy–plane which pass through p. Since this average is clearly unchanged by a rotational change of coordinates, this demonstration implies that the Laplacian $u_{xx} + u_{yy}$ retains its form under a rotation of coordinates. (A direct proof of this is requested in Problem 1.) The line through p, making an angle θ with the positive x–direction $\mathbf{i}$, is parametrized by $\mathbf{r}(t) = (x_0 + t\cos(\theta), y_0 + t\sin(\theta))$. The first derivative of u at p, along this line, is

$$\frac{d}{dt}u(\mathbf{r}(t))\Big|_{t=0} = \frac{d}{dt}u(x_0 + t\cos(\theta), y_0 + t\sin(\theta))\Big|_{t=0} = u_x(x_0, y_0)\cos(\theta) + u_y(x_0, y_0)\sin(\theta)$$

The second derivative of u at p along this line is then

$$\frac{d^2}{dt^2}u(\mathbf{r}(t))\Big|_{t=0} = u_{xx}(p)\cos^2(\theta) + 2u_{xy}(p)\cos(\theta)\sin(\theta) + u_{yy}(p)\sin^2(\theta). \tag{8}$$

As the angle θ varies, the average value of the second derivatives is

$$\frac{1}{2\pi}\int_{-\pi}^{\pi}[u_{xx}(p)\cos^2(\theta) + 2u_{xy}(p)\cos(\theta)\sin(\theta) + u_{yy}(p)\sin^2(\theta)]\,d\theta$$

$$= \frac{1}{2}\,[u_{xx}(p) + u_{yy}(p)]\,. \tag{9}$$

since $\int_{-\pi}^{\pi}\cos^2(\theta)\,d\theta = \int_{-\pi}^{\pi}\sin^2(\theta)\,d\theta = \pi$, and $\int_{-\pi}^{\pi}\sin(\theta)\cos(\theta)\,d\theta = 0$ (i.e., since $\sin(\theta)$ and $\cos(\theta)$ are orthogonal of norm–square π on the interval $[-\pi, \pi]$). Since the average (cf. (9)) of the second derivatives is the same if all of the lines are rotated about p by the same amount, we deduce that $u_{xx} + u_{yy} = u_{\bar{x}\bar{x}} + u_{\bar{y}\bar{y}}$, under a rotational change of coordinates from (x,y) to $(\bar{x},\bar{y})$. A similar argument can be used to prove the invariance of $u_{xx} + u_{yy} + u_{zz}$ under rotations about a point in space. In that case, one forms the average of the second directional derivatives over a sphere of directions represented by unit vectors.

Boundary–Value Problems for Laplace's Equation

Here we will describe two standard boundary–value problems for Laplace's equation. They are known as the Dirichlet problem and the Neumann problem. Let D be a region in the xy–plane and suppose that D is bounded by a finite number of "nice" closed curves (not part of D) all of whose points form a set which we will denote by C (cf. Figure 1).

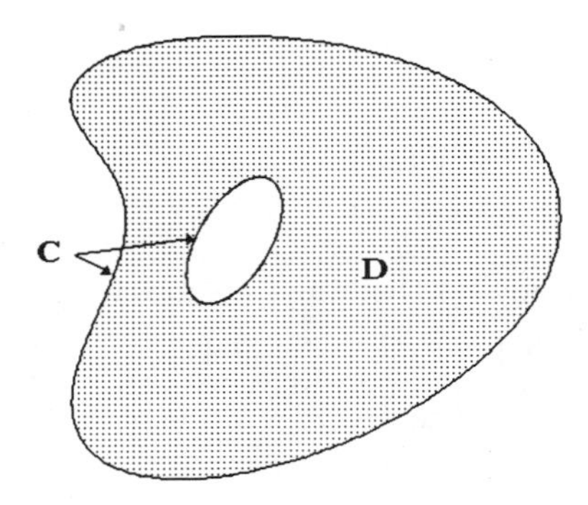

Figure 1

If we adjoin to the open set D all of the boundary points in C, then we obtain a closed set, which we denote by $D \cup C$, the union of D and C. Let f be a given continuous function defined on the boundary C. From a physical perspective, we think of D as a heat–conducting plate and f as a prescribed steady temperature distribution on the border C of the plate. Thus, the Dirichlet problem asks for a steady–state temperature (in the interior of the plate) which is induced by the prescribed temperature on the boundary C. From a mathematical point of view, the Dirichlet problem asks for a solution u of Laplace's equation on D (i.e., a harmonic function on D), such that for any point p on C, u(x,y) can be made arbitrarily close to f(p) by requiring that (x,y) (in D) be sufficiently close to p. In other words, if we define u(p) to be f(p) at each point p on C, we thereby extend the domain of u from D to $D \cup C$, and we require that the extended function u be continuous on $D \cup C$ and satisfy Laplace's equation on D. The Dirichlet problem is concisely written as

Dirichlet problem :

D.E. $u_{xx} + u_{yy} = 0$ on D

(10)

B.C. $u(p) = f(p)$ for all p on C.

Here, it is implicitly required that the function u extend continuously (via the values of f on C) to $D \cup C$. Otherwise, one could claim that by taking u to be 0 throughout D and to be f on C, the problem is solved.

Remark. In previous chapters, we required that solutions of second–order PDEs have C^2 extensions past the boundary of the domain where they are defined. However, insofar as the Dirichlet problem is concerned, *we require only that the function u itself (not its second partials) extend continuously to the boundary.* In this way, we do not need to assume that f is C^2 on C (i.e., f is only assumed to be C^0) in order to solve the Dirichlet problem. Of course, experimentally, there is no way to determine the degree of differentiability of a temperature distribution. Indeed, the question is meaningless. One can always approximate, to within experimental error, a physical temperature distribution by a C^∞ function (e.g., through any finite set of data points, one may draw a smooth curve). However, mathematics goes beyond what is relevant for concrete engineering problems, and to a small extent, so will we. □

Using the same notation as above, the Neumann problem asks for a harmonic function u on D such that, at each point p on C, the directional derivative of u in the outward normal direction $\mathbf{n}(p)$ (perpendicular to C) is equal to the value g(p), where g is some given continuous function defined on C (cf. Figure 2). (There is a difficulty, which we will not address here, concerning what is meant by the outward normal at corner points of C, if any.)

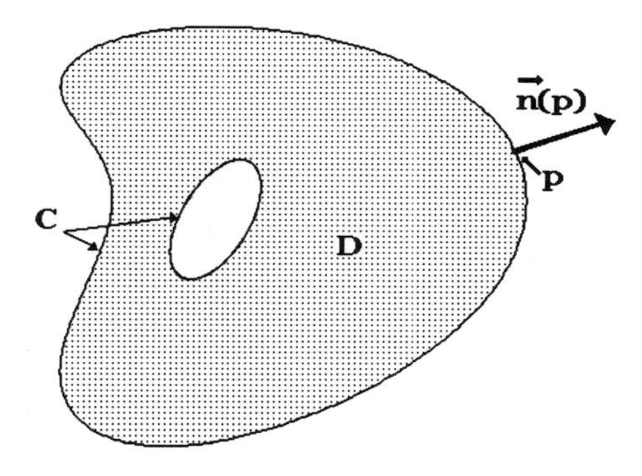

Figure 2

Concisely, we write

Neumann problem :	D.E. $u_{xx} + u_{yy} = 0$ on D	
	B.C. $\nabla u(p) \cdot \mathbf{n}(p) = g(p)$ for all p on C .	(11)

Here we have tacitly assumed that ∇u extends continuously from D to $D \cup C$. In physical terms, the normal component of the temperature gradient is given on the boundary C of a heat–conducting plate (i.e., the rate of heat loss or gain through the boundary points is prescribed), and we are to determine resulting steady–state temperature (if such exists) inside the plate. On physical grounds, in order that such a steady–state temperature distribution should exist, the net heat flux through C must be zero. Thus, we suspect (correctly) that the Neumann problem will have no solution, unless we assume that the average value of the function g on C is zero. This assumption is known as the **compatibility condition**. The analog of this compatibility condition for the one–dimensional heat flow in a rod $(0 \leq x \leq L)$ is $-u_x(0, t) + u_x(L, t) = 0$ (cf. Example 6 of Section 3.3, where there is not steady–state particular solution unless a = b).

There are other important boundary–value problems for Laplace's equation. For instance, one can combine the Dirichlet and Neumann boundary conditions to get a **B.C. of the third kind**, namely $\nabla u(p) \cdot \mathbf{n}(p) + ku(p) = h(p)$, for a given continuous function h defined on C and a constant k (which is usually positive in applications). Also, in the above problems, we can replace Laplace's equation by its inhomogeneous analog, that is, **Poisson's equation**

$$u_{xx} + u_{yy} = q(x,y). \tag{12}$$

In the context of steady–state temperatures, the function $q(x,y)$ is proportional to a time–independent, internal heat source density, perhaps due to radioactivity or microwaves. Particular solutions of Poisson's equation (12) can often be obtained by means of the integral formula

$$u_p(x,y) = \frac{1}{2\pi}\int\int_D \log[(x-\bar{x})^2 + (y-\bar{y})^2]q(\bar{x},\bar{y})\, d\bar{x}\, d\bar{y}, \tag{13}$$

which is a solution of (12) when q is a "reasonably nice" function on D. (cf. Problem 9 of Section 6.4.) Using the particular solution (13), we can reduce the boundary–value problems for Poisson's equation to corresponding problems for Laplace's equation, as the following example illustrates.

Example 3. Reduce the Dirichlet problem for Poisson's equation,

$$\begin{aligned} &\text{D.E.} \quad u_{xx} + u_{yy} = q(x,y) \text{ on } D \\ &\text{B.C.} \quad u(x,y) = g(x,y) \text{ for } (x,y) \text{ on } C, \end{aligned} \tag{14}$$

to a related Dirichlet problem for Laplace's equation.

Solution. Let $u_p(x,y)$ be a particular solution (e.g., (13)) of the D.E., and let v be a solution of the related Dirichlet problem :

$$\begin{aligned} &\text{D.E.} \quad v_{xx} + v_{yy} = 0 \text{ on } D \\ &\text{B.C.} \quad v(x,y) = g(x,y) - u_p(x,y) \text{ for } (x,y) \text{ on } C. \end{aligned} \tag{15}$$

Then, a solution of (14) is $u(x,y) = u_p(x,y) + v(x,y)$. Hence, by solving (15), we obtain a solution of (14). □

The General Solution of Laplace's Equation

Recall that the general solution of the wave equation $u_{tt} - a^2u_{xx} = 0$ was found (in Section 5.2) by factoring the wave operator, $\frac{\partial^2}{\partial t^2} - a^2\frac{\partial^2}{\partial x^2} = (\frac{\partial}{\partial t} - a\frac{\partial}{\partial x})(\frac{\partial}{\partial t} + a\frac{\partial}{\partial x})$, into two first–order operators. Then by adding the general solutions of $u_t - au_x = 0$ and $u_t + au_x = 0$, we obtained the general solution $u(x,t) = f(x + at) + g(x - at)$ of the wave equation. If we replace t by y and set $a = i = \sqrt{-1}$, then the wave equation $u_{tt} - a^2u_{xx} = 0$ becomes Laplace's equation. Thus, the general solution for the wave equation suggests that the general solution of Laplace's equation is (in some sense)

$$u(x,y) = f(x + iy) + g(x - iy). \tag{16}$$

It is rather awkward to define precisely what kind of functions f (or g) of the complex variable $z = x + iy$ (or $\bar{z} = x - iy$) should be allowed in (16). The simplest suitable functions are the power functions $f(z) = z^n$ (or $g(\bar{z}) = \bar{z}^n$) for various integers $n \geq 0$. For example, note that $z^2 = (x^2 - y^2) + i2xy$, which is a complex solution of Laplace's equation (i.e., the real and imaginary parts are harmonic). A complex solution of Laplace's equation is called a **complex harmonic function**. By using z^n for $n = 0, 1, 2, 3, \ldots$, one can generate infinitely many linearly independent complex harmonic functions (cf. Problem 5). More generally, if $f(z)$ is a power series (i.e., an infinite superposition of the functions z^n), say $f(z) = \Sigma_{n=0}^{\infty} a_n z^n$ (for complex constants $a_0, a_1, a_2, \ldots$), which converges for z in some open disk about $z = 0$, then $f(z)$ is a complex harmonic function. Such power series functions are said to be **complex analytic** (about $z = 0$). For example,

$$e^z = f(z) = \sum_{n=0}^{\infty} z^n/n! = \exp(x + iy) = e^x\cos(y) + ie^x\sin(y)$$

is harmonic and complex analytic. Power series functions of the form $\Sigma_{n=0}^{\infty} a_n\bar{z}^n$, which converge in some disk about $z = 0$, are called **conjugate–analytic** (about $\bar{z} = 0$), and they are also complex harmonic functions. With some effort, one can prove that any complex harmonic function defined about $z = 0$ can be written in the form (16) with $f(z)$ analytic and $g(\bar{z})$ conjugate–analytic. Thus, properly interpreted, (16) is a general solution of Laplace's equation. We will see more of the close connection between harmonic functions and complex–variable theory in Section 6.5, where analytic functions are used to find the streamlines of ideal fluid flows around certain obstacles, or equipotential curves about charged conductors in two–dimensional electrostatics.

Summary 6.1

1. Applications of Laplace's Equation : Solutions $u = u(x,y,z)$ of Laplace's equation, $u_{xx} + u_{yy} + u_{zz} = 0$ (or $u_{xx} + u_{yy} = 0$ when $u = u(x,y)$), can be interpreted as steady–state temperature distributions, electrostatic potentials, gravitational potentials, velocity potentials for certain fluid flows (cf. Section 6.5) and more.

2. The rotational invariance of the Laplacian : The two–dimensional Laplacian of u, namely $u_{xx} + u_{yy}$, at any point $p = (x,y)$ is twice the average of the second directional derivatives of u at p, along all lines through p. Because of this geometrical interpretation, the Laplacian of u is independent of translations or rotations of coordinates. The fact that the Laplacian Δ appears in so many applications is related to the rotational invariance of Δ.

3. Boundary–value problems for Laplace's equation : Let D be a region in the xy–plane, which is bounded by a finite number of "nice" curves (not part of D) whose union is C, the boundary of D. There are two standard types of boundary–value problems for Laplace's equation on D, the

Dirichlet problem: D.E. $u_{xx} + u_{yy} = 0$ on D

B.C. $u(p) = f(p)$ for all p on C,

and the

Neumann problem: D.E. $u_{xx} + u_{yy} = 0$ on D

$\nabla u(p) \cdot \mathbf{n}(p) = g(p)$ for all p on C ($\mathbf{n}(p) \equiv$ the normal to C at p).

Here, f and g are given continuous functions on C. The function $u(x,y)$ is required to be a C^2 solution of the D.E. in D. In the Dirichlet problem, the function u on D, along with its prescribed values $f(p)$ on C, is required to be continuous on $D \cup C$. In the Neumann problem, u_x and u_y must extend continuously to C, in such a way that the B.C. holds.

Exercises 6.1

1. Define the new coordinates in the xy–plane by

$$\bar{x} = ax + by + e \quad \text{and} \quad \bar{y} = cx + dy + f,$$

where $a, b, c, d, e,$ and f are constants, with $ad - bc \neq 0$, to ensure that the inverse transformation exists. Let $\bar{u}(\bar{x},\bar{y}) = u(x,y)$ (i.e., $\bar{u}$ is simply u in terms of the new variables).

(a) Show that if u is C^2, then

$$u_{xx} + u_{yy} = (a^2 + b^2)\bar{u}_{\bar{x}\bar{x}} + 2(ac + bd)\bar{u}_{\bar{x}\bar{y}} + (c^2 + d^2)\bar{u}_{\bar{y}\bar{y}} .$$

(b) Suppose that $(\bar{x},\bar{y})$ are the new coordinates obtained by rotating the original axes by some angle θ in the counterclockwise direction. Then verify that $a = \cos(\theta)$, $b = \sin(\theta)$, $c = -\sin(\theta)$, and $d = \cos(\theta)$. Deduce from part (a) that $u_{xx} + u_{yy} = \bar{u}_{\bar{x}\bar{x}} + \bar{u}_{\bar{y}\bar{y}}$ in this case, so that the form of Laplace's equation is retained under rotations, as well as translations of coordinate axes.

2. Let u_1 and u_2 be harmonic functions (i.e., solutions of Laplace's equation).

(a) Show that $c_1u_1 + c_2u_2$ is harmonic for any constants c_1 and c_2 .

(b) If $u(x,y)$ is harmonic, deduce that $x\,u(x,y)$ is harmonic only if $u(x,y) = ay + b$, for some constants a and b.

(c) Give an example of two harmonic functions whose product is not harmonic.

3. Use separation of variables to find all harmonic functions of the form $u(x,y) = X(x)Y(y)$. Remember to consider the cases where the separation constant is positive, negative, and zero.

4. Show that the real and imaginary parts of $(x + iy)^3$ are harmonic. (In Problem 5(d), the reader is asked to demonstrate this fact for arbitrary nonnegative integral powers of $x + iy$.)

5. Note that $\frac{\partial^2}{\partial x^2} + \frac{\partial^2}{\partial y^2} = (\frac{\partial}{\partial x} - i\frac{\partial}{\partial y})(\frac{\partial}{\partial x} + i\frac{\partial}{\partial y})$. The two factors lead us to consider the two complex, first–order equations $f_x + if_y = 0$ and $f_x - if_y = 0$.

(a) Show that the complex–valued function $f(x,y) = u(x,y) + iv(x,y)$ solves $f_x + if_y = 0$, if and only if $u(x,y)$ and $v(x,y)$ satisfy the first–order system $u_x = v_y$ and $u_y = -v_x$. These two equations are known as the **Cauchy–Riemann equations**.

(b) If u and v are C^2 functions which satisfy the Cauchy–Riemann equations, then show that u and v must be harmonic.

(c) Suppose that $U(x,y) + iV(x,y) = (x + iy)(u + iv)$, where u and v are C^2 functions which satisfy the Cauchy–Riemann equations. Show that U and V also solve the Cauchy–Riemann equations.

(d) Deduce from (b) and (c) that the real and imaginary parts of $(x + iy)^n$ are harmonic for all nonnegative integers n.

Remark. If $f = u + iv$, where u and v are C^1 functions which satisfy the Cauchy–Riemann equations (or equivalently, $f_x + if_y = 0$), then it turns out that $f(x + iy) = u(x,y) + iv(x,y)$ is a complex analytic function of the complex variable $z = x + iy$. The function $\bar{f} = u - iv$ satisfies $\bar{f}_x - i\bar{f}_y = 0$, and so $\bar{f}$ is a conjugate–analytic function of z.

6. Descibe, in terms of steady–state state temperature distributions, the difference between the physical significance of the Dirichlet problem and the Neumann problem for Laplace's equation.

7. (a) What is the physical significance of $q(x,y)$ in Poisson's equation $u_{xx} + u_{yy} = q(x,y)$ in the steady–state temperature scenario ?

(b) Give a physical explanation for why the Neumann problem

$$\text{D.E.} \quad u_{xx} + u_{yy} = q(x,y) \quad \text{for } (x,y) \text{ in } D.$$

$$\text{B.C.} \quad \nabla u(p) \cdot \mathbf{n}(p) = g(p) \quad \text{for all } p \text{ on } C,$$

for Poisson's equation, will have no solution, unless we assume the **compatibility condition**

$$\iint_D q(x,y)\, dxdy = \int_C g(p(s))\, ds,$$

where s denotes the arclength parameter along the boundary C of the region D.

6.2 The Dirichlet Problem for a Rectangle

The proof of the following uniqueness theorem will be postponed until Section 6.4 (cf. also Problem 10, for rectangles).

> **Theorem 1 (The Uniqueness Theorem for the Dirichlet Problem). There is at most one solution of the Dirichlet problem for a bounded open set D, with a continuous function F given on the boundary of D.**

Thus, knowing that any solution that we find is the only solution with the given boundary values, we now proceed to solve the Dirichlet problem in the case that D is the rectangle $0 < x < L$, $0 < y < M$. Suppose that F is 0 on three of the sides, and F equals a suitable continuous function $f(x)$ on the remaining side $y = 0$ (cf. Figure 1).

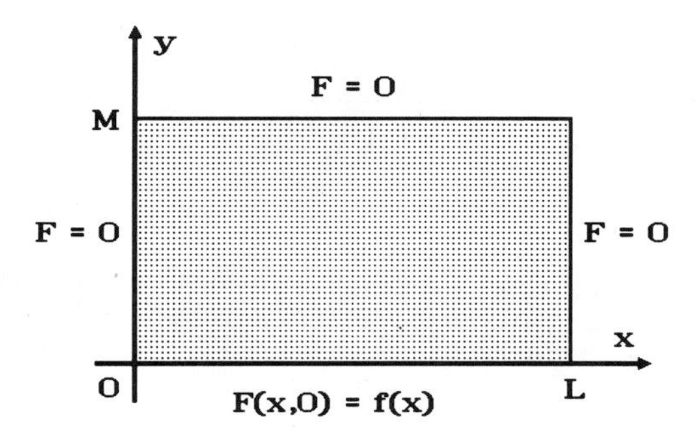

Figure 1

More precisely, we attempt to solve the following Dirichlet problem :

$$
\begin{array}{llll}
\text{D.E.} & u_{xx} + u_{yy} = 0 & 0 < x < L, & 0 < y < M \\
\text{B.C.} & \begin{cases} u(x,0) = f(x), \\ u(0,y) = 0, \end{cases} & \begin{array}{l} u(x,M) = 0 \\ u(L,y) = 0 \end{array} & \begin{array}{l} 0 \le x \le L \\ 0 \le y \le M. \end{array}
\end{array}
\qquad (*)
$$

In order that the B.C. be consistent, we must assume that $f(0) = f(L) = 0$. The product solutions $u(x,y) = X(x)Y(y)$ of the D.E. are easily determined by separation of variables. By the D.E., $X''(x)Y(y) + X(x)Y''(y) = 0$, and hence for come constant c ,

$$X''(x)/X(x) = -Y''(y)/Y(y) = c = \pm b^2 .$$

Consequently, we obtain the following results.

Case 1 $(c = -b^2 < 0)$:

$$u(x,y) = [c_1\cos(bx) + c_2\sin(bx)]\,[d_1e^{by} + d_2e^{-by}]\,. \tag{1}$$

Case 2 $(c = b^2 > 0)$:

$$u(x,y) = [c_1e^{bx} + c_2e^{-bx}]\,[d_1\cos(by) + d_2\sin(by)]\,. \tag{2}$$

Case 3 $(c = 0)$:

$$u(x,y) = (c_1 + c_2x)\,(d_1 + d_2y)\,. \tag{3}$$

One can check that the only product solutions which meet the three homogeneous B.C. of (*) are constant multiples of the following family of Case 1 product solutions :

$$u_n(x,y) = \sin(n\pi x/L)\sinh[n\pi(M - y)/L], \quad n = 1, 2, \ldots\,. \tag{4}$$

See Example 1 below regarding the Case 3 product solutions, and the related Example 7, where the Case 1 product solutions are also expressed in terms of hyperbolic functions. Recall that the hyperbolic sine and cosine are defined by

$$\sinh(z) \equiv \tfrac{1}{2}(e^z - e^{-z}) \quad \text{and} \quad \cosh(z) \equiv \tfrac{1}{2}(e^z + e^{-z}). \tag{5}$$

Since $\sinh(0) = 0$, we know that (4) meets the B.C. $u(x,M) = 0$. By the superposition principle, a more general solution of the D.E. and the homogeneous B.C. of (*) is

$$u(x,y) = \sum_{n=1}^{N} A_n \sin(n\pi x/L)\sinh[n\pi(M - y)/L]\,. \tag{6}$$

Setting $y = 0$, we obtain

$$u(x,0) = \sum_{n=1}^{N} A_n \sinh(n\pi M/L)\sin(n\pi x/L)\,. \tag{7}$$

Thus, when $f(x)$ in (*) is of the form in (7), the unique solution of problem (*) is (6). By Theorem 1 of Section 4.3, if $f(x)$ is continuous and piecewise C^1, with $f(0) = f(L) = 0$, then the partial sums $S_N(x)$ of the Fourier sine series of $f(x)$ converge uniformly to $f(x)$. Hence, in this case, to within experimental error, we may assume that $f(x)$ is of the form (7), where

$$A_n = [\sinh(n\pi M/L)]^{-1} \frac{2}{L} \int_0^L f(x)\sin(n\pi x/L)\, dx , \tag{8}$$

for $n = 1, 2, \ldots, N$. If the B.C. of $(*)$ are changed to

$$\begin{array}{lll} u(x,0) = 0, & u(x,M) = g(x) & 0 \le x \le L \\ u(0,y) = 0, & u(L,y) = 0 & 0 \le y \le M , \end{array} \tag{9}$$

then in place of the product solutions (4), we have

$$u_n(x,y) = \sin(n\pi x/L)\sinh(n\pi y/L) . \tag{10}$$

Consequently a more general solution of the D.E. and the homogeneous B.C. of (9) is

$$u(x,y) = \sum_{n=1}^{N} B_n \sin(n\pi x/L)\sinh(n\pi y/L) , \tag{11}$$

and

$$u(x,M) = \sum_{n=1}^{N} B_n \sinh(n\pi M/L)\sin(n\pi x/L) . \tag{12}$$

Assuming that $g(x)$ is adequately represented by the N–th partial sum of its Fourier sine series, we get that (11) is the harmonic function which meets the B.C. (9), where

$$B_n = [\sinh(n\pi M/L)]^{-1} \frac{2}{L} \int_0^L g(x)\sin(n\pi x/L)\, dx . \tag{13}$$

By adding solutions (6) and (11), we obtain the solution of the Dirichlet problem with B.C.

$$\begin{array}{lll} u(x,0) = f(x), & u(x,M) = g(x) & 0 \le x \le L \\ u(0,y) = 0, & u(L,y) = 0 & 0 \le y \le M, \end{array}$$

where we assume that $f(x)$ and $g(x)$ are finite Fourier sine series with at most N terms. The solution of the Dirichlet problem with B.C. of the form

$$\begin{array}{lll} u(x,0) = 0, & u(x,M) = 0 & 0 \le x \le L \\ u(0,y) = h(y), & u(L,y) = k(y) & 0 \le y \le M, \end{array}$$

can be obtained by simply switching the roles of x and y (and the limits L and M) and replacing f by h, and g by k, in the formulas (8) and (13). By the superposition principle we obtain the following result.

Theorem 2. Let a_n, b_n, c_n, and d_n be the Fourier sine coefficients (assumed to vanish for $n > N$) of $f(x)$, $g(x)$, $h(y)$, and $k(y)$. Then the solution of the Dirichlet problem

$$\begin{array}{lllll} \text{D.E.} & u_{xx} + u_{yy} = 0 & & 0 < x < L, & 0 < y < M \\ \text{B.C.} & \begin{cases} u(x,0) = f(x), \\ u(0,y) = h(y), \end{cases} & \begin{array}{l} u(x,M) = g(x) \\ u(L,y) = k(y) \end{array} & \begin{array}{l} 0 \le x \le L \\ 0 \le y \le M, \end{array} & \end{array} \tag{14}$$

is

$$\begin{aligned} u(x,y) = \sum_{n=1}^{N} \Big[& A_n \sin(n\pi x/L) \sinh[n\pi(M-y)/L] \\ & + B_n \sin(n\pi x/L) \sinh(n\pi y/L) \\ & + C_n \sin(n\pi y/M) \sinh[n\pi(L-x)/M] \\ & + D_n \sin(n\pi y/M) \sinh(n\pi x/M) \Big], \end{aligned} \tag{15}$$

where

$$A_n = a_n/\sinh(n\pi M/L) \qquad B_n = b_n/\sinh(n\pi M/L)$$

$$C_n = c_n/\sinh(n\pi L/M) \qquad D_n = d_n/\sinh(n\pi L/M).$$

Note that the solution (15) vanishes at the corners of the rectangle. This is because we have implicitly assumed that the functions f, g, h, and k vanish at the endpoints of their intervals of definition. In general, the continuous boundary data for the Dirichlet problem will not vanish at the corners. However, we can easily handle this case by subtracting, from the unknown function u, a particular solution U which has the prescribed values at the corners. The next example shows how to find U. Letting $v = u - U$, we then solve (as above) the related Dirichlet problem for v which has boundary data that vanishes at the corners (cf. Problem 4 for a specific example).

Example 1. Find the unique harmonic function of the form

$$U(x,y) = a + bx + cy + dxy, \tag{16}$$

where a, b, c and d are constants, such that $U(0,0) = A$, $U(L,0) = B$, $U(0,M) = C$ and $U(L,M) = D$, for given constants A, B, C, and D. Deduce from this result that any Case 3 (cf. (3)) product solution that satisfies the two B.C. $u(0,y) = 0$ and $u(L,y) = 0$, for $0 \le y \le L$, must be identically zero.

Solution. Note that $U(0,0) = A$ implies $a = A$. Then $U(L,0) = B$ implies $a + bL = B$ or

$b = (B - a)/L$. Similarly, $U(0,M) = C$ implies $c = (C - A)/M$. Also, $U(L,M) = D$ implies $d = (D - a - bL - cM)/(LM) = (D + A - B - C)/(LM)$. Thus, we have found the unique values for the constants a, b, c and d in terms of the given constants A, B, C and D. Observe that any Case 3 product solution u, when multiplied out, is of the form (16). The B.C. $u(0,y) = 0$ and $u(L,y) = 0$ $(0 \le y \le M)$ imply that u vanishes at the corners of the rectangle (i.e., $A = B = C = D = 0$). Hence, $u(x,y) \equiv 0$ by what we have just shown. □

Example 2. Solve the problem

$$\text{D.E.} \quad u_{xx} + u_{yy} = 0 \qquad 0 < x < \pi,\ 0 < y < \pi$$

$$\text{B.C.} \quad \begin{cases} u(x,0) = 0, & u(x,\pi) = 5\sin(2x) - 7\sin(8x) \qquad 0 \le x \le \pi \\ u(0,y) = \sin(y), & u(\pi,y) = 0 \qquad 0 \le y \le \pi. \end{cases}$$

Solution. Method 1 (By inspection). Note that the three product solutions of the D.E. which are relevant to the B.C. are $\sin(2x)\sinh(2y)$, $\sin(8x)\sinh(8y)$ and $\sin(y)\sinh(\pi - x)$. By forming a superposition, we have

$$u(x,y) = E \sin(2x)\sinh(2y) + F \sin(8x)\sinh(8y) + G \sin(y)\sinh(\pi - x).$$

The inhomogeneous B.C. dictate that $E = 5/\sinh(2\pi)$, $F = -7/\sinh(8\pi)$, and $G = 1/\sinh(\pi)$. Also, the homogeneous B.C. are met. In this method, one should carefully check that all of the B.C. are met.

Method 2 (Deriving the solution via separation of variables). In spite of the fact that separation of variables was carried out for Laplace's equation in general, some instructors will insist that this procedure be repeated in every specific problem. One then determines the product solution which meet each set of three homogeneous B.C. (There are four possibilities, depending on which B.C. is taken to be inhomogeneous. Only two of the B.C. in the above problem are inhomogeneous, whence only two possibilities need to be considered here.) By forming a superposition of the resulting product solutions, we find the solution as in method 1. This was essentially the method employed in obtaining the solution (15) of the general problem (14). Although this method 2 is cumbersome, it is one way of ensuring that the student understands how formula (15) was derived.

Method 3 (Using the derived formula (15)). Note that $b_2 = 5$, $b_8 = -7$, $c_1 = 1$ and all other Fourier sine series coefficients for the boundary functions are zero. Remember to divide by the correct factors to obtain B_2, B_8 and C_1 (i.e., E, F and G in Method 1). □

The Maximum/Minimum Principle for harmonic functions on rectangles.

In Section 6.4, we prove the Maximum/Minimum Principle for harmonic functions. In the special case of rectangles, this principle is as follows.

> The maximum (and minimum) of a function $u(x,y)$, which is continuous on a closed rectangle $D:\ 0 \le x \le L,\ 0 \le y \le M$, and which is harmonic in the interior D, must achieve its maximum and minimum on the boundary (i.e., on one of the four edges). Moreover, if the maximum or minimum is also achieved at a point in the interior, then u must be constant on D.

Actually, the principle holds for any reasonably nice, bounded (i.e., of finite extent) region with boundary. Uniqueness of solutions of the Dirichlet problem for such regions follows at once from the Maximum/Minimum Principle, since the maximum and minimum of the difference of two solutions occurs on the boundary, where the difference is zero. The next example illustrates the Maximum/Minimum Principle.

Example 3. Find the points where the harmonic function $u(x,y) = x^3 - 3xy^2 + xy - x$ achieves its maximum and minimum values in the square region $0 \leq x \leq 1,\ 0 \leq y \leq 1$.

Solution. To find the maximum and minimum for u, one would usually compute u_x and u_y, and solve the equations $u_x = 0$ and $u_y = 0$ simultaneously for x and y, in order to find the critical points in the interior of the square. Then one would have to search the boundary for competing extreme values. However, since u is harmonic ($u_{xx} = 6x$ and $u_{yy} = -6x$), the Maximum/Minimum Principle implies that we need only to consider the boundary, since no interior critical point can be a maximum or minimum, unless u is constant. On the boundary,

$$\begin{aligned} &u(x,0) = x^3 - x \qquad && u(x,1) = x^3 - 3x \qquad && 0 \leq x \leq 1 \\ &u(0,y) = 0 && u(1,y) = -3y^2 + y && 0 \leq y \leq 1 \end{aligned} \tag{17}$$

One easily finds the maxima and minima for each of the functions (17). The largest maximum occurs when $x = 1$ and $y = 1/6$, with $u(1,1/6) = 1/12$. The smallest minimum is at $(1,1)$, with $u(1,1) = -2$. The Maximum/Minimum Principle then yields $-2 \leq u(x,y) \leq 1/12$ for any point (x,y) in the square region. □

Remark (Formal Solutions). Recall that in cases where the boundary (or initial) data of a boundary–value problem cannot be expressed in terms of a finite Fourier series of the correct form, then the solution procedure leads us to an infinite series expression which is called a **formal solution** of the problem (cf. Section 4.3 for a discussion of formal solutions). Formal solutions may or may not actually converge to a solution of the problem. Nevertheless, by truncating the formal solution at a finite number of terms, one can often obtain a genuine solution of the D.E. which meets the B.C. by an experimentally allowable error. This is all that is needed in engineering applications. In Example 4, we find the formal solution of a Dirichlet problem and in Example 5, we analyze the truncation error using the Maximum/Minimum Principle. □

Example 4. Find the formal solution of the problem

$$\begin{aligned} &\text{D.E.} \quad u_{xx} + u_{yy} = 0 && 0 < x < \pi,\ 0 < y < \pi \\ &\text{B.C.} \quad \begin{cases} u(x,0) = x^3(x - \pi), & u(x,\pi) = 0 \quad 0 \leq x \leq \pi \\ u(0,y) = 0, & u(\pi,y) = 0 \quad 0 \leq y \leq \pi. \end{cases} \end{aligned} \tag{18}$$

Solution. We have found (cf. equation (4)) that the product solutions of the D.E. which satisfy the homogeneous B.C. are the multiples of $u_n(x,y) = \sin(nx)\sinh[n(\pi - y)]$, $n = 1, 2, 3, \ldots$.

Since $x^3(x - \pi)$ is not a finite sine series, we seek a formal solution of the form

$u(x,y) = \sum_{n=1}^{\infty} A_n \sin(nx)\sinh[n(\pi - y)]$, where the A_n are determined by the inhomogeneous B.C. $x^3(x - \pi) = u(x,0) = \sum_{n=1}^{\infty} A_n \sin(nx)\sinh(n\pi)$. In other words, $A_n \sinh(n\pi)$ is the Fourier sine coefficient $a_n = \frac{2}{\pi}\int_0^{\pi} x^3(x - \pi)\sin(nx)\, dx$. Applying Green's formula twice (as in Example 6 of Section 4.1, except that here we use the interval $[0,\pi]$), we obtain

$$a_n = 12\pi\left[(-1)^n n^{-3} + 4\pi^{-2}[(-1)^{n+1} + 1]n^{-5}\right], \quad n \geq 1. \tag{19}$$

Alternatively, a tedious calculation involving four integrations by parts, yields the same result. Hence, the required formal solution is

$$u(x,y) = \sum_{n=1}^{\infty} a_n \sin(nx) \frac{\sinh[n(\pi - y)]}{\sinh(n\pi)}, \tag{20}$$

where a_n is given by (19). □

Remark. Applying theorems about the validity of termwise differentiation, one can prove that the formal solution (20) is in fact the exact solution of (18). In a practical situation, the data for the B.C. can only be specified to within some experimental error, say $\epsilon > 0$. Thus, in practice, one need only determine a solution which meets the B.C. within ϵ. Even if we were to prove that the formal solution is exact, we would still have to figure out where to truncate the series in order to evaluate the solution (to within the error ϵ) at various points (x,y) in the square. Thus, for concrete applications, not only is a proof of the validity of termwise differentiation unnecessary, but also it is not sufficient. The following example illustrates an error analysis. □

Example 5. Find a value for N, such that the truncation $u_N(x,y)$, at the N–th term, of the formal solution $u(x,y)$ in (20) meets the B.C. of (18) to within an error of ϵ. Use the Maximum/Minimum Principle to analyze the difference between two truncations $u_N(x,y)$ and $u_M(x,y)$ in the square region.

Solution. Note that $u_N(x,0)$ is the N–th partial sum of the Fourier sine series of $f(x) = x^3(x - \pi)$. Since $f(x)$ is C^1 with $f(0) = f(\pi) = 0$, we know that FSS $f(x)$ converges uniformly to $f(x)$ in the interval $[0,\pi]$ by Theorem 1 of Section 4.3. More precisely, for $0 \leq x \leq \pi$, we have (where a_n is given by (19))

$$|x^3(x - \pi) - u_N(x,0)| = \left| \sum_{n=N+1}^{\infty} a_n \sin(nx) \right| \leq \sum_{n=N+1}^{\infty} |a_n|$$

$$\leq 12\pi \sum_{n=N+1}^{\infty} n^{-3}[1 + 8/(\pi^2 n^2)] \leq 24\pi \sum_{n=N+1}^{\infty} n^{-3},$$

where we have used the fact that $1 + 8/(\pi^2 n^2) < 2$, for $n \geq 1$. Since $\sum_{n=N+1}^{\infty} n^{-3} \leq \int_N^{\infty} x^{-3}\,dx = \frac{1}{2}N^{-2}$, we get $|x^3(x-\pi) - u_N(x,0)| \leq 12\pi\, N^{-2}$. Thus, it suffices to choose $N \geq (12\pi/\epsilon)^{1/2}$, in order that $u_N(x,y)$ meet the B.C. of (18) to within an error of ϵ. One would also like to know that if two solutions differ by at most ϵ on the boundary, then they differ by at most ϵ inside the square. Since the difference of two harmonic functions is harmonic, this is a direct consequence of the Maximum/Minimum Principle. For two truncations $u_N(x,y)$ and $u_M(x,y)$ of the formal solution (20), we have (for $M > N$) the following estimate of the difference on the boundary where $y = 0$ and $0 \leq x \leq \pi$:

$$|u_M(x,0) - u_N(x,0)| = \Big|\sum_{n=N+1}^{M} a_n \sin(nx)\Big| \leq \sum_{n=N+1}^{M} |a_n|$$

$$\leq 24\pi \sum_{n=N+1}^{M} n^{-3} \leq 24\pi \int_N^M x^{-3}\,dx = 12\pi\,(N^{-2} - M^{-2}).$$

On the other three edges of the boundary, both $u_M(x,y)$ and $u_N(x,y)$ are zero. Thus, by the Maximum/Minimum Principle, we have

$$|u_M(x,y) - u_N(x,y)| < 12\pi\,(N^{-2} - M^{-2}), \text{ for } 0 \leq x, y \leq \pi. \tag{21}$$

Inequality (21) can be improved by a direct estimation, without using the Maximum/Minimum Principle. Indeed, when $y > 0$, we have $2\sinh[n(\pi - y)] = \exp[n(\pi - y)] - \exp[n(y - \pi)] = e^{-ny}(e^{n\pi} - e^{2ny}e^{-n\pi}) \leq e^{-ny}(e^{n\pi} - e^{-n\pi}) = 2\,e^{-ny}\sinh(n\pi)$, whence $\sinh[n(\pi - y)]/\sinh(n\pi) \leq e^{-ny}$. Using this fact, we obtain

$$\begin{aligned} |u_M(x,y) - u_N(x,y)| &\leq \sum_{n=N+1}^{M} |a_n| \sinh[n(\pi - y)]/\sinh(n\pi) \\ &\leq \sum_{n=N+1}^{M} |a_n|\, e^{-ny} \leq e^{-(N+1)y} \sum_{n=N+1}^{M} |a_n| \\ &< 12\pi\,(N^{-2} - M^{-2})\, e^{-(N+1)y}. \end{aligned} \tag{22}$$

Thus, for $y > 0$, the estimate (22) improves (21) dramatically, since $e^{-(N+1)y}$ decreases rapidly as N increases. For example, at the center $p = (\pi/2, \pi/2)$, choosing $N = 5$ and $M = 10$, we have, by (21), $|u_{10}(p) - u_5(p)| \leq 12\pi\,(1/25 - 1/100) \approx 1.13$. However, (22) improves this bound by a factor of $e^{-3\pi} \approx 8 \times 10^{-5}$. Explicit evaluation, using a calculator, yields $u_5(p) = -1.411748...$ and $u_{10}(p) = -1.411747...$ (i.e., the difference is less than 10^{-5}). □

Example 6 (The compatibility condition for a Neumann problem). The Neumann problem for a rectangle is

$$\text{D.E.} \quad u_{xx} + u_{yy} = 0 \qquad 0 < x < L,\ 0 < y < M$$

$$\text{B.C.} \quad \begin{cases} u_y(x,0) = f(x), & u_y(x,M) = g(x) \quad 0 \le x \le L \\ u_x(0,y) = h(y), & u_x(L,y) = k(y) \quad 0 \le y \le M\,. \end{cases} \tag{23}$$

Show that problem (23) has no solution, unless the following **compatibility condition** holds

$$\int_0^L g(x)\,dx - \int_0^L f(x)\,dx + \int_0^M k(y)\,dy - \int_0^M h(y)\,dy\,, \tag{24}$$

i.e., the integral of the outward unit normal component of ∇u around the boundary is 0.

Solution. If $u(x,y)$ is a solution of (23), then

$$0 = \int_0^M \int_0^L (u_{xx} + u_{yy})\,dx\,dy = \int_0^M \int_0^L u_{xx}\,dx\,dy + \int_0^L \int_0^M u_{yy}\,dy\,dx$$

$$= \int_0^M \left[u_x(L,y) - u_x(0,y)\right] dy + \int_0^L \left[u_y(x,M) - u_y(x,0)\right] dx$$

$$= \int_0^M k(y)\,dy - \int_0^M h(y)\,dy + \int_0^L g(x)\,dx - \int_0^L f(x)\,dx\,,$$

where we have used the Fundamental Theorem of Calculus, and the fact that we may change the order of integration (cf. Appendix A.2). □

Remarks. (1) Alternatively, some readers may recognize that the compatibility condition is an immediate consequence of the following special case of Green's theorem,

$$\int_C \nabla u \cdot \mathbf{n}\,ds = \int_C u_x\,dy - u_y\,dx = \iint_R (u_{xx} + u_{yy})\,dxdy\,,$$

i.e., the flux of the gradient of u through the boundary is the integral of Δu in the interior. This result holds for domains R of finite extent which are bounded by a finite number of regular, simple closed curves C. Hence the compatibility condition is necessary in order to solve the Neumann problem for such domains.

(2) If a solution of (23) exists, it is not unique, since one can always add a constant to a solution to obtain another solution. However, in Problem 9, we lead the reader through the demonstration that any two solutions of (23) must differ by a constant. Also, while it is implicitly required that u_x and u_y be continuous on the closed rectangle, we do not demand that the second partials of u extend continuously to the closed rectangle. □

Example 7. Find the product solutions $X(x)Y(y)$ of the D.E. and homogeneous B.C. of the problem.

$$\text{D.E.} \quad u_{xx} + u_{yy} = 0 \qquad 0 < x < L\,, \quad 0 < y < M$$

$$\text{B.C.} \quad \begin{cases} u_y(x,0) = f(x), & u_y(x,M) = 0 \quad 0 \le x \le L \\ u_x(0,y) = 0, & u_x(L,y) = 0 \quad 0 \le y \le M. \end{cases} \tag{25}$$

Solution. Recall that any product solution of the D.E. must be of one of the forms (1), (2) or (3). We first show that there is no nonzero Case 2 product solution which meets the last two B.C. . In Case 2, we have $u_x(0,y) = b(c_1 - c_2)(d_1\cos(by) + d_2\sin(by))$, whence the B.C. $u_x(0,y) = 0$ yields $c_1 = c_2$ or $d_1 = d_2 = 0$. Then $0 = u_x(L,y)$ $= bc_1(e^{bL} - e^{-bL})\cdot(d_1\cos(by) + d_2\sin(by))$ implies $c_1 = 0$ or $d_1 = d_2 = 0$. Thus, in Case 2, the last two B.C. force the product solution to vanish. In Case 1, $0 = u_x(0,y)$ yields $0 = bc_2(d_1e^{by} + d_2e^{-by})$, whence $c_2 = 0$ or $d_1 = d_2 = 0$. Also, $0 = u_x(L,y)$ $= -bc_1\sin(bL)(d_1e^{by} + d_2e^{-by})$ implies that, in order to avoid a trivial solution, we must have $\sin(bL) = 0$ or $b = n\pi/L$, for integers $n \ge 1$. Also, $u_y(x,M) = 0$ implies $d_1e^{bM} - d_2e^{-bM} =$ 0. Thus, $d_1e^{by} + d_2e^{-by} = e^{-bM}(d_1e^{bM}e^{by} + d_2e^{b(M-y)}) = e^{-bM}(d_2e^{-bM}e^{by} + d_2e^{b(M-y)})$ $= 2\, e^{-bM}d_2\cosh[b(M - y)]$. Thus the Case 2 product solutions which meet the three homogeneous B.C. are the constant multiples of

$$u_n(x,y) = \cos(n\pi x/L)\cosh[n\pi(M - y)/L]\,, \quad n = 1, 2, \ldots\,. \tag{26}$$

The only Case 3 product solutions $(c_1 + c_2x)(d_1 + d_2y)$ which meet the homogeneous B.C. of (25) are constant (cf. Problem 6), and we can include these in (26) by allowing $n = 0$. □

Example 8. Solve problem (25) when $f(x) = \sum_{n=1}^{N} a_n\cos(n\pi x/L)$, a finite cosine series without a constant term $\frac{1}{2}a_0$. Why must the constant term be zero in order that a solution should exist?

Solution. Let $u(x,y) = \sum_{n=1}^{N} A_nu_n(x,y)$, where u_n is given by (26). Then, by the superposition principle, u satisfies the D.E. and the homogeneous B.C. of (25). Note that

$$u_y(x,0) = \sum_{n=1}^{N} -A_n(n\pi/L)\, \sinh(n\pi M/L)\cos(n\pi x/L)\,. \tag{27}$$

This is $f(x)$, if

$$A_n = -a_n L\,[n\pi \sinh(n\pi M/L)]^{-1}, \quad n = 1, 2, \ldots, N. \tag{28}$$

Observe that since (27) has no constant term, (27) can agree with $f(x)$, only if $f(x)$ has no constant term. Also, by Example 6, $\frac{1}{2}a_0 = \frac{1}{L}\int_0^L f(x)\,dx$ must vanish in order for a solution to exist. For each constant A_0, we obtain a solution

$$A_0 + \sum_{n=1}^{N} A_n \cos(n\pi x/L)\cosh[n\pi(M-y)/L], \tag{29}$$

where the A_n are expressed in terms of the given a_n by (28). By Problem 9, all solutions of (25) are of the form (29), when $f(x) = \sum_{n=1}^{N} a_n\cos(n\pi x/L)$. □

Summary 6.2

1. The Dirichlet problem for the rectangle : Suppose that a_n, b_n, c_n and d_n are the Fourier sine coefficients (which are assumed to vanish for $n > N$) of $f(x)$, $g(x)$, $h(y)$, and $k(y)$, respectively. Then by the superposition principle, the solution of the Dirichlet problem

$$\text{D.E.} \quad u_{xx} + u_{yy} = 0 \qquad 0 < x < L, \quad 0 < y < M$$

$$\text{B.C.} \quad \begin{cases} u(x,0) = f(x), & u(x,M) = g(x) \qquad 0 \le x \le L \\ u(0,y) = h(y), & u(L,y) = k(y) \qquad 0 \le y \le M\,, \end{cases} \tag{S1}$$

is

$$\begin{aligned} u(x,y) = \sum_{n=1}^{N} \Big[& A_n \sin(n\pi x/L) \sinh[n\pi(M-y)/L] \\ & + B_n \sin(n\pi x/L) \sinh(n\pi y/L) \\ & + C_n \sin(n\pi y/M) \sinh[n\pi(L-x)/M] \\ & + D_n \sin(n\pi y/M) \sinh(n\pi x/M) \Big]\,, \end{aligned} \tag{S2}$$

where

$$A_n = a_n/\sinh(n\pi M/L) \qquad B_n = b_n/\sinh(n\pi M/L)$$
$$C_n = c_n/\sinh(n\pi L/M) \qquad D_n = d_n/\sinh(n\pi L/M).$$

Note that the above functions $f(x)$, $g(x)$, $h(y)$, $k(y)$ all vanish at the endpoints of the intervals on which they are defined (i.e., the function which they define on the boundary of the rectangle is zero at the corners). In the event that the given function on the boundary does not vanish at the corners, one can subtract, from $u(x,y)$, a harmonic function with the given values at the corners (cf. Example 1), to obtain a related problem with a boundary function which vanishes at the corners.

2. The Maximum/Minimum Principle and the approximation of boundary functions : For rectangles, the Maximum/Minimum Principle states that the maximum (and minimum) of a function $u(x,y)$, which is continuous on a closed rectangle $D : 0 \le x \le L$, $0 \le y \le M$ and which is harmonic in the interior D, must achieve its maximum and minimum values on the boundary (i.e., on one of the four edges). As a consequence, if one uniformly approximates (say to within an error ϵ) the given boundary functions, by functions of the form in (S1), then the solution of the Dirichlet problem for the approximate boundary functions will be within ϵ of the exact solution (if it exists) throughout the interior of the rectangle. In other words, small changes in the boundary functions induce small changes in the solutions (if they exist).

3. The Neumann problem : The Neumann problem for a rectangle is

$$\text{D.E.} \quad u_{xx} + u_{yy} = 0 \qquad 0 < x < L, \quad 0 < y < M$$

$$\text{B.C.} \quad \begin{cases} u_y(x,0) = f(x), & u_y(x,M) = g(x) \quad 0 \le x \le L \\ u_x(0,y) = h(y), & u_x(L,y) = k(y) \quad 0 \le y \le M . \end{cases} \tag{S3}$$

The problem (S3) has no solution, unless the following compatibility condition holds

$$\int_0^L g(x)\,dx - \int_0^L f(x)\,dx + \int_0^M k(y)\,dy - \int_0^M h(y)\,dy = 0 \ , \tag{S4}$$

i.e., the integral of the outward unit normal component of ∇u (around the boundary) is 0. Neumann problems are considered in Examples 7 and 8.

Exercises 6.2

1. Verify that if the Case 2 product solution of $u_{xx} + u_{yy} = 0$ (cf. (2)), namely

$$u(x,y) = (c_1e^{bx} + c_2e^{-bx})(d_1\cos(by) + d_2\sin(by)) \qquad (b > 0),$$

satisfies the B.C. $u(0,y) = 0$ and $u(L,y) = 0$, then $u(x,y) \equiv 0$.

2. Solve the problem

$$\text{D.E.} \quad u_{xx} + u_{yy} = 0 \qquad 0 < x < L, \quad 0 < y < M$$

$$\text{B.C.} \quad \begin{cases} u(x,0) = f(x), & u(x,M) = g(x) \quad 0 \le x \le L \\ u(0,y) = h(y), & u(L,y) = k(y) \quad 0 \le y \le M , \end{cases}$$

(a) when $f(x) = 9\sin(8\pi x/L)$, $g(x) = 0$ and $h(y) = k(y) = 0$
(b) when $f(x) = 0$, $g(x) = \sin(\pi x/L)$ and $h(y) = k(y) = 0$
(c) when $f(x) = 9\sin(8\pi x/L)$, $g(x) = \sin(\pi x/L)$ and $h(y) = k(y) = 0$
(d) when $f(x) = g(x) = 0$ and $h(y) = 9\sin(8\pi y/M)$ and $k(y) = \sin(\pi y/M)$.

3. Solve the problem

$$\text{D.E.} \quad u_{xx} + u_{yy} = 0 \qquad 0 < x < \pi, \quad 0 < y < \pi$$

$$\text{B.C.} \quad \begin{cases} u(x,0) = \sin(x), & u(x,\pi) = \sin(x) \quad 0 \le x \le \pi \\ u(0,y) = \sin(y), & u(\pi,y) = \sin(y) \quad 0 \le y \le \pi . \end{cases}$$

4. (a) Find a function of the form $U(x,y) = a + bx + cy + dxy$, such that $U(0,0) = 0$, $U(1,0) = 1$, $U(0,1) = -1$, and $U(1,1) = 2$.

(b) Use the answer in (a) to solve the problem

$$\text{D.E.} \quad u_{xx} + u_{yy} = 0 \qquad 0 < x < 1, \quad 0 < y < 1$$

$$\text{B.C.} \quad \begin{cases} u(x,0) = 3\sin(\pi x) + x, & u(x,1) = 3x - 1 \quad 0 \le x \le 1 \\ u(0,y) = \sin(2\pi y) - y, & u(1,y) = y + 1 \quad 0 \le y \le 1 \end{cases}$$

Hint. Solve the related problem for $v(x,y) = u(x,y) - U(x,y)$.

5. (a) Show that the formal solution of the problem

$$\text{D.E.} \quad u_{xx} + u_{yy} = 0 \qquad 0 < x < \pi, \quad 0 < y < \pi$$

$$\text{B.C.} \quad \begin{cases} u(x,0) = 0, & u(x,\pi) = x(\pi - x) \quad 0 \le x \le \pi \\ u(0,y) = 0, & u(\pi,y) = 0 \quad 0 \le y \le \pi \end{cases}$$

is $$u(x,y) = \frac{\pi}{8}\sum_{k=0}^{\infty} (2k+1)^{-3}\sin[(2k+1)x]\,\frac{\sinh[(2k+1)y]}{\sinh[(2k+1)\pi]}.$$

(b) Estimate the number of terms of the formal solution which are needed to approximate the B.C. to within an error of .01.

(c) It turns out that the formal solution in part (a) is actually the exact solution of the Dirichlet problem. Will the truncation, in part (b), of the formal solution be within .01 of the exact solution throughout the interior of the square? Explain why.

6. (a) Verify that if the Case 3 (cf.(3)) product solution $u(x,y) = (c_1 + c_2x)(d_1 + d_2y)$ meets the B.C. $u_x(0,y) = 0$ and $u_y(x,M) = 0$, then $u(x,y) = c_1d_1$.

(b) Show that a (harmonic) function of the form $U(x,y) = a + bx + cy + dxy$ is a Case 3 product solution, if and only if $ad - bc = 0$.

7. (a) Find *a* solution of the Neumann problem

$$\text{D.E.} \quad u_{xx} + u_{yy} = 0 \qquad 0 < x < \pi, \quad 0 < y < \pi$$

$$\text{B.C.} \quad \begin{cases} u_y(x,0) = \cos(x) - 2\cos^2(x) + 1, & u_y(x,\pi) = 0 \quad 0 \le x \le \pi \\ u_x(0,y) = 0, & u_y(\pi,y) = 0 \quad 0 \le y \le \pi. \end{cases}$$

(b) By adding a constant, find a solution such that $u(0,0) = 0$.

8. (a) Find a formal solution of the Neumann problem

$$\text{D.E.} \quad u_{xx} + u_{yy} = 0 \qquad 0 < x < L, \quad 0 < y < M$$

$$\text{B.C.} \quad \begin{cases} u_y(x,0) = 0, & u_y(x,M) = g(x) \quad 0 \le x \le L \\ u_x(0,y) = 0, & u_x(L,y) = 0 \quad 0 \le y \le M\,, \end{cases}$$

given that $g(x)$ is continuous and $\int_0^L g(x)\,dx = 0$.

(b) Why is the assumption $\int_0^L g(x)\,dx = 0$ needed in (a) ?

9. In this problem we demonstrate that, given two solutions $u_1(x,y)$ and $u_2(x,y)$ of the Neumann problem (23), the difference $u(x,y) = u_1(x,y) - u_2(x,y)$ is constant.

(a) Explain why the difference u solves (23) with homogeneous B.C. $(f = g = h = k = 0)$.
(b) Show that $(uu_x)_x + (uu_y)_y = (u_x)^2 + (u_y)^2$.
(c) Integrate the identity in (b) over the rectangle $(0 \le x \le L, 0 \le y \le M)$ to obtain

$$\int_0^M \int_0^L \left[(u_x)^2 + (u_y)^2\right] dxdy = \int_0^M [u(x,y)\, u_x(x,y)]\Big|_{x=0}^{x=L} dy + \int_0^L [u(x,y)\, u_y(x,y)]\Big|_{y=0}^{y=M} dx\,. \quad (*)$$

(d) Use (a) and (*) to deduce that the left side of (*) is 0. Conclude that $u_x = u_y \equiv 0$, and so u must be identically constant.

10. Using the same procedure as in Problem 9, prove that the Dirichlet problem for the rectangle has at most one solution, such that u, u_x and u_y extend continuously to the boundary.

6.3. The Dirichlet Problem for Annuli and Disks

In order to solve the Dirichlet problem when the region D is an annulus or a disk, it is most natural to use polar coordinates. Polar coordinates (r,θ) of a point in the plane are related to its Cartesian coordinates (x,y) by the formulas

$$x = r\cos\theta \quad \text{and} \quad y = r\sin\theta . \tag{1}$$

For any fixed point (x,y), r equals $(x^2+y^2)^{\frac{1}{2}}$, and there are infinitely many values for θ for which equations (1) hold. For $(x,y) \neq (0,0)$, all these values of θ differ by 2π, but there is a unique value for θ which lies in the interval $(-\pi,\pi]$. (Alternatively, one could use the interval $[0,2\pi)$.) Thus, (1) defines a one–to–one correspondence between the **punctured plane** (i.e., the plane minus the origin) and the set of pairs (r,θ) such that $r > 0$ and $-\pi < \theta \leq \pi$. This correspondence fails at the origin, where $r = 0$ but θ is undetermined. Note that as a moving point (x,y) crosses the negative x–axis, θ suffers a jump of 2π. Thus, θ is not a continuous function at points on the negative x–axis. If we had restricted θ to the interval $[0,2\pi)$, then the discontinuities would occur on the positive x–axis, but there is no way of defining θ so that it is continuous everywhere on the entire punctured plane. There must always be a curve of discontinuity which issues from the origin. If the curve of discontinuity is removed from the punctured plane, then the transformation (1) will be smoothly invertible on the resulting region. For example, if the curve of discontinuity is the nonpositive x–axis and the range of θ is chosen to be $(-\pi,\pi)$, then for any point (x,y) which is not on this curve of discontinuity, we can define the (smooth) inverse transformation of (1) by

$$r = (x^2+y^2)^{\frac{1}{2}}, \qquad \theta = \begin{cases} \arccos[x/(x^2+y^2)^{\frac{1}{2}}] & y \geq 0 \\ -\arccos[x/(x^2+y^2)^{\frac{1}{2}}] & y < 0 . \end{cases}$$

Nevertheless, polar coordinates cannot be well–defined in any region which includes the origin. Thus, solutions of Laplace's equation in terms of polar coordinates must be reexamined in Cartesian coordinates to see if they are valid about the origin.

Let $u(x,y)$ be a C^2 function which is defined in a region on which polar coordinates (r,θ) have been chosen such that (1) has a smooth inverse. In terms of these polar coordinates, we let $U(r,\theta) \equiv u(x,y) = u(r\cos\theta, r\sin\theta)$. In this chapter, we refrain from the popular abuse of notation, "$u(r,\theta) = u(x,y)$". It is not true that $u_{xx} + u_{yy} = U_{rr} + U_{\theta\theta}$. Instead, we have the following result.

Proposition 1. With the above notation,

$$u_{xx} + u_{yy} = U_{rr} + r^{-1}U_r + r^{-2}U_{\theta\theta} \qquad (r > 0). \tag{2}$$

Proof. By the chain rule, we have

$$U_r = u_x x_r + u_y y_r = u_x \cos\theta + u_y \sin\theta \quad \text{and} \quad U_\theta = u_x x_\theta + u_y y_\theta = -u_x r\sin\theta + u_y r\cos\theta.$$

Hence,

$$U_{rr} = u_{xx} \cos^2\theta + 2u_{xy} \sin\theta \cos\theta + u_{yy}\sin^2\theta$$

and

$$U_{\theta\theta} = -(u_{xx} x_\theta + u_{xy} y_\theta)\, r \sin\theta - u_x\, r \cos\theta + (u_{yx}x_\theta + u_{yy} y_\theta)\, r \cos\theta - u_y\, r\sin\theta$$
$$= r^2(u_{xx} \sin^2\theta - 2\, u_{xy} \cos\theta \sin\theta + u_{yy} \cos^2\theta) - r(u_x \cos\theta + u_y \sin\theta).$$

Thus, $U_{rr} + r^{-2}\, U_{\theta\theta} = u_{xx} + u_{yy} - r^{-1}\, U_r$. □

The Dirichlet problem for the annulus

For two fixed positive numbers r_i and r_o $(r_o > r_i)$, let the annulus A be the open region between the inner circle $r = r_i$ and the outer circle $r = r_o$ (cf. Figure 1). Unlike a disk about the pole (0,0), the annulus does not include the pole, and hence we need not worry about the lack of uniqueness of polar coordinates at $r = 0$.

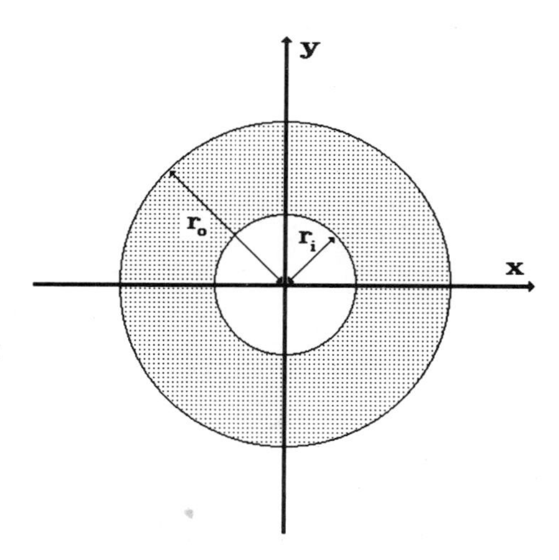

Figure 1

Given continuous functions f and g on the outer and inner circles, respectively, the Dirichlet problem asks for a harmonic function u(x,y) on A, which extends continuously to the boundary of A, so that $u = f$ on $r = r_o$ and $u = g$ on $r = r_i$. When formulating the Dirichlet problem in terms of polar coordinates where $-\pi < \theta \leq \pi$, we must ensure that the solution, u(x,y) or $U(r,\theta)$, that we obtain, does not jump when (x,y) crosses the negative x–axis. For example, by Proposition 1, the function $U(r,\theta) = \theta$ is harmonic in any open region which does not include points of the negative x–axis, but this function is not harmonic on the annulus A, since it is discontinuous on the segment of the negative x–axis between $x = -r_i$ and $x = -r_o$. One way of ensuring that a function $U(r,\theta)$ gives rise to a harmonic function on A, is to require that $U(r,\theta)$ be a C^2 function defined for $r_i < r < r_o$ and for all real θ, such that the right side of (2) is 0, and $U(r,\theta)$ is periodic in θ of period 2π (i.e., $U(r,\theta + 2\pi) = U(r,\theta)$). For example, the function $U(r,\theta) = r^2\cos\theta \sin\theta$ meets these requirements, and it yields the harmonic function $u(x,y) = xy$. With the above in mind, we formulate the Dirichlet problem for the annulus in polar coordinates, as follows:

$$\text{D.E.} \quad U_{rr} + r^{-1} U_r + r^{-2} U_{\theta\theta} = 0\,, \qquad r_i < r < r_o$$

$$\text{B.C.} \quad \begin{cases} U(r_o,\theta) = f(\theta)\,, & f(\theta + 2\pi) = f(\theta) \\ U(r_i,\theta) = g(\theta)\,, & g(\theta + 2\pi) = g(\theta) \end{cases} \tag{3}$$

$$\text{P.C.} \quad U(r,\theta + 2\pi) = U(r,\theta)\,, \quad r_i < r < r_o\,.$$

Here, θ is unrestricted (i.e., $-\infty < \theta < \infty$), and "P.C." stands for "periodicity condition". Note that the continuous functions f and g must be periodic (of period 2π), or else the P.C. cannot be met.

As usual, we first seek product solutions of the D.E. of the form $U(r,\theta) = R(r)\,T(\theta)$, using separation of variables. Inserting this form into the D.E. in (3), we obtain

$$R''(r)\,T(\theta) + r^{-1}R'(r)\,T(\theta) + r^{-2}R(r)\,T''(\theta) = 0$$

$$\frac{r^2R''(r) + rR'(r)}{R(r)} = \frac{-T''(\theta)}{T(\theta)} = c = \pm\, b^2 \qquad (b \geq 0)\,.$$

The ODE for $T(\theta)$ is $T''(\theta) \pm b^2T(\theta) = 0$. We only get periodic solutions of period 2π, when $b = n$ and $c = +b^2 = n^2$, for $n = 0, 1, 2, \ldots$. In this case, we obtain

$$T_n(\theta) = a_n\cos(n\theta) + b_n\sin(n\theta)\,, \quad n = 0, 1, 2, \ldots\,,$$

where a_n and b_n are arbitrary constants. The ODE for $R(r)$ is $r^2R''(r) + rR'(r) - n^2R(r) = 0$, which is an instance of **Euler's equation** and is solved by assuming that $R(r)$ is of the form r^m. Substituting this form into the ODE, we get $r^2m(m-1)r^{m-2} + rmr^{m-1} - n^2r^m = 0$ or $(m^2-n^2)r^m = 0$. Thus, r^m is a solution if $m = \pm n$. For $n \geq 1$, the general solution is

$$R_n(r) = c_n r^n + d_n r^{-n}, \qquad n = 1, 2, 3, \dots ,$$

and when $n = 0$,

$$R_0(r) = c_0 + d_0 \log(r),$$

(cf. Example 1 of Section 6.1). Multiplying the above expressions for $T_n(\theta)$ and $R_n(r)$, we obtain the following family of product solutions of the D.E. and P.C. of (3) :

$$\begin{aligned} U_0(r,\theta) &= a_0 + \alpha_0 \log(r) \\ U_n(r,\theta) &= (a_n r^n + \alpha_n r^{-n})\cos(n\theta) + (b_n r^n + \beta_n r^{-n})\sin(n\theta), \qquad n \geq 1. \end{aligned} \tag{4}$$

By the superposition principle, a more general solution of the D.E. and P.C. of (3) is

$$U(r,\theta) = U_0(r,\theta) + \sum_{n=1}^{N} U_n(r,\theta). \tag{5}$$

Suppose that $f(\theta)$ and $g(\theta)$ are finite Fourier series of the form

$$\begin{aligned} f(\theta) &= \tfrac{1}{2}A_0 + \sum_{n=1}^{N} A_n \cos(n\theta) + B_n \sin(n\theta) \\ g(\theta) &= \tfrac{1}{2}C_0 + \sum_{n=1}^{N} C_n \cos(n\theta) + D_n \sin(n\theta). \end{aligned} \tag{6}$$

Comparing Fourier coefficients in the equations $U(r_o,\theta) = f(\theta)$ and $U(r_i,\theta) = g(\theta)$, we obtain the following pairs of equations

$$\begin{aligned} a_0 + \alpha_0 \log(r_o) &= \tfrac{1}{2}A_0 \\ a_0 + \alpha_0 \log(r_i) &= \tfrac{1}{2}C_0 \end{aligned}$$

$$\begin{aligned} a_n r_o^{\,n} + \alpha_n r_o^{\,-n} &= A_n \qquad\qquad & b_n r_o^{\,n} + \beta_n r_o^{\,-n} &= B_n \\ a_n r_i^{\,n} + \alpha_n r_i^{\,-n} &= C_n \qquad\qquad & b_n r_i^{\,n} + \beta_n r_i^{\,-n} &= D_n , \end{aligned} \tag{7}$$

for $n = 1, 2, 3, \dots$. In the first pair, we wish to solve for a_0 and α_0 in terms of the given constants A_0 and C_0, and similarly we wish to solve the other pairs. Recall that the solution of the arbitrary system (assuming $ad - bc \neq 0$)

$$\begin{cases} ax + by = e \\ cx + dy = f \end{cases} \quad \text{is} \quad \begin{cases} x = (ed - fb)/(ad - bc) \\ y = (ad - ce)/(ad - bc). \end{cases}$$

Thus, we obtain, with $Q \equiv r_o/r_i$,

$$a_0 = \frac{\frac{1}{2} C_0 \log r_o - \frac{1}{2} A_0 \log r_i}{\log Q} \qquad \alpha_0 = \frac{\frac{1}{2} A_0 - \frac{1}{2} C_0}{\log Q}$$

$$a_n = \frac{A_n r_i^{-n} - C_n r_o^{-n}}{Q^n - Q^{-n}} \qquad \alpha_n = \frac{C_n r_o^{\,n} - A_n r_i^{\,n}}{Q^n - Q^{-n}} \tag{8}$$

$$b_n = \frac{B_n r_i^{-n} - D_n r_o^{-n}}{Q^n - Q^{-n}} \qquad \beta_n = \frac{D_n r_o^{\,n} - B_n r_i^{\,n}}{Q^n - Q^{-n}}.$$

This provides us with the constants a_n, b_n, c_n, d_n in terms of the given Fourier coefficients A_n, B_n, C_n, D_n of $f(\theta)$ and $g(\theta)$. In summary, we state the following result.

Proposition 2. The solution of (3), **where** $f(\theta)$ **and** $g(\theta)$ **are given by** (6), **is**

$$U(r,\theta) = a_0 + \alpha_0 \log r + \sum_{n=1}^{N} \{[a_n r^n + \alpha_n r^{-n}] \cos(n\theta) + [b_n r^n + \beta_n r^{-n}] \sin(n\theta)\}, \tag{9}$$

where a_n, α_n, b_n, and β_n **are defined by** (8).

Example 1 (A steady–state temperature problem for the annulus). Solve the problem

$$\begin{aligned} &\text{D.E.} \quad U_{rr} + r^{-1}U_r + r^{-2}U_{\theta\theta} = 0\,, \qquad 1 < r < 2 \\ &\text{B.C.} \quad \begin{cases} U(1,0) = 3 + 4\cos(2\theta) \\ U(2,\theta) = 5\sin(\theta) \end{cases} \\ &\text{P.C.} \quad U(r,\theta + 2\pi) = U(r,\theta). \end{aligned} \tag{10}$$

Solution. As we have seen, separation of variables and the superposition principle lead us to a solution of the form (9). We could then simply use the formulas (8) with $A_0 = 6$, $A_2 = 4$, $D_1 = 5$ and all other A_n, B_n, C_n, and D_n equal to 0. Note that $Q \equiv r_o/r_i = 2$. Rather than using the formulas (8), as indicated in Proposition 2, we could also appeal to the method by which the formulas (8) were derived. Specifically, if we equate the Fourier coefficients in the B.C. with those of $U(r,\theta)$ in (9), using $r = 1$ and $r = 2$, then we obtain (cf. (7))

$$\begin{cases} a_0 + \alpha_0 \log(1) = 3 \\ a_0 + \alpha_0 \log(2) = 0 \end{cases} \qquad \begin{cases} b_1 + \beta_1 = 0 \\ 2b_1 + \tfrac{1}{2}\beta_1 = 5 \end{cases} \qquad \begin{cases} a_2 + \alpha_2 = 4 \\ 2^2 a_2 + 2^{-2}\alpha_2 = 0 . \end{cases}$$

Solving these systems, we obtain $a_0 = 3$, $\alpha_0 = -3/\log(2)$, $b_1 = 10/3$, $\beta_1 = -10/3$, $a_2 = -4/15$, $\alpha_2 = 64/15$. All other systems in (7) have solutions zero. The solution of (8) is then

$$U(r,\theta) = 3 - 3\log(r)/\log(2) + (10r/3 - 10r^{-1}/3)\sin(\theta)$$
$$+ (-4r^2/15 + 64r^{-2}/15)\cos(2\theta) . \quad \square$$

The Dirichlet problem for the disk

The Dirichlet problem for the disk of radius r_o and center at $(0,0)$ can be expressed as

D.E.	$u_{xx} + u_{yy} = 0$	$(x^2 + y^2)^{\frac{1}{2}} < r_o$	(11)
B.C.	$U(r_o,\theta) = f(\theta)$,	$f(\theta + 2\pi) = f(\theta)$,	

where $U(r,\theta) = u(x,y) = u(r\cos\theta, r\sin\theta)$ and $f(\theta)$ is a given periodic, continuous function of period 2π.

Remark. In (11) we do not write the D.E. in terms of polar coordinates, because it would not make sense at the pole $(0,0)$ which is in the disk $(x^2 + y^2)^{\frac{1}{2}} < r_o$. One's instinct is to solve problem (11) just as in the case of the annulus, but in order to avoid the singularity at the pole in (9), set $\alpha_n = 0$ for $n \geq 0$ and $\beta_n = 0$ for $n \geq 1$. Then one determines the a_n and b_n from the B.C. that now only involves $f(\theta)$. (For the time being, we assume that $f(\theta)$ is a finite Fourier series.) While this approach does yield the solution of (11), there is a potential difficulty. Functions which appear to be C^2 in polar coordinates, might not be C^2 at $(0,0)$. For example, the function $U(r,\theta) = r$ seems nice, but in terms of Cartesian coordinates, we get

$u(x,y) = (x^2 + y^2)^{\frac{1}{2}}$, which is not C^1 at $(0,0)$, because the graph $z = u(x,y)$ is a cone with vertex at $(0,0,0)$. Also note that the functions $\cos(n\theta)$ and $\sin(n\theta)$ (for $n = 1,2, \ldots$) are not continuously definable at $(0,0)$, since each of these functions has values ranging from -1 to 1 on the rays issuing from the origin. On the other hand, $r^2\cos(2\theta) = x^2 - y^2$ is a C^∞ function. We now show that in general $r^n\cos(n\theta)$ and $r^n\sin(n\theta)$ are polynomials in x and y, and hence are C^∞. Let $z = x + iy = r\cos\theta + ir\sin\theta = re^{i\theta}$ (cf. Euler's formula (26) of Section 1.1). Then, we have **De Moivre's formula** [after Abraham De Moivre (1667–1754), French–born English mathematician and a founder of probability theory] :

$$z^n = (re^{i\theta})^n = r^n e^{in\theta} = r^n[\cos(n\theta) + i\sin(n\theta)]. \tag{12}$$

In other words, $r^n\cos(\theta)$ and $r^n\sin(n\theta)$ are the real and imaginary parts of the product $(x + iy) \ldots (x + iy)$ [n factors]. But, if this product were multiplied out, the real and imaginary parts would be n–th degree polynomials in x and y. These polynomials are harmonic. Indeed, they are the functions $r^n\cos(n\theta)$ and $r^n\sin(n\theta)$ which are harmonic for $r > 0$, and there is no difficulty at the origin, since the second partials of the polynomials are continuous at $(0,0)$. □

Proposition 3. In the Dirichlet problem (11), **if**

$$f(\theta) = \tfrac{1}{2}a_0 + \sum_{n=1}^{N} a_n\cos(n\theta) + b_n\sin(n\theta), \tag{13}$$

then the solution of (11) **is**

$$U(r,\theta) = \tfrac{1}{2}a_0 + \sum_{n=1}^{N} (r/r_0)^n[a_n\cos(n\theta) + b_n\sin(n\theta)]. \tag{14}$$

Proof. It follows from the above remark and the superposition principle, that (14) defines a harmonic function throughout the disk. Note that if we set $r = r_0$ in the right side of (14), then the result is $f(\theta)$, whence the B.C. of (11) is met. By the uniqueness theorem for the Dirichlet problem (cf. Theorem 1 of Section 6.2), (14) is the only solution of (11). □

Example 2. Solve the Dirichlet problem for the disk of radius 1:

$$\begin{aligned} &\text{D.E.} \quad u_{xx} + u_{yy} = 0 \qquad r < 1 \\ &\text{B.C.} \quad U(1,0) = -1 + 8\cos^2(\theta) \\ &\text{P.C.} \quad U(r,\theta + 2\pi) = U(r,\theta)\ . \end{aligned} \tag{15}$$

Solution. The Fourier series of $-1 + 8\cos^2(\theta)$ is $3 + 4\cos(2\theta)$, which of the same as the function of the first B.C. in Example 1. In spite of the fact that the solution of (10) in Example 1 satisfies the B.C., it is not the solution of (15), because the terms involving $\log(r)$, r^{-1}, r^{-2} are undefined at the center of the disk. Instead, we use (14) to obtain the correct solution $U(r,\theta) = 3 + 4r^2\cos(2\theta)$. □

The Poisson Integral Formula

We now derive the Poisson Integral Formula (cf. (20) below) which expresses the solution (14) in terms of an integral. We will prove that this formula yields a solution of the Dirichlet problem (11), even when the Fourier series of the continuous function $f(\theta)$ in (11) has infinitely many nonzero terms. So far, we have not proven that (14) is valid when $N = \infty$, since the superposition principle can fail for infinite linear combinations, as we have seem in previous chapters. However, as a consequence of the Poisson Integral Formula, we find that (14) is in fact valid for $r < r_0$, even if $N = \infty$. While the following informal derivation of the Poisson Integral Formula is not a substitute for a rigorous proof (cf. Theorem 1 below), this derivation explains how the formula arises.

Proceeding unrigorously, taking $N = \infty$ in (14), and expressing the Fourier coefficients a_n and b_n of f as integrals, we get

$$U(r,\theta) = \frac{1}{2\pi}\int_{-\pi}^{\pi} f(t)\,dt + \sum_{n=1}^{\infty}\left\{(r/r_0)^n\left[\frac{1}{\pi}\int_{-\pi}^{\pi} f(t)\cos(nt)dt\,\cos(n\theta) + \frac{1}{\pi}\int_{-\pi}^{\pi} f(t)\sin(nt)dt\,\sin(n\theta)\right]\right\}.$$

Combining the integrals, we have

$$\begin{aligned} U(r,\theta) &= \frac{1}{\pi}\int_{-\pi}^{\pi} f(t)\left\{\tfrac{1}{2} + \sum_{n=1}^{\infty}(r/r_0)^n\,[\cos(nt)\cos(n\theta) + \sin(nt)\sin(n\theta)]\right\}dt \\ &= \frac{1}{\pi}\int_{-\pi}^{\pi} f(t)\left\{\tfrac{1}{2} + \sum_{n=1}^{\infty}(r/r_0)^n\cos[n(\theta - t)]\right\}dt\,. \end{aligned} \tag{16}$$

We can sum up the series in (16), as follows. Let $\tau \equiv (r/r_0)[\cos(\theta - t) + i\sin(\theta - t)]$. The expression in braces in (16) is then the real part

$$\mathrm{Re}\left[\tfrac{1}{2} + \sum_{n=1}^{\infty}\tau^n\right]. \tag{17}$$

For $|\tau| < 1$, the geometric series $\sum_{n=1}^{\infty}\tau^n$ (with complex terms τ^n) converges to $\tau/(1 - \tau)$, and (17) is

$$\mathrm{Re}[\tfrac{1}{2} + \tau/(1 - \tau)] = \tfrac{1}{2}\,\mathrm{Re}[(1 + \tau)/(1 - \tau)]\,.$$

For any complex number $z = a + ib \neq 0$, we have $1/z = \bar{z}/|z|^2$, where $\bar{z} = a - ib$. Thus,

$$\frac{1+\tau}{1-\tau} = \frac{(1+\tau)(1-\bar{\tau})}{|1-\tau|^2} = \frac{1-|\tau|^2}{|1-\tau|^2} + \frac{\tau - \bar{\tau}}{|1-\tau|^2}.$$

Since the last term is purely imaginary, we obtain

$$\mathrm{Re}\left[\frac{1+\tau}{1-\tau}\right] = \frac{1-|\tau|^2}{|1-\tau|^2} = \frac{1-(r/r_o)^2}{1-(\tau+\bar{\tau})+|\tau|^2}$$

$$= \frac{r_o^2 - r^2}{r_o^2 - 2rr_o\cos(\theta - t) + r^2} \equiv P(r, r_o, \theta - t) . \tag{18}$$

The expression $P(r, r_o, \theta - t)$ is known as the **Poisson kernel**. Combining (16), (17) and (18), we get the **Poisson Integral Formula**

$$U(r,\theta) = \frac{1}{2\pi}\int_{-\pi}^{\pi} P(r, r_o, \theta - t)\, f(t)\, dt \qquad (r < r_o) . \tag{19}$$

Remark. While the above argument suggests that (19) solves the Dirichlet problem (11), it does not constitute a proof, because we have assumed that (14) is valid even when $N = \infty$, and we have interchanged the integral and the infinite sum in (16), without justification. Before giving the proof, we make a number of observations. The denominator of $P(r, r_o, \theta - t)$ in (18) is the square of the distance from (r,θ) to (r_o,t) in polar coordinates (simply compute $|re^{i\theta} - r_o e^{it}|^2$, using the formula $|z|^2 = z\bar{z}$). Thus, $P(r, r_o, t - \theta)$ is a C^∞ function of (r,θ), as long as $r < r_o$. By Leibniz's rule (cf. Appendix A.3), we are free to differentiate $U(r,\theta)$ [with respect to r and θ] as many times as we like, by differentiating $P(r, r_o, \theta - t)$ under the integral. In particular, we could check that $U(r,\theta)$ is harmonic (for $r < r_o$) by verifying that $P(r, r_o, \theta - t)$ satisfies the D.E. of (11) as a function of r and θ, for each fixed t. (Since we already know that $P(r, r_o, \theta - t)$ defines a smooth function at the pole, there is no difficulty there.) This verification is as straightforward as it is tedious, and we will leave it to the diligent reader. The fact that P is harmonic is believable since it is twice the sum (although infinite) of harmonic functions in (16). Alternatively, for those familiar with complex variable theory, we have already exhibited P as the real part of an analytic function of $\tau = (r/r_o)e^{i(\theta - t)}$, $r < r_o$ in (18), whence P must be harmonic. At any rate, the fact that P is harmonic together with

Leibniz's rule gives the result that $U(r,\theta)$ in (19) is harmonic for $r < r_0$. Thus, in order to establish the following result, we need only to check that $U(r,\theta)$ extends continuously to the given function $f(\theta)$ on the boundary $r = r_0$. □

Theorem 1 (The Poisson Integral Formula). Let $f(\theta)$ be a continuous periodic function of period 2π. Define

$$U(r,\theta) \equiv \frac{1}{2\pi}\int_{-\pi}^{\pi} \frac{(r_0^2 - r^2)\, f(t)}{r_0^2 - 2rr_0\cos(\theta - t) + r^2}\, dt \qquad (r < r_0) \tag{20}$$

and

$$U(r_0,\theta) \equiv f(\theta) \qquad (r = r_0). \tag{21}$$

Then $U(r,\theta)$ is harmonic on the open disk $r < r_0$, and continuous on the closed disk $r \leq r_0$. In other words, $U(r,\theta)$ solves the Dirichlet problem (11).

Proof. By the above remarks, $U(r,\theta)$ is harmonic in the open disk. To obtain the continuity of $U(r,\theta)$ on the closed disk, we must show that, given any boundary point (r_0,θ_0), $U(r,\theta)$ approaches $f(\theta_0)$, as (r,θ) approaches (r_0,θ_0). We first need to show that

$$\frac{1}{2\pi}\int_{-\pi}^{\pi} \frac{r_0^2 - r^2}{r_0^2 - 2rr_0\cos(\theta - t) + r^2}\, dt = 1\,, \tag{22}$$

for $r < r_0$. Observe that the integral is independent of θ, since changing θ only serves to translate the integrand (a periodic function of t of period 2π) without changing the integral. Thus, the left side of (22) is a harmonic function (by the above remark with $f(\theta) \equiv 1$) which is independent of θ, and thus must be of the form $C \log r + K$. However, the value of the left side (22) is clearly 1, when $r = 0$, and hence $C = 0$ and $K = 1$ (i.e., (22) holds for all (r,θ) with $r < r_0$). Multiplying by the constant $f(\theta_0)$ in (22), we obtain,

$$f(\theta_0) = (2\pi)^{-1}\int_{-\pi}^{\pi} P(r, r_0, \theta - t)\, f(\theta_0)\, dt \tag{23}$$

and subtracting corresponding sides of (23) from (20), we obtain

$$U(r,\theta) - f(\theta_0) = (2\pi)^{-1}\int_{-\pi}^{\pi} P(r, r_0, \theta - t)\, [f(t) - f(\theta_0)]\, dt. \tag{24}$$

To demonstrate the continuity of $U(r,\theta)$ at (r_0,θ_0), we must show that the right side of (24) can be made arbitrarily small, by requiring that (r,θ) be sufficiently close to (r_0,θ_0). The interval $[-\pi,\pi]$ of integration in (24) may be replaced by the new interval $[\theta_0-\pi,\theta_0+\pi]$ (of length 2π) without changing the value of the integral, since the integrand is periodic of period 2π. For any δ with $0<\delta<\pi$, we split the new interval into three pieces, namely

$$i_1=[\theta_0-\pi,\theta_0-\delta],\qquad i_2=[\theta_0-\delta,\theta_0+\delta],\qquad i_3=[\theta_0+\delta,\theta_0+\pi]\ .$$

We denote the integral, with respect to t, of $P(r,r_0,\theta-t)\,[f(t)-f(\theta_0)]$ over the interval i_1, by I_1, and define I_2 and I_3 similarly. To estimate I_2, let $M(\delta)$ be the maximum of $|f(t)-f(\theta_0)|$ as t ranges over the interval i_2 of length 2δ about θ_0. Since f is continuous at θ_0, $M(\delta)$ can be made arbitrarily small by choosing δ sufficiently small. We have

$$|I_2|=\left|\int_{i_2}P(r,r_0,\theta-t)\,[f(t)-f(\theta_0)]\,dt\right|\le M(\delta)\int_{i_2}P(r,r_0,\theta-t)\,dt\le M(\delta)\ ,\qquad(25)$$

where in the last inequality we have used (22) and the fact that $P(r,r_0,\theta-t)\ge 0$ (for $r<r_0$). To estimate I_1 and I_3, we use $1-\cos x=x^2(1/2-x^2/24+\ldots)\ge x^2/24$ for $|x|\le\sqrt{11}$. To prove this, note that for $x^2<12$ the series $1/2-x^2/24+x^4/720-\ldots$ is alternating with terms of decreasing absolute value, whence $1/2-x^2/24$ (which is $\ge 1/24$ for $|x|<\sqrt{11}$) underestimates the entire series. Hence, we obtain (for $|\theta-t|<\sqrt{11}$)

$$r_0^2-2r_0r\cos(\theta-t)+r^2=(r_0-r)^2+2r_0r[1-\cos(\theta-t)]\ge 2r_0r[1-\cos(\theta-t)]\ge r_0r(\theta-t)^2/12,$$

Thus, for $|\theta-t|<\sqrt{11}$ (and since $(r_0+r)/r_0<2$ for $r<r_0$)

$$P(r,r_0,\theta-t)\ \le\ \frac{12(r_0^2-r^2)}{r_0r(\theta-t)^2}\ \le\ \frac{24(r_0-r)}{r(\theta-t)^2}\ .\qquad(26)$$

Let M be the maximum of $|f(\alpha)|$ for $-\pi\le\alpha\le\pi$. (M exists since $f(x)$ is continuous on the closed interval $[-\pi,\pi]$.) Then, $|f(t)-f(\theta_0)|\le 2M$. Suppose that θ is within a distance of $\frac{1}{2}\delta$ from θ_0 (i.e., $|\theta-\theta_0|<\frac{1}{2}\delta$), where $\frac{1}{2}\delta<\sqrt{11}-\pi\approx .175$. Then, for t in i_1 or i_3, we have

$$\tfrac{1}{2}\delta\ <\ |\theta-t|\ \le\ |\theta-\theta_0|+|\theta_0-t|\ <\ \sqrt{11}-\pi+\pi\ =\ \sqrt{11}\ .$$

Thus, we may use (26) to deduce that

$$|I_1|\le\frac{1}{2\pi}\int_{i_1}P(r,r_0,\theta-t)\,|f(t)-f(\theta_0)|\,dt\ \le\ \frac{24M(r_0-r)}{r\,\delta^2/4},\qquad(27)$$

and the same bound holds for $|I_3|$. Hence, from (25) and (27), we have

$$|I_1 + I_2 + I_3| \leq M(\delta) + 192M(r_0 - r)/(r\delta^2), \tag{28}$$

Given any $\epsilon > 0$, we can choose $\delta > 0$ so small that $M(\delta) < \frac{1}{2}\epsilon$. Then, choosing θ close to θ_0 so that $|\theta - \theta_0| < \min(\frac{1}{2}\delta, \sqrt{11} - \pi)$, and choosing r close enough to r_0 so that $192(r_0 - r)/(r\delta^2) < \frac{1}{2}\epsilon$, it follows from (28) and (24) that $|U(r,\theta) - f(\theta_0)| \leq \epsilon$. In other words, we can make $U(r,\theta)$ arbitrarily close to $f(\theta_0)$, by choosing θ sufficiently close to θ_0 and r sufficiently close to (but less than) r_0. □

Even for simple functions, the integral (20) may be difficult to evaluate exactly, but by numerical integration, say by Simpson's rule, one can find an approximate value for (20) for any given (r,θ). If it is possible to represent $f(\theta)$ by a finite Fourier series, then the solution of (11) is more conveniently expressed by (14) than by (20). Indeed, in this case, (14) would have to be the result of the integration in (20), by the uniqueness theorem for the Dirichlet problem for the disk, as the following example illustrates.

Example 3. Show that for $r < 1$, we have

$$\frac{1}{2\pi}\int_{-\pi}^{\pi} \frac{(1 - r^2)\ \sin(t)}{1 - 2r\cos(\theta - t) + r^2}\, dt = r \sin\theta\ . \tag{29}$$

Solution. Both sides of (29) are solutions of the Dirichlet problem for the disk with radius $r_0 = 1$ with B.C. $f(\theta) = \sin(\theta)$. By uniqueness for the solution of this problem, the two sides must be equal for $r < 1$. Incidentally, (29) is not always valid for $r > 1$. Indeed, for $r > 1$, the left side is $-r^{-1}\sin\theta$, as is shown in Problem 13. □

The Mean–Value Theorem and the regularity of harmonic functions

We now explore some remarkable consequences of the Poisson Integral Formula.

Theorem 2 (The Mean–Value Theorem). Let u be a harmonic function on some open region R. Then the value of u at the center of any closed disk D contained in R is the average (or mean) of the values of u on the circular boundary of D.

Proof. By introducing polar coordinates with pole at the center of the disk, we may assume that D is the disk $r \leq r_0$. On the boundary of D, u is $U(r_0,\theta)$. Thus, the Poisson Integral Formula (20), with $r = 0$, yields

$$u(0,0) = U(0,\theta) = \frac{1}{2\pi}\int_{-\pi}^{\pi} P(0,r_0,\theta - t)U(r_0,t)\, dt = \frac{1}{2\pi}\int_{-\pi}^{\pi} U(r_0,t)\, dt, \tag{30}$$

since $P(0,r_0,\theta - t) = 1$ by (18). The final integral is the **mean–value** (or **average**) of u on the boundary of D. □

Although a harmonic function u on an open region R is only required (by definition) to be C^2, u is actually C^∞ on R, as the following theorem shows.

Theorem 3 (Regularity of harmonic functions). If u is harmonic on an open region R, then u is C^∞ on R.

Proof. Let p be any point in R and choose polar coordinates (r,θ) with p as the pole. Suppose that r_0 is chosen small enough so that the disk $r \le r_0$ is contained in R, and let $x = r\cos\theta$ and $y = r\sin\theta$. Now, $P(r,r_0,\theta - t)$ is infinitely differentiable when viewed as a function of (x,y,t) for $x^2 + y^2 < r_0^{\,2}$. Indeed,

$$P(r,r_0,\theta - t) = \frac{r_0^{\,2} - x^2 - y^2}{[x - r_0\cos t]^2 + [y - r_0\sin t]^2} . \tag{31}$$

Thus, it follows from repeated use of Leibniz's rule (cf. Appendix A.3) that all of the derivatives of u in the disk $r < r_0$ can be computed by differentiating under the integral in the Poisson Integral Formula. In particular, all of the partial derivatives of u with respect to x and y exist at any point p of R, and hence u is C^∞ on R. □

Theorem 4 (Infinite series solutions). Let $f(\theta)$ be a continuous periodic function of period 2π, with FS $f(\theta) = \frac{1}{2}a_0 + \sum_{n=1}^{\infty} a_n\cos(n\theta) + b_n\sin(n\theta)$. Then the the Dirichlet problem

$$\begin{aligned} &\text{D.E.} \quad u_{xx} + u_{yy} = 0 \qquad (x^2 + y^2)^{\frac{1}{2}} < r_0 \\ &\text{B.C.} \quad U(r_0,\theta) = f(\theta) \end{aligned} \tag{32}$$

has the (unique) solution $U(r,\theta)$ given by

$$U(r,\theta) = \tfrac{1}{2}a_0 + \sum_{n=1}^{\infty} (r/r_0)^n[a_n\cos(n\theta) + b_n\sin(n\theta)] \quad (r < r_0) , \tag{33}$$

while $U(r_0,\theta)$ is *defined* to be $f(\theta)$. (Note that it is not necessary to assume that FS $f(\theta) = f(\theta)$, but if this is the case, then (33) is also valid for $r = r_0$.)

Proof. By the uniqueness theorem (cf. Theorem 1 of Section 6.2) and Theorem 1, the solution $U(r,\theta)$ of problem (32) is given by (20) for $r < r_0$. Thus, we need only to show that the right sides of (20) and (33) are equal. It follows from Theorem 3 that for any fixed $r < r_0$, $U(r,\theta)$ is a C^∞ periodic function of θ. Thus, by the convergence theorems of Chapter 4 (e.g., Theorem 1 of Section 4.2), we know that for fixed $r < r_0$, the Fourier series of $U(r,\theta)$ converges to $U(r,\theta)$. Thus, it suffices to prove that the Fourier series of $U(r,\theta)$ is the right side of (33). We compute the Fourier coefficients $a_n(r)$ of $U(r,\theta)$:

$$a_n(r) = \frac{1}{2\pi}\int_{-\pi}^{\pi} U(r,\theta)\cos(n\theta)\,d\theta$$

$$= \left[\frac{1}{2\pi}\right]^2 \int_{-\pi}^{\pi}\left[\int_{-\pi}^{\pi} P(r, r_0, \theta - t)\, f(t)\,dt\right]\cos(n\theta)\,d\theta$$

$$= \left[\frac{1}{2\pi}\right]^2 \int_{-\pi}^{\pi}\left[\int_{-\pi}^{\pi} P(r, r_0, \theta - t)\cos(n\theta)\,d\theta\right] f(t)\,dt$$

$$= \frac{1}{2\pi}\int_{-\pi}^{\pi} (r/r_0)^n \cos(nt)\, f(t)\,dt = (r/r_0)^n\, a_n ,$$

where we have interchanged the order of integration (cf. Appendix A.2) and used the fact that the solution of the Dirichlet problem with boundary function $\cos(n\theta)$ is $(r/r_0)^n\cos(n\theta)$. A similar computation yields the desired result for $b_n(r)$. □

Remark. Observe that when the n–th term on the right side of (33) is expressed in terms of x and y it is an n–th degree polynomial in x and y. Thus, Theorem 4 tells us that a function, which is continuous on a closed disk and harmonic in the interior, is equal to a power series in x and y (i.e., harmonic functions are **real analytic** in x and y). (We assume that the x and y coordinates are chosen so the origin is the center, say p, of the disk.) By a simple result of advanced calculus, this power series must be the Taylor series of u at p. Consequently, if all of the partial derivatives of u vanish at p, then u must be constant throughout the disk. □

Summary 6.3

1. The Laplacian in polar coordinates : For a C^2 function $u(x,y)$ on a region such that $u(x,y) = U(r,\theta)$, in terms of polar coordinates, we have

$$U_{xx} + U_{yy} = U_{rr} + r^{-1}U_r + r^{-2}U_{\theta\theta} .$$

2. The Dirichlet Problem for the annulus : The Dirichlet problem for the annulus may be expressed in terms of polar coordinates, as follows.

$$\text{D.E.} \quad U_{rr} + r^{-1} + r^{-2}U_{\theta\theta} = 0, \qquad r_i < r < r_o$$

$$\text{B.C.} \quad \begin{cases} U(r_o, \theta) = f(\theta), & f(\theta + 2\pi) = f(\theta) \\ U(r_i, \theta) = g(\theta), & g(\theta + 2\pi) = g(\theta) \end{cases} \tag{S1}$$

$$\text{P.C.} \quad U(r,\theta + 2\pi) = U(r,\theta) .$$

Here "P.C." stands for "periodicity condition". If $f(\theta)$ and $g(\theta)$ are finite Fourier series, then the solution of (S1) is of the form (cf. Proposition 2)

$$U(r, \theta) = a_0 + \alpha_0 \log r + \sum_{n=1}^{N} U_n(r, \theta), \tag{S2}$$

where

$$U_n(r, \theta) = (a_n r^n + \alpha_n r^{-n})\cos(n\theta) + (b_n r^n + \beta_n r^{-n})\sin(n\theta), \quad n \geq 1,$$

and the coefficients a_n, b_n, α_n, and β_n are given in terms of the Fourier coefficients of $f(\theta)$ and $g(\theta)$ by the formulas in (8).

3. The Dirichlet problem for a disk : The Dirichlet problem for the disk of radius r_o and center at (0,0) can be expressed as

$$\begin{aligned} &\text{D.E.} \quad u_{xx} + u_{yy} = 0, && (x^2 + y^2)^{\frac{1}{2}} < r_o \\ &\text{B.C.} \quad U(r_o,\theta) = f(\theta), && f(\theta + 2\pi) = f(\theta), \end{aligned} \tag{S3}$$

where $U(r,\theta) = u(x,y) = u(r\cos\theta, r\sin\theta)$ an $f(\theta)$ is a given periodic, continuous function. If $f(\theta)$ is of the form

$$f(\theta) = \tfrac{1}{2}a_0 + \sum_{n=1}^{N} a_n\cos(n\theta) + b_n\sin(n\theta) , \tag{S4}$$

then the solution of (S3) is

$$U(r,\theta) = \tfrac{1}{2}a_0 + \sum_{n=1}^{N} (r/r_0)^n[a_n\cos(n\theta) + b_n\sin(n\theta)]. \qquad (S5)$$

The solution of (S3) can also be expressed by the Poisson Integral Formula, which in turn can be used to justify formula (S5) (for $r < r_0$), when $N = \infty$ and $f(\theta)$ is continuous (cf. Theorem 4).

4. The Poisson Integral Formula (Theorem 1) : Let $f(\theta)$ be a continuous periodic function of period 2π. Define $U(r_0,\theta) \equiv f(\theta)$ and set

$$U(r,\theta) = \frac{1}{2\pi}\int_{-\pi}^{\pi} \frac{(r_0^2 - r^2)\, f(t)}{r_0^2 - 2rr_0\cos(\theta - t) + r^2}\, dt \quad (r < r_0) \qquad (S6)$$

Then $U(r,\theta)$ is harmonic on the open disk $r < r_0$, and continuous on the closed disk $r \leq r_0$. In other words, $U(r,\theta)$ solves the Dirichlet problem (S3).

5. The Mean–Value Theorem (Theorem 2) : Let u be a harmonic function on some open region R. Then the value of u at the center of any closed disk D contained in R is the average (or mean) of the values of u on the circular boundary of D.

6. Regularity of harmonic functions (Theorem 3) : If u is harmonic on an open region R, then u is C^∞ on R. Indeed, u is real analytic on R.

Exercises 6.3

1. Solve the Dirichlet problem

$$\begin{aligned} &\text{D.E.} \quad U_{rr} + r^{-1}U_r + r^{-2}U_{\theta\theta} = 0\,, \qquad 1 < r < 2 \\ &\text{B.C.} \quad U(2,\theta) = f(\theta), \quad U(1,\theta) = g(\theta) \\ &\text{P.C.} \quad U(r,\theta + 2\pi) = U(r,\theta) \end{aligned}$$

for the annulus, when

(a) $f(\theta) = 1 + 2\cos\theta + \cos(2\theta)$ and $g(\theta) = \sin(2\theta)$

(b) $f(\theta) = \sin^2\theta$ and $g(\theta) = \sin\theta\cos\theta$

(c) $f(\theta) = a$ and $g(\theta) = b$, where a and b are constants

(d) $f(\theta) = a \sin(3\theta)$ and $g(\theta) = b \sin(3\theta)$, where a and b are constants

(e) $f(\theta) = 1 + 3 \cos\theta - 17 \sin(8\theta)$ and $g(\theta) = 0$.

2. (a) Show that any harmonic function, defined on the annulus $r_i < r < r_o$, which is of the form $U(r,\theta) = h(\theta)$ (i.e., not depending on r) must be a constant, i.e., independent of θ also.

(b) Suppose that the harmonic function in (a) is only required to be defined in the upper half $(0 < \theta < \pi)$ of the annulus. Show that $h(\theta)$ need not be constant in this case. If $h(0) = a$ and $h(\pi) = b$, then what is $h(\theta)$ for $0 < \theta < \pi$?

3. Solve the Dirichlet problem

$$\text{D.E.} \quad u_{xx} + u_{yy} = 0 \qquad (x^2 + y^2 < 4)$$

$$\text{B.C.} \quad U(2,\theta) = f(\theta) \qquad (f(\theta + 2\pi) = f(\theta)),$$

for the disk $r \leq 2$, in the cases where $f(\theta)$ is given in the various parts of Problem 1. Why can we not just use the answers in Problem 1 to solve the problem here ?

4. (a) Write down the solution of the problem in Problem 3, when $f(\theta) = \theta^2$ for $-\pi \leq \theta \leq \pi$, using the Poisson Integral Formula.

(b) Express the solution in (a) in terms of an infinite series of the form (33) in Theorem 4. Is this merely a formal solution, or is it an exact solution ?

(c) Estimate the number of terms needed in the series in (b) in order to approximate the exact solution throughout the disk $r \leq 2$ to within an error of .01 .

5. (a) Use (22) to compute the integral $\int_{-\pi}^{\pi} [1 - a \cos(x)]^{-1}\, dx$, where $0 < a < 1$.

(b) Use the Poisson Integral Formula to compute the integral

$$\int_{-\pi}^{\pi} f(x)\, [1 - a \cos(x)]^{-1}\, dx ,$$

when $f(x)$ is

(i) $\cos(nx)$ $n = 0, 1, 2, \ldots$ (ii) $\sin^2(x)$ (iii) $\cos^3 x$.

6. Let $U_1(r,\theta)$ and $U_2(r,\theta)$ solve the Dirichlet problem on the disk $r \leq r_o$ with $U_1(r_o,\theta) = f_1(\theta)$, and $U_2(r,\theta) = f_2(\theta)$. Without appealing to the Maximum/Minimum Principle, use the Poisson Integral Formula to deduce that if $|f_1(\theta) - f_2(\theta)| \leq \epsilon$, then $|U_1(r,\theta) - U_2(r,\theta)| \leq \epsilon$ for $r < r_o$. In other words, harmonic functions which are close on the circle will be close on the disk. **Hint.** Use (22) and the fact that $P(r,r_o,\theta-t) \geq 0$.

7. (a) Suppose that $U(r,\theta)$ is harmonic in the disk $r < r_o$, and let $V(r,\theta) \equiv U(Rr,\theta)$, where $R > 0$ is a constant. Show that V is harmonic on the disk $r < r_o/R$.
(b) Assuming that $R < 1$, use the Poisson Integral Formula and (a) to deduce that

$$U(Rr,\theta) = \frac{1}{2\pi}\int_{-\pi}^{\pi} P(r,r_o,\theta - t)\, U(Rr_o,t)\, dt.$$

8. Show that for $0 < r_1 < r_2 < r_3$, the Poisson kernel (cf. (18)) has the following "convolution" property:

$$P(r_1,r_3,\theta) = \frac{1}{2\pi}\int_{-\pi}^{\pi} P(r_1,r_2,\theta - t)\, P(r_2,r_3,t)\, dt\,.$$

Hint. The left side is a harmonic function of (r_1,θ) for $r_1 < r_3$. Apply the Poisson Integral Formula to this function on the disk $r \le r_2$.

9. A nearly flat heat conducting plate is in the shape of a disk of radius 5. Assume that the plate is insulated on the two flat faces. The boundary of the plate is given a steady temperature distribution of $f(\theta) = 10\,\theta^2$, where the central angle θ ranges from $-\pi$ to π. What is the steady–state temperature at the center of the plate ?

10. Find solutions $U(r,\theta)$ of the Neumann problem

$$\begin{aligned} &\text{D.E.} \quad u_{xx} + u_{yy} = 0 \qquad\qquad x^2 + y^2 < r_o^2 \\ &\text{B.C.} \quad U_r(r_o,\theta) = f(\theta) \\ &\text{P.C.} \quad U(r,\theta + 2\pi) = U(r,\theta)\,, \end{aligned}$$

where $f(\theta) = \frac{1}{2}a_0 + \sum_{n=1}^{N} a_n\cos(n\theta) + b_n\sin(n\theta)$ with N finite and $a_0 = 0$. Why must we assume that $a_0 = 0$?

11. To obtain the Neumann problem for the annulus $(r_i < r < r_o)$, replace U by U_r in the B.C. of (3). Then take $f(\theta)$ and $g(\theta)$ to be as in (6).

(a) Under what condition on A_0 and B_0 will solutions of this Neumann problem exist ?

(b) Under the condition in (a), find solutions of the Neumann problem for the annulus.

Hint . Much work can be avoided by noting that if $U(r,\theta)$ solves $r^{-1}(rU_r)_r + r^{-2}U_{\theta\theta} = 0$, then $V \equiv rU_r$ solves $r^{-1}(rV_r)_r + r^{-2}V_{\theta\theta} = 0$. Thus, solve the Dirichlet problem for V with B.C. $V(r_o,\theta) = r_o f(\theta)$ and $V(r_i,\theta) = r_i g(\theta)$, using the previous formulas (8), then divide by r and integrate with respect to r to obtain the desired solutions U of the Neumann problem.

(c) Let $V(r,\theta)$ be the difference of any two solutions of a fixed Neumann problem. Show that $r^{-1}[r(V\,V_r)]_r + r^{-2}(V\,V_\theta)_\theta = (V_r)^2 + r^{-2}(V_\theta)^2$, using the D.E. . Integrate this result over the annulus $r_i < r < r_o$ with respect to the area element $r dr d\theta$ to deduce that V must be constant. (cf. Problem 9 in Section 6.2).

12. (a) Show that for $r < r_o$, we have

$$\frac{r_o - r}{r_o + r} \le \frac{r_o^2 - r^2}{r_o^2 - 2rr_o\cos(\theta - t) + r^2} \le \frac{r_o + r}{r_o - r}.$$

(b) Use the result in (a) to deduce that if u is harmonic on the open disk $r < r_o$, continuous on the closed disk $r \le r_o$, and nonnegative, then **Harnack's inequality** holds:

$$\frac{r_o - r}{r_o + r}\, u(0,0) \le U(r,\theta) \le \frac{r_o + r}{r_o - r}\, u(0,0) \qquad (r < r_o) .$$

(c) Use Harnack's inequality to prove that a nonnegative harmonic function $u(x,y)$, which is defined for all (x,y), must be constant. **Hint.** Consider large r_o in (b).

(d) Use (c) to prove that the graph $z = u(x,y)$ of a nonconstant harmonic function, which is defined for all (x,y), must intersect every horizontal plane $z =$ const. (i.e., the range of u is the set of all real numbers.). In particular, if there is a constant M such that $|u(x,y)| \le M$ for all (x,y) (i.e., u is bounded), then $u \equiv$ constant.

13. (a) Show that $\frac{1}{2\pi}\int_{-\pi}^{\pi} \frac{(1 - r^2)\sin(t)}{1 - 2r\cos(\theta - t) + r^2}\, dt = -r^{-1}\sin(\theta)$ for $r > 1$.

Hint. Let $\rho = r^{-1} < 1$ and verify that $P(r,1,\theta - t) = -P(\rho,1,\theta - t)$.

(b) More generally, show that if $U(r,\theta) = \frac{1}{2\pi}\int_{-\pi}^{\pi} P(r,r_o,\theta-t)\, f(t)\, dt$ for $r < r_o$, where $f(t)$ is a continuous function, then for $r > r_o$, $\frac{1}{2\pi}\int_{-\pi}^{\pi} P(r,r_o,\theta-t)\, f(t)\, dt = -U(r_o^2 r^{-1},\theta)$.

(c) With $U(r,\theta)$ as in (b), show that $V(r,\theta) = U(r_o^2 r^{-1},\theta)$ is harmonic for $r > r_o$, and $V(r,\theta)$ extends continuously to the function $f(\theta)$ on the circle $r = r_o$. What is $\lim_{r\to\infty} V(r,\theta)$?

6.4. The Maximum Principle and Uniqueness for the Dirichlet Problem

In the previous sections, we have seen that the Maximum/Minimum Principle (or simply, Maximum Principle) and its consequence, the uniqueness theorem, are very useful results. Without uniqueness, we could not have referred to *the* solution of the Dirichlet problem, as we did many times. Also there are certain places where uniqueness is used in a more subtle fashion. For example, the proof of the mean–value theorem (Theorem 2 of Section 6.3) assumes that every harmonic function is given in a suitable disk by the Poisson Integral Formula, but this assumes uniqueness. For the same reason, in proving that harmonic functions are C^∞ (cf.Theorem 3 of Section 6.2) we have used uniqueness. In addition, the Maximum Principle can be used to estimate the difference of solutions of Dirichlet problems in terms of the difference in their boundary values. This estimation is of great importance in applications (cf. Example 5 of Section 6.2). Before we prove the Maximum Principle, we first review some essential terminology.

A subset D of the xy–plane is **open** if each point of D is the center of some disk (of positive radius) which lies entirely inside D. A point p is a **boundary point** of a subset E of the xy–plane, if every disk, centered at p, contains points that are in E and points that are not in E. A boundary point of E need not be in the set E. Indeed, an open set cannot contain any of its boundary points. A set which contains all of its boundary points is called a **closed** set. The set of all boundary points of E is called the **boundary** of E. The **closure** of a set E is the closed set $\overline{E}$ consisting of the points which are either in E or in the boundary of E. For example, if D is the open disk $r < 1$, then D is an open set, and the circle $r = 1$ is the boundary of D, while $\overline{D}$ is the closed disk $r \leq 1$. The **diameter** of the nonempty set E is the maximum distance (if such exists) between pairs of points in $\overline{E}$. Thus, the diameter of a rectangle (open or closed) is the length of a diagonal, whereas the diameter of a disk is its diameter in the usual sense. If there are points in E which are arbitrarily far apart, then the diameter of E is said to be infinite, and E is said to be **unbounded** (e.g., a half–plane, or the exterior of a circle). If the diameter of E is finite, then E is said to be **bounded**. This standard terminology can be tricky. For example, an unbounded set can include its boundary (e.g., the closed half–plane $y \geq 0$). Also, an open interval in the x–axis is not an open subset of the xy–plane. A subset E is (pathwise) **connected** if any two points in E can be joined by a continuous curve which lies totally inside E.

A fact which we will need in this section, is that a function which is continuous on a closed and bounded set achieves its maximum at some point in this set (cf. Appendix A.4). A significant improvement of this fact for harmonic functions is the **Maximum Principle**. This states that if D is a bounded open set and if u is a continuous function on $\overline{D}$ and a harmonic function on D, then u achieves its maximum on the boundary of D. It is possible that this function u will also achieve its maximum at some point of D. However a stronger version of the Maximum Principle (the **Strong Maximum Principle**) implies that in that case u must be identically constant if D is connected, as well as bounded and open. While we are concerned with the two–dimensional setting, we note that the corresponding Strong Maximum Principle in dimension one, for harmonic functions $u = f(x)$, is completely trivial. Indeed, in dimension one, Laplace's equation is $f''(x) = 0$, whence $f(x) = ax + b$. On any closed interval (a bounded, closed subset of the line) such a function achieves its maximum at an endpoint. Moreover, if f(x) also achieves its maximum inside the interval, then clearly f(x) must be constant. Of course, the situation in dimension two is more complicated. For a specific example illustrating the Maximum Principle, see Example 3 of Section 6.2. As a preliminary indication of the unlikely occurrence of a local maximum for a nonconstant harmonic function, we offer the following example.

Example 1. Let $u(x,y)$ be a harmonic function on some open set D. Show that if u has a local maximum at some point $p = (x_0,y_0)$, then all of the first and second partial derivatives of u at p must be zero (i.e., the graph is horizontal and flat "to second order" at p).

Solution. If u has a local maximum ar p, then we know that u_x and u_y vanish at p. Since the graph of u cannot be concave up in any direction, the second directional derivative of u at p in any direction $\cos\theta\,\mathbf{i} + \sin\theta\,\mathbf{j}$ must be nonpositive, i.e.,

$$\frac{d^2}{dt^2}u(x_0 + t\cos\theta,\, y_0 + t\sin\theta)\Big|_{t=0} = \cos^2\theta\, u_{xx}(p) + 2\sin\theta\cos\theta\, u_{xy}(p) + \sin^2\theta\, u_{yy}(p) \le 0. \quad (1)$$

Setting $\theta = 0$, we get $u_{xx}(p) \le 0$. Setting $\theta = \pi/2$, we get $u_{yy}(p) \le 0$. Laplace's equation, $u_{xx} + u_{yy} = 0$, then implies that $u_{xx}(p) = 0$ and $u_{yy}(p) = 0$ (Why?). Setting $\theta = \pi/4$ in (1), we get $u_{xy}(p) \le 0$, while taking $\theta = -\pi/4$ yields $-u_{xy}(p) \le 0$. Thus, $u_{xy}(p) = 0$ also. □

Theorem 1 (The Maximum/Minimum Principle). Let $u = u(x,y)$ be a continuous function on $\overline{D}$, for some open, bounded set D. If u is harmonic on D, then the maximum and the minimum values of u are achieved on the boundary of D.

Proof. We know (cf. Appendix A.4) that the maximum of u is achieved at some point in $\overline{D}$, since u is continuous on the closed, bounded set $\overline{D}$. We prove the theorem by contradiction. Suppose that the maximum is not achieved on the boundary. Then the maximum is achieved at some point (x_0,y_0) in D, say $M \equiv u(x_0,y_0) > M_b$, where M_b is the maximum of u on the boundary of D. Let

$$v(x,y) = u(x,y) + \epsilon\,[(x - x_0)^2 + (y - y_0)^2]\ , \quad (2)$$

for some constant $\epsilon > 0$. Then $v(x_0,y_0) = u(x_0,y_0) = M$, and the maximum of v on the boundary of D is at most $M_b + \epsilon d^2$, where d is the diameter of D. For ϵ sufficiently small, we have $M > M_b + \epsilon d^2$ (i.e., $0 < \epsilon < (M - M_b)/d^2$)). For such ϵ, the maximum of v cannot occur on the boundary of D, since the value M of v at (x_0,y_0) is larger than the value of v at any boundary point. There may, however, be points in D where v is greater than M. Let the maximum of v be achieved at (x_1,y_1), which (as we have just seen) must be in D. At (x_1,y_1), we must have $v_{xx} \le 0$ and $v_{yy} \le 0$, since the graph of v cannot be concave up in the x or y directions at (x_1,y_1). Thus, at (x_1,y_1), we have $v_{xx} + v_{yy} \le 0$. However, by (2), we have $v_{xx} + v_{yy} = u_{xx} + u_{yy} + 2\epsilon + 2\epsilon = 4\epsilon > 0$, where we have finally used the assumption that u is harmonic on D. Thus, we have reached a contradiction, and our original assumption, that the maximum of u is not achieved on the boundary, is false. Since the minimum of u is the negative of the maximum of the function $-u$ (also harmonic on D), the minimum of u must also be achieved on the boundary of D, by the above result for the maximum. □

Remark. Note that we have not used any of the results of the previous sections in the above proof. The uniqueness theorem below is a direct consequence of the Maximum Principle, and hence we have avoided any circular reasoning (i.e., the proof of uniqueness does not involve the use of previous results in which uniqueness was assumed). □

Theorem 2 (The uniqueness theorem for the Dirichlet problem). For some open, bounded set D, let u_1 and u_2 be continuous functions on $\overline{D}$ which are harmonic on D. If u_1 and u_2 are equal at all boundary points of D, then they are equal throughout $\overline{D}$. In other words, there is at most one solution of a Dirichlet problem for D.

Proof. The difference $v \equiv u_1 - u_2$ is a continuous function on $\overline{D}$ which is harmonic on D. Since $v \equiv 0$ on the boundary of D, the Maximum/Minimum Principle implies that $v \leq 0$ and $v \geq 0$ on $\overline{D}$. Hence, $v \equiv 0$ on $\overline{D}$, and $u_1 \equiv u_2$ on $\overline{D}$. □

We next show that if the prescribed function on the boundary, in a Dirichlet problem for a bounded open set D, is changed by at most ϵ, then the solutions (if they exist) differ by at most ϵ throughout D. Roughly speaking, solutions depend continuously on the boundary data.

Theorem 3 (Continuous dependence of solutions on boundary data). Let u_i (for $i = 1, 2$) be the solution (if it exists) of the Dirichlet problem

$$\text{D.E.} \quad \Delta u_i = 0 \qquad \text{on } D$$

$$\text{B.C.} \quad u_i = f_i \qquad \text{on } C,$$

where C is the boundary of the bounded, open set D. If M is the maximum of $|f_1 - f_2|$ on C, then $|u_1(x,y) - u_2(x,y)| \leq M$, for all (x,y) in $\overline{D}$.

Proof. Let $v \equiv u_1 - u_2$. For all (x,y) in $\overline{D}$, we have

$$-\max_C(|f_1 - f_2|) \leq \min_C(f_1 - f_2) \leq v(x,y) \leq \max_C(f_1 - f_2) \leq \max_C(|f_1 - f_2|),$$

where the first and last inequalities are clear and the middle two inequalities follow from the Maximum/Minimum Principle applied to v. Thus, $-M \leq v(x,y) \leq M$, and $|v(x,y)| \leq M$ for all (x,y) in $\overline{D}$. □

Remark. When $f_1 = f_2$, we have $M = 0$, and hence Theorem 3 yields the uniqueness theorem (Theorem 2) as a corollary. □

Example 2. Let $u(x,y)$ be any nonconstant harmonic function on the entire xy–plane. By Problem 12(d) of Section 6.3, we know that the zero level set of u, $\{ (x,y) \mid u(x,y) = 0 \}$, is nonempty. Show that this level set cannot contain any circle.

Solution. If this were possible, the harmonic function $u(x,y)$ would yield a solution of the Dirichlet problem for the disk D enclosed by the circle, where the boundary data is zero. By the uniqueness theorem, u would have to be 0 throughout the disk D. We show that $u \equiv 0$ throughout the xy–plane, as follows. Note that at the center p of D, all of the partial derivatives of u equal zero. Thus, the Taylor series of u about p would be zero. However, by the remark following Theorem 4 in Section 6.3, on any disk where u is harmonic, u is equal to its Taylor series about the center of the disk. Since u is harmonic everywhere on the plane, it is harmonic on arbitrarily large disks about p. Thus, the vanishing of the Taylor series of u about p implies that $u \equiv 0$ on the entire plane. Essentially the same argument proves that the zero set of $u(x,y)$ cannot contain the boundary of any nonempty, bounded, open set. □

Example 3. Suppose that $u(x,y)$ is a continuous function on the closed disk $r \leq 1$, and assume that u is harmonic on the open disk $r < 1$. If $u(\cos\theta,\sin\theta) \leq \sin\theta + \cos(2\theta)$, then show that we have $u(x,y) \leq y + x^2 - y^2$, for all (x,y) with $x^2 + y^2 \leq 1$.

Solution. Note that $v(x,y) \equiv y + x^2 - y^2$ is a harmonic function with $v(\cos\theta, \sin\theta) = \sin\theta + \cos(2\theta)$. By assumption, $u \leq v$ on the boundary of the disk $r \leq 1$. Thus, the maximum of the harmonic function $u - v$ on the boundary $r = 1$ must be less than or equal to zero. The Maximum Principle then implies that $u - v \leq 0$ throughout the disk, as desired. □

Example 4 (Uniqueness of the Poisson kernel). Suppose that for every continuous periodic function $f(\theta)$ of period 2π, a solution of the Dirichlet problem on the disk $r \leq r_0$, with boundary condition $U(r_0,\theta) = f(\theta)$, is given by the formula

$$U(r,\theta) = \frac{1}{2\pi}\int_{-\pi}^{\pi} Q(r,r_0,\theta-t)\, f(t)\, dt \tag{3}$$

for some function $Q(r,r_0,\theta-t)$, which is continuous and periodic in t of period 2π. Show that $Q(r,r_0,\theta-t)$ must equal the Poisson kernel $P(r,r_0,\theta-t)$, for all $r < r_0$ and all θ and t.

Solution. Since the solution U of the Dirichlet problem is unique, we know that (3) holds if Q is replaced by P. Thus, we have

$$0 = U(r,\theta) - U(r,\theta) = \frac{1}{2\pi}\int_{-\pi}^{\pi} [P(r,r_0,\theta-t) - Q(r,r_0,\theta-t)]\, f(t)\, dt.$$

For any fixed θ and $r < r_0$, let $f(t)$ be the continuous periodic function

$P(r,r_0,\theta-t) - Q(r,r_0,\theta-t)$ in this last integral. Then we get $\int_{-\pi}^{\pi} [f(t)]^2\, dt = 0$, whence $f(t) = 0$ for $-\pi \le t \le \pi$. Since $f(t)$ is periodic of period 2π, we then know that $f(t) = 0$ for all t. □

The Strong Maximum/Minimum Principle

The Strong Maximum/Minimum Principle states that if a harmonic function u on some connected, open set D achieves its maximum or minimum in D, then u must be identically constant on D. It is not necessary for D to be bounded or for u to extend continuously to the boundary of D. The goal of this subsection is to prove this result. The main tool to be used is the mean–value theorem which was established as a consequence of the Poisson Integral Formula (cf. Theorem 2 of Section 6.3). We next show that it is also possible (and instructive) to prove the mean–value theorem in an elementary fashion, without the use of this integral formula. This alternate proof is based on the following result, where we write $U(r,\theta) = u(x,y)$ (cf. Section 6.3).

Lemma 1. For any C^2 **function** $u(x,y)$ **defined on the disk** D $(r \le R)$, **we have the formula**

$$\iint_D (u_{xx} + u_{yy})\, dx\, dy = \int_0^{2\pi} U_r(R,\theta)\, R\, d\theta. \tag{4}$$

In other words, the integral of the Laplacian of u over the disk is the integral of the outward normal derivative of u along the boundary with respect to the arc length differential $ds = R\, d\theta$.

Proof. Computing the left side of (4) in terms of polar coordinates,

$$\int_0^{2\pi}\int_0^R (U_{rr} + r^{-1}U_r + r^{-2}U_{\theta\theta})\, r\, dr d\theta = \int_0^{2\pi}\int_0^R (rU_{rr} + U_r)\, dr d\theta + \int_0^{2\pi}\int_0^R r^{-1}U_{\theta\theta}\, dr d\theta$$

$$= \int_0^{2\pi}\int_0^R (r\, U_r)_r\, dr d\theta + \int_0^R\int_0^{2\pi} r^{-1}\, U_{\theta\theta}\, d\theta dr$$

$$= \int_0^{2\pi} (r\, U_r)\Big|_0^R\, d\theta + \int_0^R r^{-1}\, U_\theta(r,\theta)\Big|_0^{2\pi}\, dr\,. \tag{5}$$

But (5) is the right side of (4), since the second term in (5) is 0, due to the fact that U (and hence U_θ) is periodic of period 2π in θ. To justify the change in the order of integration (cf. Appendix A.2), we note that the apparent singularity in $r^{-1}U_{\theta\theta}$ is removable, because (by the proof of Proposition 1 in Section 6.3), $r^{-1}U_{\theta\theta}$ extends continuously to the function $-u_x(0,0)\cos\theta - u_y(0,0)\sin\theta$ on the edge $r = 0$, $0 \le \theta \le 2\pi$ of the rectangle $0 < r < R$, $0 < \theta < 2\pi$ in the $r\theta$–plane. □

An alternate proof of the mean–value theorem (Theorem 2 of Section 6.3)

In Lemma 1, if u is a harmonic function, then the left side of (4) is zero, and thus (by Leibniz's rule, in Appendix A.3)

$$0 = \int_0^{2\pi} U_r(R,\theta)\, d\theta = \frac{d}{dR}\left[\int_0^{2\pi} U(R,\theta)\, d\theta \right].$$

Hence, $\int_0^{2\pi} U(R,\theta)\, d\theta$ is a constant function of R. This function is $2\pi u(0,0)$ at $R = 0$, whence the mean–value theorem holds. □

Remark. Lemma 1 holds in much greater generality. Indeed, let D be a bounded open set with boundary C consisting of a finite number of smooth, simple, closed curves (e.g., as a slab of Swiss cheese), and let u be a C^2 function on D. Then, by using a theorem of Green, one has the following generalization of (4)

$$\iint_D \Delta u\, dxdy = \int_C \nabla u \cdot \mathbf{n}\, ds, \tag{6}$$

where **n** is the outward unit normal and ds is the element of arc length. For a rectangle, (6) was shown to hold in Example 6 of Section 6.2. As an immediate consequence of (6), we obtain the compatibility condition for the Neumann problem

$$\begin{array}{lll} \text{D.E.} & \Delta u = 0 & \text{on } D \\ \text{B.C.} & \nabla u \cdot \mathbf{n} = g & \text{on } C, \end{array}$$

namely $\int_C g\, ds = 0$. □

Example 5. Let u(x,y) be harmonic in the disk $x^2 + y^2 < r_0^2$. If u achieves its maximum at the point (0,0), then show that u must be constant throughout this disk.

Solution. By the mean–value theorem, we have (for any $r < r_0$)

$$\int_0^{2\pi} u(0,0)\, d\theta = 2\pi u(0,0) = \int_0^{2\pi} U(R,\theta)\, d\theta .$$

Subtracting, we obtain $\int_0^{2\pi} [u(0,0) - U(r,\theta)]\, d\theta = 0$. Since the integrand $u(0,0) - U(r,\theta)$ is continuous and nonnegative (by the assumption that the maximum of u is achieved at (0,0)), we have that $u(0,0) - U(r,\theta) \equiv 0$, as a function of θ. Thus, $U(r,\theta) \equiv u(0,0)$, for any $r < r_0$, and u is constant throughout the disk $r < r_0$. □

Theorem 4 (The Strong Maximum/Minimum Principle). Let u be a harmonic function on the open connected set D. Suppose that the maximum or minimum of u is achieved at some point in D. Then u must be constant throughout D.

Proof. Let p be the point in D where u achieves its maximum, say M, and let q be any other point in D. Since D is connected, we can join p to q by a curve, say with parametrization $(x(t),y(t))$, where $x(t)$ and $y(t)$ are continuous functions for $0 \le t \le 1$ with $p = (x(0),y(0))$ and $q = (x(1),y(1))$. We must prove that $u(q) = M$. Let $S = \{t \in [0,1] : u(x(t),y(t)) = M\}$. Certainly, $0 \in S$, but we need to show that $1 \in S$. Let t_0 be the smallest real number which is greater than or equal to all numbers in S. (Since S is bounded, the existence of such t_0 is guaranteed by an axiom of the real number system, **the least upper bound axiom.**) We must prove that t_0 is in S and $t_0 = 1$. For each positive integer n, there is a number t_n in S, such that $t_0 - t_n \le n^{-1}$, by the definition of t_0. By the continuity of $u(x(t),y(t))$, we have

$$u(x(t_0),y(t_0)) = \lim_{n\to\infty} u(x(t_n),y(t_n)) = M.$$

Thus, $t_0 \in S$, and so u achieves its maximum at $(x(t_0),y(t_0))$. Let D_0 be an open disk, completely contained in D, with center $(x(t_0),y(t_0))$. By Example 5, $u \equiv M$ on D_0. We now prove by contradiction that $t_0 = 1$. If $t_0 < 1$, then by the continuity of $x(t)$ and $y(t)$, we know that $(x(t_0+\delta),y(t_0+\delta))$ is in D_0 for sufficiently small $\delta > 0$. Thus, $t_0 + \delta$ is in S, contradicting the definition of t_0. Hence, we must have $t_0 = 1$, as desired. Since the minimum of u is the negative of the maximum of $-u$, we also know that if u attains its minimum in D, then u is constant throughout D. □

Remark. The Maximum/Minimum Principle (Theorem 1) is an immediate consequence of the Strong Maximum/Minimum Principle. Indeed, in the notation of Theorem 1, if the maximum (or minimum) is not achieved on the boundary of D, then u is constant by Theorem 4, whence the maximum (or minimum) is achieved everywhere in $\overline{D}$, and we have a contradiction. Also, since the above alternative proof of the mean–value theorem (used in Example 5) did not use uniqueness (Theorem 2), the proof of Theorem 1, by means of Theorem 4, is not circular. □

Example 6. Suppose that $u(x,y)$ is a continuous function defined on the closed horizontal strip $-1 \le y \le 1$, and assume that u is harmonic for $-1 < y < 1$. Assume that, for each $y \in [-1,1]$, $u(x,y)$ is periodic in x of period 2 (i.e., $u(x + 2,y) = u(x,y)$ for all x). If u is zero on the boundary lines $y = \pm 1$, then prove that u must be zero throughout the strip. Give an example which shows that the conclusion is false if the periodicity assumption is dropped.

Solution. The square $-1 \le x,\ y \le 1$ is closed and bounded. Thus, the continuous function u, when restricted to this square, achieves its maximum and minimum at points in the square (cf. Appendix A.4). Since u is assumed to be periodic of period 2 in x, the maximum and minimum values of u in any translate (in the x–direction) of the square will be the same as in the original square. Thus, the maximum and minimum values for u on the strip are achieved at infinitely many points. If any one of these points is not on the boundary lines of the strip, then (by Theorem 4) u must be constant on the strip, and hence identically zero on the strip (Why?). If the maximum and the minimum occur on the boundary lines where u is zero, then the function is obviously 0 on the strip. Thus, in any case $u \equiv 0$. Note that the harmonic function $u(x,y) = \cosh(\pi x/2)\cos(\pi y/2)$ is zero on the boundary $y = \pm 1$, but it is positive for $-1 < y < 1$. Thus, the periodicity assumption cannot be dropped. This also shows that uniqueness fails for the Dirichlet problem on the strip (Why?). □

Summary 6.4

1. The Maximum/Minimum Principle (Theorem 1) : Let $u = u(x,y)$ be a continuous function on $\overline{D}$, for some open, bounded set D. If u is harmonic on D, then the maximum and the minimum values of u are achieved on the boundary of D.

2. Uniqueness for the Dirichlet problem (Theorem 2) : For some open bounded set D, let u_1 and u_2 be continuous functions on $\overline{D}$ which are harmonic on D. If u_1 and u_2 are equal at all boundary points of D, then they are equal throughout $\overline{D}$. In other words, there is at most one solution of a Dirichlet problem for D.

3. Continuous dependence of solutions on boundary data : Let u_i (for $i = 1, 2$) be the solution (if it exists) of the Dirichlet problem

$$\text{D.E.} \quad \Delta u_i = 0 \qquad \text{on } D$$

$$\text{B.C.} \quad u_i = f_i \qquad \text{on } C,$$

where C is the boundary of the bounded, open set D. If M is the maximum of $|f_1 - f_2|$ on C, then $|u_1(x,y) - u_2(x,y)| \leq M$, for all (x,y) in $\overline{D}$.

4. The Strong Maximum/Minimum Principle (Theorem 4) : Let u be a harmonic function on the open, connected set D. Suppose that the maximum or minimum of u is achieved at some point in D. Then u must be constant throughout D.

Exercises 6.4

1. Let D be a bounded, open set in the xy–plane and let the functions u_1, u_2 and u_3 be continuous on $\overline{D}$ and harmonic on D. Show that if $u_1 \leq u_2 \leq u_3$ on the boundary of D, then $u_1 \leq u_2 \leq u_3$ throughout D.

2. Let $u(x,y)$ be a continuous function on $\overline{D}$, where D is some open, connected set. If u is harmonic on D, explain why generally it is a waste of time to locate a point where u achieves its maximum by solving the equations $u_x = 0$ and $u_y = 0$ simultaneously.

3. By considering product solutions, show that the Dirichlet problem

$$\text{D.E.} \quad u_{xx} + u_{yy} = 0 \qquad 0 < x < \pi,\ 0 < y < \infty$$

$$\text{B.C.} \quad u(x,0) = 0\ ,\ u(0,y) = 0\ ,\ u(\pi,y) = 0 \qquad 0 \le x \le \pi,\ 0 \le y < \infty\ .$$

has more than one solution. Why does this not contradict Theorem 2 ?

4. Note that $u_1(x,y) = 1 + \log(x^2 + y^2)$ and $u_2(x,y) = 1 - \log(x^2 + y^2)$ are harmonic, where defined. Moreover, these functions are equal on the circle $x^2 + y^2 = 1$, but unequal inside the circle. Why does this not contradict Theorem 2?

5. Show that any plane which is tangent to the graph of a harmonic function (defined on an open, connected set D) must intersect the graph in more than one point. **Hint.** The tangent plane is also the graph of a harmonic function. Consider the difference of these functions, and use the Strong Maximum Principle.

6. Use the mean–value theorem to show that for any real constant b with $|b| < 1$, we have $\int_0^{2\pi} \log[(1 + b\cos\theta)^2 + (b\sin\theta)^2]\, d\theta = 0$. Show that this equation cannot hold for $b > \sqrt{2}$.

7. Let $f(x)$ be a continuous function, defined for all real x, such that $|f(x)| \le M$ for some constant M.

(a) Show that the Dirichlet problem for the upper half–plane $y \ge 0$,

$$\text{D.E.} \quad u_{xx} + u_{yy} = 0 \qquad -\infty < x < \infty,\ y > 0$$

$$\text{B.C.} \quad u(x,0) = f(x) \qquad -\infty < x < \infty\ ,$$

does not have a unique solution, by considering the function $v(x,y) \equiv y$.

(b) Define $u(x,y) = \frac{1}{\pi}\int_{-\infty}^{\infty} \frac{y\, f(s)}{(x-s)^2 + y^2}\, ds$, and show (at least formally) that u is harmonic for $y > 0$. (The advanced reader may wish to carefully and repeatedly apply Leibniz's rule in Appendix A.3 to justify differentiating under the integral. First, consider (x,y) in the region $-A < x < A,\ 0 < \epsilon < y < \infty$, for positive constants A and ϵ .)

(c) For any point $(x_0,0)$ on the x–axis, show that $u(x,y)$ in (b) can be made arbitrarily close to $f(x_0)$ by taking (x,y) (for $y > 0$) sufficiently close to $(x_0,0)$.

Hint. First show that when $f(s) \equiv K$, for some constant K, the integral in (b) is identically K. By taking K to be $f(x_0)$, deduce that

$$u(x, y) - f(x_0) = \frac{1}{\pi}\int_{-\infty}^{\infty} \frac{y[f(s) - f(x_0)]}{(x-s)^2 + y^2}\, ds.$$

Split the interval of integration into three pieces, namely $(-\infty, x_0 - \delta]$, $[x_0 - \delta, x_0 + \delta]$ and $[x_0 + \delta, \infty)$, and estimate each of the resulting three integrals, as was done in the proof of the Poisson Integral Formula for the disk (cf. Theorem 1 of Section 6.3).

Remark. Part (c) establishes that a solution of the Dirichlet problem in (a) is given by the formula in (b), which is known as the **Poisson Integral Formula for the upper half–plane.**

8. For $r < 1$ and $z = re^{i\theta}$, let $U(r,\theta) = \mathrm{Im}\{[(1 + z)/(1 - z)]^2\}$.

(a) Verify that $U(r,\theta)$ is harmonic for $r < 1$.

(b) Show that $\lim_{r\to 1} U(r,\theta) = 0$ for each fixed θ.

(c) Why can we not use Theorem 1 to conclude that $U(r, \theta) \equiv 0$?

Hint. Show that $u(x,y) = -4y \dfrac{y^2 + (x^2 - 1)}{[y^2 + (x-1)^2]^2}$. Consider the limit of $u(x,y)$ as (x,y) approaches $(1,0)$ within the disk, along the line $y = x - 1$.

9. Let $q(x,y)$ be a C^2 function which is identically zero outside of some disk. Here we demonstrate that equation (*) below is a solution of Poisson's equation $u_{xx} + u_{yy} = q(x,y)$.

(a) For any fixed (x,y) in the plane, let

$$u(x,y) = \frac{1}{4\pi}\int_{-\infty}^{\infty}\int_{-\infty}^{\infty} \log[(x - s)^2 + (y - t)^2]\, q(s,t)\, ds\, dt. \qquad (*)$$

Show that $u(x,y) = \dfrac{1}{4\pi}\displaystyle\int_{-\infty}^{\infty}\int_{-\infty}^{\infty} \log[\bar{s}^2 + \bar{t}^2]\, q(x + \bar{s}, y + \bar{t})\, d\bar{s}\, d\bar{t}.$

$$= \frac{1}{2\pi}\int_{-\pi}^{\pi}\int_{0}^{\infty} q(x + r\cos\theta, y + r\sin\theta)\log(r)\, r\, dr\, d\theta, \qquad (**)$$

and use the fact that $\lim_{r\to 0^+} r\log(r) = 0$ (Why?), to deduce from Leibniz's rule that $u(x,y)'$ is C^2, and that u_{xx} and u_{yy} can be computed by differentiating under the integral (**).

(b) Let $Q(x,y,r,\theta) = q(x + r\cos\theta, y + r\sin\theta)$. Show that, for $r > 0$, we have $Q_{xx} + Q_{yy} = r^{-1}(rQ_r)_r + r^{-2}Q_{\theta\theta}$. Use this fact to deduce from (a) that $u_{xx} + u_{yy} = q(x,y)$.

Hint. When computing the integral of $[(rQ_r)_r + r^{-1}Q_{\theta\theta}]\log r$, use integration by parts to find the integral of $(rQ_r)_r \log r$ with respect to r, and use the Fundamental Theorem of Calculus and periodicity in θ to compute the integral of $r^{-1}Q_{\theta\theta}$ with respect to θ.

10. Let D be a bounded, open set with boundary C, and suppose that each Dirichlet problem for Laplace's equation with continuous boundary function has a (necessarily unique) solution. Let $q(x,y)$ be a C^2 function defined for all (x,y), such that q is zero outside of some disk which contains D. Use Problem 9 to show that the Dirichlet problem

$$\text{D.E.} \quad u_{xx} + u_{yy} = q(x,y) \qquad \text{on } D$$

$$\text{B.C.} \quad u(x,y) = f(x,y) \qquad \text{on } C$$

for Poisson's equation has a unique solution, where f is a given continuous function on the boundary C of D. **Hint.** See Example 3 of Section 6.1.

11. Let D be an open set and let $a \in D$. A **Green's function of** D **with logarithmic singularity at** a is a real-valued function $g(z;a)$ of $z = x + iy \in D$, such that

(i) $g(z;a)$ is harmonic in $D - \{a\}$,

(ii) $g(z;a) - \log|z-a|$ is harmonic in a disk about a,

and (iii) for each boundary point w of D, the limit of $g(z;a)$, as z approaches w from within D, is 0 (i.e., $g(z;a)$ continuously extends by zero values to $\overline{D} - \{a\}$).

(a) Prove that if a bounded, open set D has a Green's function $g(z;a)$, then $g(z;a)$ is unique.

(b) With D and $g(z;a)$ as in part (a) and D connected, prove that $g(z;a)$ is positive on D.

Hints. (a) Consider $h(z;a)$ satisfying (i) – (iii) and apply the Maximum/Minimum Principle to $h(z;a) - g(z;a)$. Why is this difference nonsingular ?
(b) Note that $\lim_{z\to a} g(z;a) = +\infty$, and apply the Strong Maximum Principle to $g(z;a)$ on D minus arbitrarily small disks about $z = a$.

12. Let D and C be as in Problem 10 and suppose that D has a Green's function $g(z;a)$ as in Problem 11. For $z = x + iy$ and $a = s + it$, let $G(x,y,s,t) \equiv g(z;a)$. Assume that G_{xx} and G_{yy} extend continuously to $\overline{D} \times \overline{D}$, so that Leibniz's rule may be applied when needed. Show that the solution of the Dirichlet problem in Problem 10, with $f(x,y) = 0$ on C, is

$$u(x,y) = \frac{1}{2\pi}\iint_D G(x,y,s,t)\, q(s,t)\, ds\, dt\,, \qquad (x,y) \in D\,.$$

For the verification that $u(x,y)$ approaches 0 as (x,y) approaches a point on C from within

D, the typical reader may formally take the limit under the integral. The more advanced reader may try to justify this operation.

Hint. To show that $u(x,y)$ solves Poisson's equation, write $G(x,y,s,t)$ as $\log|z-a| + \{g(z;a) - \log|z-a|\}$, split the integral into a sum of two integrals, and apply the result of Problem 9. Observe that the integral of $\log|z-a|\, q(s,t)$, with respect to $ds\, dt$, over the *exterior* of D is harmonic (Why?).

13. (a) Verify that the Green's function for the disk $r < 1$ is $g(z;a) = \log|(z-a)/(\bar{a}z-1)|$, $|a| < 1$, $z = x + iy = re^{i\theta}$.

(b) Use part (a) and Problems 11 and 12 to obtain an explicit integral formula for the solution of the Dirichlet problem (in Problem 10) for Poisson's equation when D is the disk $r < 1$, and the boundary function f is identically 0. What can be done, if the continuous boundary function f is arbitrary?

14. (a) By direct computation, verify that the Poisson kernel $P(r,1,\theta-t)$ for the unit disk is given by $P(r,1,\theta-t) = \frac{\partial}{\partial R} G(r,\theta;R,t)\Big|_{R=1}$, where $G(r,\theta;R,t) = g(re^{i\theta};Re^{it})$ is the Green's function for the disk in Problem 13.

(b) Derive formally the result of (a) by using the following Green's formula for the disk

$$\int_0^{2\pi}\int_0^1 [\Delta V(R,t)U(R,t) - V(R,t)\Delta U(R,t)]\, R\, dR\, dt = \int_0^{2\pi}\Big[V_R(1,t)U(1,t) - V(1,t)U_R(1,t)\Big]\, dt. \quad (*)$$

(The derivation of this formula is similar to the proof of Lemma 1. It only involves two simple integrations by parts and does not require the general Green's formula from vector analysis.)

Hints. Let $U(r,\theta)$ be a harmonic function with $U(1,\theta) = f(\theta)$, and let $V(R,t) = G(r,\theta;R,t)$ for fixed (r,θ). Note that $U(r,\theta) = \Delta\left[\frac{1}{2\pi}\int_0^{2\pi}\int_0^1 G(r,\theta;R,t)U(R,t)\, R\, dR\, dt\right]$, where $G(r,\theta;R,t)$ is $g(z;a)$ of Problem 13 (Why ?). Now, formally bring Δ under the integral and apply $(*)$, noting that $\Delta U(R,t) = 0$, by assumption. Finally, obtain $U(r,\theta) = \int_0^{2\pi} G_R(r,\theta;1,t)U(1,t)\, dt$, whence $G_R(r,\theta;1,t) = P(r,1,\theta-t)$, by the uniqueness of the Poisson kernel (cf. Example 4).

6.5 Complex Variable Theory with Applications

We have seen that the real and imaginary parts of $(x + iy)^n$, for $n = 0, 1, 2, \ldots$, are harmonic functions (cf. Problem 5(d) of Section 6.1, or use De Moivre's Formula (12) in Section 6.3). One can easily check that the real and imaginary parts of $e^{x+iy} = e^x\cos(y) + ie^x\sin(y)$ are also harmonic. More generally, suppose that we have a function $f(z)$ of the complex variable $z = x + iy$. As we will see shortly, a key property which guarantees that the real and imaginary parts of $f(z)$ will be harmonic in an open set D, is that $f(z)$ is **differentiable with respect to** z, in the sense that the limit

$$f'(z) = \lim_{h \to 0} \frac{f(z+h) - f(z)}{h} . \tag{1}$$

exists at each point $z = x + iy$ in D (i.e., $(x,y) \in D$), *regardless of how the complex number* h *approaches* 0. For example, the derivative of z^2 is $2z$, since

$$\lim_{h \to 0} h^{-1}[(z + h)^2 - z^2] = \lim_{h \to 0} h^{-1}[2hz + h^2] = \lim_{h \to 0} (2z + h) = 2z.$$

A function $f(z)$, such that $f'(z)$ exists for all z in the open set D, is said to be **complex analytic** (or simply **analytic** or **holomorphic**) on D. If $f(z)$ is analytic on the entire complex plane, then $f(z)$ is called an **entire function**. Examples of entire functions include polynomials $p(z) = a_0 + a_1z + a_2z^2 + \ldots + a_nz^n$, where the coefficients a_i are constants (possibly complex), and the functions e^z, $\sin z$, $\cos z$, $\sinh z$, $\cosh z$, which can be defined in terms of power series in z. The function z^{-1} is not entire, since its derivative $-z^{-2}$ does not exist at $z = 0$, but this function is holomorphic on the punctured plane $\{z : z \neq 0\}$. The following proposition exhibits a relation between analytic functions and harmonic functions.

Proposition 1 (The Cauchy–Riemann Equations). If $f(x + iy) = u(x,y) + iv(x,y)$ **is analytic on an open set** D, **then the real and imaginary parts** (u **and** v, **respectively) of** f **obey the Cauchy–Riemann equations in** D

$$u_x = v_y \qquad \text{and} \qquad u_y = -v_x . \tag{2}$$

If in addition, u **and** v **are** C^2 **on** D, **then they are harmonic on** D, **and in this case** $f'(z) = u_x + i\,v_x = v_y - i\,u_y$.

Proof. If we take h to be real in (1), then we get (for z in D)

$$f'(z) = \lim_{h \to 0} \frac{u(x+h,y) + iv(x+h,y) - [u(x,y) + iv(x,y)]}{h} = u_x + iv_x . \tag{2'}$$

If we take h to be the imaginary number ik, where k is real, then

$$f'(z) = \lim_{ik \to 0} \frac{u(x,y+k) + iv(x,y+k) - [u(x,y) + iv(x,y)]}{ik} = v_y - iu_y .$$

Equating the corresponding real and imaginary parts in these two expressions for $f'(z)$, we obtain the Cauchy–Riemann equations. If u and v are C^2, then using the Cauchy–Riemann equations, we obtain $u_{xx} + u_{yy} = v_{yx} - v_{xy} = 0$ and $v_{xx} + v_{yy} = -u_{yx} + u_{xy} = 0$, whence u and v are harmonic on D. □

Remarks. It turns out that the real and imaginary parts of an analytic function are always C^∞. Thus, our assumption that u and v are C^2 is actually unnecessary, but it is needed for the above proof. Conversely, if u and v are C^1 functions which satisfy the Cauchy–Riemann equations, then it can be proved that the function $f(z) = u + iv$ is analytic (cf. Rudin, 1987). □

The harmonic function v is called *a* **harmonic conjugate** of the harmonic function u, if $u + iv$ is analytic. For example, since $z^2 = x^2 - y^2 + i\,2xy$ is analytic, the function $2xy$ is a harmonic conjugate of $x^2 - y^2$. Since the partial derivatives of v are determined by u via the Cauchy–Riemann equations, if u has two harmonic conjugates, then these must differ by a constant. If u is harmonic in an open rectangular region R, then there is a function v, such that $u + iv$ is analytic in R (i.e., u has a harmonic conjugate v on R). This fact is a consequence of the next proposition which we will also use later in connection with fluid flow problems. Thus, any harmonic function u on R is the real part of some analytic function $u + iv$. Since $-v + iu = i(u + iv)$ is also analytic, we see that u is also the imaginary part of some analytic function on R. Note that if v is a harmonic conjugate of u, then $-u$ (*minus* u) is a harmonic conjugate of v, since $v - iu = -i(u + iv)$ is analytic. Thus, strictly speaking, one should not say that u and v are harmonic conjugates of each other, as is commonly (indeed, almost always) done.

Proposition 2. Let $P(x,y)$ **and** $Q(x,y)$ **be** C^1 **functions on an open rectangular region** R **(possibly with one or more sides of infinite length, so that** R **may be a strip). Then there is a** C^2 **function** $f(x,y)$ **on** R**, such that**

$$f_x(x,y) = P(x,y) \quad \text{and} \quad f_y(x,y) = Q(x,y), \tag{3}$$

if and only if the integrability condition $P_y \equiv Q_x$ **holds on** R.

Proof. If f satisfies (3), then $P_y = f_{xy} = f_{yx} = Q_x$. Conversely, we assume that $P_y \equiv Q_x$ on R, and construct f satisfying (3), as follows. Let (a,b) be a point in R, and let H be any C^2

function satisfying $H_x = P$ on R (e.g., $H(x,y) = \int_a^x P(s,y)\,ds$). Similarly, let $K(x,y)$ be a C^2 function on R such that $K_y = Q$ on R. Then set $u(x,y) = H(x,y) - K(x,y)$ and note that $u_{xy} = H_{xy} - K_{yx} = P_y - Q_x = 0$. Thus, u solves the PDE $u_{xy} = 0$, which has the general solution on R of the form $u(x,y) = k(x) - h(y)$, for C^2 functions h and k. Thus, $H(x,y) - K(x,y) = u(x,y) = k(x) - h(y)$. Define f by

$$f(x,y) = H(x,y) + h(y) \qquad \text{and} \qquad f(x,y) = K(x,y) + k(x)\ .$$

Since the two right–hand sides are equal, it does not matter which equation we use. From the first equation, we get that $f_x = H_x = P$, and from the second equation we get $f_y = K_y = Q$. □

Proposition 3. Any harmonic function u, **defined on an open rectangular region** R, **has a harmonic conjugate** v **defined on** R.

Proof. By the remarks before Proposition 2, we need only to show that for the given harmonic function u, we can solve the Cauchy–Riemann equations $v_x = -u_y$ and $v_y = u_x$. We apply Proposition 2 with $P = -u_y$ and $Q = u_x$. Since u is harmonic, we have the integrability condition $P_y = -u_{yy} = u_{xx} = Q_x$, and thus v exists. □

Remark. It can be shown that Propositions 2 and 3 remain true, if the rectangular region R is replaced by an open region without "holes" (i.e., any open region whose exterior is connected). Such a region in the plane is called **simply–connected**. Example 3 below shows that this hypothesis is necessary. □

Example 1. Find a harmonic conjugate of the harmonic function $u(x,y) = \sin(x)\cosh(y) + y$, defined on the whole plane.

Solution. Proceeding as in the proof of Proposition 2, we integrate the equation $v_y = u_x = \cos(x)\cosh(y)$ with respect to y, and $v_x = -u_y = -\sin(x)\sinh(y) - 1$ with respect to x. Then $v(x,y) = \cos(x)\sinh(y) + h(x)$ and $v(x,y) = \cos(x)\sinh(y) - x + k(y)$, where h and k are arbitrary C^1 functions. Comparing these two expressions for $v(x,y)$, we see that $h(x) = -x + c$ and $k(y) = c$, where c is an arbitrary real number. Thus, $v(x,y) = \cos(x)\sinh(y) - x + c$ is a harmonic conjugate of u. The associated analytic function is

$$f(z) = u + iv = [\sin(x)\cosh(y) + y] + i[\cos(x)\sinh(y) - x + c]\ ,$$

which turns out to be $\sin(z) - iz + ic$ (cf. Problem 1). □

Example 2. Show that if v is a harmonic conjugate of the harmonic function u, then at any point, the gradients ∇u and ∇u are of equal length and are perpendicular. Conclude that at a point where $\nabla u \neq \mathbf{0}$, the level curves of u and v are orthogonal.

Solution. Using the Cauchy–Riemann equations, for the analytic function $f = u + iv$, we have $|\nabla u|^2 = (u_x)^2 + (u_y)^2 = (v_y)^2 + (-v_x)^2 = |\nabla v|^2$, whence the gradients of u and v have equal lengths. Note also that the length $|f'(z)|$ of $f'(z) = u_x + iv_x$ (cf. (2′)) is the same as the length of these gradients. The dot product of the gradients is $u_x v_x + u_y v_y = v_y v_x - v_x v_y = 0$. Hence, the gradients are perpendicular. Recall that the gradient of a function at a point is perpendicular to the level curve through the point, if the gradient is nonzero. Thus, the level curves of u and v through a point z, where $f'(z) \neq 0$, are perpendicular, since the gradients of u and v are perpendicular. □

Example 3. Show that the function $u(x,y) = \frac{1}{2}\log(x^2 + y^2)$, or $U(r,\theta) = \log r$, which is harmonic on the punctured plane $(r > 0)$, has no harmonic conjugate defined on the punctured plane. However, if the negative x–axis is deleted from the punctured plane, then there is a unique harmonic conjugate v of u on this slit plane (which is simply–connected; cf. the Remark following Proposition 3), such that v is zero on the positive x–axis. Show that the analytic function $f = u + iv$ on the slit plane is an inverse of the exponential function e^z (i.e., $f(z)$ can be regarded as a log function).

Solution. Since the level curves of $u = \log r$ are circles centered at the pole, we know by Example 2 that the level curves of a harmonic conjugate v of u must be the rays issuing from the origin. (Note that the gradient of u does not vanish, and hence all level curves are indeed curves, and v cannot be constant.) Consequently, $V(r,\theta)$ must be of the form $h(\theta)$. However, the only harmonic functions of this form are the functions $V(r,\theta) = a\theta + b$ (cf. Problem 2 of Section 6.3). Since v is not constant, we have $a \neq 0$. Thus, $V(r,\theta + 2\pi) \neq V(r,\theta)$, and so v cannot be continuous on the entire punctured plane. On the slit plane, assuming that v is 0 on the positive x–axis (i.e., when $\theta = 0$), we have $b = 0$, and $V(r,\theta) = a\theta$. Since $u_x(1,0) = 1$ and $v_y(1,0) = a$, we must have $a = 1$, by the Cauchy–Riemann equation $u_x = v_y$. Thus, $V(r,\theta) = \theta$, for $-\pi < \theta < \pi$, and

$$f(z) = f(re^{i\theta}) = \log r + i\theta, \qquad r > 0\ , \quad -\pi < \theta < \pi\ .$$

Note that $\exp[f(z)] = \exp[\log r + i\theta] = \exp[\log r](\cos\theta + i\sin\theta) = re^{i\theta} = z$, whence f is the unique inverse for $\exp$ on the slit plane, which is real–valued on the positive x–axis. If we add integer multiples of $2\pi i$ to $f(z)$, we get other inverses for $\exp$ which are not real–valued on the positive x–axis. For this reason, $f(z)$ is known as the **principal branch of the multi–valued log function**, and $f(z)$ is denoted by $\mathrm{Log}(z)$. □

Conformal Mapping

A complex–valued function $f(z)$ of a complex variable z, assigns to each complex number $z = x + iy$, in the domain of f, a new complex number $w = f(z) = u(x,y) + iv(x,y)$. In other words, f sends (or maps) points of the z–plane (or xy–plane) to points of the w–plane (or uv–plane). A region in the z–plane may be mapped by $f(z)$ to a differently shaped region in the w–plane. As we will soon demonstrate, if $f(z)$ is analytic, the amount of distortion that a disk in the z–plane undergoes (when it is mapped by f to the w–plane) decreases as the disk becomes smaller, provided that $f'(z)$ does not vanish at the center z_0 of the disk. Although the circular shape of a small disk about z_0 is nearly preserved, the image will be magnified by about a factor of about $|f'(z_0)|$. For this reason, a mapping given by an analytic function is said to be a **conformal mapping** (i.e., locally shape–preserving, but not necessarily size–preserving) when $f'(z)$ is nonzero. Before continuing, we consider some examples.

Example 4. Show that the analytic function $f(z) = z^2$ maps the wedge $0 \le r \le 3$, $0 \le \theta \le \alpha$ to the wedge $0 \le r \le 9$, $0 \le \theta \le 2\alpha$.

Solution. The function $f(z) = z^2$ maps the point (x,y) to $(x^2 - y^2, 2xy)$, but it is much easier to see the mapping geometrically, by using polar coordinates. Indeed, by De Moivre's formula, we have $f(re^{i\theta}) = r^2e^{i2\theta}$, which means that the point, with polar coordinates (r,θ), is mapped to the point with polar coordinates $(r^2, 2\theta)$ (i.e., the polar angle is doubled and the distance to the pole is squared). In particular, any point in the sector $0 \le \theta \le \alpha$ gets mapped into the sector $0 \le \theta \le 2\alpha$, and the points on the boundary arc $(r = 3, 0 \le \theta \le \alpha$ get mapped to the arc $(r = 9, 0 \le \theta \le 2\alpha)$ (cf. Figure 1). □

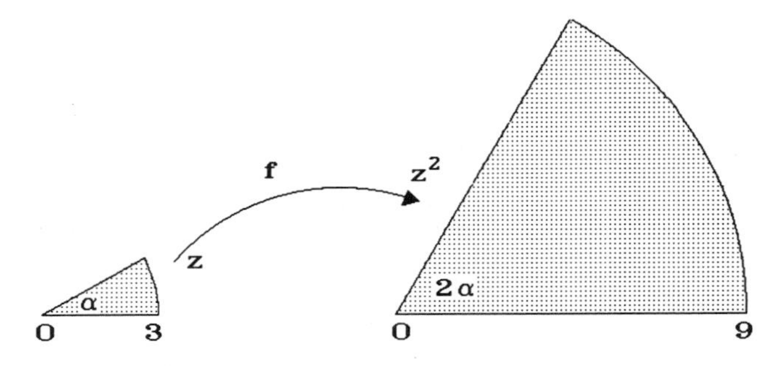

Figure 1

Example 5. Suppose that $f(z)$ is an analytic function and $f'(z_0) = Me^{i\tau} \ne 0$. As θ varies, the point $z_0 + re^{i\theta}$ $(r > 0)$ traces out a circle C of radius r about z_0. Show that for small r, the circle C is mapped by f nearly to a circle of radius $M = |f'(z_0)|r$ about $f(z_0)$ and C is rotated by τ radians.

Solution. From the definition (1) of $f'(z)$, we know that $f(z_0 + h) - f(z_0) \approx f'(z_0)h$, for small $|h|$. Taking $h = re^{i\theta}$ for small r, we then obtain

$$f(z_0 + re^{i\theta}) \approx f(z_0) + (Me^{i\tau})re^{i\theta} = f(z_0) + Mr\, e^{i(\theta+\tau)}.$$

Thus, as θ varies, we see that the circle C is mapped nearly to a circle about $f(z_0)$ of radius Mr and the circle is rotated through an angle of τ radians, since the point corresponding to θ on C is mapped nearly to the point with angle $\theta + \tau$ on the image curve. Note also that the central angle of an arc of C is nearly preserved by the mapping, which means that *conformal mappings preserve angles.* The fact that $f'(z)$ exists is crucial in this demonstration. For example, the function $g(x + iy) = 2x + iy$ is not analytic. Indeed, $g'(z)$ does not exist, because the Cauchy–Riemann equation $u_x = v_y$ is violated. Instead of mapping small circles nearly onto small circles, g maps circles onto ellipses which are twice as wide as they are tall. Also, we needed to know that $f'(z_0) \neq 0$. For instance, in Example 4, where $f(z) = z^2$ and $f'(0) = 0$, the angles at $z_0 = 0$ are not preserved, but rather they are doubled. □

One of the key properties of analytic functions f(z), or conformal mappings, which make them so useful is that they can transform a harmonic function on a region E of the w–plane into a harmonic function on the preimage D of E in the z–plane, in the sense of Proposition 4 below. In particular, if one can solve Dirichlet problems on familiar regions E (such as a disk or rectangle), one could use f to solve Dirichlet problems on an unfamiliar region D, provided f is chosen in such a way that it conformally maps D onto E. However, finding such an f explicitly can be difficult, even if one has a table or book of conformal maps.

Proposition 4. If $h(u,v)$ **is a harmonic function on an open set** E **of the** uv–**plane (i.e., the** w–**plane,** $w = u + iv$) **and if** $f(z) = u(x,y) + iv(x,y)$ **is an analytic function on the open set** D **in the** xy–**plane which maps** D **into** E, **then** $g(x,y) \equiv h(u(x,y), v(x,y))$ **defines a harmonic function** $g(x,y)$ **on** D.

Proof. Using the chain rule, $g_x = h_u u_x + h_v v_x$ and $g_{xx} = h_{uu}(u_x)^2 + 2h_{uv}u_x v_x + h_{vv}(v_x)^2$, and we have a similar expression for g_{yy}. Thus, using the solution of Example 2, we obtain

$$g_{xx} + g_{yy} = |\nabla u|^2 h_{uu} + 2\nabla u \cdot \nabla v\, h_{uv} + |\nabla v|^2 h_{vv} = |f'(z)|^2 (h_{uu} + h_{vv}) .$$

Hence, if h is harmonic, then g is harmonic. □

The next two examples illustrate how it is possible to solve boundary–value problems for Laplace's equation on a region by conformally mapping the region to a more familiar region (cf. the paragraph before Proposition 4).

Example 6. Consider a heat–conducting plate D which is the first quadrant of the xy–plane minus the quarter disk $(r < 1,\ 0 < \theta < \pi/2)$, as in Figure 2. Assume that the circular arc is insulated and the edge $y = 0$ $(x > 1)$ is held at temperature 0, while the remaining edge $x = 0$ $(y > 1)$ is held at temperature 100. Find the steady–state temperature in the plate, by conformally mapping D onto a strip by means of the analytic function $f(z) = \mathrm{Log}(z) = \log r + i\theta = u + iv = w$ (cf. Example 3).

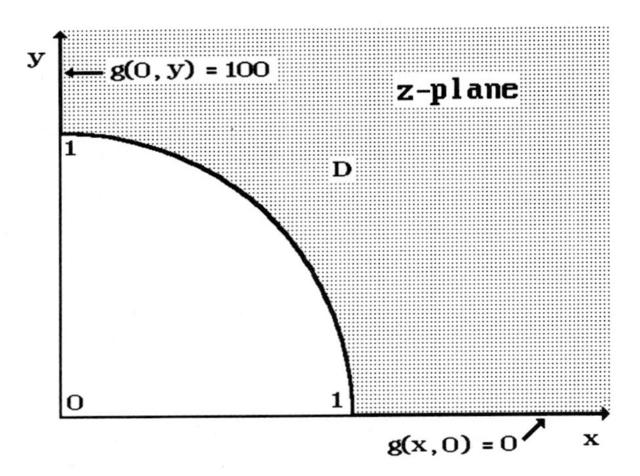

Figure 2

Solution. The region D is defined by $r > 1$ and $0 < \theta < \pi/2$. Since $u = \log r$ and $v = \theta$, the image, say E, of D under the conformal map f, is defined by $u > 0$ and $0 < v < \pi/2$, which is the strip (cf. Figure 3) in the uv–plane (or w–plane). The boundary conditions for the corresponding problem on E are also indicated in Figure 3.

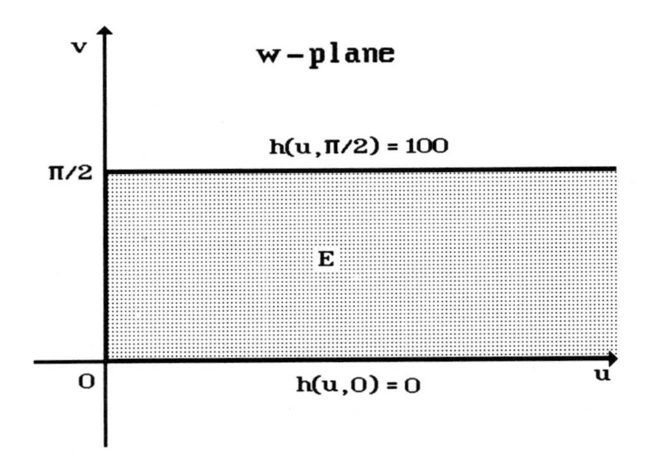

Figure 3

By inspection, we see that $h(u,v) = 200v/\pi$ is a harmonic function which solves the problem on E. Thus, in terms of polar coordinates, the solution of the original problem is $G(r,\theta) = h(U(r,\theta),V(r,\theta)) = h(\log r, \theta) = 200\theta/\pi$, or in terms of x and y, we obtain $g(x,y) = h(u(x,y),v(x,y)) = h(\frac{1}{2}\log(x^2 + y^2), \tan^{-1}(y/x)) = (200/\pi)\tan^{-1}(y/x)$. □

Example 7. Let R be a positive constant and define $F(z) = R^2z^{-1}$, for $z \neq 0$. Show that F maps the exterior $r > R$ of the circle $r = R$ onto the interior of this circle minus the pole (i.e., onto the punctured disk $0 < r < R$). Use F to solve the **exterior Dirichlet problem**

$$\begin{aligned} &\text{D.E.} \quad u_{xx} + u_{yy} = 0 \qquad x^2 + y^2 > R^2 \\ &\text{B.C.} \quad U(r,\theta) = f(\theta)\,, \end{aligned} \tag{4}$$

where $f(\theta)$ is a continuous periodic function of period 2π.

Solution. In terms of polar coordinates, $F(re^{i\theta}) = R^2(re^{i\theta})^{-1} = R^2r^{-1}e^{-i\theta}$. In other words, F maps the point (r,θ) to $(R^2/r\,,-\theta)$. If $r \geq R$, then $R^2/r \leq R$. Thus, F maps the exterior of the circle $r = R$ to the interior of that circle. Note also that the point (R,θ) on this circle is mapped to the point $(R\,,-\theta)$. Now the Poisson Integral Formula with boundary function $f(-\theta)$, yields a solution $H(r,\theta)$ [or $h(re^{i\theta})$] of the Dirichlet problem for $r < R$, namely

$$H(r,\theta) = \frac{1}{2\pi}\int_{-\pi}^{\pi} P(r,R,\theta - t)\, f(-t)\, dt\,.$$

Then we use the conformal mapping F to obtain a solution $g(re^{i\theta}) \equiv h(F(re^{i\theta})) = h(R^2r^{-1}e^{-i\theta})$ $= H(R^2/r,-\theta)$ of the given problem (4), i.e.,

$$G(r,\theta) = H(R^2/r,-\theta) = \frac{1}{2\pi}\int_{-\pi}^{\pi} P(R^2/r,R,-\theta - t)\, f(-t)\, dt = \frac{1}{2\pi}\int_{-\pi}^{\pi} P(R^2/r,R,t - \theta)\, f(t)\, dt$$

$$= \frac{1}{2\pi}\int_{-\pi}^{\pi} \frac{(R^2 - R^4r^{-2})f(t)}{R^4r^{-2} - 2R^3r^{-1}\cos(\theta - t) + R^2}\, dt = \frac{1}{2\pi}\int_{-\pi}^{\pi} \frac{(r^2 - R^2)\; f(t)}{R^2 - 2Rr\cos(\theta - t) + r^2}\, dt\,. \tag{5}$$

Note that there are many other solutions of problem (4), since we can always add linear combinations of the harmonic functions $[(r/R)^n - (R/r)^n]\sin(n\theta)$ (or use $\cos(n\theta)$), for $n = 1,2,\ldots$, which vanish on the circle $r = R$. This does not contradict the uniqueness theorem (Theorem 2 of Section 6.4), because the exterior of the circle is *not* a *bounded* open set. It is possible to show that (5) is the only solution which is bounded, in the sense that there is a constant M, such that $|G(r,\theta)| \leq M$ for all (r,θ) $r \geq 1$ (cf. Problem 6.). □

Conformal maps in fluid flow, electrostatics and heat theory

Suppose that the velocity of fluid flow at a point (x,y,z) at time t is of the form $\mathbf{v} = v_1(x,y)\,\mathbf{i} + v_2(x,y)\,\mathbf{j}$ (i.e., assume that the velocity is actually independent of z and t). Such a velocity vector field is said to describe a **steady two–dimensional fluid flow**. Henceforth, we assume that v_1 and v_2 are C^1. If the fluid is "incompressible", then the *net* amount of fluid which leaves any rectangle $(a \leq x \leq b,\ c \leq y \leq d)$ must be 0. In this case,

$$0 = \int_c^d v_1(b,y)\,dy - \int_c^d v_1(a,y)\,dy + \int_a^b v_2(x,d)\,dx - \int_a^b v_2(x,c)\,dx$$

$$= \int_c^d \int_a^b (v_1)_x\,dx\,dy + \int_a^b \int_c^d (v_2)_y\,dy\,dx = \int_c^d \int_a^b [(v_1)_x + (v_2)_y]\,dx\,dy,$$

where we have used the Fundamental Theorem of Calculus. Since the rectangle is arbitrary, we deduce that $(v_1)_x + (v_2)_y = 0$. If this equation holds for all velocity fields $\mathbf{v}$, then the fluid is said to be **incompressible**. The quantity $(v_1)_x + (v_2)_y$ is known as the **divergence** of the vector field $\mathbf{v}$. Thus, the fluid is incompressible if and only if its velocity fields are divergence–free.

If a fish swims counterclockwise along the boundary of the above rectangle, making a complete lap, the net amount of assistance that it receives from the current is

$$\int_a^b v_1(x,c)\,dx + \int_c^d v_2(b,y)\,dy - \int_a^b v_1(x,d)\,dx - \int_c^d v_2(a,y)\,dy = \int_c^d \int_a^b [(v_2)_x - (v_1)_y]\,dx\,dy\,.$$

This quantity is known as the **circulation** of the fluid flow around the rectangular loop. If the circulation is 0 around any rectangular loop, then this computation reveals that we must have $(v_2)_x - (v_1)_y = 0$. In this case, the fluid flow is said to be **irrotational**.

Proposition 5. Let $\mathbf{v} = v_1\mathbf{i} + v_2\mathbf{j}$ be a C^1 velocity vector field of an irrotational fluid flow of an incompressible fluid. Suppose that $\mathbf{v}$ is defined on a simply–connected open set R (cf. the remark following Proposition 3). Then there are C^2 functions $\Phi(x,y)$ and $\Omega(x,y)$, defined on R, such that, on R,

$$\Phi_x = v_1 \quad \text{and} \quad \Phi_y = v_2\,, \tag{6}$$

and

$$\Omega_x = -v_2 \quad \text{and} \quad \Omega_y = v_1\,. \tag{7}$$

Proof. The functions Φ and Ω exist, by Proposition 2 and the remark following Proposition 3. The integrability condition for (6) is the irrotationality condition $[(v_2)_x - (v_1)_y = 0]$, while the integrability condition for (7) is the incompressibility condition $[(v_1)_x - (-(v_2)_y) = 0]$. □

The function Φ is known as the **velocity potential** of the fluid flow, since $\nabla\Phi = \mathbf{v}$. The function Ω is known as the **stream function**, because the level curves Ω = constant are the streamlines. (Simply note that the gradient of Ω is orthogonal to $\mathbf{v}$, and so $\mathbf{v}$ must be tangent to the level curves of Ω.) Observe that (6) and (7) imply that Φ and Ω satisfy the Cauchy–Riemann equations, namely $\Phi_x = \Omega_y$ and $\Phi_y = -\Omega_x$. Thus, we have an analytic function

$$f(z) = f(x + iy) = \Phi(x,y) + i\Omega(x,y) \ , \tag{8}$$

which is known as the **complex velocity potential** of the fluid flow. Also, by Proposition 1,

$$f'(z) = \Phi_x + i\Omega_x = v_1 - iv_2 \ , \tag{9}$$

and thus

$$|f'(z)| = (v_1^2 + v_2^2)^{\frac{1}{2}} \ . \tag{10}$$

In other words, the conjugate of $f'(z)$ is the velocity vector field (regarded as a complex number) at the point (x,y) or $x + iy$, and the length of $f'(z)$ is the speed of the fluid flow at this point. Conversely, given any analytic function $f(z)$, the functions v_1 and v_2, *defined* by (9), will be the components of the velocity vector field of an irrotational fluid flow of an incompressible fluid. Indeed, by virtue of the Cauchy–Riemann equations, Φ and Ω will be solutions of the systems (6) and (7), whence the compatibility conditions of incompressibility and irrotationality must hold, by Proposition 2.

In the examples below, we examine some analytic functions and the resulting fluid flows which they define via (9). Since Ω is harmonic (by Proposition 1), one can also interpret Ω as an electrostatic potential in a region free of charge, in which case the curves Ω = constant are equipotential curves. The curves Φ = constant are then the lines of force along which charged particles will move, since the electric field is proportional to $\nabla\Omega$, and $\nabla\Omega$ is tangent to the curves Φ = constant. (Recall from Example 2 that, since Ω is a harmonic conjugate of Φ, the level curves of Φ and Ω are orthogonal, when $f'(z) \neq 0$.) If Ω is interpreted as a steady–state temperature distribution, then the level curves of Ω are curves of constant temperature (i.e., **isotherms**) and the temperature gradient $\nabla\Omega$ is tangent to the level curves of Φ which then may be interpreted as curves of heat flow. Thus, we see that any analytic function provides simultaneously a solution of several problems in at least three different contexts.

Example 8. Let $f(z) = z^2$. Sketch the level curves of the real and imaginary parts of f and interpret these curves physically.

Solution. Since $f(z) = (x^2 - y^2) + i2xy$, we have $\Phi(x,y) = \text{Re}(f(z)) = x^2 - y^2$, and $\Omega(x,y) = \text{Im}(f(z)) = 2xy$. In fluid mechanics, the streamlines (cf. Figure 4) of the fluid flow associated with f are the hyperbolas $2xy$ = constant (i.e., the level curves of the stream function Ω). The fluid velocity at the point (x,y) is the gradient of the velocity potential Φ, namely $\mathbf{v}(x,y) = 2x\mathbf{i} - 2y\mathbf{j}$, and the speed of the fluid is $2r$, where r is the distance to the origin. Note that the streamline $2xy = 0$ consists of intersecting curves (i.e., the axes). At the intersection, this streamline fails to be a "regular" curve. In general, at such a point, the velocity of the fluid flow must be 0 (or $f'(z) = 0$). Indeed, the implicit function theorem guarantees that a level curve Ω = constant will be regular at points where $\nabla\Omega \neq \mathbf{0}$. At an "irregular point" z of a streamline, we then must have $0 = |\nabla\Omega| = |\nabla\Phi| = |f'(z)|$, using Example 2. A point where the velocity is zero is a **stagnation point**. Thus, we know that irregular points on a streamline must be stagnation points. We can interpret the picture as the result of two rivers meeting head on as they approach the x–axis from above and below. By confining our attention to the first

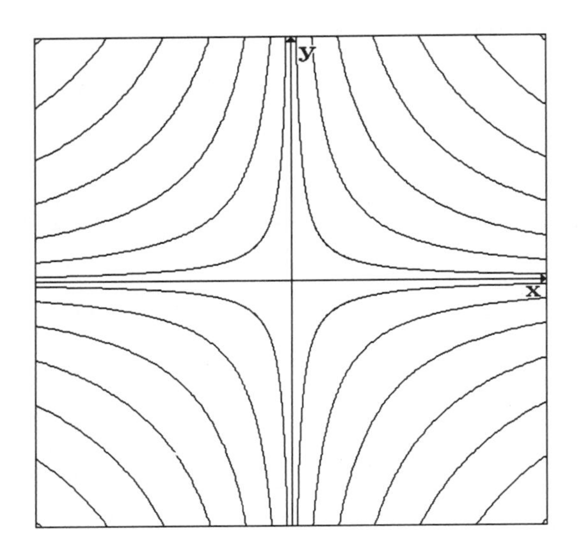

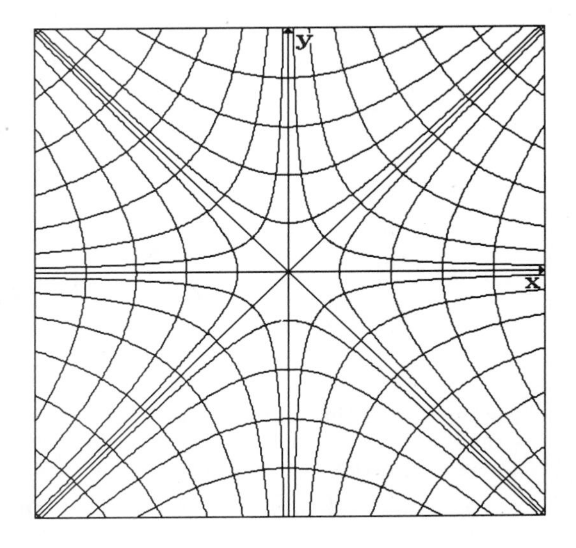

Figure 4

quadrant $(x \geq 0, y \geq 0)$, we can also think of the flow as resulting from the diversion of a stream of fluid when it meets a corner, after it settles to a steady flow. By extending the picture out of the page, the streamlines can also be interpreted as equipotential surfaces of an electrostatic potential which is produced by two distant positively charged lines (*perpendicular* to the page) at (a,a) and at (–a,–a) and two negatively charged lines at (a,–a) and (–a,a), where a is large, and we confine our attention to a small neighborhood of the origin. In electrostatic jargon, this potential is produced by a quadrapole at infinity. If we regard the positively charged lines as heat sources and the negatively charged lines as heat sinks, the equipotential surfaces are isotherms. □

Example 9. Let $f(z) = \mathrm{Log}(z) = \mathrm{Log}(re^{i\theta}) = \log r + i\theta$ for $r > 0$ and $-\pi < \theta < \pi$ (i.e., f is the principal branch of $\log z$; cf. Example 3). Analyze and interpret this function as was done in Example 8. Also, consider the related function $i\,\mathrm{Log}(z) = -\theta + i \log r$.

Solution. Here the streamlines for the fluid flow associated with $f(z)$ are the rays $\theta = \text{constant}$ (cf. Figure 5). The gradient of the velocity potential, $\log r$, is $r^{-1}\mathbf{e}_r$, where $\mathbf{e}_r$ is the unit radial vector field. Thus, the fluid appears to be emerging from a source at the pole, and the fluid slows down as it moves away from the pole. We may interpret the stream function θ as the electrostatic potential which is produced when two oppositely charged, parallel, half–planes $(x \leq 0, y = \pm\epsilon$, extending out of the page) are brought together (i.e., as $\epsilon \to 0$) along the *negative* x–axis (cf. Figure 5). If the half–planes are regarded as hot and cold objects maintained at temperatures $-\pi$ and π, then the rays $\theta = \text{constant}$ are isotherms of the resulting steady–state temperature distribution. We now consider the related function $i\,\mathrm{Log}(z) = -\theta + i \log r$. Here the streamlines are the circles $\log r = \text{constant}$, and the velocity is the gradient of $-\theta$ which is $-r^{-1}\mathbf{e}_\theta$, whence the flow is in the clockwise sense, and diminishes in speed as r increases.

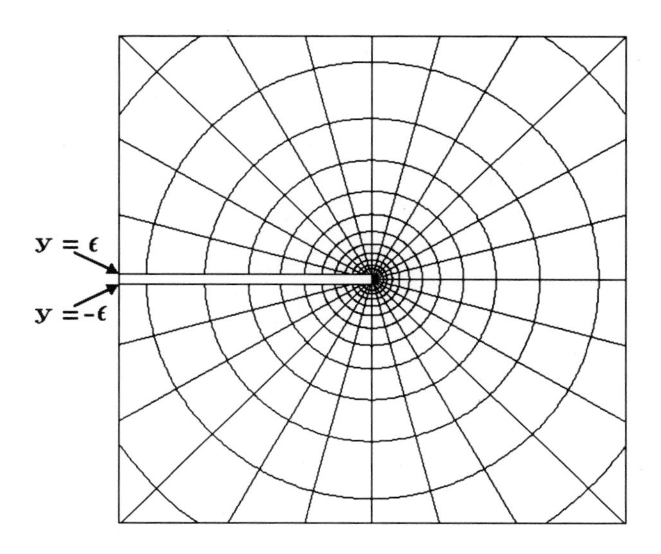

Figure 5

The electrostatic potential log r is the potential for an infinite, uniformly charged wire which is orthogonal to the page (cf. Example 1 of Section 6.1). If the wire is viewed as a very cold object, then log r may be regarded as a steady–state temperature distribution with circular isotherms. Apparently, there is a heat source at infinity. □

Example 10. In the same way as in Examples 8 and 9, supply interpretations of the function $f(z) = V_0(z + R^2z^{-1})$, where V_0 and R are positive constants and $z \neq 0$.

Solution. For $|z|$ large, we have $f(z) \approx V_0 z = V_0(x + iy)$. Thus, the velocity of the associated fluid flow is nearly $V_0\mathbf{i}$, far away from the pole. In terms of polar coordinates, we have

$$f(z) = f(re^{i\theta}) = V_0(re^{i\theta} + R^2/r\, e^{-i\theta}) = V_0(r + R^2/r)\cos\theta + iV_0(r - R^2/r)\sin\theta. \tag{11}$$

Observe that the stream function $V_0(r - R^2/r)\sin\theta$ is zero on the circle $r = R$ and also on the x–axis (except at $x = 0$). Note that this streamline has a self–intersection at $z = \pm R$. We deduce from the discussion in Example 8 that these points must be stagnation points, where the fluid velocity must be zero. With some work, one can sketch enough of the other streamlines to deduce that, outside of the circle, the function f(z) represents the steady fluid flow around a circular (or cylindrical) obstacle (cf. Figure 6). In electrostatics, the streamlines are the equipotential curves produced, when a cylindrical conductor is placed in a uniform electric field pointing in the y direction. We remark that on the surface of a conductor, electrons will arrange themselves in such a way that they cancel any tangential component of the electric field (i.e., the static electric field will be normal to the surface of the conductor, which is therefore an

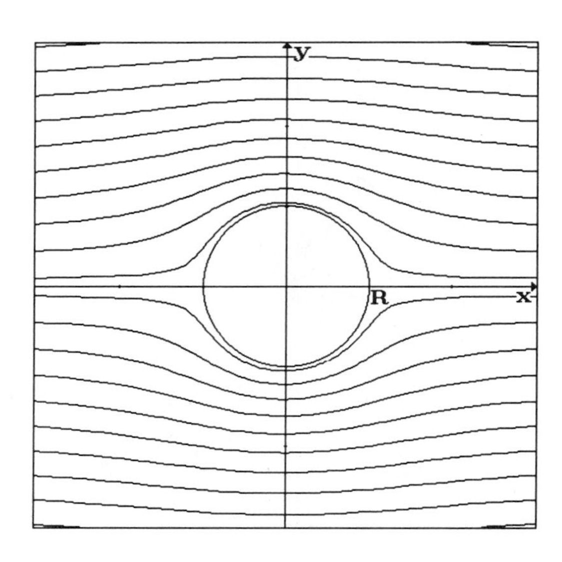

Figure 6

equipotential surface. Suppose that a disk which is maintained at temperature 0 is placed in a large heat conducting plate with a uniform temperature gradient in the y direction. The isotherms of the resulting steady–state temperature distribution will be the streamlines in Figure 6. □

Remark. In the above examples, we have essentially determined problems that a given analytic function solves. It is much harder to determine the analytic function that yields the solution to a given problem. There are catalogs of conformal maps that may help. For example, see H. Kober, *Dictionary of conformal representations*, Dover Publications, Inc., 1957. □

Summary 6.5

1. Complex analytic functions : Let $f(z)$ be a complex–valued function of the complex variable $z = x + iy$ in an open set D of the plane. The derivative of $f(z)$ with respect to z is the limit

$$f'(z) = \lim_{h \to 0} \frac{f(z + h) - f(z)}{h},$$

(if it exists) *regardless of how the complex number* h *approaches* 0. A function $f(z)$, for which $f'(z)$ exists for all $z \in D$, is said to be **complex analytic** (or simply **analytic** or **holomorphic**) on D. If $f(z)$ is analytic on the entire complex plane, then $f(z)$ is called an **entire function**.

2. Cauchy–Riemann equations (Proposition 1) : If $f(x + iy) = u(x,y) + iv(x,y)$ is analytic on an open set D, then the real and imaginary parts (u and v) of f obey the **Cauchy–Riemann equations**

$$u_x = v_y, \qquad u_y = -v_x. \tag{S1}$$

If in addition u and v are C^2 on D, then they are harmonic on D. Conversely, if u and v are C^1 and satisfy the Cauchy–Riemann equations on D, then $f = u + iv$ is analytic on D.

3. Harmonic conjugates : Let $u(x,y)$ and $v(x,y)$ be functions defined on some open set D. Then v is a **harmonic conjugate** of u on D, if $u + iv$ is analytic on D (i.e., if u and v obey the Cauchy–Riemann equations (S1)). Any two harmonic conjugates of u differ by a constant. Proposition 3 states that any harmonic function u defined on a rectangular region R has a harmonic conjugate on R. This is still true for simply–connected regions R (i.e., without holes), but Example 3 shows that the harmonic function $\log(r)$, defined on the punctured plane (i.e., with hole at (0,0)), does not have a harmonic conjugate on the punctured plane. If u and v are harmonic conjugates, then the level curves u = const. and v = const. intersect perpendicularly at any point where the common length of their gradients is not zero (cf. Example 2). In two–dimensional applications, these curves have the interpretations (depending on the context) of being electrostatic equipotential curves, isotherms of a steady–state temperature distribution, streamlines of an incompressible, irrotational fluid flow, etc. (cf. Examples 8, 9 and 10).

4. Conformal mapping : An analytic function $f(x + iy) = u(x,y) + iv(x,y)$ can be regarded as a transformation which maps the points in some region D of the xy–plane (or complex z–plane) to points in some region E in the uv–plane (or complex w–plane; $w = u + iv$). If $f'(z) \neq 0$, for all z in D, then $f(z)$ is called a **conformal mapping**, because it has the property of preserving angles or small shapes, although there is a local magnification factor equal to about $|f'(z)|$ (cf. Example 5). Proposition 4 shows that a conformal maps from D into E can be used to transfer a harmonic function on E in the w–plane to the region D in the z–plane. This is helpful in solving Dirichlet problems on unfamiliar domains D, if the image domain is familiar (cf. Examples 6 and 7). If one wishes to find a harmonic function which is constant on some curve (e.g., an isotherm, streamline, etc.), then the real (or imaginary) part of a conformal mapping which sends this curve to a vertical (or horizontal) line will be such a function. Every analytic function solves some (possibly very interesting) problem, but it is not always easy to find the appropriate function which solves a given (even simple) problem.

Exercises 6.5

1. (a) For complex $z = x + iy$, use the power series definition $\sin z = z - z^3/3! + z^5/5! + \ldots$ and Euler's formula $e^z = e^x(\cos y + i \sin y)$, to deduce that $\sin(z) = (e^{iz} - e^{-iz})/(2i) = \sin(x)\cosh(y) + i \cos(x)\sinh(y)$.

(b) Check that $\sin(z)$ is analytic by verifying that the Cauchy–Riemann equations hold for the real and imaginary parts found in part (a).

2. The argument in Example 3 shows that the harmonic function $\log r$ defined on the punctured plane, $r \neq 0$, does not have a harmonic conjugate defined on any open rectangular region which includes the pole. Why does this not violate Proposition 3?

3. (a) Show that the circle of radius $b > 0$, with center at $(1,0)$ (i.e. at $z = 1$), is traced out by $z = 1 + be^{i\theta}$, as θ varies.

(b) Show that the circle in part (a) is not mapped to a "perfect" circle by the function $f(z) = z^2$, but the image becomes more circular as $b \to 0$.

(c) Show directly that for small b, the circle is magnified by a factor of $|f'(1)|$.

4. Sketch the image of the wedge, $0 \leq r \leq 2, 0 \leq \theta \leq \pi/2$, when it is mapped into the w–plane via $w = f(z) = z^3$.

5. Use Proposition 4 to deduce that $g(x,y) = \exp[x^2 - y^2]\sin(2xy)$ is harmonic.

6. Suppose that $U(r,\theta)$ is a continuous function on the *exterior* $(r \geq R)$ of the disk $r < R$, and assume that $U(r,\theta)$ is harmonic for $r > R$, with $U(R,\theta) \equiv 0$ and $|U(r,\theta)| \leq M$, for some constant M. By completing the following steps, show that $U(r,\theta) \equiv 0$ for $r \geq R$.

(a) For any fixed $r_0 > R$, show that the function $V(r,\theta\,;\,r_0) \equiv M \log(r/R)/\log(r_0/R)$ is harmonic for $r > 0$, and is zero on the circle $r = R$ and equals M on the circle $r = r_0$.

(b) Deduce from the Maximum/Minimum Principle (or use Problem 1 of Section 6.4) that $|U(r,\theta)| \leq V(r,\theta\,;\,r_0)$ on the annulus $R \leq r \leq r_0$.

(c) For an arbitrary fixed r, take the limit of both sides of the inequality in (b) as $r_0 \to \infty$, to obtain $|U(r,\theta)| \leq 0$ (i.e., $U(r,\theta) \equiv 0$), as required.

7. Show that when the Cauchy–Riemann equations, $u_x = v_y$ and $u_y = -v_x$, are written in terms of polar coordinates, they yield the equivalent pair of equations, $U_r = r^{-1}V_\theta$ and $V_r = -r^{-1}U_\theta$, for $r > 0$.

8. (a) For any real number α define the function $z^\alpha = r^\alpha e^{i\alpha\theta}$, for $-\pi < \theta < \pi$ and $z = re^{i\theta}$. Use Problem 7 to show that z^α is an analytic function of z in the slit plane $-\pi < \theta < \pi$.

(b) Show that $z^\alpha = \exp[\alpha \operatorname{Log}(z)]$, where $\operatorname{Log}(z)$ is the principal branch of the log function defined by $\operatorname{Log}(z) = \log r + i\theta$ (cf. Example 3).

(c) Show that for $\alpha > \frac{1}{2}$, z^α maps the sector $0 < \theta < \pi/\alpha$ onto the upper half–plane $y > 0$.

(d) Roughly sketch the streamlines of the fluid flow which is associated with the analytic function $f(z) = z^\alpha$ in the sector $0 < \theta < \pi/\alpha$, $\alpha > 0$. (If this seems difficult, first consider the case $\alpha = 2$). What happens to these streamlines under the map in part (c) ?

9. Let the analytic function $f(z) = u(x,y) + i\, v(x,y)$ conformally map the open set D in the z–plane, in a one–to–one fashion, onto the open region E in the w–plane. Then there is a function $g(w) = x(u,v) + i\, y(u,v)$ which is the inverse of $f(z)$, in the sense that $w = f(z)$ if and only if $z = g(w)$. It is possible to prove that $g(w)$ is an analytic function of w, but do not bother. Instead, demonstrate that the curves $y =$ constant in D are mapped by $f(z)$ to the streamlines of the fluid flow on E associated with the analytic function g.

10. Let $f(z) = \sinh(z) = \sinh(x)\cos(y) + i \cosh(x)\sin(y)$.

(a) Show that for any nonzero c in the interval $(-\pi/2,\pi/2)$, $f(z)$ maps the horizontal line $y = c$ onto a branch of the hyperbola

$$\frac{v^2}{\sin^2 c} - \frac{u^2}{\cos^2 c} = 1 \qquad (u + iv = w = f(z))$$

(the upper or lower branch depending on whether c is positive or negative). What happens to the x–axis (i.e., $y = 0$) ?

(b) Noting that all of the hyperbolas intersect the v–axis between 1 and –1, sketch these curves, and observe that they fill the entire w–plane, except for the rays $v \geq 1$ and $v \leq -1$ on the v–axis.

(c) Deduce from (a) and (b) that $\sinh(z)$ maps the horizonal strip $-\pi/2 < y < \pi/2$ (say D) in a 1–1 fashion onto the open set E consisting of the w–plane minus the rays in (b). Thus, there is an inverse $\sinh^{-1}(w)$ on E of $\sinh(z)$ on D.

(d) Deduce from Problem 9 that the fluid flow associated with the analytic function $\sinh^{-1}(w)$ on E has the hyperbolas as streamlines. Describe the flow in words, and explain how the curves could arise electrostatically as equipotential curves, or as isotherms in a steady–state heat conduction setup.

11. (a) Describe the streamlines of fluid flow with complex velocity potential $f(z) = i(z^2 - 1)^{\frac{1}{2}}$, for $z = x + i y$ not on the segment from -1 to 1 and $x > 0$, where the square root is defined as in Problem 8. **Hint**. Consider the image of a line (say $v = \epsilon > 0$) slightly above the u–axis under the inverse transformation $g(w) = (1 - w^2)^{\frac{1}{2}}$, $w = u + iv$, $v > 0$. Is Figure 7 below relevant?

(b) What concrete physical situations will yield isotherms or equipotential curves which are the same as the streamlines found in part (a) ?

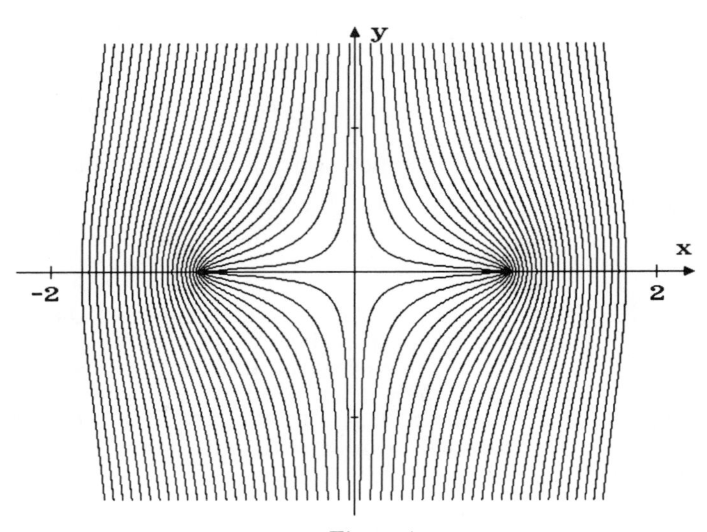

Figure 7

CHAPTER 7

FOURIER TRANSFORMS

In this chapter, we introduce the theory of Fourier transforms and use it to find solutions of PDEs on infinite domains. For example, we consider in some detail the problem of heat conduction in an infinite rod:

$$\text{D.E.}\quad u_t = ku_{xx} \qquad -\infty < x < \infty,\ t > 0,\ k > 0,$$

$$\text{I.C.}\quad u(x,0) = f(x), \tag{$*$}$$

under various assumptions on $f(x)$. The reader may wonder why such problems are of interest, since for practical purposes all domains (rods) are finite. One answer is that usually the form of the solution is easier to handle and interpret in the infinite case. This has already been seen for the wave equation, where D'Alembert's formula for the infinite string is more tractable than the Fourier series solution, which is usually a sum of infinitely many harmonics, whose convergence may be difficult to establish directly. In Chapter 5, we used D'Alembert's formula with suitable periodic initial data to solve the wave equation for a finite string with fixed or free ends, without any Fourier series (cf. Theorem 3 of Section 5.2, or Examples 1 and 2 of Section 5.3).

The analog of D'Alembert's formula for the heat equation (i.e., for problem (∗)) is the remarkable formula (for $t > 0$)

$$u(x,t) = \frac{1}{\sqrt{4\pi kt}} \int_{-\infty}^{\infty} e^{-(x-y)^2/4kt} f(y)\, dy, \tag{$**$}$$

which, as we will prove, is the only "physically reasonable" solution of (∗), under various hypotheses on $f(x)$. In particular, if $f(x)$ is continuous and periodic of period $2L$, then $u(x,t)$ given by (∗∗) provides a periodic solution to the problem of heat conduction in a circular wire of length $2L$. Note that (∗∗) is visibly simpler to deal with than the formal infinite Fourier series solutions in Section 3.4. Solution (∗∗) also has a nice interpretation as a continuous superposition (i.e., integral) of the contributions $(4\pi kt)^{-\frac{1}{2}} e^{-(x-y)^2/4kt} f(y)$ (due to heat sources at various points y) to the temperature $u(x,t)$ at x after an elapsed time t. The profiles of these contributions due to a fixed point y at various times t is illustrated in Figure 1. The area under each of these "Gaussian curves" is $f(y)$ (assumed positive here) and the height is $(4\pi kt)^{-\frac{1}{2}} f(y)$. As t increases, the influence of the initial temperature $f(y)$ at y spreads.

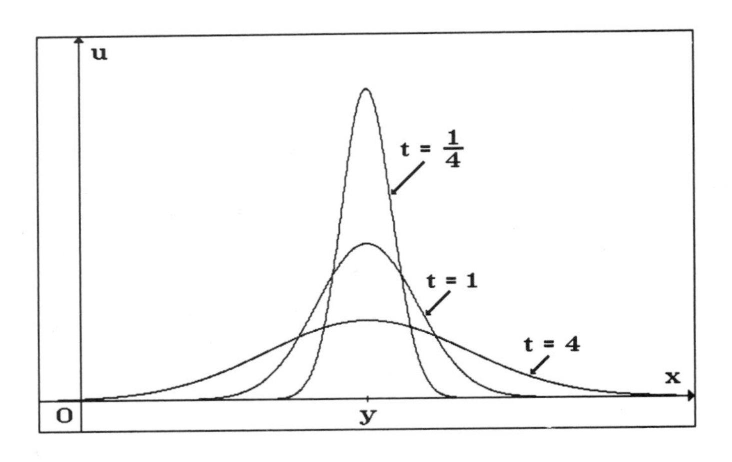

Figure 1

Formula (**) and solutions of analogous infinite problems are found by applying Fourier transform techniques in a formal, unrigorous fashion. For that reason, the solutions obtained by such methods must be verified separately anyway to make sure that they solve the given problems. Nevertheless, we will prove some properties of Fourier transforms under various assumptions (usually not the most general), so that the reader will have some basis for believing that the formal manipulations are likely to lead to a correct solution. We suggest that the reader or instructor skip the more difficult proofs during the first reading, and concentrate on the examples.

In Section 7.1, to motivate complex Fourier transforms, we first consider the complex form of Fourier series. The basic properties of Fourier transforms are covered in Section 7.2 . While we defer the proof of the Inversion Theorem to the end of the chapter, we use this theorem in Section 7.3 to prove Parseval's equality for Fourier transforms. In Section 7.4, we apply Fourier transform methods to the heat problem (*) on the infinite rod, thereby formally obtaining the formula (**), which is then rigorously shown to solve (*). We also show how Fourier transform methods can be applied to problems for the wave equation and Laplace's equation, even though we have seen that other methods suffice. In Section 7.5, using (**) together with the method of images, we answer the questions which were left unresolved in Section 4.3 concerning the validity of formal infinite series solutions for the finite rod. The method of images and (**) are also used to solve heat problems for the semi–infinite rod $(0 \leq x < \infty)$. We use (**) and the method of images for heat problems, essentially in the same way that we used D'Alembert's formula and the method of images for wave problems in Section 5.3. The Fourier sine and cosine transforms are introduced, but we do not dwell on them, since results obtained by these transforms are easily obtained by using the ordinary Fourier transform with the method of images.

Historical Remarks and Contemporary Perspectives

In the late eighteenth and early nineteenth centuries, scientists and mathematicians were impressed by the success of Fourier series methods for solving initial/boundary–value problems for finite intervals, and they sought an appropriate continuous analog of Fourier series, involving integrals instead of infinite series. This analog would enable them to give integral represention of certain functions which are not periodic. Moreover, such representations would lead to more manageable and understandable solutions in closed form (i.e., not involving infinite series; cf. (**)). The idea of the Fourier transform (or integral) was inspired by the work Pierre–Simon de Laplace (1749–1827), and is largely due to Joseph Fourier (1768–1830), Augustin–Louis Cauchy (1789–1857) and Simon D. Poisson (1781–1840). In 1811, all three presented papers orally to the Academy of Sciences of Paris, and each had the benefit of verbal accounts of the others.

The concept of a Fourier transform is a special case of the notion of an integral transform (or operator; cf. Section 1.2). Given any "reasonable" function $K(x,\xi)$, we can define the integral transform, $F(\xi)$ or $T[f](\xi)$, of a function $f(x)$ by

$$F(\xi) \equiv T[f](\xi) \equiv \int_a^b K(x,\xi)\, f(x)\, dx\,, \qquad (***)$$

where the fixed limits of integration, a and b, may be finite or infinite. Each suitable function f is transformed into a new function F or T[f], according to (***). Such transforms are linear in the sense that $T[c_1f_1 + c_2f_2] = c_1T[f_1] + c_2T[f_2]$. The transform is determined by the choice of the limits a and b, and the function $K(x,\xi)$ is known as the **kernel** of the transform. Some widely used transforms are defined by the kernels and limits in Table 1 below.

$K(x,\xi)$	Limits of integration	Name of transform
$(2\pi)^{-\frac{1}{2}}\, e^{-i\xi x}$	$a = -\infty$, $b = \infty$	Fourier
$e^{-\xi x}$	$a = 0$, $b = \infty$	Laplace
$x^{\xi-1}$	$a = 0$, $b = \infty$	Mellin
$\frac{1}{\pi}\,\frac{1}{x+\xi}$	$a = -\infty$, $b = \infty$	Hilbert
$(4\pi\xi)^{-\frac{1}{2}}\, e^{-\frac{1}{4}x^2/\xi}$	$a = -\infty$, $b = \infty$	Weierstrass

Table 1 (Some common integral transforms)

The choice of which transform to use depends on the nature of the problem at hand. The Fourier transform is helpful in solving PDEs, primarily because it converts differentiation into a simple algebraic multiplication, in the sense that $T[f'](\xi) = i\xi T[f](\xi)$ (cf. Proposition 1 of Section 7.2). Laplace transforms are ideally suited for initial–value problems for linear systems of ODEs. To solve such problems algebraically, the English electrical engineer Oliver Heaviside (1850–1925) developed his widely applied "operational calculus", based on Laplace and Fourier transforms.

As with Fourier series, Fourier transforms have many fundamental uses, apart from solving differential equations. The Fourier coefficients a_n and b_n of a function $f(x)$ on $[-L,L]$ tell us how prominently the harmonics $\sin(n\pi x/L)$ and $\cos(n\pi x/L)$ enter into the makeup of the function. For a function $f(x)$ on $(-\infty,\infty)$, the Fourier transform $F(\xi) \equiv (2\pi)^{\frac{1}{2}}\int_{-\infty}^{\infty} f(x)e^{-i\xi x}\, dx$, at a certain value ξ, tells us how prominent the harmonics $\sin(\xi x)$ and $\cos(\xi x)$ are in the makeup of the function. Often it is useful to think of an object under experimental scrutiny as an unknown function, and one primary way of learning about this object is by measuring its response to a probe (e.g., light, X rays, microwaves, electronic signals, sound waves, etc.) which is sent at a various frequencies $\xi/2\pi$. The magnitude of the response may be expected to be proportional to the prominence of the harmonic of frequency ξ in the makeup of the unknown object (i.e., the value of the Fourier transform of the unknown function, at ξ). One then hopes to approximate the unknown object from these responses. In other words, one hopes to determine a function from its Fourier transform. There is a result (the Inversion Theorem) which shows that this can be done for a large class of functions. (This is analogous to the convergence theorems for Fourier series, which reconstruct certain functions from their Fourier coefficients.) In realistic situations, the frequency response of an object may depend on the angle from which the object is probed, and so the problem can be much more complicated, and possibly a different multidimensional transform may be more suitable. For example, the Radon transform and its inversion theorem were developed by Johann Radon (1887–1956) in his 1917 paper "Über die Bestimmung von Funktionen durch ihre Integralwerte längs gewisser Manningfaltigkeiten" ([Radon], [Deans]). Although this paper was virtually unknown in applied areas before the 1970s, today the use of Radon's integral transform is fundamental in such fields as medical diagnostics, atmospheric physics, astronomy, spectroscopy, statistics, geophysics, stress analysis, etc. . Since the problems of determining objects from their frequency response (i.e., **inverse problems** or **reconstruction problems**) are nearly universal in science. integral transforms are essential tools of many trades.

7.1 Complex Fourier Series

Recall that the Fourier series of a suitable function $f(x)$ defined for $-L \le x \le L$ is

$$\text{FS } f(x) = \tfrac{1}{2}a_0 + \sum_{n=1}^{N} a_n\cos(\tfrac{n\pi x}{L}) + b_n\sin(\tfrac{n\pi x}{L}), \tag{1}$$

where

$$a_n = \frac{1}{L}\int_{-L}^{L} f(x)\cos(\tfrac{n\pi x}{L})\,dx, \quad (n = 0, 1, \ldots) \quad \text{and} \quad b_n = \frac{1}{L}\int_{-L}^{L} f(x)\sin(\tfrac{n\pi x}{L})\,dx, \quad (n = 1, 2, \ldots).$$

To obtain the complex form of the Fourier series, we use the identity

$$e^{iy} = \cos(y) + i\sin(y), \tag{2}$$

which yields

$$\cos(y) = \frac{1}{2}(e^{iy} + e^{-iy}) \quad \text{and} \quad \sin(y) = -\frac{i}{2}(e^{iy} - e^{-iy}). \tag{3}$$

Setting $y = n\pi x/L$, and replacing $\cos(n\pi x/L)$ and $\sin(n\pi x/L)$ in (1) by the corresponding formulae in (3), yields

$$\begin{aligned}\text{FS } f(x) &= \tfrac{1}{2}a_0 + \sum_{n=1}^{\infty} \left[\tfrac{1}{2}a_n(e^{in\pi x/L} + e^{-in\pi x/L}) - \tfrac{1}{2}ib_n(e^{in\pi x/L} - e^{-in\pi x/L}) \right] \\ &= \tfrac{1}{2}a_0 + \sum_{n=1}^{\infty} \left[\tfrac{1}{2}(a_n - ib_n)e^{in\pi x/L} + \tfrac{1}{2}(a_n + ib_n)e^{-in\pi x/L} \right] \\ &= \tfrac{1}{2}a_0 + \sum_{n=1}^{\infty} \tfrac{1}{2}(a_n - ib_n)e^{in\pi x/L} + \sum_{n=-\infty}^{-1} \tfrac{1}{2}(a_{-n} + ib_{-n})e^{in\pi x/L}.\end{aligned} \tag{4}$$

Let $c_0 = \tfrac{1}{2}a_0$. For $m = 1, 2, 3, \ldots$, we define

$$c_m = \tfrac{1}{2}(a_m - ib_m) = \frac{1}{2L}\int_{-L}^{L} f(x)[\cos(\tfrac{m\pi x}{L}) - i\sin(\tfrac{m\pi x}{L})]\,dx = \frac{1}{2L}\int_{-L}^{L} f(x)\,e^{-im\pi x/L}\,dx.$$

For $m = -1, -2, -3, \ldots$, we define

$$c_m = \tfrac{1}{2}(a_{-m} + ib_{-m}) = \frac{1}{2L}\int_{-L}^{L} f(x)\,[\cos(-\tfrac{m\pi x}{L}) + i\sin(-\tfrac{m\pi x}{L})]\,dx = \frac{1}{2L}\int_{-L}^{L} f(x)\,e^{-im\pi x/L}\,dx.$$

Then (4) can be written in the form

$$\text{FS}_c f(x) = \sum_{m=-\infty}^{\infty} c_m\,e^{im\pi x/L}, \quad \text{where } c_m = \frac{1}{2L}\int_{-L}^{L} f(x)\,e^{-im\pi x/L}\,dx, \quad m = 0, \pm 1, \pm 2, \ldots. \tag{5}$$

The formal expression $\text{FS}_c f(x)$ is known as the **complex Fourier series** of $f(x)$.

In the event that FS f(x) converges at x, we clearly, get the same result whether we use $FS_c f(x)$ or FS f(x), i.e.,

$$\lim_{N\to\infty} \sum_{m=-N}^{N} c_m e^{im\pi x/L} = \lim_{N\to\infty} \left[\tfrac{1}{2}a_0 + \sum_{n=1}^{N} a_n \cos(\tfrac{n\pi x}{L}) + b_n \sin(\tfrac{n\pi x}{L}) \right].$$

Thus, all of the convergence results of Chapter 4 carry over to the corresponding results for complex Fourier series. Ironically, complex Fourier series are often simpler to compute than "ordinary" Fourier series.

Example 1. Compute the complex Fourier series of the $f(x) = e^{ax}$, $-L \le x \le L$, where a is a real constant.

Solution. We have

$$c_m = \frac{1}{2L}\int_{-L}^{L} e^{ax} e^{-im\pi x/L}\, dx = \frac{1}{2L}\int_{-L}^{L} e^{(a-(im\pi/L))x}\, dx$$

$$= \frac{1}{2L}\frac{1}{a - im\pi/L} e^{(a-(im\pi/L))x}\Big|_{-L}^{L} = \frac{1}{2}\frac{1}{aL - im\pi}\left[e^{aL}e^{-im\pi} - e^{-aL}e^{im\pi}\right]$$

$$= \frac{1}{2}\frac{aL + im\pi}{(aL)^2 + (m\pi)^2}\left[e^{aL}(\cos(m\pi)-i\sin(m\pi)) - e^{-aL}(\cos(m\pi)+i\sin(m\pi))\right]$$

$$= (-1)^m \frac{1}{2}\frac{e^{aL} - e^{-aL}}{(aL)^2 + (m\pi)^2}(aL + im\pi) = (-1)^m \sinh(aL)\frac{(aL + im\pi)}{(aL)^2 + (m\pi)^2}.$$

For $m \ge 0$ the real part of c_m is $\tfrac{1}{2}a_m$ and the imaginary part is $\tfrac{1}{2}b_m$. The reader is invited to compute these ordinary Fourier coefficients directly. Of course, the complex Fourier series of

f(x) is just $\sum_{m=-\infty}^{\infty} c_m e^{im\pi x/L}$, where c_m is given above. □

The functions $e^{im\pi x/L}$ are complex–valued. The **inner product of two complex–valued functions** f(x) and g(x), piecewise continuous on [–L,L], is defined to be the complex number

$$\langle f,g\rangle \equiv \int_{-L}^{L} f(x)\overline{g(x)}\, dx, \tag{6}$$

where $\overline{g(x)}$ is the complex conjugate of g(x). Thus, if $f(x) = f_1(x) + if_2(x)$ and $g(x) = g_1(x) + ig_2(x)$, where f_1, f_2, g_1, g_2 are real–valued, then

$$<f,g> = \int_{-L}^{L} [f_1(x) + if_2(x)][g_1(x) - ig_2(x)]\, dx$$

$$= \int_{-L}^{L} [f_1(x)g_1(x)+f_2(x)g_2(x)]\, dx \;+\; i\int_{-L}^{L} [f_2(x)g_1(x)-f_1(x)g_2(x)]\, dx.$$

While in general $<f,g>$ is not real, the inner product of f with itself is real and nonnegative :

$$<f,f> = \int_{-L}^{L} f(x)\overline{f(x)}\, dx = \int_{-L}^{L} |f(x)|^2\, dx \geq 0.$$

Hence the **norm** $\|f\|$, defined as $\|f\| = \sqrt{<f,f>}$, makes sense. Also, $<f,g>$ agrees with the definition in Section 4.1, when f and g are real–valued. Using the notation $e_m(x) = e^{im\pi x/L}$,

$$<e_m,e_n> = \int_{-L}^{L} e^{im\pi x/L}\,\overline{e^{in\pi x/L}}\, dx = \int_{-L}^{L} e^{i(m-n)\pi x/L}\, dx$$

$$= \int_{-L}^{L} \left[\cos(\tfrac{(m-n)\pi x}{L}) + i\sin(\tfrac{(m-n)\pi x}{L})\right] dx = \begin{cases} 0 & \text{if } m \neq n \\ 2L & \text{if } m = n. \end{cases}$$

We say that the family $\{e_m\}$, $m = 0, \pm1, \pm2, \ldots$, forms **an orthogonal family of norm–square** $\|e_m\|^2 = 2L$. The complex Fourier coefficients c_m are essentially components of the possibly complex–valued function f with respect to this family, i.e.,

$$c_m = \frac{1}{2L}\int_{-L}^{L} f(x)\, e^{-im\pi x/L}\, dx = \frac{1}{2L}<f,e_m>.$$

If the function $f(x)$ is nice enough (e.g., if f is continuous and piecewise C^1, with $f(-L) = f(L)$), then we know (cf. Theorem 4 of Section 4.2) that the Fourier series (complex or ordinary) converges to $f(x)$, and therefore

$$f(x) = \sum_{m=-\infty}^{\infty} c_m e^{im\pi x/L} = \sum_{m=-\infty}^{\infty} \left[<f, (2L)^{-\frac{1}{2}}e_m>\, (2L)^{-\frac{1}{2}}e_m(x)\right],$$

which is the expansion of f in terms of the "orthogonal unit vectors" $(2L)^{-\frac{1}{2}}e_m$. Parseval's equality (cf. Section 4.2) assumes a pleasant form in terms of the complex Fourier coefficients :

$$\|f\|^2 = \int_{-L}^{L} |f(x)|^2\,dx = \sum_{m=-\infty}^{\infty} |\langle f, (2L)^{-\frac{1}{2}} e_m\rangle|^2$$
$$= 2L \sum_{m=-\infty}^{\infty} |\tfrac{1}{2L}\langle f, e_m\rangle|^2 = 2L \sum_{m=-\infty}^{\infty} |c_m|^2 . \tag{7}$$

It turns out that this result holds for any function (possibly complex–valued) $f(x)$ for which $\int_{-\infty}^{\infty} |f(x)|^2\,dx < \infty$. When $f(x)$ is real–valued, the result follows from Parseval's equality of Section 4.2, since $|c_m|^2 = (a_{|m|}^2 + b_{|m|}^2)/4$ for $m = \pm 1, \pm 2, \ldots$ and $|c_0|^2 = a_0^2/4$. For convenience, we state the following complex form of Parseval's equality.

Theorem (Parseval's equality). Let $f(x)$ **be a real or complex–valued function defined on** $[-L,L]$. **If** $\|f\|^2 < \infty$, **then**

$$\|f\|^2 = \int_{-L}^{L} |f(x)|^2\,dx = 2L \sum_{m=-\infty}^{\infty} |c_m|^2, \quad \textbf{where} \quad c_m = \frac{1}{2L}\int_{-L}^{L} f(x)\, e^{-im\pi x/L}\,dx,$$

$m = 0, \pm 1, \pm 2, \ldots,$ **are the complex Fourier coefficients of** $f(x)$.

Example 2. What does Parseval's equality say if $f(x) = e^{ax}$, $-L \le x \le L$ ($a \ne 0$ and real)?

Solution. We have

$$\int_{-L}^{L} |f(x)|^2\,dx = \int_{-L}^{L} e^{2ax}\,dx = \frac{1}{2a} e^{2ax}\Big|_{-L}^{L} = \frac{1}{2a}(e^{2aL} - e^{-2aL})$$
$$= \frac{1}{2a}(e^{aL} + e^{-aL})(e^{aL} - e^{-aL}) = \frac{2}{a}\cosh(aL)\sinh(aL).$$

From Example 1, $|c_m|^2 = \dfrac{\sinh^2(aL)}{a^2L^2 + m^2\pi^2}$, so that Parseval's equality yields

$$\frac{2}{a}\cosh(aL)\sinh(aL) = 2L\sinh^2(aL) \sum_{m=-\infty}^{\infty} \frac{1}{a^2L^2 + m^2\pi^2} .$$

For $L = \pi$, we obtain $\displaystyle\sum_{m=-\infty}^{\infty} \frac{1}{a^2 + m^2} = \frac{\pi\cosh(a\pi)}{a\sinh(a\pi)} = \frac{\pi}{a}\coth(a\pi).$ □

The following definition will be useful in the sequel.

Definition. A real or complex–valued function defined on $(-\infty,\infty)$ is said to be **absolutely integrable** on $(-\infty,\infty)$, if $\int_{-R}^{R} |f(x)|\, dx$ exists for all $R > 0$, and

$$\int_{-\infty}^{\infty} |f(x)|\, dx \equiv \lim_{R\to\infty} \int_{-R}^{R} |f(x)|\, dx < \infty .$$

For example, $(1+x^2)^{-1}$ is absolutely integrable, since $\int_{-\infty}^{\infty} (1+x^2)^{-1} dx = \pi < \infty$, but $x/(1+x^2)$ is not absolutely integrable. Indeed, while $\int_{-R}^{R} x/(1+x^2)\, dx = 0$,

$$\int_{-\infty}^{\infty} |x/(1+x^2)|\, dx = \lim_{R\to\infty} \int_{-R}^{R} |x/(1+x^2)|\, dx = \lim_{R\to\infty} \int_{0}^{R} 2x/(1+x^2)\, dx = \lim_{R\to\infty} \log(1+R^2) = \infty .$$

The Fourier transform

Let $f(x)$ be a real or complex–valued function of the real variable x $(-\infty < x < \infty)$. The **Fourier transform** of $f(x)$ is the function $\hat{f}(\xi)$ of the *real* variable ξ $(-\infty < \xi < \infty)$ defined by

$$\hat{f}(\xi) \equiv \frac{1}{\sqrt{2\pi}} \int_{-\infty}^{\infty} f(x)\, e^{-i\xi x}\, dx \equiv \lim_{R\to\infty} \int_{-R}^{R} f(x)\, e^{-i\xi x}\, d'x , \tag{8}$$

where

$$d'x \equiv (2\pi)^{-\frac{1}{2}} dx, \text{ when the limit (8) exists.}$$

The somewhat awkward factor of $(2\pi)^{-\frac{1}{2}}$ (which we disguise with the notation $d'x = (2\pi)^{-\frac{1}{2}} dx$) is sometimes omitted in other books. With our notation, we will find that, for "nice" functions $f(x)$, the following **Parseval's equality** (for Fourier transforms) holds

$$\int_{-\infty}^{\infty} |\hat{f}(\xi)|^2\, d\xi = \int_{-\infty}^{\infty} |f(x)|^2\, dx ,$$

(cf. Section 7.3). If the factor $(2\pi)^{-\frac{1}{2}}$ in (8) is omitted in the definition of $\hat{f}(\xi)$, then a factor of $(2\pi)^{-1}$ must be inserted on the left–hand side of Parseval's equality. Note that in definition (8), we implicitly assume that the integral $\int_{-R}^{R} f(x)e^{-i\xi x}\, dx$ exists for all real R, *but we do not assume that $f(x)$ is absolutely integrable.* If the limit in (8) exists, then it is called the

Cauchy principal value of the improper integral $\int_{-\infty}^{\infty} f(x)e^{-i\xi x}dx$. The Cauchy principal value can exist, even if $f(x)$ is not absolutely integrable.

If $\xi = m\pi/L$, then $\hat{f}(\xi)$ looks much like a complex Fourier coefficient. Indeed suppose that $f_1(x)$ is a "nice" (e.g., piecewise continuous) function defined for $-L \leq x \leq L$. Extend $f_1(x)$ to a function defined for all x, by setting

$$f(x) = \begin{cases} f_1(x) & \text{for } |x| \leq L \\ 0 & \text{for } |x| > L \ . \end{cases}$$

The complex Fourier coefficients of $f_1(x)$ are given by

$$\begin{aligned} c_m = \frac{1}{2L}\int_{-L}^{L} f_1(x)\, e^{-im\pi x/L}\, dx &= \frac{1}{2L}\int_{-\infty}^{\infty} f(x)\, e^{-i(m\pi/L)x}\, dx \\ &= \frac{\sqrt{2\pi}}{2L}\int_{-\infty}^{\infty} f(x)\, e^{-i(m\pi/L)x}\, d'x = \frac{\sqrt{2\pi}}{2L}\hat{f}(m\pi/L). \end{aligned}$$

Thus, the c_m for $f_1(x)$ are obtained essentially by evaluating the Fourier transform $\hat{f}(\xi)$ of the extended function $f(x)$ at the points $\xi = m\pi/L$, $m = 0, \pm1, \pm2, \ldots$. We will exploit this fact later (cf. Section 7.3), but for the remainder of this section, we consider some examples.

Example 3. Compute the Fourier transform of $f(x) = e^{-a|x|}$, where $a > 0$ and $-\infty < x < \infty$.

Solution.

$$\begin{aligned} \hat{f}(\xi) &= \int_{-\infty}^{\infty} e^{-a|x|}e^{-i\xi x}\, d'x = \int_{0}^{\infty} e^{-ax}e^{-i\xi x}\, d'x + \int_{-\infty}^{0} e^{ax}e^{-i\xi x}\, d'x \\ &= \int_{0}^{\infty} e^{-(a+i\xi)x}\, d'x + \int_{0}^{\infty} e^{-(a-i\xi)u}\, d'u \qquad (\text{where } u = -x) \\ &= \frac{1}{\sqrt{2\pi}}\left[\frac{e^{-(a+i\xi)x}}{-(a+i\xi)}\Big|_0^{\infty} + \frac{e^{-(a-i\xi)x}}{-(a-i\xi)}\Big|_0^{\infty}\right] = \frac{1}{\sqrt{2\pi}}\left[\frac{1}{a+i\xi} + \frac{1}{a-i\xi}\right] = \frac{1}{\sqrt{2\pi}}\frac{2a}{a^2+\xi^2}. \end{aligned}$$ □

Example 4. Find the Fourier transform of the function

$$f(x) = \begin{cases} 1 & \text{for } |x| \leq L \\ 0 & \text{for } |x| > L \ . \end{cases}$$

Solution.

$$\hat{f}(\xi) = \int_{-\infty}^{\infty} f(x)\, e^{-i\xi x}\, d'x = \int_{-L}^{L} e^{-i\xi x}\, d'x = \frac{1}{\sqrt{2\pi}} \frac{e^{-i\xi x}}{-i\xi}\Big|_{-L}^{L}$$

$$= \frac{1}{\sqrt{2\pi}} \frac{e^{-i\xi L} - e^{i\xi L}}{-i\xi} = \frac{1}{\sqrt{2\pi}} \frac{2\sin(\xi L)}{\xi}.$$

Note that even though $f(x)$ vanishes for x outside the interval $[-L,L]$, the same is not true of $\hat{f}(\xi)$. In general, it can be shown that if *both* f and $\hat{f}$ vanish outside $[-L,L]$, then $f \equiv 0$. □

Example 5. By computing the Fourier transform of the function

$$f(x) = \begin{cases} 1 & \text{for } 0 \le x \le L \\ 0 & \text{otherwise}, \end{cases}$$

show that the Fourier transform of a real–valued function need not be real–valued itself.

Solution.

$$\hat{f}(\xi) = \int_{0}^{L} e^{-i\xi x}\, d'x = \frac{1}{\sqrt{2\pi}} \frac{e^{-i\xi x}}{-i\xi}\Big|_{0}^{L} = \frac{1}{\sqrt{2\pi}} \frac{e^{-i\xi L} - 1}{-i\xi}$$

$$= \frac{1}{\sqrt{2\pi}} \frac{\cos(\xi L) - i\sin(\xi L) - 1}{-i\xi} = \frac{1}{\sqrt{2\pi}} \frac{\sin(\xi L) - i(1 - \cos(\xi L))}{\xi}. \quad □$$

Example 6. Let $f(x) = e^{-ax^2/2}$, $a > 0$, $-\infty < x < \infty$. Show that

$$\hat{f}(\xi) = \int_{-\infty}^{\infty} e^{-ax^2/2 - i\xi x}\, d'x = \frac{1}{\sqrt{a}} e^{-\xi^2/2a}.$$

Solution. Completing the square, the exponent in the integrand equals $-\frac{a}{2}(x + i\frac{\xi}{a})^2 - \frac{\xi^2}{2a}$. Thus,

$$\hat{f}(\xi) = e^{-\xi^2/2a} \int_{-\infty}^{\infty} e^{-a(x + i\xi/a)^2/2}\, d'x \equiv e^{-\xi^2/2a}\, I(\xi).$$

We will now show that $I(\xi)$ is actually a constant independent of ξ. Indeed, by differentiating under the integral (cf. Leibniz's rule, Appendix A.3) we find that

$$I'(\xi) = \int_{-\infty}^{\infty} \frac{d}{d\xi} e^{-a(x + i\xi/a)^2/2}\, d'x = \int_{-\infty}^{\infty} e^{-a(x + i\xi/a)^2/2} [-a(x + i\tfrac{\xi}{a})]\, (\tfrac{i}{a})\, d'x$$

$$= \frac{i}{a\sqrt{2\pi}} e^{-a(x + i\xi/a)^2/2}\Big|_{-\infty}^{\infty} = 0,$$

i.e.,

$$I(\xi) = I(0) = \int_{-\infty}^{\infty} e^{-ax^2/2}\, d'x.$$

To compute $I(0)$, we resort to an ingenious trick (involving polar coordinates), which the reader may have seen before:

$$[I(0)]^2 = \left[\int_{-\infty}^{\infty} e^{-ax^2/2}\, d'x\right]\left[\int_{-\infty}^{\infty} e^{-ay^2/2}\, d'y\right] = \frac{1}{2\pi}\int_{-\infty}^{\infty}\int_{-\infty}^{\infty} e^{-a(x^2+y^2)/2}\, dx\, dy$$

$$= \frac{1}{2\pi}\int_0^{2\pi}\int_0^{\infty} e^{-ar^2/2}\, r\, dr d\theta = \left.\frac{e^{-ar^2/2}}{-a}\right|_0^{\infty} = \frac{1}{a},$$

whence $I(0) = \dfrac{1}{\sqrt{a}}$. Thus, we obtain

$$\hat{f}(\xi) = e^{-\xi^2/2a}\int_{-\infty}^{\infty} e^{-a(x+i\xi/a)^2/2}\, d'x = e^{-\xi^2/2a}\, I(\xi) = e^{-\xi^2/2a}\, I(0) = \frac{1}{\sqrt{a}}\, e^{-\xi^2/2a}.$$

Hence, if $f(x) = e^{-ax^2/2}$, then $\hat{f}(\xi) = \dfrac{1}{\sqrt{a}}\, e^{-\xi^2/2a}$.

When $a = 1$, note that f and $\hat{f}$ turn out to be the same function. There are infinitely many linearly independent functions with this property (cf. Problems 13 and 14 of Exercises 7.2). □

Remark. Appendix A.6 contains a table of Fourier transforms.

Summary 7.1

1. Complex Fourier series : Let $f(x)$ be a function defined on $[-L,L]$, then the **complex Fourier series** of $f(x)$ is the expression

$$FS_c\, f(x) \equiv \sum_{m=-\infty}^{\infty} c_m e^{imx} \;, \quad \text{where} \quad c_m = \frac{1}{2L}\int_{-L}^{L} f(x)\, e^{-im\pi x/L}\, dx, \quad m = 0, \pm 1, \pm 2, \dots , \tag{S1}$$

provided that all of these integrals exist. Since the partial sum of $FS_c\, f(x)$ from $-N$ to N is the same as the partial sum of the ordinary Fourier series $FS\, f(x)$ from 0 to N, all of the convergence results in Section 4.2 also hold for complex Fourier series. In terms of the complex Fourier coefficients, Parseval's equality becomes

$$\int_{-L}^{L} |f(x)|^2\, dx = 2L \sum_{m=-\infty}^{\infty} |c_m|^2 . \tag{S2}$$

The left–hand side of (S2) is the norm–square $\|f\|^2$ and the right–hand side is proportional to the sum of the squares of the moduli of the components $<f,e_m>$ of f, relative to the orthogonal family $e_m(x) \equiv e^{im\pi x/L}$ $(m = 0, \pm 1, \pm 2, \dots)$ of norm–square $2L$. Note that $c_m = \frac{1}{2L}<f,e_m>$.

2. Fourier transforms : Let $f(x)$ be a real or complex–valued function of the real variable x $(-\infty < x < \infty)$. The **Fourier transform** of $f(x)$ is the function $\hat{f}(\xi)$ defined by

$$\hat{f}(\xi) \equiv \frac{1}{\sqrt{2\pi}}\int_{-\infty}^{\infty} f(x)\, e^{-i\xi x}\, dx \equiv \lim_{R\to\infty} \int_{-R}^{R} f(x)\, e^{-i\xi x}\, d'x , \tag{S3}$$

where $dx' = dx/\sqrt{2\pi}$, when this limit (the Cauchy principal value) exists. In Section 7.3, we show that for "nice" functions $f(x)$, the following **Parseval's equality** (for Fourier transforms) holds

$$\int_{-\infty}^{\infty} |\hat{f}(\xi)|^2\, d\xi = \int_{-\infty}^{\infty} |f(x)|^2\, dx . \tag{S4}$$

Exercises 7.1

1. Compute the complex Fourier series for each of the following functions defined on $[-L,L]$:

(a) $f(x) = x$ (b) $f(x) = x^2$ (c) $f(x) = L - |x|$

(d) $f(x) = \dfrac{x}{|x|}(L - |x|)$ (e) $f(x) = x^N e^{ax}$, $N = 1, 2, \dots$ (f) $f(x) = e^{ax}\cos(bx)$.

2. (a) Check that $\sum_{n=-N}^{N} e^{in\theta} = 1 + 2\cos(\theta) + 2\cos(2\theta) + \dots + 2\cos(N\theta)$.

(b) Use the result $\sum_{n=0}^{N} z^n = \frac{1 - z^{N+1}}{1 - z}$ (valid for any complex number $z \neq 1$), to show that

$$\sum_{n=-N}^{N} e^{in\theta} = \frac{\sin[(N + \frac{1}{2})\theta]}{\sin(\theta/2)}.$$

(c) Conclude that $1/2 + \cos(\theta) + \cos(2\theta) + \dots + \cos(N\theta) = \frac{\sin[(N + \frac{1}{2})\theta]}{2\sin(\theta/2)}$, as was proved differently in Section 4.2. (Recall that this is the Dirichlet kernel.)

3. Find the Fourier transforms of the following functions :

(a) $f(x) = \begin{cases} L - |x| & |x| < L, \\ 0 & |x| \geq L \end{cases}$ (b) $f(x) = e^{-a|x|}\cos(bx)$, $a, b > 0$

(c) $f(x) = e^{-ax^2/2}\sin(bx)$, $a > 0$ (d) $f(x) = xe^{-ax^2/2}$ $\left[= -\frac{1}{a}\frac{d}{dx}e^{-ax^2/2}\right]$.

4. (a) Check that the Fourier transform is linear, i.e., for any complex numbers a and b, $(af + bg)\hat{}(\xi) = a\hat{f}(\xi) + b\hat{g}(\xi)$.

(b) Verify that, for any real number c, $[f(x+c)]\hat{}(\xi) = e^{ic\xi}\hat{f}(\xi)$ and $[e^{icx}f(x)]\hat{}(\xi) = \hat{f}(\xi-c)$.

5. Let $f(x)$ be a continuous, absolutely integrable function defined on $(-\infty,\infty)$. Show that if $f(x)$ is periodic with period $2L$, then $f(x)$ must be identically zero.

6. Prove the following complex form of Bessel's inequality. Let $f(x)$ be a complex-valued function defined on $[-L,L]$. Suppose that $|f(x)|^2$ is integrable on $[-L,L]$. Prove that

$$\sum_{m=-\infty}^{\infty} |c_m|^2 \leq \frac{1}{2L}\int_{-L}^{L} |f(x)|^2\, dx, \quad \text{where } c_m = \frac{1}{2L}\int_{-L}^{L} f(x)e^{-im\pi x/L}\, dx, \quad m = 0, \pm 1, \pm 2, \dots.$$

Hint. Let $S_N(x) = \sum_{m=-N}^{N} c_m e^{im\pi x/L}$ denote the N-th partial sum of $FS_c\, f(x)$, and consider the **mean square error** σ_N^2 defined by $\sigma_N^2 = \frac{1}{2L}\int_{-L}^{L} |f(x) - S_N(x)|^2\, dx$. Now use the fact that $|f(x) - S_N(x)|^2 = [f(x) - S_N(x)][\overline{f(x)} - \overline{S_N(x)}]$, where $\overline{f(x)}$ denotes the complex conjugate of $f(x)$. The rest of the argument is, almost *verbatim*, the same as in the real case (see Section 4.2).

7. Let $f(x) = e^{i\beta x}$, for $-\pi \le x \le \pi$, where β is a real number, but not an integer. Use Parseval's equality for complex Fourier series to show that $\sum_{m=-\infty}^{\infty} \frac{\sin^2(\pi(\beta - m))}{(\beta - m)^2} = \pi^2$.

8. Let $f(x) = \pi\cos(\alpha x)$ for $-\pi \le x \le \pi$, where α is not an integer. Derive the following formula due to Euler

$$1 + 2\alpha^2 \sum_{k=1}^{\infty} \frac{(-1)^k}{\alpha^2 - k^2} = \frac{\alpha\pi}{\sin(\alpha\pi)}.$$

Hint. First find the complex form of the Fourier series of $f(x)$.

9. Show that the functions

$$f(x) = \begin{cases} 0 & \text{if } |x| > 1 \\ 2 & \text{if } |x| = 1 \\ 1 & \text{if } |x| < 1 \end{cases} \quad \text{and} \quad g(x) = \begin{cases} 0 & \text{if } x < -1 \\ 1 & \text{if } -1 \le x < 1 \\ 0 & \text{if } x \ge 1 \end{cases}$$

have the same Fourier transform. (Note that f and g differ only at a few points).

10. (**The Riemann–Lebesgue lemma for Fourier transforms**) Let the complex–valued function $f(x)$ be absolutely integrable on $(-\infty,\infty)$ (i.e., $\int_{-\infty}^{\infty} |f(x)|\, dx < \infty$). In what follows, you may use the standard result $\left| \int_a^b f(x)\, dx \right| \le \int_a^b |f(x)|\, dx$.

(a) Show that $\hat{f}(\xi)$ exists, i.e., $\lim_{R\to\infty} \frac{1}{\sqrt{2\pi}} \int_{-R}^{R} f(x) e^{-i\xi x}\, dx$, exists for any real ξ.
Hint. Show that the integrals, from $\pm R$ to $\pm\infty$, tend to 0 as $R \to \infty$.

(b) Show that $|\hat{f}(\xi)| \le \frac{1}{\sqrt{2\pi}} \int_{-\infty}^{\infty} |f(x)|\, dx$ (i.e., $|\hat{f}(\xi)|$ is bounded).

(c) Assume that $\lim_{a\to 0} \int_{-\infty}^{\infty} |f(x) - f(x+a)|\, dx = 0$. (This holds if $f(x)$ is absolutely integrable and piecewise continuous, by using the dominated convergence theorem in Appendix A.3 .)
Show that $\lim_{\xi\to\pm\infty} |\hat{f}(\xi)| = 0$. **Hint**. First show that $\hat{f}(\xi) = \frac{1}{\sqrt{2\pi}} \int_{-\infty}^{\infty} -f(x)\, e^{-i\xi(x - \pi/\xi)}\, dx$ $= \frac{1}{\sqrt{2\pi}} \int_{-\infty}^{\infty} -f(y + \frac{\pi}{\xi})\, e^{-i\xi y}\, dy$. Then note that $2\hat{f}(\xi) = \frac{1}{\sqrt{2\pi}} \int_{-\infty}^{\infty} [f(x) - f(x + \frac{\pi}{\xi})]\, e^{-i\xi x}\, dx$.

11. Give an example of a C^∞ function $f(x)$ defined on $(-\infty,\infty)$ for which $\lim_{R\to\infty} \int_{-R}^{R} |f(x)|\, dx$ does not exist, but for which $\lim_{R\to\infty} \int_{-R}^{R} f(x)\, dx$ does exist.

12. Let $f(x) = \begin{cases} 1 & \text{if } n < x < n + n^{-2}, \quad n = \pm 1, \pm 2, \pm 3, \ldots, \\ 0 & \text{otherwise.} \end{cases}$

(a) Draw the graph of $f(x)$ for $-4 \le x \le 4$.

(b) Verify that $\int_{-\infty}^{\infty} |f(x)|\, dx = 2\sum_{n=1}^{\infty} 1/n^2 < \infty$.

(c) Show that the limit of $f(x)$ as $x \to \infty$ does not exist. In particular, $f(x)$ does not tend to zero as $x \to \infty$. (In contrast, we know that the terms of a convergent infinite series tend to zero).

(d) Construct a *strictly* positive function $g(x)$ which is absolutely integrable on $(-\infty,\infty)$, and such that $\lim_{x\to\infty} g(x)$ does not exist and $g(x)$ is unbounded. **Hint.** In the definition of $f(x)$, consider the effect of replacing 1 by n, n^{-2} by $|n|^{-3}$, and 0 by $(1+x^2)^{-1}$.

7.2 Basic Properties of Fourier Transforms

In this section, we establish several formulas which are useful in working with Fourier transforms. These formulas are valid for functions which (along with their derivatives) decay (i.e., tend to 0 as $|x| \to \infty$) sufficiently fast. More precisely, we state the following definition.

Definition. A function $f(x)$ $(-\infty < x < \infty)$ has **decay order** (m,n), where m and n are nonnegative integers, if $f(x)$ is C^m and if there is a constant $K > 0$ such that, for all x with $|x| > 1$,

$$|f(x)| + |f'(x)| + \dots + |f^{(m)}(x)| \le K|x|^{-n}. \qquad (1)$$

Remark. If $m = 0$, then $f(x)$ is assumed to be a continuous function on $(-\infty,\infty)$. Note also that the constant K in (1) depends on f, m and n. In other words, f and its first m derivatives decay at least as fast as $K|x|^{-n}$ as $|x| \to \infty$. Obviously, if f has decay order (m,n), then it also has decay order (m',n') for $0 \le m' \le m$ and $0 \le n' \le n$. A function which has decay order (m,n) for all $m, n \ge 0$ is called **rapidly decreasing**. Clearly, any C^∞ function that is identically zero outside a closed interval is rapidly decreasing. The following example shows that there are rapidly decreasing functions that are never zero. □

Example 1. Show that e^{-x^2} is rapidly decreasing.

Solution. For any $k \ge 0$ and for any real x, the following estimate holds :

$$e^{-x^2} = \left[e^{x^2}\right]^{-1} = \frac{1}{1 + x^2 + \frac{1}{2!}x^4 + \dots + \frac{1}{k!}x^{2k} + \dots} \le \frac{1}{\frac{1}{k!}x^{2k}} = k!\,x^{-2k}.$$

Thus, e^{-x^2} decays faster than $k!|x|^{-2k}$ for any $k \ge 0$. The m–th derivative of e^{-x^2} is of the form $p_m(x)e^{-x^2}$, where $p_m(x)$ is an m–th degree polynomial. There is a constant d_m, such that $|p_m(x)| \le d_m|x|^m$ for $|x| \ge 1$ (cf. Problem 9). Thus, if $f(x) = e^{-x^2}$, then for $|x| \ge 1$

$$|f(x)| + |f'(x)| + \dots + |f^{(m)}(x)| \le (d_0 + d_1|x| + \dots + d_m|x|^m)e^{-x^2}$$
$$\le (d_0 + \dots + d_m)|x|^m e^{-x^2} \le k!(d_0 + \dots + d_m)|x|^{m-2k}.$$

Given *any* (m,n) (with $m, n \ge 0$), we can choose k so large that $2k - m \ge n$. Then the above inequalities show that $f(x)$ has decay order (m,n), whence $f(x)$ is rapidly decreasing (note that the derivatives $f^{(m)}(x) = p_m(x)e^{-x^2}$ are all continuous, as required). The function $e^{-|x|}$ is *not* rapidly decreasing, since its derivatives do not exist at $x = 0$, even though it has the required behavior for $|x| \geq 1$. □

From the differential equation viewpoint, perhaps the single most important property of Fourier transforms is that they convert differentiation into multiplication by the function $i\xi$, in the sense of the next proposition. As we will see, partial differential equations (at least those with constant coefficients) are then transformed into simpler ordinary differential equations.

Proposition 1. Let $f(x)$ **have decay order** $(1,2)$**, i.e.,** f **is** C^1 **and** $|f(x)| + |f'(x)| \leq K|x|^{-2}$, **for** $|x| \geq 1$ **and some constant** $K > 0$. **Then, for all real** ξ**, we have**

$$\left[\frac{df}{dx}\right]^{\wedge}(\xi) = i\xi\,\hat{f}(\xi)\,. \tag{2}$$

Proof. In order to show that $\hat{f}(\xi)$ exists, it suffices to show that $f(x)$ is absolutely integrable (cf. Problem 10 of Exercises 7.1). We have

$$\int_{-\infty}^{\infty} |f(x)|\;dx = \int_{-1}^{1} |f(x)|\;dx + \int_{|x|>1} |f(x)|\;dx$$

$$\leq \int_{-1}^{1} |f(x)|\;dx + 2\int_{1}^{\infty} Kx^{-2}\;dx = \int_{-1}^{1} |f(x)|\;dx + 2K < \infty\,.$$

The same argument shows that $[f'(x)]^{\wedge}(\xi)$ exists. Using integration by parts and the fact that $\lim_{x\to\pm\infty} f(x) = 0$, we obtain (recall $d'x \equiv (2\pi)^{-\frac{1}{2}}\,dx$, as in (8) of Section 7.1)

$$[f'(x)]^{\wedge}(\xi) = \int_{-\infty}^{\infty} f'(x)\,e^{-i\xi x}\;d'x$$

$$= (2\pi)^{-\frac{1}{2}}\,f(x)\,e^{-i\xi x}\Big|_{-\infty}^{\infty} - \int_{-\infty}^{\infty} f(x)(-i\xi)e^{-i\xi x}\;d'x = (i\xi)\int_{-\infty}^{\infty} f(x)\,e^{-i\xi x}\;d'x = i\xi\,\hat{f}(\xi). \quad □$$

Corollary 1. If $f(x)$ **has decay order** $(m,2)$**, then for all real** ξ,

$$[f^{(m)}(x)]^{\wedge}(\xi) = i^m\xi^m\,\hat{f}(\xi). \tag{3}$$

Proof. Since $f(x)$ has decay order $(m,2)$ we have that $f'(x)$ is of decay order $(m-1,2)$, $f''(x)$ is of decay order $(m-2,2)$, ..., and $f^{(m-1)}(x)$ is of decay order $(1,2)$. In particular, all of these derivatives are of decay order $(1,2)$, and we may repeatedly apply Proposition 1 to obtain

$$[f^{(m)}(x)]\hat{}\,(\xi) = [(f^{(m-1)})'(x)]\hat{}\,(\xi) = i\xi[f^{(m-1)}(x)]\hat{}\,(\xi)$$

$$= i\xi[(f^{(m-2)})'(x)]\hat{}\,(\xi) = (i\xi)^2[f^{(m-2)}(x)]\hat{}\,(\xi) = \ldots = (i\xi)^m\,\hat{f}(\xi). \quad \square$$

Proposition 2. Suppose $f(x)$ **has decay order** $(0,3)$ **, i.e.,** $f(x)$ **is continuous and** $|f(x)| \le K|x|^{-3}$ **for** $|x| \ge 1,\ K > 0$ **. Then, for all real** ξ

$$i\,\frac{d\hat{f}}{d\xi}(\xi) = [xf(x)]\hat{}\,(\xi)\,. \tag{4}$$

Proof. Since $f(x)$ has decay order $(0,3)$, both $f(x)\,e^{-i\xi x}$ and its derivative with respect to ξ, namely, $-ixf(x)e^{-i\xi x}$, are absolutely integrable and continuous. Thus, Leibniz's rule (cf. Appendix A.3) yields

$$i\,\frac{d\hat{f}}{d\xi}(\xi) \;=\; i\,\frac{d}{d\xi}\int_{-\infty}^{\infty} f(x)\,e^{-i\xi x}\,d'x \;=\; \int_{-\infty}^{\infty} x\,f(x)\,e^{-i\xi x}\,d'x\,. \quad \square$$

Repeated application of this result, as in the proof of Corollary 1, yields the following

Corollary 2. If $f(x)$ **has decay order** $(0,n+2)$**, then for all real** ξ

$$i^n\,\frac{d^n\hat{f}}{d\xi^n}(\xi) = [x^n\,f(x)]\hat{}\,(\xi). \tag{5}$$

In particular, both sides of equation (5) **exist !**

Theorem 1. If $f(x)$ **is rapidly decreasing, then so is** $\hat{f}(\xi)$**. More precisely, if** $f(x)$ **has decay order** $(m,n+2)$**, then** $\hat{f}(\xi)$ **has decay order** (n,m) **for any** $m,\ n \ge 0$.

Proof. Assume $f(x)$ has decay order $(m,n+2)$. Using the Corollary 1 and Corollary 2, we have (for integers α and β with $0 \le \alpha \le m,\ 0 \le \beta \le n$),

$$i^{\alpha+\beta}\,\xi^{\alpha}\,(\hat{f})^{(\beta)}(\xi) = i^{\alpha}\,\xi^{\alpha}\,[x^{\beta}\,f(x)]\hat{}\,(\xi) = \left[\frac{d^{\alpha}}{dx^{\alpha}}\,x^{\beta}f(x)\right]\hat{}\,(\xi) = \int_{-\infty}^{\infty}\left[\frac{d^{\alpha}}{dx^{\alpha}}\,x^{\beta}f(x)\right]\,e^{-i\xi x}\,d'x\,.$$

Thus,

$$|\xi|^{\alpha}\,|(\hat{f})^{(\beta)}(\xi)| \le \int_{-\infty}^{\infty}\left|\frac{d^{\alpha}}{dx^{\alpha}}\,x^{\beta}f(x)\right|\,d'x. \tag{6}$$

Note that $\frac{d^\alpha}{dx^\alpha}\left[x^\beta f(x)\right]$ is a sum of terms of the form $K\,x^b\,f^{(a)}(x)$, where K is a constant and $0 \le a \le \alpha \le m$ and $0 \le b \le \beta \le n$. Hence, for some constant C and $|x| > 1$,

$$\left|\frac{d^\alpha}{dx^\alpha}x^\beta f(x)\right| \le C\,|x|^n\left[|f(x)| + \dots + |f^{(m)}(x)|\right] \le CK\,|x|^n|x|^{-(n+2)} = CK\,|x|^{-2}.$$

Therefore, the integral in (6) is finite, and we find that for an appropriate constant A, $|(\hat{f})^{(\beta)}(\xi)| \le A|\xi|^{-\alpha}$ for all α and β with $0 \le \alpha \le m$ and $0 \le \beta \le n$. In particular, we can take $\alpha = m$, and add up these inequalities for $\beta = 0, \dots, n$ in order to obtain

$$|\hat{f}(\xi)| + |(\hat{f})'(\xi)| + \dots + |(\hat{f})^{(n)}(\xi)| \le B\,|\xi|^{-m},$$

where B is some constant. We still need to show that $(\hat{f})^{(n)}(\xi)$ is continuous, i.e., for any ξ_0 we need to show that $(\hat{f})^{(n)}(\xi) \to (\hat{f})^{(n)}(\xi_0)$ as $\xi \to \xi_0$. In order to prove this, we write

$$\begin{aligned}|(\hat{f})^{(n)}(\xi) - (\hat{f})^{(n)}(\xi_0)| &\le \int_{-\infty}^{\infty} |x|^n|f(x)|\;|e^{-i\xi x} - e^{-i\xi_0 x}|\;d'x \\ &= \left[\int_{|x|>R} + \int_{|x|\le R}\right] |x|^n|f(x)|\;|e^{-i\xi x} - e^{-i\xi_0 x}|\;d'x, \qquad (*)\end{aligned}$$

where $R \ge 1$ is a positive constant. Since $|f(x)| < K\,|x|^{-n-2}$, $|x| \ge 1$, thus for $R \ge 1$

$$\int_{|x|>R} |x|^n|f(x)|\;|e^{-i\xi x} - e^{-i\xi_0 x}|\;d'x \le 2\int_R^\infty Kx^{-2}\cdot 2\,d'x = 4(2\pi)^{-\frac{1}{2}}K\,R^{-1},$$

which can be made as small as desired by taking R large enough. Once R is chosen, the second integral in $(*)$ can be made as small as desired by taking ξ close enough to ξ_0. □

Remark. We have seen that the Fourier transform of the rapidly decreasing function $e^{-\frac{1}{2}ax^2}$ $(a > 0)$ is $a^{-\frac{1}{2}}\,e^{-\frac{1}{2}a\xi^2}$, which is also rapidly decreasing. In all of the other examples computed so far, $\hat{f}(\xi)$ is not rapidly decreasing because $f(x)$ failed to be differentiable. In general, the nonexistence of $f'(x)$ (even at a single point) slows the decay of $\hat{f}(\xi)$. For instance, in Example 3 of Section 7.1, we found that the Fourier transform of $e^{-a|x|}$ is $\frac{1}{\sqrt{2\pi}}\frac{2a}{a^2 + \xi^2}$ which is $\approx \sqrt{2/\pi}\;a\xi^{-2}$ for large ξ, even though $e^{-a|x|}$ decays faster than $|x|^{-n}$ for any integer $n > 0$, as $|x| \to \infty$. The fact that $e^{-a|x|}$ is not C^1 at $x = 0$, is responsible for the relative slowness in the decay of its transform. Observe that Theorem 1 roughly says that the smoothness of $f(x)$ governs the rate at which $|\hat{f}(\xi)|$ approaches zero as $|\xi| \to \infty$, whereas the rate of decay of $f(x)$ as $|x| \to \infty$ governs the smoothness of $\hat{f}(\xi)$. □

Example 2. Find a function $f(x)$ which is *not* rapidly decreasing, but is such that $|\hat{f}(\xi)| \leq K_m |\xi|^{-m}$ for all $|\xi| \geq 1$ and $m > 0$ (i.e., $\hat{f}(\xi)$ has decay order $(0,m)$ for all $m > 0$).

Solution. Theorem 1 says that $\hat{f}(\xi)$ will have decay order $(0,m)$ for all $m \geq 0$, if $f(x)$ has decay order $(m,2)$ for all $m \geq 0$ (i.e., $f(x)$ is a C^∞ function whose derivatives [and itself] all decay at least as fast as $K|x|^{-2}$ as $|x| \to \infty$). A function with these properties is $f(x) = (1+x^2)^{-1}$, and it is also not rapidly decreasing. We do not have to compute $\hat{f}(\xi)$ to verify that it has decay order $(0,m)$ for all $m > 0$, although a computation in Example 1 of Section 7.3 shows that $\hat{f}(\xi) = \sqrt{\pi/2}\, e^{-|\xi|}$, which does in fact have decay order $(0,m)$ for all $m > 0$. □

Remark. A beautiful result which supplements Theorem 1 is due to Godfrey H. Hardy (1877–1947), one of the greatest analysts of this century. [*G.H. Hardy , A theorem concerning Fourier transforms, J. London Math. Soc. , 8 (1933) , 227–231.*] Hardy's result is as follows.

> **Let** $f(x)$ **be defined on** $(-\infty,\infty)$. **Suppose that**
>
> $$|f(x)| \leq K_1 |x|^n e^{-\frac{1}{2}x^2}, \ \text{for } |x| \geq 1, \quad \text{and} \quad |\hat{f}(\xi)| \leq K_2 |\xi|^n e^{-\frac{1}{2}\xi^2} \ \text{for } |\xi| \geq 1,$$
>
> **where** n **is a nonnegative integer and** K_1, K_2 **are positive constants. Then**
>
> $f(x) = p(x)\, e^{-\frac{1}{2}x^2}$, **where** $p(x)$ **is a polynomial of degree not exceeding** n. □

Convolutions and their Fourier transforms

Often one wants to find a function $h(x)$ whose Fourier transform is the product $\hat{f}(\xi)\hat{g}(\xi)$ of the transforms of two functions $f(x)$ and $g(x)$. It turns out that $h(x)$ is *not* $f(x)g(x)$ in general. In order to find the function $h(x)$, we carry out the following computation, assuming that $f(x)$ and $g(x)$ have decay order $(0,2)$ to justify the interchange of order of integration (cf. Appendix A.2) and to ensure the existence of all of the integrals :

$$\begin{aligned}
\hat{h}(\xi) = \hat{f}(\xi)\hat{g}(\xi) &= \left[\int_{-\infty}^{\infty} f(x)\, e^{-i\xi x}\, d'x\right]\left[\int_{-\infty}^{\infty} g(y)\, e^{-i\xi y}\, d'y\right] \\
&= \int_{-\infty}^{\infty}\int_{-\infty}^{\infty} f(x)g(y)\, e^{-i\xi(x+y)}\, d'x\, d'y \qquad (z = x + y) \\
&= \int_{-\infty}^{\infty}\int_{-\infty}^{\infty} f(z-y)g(y)\, e^{-i\xi z}\, d'z\, d'y = \int_{-\infty}^{\infty}\int_{-\infty}^{\infty} f(z-y)g(y)\, e^{-i\xi z}\, d'y\, d'z \\
&= \int_{-\infty}^{\infty} e^{-i\xi z}\left[\int_{-\infty}^{\infty} f(z-y)g(y)\, d'y\right] d'z.
\end{aligned}$$

Replacing the dummy variable of integration z by x we see that this final integral is the Fourier transform of the function

$$h(x) = \int_{-\infty}^{\infty} f(x-y)g(y)\, d'y = \frac{1}{\sqrt{2\pi}} \int_{-\infty}^{\infty} f(x-y)g(y)\, dy,$$

which is a continuous superposition of the translates $f(x-y)$ of the function $f(x)$ with respect to a "weight" $g(y)$.

Definition. The **convolution** of the function f with g is the function $f*g$, which is defined by

$$(f*g)(x) \equiv \int_{-\infty}^{\infty} f(x-y)g(y)\, dy\ , \tag{7}$$

provided the integral exists for each x (e.g., if f is bounded and g is absolutely integrable).

Thus our computation above shows that if $f(x)$ and $g(x)$ both have decay order (0,2), then the function with Fourier transform $\hat{f}(\xi)\hat{g}(\xi)$ is $(2\pi)^{-\frac{1}{2}}(f*g)(x)$. In fact, the above computation remains valid if we merely assume that f and g are *piecewise* continuous with $|f(x)|, |g(x)| \le c|x|^{-2}$. Hence we have the following result.

Theorem 2 (The Convolution Theorem). Let $f(x)$ **and** $g(x)$ **be piecewise continuous with** $|f(x)|, |g(x)| \le \text{const.}|x|^{-2}$, $|x| \ge 1$. **Then**

$$\hat{f}(\xi)\hat{g}(\xi) = \frac{1}{\sqrt{2\pi}} (f*g)\hat{\ }(\xi). \tag{8}$$

To illustrate the use of this result and to motivate the Inversion Theorem to be introduced in Section 7.3, we offer the following simple example from ordinary differential equations. More involved examples for PDEs will be presented later in Sections 7.4 and 7.5 .

Example 3. For a given constant k and continuous function $f(x)$, consider the ODE

$$\frac{dy}{dx} + ky = f(x)\ . \tag{9}$$

Find a solution of this ODE by formally taking the Fourier transform of both sides of (9).

Solution. We know from Section 1.1 that the general solution of this problem is given by

$$y(x) = e^{-kx} \int_a^x e^{kt} f(t)\, dt + ce^{-kx}, \tag{10}$$

where a and c are arbitrary constants. For this ODE there is no advantage in using Fourier transform techniques, but it is instructive to see what the technique looks like in this familiar setting. In order to take the Fourier transform of both sides of the ODE (9), we make the assumptions that the solution $y(x)$ exists and is of decay order (1,2). We also assume that $\hat{f}(\xi)$ exists. Taking the Fourier transform of both sides of (9) and using Proposition 1, we get

$$i\xi\, \hat{y}(\xi) + k\hat{y}(\xi) = \hat{f}(\xi) \quad \text{or} \quad \hat{y}(\xi) = \frac{\hat{f}(\xi)}{k + i\xi}.$$

In practice, one often does not know whether the above assumptions hold. Therefore, after the final result is obtained below, it must be checked to ensure that it exists and satisfies (9). At any rate, from Example 3 of Section 7.1 we know that

$$\text{if} \quad g(x) = \begin{cases} 0 & x < 0 \\ e^{-kx} & x \geq 0, \end{cases} \quad \text{then} \quad \hat{g}(\xi) = \frac{1}{\sqrt{2\pi}\,(k + i\xi)}.$$

(Simply disregard the second integral in Example 3 of Section 7.1 .) Thus, by Theorem 2 we have

$$\hat{y}(\xi) = \sqrt{2\pi}\, \hat{g}(\xi)\hat{f}(\xi) = \sqrt{2\pi} \frac{1}{\sqrt{2\pi}} (g*f)\hat{\ }(\xi) = (g*f)\hat{\ }(\xi).$$

This requires us to assume that $f(x)$ is piecewise continuous with $|f(x)| \leq \text{const.}|x|^{-2}$, but we are going to check our answer anyway. To conclude that $y = f*g$, we need the Inversion Theorem and more assumptions. Nevertheless, we have been led to the *hypothetical* solution

$$y(x) = (g*f)(x) = \int_{-\infty}^{\infty} g(x-t) f(t)\, dt = \int_{-\infty}^{x} e^{-k(x-t)} f(t)\, dt = e^{-kx} \int_{-\infty}^{x} e^{kt} f(t)\, dt.$$

Now, $y(x)$ exists whenever $\int_{-\infty}^{x} e^{kt} f(t)\, dt$ exists. For example, this would be the case, if $|f(t)| \leq \text{const.}e^{-k't}$ for some $k' < k$ and all $t < 0$. Then by the product rule

$$y'(x) = -ke^{-kx} \int_{-\infty}^{x} e^{kt} f(t)\, dt + e^{-kx} e^{kx} f(x),$$

and thus $y'(x) + ky(x) = f(x)$, i.e., the hypothetical solution is an actual solution under suitable assumptions on $f(x)$. More precisely, we have shown

If $f(x)$ **is continuous and if** $\int_{-\infty}^{A} e^{kt}|f(t)|\,dt$ **exists for some real number** A**, then**

$$y(x) = \int_{-\infty}^{x} e^{-k(x-t)} f(t)\,dt, \qquad -\infty < x < \infty, \tag{11}$$

is a solution of $y' + ky = f(x)$.

Obviously, $y(x)$ in (11) is not the only solution of (9); all other solutions are obtained by adding the solutions Ce^{-kx} of the associated homogeneous equation. This illustrates the fact that the hypothetical solution given by transform methods is typically *not* the most general one. The reason is that the transform method presupposes that the solution has a certain decay rate. Adding Ce^{-kx} to $y(x)$ may destroy any decay property that $y(x)$ might have had, since $e^{-kx} \to \infty$ as $x \to -\infty$, if $k > 0$. However, in applications, one often just wants solutions that decay in some sense. If there are any such solutions, the method of Fourier transforms will most likely provide them and not any extraneous non–decaying solutions. For example, if $f(x)$ is continuous and vanishes outside a finite interval, then (11) is the only solution of (9) that is bounded (i.e., $|y(x)| <$ const.). On the other hand, if one is not careful, the solution (11) might be overlooked if one uses (10). □

Example 4. Let $f(x)$ be a rapidly decreasing function defined on $(-\infty,\infty)$, and let $F(\xi) \equiv 2\int_0^{\infty} f(x)\cos(\xi x)\,d'x$. [This is the "Fourier cosine transform" of the restriction of $f(x)$ to $(0,\infty)$ (cf. Section 7.5).] Show that if the graph of $F(\xi)$ crosses the ξ–axis at points arbitrarily far from $\xi = 0$, then all of the odd order derivatives $f^{(2k+1)}(0)$ are zero. In particular, if $f(x) = \sum_{n=0}^{\infty} \frac{f^{(n)}(0)}{n!} x^n$ (i.e., if $f(x)$ is given by its Taylor series), then $f(x)$ must be even.

Solution. By repeated application Green's formula (or by twice as many integrations by parts),

$$\sqrt{\pi/2}\,F(\xi) = \int_0^{\infty} f(x)\cos(\xi x)\,dx = -\xi^{-2}\left[\left[-f(x)\xi\sin(\xi x) - f'(x)\cos(\xi x)\right]\Big|_0^{\infty} + \int_0^{\infty} f''(x)\cos(\xi x)dx\right]$$

$$= \xi^{-2} f'(0) - \xi^{-2}\int_0^{\infty} f''(x)\cos(\xi x)\,dx = \xi^{-2}f'(0) - \xi^{-2}\left[\xi^{-2} f^{(3)}(0) - \xi^{-2}\int_0^{\infty} f^{(4)}(x)\cos(\xi x)\,dx\right]$$

$$= \ldots = \sum_{k=0}^{p}\left[(-1)^k \xi^{-2k-2} f^{(2k+1)}(0)\right] - (-1)^p \xi^{-2p-2}\int_0^{\infty} f^{(2p+2)}(x)\cos(\xi x)\,dx\,.$$

Seeking a contradiction, suppose that $f^{(2p-1)}(0)$ is the first nonvanishing odd order derivative. Then, the above becomes

$$\sqrt{\pi/2}\, F(\xi) = (-1)^{p-1} \xi^{-2p} f^{(2p-1)}(0) + (-1)^{p} \xi^{-2p-2} \left[f^{(2p+1)}(0) - \int_0^\infty f^{(2p+2)}(x) \cos(\xi x)\, dx \right].$$

Since $f(x)$ is rapidly decreasing, the term in large parentheses is a bounded function of ξ. Thus, $\lim_{|\xi| \to \infty} \sqrt{\pi/2}\, \xi^{2p} F(\xi) = (-1)^{p-1} f^{(2p-1)}(0) \neq 0$. However, since it is assumed that $F(\xi) = 0$ for infinitely many values of ξ as $|\xi|$ tends to infinity, this limit (if it exists) must be zero. □

Summary 7.2

1. **Decay orders** : A function $f(x)$ $(-\infty < x < \infty)$ has **decay order** (m,n), where m and n are nonnegative integers, if $f(x)$ is C^m and if there is a constant K such that

$$|f(x)| + |f'(x)| + ... + |f^{(m)}(x)| \leq K|x|^{-n}, \qquad \text{(S1)}$$

for all x with $|x| \geq 1$. A function with decay order (m,n) for all $m, n \geq 0$ is said to be **rapidly decreasing**. Theorem 1 states that if $f(x)$ has decay order $(m,n+2)$, then $\hat{f}(\xi)$ has decay order (n,m) for any $m, n \geq 0$. In particular, the Fourier transform of a rapidly decreasing function is also rapidly decreasing.

2. **Properties of the Fourier transform** : $\hat{f}(\xi) = \frac{1}{\sqrt{2\pi}} \int_{-\infty}^{\infty} f(x)\, e^{-i\xi x}\, dx$). We list the properties shown in this section and specify the places where precise conditions for them are stated.

A. $\left[\frac{df}{dx}\right]\hat{}\,(\xi) = i\xi\, \hat{f}(\xi)$ (Proposition 1), B. $[f^{(m)}(x)]\hat{}\,(\xi) = i^m\, \xi^m\, \hat{f}(\xi)$ (Corollary 1),

C. $i\,\frac{d\hat{f}}{d\xi}(\xi) = [x\, f(x)]\hat{}\,(\xi)$ (Proposition 2), D. $i^n\, \frac{d^n\hat{f}}{d\xi^n}(\xi) = [x^n\, f(x)]\hat{}\,(\xi)$, (Corollary 2),

E. $(e^{-\frac{1}{2}ax^2})\hat{}\,(\xi) = \frac{1}{\sqrt{a}}\, e^{-\frac{1}{2}a^{-1}\xi^2}$ $(a > 0)$, F. $\hat{f}(\xi)\hat{g}(\xi) = \frac{1}{\sqrt{2\pi}}(f*g)\hat{}\,(\xi)$, (Theorem 2),

where $(f*g)(x) \equiv \int_{-\infty}^{\infty} f(x-y)g(y)\, dy$ is the **convolution** of f and g.
A more complete list of properties, and specific Fourier transforms, is given in Appendix A.6 .

Exercises 7.2

1. Show that if $f(x)$ is absolutely integrable on $(-\infty,\infty)$, then

(a) $[e^{ibx}f(ax)]\hat{}\,(\xi) = \frac{1}{a}\, \hat{f}((\xi-b)/a)$, a, b real, $a \neq 0$

(b) $[f(\frac{x}{a} + b)]\hat{}\,(\xi) = ae^{iab\xi}\, \hat{f}(a\xi)$, a, b real, $a \neq 0$.

2. Use the result of Example 6 of Section 7.1 and Corollary 2 to find the Fourier transform of each of the following functions :

(a) $f(x) = x^2e^{-\frac{1}{2}x^2}$ (b) $g(x) = x^4e^{-\frac{1}{2}ax^2}$, $a > 0$.

3. Show that if $f(x)$ is defined on $(-\infty,\infty)$ and has decay order $(1,2)$ (i.e., f is C^2 and $|f(x)| + |f'(x)| \le K|x|^{-2}$ for $|x| \ge 1$, for some constant $K > 0$), then $[f'(x)]\hat{}\,(\xi)$ exists. **Hint.** See the proof of Proposition 1.

4. Show that if $f(x)$ is defined on $(-\infty,\infty)$ and has decay order $(m,2)$, $m \ge 1$, then $f'(x)$ has decay order $(m-1,2)$.

5. Show that if $f(x)$ $(-\infty < x < \infty)$ has decay order (m,n), then $f(x)$ also has decay order (m',n'), where $0 \le m' \le m$ and $0 \le n' \le n$.

6. Prove that if $f(x)$ $(-\infty < x < \infty)$ has decay order $(0,n+2)$, then, for all real ξ,

$$i^n \frac{d^n}{d\xi^n} \hat{f}(\xi) = [x^n f(x)]\hat{}\,(\xi) .$$

7. Find the Fourier transform of

$$f(x) = \begin{cases} x^2 e^{-x} & \text{if } x \ge 0 \\ 0 & \text{if } x < 0 . \end{cases}$$

8. Consider Example 3 in the text with $k = -1$ and $f(x) \equiv 1$. Then, the problem is to solve the very simple ODE $y'(x) - y(x) = 1$. Is it possible to find any solutions of this ODE using Fourier transforms ? **Hint.** See the remarks following equation (11).

9. Show that if $p_m(x) = a_0 + a_1 x + \dots + a_m x^m$, $a_m \ne 0$, $m \ge 1$, is an m–th degree polynomial, then $|p_m(x)| \le d_m |x|^m$, for $|x| \ge 1$ and for some positive constant d_m.

10. (a) Consider the ODE $y''(x) - y(x) = f(x)$. Show that a formal use of Fourier transforms and the Convolution Theorem leads to a formal solution $y(x) = -\frac{1}{2}\int_{-\infty}^{\infty} e^{-|x-y|} f(y)\, dy$.

Hint. By Example 3 of Section 7.1, the transform of $e^{-|x|}$ is $(2\pi)^{-\frac{1}{2}}\, 2\,(1 + \xi^2)^{-1}$.

(b) By writing $y(x) = -\frac{1}{2}\Big[\int_{-\infty}^{x} e^{y-x} f(y)\, dy + \int_{x}^{\infty} e^{x-y} f(y)\, dy\Big]$ and applying the general version of Leibniz's rule (cf. (7) of Appendix A.3), show that the formal solution is valid if $f(x)$ is continuous and absolutely integrable or bounded.

11. Verify the Convolution Theorem in the following cases:

(a) $f(x) = g(x) = e^{-\frac{1}{2}x^2}$ (b) $f(x) = g(x) = \begin{cases} 1 & \text{for } |x| \le 1 \\ 0 & \text{for } |x| > 1 \end{cases}$.

Hint (for (b)). $(f*g)(x) = \begin{cases} 0 & \text{for } |x| > 2 \\ 2-|x| & \text{for } |x| \le 2 \end{cases}$.

12. Let $f(x)$, $g(x)$ and $h(x)$ be rapidly decreasing functions and let α be a constant. Prove

(a) $f*(\alpha g) = (\alpha f)*g = \alpha(f*g)$ (b) $f*(g+h) = (f*g) + (f*h)$ (c) $f*g = g*f$.

13. (a) In the computations for Example 6 of 7.1, we found that $\int_{-\infty}^{\infty} e^{-\frac{1}{2}ax^2}\,dx = \sqrt{2\pi/a}$.

Setting $a = 2b$, we then get $\int_{-\infty}^{\infty} e^{-bx^2}\,dx = \sqrt{\pi/b}$. Differentiate this formula n times with respect to b and obtain $\int_{-\infty}^{\infty} x^{2n} e^{-bx^2}\,dx = \frac{\sqrt{\pi}}{\sqrt{b}}\frac{(2n)!}{(4b)^n n!}$.

(b) Let $f_m(x) = x^m e^{-\frac{1}{2}x^2}$. Show that this is a rapidly decreasing function for $m = 0, 1, 2, \ldots$.

(c) Recalling that $\langle f,g\rangle = \int_{-\infty}^{\infty} f(x)\overline{g(x)}\,dx$, prove (by using the result of (a) with $b = 1$) that

$$\langle f_m,f_n\rangle = \begin{cases} \dfrac{\sqrt{\pi}\,(m+n)!}{2^{m+n}\left[\frac{m+n}{2}\right]!} & \text{if } m+n \text{ is even} \\ 0 & \text{if } m+n \text{ is odd.} \end{cases}$$

14. (a) By part (c) of Problem 13, we know that $f_0, f_1, f_2, \ldots$ are not mutually orthogonal (e.g., $\langle f_0,f_2\rangle \ne 0$). Find a constant c, such that $(x^2 + c)e^{-\frac{1}{2}x^2}$ is orthogonal to f_0 and f_1.

(b) Let V_n be the set of all linear combinations of $f_0, f_1, \ldots, f_n$. Since V_{n-1} is an $(n-1)$-dimensional subspace of the n-dimensional space V_n, there should be exactly one function of the form $f_n + a_{n-1}f_{n-1} + \ldots + a_0f_0$, which is orthogonal to $f_0, f_1, \ldots, f_{n-1}$ (i.e., to all functions in V_{n-1}). Use integration by parts to show that this function is

$$k_n(x) = (-\tfrac{1}{2})^n e^{\frac{1}{2}x^2} \frac{d^n}{dx^n}(e^{-x^2}).$$

Compute $k_1(x)$, $k_2(x)$ and $k_3(x)$. Note that the factor $(-\frac{1}{2})^n$ ensures that $k_n(x)$ will be of the form $p_n(x)e^{-\frac{1}{2}x^2}$, where $p_n(x)$ is an n–th degree *monic* polynomial (i.e., the coefficient of x^n is one).

(c) Using part (b), show that the functions $k_n(x)$, $n = 0, 1, 2, \ldots$, are mutually orthogonal. Also, prove that $\|k_n\|^2 = \langle f_n,k_n\rangle = 2^{-n}n!\sqrt{\pi}$.

(d) (i) Check that $k_n'(x) = xk_n(x) - 2k_{n+1}(x)$ (i.e., $k_{n+1}(x) = \frac{1}{2}[xk_n(x) - k_n'(x)]$).
(ii) Verify also that $xk_n(x) + k_n'(x) = n\,k_{n-1}(x)$, $n = 0, 1, 2, \ldots$.

Hint. For (ii), show that $xk_n(x) + k_n'(x) = p_n'(x)e^{-\frac{1}{2}x^2}$, where $k_n(x) = p_n(x)e^{-\frac{1}{2}x^2}$ as in part (b). Then use integration by parts to show that $xk_n(x) + k_n'(x)$ is orthogonal to $f_m(x)$ for $m = 0, \ldots, n-2$. Conclude form part (b) that $xk_n(x) + k_n'(x)$ is proportional to $k_{n-1}(x)$.

(e) Using (i) of part (d) and Propositions 1 and 2, verify that $\hat{k}_{n+1}(\xi) = \frac{i}{2}[\hat{k}_n'(\xi) - \xi\hat{k}_n(\xi)]$, and hence that $i^{n+1}\,\hat{k}_{n+1}(\xi) = \frac{1}{2}[\xi\, i^n\,\hat{k}_n(\xi) - i^n\,\hat{k}_n'(\xi)]$.

(f) Since $\hat{k}_0(\xi) = e^{-\frac{1}{2}\xi^2} = k_0(\xi)$, from Example 6 of Section 7.1, and since the sequences of functions $i^n\hat{k}_n(\xi)$ and $k_n(\xi)$ obey the same recursion formula [see (d) and (e)], deduce that $k_n(\xi) = i^n\hat{k}_n(\xi)$ or equivalently that $\hat{k}_n(\xi) = (-i)^n k_n(\xi)$. In particular k_0, k_4, k_8, ... are all equal to their Fourier transforms. In the terminology of linear algebra, the Hermite functions are eigenfunctions of the Fourier transform operator, with eigenvalues $1, i, -1, -i$.

Remark. The functions $k_n(x)$ are proportional to the **Hermite functions**, while the $p_n(x)$ are proportional to the **Hermite polynomials**. These functions arise in the study of the quantum mechanical harmonic oscillator which is studied in Section 9.5 (cf. also Problem 16 below). It follows from the next problem and the Inversion Theorem (cf. Section 7.3), that the Hermite functions form a complete orthogonal family of functions in the space of functions of decay order (0,2), in the sense that there is no nonzero function of decay order (0,2) which is orthogonal to all of the Hermite functions.

15. Let $f(x)$ be a function of decay order (0,2), and let $h(x) = e^{-\frac{1}{2}x^2}$. Justify the steps in the following computation, and use the Convolution Theorem to conclude that if $f(x)$ is orthogonal to all functions of the form $x^n e^{-\frac{1}{2}x^2}$ (or equivalently $k_n(x)$), then $\hat{f}(\xi) \equiv 0$.

$$(f*h)(x) = \int_{-\infty}^{\infty} f(y)e^{-\frac{1}{2}(x-y)^2}\,dy = e^{-\frac{1}{2}x^2}\int_{-\infty}^{\infty}\left[\sum_{n=0}^{\infty} f(y)\,\frac{x^n y^n}{n!}\,e^{-\frac{1}{2}y^2}\right]dy$$
$$= e^{-\frac{1}{2}x^2}\sum_{n=0}^{\infty}\left[\frac{x^n}{n!}\int_{-\infty}^{\infty} f(y)\,y^n\,e^{-\frac{1}{2}y^2}\,dy\right].$$

Hint. To justify the interchange of the sum and the integral, it suffices (cf. the Dominated Convergence Theorem in Appendix A.3) to find $g(y)$, such that $\sum_{n=0}^{\infty} |f(y)|\,\frac{x^n y^n}{n!}\,e^{-\frac{1}{2}y^2} \le g(y)$ and $\int_{-\infty}^{\infty} g(y)\,dy < \infty$. However, $f(y) \le M$, and so $\sum_{n=0}^{\infty} |f(y)|\,\frac{x^n y^n}{n!}\,e^{-\frac{1}{2}y^2} \le M\sum_{n=0}^{\infty}\frac{x^n y^n}{n!}\,e^{-\frac{1}{2}y^2}$ $= M\,e^{xy-\frac{1}{2}y^2}$ whose integral with respect to y is finite, for each fixed x (Why ?).

16. Let H be the differential operator defined by

$$[Hf](x) \equiv \tfrac{1}{2}\,[-f''(x) + x^2 f(x)]\,.$$

In quantum mechanics (cf. Section 9.5), H is essentially the energy operator for the harmonic oscillator. The eigenvalues of H are those constants λ such that Schrödinger's equation $Hf = \lambda f$ holds for some nonzero function f (called an eigenfunction of H; cf. Section 4.4).

(a) From the particular form of H, verify that for any rapidly decreasing f, we have $\widehat{[Hf]}(\xi) = -\frac{1}{2}[\hat{f}''(\xi) + \xi^2\,\hat{f}(\xi)] = -[H(\hat{f})](\xi)$. Conclude that if f is an eigenfunction of H, then $\hat{f}$ is also an eigenfunction of H with the same eigenvalue.

(b) Verify that the Hermite function $k_n(x)$ defined in Problem 14(b) is an eigenfunction of H with eigenvalue $n + \frac{1}{2}$. (In appropriate units, these eigenvalues represent the energy levels of the quantum–mechanical harmonic oscillator). **Hint.** First show that for any C^2 function $f(x)$, we have $\frac{1}{2}\,[-f''(x) + x^2 f(x)] = \frac{1}{2}\left[-\frac{d}{dx} + x\right]\left[\frac{d}{dx} + x\right][f(x)] + \frac{1}{2}\,f(x)$. Consider $f(x) = k_n(x)$, and use (i) and (ii) of part (d) of Problem 14.

(c) Check that for any function of the form $q_n(x)e^{-\frac{1}{2}x^2}$, where $q_n(x)$ is a polynomial of degree n, we have that $H[q_n(x)e^{-\frac{1}{2}x^2}]$ is another function of the same form.

Remark. From the observation of part (c), we might have guessed that the Hermite functions k_n are eigenfunctions of H (cf. Problem 14(f)). Indeed, for those who are familiar with linear algebra, H is a symmetric (by Green's formula) operator on the space V_n of Problem 14(b) and H leaves V_{n-1} invariant. It then follows that any vector of V_n orthogonal to V_{n-1} (say $k_n(x)$) must be mapped by H onto a multiple of itself (i.e., k_n *must* be an eigenfunction of H).

17. Let $H_n(x) = [(-1)^n/n!]e^{x^2} \frac{d^n}{dx^n}(e^{-x^2})$. Note that $H_n(x)$ is a multiple of the polynomial $p_n(x)$ of Problem 14(b), i.e., $H_n(x)$ is proportional a Hermite polynomial. Let z be another variable and let D denote the operator $\frac{d}{dx}$.

(a) Show that $\sum_{n=0}^{\infty} H_n(x)z^n = e^{x^2}\sum_{n=0}^{\infty} \frac{D^n(e^{-x^2})}{n!}(-z)^n = e^{x^2}e^{-(x-z)^2} = e^{2xz-z^2}$.

(b) Define $u(x,t,z)$ and $K_n(x,t)$ by $u(x,t,z) = e^{2xz+tz^2} = \sum_{n=0}^{\infty} K_n(x,t)z^n$. Show that $K_n(x,t)$ solves the heat equation $v_t = \frac{1}{4}v_{xx}$. **Hint.** For each z the function $u(x,t,z) = e^{2xz+tz^2}$ satisfies this heat equation. Differentiate both sides of $u_t = \frac{1}{4}u_{xx}$ with respect to z n–times, and evaluate the result at $z = 0$.

Remark. Note that $K_n(x,-1) = H_n(x)$, while $K_n(x,0) = 2^n x^n/n!$. Thus, curiously, running time backward by one unit converts monomials to Hermite polynomials.

18. **(The Poisson Summation Formula).** Let $f(x)$ be any function, defined for $-\infty < x < \infty$, such that for each x, the sum $F(x) = \sum_{k=-\infty}^{\infty} f(x + 2kL)$ converges (absolutely) to a continuous and piecewise C^1 function $F(x)$. Assume that this convergence is uniform for $-L \le x \le L$.

(a) Show that $F(x)$ is periodic of period $2L$, and equals its Fourier series.

(b) Set $x = 0$ in the complex Fourier series for $F(x)$, to obtain the Poisson Summation Formula:

$$\sum_{k=-\infty}^{\infty} f(2kL) = F(0) = \sum_{m=-\infty}^{\infty} c_m = \frac{\sqrt{2\pi}}{2L}\sum_{m=-\infty}^{\infty} \hat{f}(m\pi/L). \qquad (*)$$

Pay particular attention to the last equality.

(c) Show that the hypotheses for $f(x)$ are met if $f(x)$ is of decay order $(1,2)$ (i.e., $f(x)$ is C^1 and $|f(x)| + |f'(x)| \le K|x|^{-2}$ for all $|x| > 1$ and some K).

19. Use formula $(*)$ of Problem 18 to find a simple expression for the following infinite series :

(a) $\sum_{k=1}^{\infty} \sin(ak)/k$ (b) $\sum_{k=1}^{\infty} [\sin(ak)/k]^2$ (c) $\sum_{k=0}^{\infty} 1/(a^2+k^2)$ $(a \ne 0)$.

20. (a) By taking $f(x) = (4\pi t)^{-\frac{1}{2}} e^{-x^2/4t}$, $t > 0$, in (*) of Problem 18, prove that

$$(4\pi t)^{-\frac{1}{2}} \sum_{k=-\infty}^{\infty} e^{-L^2k^2/t} = \frac{1}{2L} \sum_{m=-\infty}^{\infty} e^{-tm^2\pi^2/L^2}. \qquad (**)$$

The famous **Jacobi theta function** is defined by $\theta(t) = \sum_{n=-\infty}^{\infty} e^{-\pi n^2 t}$. By choosing $L = \sqrt{\pi}$ in (**), verify that $\theta(1/t) = \sqrt{t}\,\theta(t)$.

(b) Let $\omega(t) = \frac{1}{4} e^t \sum_{n=-\infty}^{\infty} e^{-\pi n^2 e^{4t}}$. Use the result of part (a) to verify the remarkable fact that $\omega(t)$ is an even function of t.

Remark. Let $\phi(t) = \omega''(t) - \omega(t)$. Then by part (b), $\phi(t)$ is also an even function. One of the most celebrated open problem in mathematics, known as the **Riemann Hypothesis**, is equivalent to the assertion that the Fourier cosine transform of $\phi(t)$, i.e.,

$$F(x) = \int_0^{\infty} \phi(t)\cos(xt)\, dt,$$

has only real zeros. The famous German mathematician, Georg Friedrich Bernhard Riemann (1826–1866) stated this conjecture in 1859 in his epoch–making 8–page paper entitled *Über die Anzahl der Primzahlen unter einer gegebenen Grösse* ("On the Number of Primes Less Than a Given Magnitude") (cf. also [H. M. Edwards, 1974], [PIlya, 1974]).

21. (a) Use Fubini's theorem (cf. Appendix A.2) to deduce that the convolution $(f*g)(x)$ of two absolutely integrable functions $f(x)$ and $g(x)$ is also absolutely integrable (i.e., $\int_{-\infty}^{\infty} |(f*g)(x)|\, dx \le \int_{-\infty}^{\infty}\int_{-\infty}^{\infty} |f(x-y)g(y)|\, dy\, dx < \infty$).

(b) Deduce from part (a) and the Convolution Theorem that the product of the Fourier transforms of two absolutely integrable is also the Fourier transform of an absolutely integrable function. You may use the fact that the Convolution Theorem holds for any two absolutely integrable functions.

(c) Deduce from (b) that if $p(z)$ is any polynomial such that $p(0) = 0$, then $p(\hat{f}(\xi))$ is the Fourier transform of an absolutely–integrable function, provided that $f(x)$ is absolutely integrable.

(d) Why is the condition $p(0) = 0$ necessary in part (c) ? **Hint**. For any absolutely integrable function $h(x)$, we have $\lim_{\xi\to\pm\infty} |\hat{h}(\xi)| = 0$ by Problem 10(c) of Section 7.1.

7.3 The Inversion Theorem and Parseval's Equality

One property of Fourier transforms which is essential, both in applications (cf. remarks preceding Section 7.1) and in pure mathematics, is that under mild assumptions on f and g, if $\hat{f}(\xi) = \hat{g}(\xi)$, then $f(x) = g(x)$. In other words, we can recover a suitable function $f(x)$ from its transform $\hat{f}(\xi)$. At the end of this chapter (cf. Supplement), we prove

The Inversion Theorem. Let $f(x)$ be a piecewise C^1 function (i.e., $f(x)$ and $f'(x)$ are piecewise continuous functions in any finite interval), such that $\int_{-\infty}^{\infty} |f(x)|\, dx < \infty$. Then for each real x,

$$\frac{f(x^+) + f(x^-)}{2} = \frac{1}{\sqrt{2\pi}} \int_{-\infty}^{\infty} \hat{f}(\xi)\, e^{i\xi x}\, d\xi . \qquad (1)$$

In particular, if $f(x)$ is *also* assumed to be continuous, then $f(x)$ is determined by $\hat{f}(\xi)$ by the inversion formula (1). The right side of (1) looks very much like the Fourier transform of $\hat{f}(\xi)$, except that there is no minus sign in front of $i\xi x$. Since the integral essentially gives $f(x)$ back again, the right side is the inverse Fourier transform of $\hat{f}(\xi)$. More precisely, we have :

Definition. The **inverse Fourier transform** $\check{g}(x)$ of a function $g(\xi)$ is defined by

$$\check{g}(x) = \int_{-\infty}^{\infty} g(\xi)\, e^{i\xi x}\, d'\xi = \lim_{R\to\infty} \int_{-R}^{R} g(\xi)\, e^{i\xi x}\, d'\xi,$$

whenever this limit exists, and where $d'\xi = (2\pi)^{-\frac{1}{2}} d\xi$.

For a continuous piecewise C^1 function $f(x)$ with $\int_{-\infty}^{\infty} |f(x)|\, dx < \infty$, the Inversion Theorem simply says that $[\hat{f}(\xi)]\check{}\,(x) = f(x)$.

Remark. The Fourier transform of an absolutely integrable function need not be absolutely integrable. Indeed, from Example 4 of Section 7.1 (with $L = 1$) we know that the Fourier transform of the absolutely integrable function

$$f(x) = \begin{cases} 1 \text{ if } |x| \le 1 \\ 0 \text{ if } |x| > 1 \end{cases} \qquad \text{is} \qquad \hat{f}(\xi) = \frac{2\sin(\xi)}{\xi} .$$

It can be shown (cf. Problem 4) that $\hat{f}(\xi)$ is *not* absolutely integrable. For this reason the integral (1) in the Inversion Theorem is to be interpreted as a Cauchy principal value, that is,

$$\int_{-\infty}^{\infty} \hat{f}(\xi)\, e^{i\xi x}\, d'\xi = \lim_{R\to\infty} \frac{1}{\sqrt{2\pi}} \int_{-R}^{R} \hat{f}(\xi)\, e^{i\xi x}\, d\xi . \quad \square$$

Example 1. Compute the inverse Fourier transform of the function $g(\xi) = (a^2 + \xi^2)^{-1}$ $(a > 0)$.

Solution. If we had to do this directly, we would need to compute the integral

$$\check{g}(x) = \int_{-\infty}^{\infty} \frac{e^{i\xi x}}{a^2 + \xi^2}\, d'\xi.$$

While this is not hard to do using complex contour integration, we will proceed indirectly by using the Inversion Theorem. By Example 3 of Section 7.1, we know that the function whose Fourier transform is $g(\xi)$, is $\sqrt{2\pi}\,(2a)^{-1}\, e^{-a|x|}$. Thus, the Inversion Theorem tells us immediately that

$$\int_{-\infty}^{\infty} \frac{e^{i\xi x}}{a^2 + \xi^2}\, d'\xi = \check{g}(x) = \frac{\sqrt{2\pi}}{2a}\, e^{-a|x|} . \quad \square \tag{2}$$

Remark. Suppose that we make the change of variables $\xi \to x$ and $x \to -\xi$ in formula (2). Then

$$\int_{-\infty}^{\infty} \frac{e^{-i\xi x}}{a^2 + x^2}\, d'x = \frac{\sqrt{2\pi}}{2a}\, e^{-a|\xi|} . \tag{2'}$$

Thus, we have found the Fourier transform of $(a^2 + x^2)^{-1}$ as well ! The same argument shows that, in general, $\hat{g}(\xi) = \check{g}(-\xi)$. Indeed, we have

$$\hat{g}(\xi) = \int_{-\infty}^{\infty} g(x)\, e^{-i\xi x}\, d'x = \int_{-\infty}^{\infty} g(x)\, e^{i(-\xi)u}\, d'x = \check{g}(-\xi) . \tag{3}$$

In other words, the graph of the Fourier transform of a function is just the reflection in the vertical axis of the graph of the inverse Fourier transform of the function. Letting $g = \hat{f}$ for a suitable function f, (3) yields $(\hat{f})^{\hat{}}(\xi) = (\hat{f})^{\check{}}(-\xi) = f(-\xi)$ by the Inversion Theorem (i.e., applying the Fourier transform twice to a function has the effect of reflecting the graph of the function). Applying the Fourier transform four times will give the original function back again (i.e., $f^{\hat{}\hat{}\hat{}\hat{}} = f$; why ?). $\square$

The relation of Fourier transforms to complex Fourier series

The inversion formula bears a resemblance to the convergence results for Fourier series (cf. Section 4.2). Indeed, one can use complex Fourier series to prove the inversion formula (1), when

$f(x)$ is C^2 and is zero outside of a finite interval. To do this, let f_L be the function f restricted to the interval $[-L,L]$. For any value x_0 choose L so large that $-L < x_0 < L$ and $f(x) = 0$ for $|x| > L$. In terms of complex Fourier series, Theorem 2 of Section 4.2 yields

$$f(x_0) = f_L(x_0) = \sum_{m=-\infty}^{\infty} c_m e^{im\pi x_0/L}, \quad \text{where} \quad c_m = \frac{1}{2L}\int_{-L}^{L} f_L(x)\, e^{-im\pi x/L}\, dx\,.$$

Since $f = f_L$ inside $[-L,L]$ and $f \equiv 0$ outside $[-L,L]$, we have

$$c_m = \frac{1}{2L}\int_{-\infty}^{\infty} f(x)\, e^{-im\pi x/L}\, dx = \frac{\sqrt{2\pi}}{2L}\int_{-\infty}^{\infty} f(x)\, e^{-im\pi x/L}\, d'x = \frac{\sqrt{2\pi}}{2L}\hat{f}\left[\frac{m\pi}{L}\right]. \tag{4}$$

Thus,

$$f(x_0) = \sum_{m=-\infty}^{\infty} \frac{\sqrt{2\pi}}{2L}\hat{f}\left[\frac{m\pi}{L}\right] e^{im\pi x_0/L} = \frac{1}{\sqrt{2\pi}}\sum_{m=-\infty}^{\infty} \hat{f}\left[\frac{m\pi}{L}\right] e^{im\pi x_0/L} \cdot \frac{\pi}{L} \tag{5}$$

Since the extreme left side of (5) is a constant, independent of L, the sum on the extreme right must also be independent of L. However, this sum is a Riemann sum for the integral

$$\int_{-\infty}^{\infty} \hat{f}(\xi)\, e^{i\xi x_0}\, d'\xi\,, \tag{6}$$

where the intervals of the partition have length $\Delta\xi = \pi/L$. Under the assumption that f is C^2, we know from Theorem 1 of Section 7.2 that $\hat{f}(\xi)$ is continuous (indeed, C^∞) and $|\hat{f}(\xi)| \le C|\xi|^{-2}$ for $|\xi| \ge 1$. Thus, (6) exists and the Riemann sums (5) will approach (6) as L tends to infinity (i.e., as $\Delta\xi \to 0$). However, we have observed that the sums in (5) are independent of L and hence each sum must equal (6). Thus as desired, in the limit, (5) becomes

$$f(x_0) = \int_{-\infty}^{\infty} \hat{f}(\xi)\, e^{i\xi x_0}\, d'x\,,$$

valid for all x_0, provided that $f(x)$ is a C^2 function vanishing outside a finite interval. To get the more general Inversion Theorem we need a different argument (cf. the Supplement at the end of this chapter). However, since we will only use the Inversion Theorem to obtain hypothetical solutions of PDEs (which are verified separately) there is no great need for this generality.

It is also possible to use Parseval's equality for complex Fourier series to obtain Parseval's equality (cf. (7) below) for C^1 functions $f(x)$ and $g(x)$ which are zero outside of a finite interval (cf. Problem 12). Instead, we will prove the following more general form of Parseval's equality by using the Inversion Theorem :

Parseval's Equality. If $f(x)$, $\hat{f}(\xi)$ and $g(x)$ **are absolutely integrable on** $(-\infty,\infty)$ **and** $f(x)$ **is piecewise** C^1 **on** $(-\infty,\infty)$, **then**

$$\int_{-\infty}^{\infty} f(x)\overline{g(x)}\,dx = \int_{-\infty}^{\infty} \hat{f}(\xi)\overline{\hat{g}(\xi)}\,d\xi\,. \tag{7}$$

Proof. We use the hypotheses to apply the Inversion Theorem to $f(x)$ and to justify the interchange of the order of integration (Appendix A.2) in the following computation :

$$\int_{-\infty}^{\infty} f(x)\overline{g(x)}\,dx = \int_{-\infty}^{\infty}\left[\int_{-\infty}^{\infty} \hat{f}(\xi)\,e^{i\xi x}\,d'\xi\right]\overline{g(x)}\,dx \qquad \text{(Inversion Theorem)}$$

$$= \int_{-\infty}^{\infty}\int_{-\infty}^{\infty} \hat{f}(\xi)\overline{g(x)}\,e^{i\xi x}\,d'\xi\,dx = \int_{-\infty}^{\infty}\int_{-\infty}^{\infty} \hat{f}(\xi)\overline{g(x)}\,e^{i\xi x}\,d'x\,d\xi \qquad \text{(Appendix A.2)}$$

$$= \int_{-\infty}^{\infty} \hat{f}(\xi)\overline{\left[\int_{-\infty}^{\infty} g(x)e^{-i\xi x}d'x\right]}\,d\xi = \int_{-\infty}^{\infty} \hat{f}(\xi)\overline{\hat{g}(\xi)}\,d\xi\,.$$

Observe that the inverse transform in the right side of the first equation can differ from $f(x)$ at a finite number of points in each finite interval, but this does not affect the value of the integral. □

Remarks. When $g = f$, Parseval's equality yields $\|f\| = \|\hat{f}\|$ for any rapidly decreasing function f, where $\|f\| = \left[\int_{-\infty}^{\infty} |f(x)|^2\,dx\right]^{\frac{1}{2}}$. This means that the Fourier transform is a "rotation" of the vector space of rapidly decreasing functions, since lengths of functions are unchanged. This fact can be used to extend the notion of Fourier transform to more general classes of functions (e.g., the square–integrable functions for which $\|f\| < \infty$, but which may not be absolutely integrable, such as $x/(1+x^2)$). In ordinary space, a linear transformation which does not change the distance of any point to the origin must be a rotation possibly followed by a reflection. In the infinite dimensional space of rapidly decreasing functions, we have shown (cf. Problem 14 of Section 7.2) that there is an orthogonal family $k_0(x), k_1(x), k_2(x), \ldots$ of Hermite functions such that $\hat{k}_n(\xi) = i^n k_n(\xi)$ (i.e., the values of k_n are rotated by $n\pi/2$ in the complex plane). □

Additional examples

Example 2. Find the Fourier transform of $h(x) = x^n e^{-\frac{1}{2}x^2}$, for $n = 0, 1, 2, \ldots$.

Solution. A direct computation of $\hat{h}(\xi)$, using the definition

$$\hat{h}(\xi) = \int_{-\infty}^{\infty} x^n e^{-\frac{1}{2}x^2}\,e^{-i\xi x}\,d'x\,,$$

is not very easy. One could differentiate $e^{-\frac{1}{2}\xi^2} = \int_{-\infty}^{\infty} e^{-\frac{1}{2}x^2} e^{-i\xi x}\, d'x$ n–times with respect to ξ to obtain the ensuing result. Instead, our strategy will be to use the relation $\hat{h}(\xi) = \check{h}(-\xi)$ (cf. (3)). In other words, we compute the inverse transform $\check{h}(x)$ of $h(\xi) = \xi^n e^{-\frac{1}{2}\xi^2}$, and then replace x in $\check{h}(x)$ by $-\xi$ to obtain $\check{h}(\xi)$ (cf. (3)). To compute $\check{h}(x)$, we use the formula $\xi^n \hat{f}(\xi) = i^{-n} [f^{(n)}(x)]\hat{\ }(\xi)$ (cf. Corollary 1 of Section 7.2) with $f(x) = e^{-\frac{1}{2}x^2}$, so that $\hat{f}(\xi) = e^{-\frac{1}{2}\xi^2}$. Then by the Inversion Theorem, we obtain

$$\check{h}(x) = [\xi^n \hat{f}(\xi)]\check{\ }(x) = [i^{-n}[f^{(n)}(x)]\hat{\ }]\check{\ }(x) = i^{-n} f^{(n)}(x) = i^{-m} \frac{d^n}{dx^n}\left[e^{-\frac{1}{2}x^2}\right].$$

Thus,

$$\hat{h}(\xi) = \check{h}(-\xi) = i^{-n}(-1)^n \frac{d^n}{d\xi^m}\left[e^{-\frac{1}{2}\xi^2}\right] = i^n \frac{d^n}{d\xi^n}\left[e^{-\frac{1}{2}\xi^2}\right] = i^n p_n(\xi)\, e^{-\frac{1}{2}\xi^2},$$

where $p_n(\xi)$ is a polynomial of degree n. We have $p_0(\xi) = 1$, $p_1(\xi) = -\xi$, $p_2(\xi) = \xi^2 - 1$, and in general we can compute inductively : $p_{n+1}(\xi) = -\xi p_n(\xi) + p_n{}'(\xi)$. □

Example 3. Compute $I \equiv \int_{-\infty}^{\infty} \frac{x^2}{(x^2+1)^4}\, dx$, using Parseval's equality and (2′).

Solution. Let $g(x) = \frac{1}{2}(x^2+1)^{-1}$. By Parseval's equality and Proposition 1 of Section 7.2 ,

$$I = \int_{-\infty}^{\infty} |g'(x)|^2\, dx = \int_{-\infty}^{\infty} |(g')\hat{\ }(\xi)|^2\, d\xi = \int_{-\infty}^{\infty} \xi^2\, |\hat{g}(\xi)|^2\, d\xi = \int_{-\infty}^{\infty} \tfrac{1}{8}\,\pi\xi^2 e^{-2|\xi|}\, d\xi$$

$$= \frac{\pi}{4}\int_0^{\infty} \xi^2 e^{-2\xi}\, d\xi = \frac{\pi}{4}\left[-\tfrac{1}{2}\xi^2 e^{-2\xi}\Big|_0^{\infty} + \int_0^{\infty} \xi e^{-2\xi} d\xi\right] = \frac{\pi}{4}\left[-\tfrac{1}{2}\xi e^{-2\xi}\Big|_0^{\infty} + \tfrac{1}{2}\int_0^{\infty} e^{-2\xi} d\xi\right] = \frac{\pi}{16}. \quad □$$

Example 4. By means of *formal* calculations express the Fourier transform of

$$h(x) = \int_{-\infty}^{\infty} (x-s)\, f(x-s)\, f'(s)\, ds,$$

in terms of $\hat{f}(\xi)$ and $\frac{d\hat{f}}{d\xi}(\xi)$. State conditions under which the calculations are valid.

Solution. Let $g(x) = xf(x)$. Then $h(x) = (g*f')(x)$ and hence by the Convolution Theorem

$$\hat{h}(\xi) = (g*f')\hat{\ }(\xi) = \sqrt{2\pi}\, \hat{g}(\xi)[f'(x)]\hat{\ }(\xi).$$

Now, by Proposition 1 and Proposition 2 of Section 7.2

$$\hat{g}(\xi) = i\,\frac{d\hat{f}}{d\xi}(\xi) \quad \text{and} \quad [f'(x)]\hat{\ }(\xi) = i\xi\,\hat{f}(\xi),$$

whence

$$\hat{h}(\xi) = -\sqrt{2\pi}\,\xi\,\hat{f}(\xi)\,\frac{d\hat{f}}{d\xi}(\xi).$$

Propositions 1 and 2 require that $f(x)$ be of decay order (1,2) and (0,3) respectively, and thus both propositions can be used if $f(x)$ is of decay order (1,3). Then $xf(x)$ and $f'(x)$ will satisfy the conditions in Theorem 2 (the Convolution Theorem), as well. Thus, it suffices to take $f(x)$ to be of decay order (1,3). □

The Convolution Theorem and the Inversion Theorem can be used to solve certain integral equations, as the following example illustrates.

Example 5. Solve for $g(x)$ in the integral equation

$$\int_{-\infty}^{\infty} \frac{g(s)}{(x-s)^2 + b^2}\,ds = \frac{1}{x^2 + a^2} \quad (a > b > 0\,), \tag{8}$$

You may assume that $\int_{-\infty}^{\infty} |g(x)|\,dx < \infty$ and that g is C^1.

Solution. If we set $f(x) = \dfrac{1}{x^2 + b^2}$, then (8) becomes $(f*g)(x) = \dfrac{1}{x^2 + a^2}$. Hence, by the Convolution Theorem,

$$\sqrt{2\pi}\,\hat{f}(\xi)\hat{g}(\xi) = \left[\frac{1}{x^2 + a^2}\right]\hat{\ }(\xi)\,. \tag{9}$$

Now, by Example 1, we have that

$$\left[\frac{1}{x^2 + a^2}\right]\hat{\ }(\xi) = \frac{\sqrt{2\pi}}{2a}\,e^{-a|\xi|}\,. \tag{10}$$

Using (10) (with b replacing a) to evaluate $\hat{f}(\xi)$, (9) becomes $\hat{g}(\xi) = \dfrac{b}{a\sqrt{2\pi}}\,e^{-(a-b)|\xi|}$. Applying (10) again (with $a-b$ replacing a) and using the Inversion Theorem, we obtain

$$g(x) = \frac{b(a-b)}{a\pi(x^2 + (a-b)^2)}\,. \quad \square$$

The relation between the Fourier transform and the Laplace transform

Let $f(t)$ be a function which is zero for $t < 0$. Then

$$\sqrt{2\pi}\,\hat{f}(\xi) = \int_{-\infty}^{\infty} f(t)\, e^{-i\xi t}\, dt = \int_{0}^{\infty} f(t)\, e^{-i\xi t}\, dt.$$

Although we have previously taken ξ to be real, if we let $\xi = -is$, then we have

$$\sqrt{2\pi}\,\hat{f}(-is) = \int_{0}^{\infty} e^{-st}\, f(t)\, dt\,. \tag{11}$$

The right–hand side of (11) is called the **Laplace transform** of $f(t)$, and is denoted by

$$L\, f(s) \equiv \int_{0}^{\infty} e^{-st}\, f(t)\, dt = \sqrt{2\pi}\,\hat{f}(-is). \tag{12}$$

Thus, (12) indicates that the Laplace transform of f at s is essentially the Fourier transform of f evaluated at $-is$. Formula (12) explains the many properties that these transforms have in common. The decaying factor e^{-st} $(s > 0)$ in (12) enables one to apply the Laplace transform to a wide variety of functions which need not decay as $t \to \infty$. Indeed, the Laplace transform will be defined for s sufficiently large, if $f(t) \leq Ke^{at}$, for some constants K and a. Thus, one may confidently apply Laplace transform techniques to solve systems of ODEs with constant coefficients, since the solutions are already known to grow at most exponentially. Moreover, initial conditions for such systems can be nicely encoded in the resulting algebraic problem. However, for functions which are not identically zero when $t < 0$, the Laplace transform ignores the values of $f(t)$ for $t < 0$ (Why ?), and hence such functions cannot be completely recovered from their Laplace transforms, in general. Thus, in spite of the more stringent decay requirements, the Fourier transform is usually preferable to the Laplace transform, when one wishes to transform functions of an *unrestricted* variable (e.g., $f(x)$, $-\infty < x < \infty$).

Summary 7.3

1. The Inversion Theorem : Let $f(x)$ be a piecewise C^1 function (i.e., $f(x)$ and $f'(x)$ are piecewise continuous functions in any finite interval), with $\int_{-\infty}^{\infty} |f(x)|\, dx < \infty$. Then for all real x,

$$\frac{f(x^+) + f(x^-)}{2} = \frac{1}{\sqrt{2\pi}} \int_{-\infty}^{\infty} \hat{f}(\xi)\, e^{i\xi x}\, d\xi . \tag{S1}$$

A proof of the Inversion Theorem is in the Supplement at the end of this chapter, but in this section, it was shown that in the case, where $f(x)$ is C^2 and is 0 outside of a finite interval, the result can be proved by using a convergence theorem (Theorem 2 of Section 4.2) for Fourier series.

2. The inverse Fourier transform : The inverse Fourier transform of a function $g(\xi)$ is

$$\check{g}(x) = \int_{-\infty}^{\infty} g(\xi)\, e^{i\xi x}\, d'\xi = \lim_{R\to\infty} \int_{-R}^{R} g(\xi)\, e^{i\xi x}\, d'\xi,$$

whenever this limit exists, and where $d'\xi = (2\pi)^{-\frac{1}{2}} d\xi$. The Inversion Theorem implies that for suitable $f(x)$ (i.e., continuous, piecewise C^1 and absolutely integrable), we have $(\hat{f})\check{\ }(x) = f(x)$. The graph of the Fourier transform of a function is the reflection in the vertical axis of the graph of the inverse Fourier transform of the function (i.e., $\hat{g}(\xi) = \check{g}(-\xi)$). In particular, for suitable functions $f(x)$, with $g = \hat{f}$, we obtain $(\hat{f})\hat{\ }(\xi) = f(-\xi)$. This makes it easy to compute the Fourier transform of a function which is the Fourier transform of some initial suitable function (cf. Example 1). Also, it follows that $f^{\hat{}\hat{}\hat{}\hat{}} = f$ for suitable f.

3. Parseval's equality : **If** $f(x)$, $\hat{f}(\xi)$ and $g(x)$ **are absolutely integrable on** $(-\infty,\infty)$ **and** $f(x)$ **is piecewise** C^1 **on** $(-\infty,\infty)$, **then**

$$\int_{-\infty}^{\infty} f(x)\overline{g(x)}\, dx = \int_{-\infty}^{\infty} \hat{f}(\xi)\overline{\hat{g}(\xi)}\, d\xi . \tag{S2}$$

Exercises 7.3

1. Find the inverse transform of each of the following functions :

(a) $g(\xi) = (1/\sqrt{a})e^{-\frac{1}{2}\xi^2/a}$, $a > 0$ (b) $g(\xi) = i(\xi\sqrt{2\pi})^{-1}(e^{i\xi} - 1)$ (c) $g(\xi) = \sqrt{2/\pi}\,[\sin(\xi)/\xi]$.

Hint. Review the examples of Section 7.1.

2. Prove that the convolution of two rapidly decreasing functions (cf. Section 7.2) is a rapidly decreasing function.

3. Let n be a positive integer and let $h(x) = \int_{-\infty}^{\infty} (x - s)^n f(x-s) f'(s)\, ds$.

Use formal calculations to express the Fourier transform of $h(x)$ in terms of $\hat{f}(\xi)$ and $\hat{f}^{(n)}(\xi)$.
Hint. See Example 4.

4. Let $f(x) = \begin{cases} \dfrac{\sin(x)}{x} & \text{if } x \neq 0 \\ 1 & \text{if } x = 0 \end{cases}$. Prove that $\lim_{R\to\infty} \int_0^R |f(x)|\, dx$ does not exist.

Hint. If $n\pi \le x \le (n+1)\pi$, $(n = 1,2,\ldots)$, then $\frac{1}{x} \ge \frac{1}{(n+1)\pi}$ and so

$$\int_{n\pi}^{(n+1)\pi} \left|\frac{\sin(x)}{x}\right| dx \ge \frac{1}{(n+1)\pi} \int_{n\pi}^{(n+1)\pi} |\sin(x)|\, dx = \frac{2}{(n+1)\pi} .$$

5. Use the elementary formula $\int_0^{\infty} x^n e^{-ax}\, dx = \frac{n!}{a^{n+1}}$ $(n = 0, 1, 2,\ldots;\ \mathrm{Re}(a) > 0)$ to find the Fourier transforms of the following functions

(a) $f_n(x) = [(x + |x|)/2]^n e^{-x}$, $n = 0, 1, 2,\ldots$.

(b) $h_n(x) = [(x + |x|)/2]^n \sin(x)\, e^{-x}$, $n = 0, 1, 2,\ldots$.

Hint. For part (b), use the formula $2i\cdot\sin x = e^{ix} - e^{-ix}$.

6. (a) Solve the integral equation $\int_{-\infty}^{\infty} e^{-\frac{1}{2}b(x-y)^2} g(y)\, dy = e^{-\frac{1}{2}ax^2}$, where $0 < a < b$.

(b) Why can there be no absolutely integrable, piecewise continuous function which solves this equation when $a \ge b$? **Hint.** See Problem 10 of Section 7.1.

7. Use Parseval's equality to evaluate the following definite integrals, where $a, b > 0$:

(a) $\displaystyle\int_{-\infty}^{\infty} \frac{1}{(a^2 + x^2)^2}\, dx$ (b) $\displaystyle\int_{-\infty}^{\infty} \left[\frac{\sin(ax)}{x}\right]^2 dx$

(c) $\displaystyle\int_{-\infty}^{\infty} \frac{\sin(bx)}{x(a^2 + x^2)}\, dx$ (d) $\displaystyle\int_{-\infty}^{\infty} \frac{e^{-ax^2}}{(b^2 + x^2)}\, dx.$

Remark. Note that Parseval's equality, as we have stated it, does not apply to part (b) (Why not ?), even though the correct result is obtained anyway. Try doing part (b) rigorously, by using (at $x = 0$) the result $(f*g)(x) = \int_{-\infty}^{\infty} \hat{f}(\xi)\hat{g}(\xi)\, e^{ix\xi}\, d\xi$, which is valid when $f(x)$ and $g(x)$ are piecewise continuous and absolutely integrable (so that the Convolution Theorem of Section 7.2 may be applied) *and* when $(f*g)(x)$ is continuous, piecewise C^1 and absolutely integrable (for the Inversion Theorem).

8. Let $u(x,t) = \dfrac{1}{\sqrt{2\pi}} \displaystyle\int_{-\infty}^{\infty} \left[e^{-k\xi^2 t} \int_0^t e^{k\xi^2 s}\, \hat{q}(\xi,s)\, ds \right] e^{i\xi x}\, d\xi$.

Show that $u(x,t) = \dfrac{1}{\sqrt{4\pi k}} \displaystyle\int_{-\infty}^{\infty} \left[\int_0^t \frac{q(y,s)}{\sqrt{t-s}}\, e^{-(y-x)^2/(4k(t-s))}\, ds \right] dy$, assuming that $q(y,s)$ is nice enough (e.g., bounded and continuous) to permit the interchange of order of integration.

9. Find a function $f(x)$ which solves the integral equation

$$f(x) + \int_0^{\infty} f(x-t)\, e^{-t}\, dt = \frac{1}{1 + x^2}. \qquad (*)$$

Hint. Let $g(t) = e^{-t}$ for $t \geq 0$, and 0 otherwise. Note that the integral in $(*)$ is the convolution $(f*g)(x)$.

10. Evaluate the integral $f(x) = \sqrt{2/\pi} \displaystyle\int_{-\infty}^{\infty} (1/\xi^2)(1 - \cos(\xi))\, e^{i\xi x}\, d\xi$.

Hint. Use the Inversion Theorem and Problem 3 of Section 7.1.

11. Evaluate the integral $f(x) \equiv \dfrac{1}{\sqrt{2\pi}} \displaystyle\int_{-\infty}^{\infty} \xi \sin(\xi x) e^{-\frac{1}{2}\xi^2}\, d\xi$, by considering the effect of replacing $\sin(\xi x)$ by $e^{i\xi x}$.

12. In the case where $f(x)$ is C^1 on $(-\infty,\infty)$ and is 0 outside of some finite interval, say $[-L,L]$, show that Parseval's equality $\int_{-\infty}^{\infty} |f(x)|^2\, dx = \int_{-\infty}^{\infty} |\hat{f}(\xi)|^2\, d\xi$ follows from Parseval's equality for Fourier series in the complex form (cf. Section 7.1).

Hint. Write the complex Fourier coefficients of $f(x)$ on $[-L,L]$ in terms of the Fourier transform of $f(x)$, as was done in the Fourier series derivation of the Inversion Theorem. Then apply Parseval's equality for complex Fourier series, and take a limit as $L \to \infty$ to obtain an integral from a Riemann sum (cf. (4), (5), and (6)).

13. Suppose that $f(x)$ is absolutely integrable and piecewise C^1 and $\hat{f}(\xi) = \alpha f(\xi)$ for some complex constant α. Show that α must be 1, i, -1, or $-i$. Also show that $f(x)$ is even if $\alpha = \pm 1$, and $f(x)$ is odd if $\alpha = \pm i$. Note that there are many examples of such functions (cf. the Hermite functions in Problem 14 of Section 7.2). **Hint.** First observe that $\alpha^4 = 1$ (Why ?).

14. For real $x \neq n$, define $w_n(x) = \dfrac{\sin[\pi(x-n)]}{\pi(x-n)}$, $n = 0, \pm 1, \pm 2, \ldots$, and $w_n(n) \equiv 1$.

Let $f_n(x) \equiv \begin{cases} e^{inx}/\sqrt{2\pi} & -\pi \leq x \leq \pi \\ 0 & \text{otherwise} \end{cases}$.

(a) Using Example 4 of Section 7.1, show that $\hat{f}_n(\xi) = w_n(\xi)$.

(b) Show that $\int_{-\infty}^{\infty} w_n(\xi) w_m(\xi)\, d\xi = \begin{cases} 0 & \text{for } m \neq n \\ 1 & \text{for } m = n \end{cases}$ by a formal application of Parseval's equality, or rigorously, by using the Convolution Theorem (cf. the remark following Problem 7(b)). Thus, the functions $w_n(\xi)$ form an orthogonal family of norm–square 1 on $(-\infty,\infty)$.

(c) Suppose that for some complex constants c_n $(-N < n < N)$, $g(x) = \sum_{n=-N}^{N} c_n f_n(x)$. Show that $\int_{-\infty}^{\infty} |g(x)|^2\, dx = \sum_{n=-N}^{N} |c_n|^2 = \int_{-\infty}^{\infty} |\hat{g}(\xi)|^2\, d\xi$, without using Parseval's equality.

(d) For $g(x)$ as in part (c), show that $\hat{g}(t) = \int_{-\infty}^{\infty} \hat{g}(\xi) \dfrac{\sin(\pi(\xi-t))}{\pi(\xi-t)}\, d\xi$.

Remark. Functions of the form $\hat{g}(\xi) = \sum_{n=-\infty}^{\infty} c_n w_n(\xi)$, with $\sum_{n=-\infty}^{\infty} |c_n|^2 < \infty$, are known as **Paley–Wiener functions**, and the series is known as the **cardinal series** for the function $\hat{g}(\xi)$. Such series play a central role in the Whittaker–Shannon sampling theory of band–limited signals (cf. [J.R. Higgins, *Five short stories about the cardinal series*, Bull. Amer. Math. Soc. (New Series) 12 (1985), 45–89]). The result in part (c) implies that $w(x,t) \equiv \dfrac{\sin(\pi(x-t))}{\pi(x-t)}$ is a **reproducing kernel** for the class of Paley–Wiener functions.

7.4 Fourier Transform Methods for PDEs

In this section we use Fourier transform methods to derive hypothetical solutions for boundary–value problems for PDEs on infinite domains. Since one must always separately verify that these hypothetical solutions are actual solutions, there is absolutely no need to justify any steps in the derivation of the hypothetical solution. Also, any attempt to justify the Fourier transform methods in the derivation of these hypothetical solutions is doomed to failure, because these methods presuppose that the solution not only exists (an unwarranted assumption in itself), but also that the solution decays rapidly enough so that its transform exists. The Fourier transform methods simply provide us with a hypothetical solution (to be justified by other means) which we might never have guessed without them (cf. Example 3 of Section 7.2).

The heat problem for the infinite rod

Here we will prove that if the function $f(x)$ is continuous and either absolutely integrable (i.e, $\int_{-\infty}^{\infty} |f(x)|\, dx < \infty$) or bounded (i.e., $|f(x)| \leq M$ for all x), then the following initial–value problem has a solution $u(x,t)$ which is continuous throughout the half–plane $t \geq 0, -\infty < x < \infty$.

$$\begin{array}{lll} \text{D.E.} & u_t = ku_{xx} & -\infty < x < \infty,\ t > 0 \\ \text{I.C.} & u(x,0) = f(x) & (t = 0). \end{array} \tag{1}$$

This solution is given by the formulas

$$u(x,t) = \frac{1}{\sqrt{4\pi kt}} \int_{-\infty}^{\infty} e^{-(x-y)^2/4kt} f(y)\, dy \quad (\text{for } t > 0) \quad \text{and} \quad u(x,0) = f(x) \quad (\text{for } t = 0). \tag{2}$$

Note that if we define $u(x,t) \equiv 0$ for $t > 0$, $-\infty < x < \infty$, and $u(x,0) = f(x)$ for $t = 0$, then we also have a solution of problem (1). However, this solution is not continuous at points on the x–axis unless $f(x) \equiv 0$, i.e., we do not have the **continuity condition**

$$\lim_{(x,t) \to (x_0, 0^+)} u(x,t) = f(x_0), \tag{3}$$

for all x_0, unless $f(x) \equiv 0$. Solutions of the D.E. which do not satisfy the continuity condition (3) are not of any obvious value. This is why it is important to eventually prove that (2) does satisfy (3). Since the partial derivatives with respect to x and t (*not* y) of the integrand of (2) are absolutely integrable with respect to y, Leibniz's rule (cf. Appendix A.3) tells us that $u(x,t)$ in

(2) is actually C^∞ for $t > 0$, $-\infty < x < \infty$, even though $f(x)$ need not be C^1. In other words, the temperature distribution smooths out in an instant. In the event that $\int_{-\infty}^{\infty} |f(x)|\, dx < \infty$ and $\lim_{x \to \pm\infty} f(x) = 0$, we will prove that for any $T > 0$, $u(x,t)$, defined by (2), satisfies

$$\lim_{x \to \pm\infty} \max_{0 \le t \le T} |u(x,t)| = 0 . \tag{4}$$

Moreover, we will apply the Maximum Principle to deduce that $u(x,t)$ is the unique solution of (1) with property (4). Without some additional condition, such as (4), there is no guarantee that there is only one solution of (1). Indeed, as the next example shows, there are continuous *nonzero* solutions of (1), even when $f(x) \equiv 0$, but they do not satisfy (4).

Example 1. In 1935, the Russian mathematician A.N. Tychonov demonstrated that the problem

$$\text{D.E.}\quad u_t = u_{xx} \qquad -\infty < x < \infty,\ t > 0$$

$$\text{I.C.}\quad u(x,0) = 0 ,$$

for the infinite rod with a zero initial temperature, has a solution other than the obvious trivial solution $u(x,t) \equiv 0$ (i.e., uniqueness fails !). We sketch the construction of Tychonov's solution.

Construction. Let $f(t) = \begin{cases} e^{-1/t^2} & \text{if } t \neq 0 \\ 0 & \text{if } t = 0 \end{cases}$,

and let
$$u(x,t) = \sum_{n=0}^{\infty} f^{(n)}(t) \frac{x^{2n}}{(2n)!}, \qquad -\infty < x < \infty,\ t \ge 0 .$$

We may differentiate the above series term–by–term . (For a justification, see D.V. Widder, *The Heat Equation*, Academic Press, New York (1975), Chapter III.) Thus,

$$u_t = \sum_{n=0}^{\infty} f^{(n+1)}(t) \frac{x^{2n}}{(2n)!} = \sum_{m=1}^{\infty} f^{(m)}(t) \frac{x^{2(m-1)}}{(2(m-1))!} = \sum_{n=1}^{\infty} f^{(n)}(t) \frac{x^{2n-2}}{(2n-2)!} = u_{xx} ,$$

where we have made a change of index $m = n+1$ to get the second sum, and then set $n = m$ to obtain the third sum. Since $f(0) = 0$ and $f^{(n)}(0) = 0$ for $n = 1, 2, 3, \ldots$ (cf. Problem 1), $u(x,0) = 0$, and with some effort, one could verify the continuity condition (3). But $u(x,t)$ is not identically zero, since $u(0,t) = f(t)$ and $f(t) > 0$ for all $t \neq 0$. □

Obtaining the solution (2) of problem (1) by Fourier transform methods

Having seen the virtues of the formula

$$u(x,t) = \frac{1}{\sqrt{4\pi kt}} \int_{-\infty}^{\infty} e^{-(x-y)^2/4kt} f(y)\, dy \quad (t > 0)\ , \tag{5}$$

the reader no doubt wonders how one arrives at this formula in the first place. This may be done by a purely formal application of the properties of the Fourier transform that we have developed in the previous sections. Fourier methods presuppose that $u(x,t)$ exists and (together with $f(x)$) possesses a suitable decay order. They do not constitute any proof, but, as we will show below, they provide us with the verifiable hypothetical solution given by formula (5).

Upon taking the Fourier transform of both sides of the D.E. in (1) with respect to the variable x, keeping t fixed, we obtain (formally)

$$\int_{-\infty}^{\infty} u_t(x,t)\, e^{-i\xi x}\, d'x = \int_{-\infty}^{\infty} k u_{xx}(x,t)\, e^{-i\xi x}\, d'x\ .$$

Using Corollary 1 of Section 7.2 (formally, of course), we get

$$\hat{u}_t(\xi,t) = k(i\xi)^2\, \hat{u}(\xi,t) = -k\xi^2\, \hat{u}(\xi,t). \tag{6}$$

Note that (6) is a first–order ODE in t for each fixed ξ. The solutions of (6) are of the form $\hat{u}(\xi,t) = F(\xi)\, e^{-k\xi^2 t}$, where $F(\xi)$ can be determined by setting $t = 0$:

$$F(\xi) = \hat{u}(\xi,0) = \int_{-\infty}^{\infty} u(x,0)\, e^{-i\xi x}\, d'x = \int_{-\infty}^{\infty} f(x)\, e^{-i\xi x}\, d'x.$$

Thus, $F(\xi) = \hat{f}(\xi)$, and so $\hat{u}(\xi,t) = \hat{f}(\xi)\, e^{-k\xi^2 t}$. We found in Example 6 of Section 7.1 that

$$\left[\sqrt{a}\, e^{-\frac{1}{2}ax^2}\right]\hat{}\,(\xi) = e^{-\frac{1}{2}\xi^2/a}\ ,$$

for any constant $a > 0$. Thus, setting $a = 1/2kt,\ t > 0$, we get

$$\left[\frac{1}{\sqrt{2kt}}\, e^{-x^2/4kt}\right]\hat{}\,(\xi) = e^{-k\xi^2 t}.$$

If $g(x) \equiv (2kt)^{-\frac{1}{2}}\, e^{-x^2/4kt}$, then by a formal application of the Convolution Theorem,

$$\hat{u}(\xi,t) = \hat{f}(\xi)\hat{g}(\xi) = \frac{1}{\sqrt{2\pi}}[f*g]^{\wedge}(\xi) .$$

Taking the inverse transform of both sides, with t fixed, we arrive at formula (5) :

$$u(x,t) = \frac{1}{\sqrt{2\pi}}(f*g)(x) = \frac{1}{\sqrt{2\pi}}\int_{-\infty}^{\infty} f(y)\, g(x-y)\, dy = \frac{1}{\sqrt{4\pi kt}}\int_{-\infty}^{\infty} e^{-(x-y)^2/4kt} f(y)\, dy .$$

Lest the reader be overjoyed by these formal manipulations, we again remark that they do not prove that (2) is the solution of (1). They do not even prove that a solution exists! We now go on to prove that (2) is in fact a solution of problem (1), which meets the continuity condition (3).

The proof of the validity of solution (2) of problem (1)

We first prove that (2) (or (5)) is a solution of the D.E. $u_t = ku_{xx}$ when $t > 0$ and $f(x)$ is bounded or absolutely integrable. We can rewrite the formula (5) in the form,

$$u(x,t) = \int_{-\infty}^{\infty} H(x-y,t)\, f(y)\, dy , \tag{7}$$

where

$$H(x-y,t) \equiv \frac{1}{\sqrt{4\pi kt}} e^{-(x-y)^2/4kt} \qquad (t > 0) . \tag{8}$$

Definition. The function $H(x-y,t)$ in (8) is known as the **heat kernel** (or **source solution** or **fundamental solution**) of the heat equation for the infinite rod.

We have $H_t = kH_{xx}$, as can be verified directly (cf. Example 1 of Section 3.1), and intuitively $H(x-y,t)$ represents the temperature distribution at position x at time t due to a concentrated initial temperature source at position y (cf. the Figure 1 in the introduction to this Chapter). Since $H_t = kH_{xx}$, it will follow that (7) is a solution of $u_t = ku_{xx}$ for $t > 0$, if we can justify differentiating (7) with respect to x and t under the integral. Fortunately, for $t > 0$, all of the partial derivatives with respect to x and t of the integrand of (2) are continuous and absolutely integrable with respect to y (cf. Example 1 of Section 7.2). This is clear, if $f(y)$ is continuous and absolutely integrable or bounded. (Indeed, due to the rapid decay of $H(x-y,t)$ in the variable y, we could have only required that $|f(y)| < Ce^{|y|}$.) Hence, for $t > 0$, we may freely differentiate $u(x,t)$ in (7) according to Leibniz's rule (cf. Appendix A.3). Thus, $u(x,t)$ in (7), satisfies the D.E. . Note that $f(y)$ is constant with respect to x and t, so that no assumption about the differentiability of $f(y)$ is needed.

It remains to show that the continuity condition (3) holds. We know that, for $t > 0$,

$$\int_{-\infty}^{\infty} H(x-y,t)\, dy = \frac{1}{\sqrt{4\pi kt}} \int_{-\infty}^{\infty} e^{-(x-y)^2/4kt}\, dy = 1, \tag{9}$$

by the result $\int_{-\infty}^{\infty} e^{-\frac{1}{2}az^2}\, dz = \sqrt{2\pi/a}$, proved in Example 6 of Section 7.1 . (Take $a = (2kt)^{-1}$ and $z = x-y$). Consider $u(x,t)$ given by (7). Then by (9), we have

$$\begin{aligned} |u(x,t) - f(x_0)| &= \left| \int_{-\infty}^{\infty} H(x-y,t)f(y)\, dy - f(x_0) \int_{-\infty}^{\infty} H(x-y,t)\, dy \right| \\ &= \left| \int_{-\infty}^{\infty} H(x-y,t)[f(y) - f(x_0)]\, dy \right| \le \int_{-\infty}^{\infty} H(x-y,t)\,|f(y) - f(x_0)|\, dy . \end{aligned} \tag{10}$$

To prove the continuity condition (3), we must show that $|u(x,t) - f(x_0)|$ (or the last integral in (10)) can be made as small as desired by taking (x,t) (in the half–plane $t > 0$) sufficiently close to $(x_0,0)$. In order to do this, we split the last integral in (10) into three pieces :

$$\int_{-\infty}^{x-\delta} H(x-y,t)\,|f(y)-f(x_0)|\, dy + \int_{x-\delta}^{x+\delta} H(x-y,t)\,|f(y)-f(x_0)|\, dy + \int_{x+\delta}^{\infty} H(x-y,t)\,|f(y)-f(x_0)|\, dy ,$$

where δ is a positive constant. Let these integrals, from left to right, be denoted by I_1 , I_2 and I_3. Since $f(y)$ is continuous, for any given $\epsilon > 0$, we can guarantee that $|f(y) - f(x_0)| < \epsilon$ in I_2 , if x is sufficiently close to x_0 and δ is sufficiently small. Then

$$I_2 = \int_{x-\delta}^{x+\delta} H(x-y,t)\,|f(y)-f(x_0)|\, dy \le \int_{x-\delta}^{x+\delta} H(x-y,t)\, \epsilon\, dy \le \epsilon \int_{-\infty}^{\infty} H(x-y,t)\, dy = \epsilon .$$

In other words, we can make I_2 as small as we like by choosing x sufficiently close to x_0 and δ small enough. We now show that regardless of the choice of $\delta > 0$, the integrals I_1 and I_3 can be made as small as desired, by taking $t > 0$ sufficiently small. We work only with I_3 , since I_1 is handled in the same way. We have

$$\begin{aligned} I_3 &= \int_{x+\delta}^{\infty} H(x-y,t)\, |f(y)-f(x_0)|\, dy = \frac{1}{\sqrt{4\pi kt}} \int_{x+\delta}^{\infty} e^{-(x-y)^2/4kt}\, |f(y) - f(x_0)|\, dy \\ &\le \frac{1}{\sqrt{4\pi kt}} \int_{x+\delta}^{\infty} e^{-(x-y)^2/4kt}\, |f(y)|\, dy + \frac{1}{\sqrt{4\pi kt}} \int_{x+\delta}^{\infty} e^{-(x-y)^2/4kt}\, |f(x_0)|\, dy \\ &= \frac{1}{\sqrt{4\pi kt}} \int_{\delta}^{\infty} e^{-z^2/4kt}\, |f(x+z)|\, dz + \frac{1}{\sqrt{4\pi kt}} \int_{\delta}^{\infty} e^{-z^2/4kt}\, |f(x_0)|\, dz . \end{aligned} \tag{11}$$

When $f(x)$ is bounded, say $|f(x)| \leq M$, then each of the last integrals in (11) is no greater than

$$\frac{M}{\sqrt{4\pi kt}} \int_{\delta}^{\infty} e^{-z^2/4kt}\, dz \leq \frac{M}{\sqrt{4\pi kt}} \int_{\delta}^{\infty} e^{-z\delta/4kt}\, dz = \frac{M4kt}{\delta\sqrt{4\pi kt}} e^{-\delta^2/4kt},$$

which can be made as small as desired by taking t sufficiently small and positive. When $f(y)$ is not necessarily bounded, but absolutely integrable, the second integral in (11) is treated as in the bounded case, but the first integral in (11) is estimated as follows :

$$\frac{1}{\sqrt{4\pi kt}} \int_{\delta}^{\infty} e^{-z^2/4kt} |f(x+z)|\, dz \leq \frac{e^{-\delta^2/4kt}}{\sqrt{4\pi kt}} \int_{\delta}^{\infty} |f(x+z)|\, dz \leq \frac{t^{-\frac{1}{2}}}{\sqrt{4\pi k}\ \exp(\delta^2/4kt)} \int_{-\infty}^{\infty} |f(y)|\, dy .$$

A single application of L'Hospital's rule reveals that the limit of the last quotient tends to 0, as $t \to 0^+$, while the last integral is finite since $f(x)$ is absolutely integrable. Thus, whether $f(x)$ is bounded or absolutely integrable, for any positive value of δ, we can make I_3 (and similarly I_1) arbitrarily small by taking t sufficiently small and positive. Recall that we already showed that the continuity of $f(x)$ ensures that I_2 can be made arbitrarily small for x sufficiently close to x_0 and δ sufficiently small. As we have noted, these facts and (10) ensure the continuity condition (3). In summary, we have established the following key result.

Theorem 1. Let $f(x)$, $-\infty < x < \infty$, **be continuous and either bounded or absolutely integrable. Then the function** $u(x,t)$, **defined by**

$$u(x,t) = \frac{1}{\sqrt{4\pi kt}} \int_{-\infty}^{\infty} e^{-(x-y)^2/4kt} f(y)\, dy \quad (\text{for } t > 0) \quad \text{and} \quad u(x,0) = f(x) \quad (\text{for } t = 0)\,, \tag{12}$$

is C^∞ **in the domain** $\{(x,t):\ t > 0,\ -\infty < x < \infty\}$, **continuous in** $\{(x,t):\ t \geq 0,\ -\infty < x < \infty\}$, **and satisfies** $u_t = ku_{xx}$ **for** $t > 0$, $-\infty < x < \infty$. **In particular,** (12) **solves the problem**

$$\begin{aligned} &\text{D.E.} \quad u_t = ku_{xx} \qquad -\infty < x < \infty\,,\ t > 0 \\ &\text{I.C.} \quad u(x,0) = f(x)\,, \end{aligned} \tag{13}$$

together with the continuity condition $u(x,t) \to f(x_0)$ as $(x,t) \to (x_0, 0^+)$.

In the next example, we illustrate the use of Theorem 1 and the fact that the evaluation of the integral in (12) may be difficult even in the case when $f(x)$ is a simple function. The reader who has access to symbolic integration on a computer (e.g., Macsyma, Maple, Derive or Mathematica) can verify the result (14) given below.

Example 2. Solve problem (13) in the case where $f(x) = e^{-\alpha x^2}\cos(\beta x)$, $\alpha \geq 0$.

Solution. Since $f(x)$ is bounded and continuous, one solution of (13) is (for $t > 0$)

$$u(x,t) = \frac{1}{\sqrt{4\pi kt}}\int_{-\infty}^{\infty} e^{-(x-y)^2/4kt}\, e^{-\alpha y^2}\cos(\beta y)\, dy = \frac{1}{\sqrt{4\pi kt}}\,\mathrm{Re}\left[\int_{-\infty}^{\infty} e^{-(x-y)^2/4kt}\, e^{-\alpha y^2+i\beta y}\, dy\right].$$

The last integral is evaluated, as follows. We combine the exponents and complete the square :

$$e^{-(x-y)^2/4kt}\, e^{-\alpha y^2+i\beta y} = \exp\left[-\tfrac{1}{2}\delta[y^2 - 2\delta^{-1}(\tfrac{x}{2kt}+i\beta)y] - x^2/4kt\right] \quad (\text{where } \delta \equiv \tfrac{1}{2kt} + 2\alpha)$$

$$= \exp\left[-\tfrac{1}{2}\delta\cdot[y - \delta^{-1}(\tfrac{x}{2kt}+i\beta)]^2 + \tfrac{1}{2}\delta\,\delta^{-2}(\tfrac{x}{2kt}+i\beta)^2 - x^2/4kt\right].$$

For fixed x, let $\gamma \equiv \delta^{-1}(\frac{x}{2kt}+i\beta)$. Then the real part of this last expression is

$$e^{-x^2/4kt}\mathrm{Re}\left[e^{-\frac{1}{2}\delta\gamma^2}\right]\mathrm{Re}\left[e^{-\frac{1}{2}\delta(y-\gamma)^2}\right] = \exp\left[-x^2/4kt+\tfrac{1}{2}\delta\cdot\mathrm{Re}[\gamma^2]\right]\,\cos(\tfrac{1}{2}\delta\cdot\mathrm{Im}[\gamma^2])\,\mathrm{Re}\left[e^{-\frac{1}{2}\delta(y-\gamma)^2}\right].$$

Note that only the last factor involves y, and by the solution of Example 6 of Section 7.1, $\int_{-\infty}^{\infty} e^{-\frac{1}{2}\delta(y-\gamma)^2}dy = \sqrt{2\pi/\delta} = \mathrm{Re}\left[\int_{-\infty}^{\infty} e^{-\frac{1}{2}\delta(y-\gamma)^2}dy\right]$, even when γ is a complex constant. Thus,

$$u(x,t) = \frac{1}{\sqrt{4\pi kt}}\exp\left[-x^2/4kt + \tfrac{1}{2}\delta^{-1}[(x/2kt)^2 - \beta^2]\right]\,\cos(\tfrac{\beta x}{2\delta kt})\,\sqrt{2\pi/\delta}$$

$$= \frac{1}{\sqrt{1+4kt\alpha}}\exp\left[\frac{-1}{1+4\alpha kt}(\alpha x^2 + \beta^2 kt)\right]\,\cos(\tfrac{\beta x}{1+4\alpha kt}). \tag{14}$$

Thus $u(x,0) = e^{-\alpha x^2}\cos(\beta x)$. If $\alpha = 0$, (14) is the product solution $e^{-\beta^2 kt}\cos(\beta x)$ of the D.E. . □

Remark. Observe that (14) is *not* the only solution of the problem, because one can add to (14) any constant multiple of Tychonov's solution in Example 1. In order to achieve uniqueness, one can impose some restriction on the behavior of solutions $u(x,t)$ for $|x|$ large (i.e., some boundary conditions at $\pm\infty$ are needed). From a physical perspective, one attractive condition is that the desired solution tend to zero as $|x| \to \infty$. In this case, we have the following result. □

Theorem 2 (Uniqueness Theorem). Suppose that $f(x)$ is continuous and absolutely integrable, for $-\infty < x < \infty$, and $f(x) \to 0$ as $x \to \pm\infty$. Then the function $u(x,t)$ defined by

$$u(x,t) = \frac{1}{\sqrt{4\pi kt}} \int_{-\infty}^{\infty} e^{-(x-y)^2/4kt} f(y)\, dy \quad (\text{for } t > 0) \quad \textbf{and} \quad u(x,0) = f(x) \quad (\text{for } t = 0), \tag{15}$$

is the *unique*, continuous (for $t \geq 0$) solution of the problem

$$\begin{aligned} &\text{D.E.} \quad u_t = ku_{xx} \qquad -\infty < x < \infty,\ t > 0 \\ &\text{I.C.} \quad u(x,0) = f(x) \\ &\text{"B.C."} \quad \lim_{x\to\pm\infty} \max_{0\leq t\leq T} |u(x,t)| = 0, \quad \text{for all } T > 0. \end{aligned} \tag{16}$$

Proof. If there are two solutions, say u_1 and u_2 of (16), then we would have for each small $\epsilon > 0$, and $x_0 > 0$ sufficiently large,

$$|u_1(\pm x_0,t) - u_2(\pm x_0,t)| \leq \epsilon, \qquad 0 \leq t \leq T,$$

$$|u_1(x,0) - u_2(x,0)| = 0 \leq \epsilon, \quad -x_0 \leq x \leq x_0\,.$$

Then, as a consequence of the Maximum Principle (cf. Theorem 3 of Section 3.2), we have $|u_1(x,t) - u_2(x,t)| \leq \epsilon$ in the rectangle $-x_0 \leq x \leq x_0$, $0 \leq t \leq T$. Thus, given any (x,t) in the half–plane $t > 0$, we can deduce that $|u_1(x,t) - u_2(x,t)|$ is smaller than any positive number ϵ, by choosing $T > t$ and x_0 sufficiently large. Thus, $u_1(x,t) = u_2(x,t)$. Hence, we have shown that if (16) has a solution, then there is only one solution. It remains to show that (15) is this solution. By Theorem 1, we need only to prove that (15) satisfies the "B.C." in (16).

Using (9) to get the second inequality below, for $t > 0$, we have

$$\begin{aligned} |u(x,t)| &\leq \frac{1}{\sqrt{4\pi kt}} \int_{|y-x|\leq A} e^{-(x-y)^2/4kt} |f(y)|\, dy + \frac{1}{\sqrt{4\pi kt}} \int_{|y-x|>A} e^{-(x-y)^2/4kt} |f(y)|\, dy \\ &\leq \max_{x-A\leq y\leq x+A} |f(y)| + \frac{1}{\sqrt{4\pi kt}} e^{-A^2/4kt} \int_{-\infty}^{\infty} |f(y)|\, dy. \end{aligned} \tag{17}$$

For A sufficiently large, the second term can be made arbitrarily small, since we can show that the absolute maximum of the function $(4\pi kt)^{-\frac{1}{2}} e^{-A^2/4kt}$ is $(2\pi e)^{-\frac{1}{2}} A^{-1}$, as follows :

$$\frac{d}{dt}\left[\frac{1}{\sqrt{4\pi kt}} e^{-A^2/4kt}\right] = \frac{1}{(4\pi kt)^{3/2}} e^{-A^2/4kt} \left[\frac{2\pi k}{t}\right]\left[\frac{A^2}{2k} - t\right]$$

is negative for $t > A^2/2k$, and positive for $0 < t < A^2/2k$. The maximum of $(4\pi kt)^{-\frac{1}{2}}e^{-A^2/4kt}$ (at $t = A^2/2k$) is $(2\pi e)^{-\frac{1}{2}}A^{-1}$. Thus, for A sufficiently large, we can make the second term in (17) as small as desired, and then for $|x|$ sufficiently large (and A fixed) the first term goes to 0 as $x \to \pm\infty$, since $\lim_{x\to\pm\infty} f(x) = 0$. Thus, the "B.C." in (16) holds for (15). □

Remark. Theorem 2 implies that the solution (14), with $\alpha > 0$, is the only solution which meets the condition $\lim_{x\to\pm\infty} \max_{0\le t\le T} |u(x,t)| = 0$, but when $\alpha = 0$, Theorem 2 does not apply since $\cos(\beta x)$ is not absolutely integrable, nor does it tend to zero as $x \to \pm\infty$. However, $e^{-\beta^2 kt}\cos(\beta x)$ is the unique solution which is *periodic* in x of period $2\pi/\beta$. Indeed, such a solution defines a solution on a circular wire, and we already have proved uniqueness in that case (cf. Example 1 of Section 3.2). □

Example 3. Find a formal solution of

$$\begin{aligned} &\text{D.E. } u_t = ku_{xx} + q(x,t) \qquad -\infty < x < \infty,\ t > 0 \\ &\text{I.C. } u(x,0) = 0. \end{aligned} \tag{18}$$

Solution. The following formal manipulations serve only to provide a hypothetical solution whose validity could be justified directly, under certain assumptions concerning the source term $q(x,t)$. We first take the Fourier transform of both sides of the D.E. with respect to x. Thus, we obtain

$$\hat{u}_t(\xi,t) = k(i\xi)^2\,\hat{u}(\xi,t) + \hat{q}(\xi,t)\,, \quad \text{or} \quad \hat{u}_t(\xi,t) + k\xi^2\,\hat{u}(\xi,t) = \hat{q}(\xi,t). \tag{19}$$

Note that equation (19) is a first–order linear ODE in t. Thus, if we multiply (19) by the integrating factor $e^{k\xi^2 t}$, then we obtain

$$\frac{\partial}{\partial t}[\hat{u}(\xi,t)\,e^{k\xi^2 t}] = e^{k\xi^2 t}\,\hat{q}(\xi,t)\,, \quad \text{or} \quad \hat{u}(\xi,t)\,e^{k\xi^2 t} = \int_0^t e^{k\xi^2 s}\,\hat{q}(\xi,s)\,ds + F(\xi).$$

But by the I.C. $u(x,0) = 0$, $\hat{u}(\xi,0) = F(\xi) = \frac{1}{\sqrt{2\pi}}\int_{-\infty}^{\infty} u(x,0)\,e^{-i\xi x}\,dx = 0$. Hence,

$$\hat{u}(\xi,t) = e^{-k\xi^2 t}\int_0^t e^{k\xi^2 s}\,\hat{q}(\xi,s)\,ds,$$

and a formal application of the Inversion Theorem yields

$$u(x,t) = \frac{1}{\sqrt{2\pi}}\int_{-\infty}^{\infty} \hat{u}(\xi,t)\,e^{i\xi x}\,d\xi = \frac{1}{\sqrt{2\pi}}\int_{-\infty}^{\infty}\left[e^{-k\xi^2 t}\int_0^t e^{k\xi^2 s}\,\hat{q}(\xi,s)\,ds\right]e^{i\xi x}\,d\xi$$

$$= \frac{1}{\sqrt{4\pi k}} \int_{-\infty}^{\infty} \left[\int_0^t \frac{q(y,s)}{\sqrt{t - s}} e^{-(y-x)^2/(4k(t-s))} \, ds \right] dy ,$$

cf. Problem 8 of Section 7.3 . The same hypothetical solution can also be obtained by a formal application of Duhamel's principle (cf. Problem 10). □

Example 4. Suppose $f(x)$ is a continuous, odd function defined on $(-\infty,\infty)$. If $f(x)$ is absolutely integrable on $(-\infty,\infty)$ and if $f(x) \to 0$ as $x \to \pm\infty$, show that the unique continuous solution of the following problem is also odd in the variable x :

$$\begin{aligned} &\text{D.E. } u_t = ku_{xx} && -\infty < x < \infty, \ t > 0 \\ &\text{I.C. } u(x,0) = f(x) && \qquad\qquad (20) \\ &\text{"B.C." } \lim_{x \to \pm\infty} \max_{0 \le t \le T} u(x,t) = 0, && \text{for all } T > 0 . \end{aligned}$$

Show that the conclusion is false if the "B.C." is deleted.

Solution. Method 1. By Theorem 2, the unique continuous solution of problem (20) is given by

$$u(x,t) = \frac{1}{\sqrt{4\pi kt}} \int_{-\infty}^{\infty} e^{-(x-y)^2/4kt} f(y) \, dy \ \ (\text{for } t > 0) \quad \text{and} \quad u(x,0) = f(x) \ \ (\text{for } t = 0).$$

We are given that $u(x,0)$ is an odd function. If $t > 0$, then letting $z = -y$,

$$u(-x,t) = \frac{1}{\sqrt{4\pi kt}} \int_{-\infty}^{\infty} e^{-(x+y)^2/4kt} f(y) \, dy = -\frac{1}{\sqrt{4\pi kt}} \int_{\infty}^{-\infty} e^{-(x-z)^2/4kt} f(-z) \, dz = -u(x,t) ,$$

since $f(-z) = -f(z)$. Thus, $u(x,t)$ is an odd function of x.

Method 2. Observe that $-u(-x,t)$ also solves (20) and use uniqueness (cf. Theorem 2).

Without the "B.C.", we can add the (even) Tychonov solution to the above odd solution, thereby obtaining a solution which is not odd. □

Remark. Without the "B.C.", we can also obtain more than one odd solution of (20), because we can differentiate Tychonov's solution with respect to x and get an odd solution with zero initial data which can be added to the above standard solution to obtain another odd solution. □

Example 5. Let $f(x)$ be continuous and absolutely integrable on $(-\infty,\infty)$, and let

$$u(x,t) = \frac{1}{\sqrt{4\pi kt}} \int_{-\infty}^{\infty} e^{-(x-y)^2/4kt} f(y) \, dy \qquad (t > 0) . \tag{21}$$

Verify the following relation and give a physical interpretation of it :

$$\int_{-\infty}^{\infty} u(x,t) \, dx = \int_{-\infty}^{\infty} f(y) \, dy . \tag{22}$$

Solution. Since the integrand of (21) is absolutely integrable over the xy–plane, in the following calculations we may interchange the order of integration (cf. Appendix A.2).

$$\int_{-\infty}^{\infty} u(x,t)\,dx = \int_{-\infty}^{\infty}\left[\frac{1}{\sqrt{4\pi kt}}\int_{-\infty}^{\infty} e^{-(x-y)^2/4kt} f(y)\,dy\right] dx$$
$$= \int_{-\infty}^{\infty}\frac{1}{\sqrt{4\pi kt}}\left[\int_{-\infty}^{\infty} e^{-(x-y)^2/4kt}\,dx\right] f(y)\,dy = \int_{-\infty}^{\infty} f(y)dy\ ,$$

where we have used the fact that $\int_{-\infty}^{\infty} H(x-y,t)\,dy = 1$ (cf. (9)). Equation (22) may be regarded as a special case of the law of conservation of energy. Indeed, at any time $t > 0$, the left side of (22) is (in appropriate units) the total heat energy in the rod at time t, while the right side of (22) is the initial total heat energy. □

Example 6. For $t > 0$, consider the fundamental solution (cf. (8))

$$H(x,t) \equiv \frac{1}{\sqrt{4\pi kt}}\, e^{-x^2/4kt} \tag{23}$$

of the heat equation $u_t = ku_{xx}$. For *fixed* t, we denote the function $H(x,t)$ of x by ${}_tH(x)$, so as not to confuse it with the partial derivative H_t . Show that

$$({}_tH * {}_tH)(x) = {}_{2t}H(t)\ , \qquad (t > 0). \tag{24}$$

Solution. According to the definition of convolution (cf. Section 7.2), for $t > 0$

$$({}_tH * {}_tH)(x) = \int_{-\infty}^{\infty} {}_tH(x-y)\, {}_tH(y)\,dy = \int_{-\infty}^{\infty} \frac{1}{\sqrt{4\pi kt}}\, e^{-(x-y)^2/4kt}\, \frac{1}{\sqrt{4\pi kt}}\, e^{-y^2/4kt}\,dy$$
$$= \frac{1}{4\pi kt}\, e^{-x^2/8kt} \int_{-\infty}^{\infty} e^{-(y-\frac{1}{2}x)^2/2kt}\,dy \ = \frac{1}{4\pi kt}\, e^{-x^2/8kt}\sqrt{2\pi kt}$$
$$= \frac{1}{\sqrt{8\pi kt}}\, e^{-x^2/8kt} \ = {}_{2t}H(x)\ ,$$

where we have used (9), with x replaced by $\frac{1}{2}x$. In Problem 8, a different (and easier) approach, which uses the Convolution Theorem, yields the more general result that ${}_sH * {}_tH = {}_{(s+t)}H$, for all positive s and t. This other approach is easier, because it is much simpler to multiply Fourier transforms of functions than to convolve the functions. □

Remark. Solution (21) can be written as $u(x,t) = ({}_tH * f)(x)$ (Why ?). For $s > 0$, let $g(x) = u(x,s)$. Since $u(x,s+t)$ is the solution at time t with initial temperature $g(x)$,

$$({}_{(s+t)}H*f)(x) = u(x,t+s) = ({}_tH*g)(x) = ({}_tH*({}_sH*f))(x) = (({}_sH*{}_tH)*f)(x) .$$

Comparing the extreme left with the extreme right, it is not surprising that ${}_sH*{}_tH = {}_{(s+t)}H$. Indeed, this argument shows that one can expect this convolution property of fundamental solutions of equations with translational invariance with respect to a variable. □

Example 7. Find a continuous solution of the problem

$$\begin{aligned} &\text{D.E.} \quad u_t - ku_{xx} = -\lambda u \qquad -\infty < x < \infty,\ t > 0 \\ &\text{I.C.} \quad u(x,0) = f(x) , \end{aligned} \tag{25}$$

where λ is a constant and $f(x)$ is a continuous, bounded function. Give a physical interpretation of the D.E. in (25).

Solution. Formally applying the Fourier transform to the D.E., we obtain

$$\hat{u}_t(\xi,t) + (k\xi^2 + \lambda)\hat{u}(\xi,t) = 0$$

Thus, formally $\hat{u}(\xi,t) = e^{-\lambda t}\hat{f}(\xi)e^{-k\xi^2 t}$ (Why ?). Since $e^{-\lambda t}$ does not involve ξ , we have

$$u(x,t) = [\hat{u}(\xi,t)]\check{}(x) = e^{-\lambda t}[\hat{f}(\xi)e^{-k\xi^2 t}] = e^{-\lambda t}\frac{1}{\sqrt{4\pi kt}}\int_{-\infty}^{\infty} e^{-(x-y)^2/4kt}f(y)\,dy .$$

by a formal use of the Convolution Theorem. Using Theorem 1, it is easy to check that this function solves the problem. Indeed, note that $u(x,t) = e^{-\lambda t}v(x,t)$, where $v_t = kv_{xx}$. Thus, $u_t = -\lambda e^{-\lambda t}\, v + e^{-\lambda t}v_t = -\lambda u + ku_{xx}$, as required. Moreover, $u(x,0) = e^{-\lambda 0}v(x,0) = f(x)$.

If $\lambda > 0$, then the presence of the term $-\lambda u$ in the D.E. might arise from a heat flux through the lateral surface of the rod, say because of faulty insulation (cf. Problem 4 of Section 3.1). If $\lambda < 0$, then the term $-\lambda u$ represents a heat source whose strength is proportional to the temperature, say due to a chemical or nuclear reaction. □

The derivation of D'Alembert's formula using Fourier transform methods

Although the treatment of the wave equation in Chapter 5 is adequate as it stands, it is instructive to obtain D'Alembert's formula by Fourier transform methods. Consider the problem

$$\begin{aligned} &\text{D.E.} \quad u_{tt} = a^2u_{xx} \qquad -\infty < x, t < \infty \\ &\text{I.C.} \quad u(x,0) = f(x) ,\ u_t(x,0) = g(x). \end{aligned} \tag{26}$$

We proceed formally again. The justification of the end result was already done in Chapter 5. Taking the Fourier transform of both sides of the D.E. with respect to x we get

$$\hat{u}_{tt}(\xi,t) = a^2(i\xi)^2\,\hat{u}(\xi,t) = -(a\xi)^2\,\hat{u}(\xi,t),$$

whence

$$\hat{u}(\xi,t) = c_1(\xi)\cos(a\xi t) + c_2(\xi)\sin(a\xi t) \quad \text{and} \quad \hat{u}_t(\xi,t) = -a\xi\, c_1(\xi)\sin(a\xi t) + a\xi\, c_2(\xi)\cos(a\xi t).$$

Now, upon applying Fourier transforms to the I.C., we get

$$\hat{f}(\xi) = \hat{u}(\xi,0) = c_1(\xi) \quad \text{and} \quad \hat{g}(\xi) = \hat{u}_t(\xi,0) = a\xi\, c_2(\xi).$$

Thus,

$$\hat{u}(\xi,t) = \hat{f}(\xi)\cos(a\xi t) + \hat{g}(\xi)\,\frac{\sin(a\xi)}{a\xi}. \tag{27}$$

We would like to apply the Convolution Theorem at this point, but we have not found any function whose transform is $\cos(a\xi t)$. In fact, there is no "ordinary" function with this property. Instead, we formally take the inverse transform of $\hat{f}(\xi)\cos(a\xi t)$:

$$[\hat{f}(\xi)\cos(a\xi t)]\check{}\,(x) = \int_{-\infty}^{\infty}\hat{f}(\xi)\cos(a\xi t)\,e^{i\xi x}\,d'\xi = \frac{1}{2}\int_{-\infty}^{\infty}\hat{f}(\xi)\,(e^{ia\xi t} + e^{-ia\xi t})\,e^{i\xi x}\,d'\xi$$

$$= \frac{1}{2}\int_{-\infty}^{\infty}\hat{f}(\xi)\,e^{i(x+at)\xi}\,d'\xi + \frac{1}{2}\int_{-\infty}^{\infty}\hat{f}(\xi)\,e^{i(x-at)\xi}\,d'\xi = \tfrac{1}{2}[f(x+at) + f(x-at)]. \tag{28}$$

To invert the second term in (27), we may use the Convolution Theorem, since we know from Example 4 of Section 7.1 that

$$\text{if} \quad h(x) = \begin{cases} 1/2a & |x| \le at \\ 0 & |x| > at \end{cases}, \quad \text{then} \quad \hat{h}(\xi) = \frac{1}{\sqrt{2\pi}}\,\frac{\sin(a\xi)}{a\xi}. \tag{29}$$

The Convolution Theorem and (29) then yield

$$\left[\hat{g}(\xi)\,\frac{\sin(a\xi)}{a\xi}\right]\check{}\,(x) = \int_{-\infty}^{\infty} g(y)\,h(x-y)\,dy = \frac{1}{2a}\int_{x-at}^{x+at} g(y)\,dy. \tag{30}$$

Adding (28) and (30) produces (but does not prove) D'Alembert's formula:

$$u(x,t) = \tfrac{1}{2}[f(x+at) + f(x-at)] + \frac{1}{2a}\int_{x-at}^{x+at} g(y)\,dy. \tag{31}$$

There is another way of producing (31) that involves the **Dirac delta "function"** $\delta(x)$. This is *not* a function in the usual sense, since it is declared to have the property that for any function $f(x)$,

$$\int_{-\infty}^{\infty} \delta(x)f(x)\,dx = f(0). \tag{32}$$

If $\delta(x)$ were a usual function, then changing $f(x)$ to a different value at the single point $x = 0$ would not change the left–hand side of (32), but the right–hand side would be changed. In particular, the left side of (32) cannot be defined the way integrals are usually defined (e.g., as limits of Riemann sums), because the integrand is not an ordinary function. Instead, one way to view the left–hand side of (32) is that it is merely a fancy way of writing $f(0)$. In other words, (32) is the *definition* of $\int_{-\infty}^{\infty} \delta(x)f(x)\,dx$. Similarly, we define

$$\int_{-\infty}^{\infty} \delta(x-c)f(x)\,dx = \int_{-\infty}^{\infty} \delta(c-x)f(x)\,dx = f(c)\ , \tag{33}$$

for any real number c. The Fourier transform of $\delta(x-c)$ is then (by (33))

$$\int_{-\infty}^{\infty} \delta(x-c)\, e^{-ix\xi}\, d'x \;=\; \frac{1}{\sqrt{2\pi}}\, e^{-ic\xi}\,. \tag{34}$$

Note that we then have (declaring that linearity is to hold)

$$\frac{1}{2}\int_{-\infty}^{\infty} [\delta(x+c) + \delta(x-c)]e^{-ix\xi}\, d'x \;=\; \frac{1}{2\sqrt{2\pi}}\,(e^{ic\xi} + e^{-ic\xi}) \;=\; \frac{1}{\sqrt{2\pi}}\cos(c\xi).$$

In other words, $\cos(c\xi)$ is the Fourier transform of $\frac{1}{2}[\delta(x+c) + \delta(x-c)]$. Although the Convolution Theorem was only proved in the context of certain nice ordinary functions, we observe that since

$$\hat{f}(\xi)\cos(a\xi t) = \frac{1}{\sqrt{2\pi}}\,\hat{f}(\xi)\,\tfrac{1}{2}[\delta(x-at) + \delta(x+at)]^{\wedge}(\xi),$$

we ought to have

$$[\hat{f}(\xi)\cos(a\xi t)]^{\vee}(x) = \Big[f * \tfrac{1}{2}[\delta(x-at) + \delta(x+at)]\Big](x)$$

$$= \int_{-\infty}^{\infty} f(y)\,\tfrac{1}{2}[\delta(x-y-at) + \delta(x-y+at)]\,dy \;=\; \tfrac{1}{2}[f(x-at) + f(x+at)]\,.$$

Again, this does not justify the end result which was separately proven in Chapter 5. Note that the Dirac delta "function" allows us to write the solution (31) as a single "integral" as was done for the heat equation (cf. (12)), namely

$$u(x,t) = \int_{-\infty}^{\infty} [k(x-y,t)f(y) + h(x-y,t)g(y)]\, dy, \tag{35}$$

where

$$k(z,t) \equiv \tfrac{1}{2}[\delta(z-at) + \delta(z+at)] \quad \text{and} \quad h(z,t) \equiv \begin{cases} \dfrac{1}{2a} & |z| \le at \\ 0 & \text{otherwise.} \end{cases}$$

The Dirac delta "function" belongs to a class of objects known as generalized functions or distributions. The language of distributions is quite convenient for the treatment of situations where the quantities involved are too singular to be represented by ordinary functions. This happens, for example, when a force, which is applied for an infinitesimal time, yields a finite change in momentum. Such a force is called an impulse and it is represented by a delta "function", as engineering students will quickly point out.

Example 8. By formal calculations find an integral representation for the solution of

$$\begin{aligned} &\text{D.E.} \quad u_{tt} = a^2 u_{xx} - u \qquad -\infty < x,\, t < \infty \\ &\text{I.C.} \quad u(x,0) = e^{-\frac{1}{2}x^2}, \quad u_t(x,0) = 0\,. \end{aligned} \tag{36}$$

Solution. If we take the Fourier transform of both sides of the D.E. with respect to x, we obtain

$$\hat{u}_{tt}(\xi,t) = a^2(i\xi)^2\,\hat{u}(\xi,t) - \hat{u}(\xi,t) \quad \text{or} \quad \hat{u}_{tt}(\xi,t) + (1 + a^2\xi^2)\,\hat{u}(\xi,t) = 0\,. \tag{37}$$

Since (37) is a second–order, linear homogeneous ODE in t, with constant coefficients, we find that the general solution of (37) is given by

$$\hat{u}(\xi,t) = c_1(\xi)\cos(t\sqrt{1+a^2\xi^2}) + c_2(\xi)\sin(t\sqrt{1+a^2\xi^2})\,,$$

where $c_1(\xi)$ and $c_2(\xi)$ are arbitrary functions. Moreover,

$$\hat{u}_t(\xi,t) = \sqrt{1+a^2\xi^2}\left[-c_1(\xi)\sin(t\sqrt{1+a^2\xi^2}) + c_2(\xi)\cos(t\sqrt{1+a^2\xi^2})\right].$$

Setting $t = 0$ in the expressions for $\hat{u}(\xi,t)$ and $\hat{u}_t(\xi,t)$ and using the I.C., we find that

$$c_1(\xi) = \hat{u}(\xi,0) = \frac{1}{\sqrt{2\pi}}\int_{-\infty}^{\infty} e^{-\frac{1}{2}x^2}\, e^{-i\xi x}\, dx = e^{-\frac{1}{2}\xi^2}$$

and

$$\sqrt{1+a^2\xi^2}\; c_2(\xi) = \hat{u}_t(\xi,0) = \frac{1}{\sqrt{2\pi}}\int_{-\infty}^{\infty} u_t(x,0)\, e^{-i\xi x}\, dx = 0\ ,$$

where we have used the result of Example 6 of Section 7.1. Hence,

$$\hat{u}(\xi,t) = e^{-\frac{1}{2}\xi^2}\cos(t\sqrt{1+a^2\xi^2}).$$

Finally, a formal application of the Inversion Theorem yields the integral representation

$$u(x,t) \;=\; \frac{1}{\sqrt{2\pi}}\int_{-\infty}^{\infty} \hat{u}(\xi,t)e^{i\xi x}\, d\xi \;=\; \frac{2}{\sqrt{2\pi}}\int_0^{\infty} e^{-\frac{1}{2}\xi^2}\cos(t\sqrt{1+a^2\xi^2})\cos(\xi x)\, d\xi\ , \tag{38}$$

since $\hat{u}(\xi,t)$ is an even function of ξ. All of the partial derivatives of the integrand of (38) with respect to x and t are absolutely integrable with respect to ξ. Thus, (38) can be shown to be a valid solution of (36), by differentiating under the integral (cf. Appendix A.3). □

Remark. The Fourier transform method may not be applicable (not even formally). Consider

$$\text{D.E.}\quad u_{tt} = u_{xx} + 2x \qquad -\infty < x < \infty,\ \ t > 0,$$

$$\text{I.C.}\quad u(x,0) = x^2,\quad u_t(x,0) = 0\ .$$

In Theorem 1 of Section 5.3, we found that a straightforward application of Duhamel's principle yields the solution $u(x,t) = x^2 + t^2 + xt^2$. On the other hand, it is clear that we cannot take the Fourier transform of both sides of the D.E. with respect to x, since the Fourier transform of $h(x) = x$ does not exist. Nor does the Fourier transform of $f(x) = x^2$ exist. □

Laplace's equation in a half–plane

The following result (save uniqueness) is a consequence of Problem 7 of Section 6.4.

Theorem 3. Let $f(x)$, $-\infty < x < \infty$, **be a bounded, continuous function. Then the problem**

$$\begin{aligned} &\text{D.E.}\quad u_{xx} + u_{yy} = 0 \qquad -\infty < x < \infty,\ \ y > 0 \\ &\text{B.C.}\quad u(x,0) = f(x) \qquad -\infty < x < \infty\ . \end{aligned} \tag{39}$$

has a unique bounded and continuous solution on the half–plane $y \geq 0$, **namely**

$$u(x,y) \;=\; \frac{1}{\pi}\int_{-\infty}^{\infty} \frac{y}{y^2 + (x-s)^2}\, f(s)\, ds\ . \tag{40}$$

Remark. Formula (40) is known as **Poisson's integral formula for the upper half–plane.**

Formula (40) can be found using Fourier transforms. Proceeding formally, we take the Fourier transform of the D.E. with respect to x, treating y as a constant :

$$(i\xi)^2 \hat{u}(\xi,y) + \hat{u}_{yy}(\xi,y) = 0 .$$

Treating ξ as a constant, the general solution of this "ODE" is

$$\hat{u}(\xi,y) = c_1(\xi)e^{\xi y} + c_2(\xi)\, e^{-\xi y} . \tag{41}$$

If $\hat{u}(\xi,y)$ grows exponentially as $y \to \infty$, we do not expect that $u(x,y)$ will be bounded as $y \to \infty$. Thus, we impose on $\hat{u}(\xi,y)$ the conditions that

$$c_1(\xi) = 0, \text{ for } \xi > 0 \quad \text{and} \quad c_2(\xi) = 0, \text{ for } \xi < 0 .$$

Then (41) is of the form

$$c_1(\xi)e^{-|\xi|y} + c_2(\xi)e^{-|\xi|y} = c(\xi)e^{-|\xi|y},$$

where $c(\xi) = c_1(\xi) + c_2(\xi)$. We use the B.C. to get $\hat{f}(\xi) = \hat{u}(\xi,0) = c(\xi)$. Thus,

$$u(\xi,y) = \hat{f}(\xi)e^{-|\xi|y} .$$

By Example 3 of Section 7.1 , we have (for fixed y)

$$e^{-y|\xi|} = \left[\frac{1}{\sqrt{2\pi}}\frac{2\,y}{y^2 + x^2}\right]^{\wedge}(\xi) .$$

The Convolution Theorem then yields the desired formula (40). Again, this is not a proof that (40) is a continuous solution of problem (39) for $y \geq 0$, but we at least have a hypothetical formula that may be verified, as in Problem 7 in Section 6.4. □

Example 9. Solve the problem

$$\text{D.E.} \quad u_{xx} + u_{yy} = 0 \qquad -\infty < x < \infty, \ y > 0$$

$$\text{B.C.} \quad u(x,0^+) = f(x) = \begin{cases} 0 & \text{if} \quad x < -1 \\ 1 & \text{if} \ -1 < x < 1 \\ 2 & \text{if} \quad x > 1 . \end{cases}$$

Solution. In this example, the boundary function $f(x)$ has some jump discontinuities. Nevertheless, if we use the formula (40), we obtain

$$u(x,y) = \frac{1}{\pi}\int_{-\infty}^{\infty} \frac{y}{y^2 + (x-s)^2} f(s)\, ds.$$

$$= \frac{1}{\pi}\int_{-\infty}^{-1} \frac{0\cdot y}{y^2+(x-s)^2}\, ds + \frac{1}{\pi}\int_{-1}^{1} \frac{y}{y^2+(x-s)^2}\, ds + \frac{1}{\pi}\int_{1}^{\infty} \frac{2y}{y^2+(x-s)^2}\, ds$$

$$= \frac{1}{\pi}\left[\arctan(\tfrac{x+1}{y}) + \arctan(\tfrac{x-1}{y})\right] + 1\,,$$

which is a harmonic function, for $y > 0$. Recall that $v(x,y) \equiv \arctan(y/x) = \theta$ is harmonic for $x > 0$ by Laplace's equation in polar coordinates (cf. Proposition 1 of Section 6.3). Thus, $v(y,x+1)$ and $v(y,x-1)$ will be harmonic for $y > 0$.) Moreover, we can easily check that

$$\lim_{y\to 0+} u(x,y) = \begin{cases} 0 & \text{if} \quad x < -1 \\ 1 & \text{if} \ -1 < x < 1 \\ 2 & \text{if} \quad x > 1\ . \end{cases}$$

Observe also that

$$\lim_{y\to 0+} u(-1,y) = -\tfrac{1}{2} + 1 \quad \text{and} \quad \lim_{y\to 0+} u(1,y) = \tfrac{1}{2} + 1.$$

One should note that $u(x,y)$ does not approach the same value regardless of how (x,y) approaches $(-1,0)$ (or $(1,0)$) in the upper half–plane. For example, let α be an arbitrary angle between $-\pi/2$ and $\pi/2$. If (x,y) approaches $(-1,0)$ along the ray $(x+1) = \tan(\alpha)\, y$ $(y > 0)$, then $u(x,y) = (\alpha + \arctan(\frac{x-1}{y}))\pi^{-1} + 1$ approaches $(\alpha - \pi/2)\pi^{-1} + 1 = (\alpha + \pi/2)\pi^{-1}$, which ranges from 0 to 1, as α ranges from $-\pi/2$ to $\pi/2$. □

Example 10. Consider the problem

$$\text{D.E.}\quad u_{xx} + u_{yy} = 0 \qquad -\infty < x < \infty,\ y > 0$$

$$\text{B.C.}\quad u(x,0) = e^{-x^2}\,.$$

If $u(x,y)$ is the unique solution (40) which is continuous and bounded, then show that

$$\int_{-\infty}^{\infty} u(x,y)\, dx = \sqrt{\pi}\,, \quad \text{for each } y \geq 0. \tag{42}$$

Solution. Since e^{-x^2} is bounded and continuous, formula (40) yields (for $y > 0$)

$$u(x,y) = \frac{1}{\pi}\int_{-\infty}^{\infty} \frac{y}{y^2 + (x-s)^2} e^{-s^2}\, ds. \tag{43}$$

If we integrate both sides of (43) with respect to x, then we obtain

$$\int_{-\infty}^{\infty} u(x,y)\, dx = \int_{-\infty}^{\infty} \frac{1}{\pi} \int_{-\infty}^{\infty} \frac{y}{y^2 + (x-s)^2} e^{-s^2} ds\, dx. \tag{44}$$

Since the integrand is absolutely integrable for $y > 0$, we can interchange the order of integration. Using $\int_{-\infty}^{\infty} e^{-s^2} ds = \sqrt{\pi}$ (cf. Example 6 of Section 7.1), (44) then yields (42) :

$$\int_{-\infty}^{\infty} u(x,y)\, dx = \int_{-\infty}^{\infty} \frac{y}{\pi} \left[\int_{-\infty}^{\infty} \frac{dx}{y^2 + (x-s)^2} \right] e^{-s^2} ds = \frac{y}{\pi} \frac{\pi}{y} \int_{-\infty}^{\infty} e^{-s^2} ds = \sqrt{\pi}.$$

Since $\int_{-\infty}^{\infty} u(x,0)\, dx = \int_{-\infty}^{\infty} e^{-x^2} ds = \sqrt{\pi}$, (42) also holds for $y = 0$. □

Remark. For the B.C. $u(x,0) = f(x)$, where $f(x)$ is continuous and absolutely integrable, the same argument (cf. Problem 13) yields the result $\int_{-\infty}^{\infty} u(x,y)\, dx = \int_{-\infty}^{\infty} f(x)\, dx$. □

Summary 7.4

1. Fourier transform methods for PDEs : In general, for PDEs in which at least one variable is unrestricted, Fourier transforms may be used to find a hypothetical solution which must be verified by other means. This verification is necessary, because when the Fourier transform is applied to a PDE, one is already assuming not only that a solution exists, but that it has all of the properties which are needed in order to apply the results in Sections 7.2 and 7.3. The success of the method is largely due to the fact that the Fourier transform with respect to a variable, say x, converts each partial derivative with respect to that variable into an algebraic multiplication by $i\xi$. In our examples, the PDEs were for unknown functions of two variables, and so these PDEs became ODEs in the remaining variable. In higher dimensions (e.g., for $u_t = u_{xx} + u_{yy}$) one has to take the Fourier transform in more than one variable to obtain an ODE (cf. Section 9.1). It frequently happens that when one checks a hypothetical solution, it turns out to be valid in greater generality than the Fourier methods presuppose. For example, formula (S2) below yields a valid solution even when $\hat{f}(\xi)$ does not exist (e.g., when $f(x) = \cos(x)$ or even x^2).

2. The heat problem for the infinite rod (Theorem 1) : Let $f(x)$, $-\infty < x < \infty$, **be continuous and either bounded or absolutely integrable. Then the function** $u(x,t)$, **defined by**

$$u(x,t) = \frac{1}{\sqrt{4\pi kt}} \int_{-\infty}^{\infty} e^{-(x-y)^2/4kt} f(y)\, dy \quad (\text{for } t > 0) \quad \text{and} \quad u(x,0) = f(x) \quad (\text{for } t = 0)\,, \tag{S1}$$

is C^∞ **in the domain** $\{(x,t)\colon t > 0,\ -\infty < x < \infty\}$, **continuous in** $\{(x,t)\colon t \geq 0,\ -\infty < x < \infty\}$, **and satisfies** $u_t = ku_{xx}$ **for** $t > 0$, $-\infty < x < \infty$. **In particular,** (12) **satisfies the problem**

$$\begin{aligned} &\text{D.E.} \quad u_t = ku_{xx} \quad -\infty < x < \infty\,,\ t > 0 \\ &\text{I.C.} \quad u(x,0) = f(x) \end{aligned} \tag{S2}$$

together with the continuity condition (3) (i.e., $\lim_{(x,t) \to (x_0,0^+)} u(x,t) = u(x_0,0)$).

3. Uniqueness for the heat problem on the infinite rod : Without further assumptions solutions of problem (S1) are never unique, because one can always add Tychonov's solution (cf. Example 1) to obtain other solutions. If we adjoin to problem (S1) a "boundary condition" at infinity,

$$\text{"B.C."} \quad \lim_{x \to \pm\infty} \max_{0 \leq t \leq T} |u(x,t)| = 0\,, \quad \text{for all } T > 0\,, \tag{S3}$$

then (S2) is the unique solution of (S1) with (S3), provided that $f(x)$ is continuous, absolutely integrable and $\lim_{x \to \pm\infty} f(x) = 0$ (cf. Theorem 2).

4. Fundamental solution : The heat kernel, source solution or fundamental solution of $u_t = ku_{xx}$ is

$$H(x-y,t) = \frac{1}{\sqrt{4\pi kt}} e^{-(x-y)^2/4kt}, \quad \text{for } t > 0 .$$

One can think of $H(x-y,t)$ as the temperature at x at time t due to an initial concentrated heat source at y. Defining ${}_tH(x) \equiv H(x,t)$, we can write (S2) as $u(x,t) = ({}_tH*f)(x)$ (i.e., the convolution of the heat kernel with the initial temperature. For positive s and t, it is easily shown that ${}_tH*{}_sH = {}_{(s+t)}H$, by applying the Convolution Theorem (cf. Problem 8 and the Remark following Example 5).

5. D'Alembert's formula and the Poisson integral formula for the upper half–plane : One can also use Fourier transform techniques to obtain D'Alembert's formula for the wave problem on the infinite string and the Poisson integral formula for the upper half–plane :

$$u(x,t) = \tfrac{1}{2}[f(x+at)+f(x-at)] + \frac{1}{2a}\int_{x-at}^{x+at} g(r)dr \quad \text{and} \quad u(x,y) = \frac{1}{\pi}\int_{-\infty}^{\infty} \frac{y}{y^2 + (x-s)^2} f(s)\, ds \quad (y > 0).$$

With appropriate assumptions on the functions f and g, the verification that these formulas provide solutions of the $u_{tt} = a^2u_{xx}$ and $u_{xx} + u_{yy} = 0$ was carried out in Section 5.2 and in Problem 7 in Section 6.4 , respectively.

Exercises 7.4

1. (a) By means of formal calculations verify that if

$$u(x,t) = \sum_{n=0}^{\infty} f^{(n)}(t) \frac{x^{2n}}{(2n)!}, \quad -\infty < x < \infty, \ t \geq 0, \ \text{where } f(t) = \begin{cases} e^{-1/t^2} & \text{if } t \neq 0 \\ 0 & \text{if } t = 0 \end{cases},$$

then $u(x,t)$ satisfies the heat equation $u_t = u_{xx}$.

(b) Prove that $f^{(n)}(0) = 0$, for $n = 1, 2, \ldots$.

Hint. Consider the limit of the Newton quotient $[f(t) - f(0)]/t$, as $t \to 0$.

2. Solve the problem

D.E. $u_t = ku_{xx} + (2kt)^{-\frac{1}{2}}e^{-x^2/4kt} \quad -\infty < x < \infty, \ t > 0.$

I.C. $u(x,0) = 0$.

3. Let $f(y)$ be a piecewise continuous, bounded function (i.e., $|f(x)| \leq M$, for some constant M).

(a) Show that $u(x,t) = \dfrac{1}{\sqrt{4\pi kt}} \displaystyle\int_{-\infty}^{\infty} e^{-(x-y)^2/4kt} f(y)\, dy = \dfrac{1}{\sqrt{\pi}} \int_{-\infty}^{\infty} e^{-s^2} f(x + 2s\sqrt{kt})\, ds \quad (t > 0).$

(b) Use the result of part (a) and the Dominated Convergence Theorem of Appendix A.3, to prove that if f is continuous at x_0, then $u(x,t) \to f(x_0)$ as $(x,t) \to (x_0, 0^+)$.

4. Suppose that $f(x)$ is a continuous, even function defined on $(-\infty,\infty)$. If $f(x)$ is absolutely integrable on $(-\infty,\infty)$ and if $f(x) \to 0$ as $x \to \pm\infty$, show that the unique continuous solution $u(x,t)$ of the following problem is also an *even* function of x:

$$\text{D.E.} \quad u_t = ku_{xx} \qquad -\infty < x < \infty,\ t > 0$$

$$\text{I.C.} \quad u(x,0) = f(x)$$

$$\text{"B.C."} \quad \lim_{x \to \pm\infty} \max_{0 \leq t \leq T} |u(x,t)| = 0, \quad \text{for all } T > 0.$$

5. Let $F(x) = (2\pi)^{-\frac{1}{2}} \displaystyle\int_{-\infty}^{x} e^{-\frac{1}{2}s^2}\, ds$ denote the **normal distribution function.** Verify that $F(-\infty) = 0$, $F(\infty) = 1$, and $F(-x) = 1 - F(x)$.

6. Let $u(x,t) = (4\pi kt)^{-\frac{1}{2}} \displaystyle\int_{-\infty}^{\infty} e^{-(x-y)^2/4kt} f(y)\, dy$, $t > 0$, where $f(y) = 1$ if $a \leq y \leq b$ and $f(y) = 0$ otherwise. Express $u(x,t)$ in terms of the normal distribution function defined in Problem 5.

7. (a) Show that $g(x)$ is a rapidly decreasing function if and only if $\hat{g}(\xi)$ is a rapidly decreasing function. **Hint.** Use Theorem 1 of Section 7.2, and show that $(\hat{g})\hat{}\,(x) = g(-x)$.

(b) For $t > 0$, let $u(x,t) = \dfrac{1}{\sqrt{4\pi kt}} \displaystyle\int_{-\infty}^{\infty} e^{(x-y)^2/4kt} f(y)\, dy$, where $f(x)$ is a continuous function which is identically zero outside some finite interval. For each fixed $t > 0$, show that $u(x,t)$ is a rapidly decreasing function of x, even though $f(x)$ might not be C^1.
Hint. Use the Convolution Theorem and Theorem 1 of Section 7.2, and use part (a).

(c) Suppose that in part (b), we assume that $f(x)$ is bounded, continuous and absolutely integrable, but not that it vanishes outside of some finite interval. Give an example which shows that $u(x,t)$ is not necessarily rapidly decreasing in x, for fixed $t > 0$.

8. As in Example 6, let ${}_tH(x) = \frac{1}{\sqrt{4\pi kt}} e^{-x^2/4kt}$, $k, t > 0$. Use the Convolution Theorem and the Inversion Theorem to prove that ${}_sH * {}_tH = {}_{(s+t)}H$, for all $s,t > 0$. Give a physical interpretation of this result. Is it possible to get the same result via the method of Example 6 ?

9. Let $P_y(x) = \frac{y}{\pi(x^2 + y^2)}$ for $y > 0$. Use the method of Problem 8 to prove that $P_y * P_z = P_{y+z}$ for $y, z > 0$.

10. Obtain the formal solution of the inhomogeneous heat problem of Example 3, through a direct application of Duhamel's principle (cf. Section 3.4).

11. Use the Fourier transform method (formally) to solve the following wave problem :

$$\text{D.E. } u_{tt} = a^2 u_{xx} \qquad -\infty < x, t < \infty, \; a \neq 0$$

$$\text{I.C. } u(x,0) = 2\sin(x)/(1+x^2), \; u_t(x,0) = 0 .$$

Hint. It is not necessary to explicitly compute the Fourier Transform of $f(x) \equiv 2\sin(x)/(1+x^2)$, but note that $\cos(a\xi t) = \frac{1}{2}(e^{ia\xi} + e^{-ia\xi})$.

12. (a) Consider the ODE $y''(t) + \omega^2 y(t) = f(t)$ $(\omega \neq 0)$, with initial conditions $y(0) = 0$, $y'(0) = 0$. Use Leibniz's rule (cf. Appendix A.3) to verify that if $f(t)$ is continuous, then the solution of this problem is $y(t) = \frac{1}{\omega} \int_0^t f(s)\sin[\omega(t-s)]\, ds$. Show that Duhamel's principle yields a quick derivation of this solution (cf. the argument leading up to (25) in Section 5.3).

(b) Use part (a) to solve the inhomogeneous wave equation with homogeneous I.C.:

$$\text{D.E. } u_{tt} = a^2 u_{xx} + h(x,t) , \qquad -\infty < x < \infty, \; t > 0$$

$$\text{I.C. } u(x,0) = 0, \; u_t(x,0) = 0 ,$$

by using formal Fourier transform methods. Show that if $h(x,t)$ is C^1, then the "solution" is C^2 for $-\infty < x < \infty$, $t > 0$, and satisfies the D.E.and I.C.. Verify that $\lim_{t \to 0+} u(x,t) = 0$.

13. Let $u(x,y)$ $(y \geq 0)$ be the unique continuous, bounded solution of the problem

$$\text{D.E. } u_{xx} + u_{yy} = 0 \qquad -\infty < x < \infty,\ y > 0$$
$$\text{B.C. } u(x,0) = f(x)\ ,$$

where $f(x)$ is continuous, bounded, and $\int_{-\infty}^{\infty} |f(x)|\, dx < \infty$. Show that $\int_{-\infty}^{\infty} u(x,y)dx = \int_{-\infty}^{\infty} f(x)\, dx$.

14. Solve the Dirichlet problem

$$\text{D.E. } u_{xx} + u_{yy} = 0 \qquad -\infty < x < \infty,\ y > 0$$
$$\text{B.C. } u(x,0^+) = f(x) = \begin{cases} 1 & \text{if } x > 0 \\ -1 & \text{if } x < 0\ . \end{cases}$$

15. For complex $z = x + iy$ and $a = s + it$, let $g(z;a) = \log|z-a| - \log|z-\bar{a}|$. Show that

$$-\frac{\partial}{\partial t} g(z;a)\bigg|_{t=0} = \frac{2y}{y^2 + (x-s)^2}\ . \qquad (*)$$

Remark. The function g is a **Green's function** for the upper half–plane $y > 0$ (cf. Problem 11 of Section 6.4). It may be interpreted as the potential at z caused by oppositely charged particles at the points a and $\bar{a}$ in the xy–plane. Note that $g(z;a) = 0$ when $y = 0$. The result $(*)$ is analogous to the result proved in Problem 14 of Section 6.4. In both cases, the Poisson kernel for the Dirichlet problem is the outward normal derivative (in the a variable) of the Green's function, evaluated on the boundary. This result holds for suitably nice regions, and we show it in Section 9.6 (at least formally) for compact manifolds with boundary .

16. Let $u(x,t) = \frac{1}{\sqrt{4\pi kt}} \int_{-\infty}^{\infty} e^{-(x-y)^2/4kt} f(y)\, dy$, $t > 0$, where $f(y) = \begin{cases} a & \text{if } y < 0 \\ b & \text{if } y > 0 \end{cases}$

and a and b are nonzero constants. Show that $u(x,t)$ can be expressed in the form

$$u(x,t) = \tfrac{1}{2}(a+b) + \frac{(b-a)}{\sqrt{\pi}} \int_0^{x/\sqrt{4kt}} e^{-s^2}\, ds, \quad t > 0 \text{ (cf. Problem 9 of Section 3.1).}$$

17. (a) In Problem 16, choose $a = 10$ and $b = 0$. Show that for any $t > 0$ (no matter how small), we have that $u(x,t) > 0$ for any x (no matter how large). Conclude that the heat equation implies (contrary to relativity theory) that heat diffuses with infinite speed.

(b) Suppose that $f(x)$ is bounded, continuous and nonnegative for all real x, and that $f(x) > 0$ somewhere (possibly only in some small interval). Show that for any positive value of t, $u(x,t) \equiv \frac{1}{\sqrt{4\pi kt}} \int_{-\infty}^{\infty} e^{-(x-y)^2/4kt} f(y)\, dy$ is positive for *all* x.

7.5 Applications to Problems on Finite and Semi–Infinite Intervals

In this section we will solve some problems where x is restricted to a finite or semi–infinite interval (e.g., $0 \le x \le L$ or $0 \le x < \infty$). This is done using the method of images, whereby such problems will be extended appropriately (depending on the B.C.) to related problems where x is unrestricted. In this way, we can use the previous results for unrestricted x which were obtained by Fourier transform methods or by other means (e.g., as with D'Alembert's formula). Since the method of images was already used extensively for wave problems in Chapter 5, we will concentrate on heat problems, and Dirichlet/Neumann problems for Laplace's equation.

We begin by settling the theoretical question (left dangling in Chapter 3) concerning the existence and uniqueness of solutions of the finite interval problem

$$\begin{aligned} &\text{D.E. } u_t = k u_{xx} \qquad 0 \le x \le L,\ t > 0 \\ &\text{B.C. } u(0,t) = 0\,, \quad u(L,t) = 0 \\ &\text{I.C. } u(x,0) = f(x)\,, \end{aligned} \tag{1}$$

where $f(x)$ is continuous and $f(0) = f(L) = 0$.

Remark. Note that here the D.E. is not required to hold when $t = 0$, since we do not want to insist that $f(x)$ is C^2. However, we do require that $u(x,t)$ be continuous for $t \ge 0$ and $0 \le x \le L$. This means that $u(x,t)$ not only must satisfy the D.E. for $t > 0$, but also

$$\lim_{(x,t)\to(x_0,0^+)} u(x,t) = u(x_0,0)\,. \quad \square \tag{2}$$

Theorem 1. Let $\tilde{f}_o(x)$, $-\infty < x < \infty$, be the periodic extension of the odd extension $f_o(x)$, $-L \le x \le L$, of $f(x)$. Then the only solution of problem (1), which meets (2), is

$$u(x,t) = \frac{1}{\sqrt{4\pi kt}} \int_{-\infty}^{\infty} e^{-(x-y)^2/4kt}\, \tilde{f}_o(y)\, dy \ \ (\text{for } t > 0) \ \ \textbf{and} \ \ u(x,0) = f(x) \ \ (\text{for } t = 0). \tag{3}$$

Proof. We already know from Theorem 1 of Section 7.4 that the function u defined by (3) is continuous and satisfies the D.E. for $0 \le x \le L$ and $t > 0$. Since $u(0,t)$ is the integral of an odd function, it is 0. Moreover, since $\tilde{f}_o(y)$ is also odd about $x = L$ (i.e., $\tilde{f}_o(L-y) = -\tilde{f}_o(L+y)$), the other B.C. $u(L,t) = 0$ is also met. Uniqueness was proved in Theorem 1 of Section 3.2. $\square$

The following result implies that, for problem (1), the formal infinite series solution (which was not proven to be generally valid in Chapter 3) is in fact equal to the genuine solution (3) when $t > 0$ (i.e., the formal solution is valid for $t > 0$ and satisfies the condition (2)).

Theorem 2. The solution (3) **of** (1) **is equal (for** $t > 0$**) to the convergent series**

$$u(x,t) = \sum_{n=1}^{\infty} b_n e^{-n^2\pi^2 kt/L^2} \sin(\tfrac{n\pi x}{L}), \quad (t > 0), \tag{4}$$

where

$$b_n = \frac{2}{L}\int_0^L f(x)\sin(\tfrac{n\pi x}{L})\,dx, \quad n = 1, 2, 3, \ldots .$$

Proof. We know from Theorem 1 of Section 7.4 and by inspection that $u(x,t)$ in (3) is a C^∞ odd periodic function (of period 2L) of x, for each $t > 0$. Thus, by Theorem 1 of Section 4.3, we know that $u(x,t)$ of (3) is equal to the Fourier sine series of its restriction to $[0,L]$, for each $t > 0$:

$$u(x,t) = \sum_{n=1}^{\infty} B_n(t)\sin(\tfrac{n\pi x}{L}), \text{ where } B_n(t) = \frac{2}{L}\int_0^L u(x,t)\sin(\tfrac{n\pi x}{L})\,dx .$$

We may differentiate under the integral $B_n(t)$ (by Leibniz's rule, since $u(x,t)$ is C^∞ for $t > 0$) :

$$B_n'(t) = \frac{2}{L}\int_0^L u_t(x,t)\sin(\tfrac{n\pi x}{L})\,dx = \frac{2}{L}\int_0^L k u_{xx}(x,t)\sin(\tfrac{n\pi x}{L})\,dx$$

$$= -k(\tfrac{n\pi}{L})^2 \frac{2}{L}\int_0^L u(x,t)\sin(\tfrac{n\pi x}{L})\,dx = -k(\tfrac{n\pi}{L})^2 B_n(t) ,$$

where we have used Green's formula. Hence, $B_n(t) = c_n e^{-kn^2\pi^2 t/L^2}$ for some constant c_n , and it remains to show that $c_n = b_n$. First,

$$c_n = \lim_{t\to 0+} B_n(t) = \lim_{t\to 0+} \frac{2}{L}\int_0^L u(x,t)\sin(\tfrac{n\pi x}{L})\,dx$$

$$= \lim_{t\to 0+} \frac{2}{L}\frac{1}{\sqrt{4\pi kt}}\int_{-\infty}^{\infty}\left[\int_0^L e^{-(x-y)^2/4kt}\sin(\tfrac{n\pi x}{L})\,\tilde{f}_o(y)\,dx\right]dy \quad \text{(cf. Appendix A.2)}$$

$$= \lim_{t\to 0+} \frac{2}{L}\frac{1}{\sqrt{4\pi kt}}\int_{-\infty}^{\infty} e^{-(0-z)^2/4kt}\left[\int_0^L \sin(\tfrac{n\pi x}{L})\,\tilde{f}_o(x+z)\,dx\right]dz \qquad (y = z-x) .$$

Now we use the fact that for a continuous, bounded function $h(z)$,

$$\lim_{t\to 0+} \frac{1}{\sqrt{4\pi kt}} \int_{-\infty}^{\infty} e^{-(x_0-z)^2/4kt}\, h(z)\, dz = h(x_0)\,, \tag{5}$$

which is implicit in Theorem 1 of Section 7.4 (Why ?). Note that $\frac{2}{L}\int_0^L \sin(\frac{n\pi x}{L})\, \tilde{f}_o(x+z)\, dx$ is a continuous, bounded function of z (Why ?). Letting $h(z)$ in (5) be this function, we obtain

$$c_n = \lim_{t\to 0+} \frac{2}{L} \frac{1}{\sqrt{4\pi kt}} \int_{-\infty}^{\infty} e^{-(0-z)^2/4kt} \left[\int_0^L \sin(\tfrac{n\pi x}{L})\, \tilde{f}_o(x+z)\, dx\right] dz$$

$$= h(0) = \frac{2}{L}\int_0^L \sin(\tfrac{n\pi x}{L})\, \tilde{f}_o(x+0)\, dx = \frac{2}{L}\int_0^L \sin(\tfrac{n\pi x}{L})\, f(x)\, dx = b_n\,. \quad \square$$

Note. Formula (3) may not be valid if $t = 0$, since $f(x)$ may not equal its Fourier series. If $f(x)$ is continuous and piecewise C^1, with $f(0) = f(L) = 0$, then (4) is valid, even for $t = 0$, by Theorem 1 of Section 4.3. □

We can also justify the existence, uniqueness and Fourier series representations of solutions of the heat equation on a finite interval with other types of B.C. and continuous initial data. Rather than doing this for all the standard B.C., we will just consider the case of the circular wire which is modeled by the problem :

$$\begin{array}{lll} \text{D.E.} & u_t = ku_{xx} & -\infty < x < \infty,\ t > 0 \\ \text{"B.C."} & u(x+2L,t) = u(x,t) & \\ \text{I.C.} & u(x,0) = f(x), & \end{array} \tag{6}$$

where $f(x)$ is any continuous periodic function of period $2L$. Of course, periodic functions of period $2L$ correspond to well–defined functions on a circle of circumference $2L$. In the following solution (7), we could replace the complex Fourier series by "ordinary" Fourier series, if desired.

Theorem 3. The unique continuous solution of problem (6) is given by the following equivalent formulas, for $t > 0$,

$$u(x,t) = \begin{cases} \dfrac{1}{\sqrt{4\pi kt}} \displaystyle\int_{-\infty}^{\infty} e^{-(x-y)^2/4kt}\, f(y)\, dy \\ \displaystyle\sum_{n=-\infty}^{\infty} c_n\, e^{-n^2\pi^2 kt/L^2}\, e^{-in\pi x/L}, \end{cases} \quad (t > 0) \tag{7}$$

where

$$c_n = \frac{1}{2L}\int_{-L}^{L} f(x)\, e^{-in\pi x/L}\, dx\,.$$

Proof. We know from Theorem 1 of Section 7.4 that the integral in (7) is C^∞ for $t > 0$, and solves the D.E.. It is also periodic. Indeed, replacing x by $x + 2L$ in the integral and changing variables by the formula $z = y - 2L$, we obtain the same result since $f(y) = f(z+2L) = f(z)$. Thus, the first expression of (7) is equal to its Fourier series. Now, using the same calculation as in the proof of Theorem 2, we can establish that this Fourier series is the same as the infinite sum in (7). Again, a separate formula is needed for $t = 0$, unless $f(x)$ is continuous and piecewise C^1. Uniqueness was proved in Example 1 of Section 3.2. □

Miscellaneous solved heat problems for the semi–infinite rod

Example 1. Solve the problem

$$\begin{aligned} &\text{D.E.} \quad u_t = ku_{xx} \qquad 0 < x < \infty,\ t > 0 \\ &\text{B.C.} \quad u(0^+,t) = 0 \\ &\text{I.C.} \quad u(x,0^+) = f(x)\ , \end{aligned} \tag{8}$$

where $f(x)$ is continuous and absolutely integrable on $[0,\infty)$, $f(0) = 0$. Is the solution unique ?

Solution. Let $f_o(x)$ denote the odd extension of $f(x)$ (i.e., $f_o(x) = -f(-x)$, for $x \leq 0$). Note that $f_o(x)$ is continuous on $(-\infty,\infty)$, since it is assumed that $f(0) = 0$. Moreover, it is easy to see that $f_o(x)$ is absolutely integrable on $(-\infty,\infty)$ and that $f_o(x) \to 0$ as $x \to \pm\infty$. Thus, by Theorem 1 of Section 7.4 we know that the function

$$u(x,t) = \frac{1}{\sqrt{4\pi kt}} \int_{-\infty}^{\infty} e^{-(x-y)^2/4kt} f_o(y)\, dy \ \ (\text{for } t > 0) \quad \text{and} \quad u(x,0) = f_o(x) \ \ (\text{for } t = 0), \tag{9}$$

is a continuous solution of the problem

$$\begin{aligned} &\text{D.E.} \quad u_t = ku_{xx} \qquad -\infty < x < \infty,\ t > 0 \\ &\text{I.C.} \quad u(x,0) = f_o(x)\ . \end{aligned} \tag{10}$$

Since $u(0,t) = 0$ by inspection of (9), $u(x,t)$ will satisfy not only the D.E. and I.C., but also the B.C. of problem (8). It is perhaps more satisfactory to express $u(x,t)$ in terms of $f(y)$ rather than in terms of $f_o(y)$. For this reason, we rewrite the integral (9) in the form

$$u(x,t) = \frac{1}{\sqrt{4\pi kt}} \left\{ \int_{-\infty}^{0} e^{-(x-y)^2/4kt}[-f(-y)]dy + \int_{0}^{\infty} e^{-(x-y)^2/4kt} f(y)dy \right\}.$$

Now, if we use the change of variables $y = -s$ in the first integral, interchange the limits of integration, and replace the dummy variable s by y, then we obtain that

$$u(x,t) = \frac{1}{\sqrt{4\pi kt}} \int_0^{\infty} \left[e^{-(x-y)^2/4kt} - e^{-(x+y)^2/4kt} \right] f(y)\, dy \,. \tag{11}$$

This represents the solution as a continuous superposition of fundamental solutions due to concentrated sources of strength $f(y)$ at y, and of strength $-f(y)$ at $-y$. Although the solution (11) is the one which is most appropriate for applications, it is *not* unique (cf. the Remark following Example 4 of Section 7.4), because no "B.C." at $x = \infty$ was imposed. □

Example 2. Attempt to solve the problem posed in Example 1, by using the even extension $f_e(x)$ of $f(x)$ (i.e., $f_e(x) = f(-x)$ for $x < 0$).

Solution. Once again it is easy to verify that $f_e(x)$ is continuous and absolutely integrable on $(-\infty,\infty)$. Hence, by Theorem 1 of Section 7.4, the function given by

$$u(x,t) = \frac{1}{\sqrt{4\pi kt}} \int_{-\infty}^{\infty} e^{-(x-y)^2/4kt} f_e(y)\, dy \quad (\text{for } t > 0) \quad \text{and} \quad u(x,0) = f_e(x) \quad (\text{for } t = 0) \,,$$

is a continuous solution of the problem

$$\begin{aligned} &\text{D.E.} \quad u_t = ku_{xx} \qquad -\infty < x < \infty, \ t > 0, \\ &\text{I.C.} \quad u(x,0^+) = f_e(x) \,, \end{aligned}$$

and *a fortiori* $u(x,t)$ will satisfy the D.E. and the I.C. of the given problem (8) of Example 1. Now, since $f_e(x)$ is even, we can express $u(x,t)$ in the form

$$u(x,t) = \frac{1}{\sqrt{4\pi kt}} \int_0^{\infty} \left[e^{-(x-y)^2/4kt} + e^{-(x+y)^2/4kt} \right] f(y)\, dy \,,$$

but $u(x,t)$ does not generally meet the B.C. (e.g., $u(0^+,t) > 0$, if $f(x) > 0$ for $x > 0$). □

Example 3. Solve the problem

$$\begin{aligned} &\text{D.E.} \quad u_t = ku_{xx} \qquad 0 < x < \infty, \ t > 0 \\ &\text{B.C.} \quad u_x(0^+,t) = 0 \\ &\text{I.C.} \quad u(x,0^+) = f(x) \,, \end{aligned}$$

where $f(x)$ is continuous and absolutely integrable on $[0,\infty)$.

Solution. For this problem, we take the even extension $f_e(x)$ of $f(x)$. Then, as in Example 1,

$$u(x,t) = \frac{1}{\sqrt{4\pi kt}} \int_0^\infty \left[e^{-(x-y)^2/4kt} + e^{-(x+y)^2/4kt} \right] f(y)\, dy.$$

In Problem 3, the reader is asked to supply the details of the verification of this solution. □

Remark. It appears that the technique of odd or even extension (i.e., the method of images), which was used in Example 1 and Example 3, is limited to problems with homogeneous B.C. . However, this method can also be used to solve the problem of the semi–infinite rod with a time–dependent temperature prescribed at the end $x = 0$ (cf., Example 5 below). First, we need a result (Example 4) concerning the Fourier transform of the second derivative of a suitable odd function with a jump discontinuity at $x = 0$. □

Example 4. Let $g(x)$ be an odd function which is C^2 on $(-\infty,\infty)$, except at $x = 0$. Assume that $g(0^+)$, $g'(0^+)$ and $g''(0^+)$ exist, and suppose that $g(x)$, $g'(x)$ and $g''(x)$ are absolutely integrable and tend to zero as $x \to \infty$. Then show that

$$(g'')\hat{}\,(\xi) = -\frac{2i}{\sqrt{2\pi}}\, \xi\, g(0^+) - \xi^2\, \hat{g}(\xi)\,. \tag{12}$$

In other words, due to the possible jump discontinuity at $x = 0$, there is a correction term which must be added to the usual expression $-\xi^2\, \hat{g}(\xi)$ (cf. Corollary 1 of Section 7.2).

Solution. Since $g''(x)$ is odd, we have

$$(g'')\hat{}\,(\xi) = \int_{-\infty}^{\infty} g''(x)\, e^{-i\xi x}\, d'x = \int_0^\infty g''(x)\, (e^{-i\xi x} - e^{i\xi x})\, d'x\,. \tag{13}$$

Using $2i\sin(\xi x) = (e^{i\xi x} - e^{-i\xi x})$ and Green's formula, we obtain

$$(g'')\hat{}\,(\xi) = -2i \int_0^\infty g''(x) \sin(\xi x)\, d'x$$

$$= \frac{-2i}{\sqrt{2\pi}} \left[g'(x)\sin(\xi x) - g(x)\xi\cos(\xi x) \right] \Big|_{0^+}^{\infty} + 2i\xi^2 \int_0^\infty g(x)\sin(\xi x)\, d'x\,,$$

which reduces to (12), since $g(x)$ is odd, and $g(x)$ and $g'(x)$ tend to zero as $x \to \infty$. □

Example 5. Use the result of Example 4 formally, to find a hypothetical solution of the problem

$$\text{D.E.}\quad u_t = u_{xx} \qquad x > 0,\ t > 0,$$

$$\text{B.C.}\quad u(0^+,t) = h(t)$$

$$\text{I.C.}\quad u(x,0^+) = 0\,,$$

where $h(t)$ is a given continuous function, for $0 \le t < \infty$, with $h(0) = 0$.

Solution. We define $u(0,t) = 0$ for each $t \geq 0$, and for each fixed t, let $v(x,t)$ be the odd extension of $u(x,t)$ in the x–variable. Note that $v(0^+,t) = h(t)$, $v(0,t) = 0$, and $v(0^-,t) = -h(t)$. If u satisfies the D.E., then $v_t = v_{xx}$, except possibly when $x = 0$. Taking the Fourier transform of both sides of the D.E. with respect to x, keeping t fixed, we formally obtain from Example 4 (with $g(x) = v(x,t)$)

$$(\hat{v})_t(\xi,t) = \int_{-\infty}^{\infty} v_t(x,t)\, e^{-i\xi x}\, d'x = \int_{-\infty}^{\infty} v_{xx}(x,t)\, e^{-i\xi x}\, d'x = -\frac{2i}{\sqrt{2\pi}}\, \xi h(t) - \xi^2\, \hat{v}(\xi,t)\ ,$$

or

$$(\hat{v})_t(\xi,t) + \xi^2\, \hat{v}(\xi,t) = \frac{-2i}{\sqrt{2\pi}}\, \xi h(t)\ . \tag{14}$$

The general solution of the first–order linear ODE (14) is

$$\hat{v}(\xi,t) = -\frac{2i}{\sqrt{2\pi}}\, \xi\, e^{-\xi^2 t} \int_0^t e^{\xi^2 s}\, h(s)\, ds + e^{-\xi^2 t}\, F(\xi)\ ,$$

and the initial condition $\hat{u}(\xi,0) = 0$ implies that $F(\xi) \equiv 0$. Hence, we have

$$\hat{v}(\xi,t) = \frac{-2i}{\sqrt{2\pi}}\, \xi \int_0^t e^{-\xi^2(t-s)}\, h(s)\, ds\ . \tag{15}$$

Formally applying the Inversion Theorem and interchanging the order of integration, etc.,

$$v(x,t) = \int_{-\infty}^{\infty} \hat{v}(\xi,t)\, e^{i\xi x}\, d'\xi = \frac{-2}{\sqrt{2\pi}} \int_0^t \left[\int_{-\infty}^{\infty} i\xi\, e^{-\xi^2(t-s)}\, e^{ix\xi}\, d'\xi \right] h(s)\, ds$$

$$= \frac{-2}{\sqrt{2\pi}} \int_0^t \frac{\partial}{\partial x} \left[e^{-\xi^2(t-s)} \right]^{\vee}(x)\, h(s)\, ds = \frac{-2}{\sqrt{2\pi}} \int_0^t \frac{\partial}{\partial x} \left[[2(t-s)]^{-\frac{1}{2}}\, e^{-x^2/4(t-s)} \right] h(s)\, ds\ ,$$

where we have used Example 6 of Section 7.1. Simplifying the last expression and recalling that $v(x,t) = u(x,t)$ for $x > 0$, we obtain the hypothetical solution

$$u(x,t) = \frac{x}{2\sqrt{\pi}} \int_0^t (t-s)^{-\frac{3}{2}}\, e^{-x^2/4(t-s)}\, h(s)\, ds\ , \quad x,\ t > 0. \tag{16}$$

Again, we have merely shown that *if* a solution of the problem exists *and* has all of the properties which are needed in order to justify the above formal manipulations, then the solution is given by (16). Nevertheless, one could check directly that (16) solves the D.E., that $u(x,0^+) = 0$, and (as is done in the next example) that $u(0^+,t) = 0$. □

Example 6. For $u(x,t)$ given by (16), show that, despite appearances, we have $u(0^+,t) = h(t)$ (*not* necessarily 0), for $t > 0$. Use the fact that

$$\int_0^\infty y^{-\frac{1}{2}}\, e^{-ay}\, dy = \sqrt{\pi/a}\ , \tag{17}$$

which may be easily verified by the change of variable $y = \frac{1}{2}x^2$.

Solution. In (16), change the variable of integration from s to $y = x^2/(t-s)$, so that $dy = x^2/(t-s)^2\, ds$. We then obtain (cf. Appendix A.3)

$$u(0^+,t) = \lim_{x\to 0^+} \frac{1}{2\sqrt{\pi}} \int_{x^2/t}^\infty y^{-\frac{1}{2}}\, e^{-y/4}\, h(t - \tfrac{x^2}{y})\, dy = \frac{h(t)}{2\sqrt{\pi}} \int_0^\infty y^{-\frac{1}{2}}\, e^{-y/4}\, dy\ ,$$

which is $h(t)$ by (17) with $a = \frac{1}{4}$. □

Example 7. Assuming the validity of the solutions found in Examples 1 and 5, solve the problem

$$\begin{aligned} &\text{D.E.}\quad u_t = u_{xx} \qquad x > 0,\ t > 0 \\ &\text{B.C.}\quad u(0^+,t) = h(t) \\ &\text{I.C.}\quad u(x,0^+) = f(x)\ , \end{aligned} \tag{18}$$

where $h(t)$ and $f(x)$ are continuous, and $f(x)$ is absolutely integrable.

Solution. Since the D.E., B.C. and I.C are linear, the superposition principle implies that a solution of (18) is simply the sum of the solutions which were found in Examples 1 and 5 :

$$u(x,t) = \frac{1}{\sqrt{4\pi t}} \int_0^\infty \left[e^{-(x-y)^2/4t} - e^{-(x+y)^2/4t}\right] f(y)\, dy + \frac{x}{\sqrt{4\pi}} \int_0^t e^{-x^2/4(t-y)} (t-y)^{-\frac{3}{2}}\, h(y)\, dy. \quad □$$

Example 8. Give a physical interpretation of the problem

$$\begin{aligned} &\text{D.E.}\quad u_t = ku_{xx} \qquad x > 0,\ t > 0 \\ &\text{B.C.}\quad u_x(0^+,t) = au(0^+,t) \\ &\text{I.C.}\quad u(x,0^+) = g(x)\ , \end{aligned} \tag{19}$$

where a is a positive constant and $f(x)$ is a given continuous function. Find a hypothetical solution and discuss the circumstances under which it is valid.

Solution. We have a semi–infinite rod, whose initial temperature at each cross–section x is given by the function $g(x)$. The B.C. means that heat flows through the end of the rod at a rate which is proportional to the temperature $u(0^+,t)$ at the end. This boundary condition often occurs in practice when there is faulty insulation or imperfect thermal contact at the end (cf. the discussion near the end of Section 3.3). To find a hypothetical solution, we again make all of the assumptions about a solution $u(x,t)$ of (19) (including existence) which are necessary in order that we may carry out the following steps. Let $v(x,t) = u_x(x,t) - a\,u(x,t)$. Then the B.C. of problem (19) becomes $v(0^+,t) = 0$, which was considered in Example 1. Assuming that u is C^3 and g is C^1, we have that v is C^2 and v solves the problem

$$\begin{aligned} &\text{D.E.}\quad v_t = kv_{xx} \qquad x > 0,\ t > 0 \\ &\text{B.C.}\quad v(0^+,t) = 0 \\ &\text{I.C.}\quad v(x,0^+) = g'(x) - ag(x) \equiv f(x)\,. \end{aligned} \tag{20}$$

(Note that $v_t = u_{xt} - au_t = k(u_{xxx} - au_{xx}) = kv_{xx}$.) By Example 1, the solution of (20) is

$$v(x,t) = \int_0^\infty B(x,y,t)\, f(y)\, dy\,, \tag{21}$$

where

$$B(x,y,t) \equiv \frac{1}{\sqrt{4\pi kt}}\left[e^{-(x-y)^2/4kt} - e^{-(x+y)^2/4kt} \right].$$

For v given by (21), we can solve the first–order "ODE" $u_x - au = v$ for u, and obtain

$$u(x,t) = -e^{ax}\int_x^\infty e^{-ar}\, v(r,t)\, dr\,, \tag{22}$$

which is not the most general solution, but it is the only one which might not grow rapidly with increasing x. From (21) and (22), we have (formally interchanging the order of integration first)

$$\begin{aligned} u(x,t) &= -e^{ax}\int_0^\infty\int_x^\infty e^{-ar}\, B(r,y,t)\, dr\, f(y)\, dy \\ &= -e^{ax}\int_0^\infty\int_x^\infty e^{-ar}\, B(r,y,t)\, dr\, e^{ay}\frac{d}{dy}\left[e^{-ay}\, g(y)\right] dy \\ &= e^{ax}\int_0^\infty e^{-ay}\frac{d}{dy}\left[e^{ay}\int_x^\infty e^{-ar} B(r,y,t)\, dr \right] g(y)\, dy\,, \end{aligned}$$

where we have integrated by parts with respect to y, noting that $B(r,0,t) = B(r,\infty,t) = 0$ and assuming that $e^{-ay}g(y)$ is bounded for $y \geq 0$. Carrying out the differentiation, we have

$$u(x,t) = \int_0^\infty \left[\int_x^\infty e^{a(x-r)}[B_y(r,y,t) + aB(r,y,t)]\, dr \right] g(y)\, dy\ , \tag{23}$$

which is a hypothetical solution of (19). For $t > 0$, one can check that the inner integral (say $I(x,y,t)$), and all of its partial derivatives with respect to x and t, exist and are rapidly decreasing functions of y. In what follows, let $g(x)$ be continuous for $x > 0$, and assume that for all sufficiently large x, $|g(x)| < x^p$ for some positive constant p. Then Leibniz's rule can be repeatedly applied in order to deduce that for $t > 0$, (23) defines a C^∞ function $u(x,t)$. Moreover, by using Green's formula, one can verify directly that $I_t(x,y,t) = kI_{xx}(x,y,t)$, whence (19) satisfies the D.E. $u_t = ku_{xx}$. One can check explicitly that for $t > 0$, $u_x(0,t) = au(0,t)$, by applying Leibniz's rule and using the fact that $B(0,y,t) = B_y(0,y,t) = 0$. Finally, one could, with considerable effort, show that for $x > 0$, $u(x,0^+) = g(x)$. (To begin, one can show that for any *positive* x, $\lim_{t\downarrow 0} \int_0^\infty I(x,y,t)\, dy = 1$.) Thus, regardless of what assumptions were made during the derivation of the hypothetical solution, (23) gives us a solution of problem (19), when $g(x)$ is assumed to be continuous and grows no faster than some polynomial as $x \to \infty$. □

Fourier sine and cosine transforms

When solving problems where x is restricted to the semi–infinite interval $[0,\infty)$, as an alternative to the method of images, one can utilize the Fourier sine and cosine transforms to be defined below. The Fourier sine and cosine transforms bear the same relation to the ordinary Fourier transform as the Fourier sine and cosine series bear to the ordinary Fourier series.

Definition. Let $f(x)$ be a function defined for $0 \le x < \infty$. Then the **Fourier sine transform** of $f(x)$ is

$$\hat{f}_s(\xi) \;=\; \sqrt{2/\pi} \int_0^\infty f(x) \sin(\xi x)\, dx\ ,$$

and the **Fourier cosine transform** of $f(x)$ is

$$\hat{f}_c(\xi) \;=\; \sqrt{2/\pi} \int_0^\infty f(x) \cos(\xi x)\, dx\ ,$$

provided these integrals exist.

When $f(x)$ is real–valued and $f_e(x)$ and $f_o(x)$ denote the even and odd extensions of $f(x)$,

$$\hat{f}_c(\xi) = (f_e)\hat{\ }(\xi) \quad \text{and} \quad -i\hat{f}_s(\xi) = (f_o)\hat{\ }(\xi).$$

Thus anything that can be done using the sine and cosine transforms, can also be done with the usual Fourier transform by considering extensions. Just as the theory of Fourier sine and cosine series reduces to the theory of ordinary Fourier series, the theory of Fourier sine and cosine transforms reduces to that of the ordinary Fourier transform. In particular, one can use the Inversion Theorem for the usual Fourier transform to establish the related inversion formulas for these one–sided transforms (for $x > 0$)

$$\frac{f(x+) + f(x-)}{2} = \sqrt{2/\pi} \int_0^\infty \hat{f}_s(\xi) \sin(\xi x) \, d\xi \tag{24}$$

and

$$\frac{f(x+) + f(x-)}{2} = \sqrt{2/\pi} \int_0^\infty \hat{f}_c(\xi) \cos(\xi x) \, d\xi , \tag{25}$$

where it is assumed that $f(x)$ is piecewise C^1 and absolutely integrable on $[0,\infty)$.

In the next example, we illustrate the use of formula (25) in a problem for Laplace's equation on the quadrant, $0 < x < \infty$ and $0 < y < \infty$. However, one can also get the solution immediately by evenly extending the problem to the upper half–plane, and using Poisson's integral formula (cf. (40) of Section 7.4 and replace $f(x)$ by $f_e(x)$), which we found via the usual Fourier transform methods.

Example 9. Find a hypothetical solution of the problem

$$\text{D.E.} \quad u_{xx} + u_{yy} = 0 \qquad 0 < x < \infty, \ 0 < y < \infty$$

$$\text{B.C.} \quad \begin{cases} u_x(0^+,y) = 0 \\ u(x,0^+) = f(x) , \end{cases}$$

where it is assumed that $u(x,y)$ is bounded and $u(\infty,y) = u_x(\infty,y) = 0, \ 0 < y < \infty$.

Solution. Proceeding formally, we take the Fourier cosine transform of the D.E. with respect to x , keeping y fixed. The choice of the Fourier cosine transform here is suggested by the B.C.. (The reader may consider the reason why the Fourier sine transform is unsuitable.) We obtain

$$(\hat{u}_{yy})_c(\xi,y) = \sqrt{2/\pi} \int_0^\infty u_{yy}(x,y) \cos(\xi x) \, dx = -\sqrt{2/\pi} \int_0^\infty u_{xx}(x,y) \cos(\xi x) \, dx ,$$

where $(\hat{u}_{yy})_c(\xi,y)$ denotes the Fourier cosine transform of u_{yy} . Integrating by parts twice,

$$-\sqrt{2/\pi} \int_0^\infty u_{xx}(x,y)\cos(\xi x) \, dx = \Big[-\sqrt{2/\pi}\, u_x(x,y)\cos(\xi x)\Big]_{x=0}^\infty - \xi\sqrt{2/\pi} \int_0^\infty u_x(x,y)\sin(\xi x) \, dx$$

$$= \Big[-\xi\sqrt{2/\pi}\, u(x,y)\sin(\xi x)\Big]_{x=0}^\infty + \xi^2\sqrt{2/\pi} \int_0^\infty u(x,y)\cos(\xi x) \, dx = \xi^2 \, \hat{u}_c(\xi,y),$$

where we have used the B.C. and the assumption that $u(\infty,y) = u_x(\infty,y) = 0$. Thus,

$$(\hat{u}_{yy})_c(\xi,y) - \xi^2 \hat{u}_c(\xi,y) = 0 ,$$

and the general solution of this ODE in y is $\hat{u}_c(\xi,y) = c_1(\xi)e^{\xi y} + c_2(\xi)e^{-\xi y}$, where $c_1(\xi)$ and $c_2(\xi)$ are arbitrary functions of ξ. Since the given problem requires that $u(x,y)$ be bounded, we assume that $c_1(\xi) = 0$ for $\xi > 0$ and $c_2(\xi) = 0$ for $\xi < 0$. Thus, $\hat{u}_c(\xi,y) = c(\xi)e^{-|\xi|y}$, where $c(\xi) = c_1(\xi) + c_2(\xi)$. In particular, $\hat{u}_c(\xi,0) = c(\xi)$. It follows from the second B.C. that

$$c(\xi) = \hat{u}_c(\xi,0) = \sqrt{2/\pi}\int_0^\infty u(x,0)\cos(\xi x)\,dx = \sqrt{2/\pi}\int_0^\infty f(x)\cos(\xi x)\,dx = \hat{f}_c(\xi)$$

Thus, $\hat{u}_c(\xi,y) = \hat{f}_c(\xi)\, e^{-|\xi|y}$, and by the Inversion Theorem for one–sided transforms (cf. (25))

$$u(x,y) = \sqrt{2/\pi}\int_0^\infty \hat{f}_c(\xi)\, e^{-|\xi|y} \cos(\xi x)\, d\xi = \frac{2}{\pi}\int_0^\infty \left[\int_0^\infty f(s)\cos(\xi s)\, ds\right] e^{-\xi y}\cos(\xi x)\, d\xi.$$

If the order of integration is switched, then the integral with respect to ξ can be explicitly computed and the resulting solution will be the same as the solution obtained by using the even extension of $f(x)$ for $f(s)$ in the Poisson integral formula (40) in Section 7.4. □

Summary 7.5

1. **Heat problems for finite or semi–infinite rods** : Solutions of heat problems for finite or semi–infinite rods can be obtained, by the method of images, from the solution of Theorem 1 in Section 7.4 for the infinite rod. This is done via the method of images, in the same way that solutions of finite string problems were obtained from D'Alembert's formula for the infinite string in Chapter 5. In the case where each end is insulated or maintained at 0, the extension of the initial temperature to the infinite rod must be chosen so that it is odd about each end which is held at 0 and even about each end which is insulated. In Example 5, the heat problem with a prescribed time–dependent temperature at the end is solved by the method of images, which shows that the method of images is not limited to problems with homogeneous B.C. .

2. **The case of the finite rod with ends held at 0 (Theorem 1)** : Consider the finite rod problem

$$\begin{aligned} &\text{D.E.} \quad u_t = ku_{xx} \qquad 0 \le x \le L,\ t > 0, \\ &\text{B.C.} \quad u(0,t) = 0, \quad u(L,t) = 0 \\ &\text{I.C.} \quad u(x,0) = f(x)\ , \end{aligned} \tag{S1}$$

where $f(x)$ is continuous and $f(0) = f(L) = 0$, and we require that $u(x,t)$ be continuous for $t \ge 0$ and $0 \le x \le L$. This means that $u(x,t)$, not only must satisfy the D.E. for $t > 0$, but also

$$\lim_{(x,t)\to(x_0,0^+)} u(x,t) = u(x_0,0)\ . \tag{S2}$$

We have :

Theorem 1. Let $\tilde{f}_o(x)$, $-\infty < x < \infty$, **be the periodic extension of the odd extension** $f_o(x)$, $-L \le x \le L$, **of** $f(x)$. **Then the only solution of problem (S1), which meets (S2), is**

$$u(x,t) = \frac{1}{\sqrt{4\pi kt}} \int_{-\infty}^{\infty} e^{-(x-y)^2/4kt}\, \tilde{f}_o(y)\, dy \quad (\text{for } t > 0) \quad \text{and} \quad u(x,0) = f(x) \quad (\text{for } t = 0). \tag{S3}$$

3. **The validity of formal solutions** : Integral formulas such as (3) are helpful in establishing the validity of the formal infinite Fourier series solutions heat problems on finite rods which were discussed in Section 4.3. The integral formulas show that the solutions exist and are actually C^∞ for $t > 0$, even if the continuous initial temperature distribution has corners. The proof of the validity of the formal solution consists of showing that for each $t > 0$, the Fourier series of the integral formula solution is the same as the formal solution (cf. Theorem 2 and its proof). We do *not* verify the formal solution by means of the superposition principle, which we have seen can fail for infinite series. When $t = 0$, the formal solution may not converge to $f(x)$, but the continuity condition (S2) holds, since (S2) holds for the equivalent integral solution.

4. Fourier sine and cosine transforms : Let $f(x)$ be function defined for $0 \le x < \infty$. Then the **Fourier sine transform** and the **Fourier cosine transform** of $f(x)$ are given by

$$\hat{f}_s(\xi) = \sqrt{2/\pi}\int_0^\infty f(x)\sin(\xi x)\,dx \quad \text{and} \quad \hat{f}_c(\xi) = \sqrt{2/\pi}\int_0^\infty f(x)\cos(\xi x)\,dx\,,$$

respectively, provided these integrals exist. When $f(x)$ is real–valued,

$$\hat{f}_c(\xi) = (f_e)\hat{\ }(\xi) \quad \text{and} \quad -i\hat{f}_s(\xi) = (f_o)\hat{\ }(\xi).$$

One can use the Inversion Theorem for the usual Fourier transform to establish the related inversion formulas for these one–sided transforms (for $x > 0$)

$$\frac{f(x+) + f(x-)}{2} = \sqrt{2/\pi}\int_0^\infty \hat{f}_s(\xi)\sin(\xi x)\,d\xi \quad \text{and} \quad \frac{f(x+) + f(x-)}{2} = \sqrt{2/\pi}\int_0^\infty \hat{f}_c(\xi)\cos(\xi x)\,d\xi\,,$$

where it is assumed that $f(x)$ is piecewise C^1 and absolutely integrable on $[0,\infty)$.
Example 9 illustrates the use of the Fourier cosine transform in solving a Dirichlet/Neumann problem in a quadrant, although a direct application of the method of images is simpler.

Exercises 7.5

1. Find a solution of the problem

$$\begin{aligned} &\text{D.E.} \quad u_t = ku_{xx}\,, \qquad x > \infty,\ t > 0, \\ &\text{B.C.} \quad u_x(0^+,t) = 0 \\ &\text{I.C.} \quad u(x,0^+) = f(x)\,, \end{aligned}$$

where $f(x)$ is continuous and absolutely integrable on $[0,\infty)$ and $f(x) \to 0$ as $x \to \infty$.

2. By means of *formal* calculations find a hypothetical solution of the problem

$$\begin{aligned} &\text{D.E.} \quad u_t = u_{xx}\,, \qquad x > 0,\ t > 0 \\ &\text{B.C.} \quad u_x(0^+,t) = h(t) \\ &\text{I.C.} \quad u(x,0^+) = 0\,, \end{aligned}$$

where $h(t)$ is a suitable continuous function.
Hint. Seek a solution $u(x,t)$ which is an even function of x (cf. Examples 4, 5 and 6).

3. Find a hypothetical solution of the problem

$$\text{D.E.}\quad u_t = u_{xx} \qquad 0 < x < \infty,\ t > 0$$
$$\text{B.C.}\quad u_x(0^+,t) = h(t)$$
$$\text{I.C.}\quad u(x,0^+) = f(x)\ ,$$

where $h(t)$ and $f(x)$ are suitable continuous functions.

4. (a) Using Fourier transform methods (formally), find a hypothetical solution of the inhomogeneous heat equation with homogeneous I.C. and B.C. :

$$\text{D.E.}\quad u_t = ku_{xx} + h(x,t) \qquad 0 < x < \infty,\ t > 0$$
$$\text{B.C.}\quad u(0^+,t) = 0$$
$$\text{I.C.}\quad u(x,0^+) = 0\ .$$

(b) Show that the hypothetical solution found in part (a) can also be obtained directly by a formal use of Duhamel's principle. **Hint.** See Section 3.4 for Duhamel's Principle.

5. Use both the method of images and Fourier sine transforms to find a solution of

$$\text{D.E.}\quad u_{xx} + u_{yy} = 0 \qquad 0 < x,\ y < \infty$$
$$\text{B.C.}\quad \begin{cases} u(0^+,y) = 0 \\ u(x,0^+) = f(x)\ , \end{cases}$$

where it is assumed that $u(x,y)$ is bounded and $u(\infty,y) = u_x(\infty,y) = 0,\ \ 0 \le y < \infty$. Show that the two solutions are actually the same.

6. (a) As in Example 5, find an integral formula for a hypothetical solution of the *wave* problem

$$\text{D.E.}\quad u_{tt} = u_{xx} \qquad 0 < x < \infty,\ t > 0$$
$$\text{B.C.}\quad u(0^+,t) = h(t)$$
$$\text{I.C.}\quad u(x,0^+) = u_t(x,0^+) = 0\ ,$$

where it assumed that $u(\infty,t) = u_x(\infty,t) = 0,\ \ t > 0$.

You may use the fact that the solution of the ODE $y''(t) + \xi^2 y(t) = g(t)$ with $y(0) = y'(0) = 0$, is given by $y(t) = \int_0^t g(s)\frac{\sin[\xi(t-s)]}{\xi}\,ds$ (cf. Problem 12 of Section 7.4).

(b) By assuming a solution of the form $u(x,t) = F(x+t) + G(x-t)$, arrive at a simpler hypothetical solution: $u(x,t) = h(t-x)$ if $x < t$, and $u(x,t) = 0$ if $x \geq t$. Under what circumstances will this solution be a C^2 solution of the D.E. for $t,x > 0$?

Hint. Note that from the I.C., we know that $F(x) + G(x) = 0$ and $F'(x) - G'(x) = 0$ for $x > 0$. Thus, $u(x,t) = 0$ for $0 < t < x$. Now, use the B.C. $h(t) = u(0,t) = G(-t)$ to determine $G(s)$ for $s < 0$, thereby obtaining $u(x,t)$ for $0 < x < t$.

(c) Evaluate the answer in (a) explicitly, when $h(t) \equiv 1$ for $t > 0$.

7. In Problem 6, replace the B.C. by $u_x(0^+,t) = h(t)$, $t > 0$. Find the hypothetical solution $u(x,t)$ of the resulting problem by the method of your choice. Under what assumptions on $h(t)$ is $u(x,t)$ a C^2 solution of the D.E. for $x > 0$ and $t > 0$?

8. (a) Solve the following Dirichlet problem for the quarter–plane, where $g(y)$ and $f(x)$ are given bounded, continuous functions with $g(0) = f(0) = 0$. Is there only one solution ?

$$\text{D.E.} \quad u_{xx} + u_{yy} = 0\,, \qquad 0 < x,\, y < \infty$$

$$\text{B.C.} \quad \begin{cases} u(0^+,y) = g(y) \\ \\ u(x,0^+) = f(x)\,, \end{cases}$$

(b) Solve the problem in part (a) by converting the problem to a problem in the upper half–plane by means of the conformal mapping $F(z) = z^2$ of the quarter–plane to the half–plane.

9. (a) Use Fourier transforms to find a hypothetical solution $u(x,t)$ of the following problem, where b is a constant :

$$\text{D.E.} \quad u_t + bu_x = h(x,t) \qquad -\infty < x,\, t < \infty$$

$$\text{I.C.} \quad u(x,0) = f(x)\,.$$

(b) Under what assumptions on $h(x,t)$ and $f(x)$ is the solution found in part (a) a C^1 solution of the D.E. ?

Supplement (The Inversion Theorem)

Let $f(x)$, $-\infty < x < \infty$, be a piecewise C^1 function, in the sense that $f'(x)$ exists and is continuous in every finite interval except at a finite number of points in the interval where $f'(x)$ suffers a finite jump. For such a function, the following limits exist at each point x_0 :

$$f(x_0^+) = \lim_{x\downarrow x_0} f(x) \quad \text{and} \quad f(x_0^-) = \lim_{x\uparrow x_0} f(x) ,$$

$$f'(x_0^+) = \lim_{x\downarrow x_0} \frac{f(x) - f(x_0^+)}{x - x_0} \quad \text{and} \quad f'(x_0^-) = \lim_{x\uparrow x_0} \frac{f(x) - f(x_0^-)}{x - x_0} . \tag{1}$$

The Inversion Theorem. Let $f(x)$, $-\infty < x < \infty$, **be piecewise** C^1 **and absolutely integrable** (i.e., $\int_{-\infty}^{\infty} |f(x)|\, dx < \infty$). **Then for any** x_0,

$$\tfrac{1}{2}[f(x_0^+) + f(x_0^-)] = \int_{-\infty}^{\infty} \hat{f}(\xi)\, e^{i\xi x_0}\, d'\xi \equiv \lim_{R\to\infty} \int_{-R}^{R} \hat{f}(\xi) e^{i\xi x_0}\, d'\xi . \tag{2}$$

Proof. The right side of (2) is

$$\lim_{R\to\infty} \int_{-R}^{R} \left[\int_{-\infty}^{\infty} f(x)\, e^{-i\xi x}\, d'x\right] e^{+i\xi x_0}\, d'\xi = \lim_{R\to\infty} \int_{-\infty}^{\infty} f(x) \left[\int_{-R}^{R} e^{-i\xi(x-x_0)}\, d'\xi\right] d'x , \tag{3}$$

where the interchange in the order of integration is permitted, since $\int_{-\infty}^{\infty} |f(x)|\, dx < \infty$ and $\int_{-R}^{R} |e^{-i\xi(x-x_0)}|\, d\xi = 2R < \infty$ (cf. Appendix A.2). Now for $x \neq x_0$,

$$\int_{-R}^{R} e^{-i\xi(x-x_0)}\, d\xi = \frac{e^{iR(x-x_0)} - e^{-iR(x-x_0)}}{i(x-x_0)} = \frac{2\sin R(x-x_0)}{x-x_0} .$$

and the left–hand side is $2R$ when $x = x_0$. With $d'\xi d'x \equiv \frac{1}{2\pi} d\xi dx$, (3) becomes

$$\lim_{R\to\infty} \left[\frac{1}{\pi} \int_{x_0}^{\infty} f(x) \frac{\sin R(x-x_0)}{x-x_0}\, dx + \frac{1}{\pi} \int_{-\infty}^{x_0} f(x) \frac{\sin R(x-x_0)}{x-x_0}\, dx \right].$$

We then only need to show that the limit of the first term is $\frac{1}{2}f(x_0^+)$ and the limit of the second term is $\frac{1}{2}f(x_0^-)$. As the proofs are similar, we just prove

$$\lim_{R\to\infty} \frac{1}{\pi}\int_{x_0}^{\infty} f(x)\,\frac{\sin R(x-x_0)}{x-x_0}\,dx = \tfrac{1}{2}\,f(x_0^+). \tag{4}$$

By the change of variable $(z = x - x_0)$ the integral in (4) becomes

$$\frac{1}{\pi}\int_0^{\infty} f(x_0+z)\,\frac{\sin(Rz)}{z}\,dz. \tag{5}$$

We will prove in the Lemma below that $\frac{1}{\pi}\int_0^{\infty}\frac{\sin(Rz)}{z}\,dz = \frac{1}{2}$. Multiplying this result by $f(x_0^+)$,

$$\frac{1}{\pi}\int_0^{\infty} f(x_0{}^+)\,\frac{\sin(Rz)}{z}\,dz = \frac{1}{2}\,f(x_0^+)\,.$$

Thus,

$$\left|\frac{1}{\pi}\int_0^{\infty} f(x_0+z)\,\frac{\sin(Rz)}{z}\,dz - \frac{1}{2}\,f(x_0^+)\right| = \left|\frac{1}{\pi}\int_0^{\infty}[f(x_0+z) - f(x_0^+)]\,\frac{\sin(Rz)}{z}\,dz\,\right| \equiv I. \tag{6}$$

We need to show that I in (6) approaches 0 as $R\to\infty$. For any constant $A > 1$, we have

$$I \le \left|\frac{1}{\pi}\int_0^{A}\frac{f(x_0+z) - f(x_0^+)}{z}\sin(Rz)\,dz\,\right| \tag{7}$$

$$+\frac{1}{\pi}\int_A^{\infty}|f(x_0+z)|\,\left|\frac{\sin(Rz)}{z}\right|\,dz \tag{8}$$

$$+\frac{1}{\pi}\,|f(x_0^+)|\,\left|\int_A^{\infty}\frac{\sin(Rz)}{z}\,dz\right|. \tag{9}$$

Let J and K denote the expressions (8) and (9), respectively. Since $z \ge A > 1$, we have $|\sin(Rz)/z| \le 1$ in (8), whence

$$J < \frac{1}{\pi}\int_A^{\infty}|f(x_0+z)|\,dz = \frac{1}{\pi}\int_{A+x_0}^{\infty}|f(y)|\,dy \to 0\,,$$

as $A\to\infty$, because we assumed that $\int_{-\infty}^{\infty}|f(x)|\,dx < \infty$. Thus, we can make J arbitrarily small by choosing A large enough. Now, by the change of variables $x = Rz$,

$$K = \frac{1}{\pi}\,|f(x_0^+)|\,\left|\int_{AR}^{\infty}\frac{\sin(x)}{x}\,dx\,\right|,$$

and hence, for any value of $A > 1$, we can make K as small as desired by choosing R large enough (Why ? ; cf. the Lemma below). Finally, since the limits in (1) exist, the Riemann—Lebesgue Lemma of Section 4.2 implies that (7) tends to 0 as $R \to \infty$, and thus the proof of (4) is complete (cf. also Problem 10 of Section 7.1). Since the result for x_0^- corresponding to (4) is proved in the same way, the desired formula (2) holds. □

Lemma.
$$\int_0^\infty \frac{\sin(x)}{x}\,dx \equiv \lim_{R\to\infty}\int_0^R \frac{\sin(x)}{x}\,dx = \frac{\pi}{2}. \tag{10}$$

Proof. Note that for any $x > 0$, $\int_0^\infty e^{-\lambda x}\,d\lambda = \frac{1}{x}$. Thus, we have for $0 < \epsilon < R$

$$\int_\epsilon^R \frac{\sin(x)}{x}\,dx = \int_\epsilon^R\int_0^\infty e^{-\lambda x}\sin(x)\,d\lambda dx = \int_0^\infty\int_\epsilon^R e^{-\lambda x}\sin(x)\,dxd\lambda \tag{11}$$

where the interchange of order of integration is allowed, since $|e^{-\lambda x}\sin(x)| \le e^{-\lambda x}$ and $e^{-\lambda x}$ is integrable in the strip $\lambda \ge 0$, $\epsilon \le x \le R$. Now,

$$\int_\epsilon^R e^{-\lambda x}\sin(x)\,dx = \mathrm{Im}\Big[\int_\epsilon^R e^{(i-\lambda)x}\,dx\Big] = \mathrm{Im}\left[\frac{e^{(i-\lambda)x}}{i-\lambda}\Big|_\epsilon^R\right]$$
$$= \mathrm{Im}\left[\frac{e^{(i-\lambda)R} - e^{(i-\lambda)\epsilon}}{i-\lambda}\right] = \mathrm{Im}\left[\frac{(i+\lambda)[e^{(i-\lambda)\epsilon} - e^{(i-\lambda)R}]}{1+\lambda^2}\right]. \tag{12}$$

Note that

$$\lim_{R\to\infty}\left|\int_0^\infty\int_R^\infty e^{-\lambda x}\sin(x)\,dxd\lambda\right| = \lim_{R\to\infty}\int_0^\infty\left|\mathrm{Im}\left[\frac{(i+\lambda)e^{(i-\lambda)R}}{1+\lambda^2}\right]\right|d\lambda \le \lim_{R\to\infty}\int_0^\infty e^{-\lambda R}d\lambda = \lim_{R\to\infty}\frac{1}{R}.$$

Thus, since this limit is zero, we have

$$\int_\epsilon^\infty \frac{\sin(x)}{x}\,dx = \lim_{R\to\infty}\Big[\int_0^\infty\int_\epsilon^\infty e^{-\lambda x}\sin(x)\,dx - \int_R^\infty e^{-\lambda x}\sin(x)\,dx\,d\lambda\Big]$$
$$= \int_0^\infty\int_\epsilon^\infty e^{-\lambda x}\sin(x)\,dx\,d\lambda = \int_0^\infty \mathrm{Im}\left[\frac{(i+\lambda)e^{(i-\lambda)\epsilon}}{1+\lambda^2}\right]d\lambda$$
$$= \int_0^\infty \frac{e^{-\lambda\epsilon}\cos(\epsilon)}{1+\lambda^2}\,d\lambda + \int_0^\infty \frac{e^{-\lambda\epsilon}\lambda\sin(\epsilon)}{1+\lambda^2}\,d\lambda. \tag{13}$$

Now,

$$\int_0^\infty \frac{\sin(x)}{x}\,dx = \lim_{\epsilon\to 0}\int_\epsilon^\infty \frac{\sin(x)}{x}\,dx = \lim_{\epsilon\to 0}\int_0^\infty \frac{e^{-\lambda\epsilon}\cos(\epsilon)}{1+\lambda^2}\,d\lambda + \lim_{\epsilon\to 0}\int_0^\infty \frac{e^{-\lambda\epsilon}\lambda\sin(\epsilon)}{1+\lambda^2}\,d\lambda. \tag{14}$$

Since $\lambda e^{-\lambda\epsilon} \le \frac{1}{\epsilon e}$, we have, for any constant $a > 0$,

$$\lim_{\epsilon\to 0}\left|\int_0^\infty \frac{e^{-\lambda\epsilon}\lambda\sin(\epsilon)}{1+\lambda^2}\,d\lambda\right| \le \lim_{\epsilon\to 0}\left[\left|\int_0^a \frac{e^{-\lambda\epsilon}\lambda\sin(\epsilon)}{1+\lambda^2}\,d\lambda\right|\right] + \lim_{\epsilon\to 0}\left|\int_a^\infty \frac{e^{-\lambda\epsilon}\lambda\sin(\epsilon)}{1+\lambda^2}\,d\lambda\right| .$$

$$\le \lim_{\epsilon\to 0}\sin(\epsilon)\int_0^a \frac{\lambda}{1+\lambda^2}\,d\lambda + \lim_{\epsilon\to 0}\frac{\sin(\epsilon)}{\epsilon}\,e^{-1}\,[\tfrac{\pi}{2}-\arctan(a)] = e^{-1}\,[\tfrac{\pi}{2}-\arctan(a)].$$

Since a can be arbitrarily large, the final limit in (14) is zero. It remains to show that

$$\lim_{\epsilon\to 0}\int_0^\infty \frac{e^{-\lambda\epsilon}\cos(\epsilon)}{1+\lambda^2}\,d\lambda = \frac{\pi}{2} \quad \text{or} \quad \lim_{\epsilon\to 0}\int_0^\infty \frac{1 - e^{-\lambda\epsilon}\cos(\epsilon)}{1+\lambda^2}\,d\lambda = 0 .$$

To this end,

$$\lim_{\epsilon\to 0}\int_0^\infty \frac{1 - e^{-\lambda\epsilon}\cos(\epsilon)}{1+\lambda^2}\,d\lambda = \lim_{\epsilon\to 0}\int_0^a \frac{1 - e^{-\lambda\epsilon}\cos(\epsilon)}{1+\lambda^2}\,d\lambda + \int_a^\infty \frac{1 - e^{-\lambda\epsilon}\cos(\epsilon)}{1+\lambda^2}\,d\lambda$$

$$\le \lim_{\epsilon\to 0}\,[1-\cos(\epsilon)]\int_0^a \frac{1}{1+\lambda^2}\,d\lambda + \int_a^\infty \frac{1}{1+\lambda^2}\,d\lambda = \frac{\pi}{2} - \arctan(a) ,$$

Since a can be arbitrarily large, the limit is 0. □

Remark. A formal calculation yields the Lemma quickly, if one is willing to accept the validity of switching the order of integration in the following computation :

$$\int_0^\infty \frac{\sin(x)}{x}\,dx = \int_0^\infty\int_0^\infty e^{-\lambda x}\sin(x)\,d\lambda dx = \int_0^\infty\int_0^\infty e^{-\lambda x}\sin(x)\,dx d\lambda = \int_0^\infty \frac{1}{1+\lambda^2}\,d\lambda = \frac{\pi}{2} .$$

Observe that $\lambda^2\int_0^\infty e^{-\lambda x}\sin(x)\,dx + \int_0^\infty e^{-\lambda x}\sin(x)\,dx = 1 \quad (\lambda > 0)$ is immediate from Green's formula (cf. Section 4.1), which gives us the third equality in the above computation. Thus, the complexity of the proof of the Lemma is due only to the need to justify the switching of the order of integration. Even for those who are familiar with the theory of Lebesgue integration (or alternatively, with the material in Appendices 2 and 3), the justification is not immediate, because $\exp(-\lambda x)$ is not integrable on the quarter plane $0 \le \lambda, x < \infty$. Indeed, using polar coordinates,

$$\int_0^\infty\int_0^\infty e^{-\lambda x}\,dx\,d\lambda = \int_0^{\frac{1}{2}\pi}\int_0^\infty e^{-\frac{1}{2}r^2\sin(2\theta)}\,r\,dr\,d\theta$$

$$= \int_0^{\frac{1}{2}\pi}\sin(2\theta)^{-1}\,d\theta = \lim_{\epsilon\to 0}\tfrac{1}{2}\log(\tan\theta)\Big|_{\epsilon}^{\frac{1}{2}\pi-\epsilon} = \infty .$$

However, $e^{-\lambda x}\sin(x)$ *is* absolutely integrable on the strip $0 \le x \le R$, $0 \le \lambda < \infty$, since

$$\int_0^R \int_0^\infty e^{-\lambda x} |\sin(x)| \, d\lambda \, dx \le \int_0^R \int_0^\infty x \, e^{-\lambda x} \, d\lambda \, dx = R .$$

Then, Fubini's theorem (cf. Appendix 2) validates the interchange of the order of integration in the following computation

$$\lim_{R\to\infty} \int_0^R \frac{\sin(x)}{x} dx = \lim_{R\to\infty} \int_0^R \int_0^\infty e^{-\lambda x} \sin(x) \, d\lambda \, dx = \lim_{R\to\infty} \int_0^\infty \int_0^R e^{-\lambda x} \sin(x) \, dx \, d\lambda$$

$$= \lim_{R\to\infty} \int_0^\infty \frac{1}{\lambda^2 + 1} - \frac{e^{-R\lambda}[\lambda \sin(R) + \cos(R)]}{\lambda^2 + 1} d\lambda = \int_0^\infty \frac{1}{\lambda^2 + 1} d\lambda = \frac{\pi}{2},$$

where the limit may be taken under the integral by the Dominated Convergence Theorem (cf. Appendix 3). Although longer, the proof given before this remark has the merit of being independent of the material in Appendices 2 and 3. Finally, we mention that the Lemma can also be proven using complex contour integration (should the reader know about this), which ultimately reduces the integral to half the length of a unit semicircle (i.e., $\pi/2$). □

CHAPTER 8

NUMERICAL SOLUTIONS OF PDEs – AN INTRODUCTION

The majority of practical problems involving PDEs cannot be solved by analytic methods. Therefore it is expedient to study numerical methods, which lead to approximate solutions of PDEs. Today, with the advances in computer technology, increased speed and storage capacity, the implementation of computer programs for approximating the solutions of PDEs are becoming more and more accessible. In many introductory texts, the student is confronted with several approximation or iteration schemes, and then the student is simply asked to write computer programs to implement these schemes. While in some of the exercises we will also encourage the student to write computer programs, the primary goal of this chapter is to provide an understandable and precise introduction to the numerical solution of PDEs, as opposed to an encyclopedic cookbook of algorithms without explanations.

In Section 8.1 we introduce Landau's big O–notation and we use Taylor's theorem to approximate the derivatives by finite differences. In Section 8.2 we will study the explicit difference method, using the heat equation as a model. In general, the method of finite differences converts PDEs into a system of algebraic equations. Since we do not assume familiarity with the rudiments of numerical analysis, we discuss only briefly in Section 8.4 the classical iterative methods of Jacobi, Gauss–Seidel etc., for solving such systems. In Section 8.2 we examine the nature of the discretization error (the difference between the exact solution and the numerical solution), and we will prove a convergence theorem for the explicit difference method (cf. Theorem 1, Section 8.2). In Section 8.3 we introduce some basic difference equations to study the propagation of round–off errors and to handle the problem of determining the mesh size which yields the greatest accuracy, in the presence of round–off errors. In Section 8.4 we provide an overview of some other numerical methods for PDEs, and we cite some references, where the reader can find applications, proofs, and further results.

8.1 The O Symbol and Approximations of Derivatives

In the sequel we will find it convenient to adopt the O–notation (read big "oh"), introduced by the German mathematician E. Landau (1877 – 1938) in the 1920's.

Definition. Let $f(x)$ and $g(x)$ be two functions defined on some open interval I. Suppose that x_0 is in I. Then

$$f(x) = O(g(x)), \qquad (x \to x_0),$$

means that there is a positive constant K, such that $|f(x)| \leq K|g(x)|$ for all values of x in some neighborhood of the point x_0.

We will often write equations such as

$$f(x) = 1 + x + O(x^2), \quad (x \to 0). \tag{1}$$

This means that

$$f(x) - (1 + x) = O(x^2), \quad (x \to 0).$$

That is, (1) means that there is a positive constant K such that $|f(x) - (1 + x)| \leq Kx^2$, for $|x| < \epsilon$, for some $\epsilon > 0$. Intuitively speaking, (1) says that the function $f(x)$ is close to $(1 + x)$ in a sufficiently small neighborhood of the origin. Moreover, $f(x) - 1 - x$ tends to zero at least as rapidly as Kx^2 does. The following example illustrates the use of this notation.

Example 1. Verify that $\sin(x) = x + O(x^3), \quad (x \to 0)$.

Solution. Consider the Taylor series expansion of $\sin(x)$ about $x = 0$:

$$\sin(x) = x - \frac{x^3}{3!} + \frac{x^5}{5!} - \frac{x^7}{7!} + \ldots = x + x^3\left[-\frac{1}{3!} + \frac{x^2}{5!} - \ldots\right] = x + x^3 g(x) . \tag{2}$$

The series $g(x)$ in (2) is alternating, with decreasing terms for $|x| \leq 1$. Thus, $|g(x)| \leq \frac{1}{6}$, and

$$|\sin(x) - x| \leq \frac{1}{6}|x|^3 ,$$

for all x in $[-1,1]$. Hence, we have shown that $\sin(x) - x = O(x^3)$, $(x \to 0)$, or $\sin(x) = x + O(x^3)$, $(x \to 0)$. □

The above example not only shows that $\sin(x) \to 0$ as $x \to 0$, but also that the magnitude of the error $|\sin(x) - x|$ in the approximation $\sin(x) \approx x$ behaves like $|x|^3$ (or more precisely $\frac{1}{6}|x|^3$), for x sufficiently close to zero (cf. Figure 1 and Table 1).

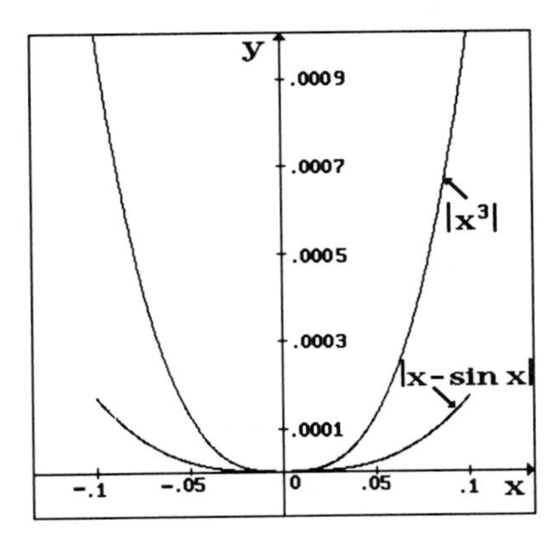

Figure 1

x	$\lvert\sin(x) - x\rvert$	$\lvert x\rvert^3/6$
0.0000	0	0
0.0015	$5.647\ldots\times10^{-10}$	$5.625\ldots\times10^{-10}$
0.0030	$4.501\ldots\times10^{-9}$	$4.500\ldots\times10^{-9}$
0.0045	$1.518\ldots\times10^{-8}$	$1.518\ldots\times10^{-8}$

Table 1

Approximation of derivatives by finite differences

The difference methods used to approximate the solutions to initial/boundary–value problems vary with the type of PDEs and with the types of side conditions prescribed. Nevertheless, the central feature of this method is based on the approximation of the derivatives by "finite differences", which we describe below. The main mathematical tool employed in forming difference approximations to derivatives is **Taylor's theorem** (named after the English mathematician Brook Taylor, 1685–1731). If $f(x)$ is a C^n function on a nontrivial, closed interval I, and x and $x + \Delta x$ are in I, then Taylor's theorem asserts that

$$f(x+\Delta x) = f(x) + f'(x)\frac{\Delta x}{1!} + f''(x)\frac{(\Delta x)^2}{2!} + \dots + f^{(n-1)}(x)\frac{(\Delta x)^{n-1}}{(n-1)!} + R_n(\Delta x), \qquad (3)$$

where

$$R_n(\Delta x) = f^{(n)}(c)\frac{(\Delta x)^n}{n!} = O((\Delta x)^n), \qquad (\Delta x \to 0) \qquad (4)$$

and where c is a number between x and $x + \Delta x$. The last equality holds, since $|f^{(n)}(y)|$ is bounded as y ranges over I. The term $R_n(\Delta x)$, given by (4), is known as **Lagrange's form of the remainder**. Thus, if $n = 2$, then from equation (3) we obtain

$$f'(x) = \frac{f(x+\Delta x) - f(x)}{\Delta x} - \frac{R_2(\Delta x)}{\Delta x} \quad (\text{for } \Delta x \neq 0), \qquad (5)$$

In particular, since $f''(x)$ is bounded in a neighborhood of x, then by (4) we have

$$f'(x) = \frac{f(x+\Delta x) - f(x)}{\Delta x} + O(\Delta x), \qquad (\Delta x \to 0). \qquad (6)$$

In the difference calculus, the expression

$$\Delta f = f(x+\Delta x) - f(x) \qquad (\Delta x > 0) \qquad (7)$$

is called the **forward difference of** $f(x)$ **with step size** Δx. In numerical analysis, $\Delta f/\Delta x$ is termed as a **first–order approximation** to $f'(x)$ as $\Delta x \to 0$. In addition, the term $O(\Delta x)$ in equation (6) is called the **truncation error** in the approximation of $f'(x)$.

There are several alternatives to the difference approximation of $f'(x)$. For example, with the above notation, we have for a sufficiently differentiable function $f(x)$, the following formulae:

$$f'(x) = \frac{f(x) - f(x-\Delta x)}{\Delta x} + O(\Delta x), \qquad (\Delta x \to 0), \qquad (8)$$

and

$$f'(x) = \frac{f(x+\Delta x) - f(x-\Delta x)}{2\Delta x} + O((\Delta x)^2) \qquad (\Delta x \to 0). \qquad (9)$$

In particular, (9) holds if f is a C^3 function (cf. Problem 6). In formulae (8) and (9)

$$\nabla f(x) = f(x) - f(x-\Delta x), \qquad \delta f(x) = f(x+\Delta x) - f(x-\Delta x), \qquad (\Delta x > 0)$$

are respectively the **backward difference** and the **central difference** of $f(x)$ at x with step size Δx. The central difference approximation of $f'(x)$ at x with step size Δx (cf. (9)) is

$$A(x) = \frac{f(x+\Delta x) - f(x-\Delta x)}{2\Delta x}. \tag{10}$$

In geometric terms, $A(x)$ approximates the slope of the tangent line at T by the slope of the chord $\overline{PQ}$ (cf. Figure 2).

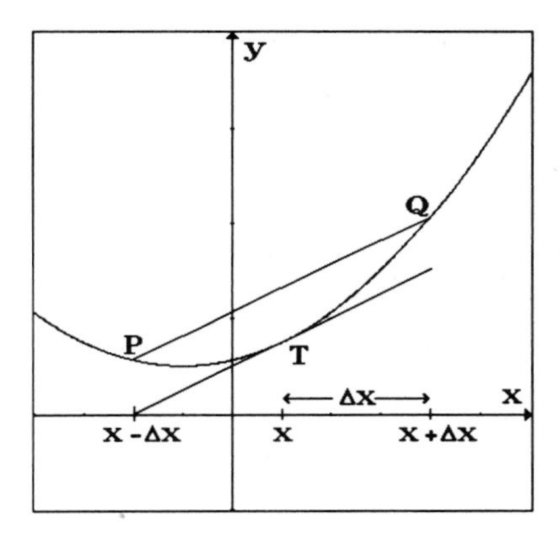

Figure 2

Example 2. Verify that

$$f'(x) = \frac{f(x+\Delta x) - f(x-\Delta x)}{2\Delta x} + O(\Delta x), \qquad (\Delta x \to 0),$$

where $f(x)$ is a C^2 function in some open interval containing the point x.

Solution. By Taylor's theorem we have, with $\Delta x > 0$ sufficiently small, that

$$f(x+\Delta x) = f(x) + f'(x)\Delta x + R_2(\Delta x) = f(x) + f'(x)\Delta x + O((\Delta x)^2) \tag{11}$$

and similarly

$$f(x-\Delta x) = f(x) - f'(x)\Delta x + O((\Delta x)^2). \tag{12}$$

Subtracting (12) from (11) yields (cf. Remark below)

$$f(x+\Delta x) - f(x-\Delta x) = 2f'(x)\Delta x + O((\Delta x)^2), \qquad (\Delta x \to 0),$$

or

$$f'(x) = \frac{f(x+\Delta x) - f(x-\Delta x)}{2\Delta x} + O(\Delta x), \qquad (\Delta x \to 0). \qquad \square$$

Remark. In the solution of Example 2, we have used the following general fact.

$$\text{If} \quad h(\Delta x) = O(g(\Delta x)) \quad \text{and} \quad k(\Delta x) = O(g(\Delta x)), \quad (\Delta x \to 0), \tag{13}$$

where $g(\Delta x)$ is a continuous function of Δx in $I \equiv [-\epsilon,\epsilon]$ $(\epsilon > 0)$, then

$$h(\Delta x) + k(\Delta x) = O(g(\Delta x)), \qquad (\Delta x \to 0). \tag{14}$$

Indeed, by (13) and the definition of the O–notation, there are constants H and K such that

$$|h(\Delta x)| \le H|g(\Delta x)| \quad \text{and} \quad |k(\Delta x)| \le K|g(\Delta x)| \tag{15}$$

hold in $[-\delta,\delta]$ $(0 < \delta \le \epsilon)$. By (15) and the triangle inequality, (14) holds, that is,

$$|h(\Delta x) + k(\Delta x)| \le |h(\Delta x)| + |k(\Delta x)| \le (H + K)|g(\Delta x)|, \quad \text{for all } \Delta x \text{ in } [-\delta,\delta]. \ \square$$

We next turn to the difference approximations of partial derivatives of a sufficiently differentiable function $u(x,y)$. Suppose that, as $\Delta x \to 0$,

$$u(x+\Delta x,y) = u(x,y) + \Delta x \frac{\partial u}{\partial x}(x,y) + \frac{(\Delta x)^2}{2!}\frac{\partial^2 u}{\partial x^2}(x,y) + \frac{(\Delta x)^3}{3!}\frac{\partial^3 u}{\partial x^3}(x,y) + O((\Delta x)^4). \tag{16}$$

Then, as before, we obtain for $\Delta x > 0$ the following finite difference approximations :

the **forward difference approximation**

$$\frac{\partial u}{\partial x}(x,y) = \frac{u(x+\Delta x,y) - u(x,y)}{\Delta x} + O(\Delta x), \qquad (\Delta x \to 0), \tag{17}$$

the **backward difference approximation**

$$\frac{\partial u}{\partial x}(x,y) = \frac{u(x,y) - u(x-\Delta x,y)}{\Delta x} + O(\Delta x), \qquad (\Delta x \to 0), \tag{18}$$

and the **central difference approximation**

$$\frac{\partial u}{\partial x}(x,y) = \frac{u(x+\Delta x,y) - u(x-\Delta x,y)}{2\Delta x} + O((\Delta x)^2), \qquad (\Delta x \to 0). \tag{19}$$

Example 3. Suppose that $u(x,y)$ is C^4 in a neighborhood of the point (x,y). Obtain a second–order difference approximation for $u_{xx}(x,y)$.

Solution. Replacing Δx by $-\Delta x$ in (16) yields (as $\Delta x \to 0$)

$$u(x-\Delta x,y) = u(x,y) - \Delta x \frac{\partial u}{\partial x}(x,y) + \frac{(\Delta x)^2}{2!}\frac{\partial^2 u}{\partial x^2}(x,y) - \frac{(\Delta x)^3}{3!}\frac{\partial^3 u}{\partial x^3}(x,y) + O((\Delta x)^4). \tag{20}$$

If we add (16) to (20) and rearrange the terms, we obtain the desired difference approximation,

$$\frac{\partial^2 u}{\partial x^2}(x,y) = \frac{u(x+\Delta x,y) - 2u(x,y) + u(x-\Delta x,y)}{(\Delta x)^2} + O((\Delta x)^2), \text{ as } \Delta x \to 0. \quad \square \tag{21}$$

Grids and approximations for partial derivatives

We will often use the following notation. Consider the rectangle $W = \{(x,t): 0 \le x \le L, 0 \le t \le T\}$. Let M and N be two positive integers and define $\Delta x = \frac{L}{M}$ and $\Delta t = \frac{T}{N}$. We next subdivide the rectangle W into congruent rectangles, whose sides are parallel to the coordinate axes and have lengths Δx and Δt (cf. Figure 3).

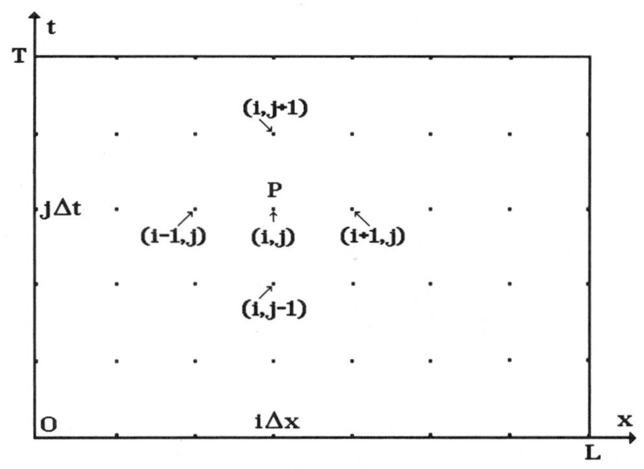

Figure 3

The representative grid point P has coordinates $(i\Delta x, j\Delta t)$, where $0 \le i \le M$ and $0 \le j \le N$. For the sake of simplicity we will also refer to the grid point P as the grid point (i,j). Let $u = u(x,y)$ be a function of two independent variables defined on the rectangle W. The value of u at $(i\Delta x, j\Delta t)$ will be denoted by

$$u_{i,j} = u(i\Delta x, j\Delta t). \tag{22}$$

If we replace in the formulas (17), (18) and (19) x by $i\Delta x$ and y by $j\Delta t$, then we obtain, for a sufficiently differentiable function u, the following difference approximations :

$$\left[\frac{\partial u}{\partial x}\right]_{(i,j)} = \frac{u_{i+1,j} - u_{i,j}}{\Delta x} + O(\Delta x), \qquad (\Delta x \to 0), \tag{23}$$

$$\left[\frac{\partial u}{\partial x}\right]_{(i,j)} = \frac{u_{i,j} - u_{i-1,j}}{\Delta x} + O(\Delta x), \qquad (\Delta x \to 0), \tag{24}$$

$$\left[\frac{\partial u}{\partial x}\right]_{(i,j)} = \frac{u_{i+1,j} - u_{i-1,j}}{2\Delta x} + O((\Delta x)^2), \qquad (\Delta x \to 0), \tag{25}$$

where $\left[\frac{\partial u}{\partial x}\right]_{(i,j)} \equiv u_x(i\Delta x, j\Delta t)$ is the value of u_x at the grid point (i,j).

Example 4. If $u(x,y)$ is C^k, for k large enough, obtain the following difference approximations :

$$\left[\frac{\partial u}{\partial t}\right]_{(i,j)} = \frac{u_{i,j+1} - u_{i,j}}{\Delta t} + O(\Delta t), \qquad (\Delta t \to 0), \qquad (26)$$

$$\left[\frac{\partial^2 u}{\partial x^2}\right]_{(i,j)} = \frac{u_{i+1,j} - 2u_{i,j} + u_{i-1,j}}{(\Delta x)^2} + O((\Delta x)^2), \qquad (\Delta x \to 0), \qquad (27)$$

$$\left[\frac{\partial^2 u}{\partial t^2}\right]_{(i,j)} = \frac{u_{i,j+1} - 2u_{i,j} + u_{i,j-1}}{(\Delta t)^2} + O((\Delta t)^2), \qquad (\Delta t \to 0), \qquad (28)$$

and (as $\Delta x, \Delta t \to 0$)

$$\left[\frac{\partial^2 u}{\partial x \partial t}\right]_{(i,j)} = \frac{u_{i+1,j+1} - u_{i-1,j+1} - u_{i+1,j-1} + u_{i-1,j-1}}{4\Delta x \Delta t} + O[(|\Delta x| + |\Delta t|)^2], \qquad (29)$$

Solution. Observe that formula (26) is similar to (23) and that using the notation in (22), formulae (27) and (28) follow from (21). In Problem 4, formula (29) is derived. □

Remark. One can also establish the above difference approximations with the aid of Taylor's theorem for functions of two variables. For the reader's convenience, we state this theorem. For a proof, we refer the reader to any standard calculus text. □

Theorem 1 (Taylor's theorem for functions of two variables). Let $u(x,y)$ be a C^n function in some disk D in the xy–plane. Let (x,y) and $(x+\Delta x, y+\Delta y)$ be interior points of D. Then

$$u(x+\Delta x, y+\Delta y) = P_{n-1}(\Delta x, \Delta y) + R_n(\Delta x, \Delta y),$$

where

$$P_{n-1}(\Delta x, \Delta y) = u(x,y) + [\Delta x\, u_x(x,y) + \Delta y\, u_y(x,y)]$$
$$+ \frac{1}{2!}[(\Delta x)^2 u_{xx}(x,y) + 2\Delta x \Delta y\, u_{xy}(x,y) + (\Delta y)^2 u_{yy}(x,y)] + \ldots$$
$$+ \sum_{k=0}^{n-1} \frac{1}{k!(n-1-k)!} (\Delta x)^{n-1-k} (\Delta y)^k \frac{\partial^{n-1} u}{\partial x^{n-1-k} \partial y^k}(x,y)$$

and (for some ξ in $(0,1)$) as $\Delta x, \Delta y \to 0$,

$$R_n(\Delta x, \Delta y) = \sum_{k=0}^{n} \frac{1}{k!(n-k)!} (\Delta x)^{n-k} (\Delta y)^k \frac{\partial^n u}{\partial x^{n-k} \partial y^k}(x+\xi\Delta x, y+\xi\Delta y) = O[(|\Delta x| + |\Delta y|)^n] .$$

Summary 8.1

1. O–notation : Let f(x), g(x) and h(x) be functions defined on some open interval I containing a point x_0. We say that $f(x) = h(x) + O(g(x))$ as $x \to x_0$, if there is a constant K such that $|f(x) - h(x)| \le K|g(x)|$ for all x in some open subinterval of I containing x_0.

2. Taylor's theorem : Let f(x) be a C^n function defined on some nontrivial closed interval I, and suppose that x and $x + \Delta x$ are in I. Then

$$f(x+\Delta x) = f(x) + f'(x)\frac{\Delta x}{1!} + f''(x)\frac{(\Delta x)^2}{2!} + \dots + f^{(n-1)}(x)\frac{(\Delta x)^{n-1}}{(n-1)!} + R_n(\Delta x),$$

where

$$R_n(\Delta x) = f^{(n)}(c)\frac{(\Delta x)^n}{n!} = O((\Delta x)^n).$$

The last equality holds, because $|f^{(n)}(y)|$ is bounded as y ranges over the closed interval I. There is also a multidimensional version of Taylor's theorem (e.g., cf. Theorem 1).

3. Differences : The following expressions are respectively known as the forward, backward and central differences of f(x) with step size $\Delta x > 0$:

$$\Delta f(x) = f(x+\Delta x) - f(x), \quad \nabla f(x) = f(x-\Delta x) - f(x), \quad \delta f(x) = f(x+\Delta x) - f(x-\Delta x)$$

For sufficiently differentiable f(x), we have, as $\Delta x \to 0$,

$$f'(x) = \frac{\Delta f(x)}{\Delta x} + O(\Delta x), \quad f'(x) = \frac{\nabla f(x)}{\Delta x} + O(\Delta x), \quad f'(x) = \frac{\delta f(x)}{2\Delta x} + O((\Delta x)^2).$$

The first two of these equations hold if f(x) is C^2 and the third holds if f(x) is C^3.

4. Grids and approximations for partial derivatives : Let u(x,y) be defined for $0 \le x \le L$ and $0 \le t \le T$. For positive integers M and N, let $\Delta x = L/M$ and $\Delta t = T/N$, and define $u_{i,j} = u(i\Delta x, j\Delta y)$, for integers $0 \le i \le M$ and $0 \le j \le N$ (i.e., $u_{i,j}$ is the value of u at the grid point $(i\Delta x, j\Delta t)$). If u(x,y) is C^k for large enough k, we have the following, as $\Delta x \to 0$ and/or $\Delta t \to 0$:

$$\left[\frac{\partial u}{\partial x}\right]_{(i,j)} = \frac{u_{i+1,j} - u_{i,j}}{\Delta x} + O(\Delta x), \quad \left[\frac{\partial u}{\partial x}\right]_{(i,j)} = \frac{u_{i+1,j} - u_{i-1,j}}{2\Delta x} + O((\Delta x)^2),$$

$$\left[\frac{\partial u}{\partial t}\right]_{(i,j)} = \frac{u_{i,j+1} - u_{i,j}}{\Delta t} + O(\Delta t), \quad \left[\frac{\partial u}{\partial t}\right]_{(i,j)} = \frac{u_{i,j+1} - u_{i,j-1}}{2\Delta t} + O((\Delta t)^2),$$

$$\left[\frac{\partial^2 u}{\partial x^2}\right]_{(i,j)} = \frac{u_{i+1,j} - 2u_{i,j} + u_{i-1,j}}{(\Delta x)^2} + O((\Delta x)^2), \quad \left[\frac{\partial^2 u}{\partial t^2}\right]_{(i,j)} = \frac{u_{i,j+1} - 2u_{i,j} + u_{i,j-1}}{(\Delta t)^2} + O((\Delta t)^2),$$

$$\left[\frac{\partial^2 u}{\partial x \partial t}\right]_{(i,j)} = \frac{u_{i+1,j+1} - u_{i-1,j+1} - u_{i+1,j-1} + u_{i-1,j-1}}{4\Delta x \Delta t} + O[(|\Delta x| + |\Delta t|)^2].$$

Exercises 8.1

1. Verify each of the following statements.

(a) $\sin(x) = O(1), \quad (x \to \infty)$,

(b) $\cos(x) = 1 + O(x^2), \quad (x \to 0)$,

(c) $e^x = 1 + x + x^2/2! + O(x^3), \quad (x \to 0)$.

2. Let I be an open interval containing the point x. Verify each of the following statements.

(a) $f'(x) = \dfrac{[f(x) - f(x-\Delta x)]}{\Delta x} + O(\Delta x), \quad (\Delta x \to 0)$, where f is a C^2 function on I.

(b) $f''(x) = \dfrac{[f(x+\Delta x) - 2f(x) + f(x-\Delta x)]}{(\Delta x)^2} + O((\Delta x)^2), \quad (\Delta x \to 0)$, where f is C^4 on I.

3. If $u = u(x,y)$ is C^4 in a neighborhood of a point (x,y), show that

(a) $u_{yy}(x,y) = \dfrac{u(x,y+\Delta y) - 2u(x,y) + u(x,y-\Delta y)}{(\Delta y)^2} + O((\Delta y)^2)$, as $\Delta y \to 0$.

(b) As $\Delta x, \Delta y \to 0$, $\quad 4\Delta x\Delta y \cdot u_{xy}(x,y)$

$$= u(x+\Delta x,y+\Delta y) - u(x-\Delta x,y+\Delta y) - u(x+\Delta x,y-\Delta y) + u(x-\Delta x,y-\Delta y) + O((|\Delta x|+|\Delta y|)^4).$$

4. Let $u = u(x,t)$ be a C^5 function. Show that, as $\Delta x, \Delta t \to 0$,

$$\left[\frac{\partial^2 u}{\partial x \partial t}\right]_{(i,j)} = \frac{u_{i+1,j+1} - u_{i-1,j+1} - u_{i+1,j-1} + u_{i-1,j-1}}{4\Delta x \Delta t} + O[(|\Delta x|+|\Delta t|)^2].$$

Hint. Use Taylor's theorem for functions of two variables (Theorem 1), to obtain the result analogous to 3(b), but write the remainder explicitly in the form $A(\Delta x)^4 + B(\Delta x)^3(\Delta t) + \ldots + E(\Delta t)^4$. Then apply the mean value (or Taylor's) theorem to the C^1 functions A and E to obtain $A = O(\Delta t)$ and $E = O(\Delta x)$, thereby justifying the division by $\Delta x \Delta t$.

5. Let $f(x)$ and $g(x)$ be two continuous functions, defined on an open interval I containing the point $x = 0$.

(a) Show that if $f(x) = O(x)$ and $g(x) = O(x^2)$ as $x \to 0$, then $f(x) + g(x) = O(x)$, as $x \to 0$.

(b) Find some functions f and g as in (a), for which $f(x) + g(x) \neq O(x^2)$ as $x \to 0$.

6. (Higher–order approximations). Let f(x) be a C^3 function. Use Taylor's theorem to show that

(a) $f'(x) = \dfrac{f(x+\Delta x) - f(x-\Delta x)}{2\Delta x} + O((\Delta x)^2)$, $(\Delta x \to 0)$. Compare this result with Example 2.

(b) $f'(x) = \dfrac{-f(x+2\Delta x) + 4f(x+\Delta x) - 3f(x)}{2\Delta x} + O((\Delta x)^2)$, $(\Delta x \to 0)$.

7. Let f(x) be a C^5 function. Show that

$$f'(x) = \frac{f(x-2\Delta x) - 8f(x-\Delta x) + 8f(x+\Delta x) - f(x+2\Delta x)}{12\Delta x} + O((\Delta x)^4), \ (\Delta x \to 0).$$

8. (The **Richardson extrapolation procedure** for higher–order approximations). Let f(x) be C^6.

(a) Show that $\dfrac{f(x+\Delta x) - 2f(x) + f(x-\Delta x)}{(\Delta x)^2} = f''(x) + \dfrac{(\Delta x)^2}{12} f^{(4)}(x) + O((\Delta x)^4)$, $(\Delta x \to 0)$.

(b) Now replace in part (a), Δx by $s\Delta x$, where s is any real number, $s \neq \pm 1, 0$. Show that

$$\frac{f(x+s\Delta x) - 2f(x) + f(x-s\Delta x)}{s^2(\Delta x)^2} = f''(x) + \frac{s^2(\Delta x)^2}{12} f^{(4)}(x) + O((\Delta x)^4), \quad (\Delta x \to 0).$$

(c) Use (a) and (b) to eliminate the term involving $(\Delta x)^2$, and deduce that (as $\Delta x \to 0$)

$$f''(x) = \frac{1}{(\Delta x)^2 s^2 (s^2-1)} \{-f(x+s\Delta x) + s^4 f(x+\Delta x) - (2s^4-2)f(x) + s^4 f(x-\Delta x) - f(x-s\Delta x)\} + O((\Delta x)^4).$$

9. (**Forward difference operator and derivatives**). Three basic operators of the difference calculus are defined as follows. The **forward difference operator** Δ_h is defined as

$$\Delta_h f(x) = f(x+h) - f(x).$$

The **shift operator**, E_h, is defined by $E_h f(x) = f(x+h)$ and the **identity**, I, is given by $If(x) = f(x)$. The higher–order differences are defined by

$$\Delta_h^n f(x) = \Delta_h(\Delta_h^{n-1} f(x)), \ n = 1, 2, \ldots, \quad \text{where } \Delta^0 f(x) = If(x).$$

(a) Verify that $\Delta_h^2 = E_h^2 - 2E_h + I = (E_h - I)^2$ and

$$\Delta_h^n f(x) = \sum_{k=0}^{n} \begin{bmatrix} n \\ k \end{bmatrix} (-1)^k f(x+(n-k)h), \text{ where } \begin{bmatrix} n \\ k \end{bmatrix} = \frac{n!}{k!(n-k)!}.$$

(b) Verify that $f(x+nh) = \sum_{k=0}^{n} \begin{bmatrix} n \\ k \end{bmatrix} \Delta_h^k f(x)$.

(c) If $p(x)$ is a polynomial of degree n, show that $\Delta_h p(x)$ is a polynomial of degree less than n, and so conclude that $\Delta_h^{n+1} p(x) = 0$ (i.e., the forward difference operator behaves like the differentation operator $\frac{d}{dx}$, in that it also reduces that degree of a polynomial).

10. (**Noncentral difference quotients**). We know that (check) if $f(x)$ is defined in a neighborhood of the origin, and if $f(x)$ is differentiable at the origin, then $\lim_{\Delta x \to 0} \frac{f(\Delta x) - f(-\Delta x)}{2\Delta x}$ exists and equals $f'(0)$. Show that the converse is not true in general. **Hint.** Consider $|x|$.

Remark. It can be proved (cf. [P.P.B. Eggermont]) that if $f(x)$ is defined and continuous in a neighborhood of the origin, and for some constant $a \neq \pm 1$, $\lim_{\Delta x \to 0} \frac{f(\Delta x) - f(a\Delta x)}{(1-a)\Delta x}$ exists, then $f'(0)$ exists and is equal to this limit. Problem 10 shows that the result may fail when $a = -1$.

11. Suppose that $f(x)$ is C^n in a neighborhood of the origin. Prove that

$$f^{(n)}(0) = \lim_{\Delta x \to 0} \left[\frac{1}{(\Delta x)^n} \sum_{k=0}^{n} (-1)^k \frac{n!}{(n-k)!k!} f\left(\frac{n-2k}{2} \cdot \Delta x\right) \right] \qquad (*)$$

Remark. For other related formulas see [J.B. Wilker] and [G. Birkhoff and Gian–Rota, p.203].

8.2 The Explicit Difference Method and the Heat Equation

Consider the following heat problem for the temperature $u = u(x,t)$ in a rod of length L :

$$
\begin{array}{lll}
\text{D.E.} & u_t = u_{xx}, & 0 \le x \le L,\ 0 \le t \le T, \\
\text{B.C.} & u(0,t) = A(t),\ u(L,t) = B(t), & \\
\text{I.C.} & u(x,0) = f(x)\ . &
\end{array}
\tag{1}
$$

Our goal in this section is to describe one of the simplest difference techniques, known as the **explicit difference method** for solving an initial/boundary–value problem of the form (1). To this end, we first introduce a finite grid on the rectangle

$$W = \{(x,t):\ 0 \le x \le L,\ 0 \le t \le T\}.$$

Let M and N be two positive integers and set

$$\Delta x = L/M \quad \text{and} \quad \Delta t = T/N.$$

If $u = u(x,t)$ is C^4 on W, then by Example 4 of Section 8.1,

$$\left[\frac{\partial^2 u}{\partial x^2}\right](i,j) = \frac{u_{i+1,j} - 2u_{i,j} + u_{i-1,j}}{(\Delta x)^2} + O((\Delta x)^2),\ (\Delta x \to 0), \tag{2}$$

and

$$\left[\frac{\partial u}{\partial t}\right](i,j) = \frac{u_{i,j+1} - u_{i,j}}{\Delta t} + O(\Delta t),\quad (\Delta t \to 0), \tag{3}$$

where

$$u_{i,j} = u(i\Delta x, j\Delta t)\ ,\ \left[\frac{\partial^2 u}{\partial x^2}\right](i,j) = u_{xx}(i\Delta x, j\Delta t)\ ,\ \left[\frac{\partial u}{\partial t}\right](i,j) = u_t(i\Delta x, j\Delta t)\ .$$

There are many other choices for the difference approximations of the derivatives u_{xx} and u_t at the grid point (i,j). Equations (2) and (3) involve the forward difference approximations to u_{xx} and u_t at (i,j), which are commonly used in the explicit difference method. The D.E. yields

$$\left[\frac{\partial u}{\partial t}\right](i,j) = \left[\frac{\partial^2 u}{\partial x^2}\right](i,j)\ ,\ \text{ for } i = 1, \dots, M-1 \text{ and } j = 1, \dots, N-1\ , \tag{4}$$

and using (2) and (3) we have

$$\frac{u_{i,j+1} - u_{i,j}}{\Delta t} + O(\Delta t) = \frac{u_{i+1,j} - 2u_{i,j} + u_{i-1,j}}{(\Delta x)^2} + O((\Delta x)^2),\ (\Delta x, \Delta t \to 0)\ . \tag{5}$$

Equation (5) suggests that, upon neglecting the truncation errors $O(\Delta t)$ and $O((\Delta x)^2)$, we consider an approximation $v_{i,j}$ to $u_{i,j}$ which satisfies the system of difference equations

$$\frac{v_{i,j+1} - v_{i,j}}{\Delta t} = \frac{v_{i+1,j} - 2v_{i,j} + v_{i-1,j}}{(\Delta x)^2}, \quad i = 1, ..., M-1, \quad j = 0, ..., N-1.$$

Let
$$\lambda \equiv \frac{\Delta t}{(\Delta x)^2} = \frac{TM^2}{L^2 N}. \tag{6}$$

Then (5) can be written in the form (where $i = 1, ..., M-1$, $j = 0, ..., N-1$)

$$v_{i,j+1} = \lambda v_{i+1,j} + (1-2\lambda)v_{i,j} + \lambda v_{i-1,j}. \tag{7}$$

The **local discretization error at** (i,j) is defined to be the difference

$$w_{i,j} = u_{i,j} - v_{i,j}. \tag{8}$$

We say that the method used to produce the approximation $v_{i,j}$ **converges**, if $\max |w_{i,j}| \to 0$ as $\Delta x \to 0$ (and $\Delta t = \lambda(\Delta x)^2 \to 0$), where the maximum is taken over all of the grid points (i,j) in the rectangle W. We will see below (cf. Theorem 1) that if u is C^4, and $0 < \lambda \leq 1/2$, then the explicit difference method converges.

Taking into account the B.C. and the I.C. in problem (1), we have

$$v_{0,j} = u(0,j\Delta t) = A(j\Delta t), \quad v_{M,j} = u(L,j\Delta t) = B(j\Delta t), \quad j = 0, ..., N$$

and

$$v_{i,0} = u(i\Delta x,0) = f(i\Delta x), \quad i = 0, ..., M. \tag{9}$$

We assume that $A(0) = f(0)$ and $B(0) = f(L)$ (Otherwise, problem (1) has no solution. Why ?). Then, each of the quantities $v_{0,0}$ and $v_{M,0}$ is consistently defined in (9).

In summarizing the foregoing discussion, the explicit difference method for solving the given problem (1), consists in solving the system of difference equations

$$v_{i,j+1} = \lambda v_{i+1,j} + (1-2\lambda)v_{i,j} + \lambda v_{i-1,j}, \tag{10}$$

$i = 1, ..., M-1$, $j = 0, ..., N-1$, subject to the side conditions (9). It is not hard to show that the system of equations (10), subject to (9), has a unique solution. That is, if $v_{i,j}$ and $\bar{v}_{i,j}$, both satisfy (10), subject to (9), then $v_{i,j} - \bar{v}_{i,j} = 0$ ($i = 1, ..., M-1$, $j = 1, ..., N$). Indeed, (10) gives an explicit formula for the unknown temperature $v_{i,j+1}$ at the grid point $(i,j+1)$ in terms of the previous temperatures along the j–th "time row". Thus, if $j = 0$, then (10) and (9) yield

$$v_{i,1} = \lambda v_{i+1,0} + (1-2\lambda)v_{i,0} + \lambda v_{i-1,0} = \lambda f((i+1)\Delta x) + (1-2\lambda)f(i\Delta x) + \lambda f((i-1)\Delta x),$$

and we leave the rest of the argument to the reader (cf. Problem 4).

We will next illustrate the explicit difference method by some examples, which do not require the use of a calculator or a computer.

Example 1. Consider

D.E. $u_t = u_{xx}$, $\quad 0 \le x \le 5$, $0 \le t \le 0.5$,

B.C. $u(0,t) = 2t$, $u(5,t) = 25 + 2t$,

I.C. $u(x,0) = x^2$.

Use the explicit difference method with grid spacings $\Delta x = 1$ and $\Delta t = 0.1$ to find approximations for (a) $u(1, 0.1)$ and (b) $u(3, 0.5)$, with all calculations to be done by hand.

Solution. In this problem $L = 5$, $T = 0.5$, and so $M = L/\Delta x = 5$ and $N = T/\Delta t = 0.5/0.1 = 5$. Moreover, since $\lambda = \Delta t/(\Delta x)^2 = 0.1$, the discretization (10) of the D.E. is

$$v_{i,j+1} = (0.1)v_{i+1,j} + (0.8)v_{i,j} + (0.1)v_{i-1,j}, \tag{11}$$

$i = 1, 2, 3, 4$ and $j = 0, 1, 2, 3, 4$. The B.C. and I.C. yield, by (9), the side conditions

$$v_{0,j} = 2j\Delta t = (0.2)j, \quad v_{5,j} = 25 + 2j\Delta t = 25 + (0.2)j, \quad j = 0, ..., 5,$$

and

$$v_{i,0} = (i\Delta x)^2 = i^2, \quad i = 0, ..., 5. \tag{12}$$

(a) Since $u(1, 0.1) = u(\Delta x,\Delta t) = u_{1,1}$, we are required to find $v_{1,1}$. Thus, if we set $i = 1$ and $j = 0$ in (11), we obtain the desired approximation

$$v_{1,1} = (0.1)v_{2,0} + (0.8)v_{1,0} + (0.1)v_{0,0} = 1.2 .$$

(b) Since $u(3, 0.5) = u(3\Delta x,5\Delta t) = u_{3,5}$, the problem here is to find $v_{3,5}$. By repeated application of (11), using (12), we find that $v_{3,5} = 10$. In Table 1 we provide the values of $v_{i,j}$ for $i = 0, 1,..., 5$ and $j = 0, 1,..., 5$.

In the numerical solution of PDEs it is of utmost importance to know how accurate the approximations $v_{i,j}$ are. In Table 2 we compare the values of $v_{i,5}$ with $u_{i,5}$, $i = 0, ..., 5$, where $u_{i,5}$ was calculated from the exact solution $u(x,t) = 2t + x^2$ of Example 1.

	t \ x	B.C. 0	1	2	3	4	B.C. 5
I.C.	0	0	1	4	9	16	25
	0.1	0.2	1.2	4.2	9.2	16.2	25.2
	0.2	0.4	1.4	4.4	9.4	16.4	25.4
	0.3	0.6	1.6	4.6	9.6	16.6	25.6
	0.4	0.8	1.8	4.8	9.8	16.8	25.8
	0.5	1	2	5	10	17	26

Table 1

The values of $v_{i,j}$ for $\lambda = 0.1$.

x	0	1	2	3	4	5	
$v_{i,5}$	1	2	5	10	17	26	approximation
$u_{i,5}$	1	2	5	10	17	26	analytic solution

Table 2

The values of $u_{i,5}$ and $v_{i,5}$ for $\lambda = 0.1$.

From Table 2 we see that our approximations $v_{i,5}$ are in complete agreement with the analytic solutions $u_{i,5}$, a rare occurrence. In this example, the lack of errors in our approximations is due to the simple choice of the B.C. and I.C. (cf. Problem 5 and Theorem 1, below). □

Example 2. Consider

$$\begin{array}{lll} \text{D.E.} & u_t = u_{xx}, & 0 < x < 4,\ t > 0, \\ \text{B.C.} & u(0,t) = 5,\ u(4,t) = 5, & t > 0, \\ \text{I.C.} & u(x,0) = 0, & 0 \le x \le 4. \end{array} \tag{13}$$

Use the explicit difference method, with $\Delta x = 1$ and $\Delta t = 0.125$, to find $v_{1,8}$, $v_{2,8}$ and $v_{3,8}$. Note that the problem has no exact continuous solution in the strip $0 \le x \le 4$, $t \ge 0$ (Why not ?).

Solution. The values $v_{1,8}$, $v_{2,8}$ and $v_{3,8}$ are in the last row of Table 3, which was constructed by solving (10) and (9), with $v_{0,0} = 0$, $v_{4,0} = 0$, $T = 1$ and $\lambda = 0.125$ (Why ?).

	t \ x	B.C. 0	1	2	3	B.C. 4
I.C	0	0	0	0	0	0*
	0.125	5	0.000...	0.000...	0.000...	5
	0.250	5	0.625...	0.000...	0.625...	5
	0.375	5	1.0938...	0.1563...	1.0938...	5
	0.500	5	1.4648...	0.3906...	1.4648...	5
	0.625	5	1.7725...	0.6592...	1.7725...	5
	0.750	5	2.0367...	0.9375...	2.0367...	5
	0.875	5	2.2697...	1.2123...	2.2697...	5
	1.000	5	2.4788...	1.4767...	2.4788...	5

Table 3

The values of $v_{i,j}$ for $\lambda = 0.125$.

* Note that the B.C. and I.C. do not match at the corners.

Remark. The diffusion of heat into the rod is evident in Table 3. The symmetry in the columns $x = 1$ and $x = 3$, is due to the symmetry of the problem about $x = 2$. Although there is no exact continuous solution which can be compared with the $v_{i,j}$, there is the formal solution

$$u(x,t) = 5 - \frac{20}{\pi} \sum_{k=0}^{\infty} \frac{1}{2k+1} \exp\left[-(2k+1)^2\pi^2 t/16\right] \sin\left[\frac{(2k+1)\pi x}{4}\right],$$

and one can check that $u(1,1) = u(3,1) \approx 2.564936$ and $u(2,1) \approx 1.572771$ (six place accuracy is obtained even when trucating the series at $k = 2$, due to the rapid exponential square decay of the terms as $k \to \infty$. The disagreement with the last row of Table 3 is nearly .1 . Better agreement is expected by choosing a value $v_{0,0} = v_{4,0}$, say C, *between* 0 and 5 (Why ?). For $C = 2.5$, $v_{1,8} = v_{3,8} \approx 2.573$ and $v_{2,8} \approx 1.601$. If $C = 2$, the results are actually better, namely, $v_{1,8} = v_{3,8} \approx 2.554$ and $v_{2,8} \approx 1.576$. □

In the next example, the B.C. arise from imperfect insulation at the ends (cf. Section 3.4). Separation of variables leads to a Sturm–Liouville problem (cf. Section 4.4), but these sections are not needed to apply the explicit difference method.

Example 3. Consider

$$\text{D.E.} \quad u_t = u_{xx}, \qquad 0 \le x \le 1, \ t \ge 0,$$

$$\text{B.C.} \quad u_x(0,t) = u(0,t), \ u_x(1,t) = -u(1,t), \tag{14}$$

$$\text{I.C.} \quad u(x,0) = 1 .$$

(a) Develop an explicit difference method for problem (14), using central difference approximations for the B.C. .

(b) Use $\Delta t = 1/400$ and $\lambda = 1/4$ to find $v_{1,2}$, that is, approximate $u(0.1, 0.005)$.

Solution. (a) We know that the explicit difference approximations $v_{i,j}$ satisfy

$$v_{i,j+1} = \lambda v_{i+1,j} + (1-2\lambda)v_{i,j} + \lambda v_{i-1,j}, \tag{15}$$

$i = 1, 2, \dots, M-1$, $j = 0, 1, \dots, N-1$, where M is determined by the formula $\Delta x = L/M$, $L = 1$, and $\lambda = \Delta t/(\Delta x)^2$. The first term on the right–hand side of

$$\left[\frac{\partial u}{\partial x}\right]_{(i,j)} = \frac{u_{i+1,j} - u_{i-1,j}}{2\Delta x} + O(\Delta x^2), \quad (\Delta x \to 0), \tag{16}$$

is the central difference approximation to u_x at the grid point (i,j), and using it in the B.C. ,

$$v_{0,j} = \frac{v_{1,j} - v_{-1,j}}{2\Delta x}, \tag{17}$$

and

$$v_{M,j} = -\left[\frac{v_{M+1,j} - v_{M-1,j}}{2\Delta x}\right], \tag{18}$$

for $j = 0, 1, 2, \dots$. Now, the terms $v_{-1,j}$ and $v_{M+1,j}$ from (17) and (18) can be eliminated with the aid of (15), as follows. By setting $i = 0$ and $i = M$ in (15), we obtain

$$v_{0,j+1} = \lambda v_{1,j} + (1-2\lambda)v_{0,j} + \lambda v_{-1,j} \tag{19}$$

and

$$v_{M,j+1} = \lambda v_{M+1,j} + (1-2\lambda)v_{M,j} + \lambda v_{M-1,j} \,. \tag{20}$$

Elimination of $v_{-1,j}$ from (17) and (19) and elimination of $v_{M+1,j}$ from (18) and (20) yield

$$v_{0,j+1} = 2\lambda v_{1,j} + (1 - 2\lambda - 2\lambda\Delta x)v_{0,j} \tag{21}$$

and

$$v_{M,j+1} = 2\lambda v_{M-1,j} + (1 - 2\lambda - 2\lambda\Delta x)v_{M,j}, \tag{22}$$

for $j = 0, 1, 2, \ldots$. Hence, the required difference approximation to (14) is given by (15), (21), (22) and $v_{i,0} = 1$, $i = 0, \ldots, M$.

(b) Since $\lambda = 1/4$ and $\Delta t = 1/400$, $\Delta x = 0.1$. We want to find $v_{1,2}$. By (15), with $i = j = 1$,

$$v_{1,2} = \frac{1}{4}v_{2,1} + \frac{1}{2}v_{1,1} + \frac{1}{4}v_{0,1}. \tag{23}$$

We will now compute $v_{0,1}$, $v_{1,1}$ and $v_{2,1}$. By (21), with $j = 0$ $(v_{0,0} = v_{1,0} = 1)$

$$v_{0,1} = \frac{1}{2}v_{1,0} + (1 - \frac{1}{2} - \frac{0.1}{2})v_{0,0} = 0.95. \tag{24}$$

By (15), with $i = 1$, $j = 0$ $(v_{0,0} = v_{1,0} = v_{2,0} = 1)$

$$v_{1,1} = \frac{1}{4}v_{2,0} + (1 - \frac{1}{2})v_{1,0} + \frac{1}{4}v_{0,0} = 1. \tag{25}$$

By (15), with $i = 2$, $j = 0$ $(v_{1,0} = v_{2,0} = v_{3,0} = 1)$

$$v_{2,1} = \frac{1}{4}v_{3,0} + (1 - \frac{1}{2})v_{2,0} + \frac{1}{4}v_{1,0} = 1. \tag{26}$$

Hence by (24), (25) and (26), we obtain for (23) $v_{1,2} = \frac{1}{4}(1) + \frac{1}{2}(1) + \frac{1}{4}(0.95) = 0.9875$. □

Remark. The need for a numerical approach for a solution of (14) becomes clear when we consider the following *formal* solution :

$$u(x,t) = \sum_{n=1}^{\infty} c_n e^{-\lambda_n t} \varphi_n(x), \tag{27}$$

where the constants $0 < \lambda_1 < \lambda_2 < \lambda_3 \ldots$ are the positive roots of the equation

$$\tan(\sqrt{\lambda}) = \frac{2\sqrt{\lambda}}{\lambda - 1}, \tag{28}$$

$$\varphi_n(x) = \cos(\sqrt{\lambda_n}\, x) + \frac{1}{\sqrt{\lambda_n}} \sin(\sqrt{\lambda_n}\, x), \tag{29}$$

and

$$c_n = \|\varphi_n\|^{-2} \int_0^1 1 \cdot \varphi_n(x)\, dx, \text{ with } \|\varphi_n\|^2 \equiv \int_0^1 \varphi_n(x)^2\, dx. \tag{30}$$

By graphing both sides of (28) as functions of $\sqrt{\lambda}$ (cf. Figure 4 of Section 3.3), one easily sees that $(n-1)\pi < \sqrt{\lambda_n} < (n-\frac{1}{2})\pi$ for $n = 1, 2, \ldots$. With considerable effort, it can be shown that

$$c_n = \begin{cases} 0 & \text{for n even} \\ \dfrac{4}{3+\lambda_n} & \text{for n odd} \end{cases} .$$

If $0 < \beta_1 < \beta_2 < \beta_3$ ... denote the positve roots of the equation $\beta\tan(\beta) = \frac{1}{2}$, then $\lambda_{2k-1} = 4\beta_k^2$, $k = 0, 1, 2,...$ (cf. Problem 3), and we may write (27) in the form

$$u(x,t) = 4\sum_{k=1}^{\infty} \frac{e^{-4\beta_k^2 t}}{3+4\beta_k^2}\left[\cos(2\beta_k x) + \frac{\sin(2\beta_k x)}{2\beta_k}\right],$$

If $u_p(x,t)$ denotes the sum of the first p terms $(p > 1)$, we have

$$|u(x,t)-u_p(x,t)| \le 2\exp(-4\beta_{p+1}^2 t)\sum_{k=p+1}^{\infty} \beta_k^{-2} \le 2e^{-4p\pi^2 t}\int_p^{\infty} [\pi(x-1)]^{-2}\,dx \le 2e^{-4p\pi^2 t}\frac{1}{\pi^2(p-1)} .$$

We want to evaluate $u(0.1,0.005)$. Since $t = 0.005$, $\exp(-4\pi^2 t) = \exp(-\pi^2/50) < 0.821$. Thus,

$$|u(0.1,0.005)-u_p(0.1,.005)| \le 2\pi^{-2}(0.821)^p/(p-1) < 0.203\,(0.821)^p/(p-1)$$

For $p = 15$, the right side is ≈ 0.00075. Thus, according to this estimate, for three–place accuracy, we need to compute more than 15 zeros of $\beta\tan(\beta) = \frac{1}{2}$, assuming that the these zeros are computed with a somewhat greater precision, say by using the Newton–Raphson method (cf. Problems 12 and 13). However, since time may be costly, the numerical approach seems more attractive. To estimate the accuracy of the numerical solution, one might keep refining the grid and computing the approximations for $u(0.1,0.005)$, until one sees sufficiently small changes in the approximation. Of course this does not *prove* that the approximation is sufficiently accurate, but in applications the chief goal is rarely the maximization of rigor. □

Convergence of the explicit difference method

One hopes that the discretization errors $w_{i,j} = u(i\Delta x, j\Delta t) - v_{i,j}$ approach zero as $\Delta x \equiv L/M$ and $\Delta t \equiv T/N$ approach zero. For the problem of a rod with temperature prescribed at the ends, this is proven below (cf. Theorem 1), under the assumptions that a C^4 solution u exists and $\Delta t/(\Delta x)^2 \le \frac{1}{2}$ is maintained. For the proof of Theorem 1, we need the following result.

Proposition 1. If $u(x,t)$ **is a** C^4 **solution of the problem**

$$\begin{aligned} &\text{D.E.} \quad u_t = u_{xx}, \qquad 0 \le x \le L,\ 0 \le t \le T, \\ &\text{B.C.} \quad u(0,t) = A(t),\ u(L,t) = B(t), \\ &\text{I.C.} \quad u(x,0) = f(x), \end{aligned} \tag{31}$$

then

$$|u_{xxxx}(x,t)| \le \max_{\substack{0\le x\le L \\ 0\le t\le T}} \{|A''(t)|, |B''(t)|, |f^{(4)}(x)|\}. \tag{32}$$

Proof. Let $w(x,t) = u_{tt}(x,t)$. Using the D.E. and the assumption that u is C^4, we have $w_t = u_{ttt} = u_{xxtt} = u_{ttxx} = w_{xx}$. Thus, w satisfies the problem

$$\text{D.E. } w_t = w_{xx}, \qquad 0 \le x \le L,\ 0 \le t \le T,$$

$$\text{B.C. } w(0,t) = A''(t),\ w(L,t) = B''(t),$$

$$\text{I.C. } w(x,0) = f^{(4)}(x),$$

since $w = u_{tt} = u_{xxt} = u_{txx} = u_{xxxx}$. By the Maximum/Minimum Principle (cf. Theorem 3 Section 3.2, with $u_1(x,t) = w(x,t)$ and $u_2 \equiv 0$), we obtain the required estimate (32). □

Theorem 1 (A Convergence Theorem for the Explicit Difference Method). Let $u(x,t)$ **be a** C^4 **solution of the problem** (31). **Let** M **and** N **be positive integers, and define** $\Delta x = L/M$, $\Delta t = T/N$ **and** $\lambda = \dfrac{\Delta t}{(\Delta x)^2}$. **Let** $v_{i,j}$ **be the solution of the system of difference equations**

$$v_{i,j+1} = \lambda v_{i+1,j} + (1-2\lambda)v_{i,j} + \lambda v_{i-1,j}, \tag{33}$$

$i = 1, \ldots, M-1$, $j = 0, \ldots, N-1$, **meeting the side conditions with** $A(0) = f(0)$ **and** $B(0) = f(L)$:

$v_{0,j} = A(j\Delta t)$, $j = 0, \ldots, N$, $v_{M,j} = B(j\Delta t)$, $j = 0, \ldots, N$, $v_{i,0} = f(i\Delta x)$, $i = 0, \ldots, M$.

For $u_{i,j} \equiv u(i\Delta x, j\Delta t)$, **let** $w_{i,j} \equiv u_{i,j} - v_{i,j}$ **denote the local discretization error. If** $\lambda \le 1/2$, **then**

$$|w_{i,j}| \le \tfrac{T}{2}K\left[\Delta t + \frac{(\Delta x)^2}{6}\right], \tag{34}$$

where

$$K = \max_{\substack{0\le x\le L \\ 0\le t\le T}} \{|A''(t)|, |B''(t)|, |f^{(4)}(x)|\}, \tag{35}$$

i.e., the discretization error tends to 0 **as** $M, N \to \infty$, **if** $\lambda \equiv TM^2/(L^2N) \le 1/2$.

Proof. Using Taylor's theorem with Lagrange's form of the remainder, we obtain

$$u_{i+1,j} = u_{i,j} + \Delta x\left[\frac{\partial u}{\partial x}\right]_{(i,j)} + \frac{(\Delta x)^2}{2!}\left[\frac{\partial^2 u}{\partial x^2}\right]_{(i,j)} + \frac{(\Delta x)^3}{3!}\left[\frac{\partial^3 u}{\partial x^3}\right]_{(i,j)} + \hat{R}_4, \tag{36}$$

$$u_{i-1,j} = u_{i,j} - \Delta x\left[\frac{\partial u}{\partial x}\right]_{(i,j)} + \frac{(\Delta x)^2}{2!}\left[\frac{\partial^2 u}{\partial x^2}\right]_{(i,j)} - \frac{(\Delta x)^3}{3!}\left[\frac{\partial^3 u}{\partial x^3}\right]_{(i,j)} + \tilde{R}_4, \tag{37}$$

and

$$u_{i,j+1} = u_{i,j} + \Delta t\left[\frac{\partial u}{\partial t}\right]_{(i,j)} + \bar{R}_2, \tag{38}$$

where

$$\hat{R}_4 = \frac{(\Delta x)^4}{4!}\hat{u}_{xxxx}, \quad \tilde{R}_4 = \frac{(\Delta x)^4}{4!}\tilde{u}_{xxxx}, \quad \text{and} \quad \bar{R}_2 = \frac{(\Delta t)^2}{2!}\bar{u}_{tt}.$$

The caret (ˆ), tilde (~) and bar (–) indicate that the respective partial derivatives are evaluated at appropriate intermediate values, given by Lagrange's form of the remainder. Note that these intermediate values depend on the grid point (i,j) under consideration. However, for notational simplicity, in our symbols for the remainders, we have *not* incorporated the letters i and j.

We add (36) to (37), and solve for $(u_{xx})_{(i,j)}$. Since $u_{xx} = u_t$, we replace $(u_t)_{(i,j)}$ in (38) by the resulting expression for $(u_{xx})_{(i,j)}$, and obtain

$$u_{i,j+1} = \lambda u_{i+1,j} + (1-2\lambda)u_{i,j} + \lambda u_{i-1,j} + E_{i,j}, \tag{39}$$

where

$$E_{i,j} = \bar{R}_2 - \lambda(\hat{R}_4 + \tilde{R}_4), \tag{40}$$

$i = 1, ..., M$, $j = 0, ..., N$. If we subtract (33) from (39) we obtain

$$w_{i,j+1} = \lambda w_{i+1,j} + (1-2\lambda)w_{i,j} + \lambda w_{i-1,j} + E_{i,j}. \tag{41}$$

Moreover, it follows from the definition of $w_{i,j}$, that $w_{i,j}$ satisfies the side conditions

$$w_{0,j} = w_{M,j} = 0, \quad j = 0, ..., N \quad \text{and} \quad w_{i,0} = 0, \quad i = 0, ..., M.$$

In order to estimate $|E_{i,j}|$, we first use Proposition 1 and (35) to get

$$|u_{xxxx}(x,t)| \leq K, \quad \text{for all } 0 \leq x \leq L, 0 \leq t \leq T. \tag{42}$$

Since $\lambda \equiv \Delta t/(\Delta x)^2$ and $u_{tt} = u_{xxxx}$, (40) and (42) yield

$$|E_{i,j}| = \left|\frac{(\Delta t)^2}{2}\bar{u}_{tt} - \frac{(\Delta t)(\Delta x)^2}{24}(\hat{u}_{xxxx}+\tilde{u}_{xxxx})\right| \leq \tfrac{1}{2}\Delta t\left[\Delta t|\bar{u}_{tt}| + \frac{(\Delta x)^2}{12}\left[|\hat{u}_{xxxx}|+|\tilde{u}_{xxxx}|\right]\right]$$

$$\leq \tfrac{1}{2}\Delta t\left[\Delta t\cdot K + \frac{(\Delta x)^2}{12}\cdot 2K\right] = \tfrac{1}{2}K\Delta t\left[\Delta t + \frac{(\Delta x)^2}{6}\right] \equiv E^*. \tag{43}$$

We still need to estimate the absolute value of the discretization error. Let

$$\omega_j \equiv \max_{0\leq i\leq M} |w_{i,j}|, \quad j = 0, ..., N.$$

Then by (41), (43), and the assumption that $0 < \lambda \leq \frac{1}{2}$, we have

$$|w_{i,j+1}| \leq \lambda\omega_j + (1-2\lambda)\omega_j + \lambda\omega_j + |E_{i,j}| \leq \omega_j + E^*. \tag{44}$$

Since inequality (44) is true for all $i = 0, \ldots, M$, iteration of (44) yields

$$\omega_{j+1} = \max_{0\leq i\leq M} |w_{i,j+1}| \leq \omega_j + E^* \leq \omega_{j-1} + 2E^* \leq \ldots \leq \omega_0 + (j+1)E^*$$

or since $\omega_0 = 0$, $\omega_j \leq jE^*$, $0 \leq j \leq N$. Since $j\Delta t \leq T$, $1 \leq j \leq N$, we obtain the desired result (34), i.e.,

$$|w_{i,j}| \leq \omega_j \leq jE^* = \tfrac{1}{2}j\Delta tK\left[\Delta t + \frac{(\Delta x)^2}{6}\right] \leq \tfrac{1}{2}KT\left[\Delta t + \frac{(\Delta x)^2}{6}\right]. \quad \square$$

Remarks. (a) Many authors estimate $|w_{i,j}|$ in terms of bounds on the derivatives of the unknown solution $u(x,t)$ of the given initial/boundary value problem. These bounds involve information about a presumably unknown solution.

(b) From (34), we have $\max\limits_{\substack{0\leq i\leq M\\0\leq j\leq N}} |w_{i,j}| = O[\Delta t + (\Delta x)^2]$, where M and N tend to infinity in such a way that $0 < \lambda = \Delta t/(\Delta x)^2 \leq \frac{1}{2}$. If $\lambda = 1/6$ is maintained as $M, N \to \infty$, then for a C^6 solution, the local discretization error has magnitude $O((\Delta x)^4)$ as $\Delta x \to 0$ (cf. Problem 7).

(c) We emphasize here that the estimate (34) of Theorem 1 is a theoretical estimate. That is, in practice we cannot assume that the computational procedure used for solving the difference equations (32) is exact. The computer can only retain a finite number of digits and consequently we must also contend with **round–off errors** (cf. Section 8.3). It is clear that the round–off error present at the j–th step of our numerical solution of the difference equations will influence the accuracy at the (j+1)–st step. Roughly speaking, the numerical method used is said to be **stable** if the errors (from whatever source) do not accumulate and destroy the given accuracy by the individual local truncation errors. While the notion of stability in the analysis of numerical solutions is extremely important, a rigorous treatment of stability is outside the scope of this text.

(d) If $u(x,t) = P''(x)t + P(x)$, where $P(x)$ is a polynomial of degree less than 4, then the right–hand side of (34) is zero (Why ?). Thus, in this case, the local discretization error is zero, and the numerical solution (without rounding off) is exact (cf. Example 1, where $P(x) = x^2$). □

In the following examples, we illustrate how the difference approximations are influenced by the value of $\lambda = \Delta t/(\Delta x)^2$ and the mesh sizes Δx and Δt.

Example 4. Consider

$$\begin{aligned} &\text{D.E.} \quad u_t = u_{xx}, \qquad 0 \leq x \leq 1,\ t \geq 0,\\ &\text{B.C.} \quad u(0,t) = 0,\ u(1,t) = 1 \\ &\text{I.C.} \quad u(x,0) = x + 1000\sin(\pi x). \end{aligned} \tag{45}$$

(a) Calculate by hand $u(0.5, 0.5)$ with $T = 0.5$, $M = 2$ and $N = 1$ ($\lambda = 2$), using the explicit difference method.

(b) Use the explicit difference method with $T = 0.5$ and $\lambda = 1/6$ to approximate $u(0.5, 0.5)$. Compare your approximations with the actual solution.

Solution.

(a) Here $\Delta x = L/M = 0.5$ and $\Delta t = T/N = 0.5$ ($\lambda = \Delta t/(\Delta x)^2 = 2$). Since $u(0.5, 0.5) = u_{1,1}$, we must find $v_{1,1}$. The system of difference equations is given by

$$v_{i,j+1} = 2v_{i+1,j} - 3v_{i,j} + 2v_{i-1,j}$$

with side conditions

$$v_{0,j} = 0\,, \quad v_{2,j} = 1\,, \quad v_{i,0} = \frac{i}{2} + 1000\sin(\tfrac{1}{2}\pi i), \qquad (i, j = 0, 1, 2, \ldots).$$

Thus, $v_{1,1} = 2 - (3)(1000.5) + 0 = -2999.5$, which is a remarkably poor approximation to $u(0.5,0.5) = 7.69188335\ldots$, computed from the solution $u(x,t) = x + 1000\,e^{-\pi^2 t}\sin(\pi x)$ of (45). □

(b) In this case, $\Delta x = 1/M$ and $\Delta t = 0.5/N$, where the positive integers M and N are such that $M^2/2N = \lambda = 1/6$, as required. Table 4 lists $v_{M/2,N}$ for various M and N.

M	N	Approximation to u(0.5,0.5)	Analytic solution
2	12	8.207346...	7.691883...
4	48	7.718356...	7.691883...
6	108	7.696947...	7.691883...
8	192	7.693468...	7.691883...

Table 4

$$\frac{M^2}{2N} = \lambda = \frac{1}{6}$$

Note that for $M = 8$, $N = 192$, $\Delta x = 0.125$ and $\Delta t = 0.0026$, our approximation agrees with the actual solution up to three places. At a grid point, the **percentage error** is the difference of the numerical and actual solutions, expressed as a percentage of the value of actual solution, e.g.,

$$\frac{v_{4,192} - u(0.5,\ 0.5)}{u(0.5,\ 0.5)} \cdot 100\ \% = 0.020\ \ldots\ \%\ . \quad \square$$

Example 5. Use the explicit method with $\lambda = 1/2$ to approximate $u(0.5, 0.5)$, where u is the solution of the problem (45) of the previous example.

Solution. As in the previous example, $L = 1$, $T = 0.5$, $\Delta x = 1/M$ and $\Delta t = 0.5/N$. Now we will select the positive integers M and N such that $\lambda = 1/2$. Table 5 provides five different approximations $v_{M/2,N}$ to u(0.5, 0.5).

M	N	Approximation to u(0.5,0.5)	Analytic solution
4	16	4.406250...	7.691883...
8	64	6.800603...	7.691883...
10	100	7.116564...	7.691883...
12	144	7.290535...	7.691883...
14	196	7.396216...	7.691883...

Table 5

$$\frac{M^2}{2N} = \lambda = \frac{1}{2}$$

We observe that the approximations are not as accurate with $\lambda = 1/2$ as the approximations with $\lambda = 1/6$, (cf. Example 4). □

The purpose of the next example is to use inequality (34) to provide an estimate of the local discretization error in a concrete setting.

Example 6. Consider

$$\begin{aligned} &\text{D.E.}\ \ u_t = u_{xx}\,, \qquad 0 \le x \le 1,\ \ t \ge 0, \\ &\text{B.C.}\ \ u(0,t) = 0,\ \ u(1,t) = 1\,, \\ &\text{I.C.}\ \ u(x,0) = x + \pi^{-4}\sin(\pi x)\,. \end{aligned} \tag{46}$$

Use the explicit method with $T = 0.5$, $\Delta x = 1/8$ and $\Delta t = 0.5/64$ to approximate u(0.5, 0.5), and use inequality (34) to estimate the local discretization error in this case.

Solution. We must find the difference approximation $v_{4,64}$, since $M = L/\Delta x = 8$ and $N = T/\Delta t = 64$. With the help of a suitable computer program, we find that $v_{4,64} = 0.500064682...$, while $u(0.5, 0.5) = 0.500073832...$, and so

$$|\, u(0.5,0.5) - v_{4,64} \,| = 0.0000091... \,. \tag{47}$$

Since $u(x,0) = f(x) = x + \pi^{-4}\sin(\pi x)$, $|f^{(4)}(x)| = |\cos(\pi x)| \le 1$. Thus, $K = 1$ in (34), and

$$|u(0.5,0.5) - v_{4,64}| = |u_{4,64} - v_{4,64}| = |w_{4,64}| \le \frac{T}{2}\left[\Delta t + \frac{(\Delta x)^2}{6}\right] = 0.0026041\,. \tag{48}$$

Thus, the local discretization error (47), in this specific calculation, is "much smaller" than the theoretical error bound provided by (34). □

Summary 8.2

1. The explicit difference method for a heat problem : Consider

$$\begin{aligned} &\text{D.E. } u_t = u_{xx}, && 0 \le x \le L,\ 0 \le t \le T, \\ &\text{B.C. } u(0,t) = A(t),\ u(L,t) = B(t), && \\ &\text{I.C. } u(x,0) = f(x) \quad (A(0) = f(0),\ B(0) = f(L)) . && \end{aligned} \tag{S1}$$

The numerical solution of this problem by the explicit difference method is as follows. For positive integers M and N, let $\Delta x = L/M$, $\Delta t = T/N$ and $\lambda = \Delta t/(\Delta x)^2$. Solve the system

$$v_{i,j+1} = \lambda v_{i+1,j} + (1-2\lambda)v_{i,j} + \lambda v_{i-1,j}, \tag{S2}$$

$(i = 1, \ldots M-1 ;\ j = 0, \ldots, N-1)$, subject to the side conditions

$$v_{0,j} = u(0,j\Delta t) = A(j\Delta t),\ v_{M,j} = u(L,j\Delta t) = B(j\Delta t),\quad j = 0, \ldots, N$$

and

$$v_{i,0} = u(i\Delta x,0) = f(i\Delta x),\ i = 0, \ldots, M. \tag{S3}$$

Ideally, the local discretization error $w_{i,j} \equiv u(i\Delta x,j\Delta t) - v_{i,j}$ is small in absolute value.

2. The error estimate (Theorem 1) : If (S1) **has a** C^4 **solution** $u(x,t)$, **then for** $0 < \lambda \le \frac{1}{2}$,

$$|u(i\Delta x,j\Delta t) - v_{i,j}| = |w_{i,j}| \le \tfrac{1}{2}KT(\Delta t + \tfrac{1}{6}(\Delta x)^2) \quad (i = 0, 1,\ldots, M ;\ j = 0, 1,\ldots,N) \tag{S4}$$

where

$$K \equiv \max_{\substack{0\le x\le L \\ 0\le t\le T}} \{|A''(t)|, |B''(t)|, |f^{(4)}(x)|\} . \tag{S5}$$

However, one must be wary of the accumulation of round–off errors in the computation of $v_{i,j}$.

Exercises 8.2

1. Consider the problem

$$\text{D.E.}\quad u_t = u_{xx}, \qquad 0 \le x \le 1,\ t \ge 0,$$

$$\text{B.C.}\quad u(0,t) = 4t,\ u(1,t) = 4t$$

$$\text{I.C.}\quad u(x,0) = 2x(x-1).$$

(a) (Calculation by hand) Use the explicit difference method with $T = 1$, $\Delta x = .5$ and $\Delta t = 1$ ($\lambda = 4$) to approximate $u(0.5, 1)$. Compare your answer with the actual solution.

(b) (Calculation with computer or calculator) Use the explicit difference method with $T = 1$, $\Delta x = 0.25$, and (i) $\lambda = 1/2$, (ii) $\lambda = 1/6$, to approximate $u(0.25, 1)$, $u(0.5, 1)$ and $u(0.75, 1)$.

2. Consider the problem

$$\text{D.E.}\quad u_t = u_{xx}, \qquad 0 \le x \le 5,\ t \ge 0,$$

$$\text{B.C.}\quad u(0,t) = 2t,\ u(5,t) = 25 + 2t,$$

$$\text{I.C.}\quad u(x,0) = x^2.$$

(a) (Calculation by hand) Use the explicit difference method with $T = 2$, $\Delta x = 1$ and $\Delta t = 2$ to approximate $u(1,2)$, $u(2,2)$, $u(3,2)$ and $u(4,2)$.

(b) (Calculation with computer or calculator) Use the explicit difference method with $T = 1$, $\Delta x = 1$ and (i) $\lambda = 1/2$ and (ii) $\lambda = 1/6$ to approximate $u(1,2)$, $u(2,2)$, $u(3,2)$ and $u(4,2)$.

3. Consider Example 3.

(a) Compute the coefficients c_n in formula (30) in terms of λ_n. Verify that $c_n = 4/(3 + \lambda_n)$ for n odd, and $c_n = 0$ for n even.

(b) Let $0 < \lambda_1 < \lambda_2 < \ldots$ be the positive roots of $\tan\sqrt{\lambda} = 2\sqrt{\lambda}/(\lambda-1)$ and let $\lambda_n = 4\alpha_n^2$ (where $\alpha_n > 0$). Show that $\tan(\alpha_n) = \frac{1}{2\alpha_n}$, if n is odd, and $\tan(\alpha_n) = -2\alpha_n$, if n is even.

(c) Let $\beta_k = \alpha_{2k-1}$ ($k = 1, 2, \ldots$), where the α_n were defined in (b). By part (b), $0 < \beta_1 < \beta_2 < \ldots$ are the positive roots of $\beta\tan\beta = \frac{1}{2}$. Show that $(k-1)\pi < \beta_k < (k-\frac{1}{2})\pi$.

(d) Let $u_4(x,t) = 4\sum_{k=1}^{4} \dfrac{e^{-4\beta_k^2 t}}{3 + 4\beta_k^2}\left[\cos(2\beta_k x) + \dfrac{\sin(2\beta_k x)}{2\beta_k}\right]$. Using the approximate values $\beta_1 = 0.653271\ldots$, $\beta_2 = 3.292310\ldots$, $\beta_3 = 6.361620\ldots$, $\beta_4 = 9.477485\ldots$, (cf. Problem 13), calculate $u_4(0.1, 0.005)$ and compare it with $v_{1,2}$ in Example 3(b).

4. (A Uniqueness Theorem) Suppose $v_{i,j}$ and $\bar{v}_{i,j}$, $i = 0, ..., M$, $j = 0, ..., N$, both satisfy (9) and (10). Show that $v_{i,j} = \bar{v}_{i,j}$ for all i and j.

5. Consider the problem

$$\text{D.E.}\quad u_t = u_{xx}\,, \qquad 0 \le x \le L,\ t \ge 0,$$
$$\text{B.C.}\quad u(0,t) = a_1 t + a_0,\ u(L,t) = b_1 t + b_0,$$
$$\text{I.C.}\quad u(x,0) = c_2 x^2 + c_1 x + c_0\,,$$

where a_0, a_1, b_0, b_1, c_0, c_1 and c_2 are constants.

(a) Under what conditions on the constants will this problem have a solution ?
(b) If the conditions on the constants are met, show that the solution must be of the form $u(x,t) = c(2t + x^2) + bx + a$.
(c) Assuming that the problem has a solution, prove that the local discretization error in the explicit difference method is zero. Is this error defined if there is no solution ?

6. Consider the problem

$$\text{D.E.}\quad u_t = u_{xx}\,, \qquad 0 \le x \le 1,\ t \ge 0,$$
$$\text{B.C.}\quad u(0,t) = 0,\ u(1,t) = 1\,,$$
$$\text{I.C.}\quad u(x,0) = x + \pi^{-4}\sin(\pi x)\,.$$

Suppose that the explicit method with $T = 0.5$, $\Delta x = 1/8$ and $\Delta t = 1/256$ yields the difference approximation $v_{4,128} = 0.500071505...$ to $u(0.5, 0.5)$. Use formula (34) to estimate the local discretization error in this case. Calculate the actual discretization error, accurate to 8 places.

7. Let $u(x,t)$ be a C^6 solution of the problem

$$\text{D.E.}\quad u_t = u_{xx}\,, \qquad 0 \le x \le L,\ 0 \le t \le T,$$
$$\text{B.C.}\quad u(0,t) = A(t),\ u(L,t) = B(t)$$
$$\text{I.C.}\quad u(x,0) = f(x)\,.$$

Prove that if $\lambda = 1/6$, then the local discretization error $|w_{i,j}|$ has order of magnitude $O((\Delta x)^4)$ as $\Delta x \to 0$ (cf. (34) for the notation).

8. (a) Write a general computer program for the explicit difference solution of the problem.

$$\text{D.E.}\quad u_t = ku_{xx} + h(x,t), \qquad 0 \le x \le L,\ 0 \le t \le T,$$
$$\text{B.C.}\quad u(0,t) = 0,\ u(L,t) = 0\,,$$
$$\text{I.C.}\quad u(x,0) = f(x)\,.$$

(b) How did you test your program ?

(c) Discuss some of the stopping procedures that one might use in such a program.

9. Consider the problem

$$\text{D.E.}\quad u_t = u_{xx} - 2u\,,\qquad 0 \le x \le 1,\ t \ge 0,$$
$$\text{B.C.}\quad u(0,t) = 0,\ u(1,t) = 0\,,$$
$$\text{I.C.}\quad u(x,0) = \sin(\pi x)\,.$$

Write a computer program to approximate, via the explicit difference method, $u(0.5,1)$ until there is no change in the third place to the right of the decimal, using $\lambda = 1/6$. Include a stopping procedure in your program, and specify the number of iterations and the values of Δt and Δx in the output. Compare your result with the exact solution $u(x,t) = e^{-(2+\pi^2)t}\sin(\pi x)$.

10. (a) Replace the D.E. in Problem 9 by $u_t = u_{xx}$ and write a program as in Problem 9 to solve this new problem (same I.C.), but with any fixed $\lambda \equiv \Delta t/(\Delta x)^2 > 0$, using

$$\frac{v_{i,j+1} - v_{i,j-1}}{2\Delta t} = \frac{v_{i+1,j} - v_{i,j+1} - v_{i,j-1} + v_{i-1,j}}{(\Delta x)^2}\,,$$

known as the **DuFort–Frankel method** (This method is unconditionally stable, in the sense that the constant $\lambda > 0$ is arbitrary.) How can the $v_{i,-1}$ be determined ?

(b) If $u(x,t)$ is any C^4 function on the rectangle $0 \le x \le L,\ 0 \le t \le T$, then show that

$$(u_{i+1,j} - u_{i,j+1} - u_{i,j-1} + u_{i-1,j})(\Delta x)^{-2} - \tfrac{1}{2}(u_{i,j+1} - u_{i,j-1})(\Delta t)^{-1}$$
$$= (u_{xx} - u_t)_{i,j} - (\Delta t/\Delta x)^2(u_{tt})_{i,j} + O((\Delta x)^2 + (\Delta t)^2 + (\Delta t)^4/(\Delta x)^2)\,.$$

Thus, if $\mu \equiv \Delta t/\Delta x$ (not λ) is held constant as Δx and $\Delta t \to 0$, we suspect that the $v_{i,j}$ in (a) will not "converge" to a solution of $u_t = u_{xx}$, but rather to a solution of $u_t + \mu^2 u_{tt} = u_{xx}$.

11. Replace the homogeneous D.E. in problem (31) for Theorem 1 by $u_t = u_{xx} + q(x,t)$. If this new problem has a C^4 solution, then prove that the discretization error $w_{i,j} \equiv v_{i,j} - u(i\Delta x, j\Delta t)$ for the explicit difference method satisfies $|w_{i,j}| \le \tfrac{1}{2}TK(\Delta t + \tfrac{1}{6}(\Delta x)^2)$ where K is now the maximum of $|u_{xxxx}|$ on the whole rectangle $0 \le x \le L$, $0 \le t \le T$, provided that $\Delta t/(\Delta x)^2 \le \tfrac{1}{2}$.

Hint. Note that in this case (33) becomes, where $q_{i,j} \equiv q(i\Delta x, j\Delta t)$,
$v_{i,j+1} = \lambda v_{i+1,j} + (1 - 2\lambda)\, v_{i,j} + \lambda v_{i-1,j} + q_{i,j}\,\Delta t$, $i = 1, \ldots M - 1$; $j = 0, \ldots N-1$.

12. (**The Newton–Raphson Method**) Let $f(x)$ be a C^2 function on the interval $[a,b]$ $(b > a)$, such that $f'(x) \neq 0$ for all x in $[a,b]$. Let $f(r) = 0$, for some unknown root r in (a,b).

(a) Let x_0 in (a,b) be an initial incorrect guess for r. If the graph of $f(x)$ is not too curved, then explain intuitively (in terms of the tangent line to the graph of $f(x)$ at x_0, why $x_1 \equiv x_0 - f(x_0)/f'(x_0)$ will be a better approximation to r (i.e. $|x_1 - r| < |x_0 - r|$).

(b) Prove that $|x_1 - r| \leq \rho |x_0 - r|$, where $\rho \equiv \frac{1}{2} M |b-a|$ and $M \equiv \max\limits_{a \leq x \leq b} |f''(x)| / \min\limits_{a \leq x \leq b} |f'(x)|$.

Hint. Use Taylor's theorem to obtain $|f(x_0) + f'(x_0)(r - x_0)| = \frac{1}{2}|f''(c)|(r-x_0)^2$, for some c between r and x_0. Then note that $f(x_0) = x_0 f'(x_0) - x_1 f'(x_0)$.

(c) Assuming that $x_n \in (a,b)$, define inductively $x_{n+1} = x_n - f(x_n)/f'(x_n)$, $n = 1, 2, \ldots$. If a and b are chosen sufficiently close (specifically, if $\rho \equiv \frac{1}{2} M |b-a| < 1$), then show that $|x_n - r| \leq \rho^n |b-a|$ (i.e., $x_n \to r$ as $n \to \infty$, and we also have an estimate of the error).

Remark. If $\min(b - r, r - a) > |x_0 - r|$ and $\rho < 1$, then it automatically follows that $x_n \in (a,b)$. In other words, the assumption $x_n \in (a,b)$ is unnecessary, if x_0 is chosen to be closer to r than either of the endpoints a and b, and if $\rho < 1$.

13. Use the Newton–Raphson method in Problem 12 to find approximations (correct to six decimal places, if you have access to a computer) of the first four positive roots β_1, β_2, β_3 and β_4, of the equation $x \tan(x) = \frac{1}{2}$. Take $f(x) = 2x \tan(x) - 1$ and take the initial approximation to β_k to be $(k-1)\pi$ if $k > 1$, and for β_1 use 0.5 (cf. Problem 3). What goes wrong if one uses 0 for the initial approximation to β_1 ?

8.3 Difference Equations and Round–off Errors

If the difference equation obtained from a given PDE is not solved exactly, then, in addition to the discretization error, we must contend with **round–off errors**. This error is naturally present in the iterative solution, since the iterations are usually only continued until no change takes place up to a certain number of digits. Mesh sizes (i.e., Δt or Δx) affect the discretization error and the round–off error in different ways. In general, the discretization error decreases with decreasing Δt, while the total round–off error tends to increase. For this reason one *cannot* assert that diminishing the mesh size Δt always improves the accuracy.

In order to gain insight into the question of round–off errors, we will find the "explicit" solution (i.e., exact solution, without any round–off error) to the (partial) difference equation

$$v_{m,n+1} = \lambda v_{m+1,n} + (1-2\lambda)v_{m,n} + \lambda v_{m-1,n}, \quad (\lambda = \frac{\Delta t}{(\Delta x)^2}) \tag{1}$$

$(m = 1,\ldots,M-1;\ n = 1,\ldots,N-1)$, with side conditions

$$v_{0,n} = A(n\Delta t),\quad v_{M,n} = B(n\Delta t),\quad v_{m,0} = f(m\Delta x) \quad (m = 0,\ldots,M;\ n = 0,\ldots,N). \tag{2}$$

In this section, we use the subscripts m and n in place of i and j (as in Section 8.2), since $i = \sqrt{-1}$ will arise. For the sake of simplicity, we assume that $A(t) = B(t) = 0$, and $f(x)$ is a linear combination of sine functions, as in Section 3.1. Our approach to solving (1) will be based on the method of "separation of variables". This method reduces (1) to two (ordinary) difference equations, which we will need to solve. The derivation of the solutions to these ordinary difference equations is analogous to the derivation of solutions to ODEs with constant coefficients (see also Section 1.1). For this reason, we omit some proofs.

The general first–order, linear, homogeneous difference equation with constant coefficients can be written in the form

$$y_{n+1} = ay_n,\quad n = 0, 1, 2, \ldots, \tag{3}$$

where a is a constant. Since $y_n = ay_{n-1} = a^2y_{n-2} = \ldots = a^ny_0$, the general solution of (3) is

$$y_n = K\,a^n,\quad n = 0, 1, 2, \ldots, \tag{4}$$

where K is the initial value y_0. If we had attempted a trial solution of the form $y = r^n$, where r is a nonzero constant, then we would have found that $r = a$ (Why ?), and the general solution (4) would be obtained by multiplying the trial solution by a constant. Note that (3) is analogous to the ODE $y'(x) = ay(x)$, whose general solution $y(x) = Ke^{ax}$ is analogous to (4).

The general second–order, linear, homogeneous difference equation with constant coefficients can be written in the form

$$y_{n+2} + a_1 y_{n+1} + a_0 y_n = 0, \quad n = 0, 1, 2, \ldots, \tag{5}$$

where a_0 and a_1 are constants, which, for convenience, will be assumed to be real. As before, we attempt a trial solution of the form $y_n = r^n$, where r is a nonzero constant. If we substitute $y_n = r^n$ into (5) we obtain the **auxiliary equation**

$$r^2 + a_1 r + a_0 = 0. \tag{6}$$

As with the ODE $y'' + a_1 y' + a_0 y = 0$, there are three cases.

Case 1. The roots r_1 and r_2 of the auxiliary equation (6) are real and distinct. For arbitrary constants c_1 and c_2, the general solution of the difference equation (5) is given by

$$y_n = c_1 r_1^n + c_2 r_2^n, \quad n = 0, 1, 2, \ldots, \tag{7}$$

Case 2. The roots r_1 and r_2 of (6) are equal. Then the general solution of (5) is

$$y_n = (c_1 + c_2 \cdot n) r_1^n, \quad n = 0, 1, 2, \ldots, \tag{8}$$

Case 3. The roots r_1 and r_2 of (6) are complex, say $r_1 = a + ib$ and $r_2 = a - ib$. Let

$$a = R\cos\theta \text{ and } b = R\sin\theta, \text{ where } R = \sqrt{a^2 + b^2}. \tag{9}$$

Then the general solution of (5) is given by

$$y_n = R^n(c_1 \cos(n\theta) + c_2 \sin(n\theta)), \quad n = 0, 1, 2, \ldots. \tag{10}$$

Example 1. (a) Find the general solution of the difference equation

$$y_{n+2} - 6y_{n+1} + 8y_n = 0\,, \qquad n = 0,\,1,\,2,\ldots\,. \tag{11}$$

(b) Find the particular solution y_n of (11), which satisfies the conditions $y_0 = 3$ and $y_1 = 2$.

Solution. (a) The auxiliary equation $r^2 - 6r + 8 = 0$ has roots $r_1 = 2$ and $r_2 = 4$. Hence, by (7), the general solution of (11) is

$$y_n = c_1 2^n + c_2 4^n\,, \qquad n = 0,\,1,\,2,\ldots\,. \tag{12}$$

(b) Using the conditions $y_0 = 3$ and $y_1 = 2$, we obtain from (12) $c_1 + c_2 = 3$ and $2c_1 + 4c_2 = 2$. Thus, $c_1 = 5$ and $c_2 = -2$, and $y_n = 5\cdot 2^n - 2\cdot 4^n$. □

Example 2. Find the general solution of the difference equation

$$y_{n+2} - 4y_{n+1} + 4y_n = 0\,, \qquad n = 0,\,1,\,2,\,\ldots\,. \tag{13}$$

Solution. The auxiliary equation $r^2 - 4r + 4 = 0$ has roots $r_1 = r_2 = 2$. Hence, by (8), the general solution of (13) is $y_n = (c_1 + c_2\cdot n)\,2^n$. □

Example 3. Find the general solution of the difference equation

$$y_{n+2} - 2y_{n+1} + 4y_n = 0\,, \qquad n = 0,\,1,\,2,\,\ldots\,. \tag{14}$$

Solution. The auxilliary equation $r^2 - 2r + 4 = 0$ has roots $1 \pm i\sqrt{3}$. In (9), $R = 2$ and $\theta = \pi/3$. Thus, by (10), the general solution of (14) is $y_n = 2^n(c_1 \cos(n\pi/3) + c_2 \sin(n\pi/3)$. □

Example 4. Find the general solution of the difference equation

$$y_{n+2} - 2\cos\alpha\, y_{n+1} + y_n = 0\,, \qquad n = 0,\,1,\,2,\,\ldots\,. \tag{15}$$

Solution. The roots of $r^2 - 2\cos\alpha\, r + 1 = 0$ are $\cos\alpha \pm i\cdot\sin\alpha$. If $\sin\alpha \neq 0$, then the roots are complex. By (9), $R = 1$, $\cos\alpha = \cos\theta$, and the general solution of (15) is

$$y_n = c_1\cos(n\alpha) + c_2\sin(n\alpha)\,, \quad \text{if } \alpha \neq 0,\,\pm\pi,\,\pm 2\pi,\,\ldots\,. \tag{16}$$

If $\sin(\alpha) = 0$, then instead of (16), the general solution of (15) is $y_n = c_1 + c_2\cdot n$ (Why ?). □

A partial difference equation

We are now ready to find the explicit solution of the following partial difference equation.

$$v_{m,n+1} = \lambda v_{m+1,n} + (1-2\lambda)v_{m,n} + \lambda v_{m-1,n}, \tag{17}$$

($m = 1, \ldots, M-1$; $n = 0, \ldots, N-1$), satisfying the side conditions

$$v_{0,n} = 0, \quad v_{M,n} = 0, \quad v_{m,0} = \sin(\frac{mp\pi}{M}) \quad (n = 0, \ldots, N ; \ m = 0, \ldots, M), \tag{18}$$

where p is a positive integer. This is the system which arises in the solution of the problem

$$\begin{aligned} &\text{D.E.} \quad u_t = u_{xx} \quad 0 \le x \le 1, \ 0 \le t \le T, \\ &\text{B.C.} \quad u(0,t) = 0 \quad u(1,t) = 0 \\ &\text{I.C.} \quad u(x,0) = \sin(p\pi x), \end{aligned} \tag{$*$}$$

when the explicit difference method is used with $\Delta x = 1/M$, $\Delta t = \lambda(\Delta x)^2 = \lambda M^{-2}$ and $T = N\Delta t$. Of course, the solution of ($*$) is $u(x,t) = e^{-\pi^2p^2t}\sin(\pi px)$, and we wish to see how closely the solution of (17) and (18) approximates this solution. We mimic the method of separation of variables for PDEs and seek product solutions of (17) of the form

$$v_{m,n} = F_m G_n, \tag{19}$$

where F_m is independent of n and G_n is independent of m. Now, (19) and (17) yield

$$F_m G_{n+1} = \lambda F_{m+1} G_n + (1-2\lambda)F_m G_n + \lambda F_{m-1} G_n,$$

or

$$\frac{G_{n+1}}{G_n} = \frac{\lambda F_{m+1} + (1-2\lambda)F_m + \lambda F_{m-1}}{F_m} = c, \tag{20}$$

where c is a constant, independent of m and n. Thus, F_m and G_n must satisfy

$$G_{n+1} = cG_n \tag{21}$$

and

$$F_{m+1} - 2[1 - \frac{1}{2\lambda}(1-c)]F_m + F_{m-1} = 0. \tag{22}$$

Note that (22) is of the form (5), and of the possible forms (7), (8) and (10), only form (10) can meet the "B.C." in (18) (Why ?). Thus, the roots of the auxiliary equation are complex, and so $|1 - \frac{1}{2\lambda}(1-c)| < 1$ (Why ?). We may set $[1 - \frac{1}{2\lambda}(1-c)] = \cos\alpha$, and (16) yields

$$F_m = c_1 \cos(m\alpha) + c_2 \sin(m\alpha) = c_2 \sin(\tfrac{m p \pi}{M}), \tag{26}$$

where we have used (18) to obtain $c_1 = 0$ and $\alpha = \frac{p\pi}{M}$. Since $[1 - \frac{1}{2\lambda}(1-c)] = \cos\alpha$ yields $c = 1 - 2\lambda(1-\cos\alpha) = 1 - 4\lambda\sin^2(\frac{1}{2}\alpha)$, the solution of (21) is given by (cf. (4))

$$G_n = Kc^n = K\left[1 - 4\lambda\sin^2(\tfrac{1}{2}\alpha)\right]^n = K\left[1 - 4\lambda\sin^2(\tfrac{p\pi}{2M})\right]^n . \tag{27}$$

Using (18), $\sin(\frac{m p \pi}{M}) = v_{m,0} = F_m G_0 = Kc_2\sin(\frac{m p \pi}{M})$. Thus, we take $Kc_2 = 1$, and

$$v_{m,n} = \left[1 - 4\lambda\sin^2(\tfrac{p\pi}{2M})\right]^n \sin(\tfrac{m p \pi}{M}), \tag{28}$$

where $0 \le m \le M$ and $0 \le n \le N$. Moreover, we know that (28) is the unique solution of the difference equation (17) satisfying the side conditions (18) (cf. Problem 4 of Exercises 8.2). The superposition principle for linear homogeneous equations then yields the following result.

Theorem 1. The unique solution of the problem

$$v_{m,n+1} = \lambda v_{m+1,n} + (1-2\lambda)v_{m,n} + \lambda v_{m-1,n}, \quad (m = 1, \ldots, M-1 \,;\, n = 0, \ldots, N-1),$$

$$v_{0,n} = 0, \quad v_{M,n} = 0, \quad (n = 0, \ldots, N), \tag{29}$$

$$v_{m,0} = \sum_{p=1}^{k} b_p \sin(\tfrac{m p \pi}{M}), \quad (m = 0, \ldots, M),$$

is

$$v_{m,n} = \sum_{p=1}^{k} b_p \left[1 - 4\lambda\sin^2(\tfrac{p\pi}{2M})\right]^n \sin(\tfrac{m p \pi}{M}). \tag{30}$$

Problem (29) is encountered in applying the explicit difference method to

$$\begin{aligned} &\text{D.E. } u_t = u_{xx}, && 0 \le x \le 1,\ 0 \le t \le T,\\ &\text{B.C. } u(0,t) = 0,\ u(1,t) = 0, && \\ &\text{I.C. } u(x,0) = \sum_{p=1}^{k} b_p \sin(\pi p x), && \end{aligned} \tag{31}$$

with $\Delta x = 1/M$, $\Delta t = \lambda(\Delta x)^2$ and $T = N\Delta t$. The solution of (31) is

$$u(x,t) = \sum_{p=1}^{k} b_p e^{-\pi^2 p^2 t} \sin(\pi p x) . \tag{32}$$

We have the following exact formula for the local discretization error at $(m\Delta x, n\Delta t)$:

$$w_{m,n} = u(m\Delta x, n\Delta t) - v_{m,n} = \sum_{p=1}^{k} b_p \left[e^{-\pi^2 p^2 n\Delta t} - [1 - 4\lambda \sin^2(\tfrac{p\pi}{2M})]^n \right] \sin(\tfrac{mp\pi}{M}) . \quad (33)$$

By Theorem 1 of Section 8.2, we know that for $0 < \lambda = \Delta t/(\Delta x)^2 = TM^2/N \leq \frac{1}{2}$,

$$\max_{\substack{0 \leq m \leq M \\ 0 \leq n \leq N}} |w_{m,n}| \leq \tfrac{1}{2} KT \left[\Delta t + \tfrac{1}{6}(\Delta x)^2 \right]$$

$$\leq \tfrac{1}{2} KT \left[\tfrac{1}{2}(\Delta x)^2 + \tfrac{1}{6}(\Delta x)^2 \right] = \tfrac{1}{3}\frac{KT}{M^2} = \tfrac{1}{3} KT(\Delta x)^2 = \frac{KT}{3\lambda}\Delta t , \quad (34)$$

where $K \equiv \max_{0 \leq x \leq 1} |u^{(4)}(x,0)| \leq \pi^4 \sum_{p=1}^{k} p^4 |b_p|$ (Why ?). For a more precise estimate,

$$\left[e^{-\pi^2 p^2 n\Delta t} - [1 - 4\lambda \sin^2(\tfrac{p\pi}{2M})]^n \right] = e^{-\pi^2 p^2 n\Delta t} \left[1 - \left[\frac{1 - 4\lambda \sin^2(\frac{1}{2}\pi p\Delta x)}{e^{-\pi^2 p^2 \lambda (\Delta x)^2}} \right]^n \right] \quad (35)$$

Let $z \equiv p\pi\Delta x$. Then,

$$\frac{1 - 4\lambda \sin^2(\frac{1}{2}\pi p\Delta x)}{e^{-\pi^2 p^2 \lambda (\Delta x)^2}} = e^{\lambda z^2} \left[1 - 2\lambda[1 - \cos(z)] \right] = [1 + \lambda z^2 + \tfrac{1}{2}\lambda^2 z^4 + \ldots][1 - \lambda z^2 + \frac{\lambda}{12} z^4 - \ldots]$$

$$= 1 + \tfrac{1}{2}\lambda \left[\tfrac{1}{6} - \lambda \right] z^4 - \frac{\lambda}{60} \left[20\lambda^2 - 5\lambda + \tfrac{1}{6} \right] z^6 + O(z^8) \qquad (z \to 0) . \quad (36)$$

Thus,

$$\left[\frac{1 - 4\lambda \sin^2(\frac{1}{2}\pi p\Delta x)}{e^{-\pi^2 p^2 \lambda (\Delta x)^2}} \right]^n = 1 + n\tfrac{1}{2}\lambda \left[\tfrac{1}{6} - \lambda \right] (p\pi\Delta x)^4 - \frac{n\lambda}{60} \left[20\lambda^2 - 5\lambda + \tfrac{1}{6} \right] (p\pi\Delta x)^6 + O((\Delta x)^8)$$

Using (33), (35) and (37), as $\Delta x \to 0$,

$$w_{m,n} = -\tfrac{1}{2} n\lambda \left[\tfrac{1}{6} - \lambda \right] (\pi\Delta x)^4 \sum_{p=1}^{k} p^4 b_p e^{-\pi^2 p^2 n\Delta t} \sin(\tfrac{mp\pi}{M}) + O((\Delta x)^6) \quad (38)$$

$$= -\tfrac{1}{2} n\Delta t \left[\tfrac{1}{6} - \lambda \right] \pi^4 (\Delta x)^2 \sum_{p=1}^{k} p^4 b_p e^{-\pi^2 p^2 n\Delta t} \sin(\tfrac{mp\pi}{M}) + n\Delta t \cdot O((\Delta x)^4) \quad (39)$$

The largest that $n\Delta t$ can be is $N\Delta t = T$, and $|\frac{1}{6} - \lambda| \le \frac{1}{3}$ for $0 \le \lambda \le \frac{1}{2}$. These inequalities show that (39) is is consistent with (34). When $\lambda = \frac{1}{6}$, we obtain $\frac{\lambda}{60}\left[20\lambda^2 - 5\lambda + \frac{1}{6}\right] = \frac{-1}{3240}$

$$w_{m,n} = -n\,\frac{1}{3240}\,(\pi\Delta x)^6 \sum_{p=1}^{k} p^6 b_p e^{-\pi^2 p^2 n\Delta t}\sin(\tfrac{mp\pi}{M}) + O((\Delta x)^8) \tag{40}$$

$$= -n\Delta t\,\frac{\pi^6}{540}\,(\Delta x)^4 \sum_{p=1}^{k} p^6 b_p e^{-\pi^2 p^2 n\Delta t}\sin(\tfrac{mp\pi}{M}) + n\Delta t\cdot O((\Delta x)^6)\,. \tag{41}$$

Round–off errors

The exact values $v_{m,n}$ in the formula (30) for the solution of the explicit difference problem may not be the values obtained by a computer or calculator. When using such devices, errors are introduced when rounding–off occurs (i.e., typically only a finite number of decimal places are retained in a computer). Let us therefore assume, for the sake of simplicity, that the net round–off errors incurred in the calculation of $v_{m,n+1}$ in terms of the previously calculated $v_{m,n}$'s is some constant, say r. We expect that, in the absence of cancellations, these round–off errors will increase as we increase the number of grid points. Let $P_{m,n}$ denote the **aggregate propagated error** at the (m,n)–th grid point. That is, if $v_{m,n}$ denotes the exact solution of our difference equation, then $\bar{v}_{m,n} \equiv v_{m,n} + P_{m,n}$ is our "observed value" at the grid point (m,n) (by the observed value, we mean the number obtained by our calculator, computer, etc.). Assuming that the round–off errors are not self–correcting (e.g., if there is a systematic error, such as always rounding down instead of up), it is reasonable to assume that $P_{m,n}$ will roughly satisfy the difference equation, with $0 < \lambda \le \frac{1}{2}$,

$$P_{m,n+1} = \lambda P_{m+1,n} + (1-2\lambda)P_{m,n} + \lambda P_{m-1,n} + r, \tag{42}$$

where r is the net round–off error introduced in passing from step n to step $n+1$. Suppose that there is no round–off error in the given initial and boundary data. Thus,

$$P_{0,n} = P_{M,n} = 0 \;\; (n = 0,\ldots,N) \quad \text{and} \quad P_{m,0} = 0 \;\; (m = 0,\ldots,M)\,. \tag{43}$$

By induction, it easily follows from (42) and (43) that $|P_{m,n}| \le n|r|$ (cf. Problem 5). Of course, $P_{m,n}$ is zero if $m = 0$ or $m = M$. Assuming a worst–case scenario for round–off errors, we have $P_{m,n} = nr$, for a given value of r, say determined empirically. The total round–off error, after N stages, is then $Nr = Tr/\Delta t$. Thus, the round–off error is inversely proportional to Δt (or $(\Delta x)^2$). However, as we will see, the leading term in the discretization error is directly proportional to Δt (or $(\Delta x)^2$) if $\lambda \neq 1/6$ (and to $(\Delta t)^2$ (or $(\Delta x)^4$), if $\lambda = 1/6$). In order to obtain an optimal result at some fixed point $(x,T) \equiv (m/M, N\Delta t) = (m/M, \lambda N/M^2)$, one should choose Δt, so that the following net sum of the discretization and round–off errors is as close to zero as possible (cf. (39) and (41)):

$$E(\Delta t) = u(m\Delta x, N\Delta t) - \bar{v}_{m,N} = u(m\Delta x, N\Delta t) - (v_{m,N} + rN) = w_{m,N}(\Delta t) - rT/\Delta t\ , \quad (44)$$

where

$$w_{m,N}(\Delta t) = \begin{cases} -\frac{T}{2}\left[\frac{1}{6\lambda} - 1\right]\sum_{p=1}^{k} e^{-\pi^2 p^2 T}(\pi p)^4\, b_p \sin\left(\frac{m\pi p}{M}\right)\Delta t + O((\Delta t)^2) & (\lambda \neq \frac{1}{6}) \\ -\frac{T}{15}\sum_{p=1}^{k} e^{-\pi^2 p^2 T}(\pi p)^6 b_p \sin\left(\frac{m p \pi}{M}\right)(\Delta t)^2 + O((\Delta t)^3) & (\lambda = \frac{1}{6})\ . \end{cases}$$

Suppose that $\lambda \neq 1/6$ and that the $O((\Delta t)^2)$ term is negligible, say because the optimal Δt is known to be much less than 1. Then dropping the $O((\Delta t)^2)$ term,

$$E(\Delta t) \approx \alpha\Delta t - rT/\Delta t\ , \quad \text{where} \quad \alpha = -\frac{T}{2}\left[\frac{1}{6\lambda} - 1\right]\sum_{p=1}^{k}(\pi p)^4 e^{-\pi^2 p^2 T}\, b_p \sin(\pi x)\ . \quad (45)$$

If α and r have the same sign (i.e., $\alpha r > 0$), then we can choose Δt so that $E(\Delta t) \approx 0$, namely

$$\alpha\Delta t - rT/\Delta t \approx 0 \quad \text{yields} \quad \Delta t \approx \sqrt{rT/\alpha} = \sqrt{|rT/\alpha|}\ . \quad (46)$$

If α and r have opposite signs, then $E(\Delta t)$ is never zero, but we can minimize it as follows

$$\frac{dE}{d(\Delta t)} \approx \alpha + rT(\Delta t)^{-2} = 0 \quad \text{yields} \quad \Delta t \approx \sqrt{-rT/\alpha} = \sqrt{|rT/\alpha|}\ . \quad (47)$$

Thus, under the above assumptions, in either case, the optimal value for Δt is given by $\Delta t \approx \sqrt{|rT/\alpha|}$. Since $\Delta t = \lambda(\Delta x)^2 = \lambda M^{-2}$, we have $M \approx \sqrt{\lambda/\Delta t}$. Thus, the optimal value for M (the number of subintervals in the x–direction) is approximately given by (for fixed x)

$$\boxed{M \approx \sqrt{\lambda/\Delta t} = \left|\frac{\alpha\lambda^2}{rT}\right|^{\frac{1}{4}} = \left|\frac{\lambda^2}{2r}\left[\frac{1}{6\lambda} - 1\right]\sum_{p=1}^{k} e^{-\pi^2 p^2 T}(\pi p)^4\, b_p \sin(\pi x)\right|^{\frac{1}{4}}.} \quad (48)$$

The optimal value of M when $\lambda = 1/6$ is the subject of Problem 6. In the following examples we test the validity of the estimate (48) in a definite problem, where $\lambda = 1/12$ and $\lambda = 1/4$. A systematic round–off error is introduced by instructing the computer to ignore all digits more than six places to the right of the decimal (e.g., replace each number α by $10^{-6}\cdot\mathrm{INT}(\alpha\cdot 10^{6})$ in all computations, where $\mathrm{INT}(x)$ is the greatest integer less than or equal to x).

Example 5. Consider

$$\text{D.E.}\quad u_t = u_{xx} \qquad 0 \le x \le 1\ ,\ 0 \le t \le 0.5$$

$$\text{B.C.}\quad u(0,t) = 0\ ,\ u(1,t) = 0$$

$$\text{I.C.}\quad u(x,0) = \sin(\pi x)\ .$$

Let $\lambda = 1/12$, $\Delta x = 1/M$ and $\Delta t = \lambda(\Delta x)^2 \equiv 1/(12M^2)$. Note that $T \equiv 0.5 = N\Delta t$ implies that $N = 6M^2$. Let $v_{m,n}$ $(0 \le m \le M\ ;\ 0 \le n \le N)$ be the exact solution (cf. (30) with $k = 1$, $p = 1$ and $b_1 = 1$) of the associated difference equations. Suppose that we are interested in the point $x = 0.5$ at time $t = 0.5$. Thus, we consider $v_{\frac{1}{2}M,N} = \left[1 - \frac{1}{3}\sin^2(\frac{\pi}{2M})\right]^N$. Assume that in the iterative computations for the numerical solution of the difference equations, a systematic round–off error is introduced by ignoring all digits past the sixth place to the right of the decimal in the computed (observed) values $\bar{v}_{m,n}$ for the $v_{m,n}$ (but $\lambda = 1/12$ is kept exact, as well as the initial values $v_{m,0}$). Use (48) to find the value of M for which $\bar{v}_{\frac{1}{2}M,N}$ is closest to $u(0.5,0.5)$. Check the result.

Solution. We use $\lambda = 1/12$, $k = p = 1$, and $b_p = 1$, note that α in (45) is given by $\alpha = -\frac{1}{4}\pi^4 e^{-\frac{1}{2}\pi^2}$, which has the same sign as $r \approx -10^{-7}$. Thus, (48) gives the optimal value

$$M \approx \left|\frac{\alpha\lambda^2}{rT}\right|^{\frac{1}{4}} = \pi\left[10^7/288\right]^{\frac{1}{4}} e^{-\pi^2/8} \approx 12.48\ldots\ .$$

For various values of M, in Table 1 below we have tabulated the observed numerical values $\bar{v}_{\frac{1}{2}M,N}$ (under the above truncation scheme), the exact solution $v_{\frac{1}{2}M,N}$ (up to ten places), and the accumulated round–off errors $\bar{v}_{\frac{1}{2}M,N} - v_{\frac{1}{2}M,N}$. The predicted optimal value of M is indeed about the best value $(M = 12)$ for which $\bar{v}_{\frac{1}{2}M,N}$ is closest to $u(0.5,0.5) = 0.0071918833\ldots$.

$\lambda = \frac{1}{12}$

M	N	$\bar{v}_{\frac{1}{2}M,N}$	$v_{\frac{1}{2}M,N}$	$\bar{v}_{\frac{1}{2}M,N} - v_{\frac{1}{2}M,N}$
2	24	.012576	.0125791152...	−.0000031...
4	96	.008182	.0076141395...	−.0000104...
6	216	.007589	.0076141395...	−.0000251...
8	384	.007378	.0074251909...	−.0000471...
10	600	.007270	.0073399752...	−.0000699...
12	864	.007193	.0072942670...	−.0001012...
14	1176	.007133	.0072669021...	−.0001339...
16	1536	.007091	.0072492196...	−.0001582...
18	1944	.007045	.0072371320...	−.0001921...
20	2400	.006997	.0072285034...	−.0002315...

Table 1 [u(0.5,0.5) = 0.0071918833...]

Example 6. Consider what happens in Example 5, when λ is changed from $1/12$ to $1/4$.

Solution. In this case, $\alpha = \frac{\pi^4}{12} e^{-\frac{1}{2}\pi^2}$ which has the opposite sign of $r \approx -10^{-7}$. Using (48), the optimal value of M at $x = 0.5$ is approximately given by

$$M \approx \left|\frac{\alpha\lambda^2}{rT}\right|^{\frac{1}{4}} = \pi\left[\frac{10^7}{96}\right]^{\frac{1}{4}} e^{-\pi^2/8} \approx 16.43\ldots .$$

In this case where α and r have opposite signs, $E(\Delta t) \approx \alpha\Delta t - Nr(\Delta t)^{-1}$ is never zero. Thus, for $M < 16$, we should observe $\bar{v}_{\frac{1}{2}M,N}$ approaching the exact value u(0.5,0.5) from below (since $\alpha > 0$), but not reaching u(0.5,0.5), and for $M > 16$, we should see $\bar{v}_{\frac{1}{2}M,N}$ moving back down, away from u(0.5,0.5). This is what we find in Table 2 below. Note that the minimum value for $E(\Delta t) \approx \alpha\Delta t - rT/\Delta t$ is $E(\sqrt{|rT/\alpha|}) \approx 2\sqrt{rT\alpha} = 0.0001080\ldots$. Adding this to $v_{8,512}$, we obtain 0.007190... , while $u(0.5,0.5) \approx 0.0071918\ldots$. Discrepancies arise not only because of the innaccurate prediction of the round–off error, but also because of the fact that we have dropped the $O((\Delta t)^2)$ term in going from (44) to (45). □

$\lambda = \frac{1}{4}$

M	N	$\bar{v}_{\frac{1}{2}M,N}$	$v_{\frac{1}{2}M,N}$	$\bar{v}_{\frac{1}{2}M,N} - v_{\frac{1}{2}M,N}$
2	8	.003906	.0039062500...	–.0000002...
4	32	.006297	.0063006033...	–.0000036...
6	72	.006784	.0067905358...	–.0000065...
8	128	.006952	.0069651178...	–.0000131...
10	200	.007026	.0070464573...	–.0000204...
12	288	.007062	.0070907817...	–.0000287...
14	392	.007079	.0071175554...	–.0000385...
16	512	.007082	.0071349516...	–.0000529...
18	648	.007082	.0071468870...	–.0000648...
20	800	.007076	.0071554286...	–.0000794...

Table 2 [u(0.5,0.5) = 0.0071918833...]

Remarks. In general, the guess for r could be wrong by an order of magnitude, and hence the predicted value of M could be wrong by a factor of $10^{\frac{1}{4}} \approx 1.78$. Thus, we have made exceptionally good guesses (but quite by accident) for r in Examples 5 and 6. The reader is asked to consider the case $\lambda = 1/6$ in Problem 6. In general, predicting round–off errors is difficult, and a probability density function of the error must be ascertained. In the above analysis, where a systematic error is assumed or introduced, one might expect the round–off error at the n–th stage to be of the form rn. If the round–off error is normally distributed with mean zero and variance σ^2, then the error at the n–th stage can be expected to be roughly in the interval $[-\sigma\sqrt{2n}, \sigma\sqrt{2n}\,]$, which is not as large as rn (for large n). However, this sort of error is worse, in the sense that it could be anywhere in this interval, and hence it is more difficult to correct, even though the correction is smaller. □

Summary 8.3

1. **Homogeneous ordinary difference equations with constant coefficieants :** The general solution of the first–order difference equation $y_{n+1} = ay_n$, $n = 0, 1, 2, \ldots$, is $y_n = K\,a^n$. The general solution of the second–order difference equation $y_{n+2} + a_1y_{n+1} + a_2y_n = 0$, $n = 0, 1, 2, \ldots$, can be found from the roots r_1 and r_2 of the auxiliary equation $r^2 + a_1r + a_2 = 0$, as follows :

$$y_n = c_1r_1^{\,n} + c_2r_2^n \qquad \text{for } r_1 \text{ and } r_2 \text{ real and distinct,}$$

$$y_n = (c_1 + c_2\cdot n)r_1^n \qquad \text{for } r_1 = r_2\,,$$

$$y_n = cR^n(c_1\cos(n\theta) + c_2\sin(n\theta)) \qquad \text{for } r_1 = R(\cos\theta + i\sin\theta)\,,\ r_2 = \bar{r}_1\,.$$

2. **An exact solution of the explicit difference scheme for a heat problem (Theorem 1) :**

The unique solution of the problem

$$\begin{aligned}
&v_{m,n+1} = \lambda v_{m+1,n} + (1-2\lambda)v_{m,n} + \lambda v_{m-1,n}, \quad (m = 1, \ldots, M-1\ ;\ n = 0, \ldots, N-1),\\
&v_{0,n} = 0, \quad v_{M,n} = 0, \quad (n = 0, \ldots, N), \qquad\qquad (S1)\\
&v_{m,0} = \sum_{p=1}^{k} b_p \sin(\tfrac{mp\pi}{M}), \quad (m = 0, \ldots, M),
\end{aligned}$$

is

$$v_{m,n} = \sum_{p=1}^{k} b_p\,[1 - 4\lambda\sin^2(\tfrac{p\pi}{2M})]^n \sin(\tfrac{mp\pi}{M}). \qquad (S2)$$

Problem (S1) is encountered in applying the explicit difference method (with $\Delta x = 1/M$, $\Delta t = \lambda(\Delta x)^2$ and $T = N\Delta t$) to the problem

$$\begin{aligned}
&\text{D.E. } u_t = u_{xx}\,, \qquad 0 \le x \le 1,\ 0 \le t \le T,\\
&\text{B.C. } u(0,t) = 0,\ u(1,t) = 0, \qquad\qquad (S3)\\
&\text{I.C. } u(x,0) = \sum_{p=1}^{k} b_p\sin(\pi px)\,.
\end{aligned}$$

3. The estimate of the local discretization error $w_{m,N} = u(m\Delta x, n\Delta t) - v_{m,N}$: The local discretization error, when the explicit difference method (S1) is applied to (S3), is

$$w_{m,N}(\Delta t) = \begin{cases} -\frac{T}{2}\left[\frac{1}{6\lambda} - 1\right]\sum_{p=1}^{k} e^{-\pi^2p^2T}(\pi p)^4\, b_p \sin\left(\frac{m\pi p}{M}\right)\Delta t + O((\Delta t)^2) & (\lambda \neq \frac{1}{6}) \\ -\frac{T}{15}\sum_{p=1}^{k} e^{-\pi^2p^2T}(\pi p)^6 b_p \sin\left(\frac{mp\pi}{M}\right)(\Delta t)^2 + O((\Delta t)^3) & (\lambda = \frac{1}{6})\,. \end{cases}$$

4. The net error in the presence of a round–off error : The round–off error $P_{m,n}$ at the grid point m,n is given by $\bar{v}_{m,n} = v_{m,n} + P_{m,n}$, where $v_{m,n}$ is the exact solution of problem (S1), and $\bar{v}_{m,n}$ is the "observed value" obtained iteratively with a computer. For a round–off error of the form $P_{n,m} = rn$, the total error at the grid point (m,N) is

$$E(\Delta t) = u(m\Delta x, T) - \bar{v}_{m,N} = \left[u(m\Delta x, T) - v_{m,N}\right] + \left[v_{m,N} - \bar{v}_{m,N}\right] = w_{m,N}(\Delta t) - rT/\Delta t$$

Note that $w_{m,N}(\Delta t) \to 0$ as $\Delta t \to 0$ (cf. Theorem 1 of Section 8.2), but $rT/\Delta t \to \pm\infty$, as $\Delta t \to 0$. Thus (cf. (47)), there is a strictly positive value of Δt (or equivalently, for $M = (\Delta x)^{-1} = \sqrt{\lambda/\Delta t)}$), such that $E(\Delta t)$ is minimal. In the case where $\lambda \neq 1/6$ and the $O((\Delta t)^2)$ term in $w_{m,N}(\Delta t)$ is negligible, the optimal value of M is

$$M \approx \sqrt{\lambda/\Delta t} = \left|\frac{\alpha\lambda^2}{rT}\right|^{\frac{1}{4}} = \left|\frac{\lambda^2}{2r}\left[\frac{1}{6\lambda} - 1\right]\sum_{p=1}^{k} e^{-\pi^2p^2T}(\pi p)^4\, b_p \sin(\pi x)\right|^{\frac{1}{4}}. \qquad \text{(S4)}$$

Thus, increasing M beyond this point will actually yield poorer results, under our assumptions. The result (S4) is illustrated in Examples 5 and 6. The case where $\lambda = 1/6$ is covered in Problem 6 . For a Gaussian random round–off error, with mean zero and variance σ^2 at each step, the error is likely to be in $[-\sigma\sqrt{2n}\,,\ \sigma\sqrt{2n}]$, after n steps, but it is hard to predict where it is in this interval for the purpose of corrections.

Exercises 8.3

1. Consider the difference equation

$$y_{n+2} + a_1 y_{n+1} + a_0 y_n = 0, \quad n = 0, 1, 2, \ldots, \qquad (*)$$

where a_0 and a_1 are real constants. Let r_1 and r_2 be the roots of the associated auxiliary equation $r^2 + a_1 r + a_0 = 0$.

(a) If r_1 and r_2 are real and distinct, verify that $y_n = c_1 r_1^n + c_2 r_2^n$, where c_1 and c_2 are constants, satisfies the difference equation (*).

(b) If $r_1 = a + ib$ and $r_2 = a - ib$, verify that $y_n = R^n(c_1 \cos(n\theta) + c_2 \sin(n\theta))$, where $R = \sqrt{a^2 + b^2}$, $a = R\cos\theta$, and $b = R\sin\theta$, satisfies the difference equation (*).

2. Find the general solution of each of the following difference equations :

(a) $y_{n+2} - 5y_{n+1} + 6y_n = 0$ (b) $y_{n+2} - 2y_{n+1} + y_n = 0$

(c) $y_{n+2} - 2y_{n+1} + 2y_n = 0$ (d) $y_{n+2} - 2\cos\theta y_{n+1} + y_n = 0$.

3. (a) Verify directly that $v_{m,n} = \left[1 - 4\lambda\sin^2(\frac{p\pi}{2M})\right]^n \sin(\frac{mp\pi}{M})$, where p, m, n and M $(m \le M)$ are positive integers and where $\lambda > 0$, satisfies the (partial) difference equation $v_{m,n+1} = \lambda v_{m+1,n} + (1-2\lambda)v_{m,n} + \lambda v_{m-1,n}$. Is the assumption $\lambda \le \frac{1}{2}$ necessary, here ?

(b) Show that in applying the explicit difference method to the problem (31), $w_{m,n} = O((\Delta x)^2)$ as $\Delta x \to 0$, provided $\Delta t/(\Delta x)^2 \le A$, for some positive constant A, say 10^6 (i.e., the restriction $\lambda \le \frac{1}{2}$ is unnecessary for such I.C. given by a finite Fourier sine series.)

4. (a) For any constants A and B , show that $\lim_{s \to 0} [1-4\lambda\sin^2(As)]^{B/s^2} = e^{-4\lambda A^2 B}$.

Hint. For $|s|$ small, take a logarithm and apply L'Hôspital's rule twice.

(b) Use (a) to show directly that the local discretization error $w_{m,n}$ (cf. (33)) tends to 0, as $\Delta x \to 0$ (or $M \to \infty$), where $\lambda = \Delta t/(\Delta x)^2$ is fixed.

5. Verify that if $P_{m,n}$ satisfies the problem

$$P_{m,n+1} = \lambda P_{m+1,n} + (1-2\lambda)P_{m,n} + \lambda P_{m-1,n} + r\,,\quad 1 \le m \le M-1\,,\ 0 \le n \le N-1$$

$$P_{0,n} = P_{M,n} = 0\,,\quad 0 \le n \le N,$$

$$P_{m,0} = 0\,,\quad 0 \le m \le M\,,$$

where r is independent of m and n, and $\lambda \le \frac{1}{2}$, then $|P_{m,n}| \le n|r|$. Is $P_{m,n} \ge 0$, if $r \ge 0$?

6. Consider the problem

D.E. $u_t = u_{xx}\,,\quad 0 \le x \le 1\,,\ 0 \le t \le 0.5$

B.C. $u(0,t) = 0\,,\ u(1,t) = 0$

I.C. $u(x,0) = \sin(\pi x)$.

As in Examples 5 and 6, we will analyze the combined effect of discretization and round–off errors, but this time when $\lambda = 1/6$. We take $\Delta x \equiv M^{-1}$ for M a positive integer, and $\Delta t = \lambda(\Delta x)^{-2} = M^{-2}/6$. With $T = 0.5$, we have $N = 0.5/\Delta t = 3M^2$.

(a) For what values of m and n is $(m\Delta x, n\Delta t) = (0.5,0.5)$. Must M be even for this ?

(b) What is the exact value of $v_{\frac{1}{2}M,N}$ for the corresponding explict difference problem ?

(c) Show that the local discretization error at $m = \frac{1}{2}M$, $n = N$ is of the form

$$w_{\frac{1}{2}M,N}(\Delta t) = -\frac{\pi^6}{30}e^{-\frac{1}{2}\pi^2}(\Delta t)^2 + O((\Delta t)^3).$$

(d) Drop the $O((\Delta t)^3)$ term in (c) and assume that the round–off error after the n–th step is of the form nr. What is the value of M for which the observed value $\overline{v}_{\frac{1}{2}M,N}$ is closest to $u(0.5,0.5)$? Consider the cases $r > 0$ and $r < 0$ separately.

(e) Impose $r \approx -10^{-7}$, by dropping (at each stage n) all digits after the sixth place to the right of the decimal in the iteratively computed values for the $v_{m,n}$. Test your answer in (d).

8.4 An Overview of Some Other Numerical Methods for PDEs (Optional)

We begin this section with a brief description of some finite difference methods for parabolic, hyperbolic and elliptic PDEs (e.g., the heat, wave and Laplace's equation, respectively). We also consider the method of lines for problems in higher dimensions, and the Rayleigh–Ritz approximation for solutions of certain linear boundary value problems. Since the the procedures we touch upon, ultimately all lead to the problem of solving systems of algebraic equations, we also discuss some iterative methods. Throughout this section we provide references, where the reader may find applications, proofs and further developments.

We will use the notation of Sections 8.1 and 8.2 . Thus, given a rectangle $W: 0 \leq x \leq L$, $0 \leq t \leq T$, and positive integers M and N, we set $\Delta x = L/M$ and $\Delta y = T/N$. For a function $u(x,y)$ on W, we set $u_{i,j} = u(i\Delta x, j\Delta t)$, for $i = 0, ..., M$ and $j = 0, ..., N$. It will be convenient to use the following standard notation for the **second (–order) central differences** :

$$\delta_x^2\, u_{i,j} \equiv u_{i+1,j} - 2\, u_{i,j} + u_{i-1,j} \qquad (i = 1, \ldots, M-1)$$

and

$$\delta_t^2\, u_{i,j} \equiv u_{i,j+1} - 2\, u_{i,j} + u_{i,j-1} \qquad (j = 1, \ldots, N-1). \tag{1}$$

Of course, these definitions make sense for any $(M+1)\times(N+1)$ array (matrix) of real numbers.

Finite difference methods

We have used the explicit difference method in the previous sections for the problem

$$\begin{aligned} &\text{D.E.} \quad u_t = u_{xx} \quad 0 \leq x \leq L\,,\ 0 \leq t \leq T \\ &\text{B.C.} \quad u(0,t) = A(t)\,,\ u(L,t) = B(t) \\ &\text{I.C.} \quad u(x,0) = f(x) \quad (A(0) = f(0)\,,\ B(0) = f(L))\,. \end{aligned} \tag{2}$$

It is straightforward to solve the system of difference equations $(i = 1,\ldots,M-1\ ;\ j = 0,\ldots,N-1)$

$$v_{i,j+1} = \lambda v_{i+1,j} + (1-2\lambda)v_{i,j} + \lambda v_{i-1,j} \qquad (\lambda \equiv \Delta t/(\Delta x)^2) \tag{3}$$

or equivalently,

$$\frac{v_{i,j+1} - v_{i,j-1}}{\Delta t} = \frac{\delta_x^2\, v_{i,j}}{(\Delta x)^2}\,, \tag{4}$$

subject to

$$v_{0,j} = A(j\Delta t)\,,\ v_{M,j} = B(j\Delta t) \ \text{ and } \ v_{i,0} = f(i\Delta x) \qquad (i = 0,\ldots,M\ ;\ j = 0,\ldots,N)\,.$$

However, there are two weaknesses in this method. First, it is only **conditionally stable**, in the sense that a condition on λ is needed (i.e., $\lambda \le \frac{1}{2}$). Second, the local discretization error $w_{i,j} \equiv u_{i,j} - v_{i,j} = O(\Delta t + (\Delta x)^2)$ as Δt and $\Delta x \to 0$, instead of $O((\Delta t)^2 + (\Delta x)^2)$, (cf. Theorem 1 of Section 8.2). These weaknesses can be overcome, if (4) is replaced by

$$\frac{v_{i,j+1} - v_{i,j-1}}{\Delta t} = \frac{\delta_x^2 v_{i,j+1} + \delta_x^2 v_{i,j}}{2(\Delta x)^2}$$

or

$$v_{i+1,j+1} - (2 + \tfrac{2}{\lambda})v_{i,j+1} + v_{i-i,j+1} = -v_{i+1,j} + (2 - \tfrac{2}{\lambda})v_{i,j} - v_{i-1,j} \qquad (5)$$

Observe that the right–hand side of (5) was found in the previous stage j, and the $v_{i,j+1}$ appear in three different equations (Why ?) involving $v_{i-2,j+1}, \dots, v_{i+2,j+1}$. Thus, the $v_{i,j+1}$ are not simply given *explicitly* in terms of the previous values as in (3), but rather one must solve a nontrivial system of linear equations. Consequently, this method, which is called the **Crank–Nicolson method**, is an example of an ***implicit* difference method** . This method does *not* require $\lambda \le \frac{1}{2}$, and its local discretization error is $O((\Delta t)^2 + (\Delta x)^2)$ (cf. [Isaacson and Keller] for precise statements and proofs).

For hyperbolic PDEs, we take as our model the wave problem (cf. Chapter 5)

D.E. $u_{tt} = a^2 u_{xx}$, $\quad 0 \le x \le L, \; 0 \le t \le T$

B.C. $u(0,t) = A(t)$, $u(L,t) = B(t)$

I.C. $u(x,0) = f(x)$, $u_t(x,0) = g(x)$.

In this case, the explicit difference method entails solving the trivial system

$$\delta_t^2 v_{i,j} = a^2 \mu^2 \delta_x^2 v_{i,j} \qquad (\mu \equiv \frac{\Delta t}{\Delta x}),$$

while for the **implicit method** we must contend with the nontrivial system

$$\delta_t^2 v_{i,j} = a^2 \mu^2 \tfrac{1}{2}(\delta_x^2 v_{i,j+1} + \delta_x^2 v_{i,j-1}) .$$

Here both of the methods have local discretization error $O((\Delta t)^2 + (\Delta x)^2)$, but the explicit method requires $a^2\mu^2 < 1$ (**the Courant–Friedrichs–Levy condition**), while no such condition is necessary for the implicit method (cf. [Isaacson and Keller] or [Mitchell). For these methods, the difference approximations to u_t in the I.C. are required to be $O((\Delta t)^2)$ approximations.

For elliptic PDEs, we consider the following Dirichlet problem for Poisson's equation.

$$\text{D.E.} \quad u_{xx} + u_{yy} = \rho(x,y), \qquad 0 \le x \le L, \ 0 \le y \le T$$

$$\text{B.C.} \quad \begin{cases} u(x,0) = f(x), \ u(x,T) = g(x) \\ u(0,y) = h(y), \ u(L,y) = k(y). \end{cases} \tag{6}$$

Again, using the second differences, we arrive at the system of equations

$$\frac{1}{(\Delta x)^2} \delta_x^2 v_{i,j} + \frac{1}{(\Delta y)^2} \delta_y^2 v_{i,j} = \rho_{i,j}, \tag{7}$$

where

$$v_{i,0} = f(i\Delta x), \ v_{i,N} = g(i\Delta x), \ v_{0,j} = h(j\Delta y), \ v_{M,j} = k(j\Delta y) \text{ and } \rho_{i,j} = \rho(i\Delta x, j\Delta y).$$

This method again has a local discretization error $O((\Delta x)^2+(\Delta y)^2)$. In the special case when $\Delta x = \Delta y$, (7) reduces to the following **five–point formula**

$$v_{i,j} = \tfrac{1}{4}(v_{i-1,j} + v_{i+1,j} + v_{i,j-1} + v_{i,j+i}) + (\Delta x)^2 \rho_{i,j},$$

(cf. [Forsythe and Wasow]). If the Neumann problem is considered, then the normal derivative is usually expressed via a higher–order approximation (cf. [Vemuri and Karplus], and for additional techniques, see [Ames]).

Remarks. (1) If the region is a disk or wedge, then it is often desirable to use polar coordinates. For complicated regions, the theory of finite elements is used (cf. [Ciarlet, Kesavan and Ranjan] or [Zienkiewicz]).
(2) The local discretization errors associated with the above finite difference methods, can be further reduced by considering higher–order Taylor series approximations (see also the **Richardson extrapolation procedure for higher–order approximations** in Problem 8 of Section 8.1). These approximations should be considered in today's computing environment with symbolic, algebraic manipulation systems and/or software, such as Macsyma, Maple, Mathematica and Reduce. □

Solutions in several dimensions

In higher dimensions a variant of the finite difference method is called the **method of lines**. The basic idea of this method is to discretize all but one of the independent variables. This then leads to a system of ODEs which can be solved by any of the standard methods, such as Runge–Kutta, Hammings, Milne or Adams–Moulton methods (for details, see [Hildebrand, 1956] or [Milne, 1953]). For instance, in problem (6), if the y–variable remains continuous and we discretize in the x–variable, then (6) yields the following system of ODEs.

$$\frac{v_{i-1}(y) - 2v_i(y) + v_{i+1}(y)}{(\Delta x)^2} + \frac{d^2 v_i}{dy^2} = \rho(i\Delta x, y)$$

$$v_i(0) = f(i\Delta x)\,,\quad v_i(L) = g(i\Delta x) \qquad (i = 0, 1, \ldots, M)\,. \tag{8}$$

Using any of the above cited methods, (8) reduces to the problem of solving a system of algebraic equations. One could also try to solve this system exactly by Laplace transform methods.

A variational approximation method

It often happens that the solution u of a boundary value problem for a PDE on a region R also minimizes a certain integral $I(v)$ over R (e.g., $I(v) \equiv \int_R (v_x)^2 + (v_y)^2\,dxdy$, in the case where the PDE is Laplace's equation), over all suitable competing functions v which satisfy the B.C., but not necessarily the D.E. . Then the boundary–value problem is equivalent to the minimization problem of finding a suitable function v which makes the integral $I(v)$ the smallest. There are usually infinitely many linearly independent competing functions which satisfy the B.C. (i.e., the set of competing functions is usually infinite–dimensional). However, the minimization problem can often be solved approximately by what is known as a **variational approximation method**. Essentially, such methods consist of first selecting a family of competing functions which depend "nicely" on a finite number of new variables, say $v(c_1,c_2,\ldots,c_n) = v(x,y;c_1,c_2,\ldots,c_n)$, where c_1, c_2, ... are the new variables. Let

$$F(c_1,c_2,\ldots,c_n) = I(v(c_1,c_2,\ldots,c_n))\,. \tag{9}$$

One then finds the minimum of the function F which depends only on the finite number of variables. The possible values for $c_1,\ldots,c_n$ are typically found by solving the system

$$\frac{\partial F}{\partial c_1}(c_1,\ldots,c_n) = 0,\quad \frac{\partial F}{\partial c_2}(c_1,\ldots,c_n) = 0\,,\ \ldots,\quad \frac{\partial F}{\partial c_n}(c_1,\ldots,c_n) = 0 \tag{10}$$

of algebraic equations and testing which of the solutions, if any, is an absolute minimum. Of course, the function in the family associated with an optimal set of values for c_1, ..., c_n will be the best function *within the family*, as far as the minimization of I is concerned, but an exact minimizing function for I may not be in the family in the first place. In order to get a reasonably good approximate solution this way, one must choose a large family, or at least a family where some of the members are expected to nearly minimize I. Thus, in this approach, it is helpful to have either some feeling for the qualitative nature of the solution, or a corresponding measure of perseverance. In the special case where the PDE and B.C. are linear and the integral

$I(v)$ is quadratic (i.e. $I(\alpha v) = \alpha^2 I(v)$ for all constants α), one can take $v(c_1,...,c_n) = g_0 + c_1 g_1 + ... + c_n g_n$, were g_0 is a guess for the solution, which at least meets the (possibly inhomogeneous) B.C. and $g_1, ..., g_n$ are functions which satisfy the homogeneous B.C. . Moreover, $F(c_1,...,c_n)$ will be a quadratic polynomial in $c_1,...,c_n$, and (10) will thus be a linear system of equations.

The above discussion makes sense for problems in which the unknown function has any finite number of independent variables. For definiteness, consider the one–dimensional problem

$$\text{D.E.} \quad -\frac{d}{dx}\left[K(x)\frac{dy}{dx} \right] + q(x)y = f(x) , \qquad 0 \le x \le 1$$

$$\text{B.C.} \quad y(0) = y(1) = 0 , \tag{11}$$

where $K(x)$ is a positive C^1 function, and $q(x)$ and $f(x)$ are continuous, with $q(x) \ge 0$. It can be proven that there is only one C^2 function $y(x)$ which solves this problem . Let

$$I(w) \equiv \int_0^1 \left[K(x)[w'(x)]^2 + q(x)[w(x)]^2 - 2f(x)w(x)\right] dx , \tag{12}$$

where w is in the set, say S, of all C^2 functions which are zero at $x = 0$ and $x = 1$. It can be shown (cf. [Schultz]) that the solution $y(x)$ of the problem (11) is the unique function for which $I(y) \le I(w)$ for all of the functions w in S. In the variational method, known as the **Rayleigh–Ritz procedure**, one selects linearly independent functions $y_1(x), y_2(x), ..., y_n(x)$ in S. A suitable family of functions is then $c_1y_1 + c_2y_2 + ... + c_ny_n$, where $c_1, ..., c_n$ are arbitrary constants. In the general spirit of variational methods, one then seeks to minimize (cf. (9))

$$F(c_1 ,..., c_n) = I(c_1y_1 + ... + c_ny_n) = \sum_{i,j=1}^n a_{ij}c_ic_j - \sum_{p=1}^n 2b_pc_p ,$$

where

$$a_{ij} = \int_0^1 K(x)y_i'(x)y_j'(x) + q(x)y_i(x)y_j(x)\, dx , \text{ and } b_p = \int_0^1 f(x)y_p(x)\, dx .$$

Differentiation of F with respect to each of $c_1 ,..., c_n$ yields the system of linear equations

$$\sum_{j=1}^n a_{ij}\, c_j = b_i \quad (i = 1, ..., n) . \tag{13}$$

If $(\hat{c}_1 ,..., \hat{c}_k)$ is a solution of the system (13), then $\hat{c}_1y_1(x) + ... + \hat{c}_ky_k(x)$ is called a **Rayleigh–Ritz approximation** to the solution of the problem (11).

Remarks. (1) The Rayleigh Ritz procedure can be extended to higher dimensions. For example, one can, replace $-\frac{d}{dx}\left[K(x)\frac{dy}{dx}\right]$ by the negative of the Laplacian $(-\Delta w)$ on a bounded region or compact manifold with boundary (cf. Chapter 9), and the lead term in $I(w)$ in (12) is then replaced by the norm–square of the gradient (i.e., $\|\nabla w\|^2$).

(2) A generalization of the Rayleigh–Ritz method, known as the **Galerkin method**, is applicable to more general boundary value problems. For an excellent treatment of variational methods, suitable for engineering and science students, consult [Schultz] (cf. also [Birkhoff and Lynch]). □

Computational linear algebra and iterative methods

We have seen above that the numerical solution of PDEs eventually becomes a problem of solving systems of algebraic equations. If the resulting system is linear, then it can be written as

$$\sum_{j=1}^{n} a_{ij}\, x_j = b_i \quad (i = 1, \ldots, n)\,, \text{ or in matrix notation,} \quad \mathbf{A}\mathbf{x} = \mathbf{b}\,. \tag{14}$$

There are two types of methods for solving such systems : **direct methods** (which yield a solution without an initial guess) and **iterative methods** (in which a sequence of approximate solutions with increasing accuracy are produced from an initial guess). If n is small, say $n < 1000$, or if $\mathbf{A}$ has a **band structure** (i.e., $a_{i,j} = 0$, if $|i-j| > k$ for some k considerably smaller than n), the direct methods of Gaussian elimination, band elimination, or matrix factorization techniques can be effectively employed. The main problem with the direct methods is that they require enormous amounts of computer memory. Thus, if the system is large, say $n > 20{,}000$ (cf. [Varga] or [D. M. Young]), then we must use some iterative methods which require much less memory. However, since the iterative methods involve more arithmetic operations, they are somewhat slower. Also, if the algebraic system is not linear, then the direct methods do not apply and again iterative methods must be considered. For large linear systems, a combination of direct and iterative methods (e.g., conjugate gradient methods) are often more efficient than either type by itself (cf. [Birkhoff and Lynch]).

Not every linear system has a solution (e.g., $x + y = 1$, $2x + 2y = 1$). If the linear system (14) has a unique solution for every possible sequence $b_1, \ldots, b_n$ then the the matrix $\mathbf{A} = [a_{ij}]$ is called **nonsingular** or **invertible**. In the case where $\mathbf{A}$ is invertible, there is a unique matrix $C = [c_{ij}]$, such that for all sequences $(b_1, \ldots, b_n)$,

$$\sum_{j=1}^{n} a_{ij}\, x_j = b_i\,, \quad \text{if and only if} \quad x_i = \sum_{j=1}^{n} c_{ij}\, b_j \qquad (i = 1,\ldots,n). \tag{15}$$

Note that once the inverse matrix is found, the solution of the problem on the left in (15) is immediately given by the equations on the right. It is common to write the matrix $\mathbf{C}$ as $\mathbf{A}^{-1}$. Direct methods can be used to find the matrix $\mathbf{A}^{-1}$ (if it exists) according to various algorithms which are best covered in a linear algebra course. However, in solving linear systems (14), one almost never computes $\mathbf{A}^{-1}$, but rather the system is put into a simple "triangular" form, such as

$$x_1 = e_1 \ , \ d_{21}x_1 + x_2 = e_2 \ , \ d_{31}x_1 + d_{32}x_2 + x_3 = e_3 \ , \ \ldots \tag{16}$$

in which each equation is easily solved by using the solution of the previous equation. If **A** is nonsingular, this form is found using algorithms which are standard in a linear algebra course. Here we will describe some iterative methods which are usually not covered in such courses.

It is good if one can easily determine that a matrix **A** is invertible, for then we at least know that the system (14) will have a solution. However, in general this may be at least as difficult as finding $\mathbf{A}^{-1}$, unless **A** has a special form. For example, in systems of difference equations for PDEs, it frequently happens that the nonzero entries of the matrix **A** for the system are near the **diagonal** formed by the entries $a_{11}, a_{22}, \ldots, a_{nn}$ (i.e., **A** has a band structure). This happens because each of the difference equations only involves the values at grid points which are close to one another. The matrix **A** is called **diagonally dominant**, if

$$\sum_{j=1,\ j\neq i}^{n} |a_{ij}| < |a_{i,i}| \ , \quad i = 1,\ldots, n \ ,$$

or equivalently, (17)

$$|a_{i1}| + |a_{i2}| + \ldots + |a_{in}| < 2|a_{ii}| \ , \quad i = 1,\ldots, n.$$

in which case **A** can be shown to be invertible.

For example, consider the heat problem

$$\begin{aligned} &\text{D.E.} \quad u_t = u_{xx} \ , \qquad 0 \le x \le L \ , \ t \ge 0 \\ &\text{B.C.} \quad u(0,t) = 0 \ , \ u(L,t) = 0 \\ &\text{I.C.} \quad u(x,0) = f(x) \ . \end{aligned} \tag{18}$$

Using the backward difference approximation to the D.E., (18) becomes

$$\begin{aligned} &(1+2\lambda)v_{i,j} - \lambda v_{i+1,j} - \lambda v_{i-1,j} = v_{i,j-1} && (1 \le i \le M-1 \ ; \ 1 \le j \le N) \\ &v_{0,j} = v_{M,j} = 0 && (j = 0,\ldots,N) \\ &v_{i,0} = f(i\Delta x) && (i = 0,\ldots,M) \ , \end{aligned} \tag{19}$$

where $\lambda \equiv \Delta t/(\Delta x)^2$. If $v_{i,j-1}$ $(i = 0,\ldots,M)$ have been determined, then according to (19), the $v_{i,j}$ are found by solving the following system

$$(1+2\lambda)v_{1,j} - \lambda v_{2,j} + 0\cdot v_{3,j} + 0\cdot v_{4,j} + \dots + 0\cdot v_{M-1,j} = v_{1,j-1}$$

$$-\lambda v_{1,j} + (1+2\lambda)v_{2,j} - \lambda v_{3,j} + 0\cdot v_{4,j} + \dots + 0\cdot v_{M-1,j} = v_{2,j-1}$$

$$\vdots \qquad\qquad \vdots$$

$$0\cdot v_{1,j} + 0\cdot v_{2,j} + \dots + 0\cdot v_{M-2,j} + (-\lambda)v_{M-1,j} + (1+2\lambda)v_{M-1,j} = v_{M-1,j-1}\,.$$

The following matrix **A** for this system is clearly diagonally dominant (and hence nonsingular) :

$$\begin{bmatrix} (1+2\lambda) & -\lambda & 0 & \cdots & & & & 0 \\ -\lambda & (1+2\lambda) & -\lambda & 0 & \cdots & & & 0 \\ 0 & & -\lambda & (1+2\lambda) & -\lambda & 0 & \cdots & 0 \\ \vdots & & & & & & & \vdots \\ & & & & & & & -\lambda \\ 0 & \cdots & & & & 0 & -\lambda & (1+2\lambda) \end{bmatrix}.$$

In iterative methods one seeks to generate an infinite sequence of vectors

$$\mathbf{x}^{(0)} = (x_1^{(0)}, x_2^{(0)}, \dots, x_n^{(0)}),\quad \mathbf{x}^{(1)} = (x_1^{(1)}, x_2^{(1)}, \dots, x_n^{(1)}),\quad \mathbf{x}^{(2)} = (x_1^{(2)}, x_2^{(2)}, \dots, x_n^{(2)}), \dots$$

which converges to the true solution vector $\mathbf{x} = (x_1, x_2, \dots, x_n)$ of $\mathbf{Ax} = \mathbf{b}$, in the sense that

$$\|\mathbf{x}^{(k)} - \mathbf{x}\|^2 \equiv \sum_{i=1}^{n} (x_i^{(k)} - x_i)^2 \to 0\,, \quad \text{as } k \to \infty \tag{20}$$

(i.e., the length or norm of the difference tends to 0). The basic steps in an iterative procedure are outlined as follows :

(a) Write the matrix **A** as a the difference of two matrices, say $\mathbf{A} = \mathbf{M} - \mathbf{N}$ (called a **splitting** of **A**), where **M** is chosen to be nonsingular matrix whose inverse is easily computed (e.g., if **M** is a diagonal matrix with nonzero diagonal entries, then $\mathbf{M}^{-1}$ is obtained by replacing each diagonal entry by its reciprocal).

(b) Select *any* initial vector $\mathbf{x}^{(0)}$.

(c) For $k = 0, 1, 2, \dots$, define inductively

$$\mathbf{Mx}^{(k+1)} = \mathbf{Nx}^{(k)} + \mathbf{b} \quad \text{or equivalently}, \quad \mathbf{x}^{(k+1)} = \mathbf{Tx}^{(k)} + \mathbf{M}^{-1}\mathbf{b}\,, \tag{21}$$

where $\mathbf{T} = \mathbf{M}^{-1}\mathbf{N}$ (matrix product) is known as the **iteration matrix**.

Theorem 1 (A Convergence Result). **Suppose that the iteration matrix T has the property, that**

$$\text{for some fixed } r < 1, \ \|\mathbf{T}\,\mathbf{y}\| \le r\|\mathbf{y}\|, \text{ for every vector } \mathbf{y} \tag{22}$$

(i.e., T contracts the lengths of all nonzero vectors). **Then the sequence** $\mathbf{x}^{(0)}, \mathbf{x}^{(1)}, \mathbf{x}^{(2)}, \ldots$ **converges to the unique solution x of** $\mathrm{Ax} = \mathrm{b}$. **In particular, A is nonsingular.**

Proof. First we establish the uniqueness of the solution. Suppose that $\mathbf{Ay} = \mathbf{b}$ and $\mathbf{Az} = \mathbf{b}$. Then $\mathbf{A}(\mathbf{y}-\mathbf{z}) = \mathbf{b} - \mathbf{b} = \mathbf{0}$. Thus, $\mathbf{0} = \mathbf{N}^{-1}(\mathbf{M} - \mathbf{N})(\mathbf{y}-\mathbf{z}) = (\mathbf{T} - \mathbf{I})(\mathbf{y}-\mathbf{z})$ or $\mathbf{T}(\mathbf{y}-\mathbf{z}) = \mathbf{y}-\mathbf{z}$. However, by property (22), $\|\mathbf{T}(\mathbf{y}-\mathbf{z})\| \le r\|(\mathbf{y}-\mathbf{z})\|$, which cannot be true unless $\mathbf{y} = \mathbf{z}$ (Why ?). For linear systems of n equations in n unknowns ($n < \infty$), the uniqueness of solutions will also ensure the existence of solutions, as is proved in linear algebra courses. (This is false if $n = \infty$.) While a modification (using the Bolzano–Weierstrass Theorem of Appendix A.4) of the following argument yields this existence fact, for simplicity, we will assume the existence of the solution $\mathbf{x}$, and prove that the sequence $\mathbf{x}^{(0)}, \mathbf{x}^{(1)}, \mathbf{x}^{(2)}, \ldots$, defined in step (c), converges to $\mathbf{x}$. Note that if $\mathbf{Ax} = \mathbf{b}$, then $\mathbf{x} - \mathbf{Tx} = \mathbf{x} - \mathbf{M}^{-1}\mathbf{Nx} = \mathbf{M}^{-1}(\mathbf{M} - \mathbf{N})\mathbf{x} = \mathbf{M}^{-1}\mathbf{Ax} = \mathbf{M}^{-1}\mathbf{b}$. Thus, $\mathbf{x} = \mathbf{Tx} + \mathbf{M}^{-1}\mathbf{b}$, and we have

$$\|\mathbf{x} - \mathbf{x}^{(k+1)}\| = \|\mathbf{Tx} + \mathbf{M}^{-1}\mathbf{b} - \mathbf{Tx}^{(k)} - \mathbf{M}^{-1}\mathbf{b}\| = \|\mathbf{T}(\mathbf{x} - \mathbf{x}^{(k)})\| \le r\|\mathbf{x} - \mathbf{x}^{(k)}\| \ \ldots$$

$$\le r^2\|\mathbf{x} - \mathbf{x}^{(k-1)}\| \le \ldots \le r^{k+1}\|\mathbf{x} - \mathbf{x}^{(0)}\|.$$

Since $r^k \to 0$, as $k \to \infty$, this shows that $\|\mathbf{x} - \mathbf{x}^{(k)}\| \to 0$, as $k \to \infty$, as desired. □

Remark. It can happen that the system $\mathbf{Ax} = \mathbf{b}$ is soluble, but property (22) does not hold. For example, consider the system $\frac{1}{2}x_1 - x_2 = 1$, $\frac{1}{2}x_2 = 0$ with $\mathbf{M} = \begin{bmatrix} \frac{1}{2} & 0 \\ 0 & \frac{1}{2} \end{bmatrix}$ $\mathbf{N} = \begin{bmatrix} 0 & 1 \\ 0 & 0 \end{bmatrix}$. Then $\mathbf{T} = \mathbf{M}^{-1}\mathbf{N} = 2\mathrm{N}$, and $\|\mathbf{Ty}\| = 2\|\mathbf{y}\|$ for $\mathbf{y} = (0,1)$, but nevertheless $\mathbf{x} = (2,0)$ is the solution. For this reason, there are properties less restrictive than (22), which are commonly used to obtain convergence. For those who have had linear algebra, one such property is that each of the (possibly complex) eigenvalues of $\mathbf{T}$ have modulus less than one (i.e., the **spectral radius** of $\mathbf{T}$ is less than 1). The above $\mathbf{T}$ has this property, since its eigenvalues are all 0. □

We now state three very useful iterative schemes, which are distinguished by the way that the splitting $\mathbf{A} = \mathbf{M} - \mathbf{N}$ is made. Suppose that $\mathbf{A}$ is a nonsingular $n \times n$ matrix with all diagonal entries $a_{ii} \ne 0$. We can uniquely write $\mathbf{A}$ as a sum

$$\mathbf{A} = \mathbf{D} - \mathbf{E} - \mathbf{F}, \text{ e.g. } \begin{bmatrix} a_{11} & a_{12} \\ a_{21} & a_{22} \end{bmatrix} = \begin{bmatrix} a_{11} & 0 \\ 0 & a_{22} \end{bmatrix} - \begin{bmatrix} 0 & -a_{12} \\ 0 & 0 \end{bmatrix} - \begin{bmatrix} 0 & 0 \\ -a_{21} & 0 \end{bmatrix} \text{ when } n = 2,$$

where $\mathbf{D}$ is a nonsingular diagonal matrix with the same diagonal entries as $\mathbf{A}$, and $\mathbf{E}$ and $\mathbf{F}$ are matrices with entries which are zero above the diagonal and zero below the diagonal, respectively. We have some standard choices

$$\mathbf{M} = \mathbf{D} \quad \text{and} \quad \mathbf{N} = \mathbf{E} + \mathbf{F} \tag{23}$$

$$\mathbf{M} = \mathbf{D} - \mathbf{E} \quad \text{and} \quad \mathbf{N} = \mathbf{F} \tag{24}$$

$$\mathbf{M} = \frac{1}{\omega}(\mathbf{D} - \omega\mathbf{E}) \quad \text{and} \quad \mathbf{N} = \frac{1}{\omega}(\omega\mathbf{F} + (1-\omega)\mathbf{D}) . \tag{25}$$

Then (23) yields the **point Jacobi iteration** , (24) yields the **point Gauss–Seidel iteration** and (25) the **point successive overrelaxation (SOR) iteration with parameter** ω (cf. [Berman and Plemmons], [Ortega and Rheinboldt], [Varga] or [D. M. Young]).

Remarks. (1) It is a consequence of the **Perron–Frobenius theory of nonnegative matrices** that in the splitting $\mathbf{A} = \mathbf{M} - \mathbf{N}$, if the entries of the matrices $\mathbf{M}^{-1}$ and $\mathbf{N}$ are nonnegative, then the spectral radius of $\mathbf{M}^{-1}\mathbf{N}$ is less than 1 if and only if $\mathbf{A}$ is nonsingular and the entries of $\mathbf{A}^{-1}$ are nonnegative.
(2) For large systems of linear equations the coefficient matrix is partitioned into square blocks and this leads to the very useful theory of **block iterative methods** (cf. [Varga]).
(3) The computer implementation of these iterative methods is straightforward. For example, with the above notation and assumptions, the point Gauss–Seidel iteration is given by

$$x_i^{(k+1)} = -\frac{1}{a_{ii}}\left[\sum_{j=1}^{i-1} a_{ij}x_j^{(k)} + \sum_{j=i+1}^{n} a_{ij}x_j^{(k)} - b_i\right], \quad (i = 1, \ldots, n) ,$$

where $x_i^{(k+1)}$ is the i–th component of $\mathbf{x}^{(k+1)}$, and the first sum is defined to be zero when $i = 1$. In practice, the iterations should be stopped at some time, say when $\|\mathbf{x}^{(k+1)} - \mathbf{x}^{(k)}\| < \epsilon$. Under the assumptions in Theorem 1, this will eventually happen for any given positive $\epsilon > 0$. □

Nonlinear equations

The finite difference methods, be they explicit (cf. Section 8.2) or implicit (cf. (5) above), are also applicable to nonlinear PDEs. Thus, for example, if the PDE is of the form

$$u_t = f(x,t,u,u_x,u_{xx}) \qquad (0 \le x \le L,\ 0 \le t \le T), \tag{26}$$

where f is a suitable function of 5 variables, then an explicit difference scheme for (26) is

$$v_{i,j+1} = v_{i,j} + \Delta t \cdot f(i\Delta x,\, j\Delta t,\, v_{i,j},\, \frac{1}{2\Delta x}(v_{i+1,j} - v_{i-1,j}),\, \frac{1}{(\Delta x)^2}\,\delta_x^2 v_{i,j}) , \tag{27}$$

$(i = 1,\ldots,M-1\ ;\ \ j = 0,\ldots,N-1)$, where $\delta_x^2 v_{i,j}$ denotes the second–order central difference (cf. (1)). Note that (27) reduces to (3) when $f \equiv u_{xx}$. While this method is easy to use, its main disadvantage is, as was noted in the beginning of this section, that it is only conditionally stable; e.g., for stability it is required that $\lambda \equiv \Delta t/\Delta x^2 \le \frac{1}{2}$ when $f \equiv u_{xx}$.

The stability limitation can be overcome by using an implicit difference method (e.g., a Crank–Nicolson type method; cf. (5)). However in this case, the algebraic problem of determining the $v_{i,j+1}$ leads, in general, to the problem of solving a nonlinear system of equations. If the nonlinear system has a solution, then we may attempt to use an iterative method (cf. [Isaacson and Keller]) to approximate the solution. One such method is a several variable analogue of the Newton–Raphson method (cf. Problem 12 of Section 8.2). The Newton–Raphson method and its variations play a central role in the study of iterative methods for solving systems of nonlinear equations (cf. [Ortega and Rheinboldt]). In its simplest form, this method may be described as follows. Using vector notation, we can write the nonlinear system as $\mathbf{F}(\mathbf{x}) = \mathbf{0}$, where

$$\mathbf{F}(\mathbf{x}) = \begin{bmatrix} f_1(\mathbf{x}) \\ \vdots \\ f_n(\mathbf{x}) \end{bmatrix}, \quad \mathbf{0} = \begin{bmatrix} 0 \\ \vdots \\ 0 \end{bmatrix}, \quad \text{and} \quad \mathbf{x} = [x_1,\dots,x_n] .$$

Now, if we let $J(\mathbf{x})$ be the Jacobian matrix of $\mathbf{F}$, i.e., $J(\mathbf{x})$ is the matrix whose (i,j)–th entry is $\partial f_i/\partial x_j$, then the n–dimensional Newton–Raphson iterative scheme becomes

$$\mathbf{x}^{(k+1)} = \mathbf{x}^{(k)} - [J(\mathbf{x}^{(k)})]^{-1}\,\mathbf{F}(\mathbf{x}^{(k)}) \qquad (k \ge 0) \tag{28}$$

where $\mathbf{x}^{(0)}$ is some initial guess. Since usually it is inefficient to compute inverses of matrices, we may wish to carry out the iteration in (28) as follows. Suppose that $\mathbf{x}^{(k)}$ has been calculated. Define $\mathbf{y}^{(k+1)} = \mathbf{x}^{(k+1)} - \mathbf{x}^{(k)}$. Then (28) becomes

$$J(\mathbf{x}^{(k)})\mathbf{y}^{(k+1)} = -\mathbf{F}(\mathbf{x}^{(k)}) . \tag{29}$$

which is a linear system of equations. We solve (29) for $\mathbf{y}^{(k+1)}$ using, say, Gaussian elimination (or perhaps an iterative method described in (21)), and then we determine $\mathbf{x}^{(k+1)}$ from the equation $\mathbf{x}^{(k+1)} = \mathbf{y}^{(k+1)} + \mathbf{x}^{(k)}$. For the error analysis, proofs of theorems on convergence and for many other techniques, we refer the interested reader to [Ortega and Rheinboldt].

CHAPTER 9

PDEs IN HIGHER DIMENSIONS

In the previous chapters, we have considered one–dimensional heat flow ($u_t = ku_{xx}$) and wave propagation ($u_{tt} = a^2u_{xx}$), and two–dimensional steady–state temperatures, electrostatics and fluid flow ($u_{xx} + u_{yy} = 0$). Since we live in a world with three spatial dimensions, it is also important to examine the corresponding results in higher dimensions, where many of the most relevant initial/boundary–value problems are posed. Since most of the techniques and central ideas in the study and application of PDEs are already manifest in the treatment of the lower–dimensional situations, the extension of what we have learned to the higher–dimensional case is not as conceptually difficult as one might think. In particular, the technique of separation of variables, the superposition principle (for homogeneous, linear equations), and certain aspects of Fourier series and transforms, all carry over to higher dimensions.

Many of the difficulties are only technical in nature, due to the variety of coordinate systems. For example, there is basically only one standard coordinate system on a line, but (in addition to Cartesian or rectangular coordinates) in the plane we have polar coordinates, and in space we have cylindrical and spherical coordinates. The choice of the system depends on the symmetry (if any) of the region under consideration. When the variables are separated, some of ODEs which result can be be difficult to solve. Indeed, some of the solutions cannot be expressed in terms of elementary functions (e.g., sines, cosines, exponentials, etc.), and new special functions, such as Bessel functions, need to be introduced. It is unfortunate that in many treatments, the main ideas behind the solution process is obscured by a focusing on the special functions instead of on the product solutions (of the PDE) which result when the solutions of the ODEs are multiplied together. The time–independent part of a product solution is usually an eigenfunction of the Laplace operator Δ for the region under consideration. The special functions serve primarily to construct the eigenfunctions which satisfy the B.C. . Once each initial condition is expressed as a linear combination of eigenfunctions (i.e., written as an eigenfunction expansion), then it is usually simple to find the solution of the initial boundary–value problem. Eigenfunction expansions (cf. Section 4.4) constitute the natural generalization of Fourier series for functions defined on smooth (possibly curved) multidimensional spaces known as manifolds, even where separation of variables not feasible, say due to a lack of symmetry.

In Section 9.1, we consider the standard higher–dimensional heat, wave and Laplace's equations in terms of rectangular coordinates, and we use multiple Fourier series and Fourier transforms to solve initial/boundary–value problems for these PDEs. In Section 9.2, we introduce the unifying concepts of eigenfunctions and eigenvalues of the Laplace operator, working primarily on rectangles. We also prove a uniform convergence theorem for eigenfunction expansions (multiple Fourier series) of suitable functions on a rectangle. Section 9.3 deals with the standard PDEs written in terms of spherical coordinates. The Laplace operator on a sphere is defined in a geometrically natural way, and its eigenfunctions (known as spherical harmonics) are introduced. In Section 9.4, we prove that the eigenfunction expansion of a C^2 function f on a sphere (the Laplace series of f) converges uniformly to f. Among other problems, we also study heat flow in a solid ball, and the wave problem for a vibrating balloon. In Section 9.5, we consider a number of special functions, such as Bessel functions and their use in solving heat problems in cylinders and in expressing the vibrational modes of a circular drum. We also solve Schrödinger's equation for the quantum–mechanical description of the energy states of the electron in a hydrogen atom. In Section 9.6, we introduce the notion of a smooth k–dimensional manifold in n–dimensional space, and define the Laplace operator on such an object. In this last section, results are stated (but not always proved) concerning the nature of the eigenfunctions and their use in constructing solutions and Green's functions for the standard PDE problems on manifolds.

9.1 Higher–Dimensional PDEs – Rectangular Coordinates

Here we reintroduce the higher–dimensional heat, wave, and Laplace's equations which are most relevant for applications. In this section, we confine ourselves to boundary–value problems on rectangular plates or solids, and hence we use rectangular coordinates. Since the techniques (separation of variables, the superposition principle, and Fourier series) are now quite familiar, we will be able to cover a lot of territory.

The heat equation for rectangular plates and solids

We found in Section 6.1 that for two–dimensional heat flow in a homogeneous medium, the only possible second–order linear PDE which the temperature $u(x,y,t)$ can obey is

$$u_t = k(u_{xx} + u_{yy}) , \tag{1}$$

where k is a positive constant (the heat diffusivity) and where we have assumed that there are no heat sources. The simplest initial/boundary–value problem for (1), arises in the case of a rectangular plate with edges maintained at temperature 0, i.e.,

$$\begin{array}{ll} \text{D.E.} & u_t = k(u_{xx} + u_{yy}) \qquad 0 \le x \le L,\ 0 \le y \le M,\ t \ge 0 \\ \text{B.C.} & \begin{cases} u(x,0,t) = 0, & u(x,M,t) = 0 \\ u(0,y,t) = 0, & u(L,y,t) = 0 \end{cases} \\ \text{I.C.} & u(x,y,0) = f(x,y) , \end{array} \tag{2}$$

where $f(x,y)$ is a continuous function which is zero on the boundary of the plate. Proceeding with separation of variables, we substitute the product $u(x,y,t) = X(x)Y(y)T(t)$ into (1), and separate t from x and y to obtain

$$\frac{T'}{kT} = \frac{X''}{X} + \frac{Y''}{Y} = b , \tag{3}$$

for some constant b. Thus, we must have

$$\frac{X''}{X} = b - \frac{Y''}{Y} = c \quad \text{or} \quad \begin{cases} X'' - cX = 0 \\ Y'' + (c-b)Y = 0 \end{cases} \tag{4}$$

for some constant c. By the B.C. $u(0,y,t) = 0$ and $u(L,y,t) = 0$, we can avoid the zero solution, only if $c = c_n \equiv -(n\pi/L)^2$ and $X(x)$ is a constant multiple of $\sin(n\pi x/L)$, $n = 1, 2, 3, \ldots$. Similarly, from the B.C. $u(x,0,t) = 0$ and $u(x,M,t) = 0$ and the equation $Y'' + (c-b)Y = 0$, we deduce that $c-b = (m\pi/M)^2$ and that $Y(y)$ must be a constant multiple of $\sin(m\pi y/M)$,

$m = 1, 2, 3, \ldots$. The possible values of $-b$, denoted by $\lambda_{n,m}$, are then given by

$$\lambda_{n,m} \equiv (m\pi/M)^2 - c_n = \pi^2[(n/L)^2 + (m/M)^2], \quad m,n = 1, 2, 3, \ldots . \tag{5}$$

From (3), we have $T' = kbT$, and thus the family of product solutions of the D.E. and B.C. of (2) consists of the constant multiples of

$$u_{n,m}(x,y,t) = e^{-\lambda_{n,m}kt}\sin(\tfrac{n\pi x}{L})\sin(\tfrac{m\pi y}{M}), \quad m,n = 1, 2, 3, \ldots . \tag{6}$$

In view of the linear, homogeneous D.E. and B.C. of (2), we may apply the superposition principle to obtain a more general solution of the D.E. and B.C. :

$$u(x,y,t) = \sum_{n,m=1}^{N} b_{n,m}\, u_{n,m}(x,y,t) = \sum_{n,m=1}^{N} b_{n,m}\, e^{-\lambda_{n,m}kt}\sin(\tfrac{n\pi x}{L})\sin(\tfrac{m\pi y}{M}), \tag{7}$$

where the integers n and m run independently from 1 to some finite N. When $t = 0$, we have

$$u(x,y,0) = \sum_{n,m=1}^{N} b_{n,m}\sin(\tfrac{n\pi x}{L})\sin(\tfrac{m\pi y}{M}) . \tag{8}$$

If the initial temperature $f(x,y)$ in (2) is of this form, then (7) is a solution of (2). Of course, not every given $f(x,y)$ will be exactly of this form. However, one might expect that a reasonably nice $f(x,y)$ could be approximated by a sum of the form (8), say within some positive experimental error. There is a Maximum Principle (stated and proved in the same way as the Maximum Principle in Section 3.2) which implies that two solutions of the D.E. and B.C. cannot differ by more than their maximum difference at $t = 0$. Hence, the error in the initial approximation will not grow as t increases. Moreover, uniqueness of the solution of (2) is ensured by the Maximum Principle. The sum (8) is known as a (finite) double Fourier sine series. More precisely, we have :

The **double Fourier sine series** of a function $f(x,y)$ on the rectangle $(0 \le x \le L, 0 \le y \le M)$ is the expression

$$\sum_{n,m=1}^{\infty} b_{n,m}\sin(\tfrac{n\pi x}{L})\sin(\tfrac{m\pi y}{M}), \tag{9}$$

where

$$b_{n,m} = \frac{4}{LM}\int_0^M\int_0^L f(x,y)\sin(\tfrac{n\pi x}{L})\sin(\tfrac{m\pi y}{M})\,dxdy, \quad m,n = 1, 2, 3, \ldots, \tag{10}$$

provided these integrals exist.

In Section 9.2, we will discuss double Fourier series and some of their convergence properties. If the double Fourier sine series of the initial temperature exists, then the *formal* solution of (2) is given by (7), with N replaced by ∞, where $b_{n,m}$ is given by (10). It is clear that this formal solution is an exact C^∞ solution, if all but a finite number of the $b_{n,m}$ are zero. It is much more difficult to show that if the function $f(x,y)$ is continuous on the closed rectangle and is zero on the edges, then the formal solution defines an actual solution of (1) for $t > 0$, which extends continuously to $f(x,y)$ as $t \to 0^+$. Recall that this type of result was proved in the one–dimensional case (cf. Theorem 1 of Section 7.5), by expressing the solution as a convolution of a suitably extended initial temperature function with the fundamental solution of the heat equation. In the case of problem (2), let $\tilde{f}_{o,o}(x,y)$ be the unique extension of $f(x,y)$ (to the whole xy–plane) which is odd and periodic in x (of period $2L$), and odd and periodic in y (of period $2M$). Then, the unique continuous solution of problem (2) can be written in the form

$$u(x,y,t) = \frac{1}{4\pi kt}\int_{-\infty}^{\infty}\int_{-\infty}^{\infty} e^{-[(x-\bar{x})^2 + (y-\bar{y})^2]/(4kt)}\, \tilde{f}_{o,o}(\bar{x},\bar{y})\, d\bar{x}d\bar{y}\,, \quad t > 0\,, \qquad (11)$$

where $u(x,y,0) \equiv f(x,y)$, when $t = 0$. Solution (11) is obtained through the application of the method of images (cf. Section 7.5), whereby one suitably extends (depending on the B.C.) the initial temperature $f(x,y)$ to a temperature $F(x,y)$, defined on the whole plane. Then

$$u(x,y,t) = \frac{1}{4\pi kt}\int_{-\infty}^{\infty}\int_{-\infty}^{\infty} e^{-[(x-\bar{x})^2 + (y-\bar{y})^2]/(4kt)}\, F(\bar{x},\bar{y})\, d\bar{x}d\bar{y} \qquad (t > 0), \qquad (12)$$

is the solution of the initial–value problem for the heat equation on the whole plane. The solution (12) is found formally by applying two–dimensional Fourier transform methods which we discuss in Section 9.2 . (Note that the square root of $4\pi kt$ does not appear in (12), since the two–dimensional source solution of the heat equation turns out to be the product of the two one–dimensional source solutions in the x and y directions.) Just as in the one–dimensional case (cf. Theorem 1 of Section 7.4), using Leibniz's rule one can prove directly that (12) is a C^∞ solution of the heat equation, for $t > 0$, provided that $F(x,y)$ is continuous and bounded (or absolutely integrable). In that case, as before, one also can show that (12) continuously extends to the initial temperature $F(x,y)$, as $t \to 0^+$ ($F(x,y) = f(x,y)$ for $0 \le x \le L,\ 0 \le y \le M$).

In choosing the extension $F(x,y)$, we use the odd extension across any edge where the solution is required to be zero by the B.C. (as in (2)), and we use the even extension across any edge which is insulated (i.e., if the partial derivative [in the direction normal to the edge] of the temperature is 0 in a B.C. ; cf. Example 2 below).

Example 1. Solve the problem

$$\text{D.E.}\quad u_t = 2(u_{xx} + u_{yy}) \qquad 0 \le x \le 3,\ 0 \le y \le 5,\ t \ge 0$$

$$\text{B.C.}\ \begin{cases} u(x,0,t) = 0, & u(x,5,t) = 0 \\ u(0,y,t) = 0, & u(3,y,t) = 0 \end{cases} \tag{13}$$

$$\text{I.C.}\quad u(x,y,0) = \cos[\pi(x+y)] - \cos[\pi(x-y)] + \sin(2\pi x)\sin(\tfrac{3\pi y}{5})\ .$$

Solution. We can use (7), provided that the I.C. can be put in the form (8). Using the formula $\cos(a+b) = \cos(a)\cos(b) - \sin(a)\sin(b)$, we obtain the desired form

$$u(x,y,0) = -2\sin(\tfrac{3\pi x}{3})\sin(\tfrac{5\pi y}{5}) + \sin(\tfrac{6\pi x}{3})\sin(\tfrac{3\pi y}{5})\ .$$

Since this is of the form (8), with $L = 3$, $M = 5$, we may use (7), with $k = 2$ to obtain

$$u(x,y,t) = -2e^{-4\pi^2 t}\sin(\pi x)\sin(\pi y) + e^{-218\pi^2 t/25}\sin(2\pi x)\sin(\tfrac{3\pi y}{5})\ .$$

Although it would be inconvenient, one could also write the solution in the form (11), with $\tilde{f}_{0,0}(x,y) = \cos(\pi(x+y)) - \cos(\pi(x-y)) + \sin(2\pi x)\sin(\frac{3\pi y}{5})$, which has the required oddness and periodicity properties. □

One can similarly handle other types of boundary conditions, where some edges are insulated and others are maintained at 0, as the next example illustrates.

Example 2. Solve

$$\text{D.E.}\quad u_t = k(u_{xx} + u_{yy}), \qquad 0 \le x \le L,\ 0 \le y \le M,\ t \ge 0$$

$$\text{B.C.}\ \begin{cases} u(x,0,t) = 0, & u_y(x,M,t) = 0 \\ u_x(0,y,t) = 0, & u_x(L,y,t) = 0 \end{cases} \tag{14}$$

$$\text{I.C.}\quad u(x,y,0) = \cos(\tfrac{2\pi x}{L})\sin(\tfrac{3\pi y}{2M})\ .$$

How might one treat the general I.C. of the form $u(x,y,0) = f(x,y)$?

Solution. Here, the edge $y = 0$ $(0 \le x \le L)$ is maintained at zero, while the other three edges are insulated. As usual, we seek the product solutions of the D.E. which satisfy the B.C., and then consider the I.C. . Separation of variables leads to equations (4) for X and Y. The B.C. $u_x(0,y,t) = 0$ and $u_x(L,y,t) = 0$ imply that $X(x)$ is a constant multiple of $\cos(\frac{n\pi x}{L})$, $n = 0, 1, 2, \ldots$. The B.C. $u(x,0,t) = 0$ and $u_y(x,M,t) = 0$ imply that $Y(y)$ is a constant multiple of $\sin\left[\frac{(m+\frac{1}{2})\pi y}{M}\right]$, $m = 0, 1, 2, \ldots$ (cf. Example 2 of Section 3.3). The family of product solutions, which satisfy the D.E. and B.C. of (10), consists of the constant multiples of

$$u_{n,m}(x,y,t) = \exp\left[-[(n/L)^2+((m+\tfrac{1}{2})/M)^2]\pi^2 kt\right]\cos(\tfrac{n\pi x}{L})\sin\left[\tfrac{(m+\frac{1}{2})\pi y}{M}\right], \quad (15)$$

$n, m = 0, 1, 2,\ldots$. In the case of a more general initial temperature $f(x,y)$, one would attempt to write $f(x,y)$ as a linear combination of the functions $u_{n,m}(x,y,0)$, and then the same linear combination of the $u_{n,m}(x,y,t)$ would give the solution for $t > 0$. For the problem at hand, observe that the I.C. is simply $u(x,y,0) = u_{2,1}(x,y,0)$. Thus,

$$u(x,y,t) = u_{2,1}(x,y,t) = \exp\left[-[(2/L)^2 + (3/2M)^2]\pi^2 kt\right]\cos(\tfrac{2\pi x}{L})\sin(\tfrac{3\pi y}{2M}) .$$

If the I.C. were $u(x,y,0) = f(x,y)$, where $f(x,y)$ is continuous on the closed rectangle and is zero on the edge $y = 0$, but is *not* necessarily a finite linear combination of the $u_{n,m}(x,y,0)$ in (15), then it can be shown that the following is a solution of the D.E. and B.C., when $t > 0$:

$$u(x,y,t) = \sum_{n,m=1}^{\infty} c_{m,n}\, u_{n,m}(x,y,t) , \quad (16)$$

where

$$c_{n,m} = \frac{4}{LM}\int_0^M\int_0^L f(x,y)\cos(\tfrac{n\pi x}{L})\sin(\tfrac{(m+\frac{1}{2})\pi y}{M})\, dxdy. \quad (17)$$

This solution continuously extends to $f(x,y)$ as $t \to 0^+$, but the B.C. involving partial derivatives will not hold, unless $f(x,y)$ satisfies these B.C. . Equivalently, we have the integral formula

$$u(x,y,t) = \frac{1}{4\pi kt}\int_{-\infty}^{\infty}\int_{-\infty}^{\infty} e^{-[(x-\bar{x})^2 + (y-\bar{y})^2]/(4kt)}\, F(\bar{x},\bar{y})\, d\bar{x}d\bar{y} , \quad (18)$$

where $F(x,y)$ is the unique extension (to the entire xy–plane) of $f(x,y)$, which is odd and periodic (of period $2L$) in x, and is periodic in y (of period $4M$), even about $y = M$ and odd about $y = 0$ (cf. Example 2 of Section 5.3, where such an extension was used). □.

It is straightforward to extend the above considerations to the case of three–dimensional heat flow. Rather than dealing with generalities, we consider an example.

Example 3. A solid cube of edge length 1, and with heat diffusivity constant k, is initially at the constant boiling temperature 100° C and is suddenly (at $t = 0$) placed in ice water at 0° C. Find the temperature at the center of the cube as a function of t, and show that this temperature is $\approx 6400\pi^{-3} e^{-3\pi^2 kt}$, for t sufficiently large. Estimate the smallest value for t such that the actual temperature at the center is within 10 % of this approximation (i.e., between $5760\pi^{-3}e^{-3\pi^2 kt}$ and $7040\pi^{-3}e^{-3\pi^2 kt}$).

Solution. Let the cube be given by $0 \le x, y, z \le 1$. In view of the B.C. $u(0,y,z) = 0$, $u(1,y,z) = 0$, $u(x,0,z) = 0, \ldots$, separation of variables leads to the following family of product solutions of the three–dimensional heat equation $u_t = k(u_{xx} + u_{yy} + u_{zz})$ for these B.C. :

$$u_{n,m,p}(x,y,z,t) = e^{-(m^2 + n^2 + p^2)\pi^2 kt}\sin(n\pi x)\sin(m\pi y)\sin(p\pi z) .$$

When x, y or z are 0 or 1, all of these solutions are 0. Thus, no superposition

$$u(x,y,z,t) = \sum_{n,m,p=1}^{\infty} B_{n,m,p}\, u_{n,m,p}(x,y,z,t)$$

of these solutions can possibly be 100 on the faces of the cube when $t = 0$. Hence, we proceed formally. Assume that strictly inside the cube we have

$$100 = \sum_{n,m,p=1}^{\infty} B_{n,m,p}\, u_{n,m,p}(x,y,z,0). \tag{19}$$

Note that the functions $u_{n,m,p}(x,y,z,0)$ are orthogonal on the cube, in the sense that

$$\int_0^1\int_0^1\int_0^1 u_{n,m,p}(x,y,z,0)u_{N,M,P}(x,y,z,0)\, dxdydz = \begin{cases} 0 & \text{if } n \ne N \text{ or } m \ne M \text{ or } p \ne P \\ \frac{1}{8} & \text{if } n = N \text{ and } m = M \text{ and } p = P . \end{cases}$$

It then formally follows from this orthogonality and (19) that

$$B_{n,m,p} = 8\int_0^1\int_0^1\int_0^1 100 \sin(n\pi x)\sin(m\pi y)\sin(p\pi z)\, dxdydz$$

$$= \frac{800}{\pi^3}\frac{1}{n}[1 - (-1)^n]\,\frac{1}{m}[1 - (-1)^m]\,\frac{1}{p}[1 - (-1)^p] = \begin{cases} \dfrac{6400}{\pi^3 nmp} & \text{if n, m and p are odd} \\ 0 & \text{otherwise} \end{cases} .$$

Thus, formally we have

$$u(x,y,z,t) = 6400\pi^{-3} \sum_{n,m,p \text{ odd}}^{\infty} \frac{1}{nmp} e^{-(m^2+n^2+p^2)\pi^2 kt} \sin(n\pi x)\sin(m\pi y)\sin(p\pi z)\,. \qquad (20)$$

Although we have derived this formula formally, it is possible to show that for $t > 0$, (20) defines a C^∞ solution of the heat equation and that $u(x,y,z,0^+) = 100$, for (x,y,z) strictly inside the cube. Evaluating the solution at the center of the cube, we have

$$u(\tfrac{1}{2},\tfrac{1}{2},\tfrac{1}{2},t) = 6400\,\pi^{-3} \sum_{n,m,p \text{ odd}}^{\infty} \frac{1}{nmp} e^{-(m^2+n^2+p^2)\pi^2 kt} (-1)^{(n+m+p-3)/2}$$

$$= 6400\,\pi^{-3} \sum_{q=0}^{\infty} (-1)^q \left[\sum_{\substack{n+m+p=2q+3 \\ n,m,p \text{ odd}}} \frac{1}{nmp} e^{-(m^2+n^2+p^2)\pi^2 kt} \right] \qquad (21)$$

The sum over q is alternating, but it is not true that the expression in brackets, say $A(q,t)$, always decreases as q increases (e.g., $A(0,0) = 1$ and $A(1,0) = 1$). (The reader may wish to decide whether $A(q,0)$ decreases with q for $q \geq 1$.) Thus, we cannot deduce that the error of a partial sum is not greater than the next term. However, shortly we will find a positive t_0, such that, for $t > t_0$, $A(q,t)$ always decreases with q, and thus the very first term, $e^{-3\pi^2 kt}$, will differ from the entire sum (21) by no more than the magnitude of the second term, $e^{-11\pi^2 kt}$, provided $t > t_0$. This second term is less than 10% of the first term for $t > \log_e(10)/(8\pi^2 k) \approx .02916/k$. Hence, in approximating (21) by using only the first term, the error committed will be less than 10% for $t > .0292/k$, *if* the value of t_0, which we now will determine, is less than $(.0292)/k$. To find t_0, note that each triplet (n,m,p), with $n+m+p = 2q+3$, gives rise to three triplets (i.e., $(n+2,m,p)$, $(n,m+2,p)$, $(n,m,p+2)$) each having the sum $2(q+1) + 3$. However, in the process, the factor $(n^2 + m^2 + p^2)$ in the exponent is increased by at least 6 (e.g., $((n+2)^2 > n^2 + 2n + 4 \geq n^2 + 6)$. Thus, taking all triplets into account, we deduce that $A(q+1,t) < 3e^{-6\pi^2 kt} A(q,t)$. Hence, the $A(q,t)$ must strictly decrease with q provided $3e^{-6\pi^2 kt} < 1$ or $t > \log_e(3)/(6\pi^2 k) \approx (.0186)/k \equiv t_0$. Thus, since $t_0 < (.0292)/k$, the temperature at the center will be within 10% of $6400\pi^{-3}e^{-3\pi^2 kt}$, provided $t > (.0292)/k$. One can check that when $t = (.0292)/k$, this approximate temperature is $\approx 87°$ C, regardless of k. □

Remark. The function (20) can be expressed in terms of an integral formula, as follows. We decompose space into cubes, with integers as coordinates of the vertices of the cubes. Let $F(x,y,z)$ be the function which is 1 inside the cube $0 < x,y,z < 1$, and which is 0 on the faces of this cube, and which is -1 strictly inside the adjacent cubes which share a common face with the original cube. Repeat this pattern, so that $F(x,y,z)$ is defined throughout space, with $F(x,y,z)$ being 1 or -1 within each cube and 0 (the average) on the common faces. By applying Fourier transform techniques, we obtain the alternate solution

$$u(x,y,z,t) = (4\pi kt)^{-\frac{3}{2}} \int_{-\infty}^{\infty} \int_{-\infty}^{\infty} \int_{-\infty}^{\infty} e^{-[(x-\bar{x})^2 + (y-\bar{y})^2 + (z-\bar{z})^2]/(4kt)}\, F(\bar{x},\bar{y},\bar{z})\, d\bar{x}\, d\bar{y}\, d\bar{z}\,. \qquad (22)$$

This form of the solution appears to be somewhat more difficult to evaluate. However, if we set $x = y = z = 1/2$, and restrict the domain of integration to the finite box B, $-1 < \bar{x}, \bar{y}, \bar{z} < 2$, (containing 27 unit cubes, with the original one at the center), one can obtain a very accurate approximation to $u(\frac{1}{2},\frac{1}{2},\frac{1}{2},t)$ for *small* $t > 0$ (i.e., when the approximation in Example 3 is bad). For small t, this new approximation will be good, because the integrand of (22) is extremely small outside of B, for $x = y = z = 1/2$ and t sufficiently small and positive. The integral over B can be evaluated numerically, or with a good table of normal distributions. □

Laplace's equation on a rectangular solid

Perhaps the simplest type of problem for Laplace's equation on a rectangular solid is

$$\text{D.E.}\quad u_{xx} + u_{yy} + u_{zz} = 0 \qquad 0 < x < L,\ 0 < y < M,\ 0 < z < P$$

$$\text{B.C.}\quad \begin{cases} u(x,y,0) = f(x,y)\,, \quad u(x,y,P) = g(x,y) \\ u(x,y,z) = 0 \quad \text{on the faces } x = 0,\ x = L,\ y = 0 \text{ and } y = M\,, \end{cases} \qquad (23)$$

where f and g are given continuous functions, on the rectangle $0 \le x \le L$, $0 \le y \le M$, which are zero on the boundary of this rectangle. We seek a (C^2) solution of the D.E. inside the rectangular solid which extends continuously to the boundary of the solid in such a way that the B.C are met. The familiar approach is to use separation of variables to find those product solutions $X(x)Y(y)Z(z)$ which meet the homogeneous B.C., and then, by the superposition principle, take a linear combination of these product solutions in an effort to meet the B.C. $u(x,y,0) = f(x,y)$ and $u(x,y,P) = g(x,y)$. Substitution of the product $X(x)Y(y)Z(z)$ into the D.E. yields $\frac{X''}{X} + \frac{Y''}{Y} + \frac{Z''}{Z} = 0$. There is no way that any of these terms can cancel with the sum of the remaining two terms, unless the term is constant (Why ?). From past experience, we know that in order that $X(x)$ meet the B.C. $X(0) = X(L) = 0$, we must have that $X(x)$ is a constant multiple of $\sin(\frac{n\pi x}{L})$, $n = 1, 2, 3, \ldots$, and similarly $Y(y)$ is a constant multiple of $\sin(\frac{m\pi y}{M})$, $m = 1, 2, 3, \ldots$. For each pair (n,m), the equation for Z is $Z'' - \pi^2[(n/L)^2 + (m/M)^2]Z = 0$. In view of the fact that the (possibly) inhomogeneous B.C. are prescribed on the faces $z = 0$ and $z = P$, a very convenient way of expressing the general solution of this equation for Z is

$$Z_{n,m}(z) = \frac{1}{\sinh(P\sqrt{\lambda_{n,m}})} \left[a_{n,m} \sinh[(P-z)\sqrt{\lambda_{n,m}}] + b_{n,m} \sinh(z\sqrt{\lambda_{n,m}}) \right],$$

where $\sqrt{\lambda_{n,m}} = \pi\sqrt{(n/L)^2 + (m/M)^2}$. Thus, a more general solution of the homogeneous D.E. and B.C. is

$$u(x,y,z) = \sum_{n,m=1}^{N} Z_{n,m}(z)\sin(\tfrac{n\pi x}{L})\sin(\tfrac{m\pi y}{M}) . \tag{24}$$

Note that on the faces $z = 0$ and $z = P$, we have

$$u(x,y,0) = \sum_{n,m=1}^{N} a_{n,m} \sin(\tfrac{n\pi x}{L})\sin(\tfrac{m\pi y}{M}) \quad \text{and} \quad u(x,y,P) = \sum_{n,m=1}^{N} b_{n,m} \sin(\tfrac{n\pi x}{L})\sin(\tfrac{m\pi y}{M}) .$$

Thus, in the event that $f(x,y)$ and $g(x,y)$ are respectively given by these finite double Fourier sine series, a solution of problem (23) is given by (24). When $f(x,y)$ or $g(x,y)$ are not finite double Fourier sine series, we have a formal solution which, with much more effort, can be shown to be valid strictly within this rectangular solid and to extend continuously to the prescribed values on the faces. There is a Maximum/Minimum Principle for the Dirichlet problem (23) (or more generally for a Dirichlet problem on any bounded open set in $\mathbb{R}^n$, $n = 1, 2, 3, \ldots$) which can be proved just as Theorem 1 in Section 6.4 was proved. Thus, a reasonable approximation of boundary data leads to a reasonable solution in the interior.

Example 4. Find the value of the constant c such that the following Neumann problem has a solution, and find such a solution. Is the solution unique ?

$$\text{D.E.} \quad u_{xx} + u_{yy} + u_{zz} = 0 \quad 0 < x, y, z < \pi$$

$$\text{B.C.} \begin{cases} u_x(0,y,z) = u_x(\pi,y,z) = 0\,, \; u_y(x,0,z) = u_y(x,\pi,z) = 0 \\ u_z(x,y,0) = 0\,, \; u_z(x,y,\pi) = c + 4\sin^2(x)\cos^2(y) \,. \end{cases} \tag{25}$$

Solution. One quickly verifies that the product solutions of the D.E. which satisfy the homogeneous B.C. are of the form

$$u_{n,m}(x,y,z) = a_{n,m} \cosh[\sqrt{n^2+m^2}\, z]\cos(nx)\cos(my) \,, \quad n,m = 0, 1, 2, \ldots .$$

We apply the superposition principle to form the more general solution

$$u(x,y,z) = \sum_{n,m=0}^{N} a_{n,m} \cosh[\sqrt{n^2+m^2}\, z]\cos(nx)\cos(my) \tag{26}$$

for which

$$u_z(x,y,\pi) = \sum_{n,m=0}^{N} a_{n,m} \sqrt{n^2+m^2} \sinh[\sqrt{n^2+m^2}\,\pi]\cos(nx)\cos(my) . \qquad (27)$$

Observe that this last expression is a finite double cosine series which has a zero constant term (i.e., $\sqrt{0^2+0^2} = 0$). The inhomogeneous B.C. in (25) can be put in the form of a finite double cosine series :

$$u_z(x,y,\pi) = (c + 1) - \cos(2x) + \cos(2y) - \cos(2x)\cos(2y) . \qquad (28)$$

If $c = -1$, then we can match the coefficients of (27) and (28) to obtain a solution

$$u(x,y,z) = \frac{\cosh(2z)[\cos(2y) - \cos(2x)]}{2\sinh(2\pi)} - \frac{\cosh(z\sqrt{8})\cos(2x)\cos(2y)}{\sqrt{8}\sinh(\pi\sqrt{8})} .$$

We may add any constant to this solution to obtain another solution. In order to understand why we must choose $c = -1$, recall that $u(x,y,z)$ can be interpreted as a steady–state temperature function, which can exist only when the net flux of heat energy through the faces is 0. Thus, the integral of $c + 4\sin^2(x)\cos^2(y)$ over the face $z = \pi$ must be zero, and so $c = -1$. □

The wave equation on a rectangle

Let $u(x,y,t)$ be the transverse displacement (in the z–direction) of a homogeneous membrane which is attached to the boundary of the rectangle $0 \le x \le L$, $0 \le y \le M$. Using the same argument as in Section 6.1, it is possible to deduce that the only linear second–order PDE which governs the displacement, is the two–dimensional wave equation $u_{tt} = a^2(u_{xx} + u_{yy})$, where a is the constant speed at which disturbances spread in the membrane. The appropriate initial/boundary–value problem for $u(x,y,t)$ is

$$\begin{array}{ll} \text{D.E.} & u_{tt} = a^2(u_{xx} + u_{yy}) \quad 0 \le x \le L,\ 0 \le y \le M,\ -\infty < t < \infty \\ \text{B.C.} & \begin{cases} u(x,0,t) = 0, & u(x,M,t) = 0 \\ u(0,y,t) = 0, & u(L,y,t) = 0 \end{cases} \\ \text{I.C.} & u(x,y,0) = f(x,y),\quad u_t(x,y,0) = g(x,y) . \end{array} \qquad (29)$$

The procedure for separation of variables is similar to the case of the two–dimensional heat problem (2), except that, in place of $T' - bkT = 0$, we have $T'' - ba^2T = 0$. As before, the permissible values of $-b$ are the $\lambda_{n,m}$ of (5), and thus, we obtain the set of product solutions

$$u_{n,m}(x,y,t) = \left[a_{n,m}\cos(at\sqrt{\lambda_{n,m}}) + b_{n,m}\sin(at\sqrt{\lambda_{n,m}})\right] \sin(\frac{n\pi x}{L})\sin(\frac{m\pi y}{M}) \,, \tag{30}$$

where

$$\lambda_{n,m} = \pi^2[(n/L)^2 + (m/M)^2] \,, \quad n,\, m = 1,\, 2,\, 3 \,\dots \,. \tag{31}$$

If the initial displacement and velocity are finite double Fourier sine series

$$f(x,y) = \sum_{n,m=1}^{N} A_{n,m} \sin(\frac{n\pi x}{L})\sin(\frac{m\pi y}{M}) \,, \qquad g(x,y) = \sum_{n,m=1}^{N} B_{n,m} \sin(\frac{n\pi x}{L})\sin(\frac{m\pi y}{M}) \,,$$

then the solution of problem (29) is

$$u(x,y,t) = \sum_{n,m=1}^{N} u_{n,m}(x,y,t), \quad \text{where} \quad a_{n,m} = A_{n,m} \quad \text{and} \quad b_{n,m} = \frac{B_{n,m}}{a\sqrt{\lambda_{n,m}}} \,. \tag{32}$$

The function $u_{n,m}(x,y,t)$, defined by (30), is known as the **(n,m)–th harmonic** of the rectangular drum. The **frequency** of this harmonic is the number of oscillations per unit time that it executes, namely

$$\nu_{n,m} = \frac{a}{2\pi}\sqrt{\lambda_{n,m}} = \frac{a}{2}\,[(n/L)^2 + (m/M)^2]^{\frac{1}{2}} \,. \tag{33}$$

Example 5. Show that if $(L/M)^2$ is not rational, then no two of the harmonics (30) can have the same frequency. For the square drum, show that there are infinitely many pairs of harmonics which have the same frequency.

Solution. Suppose that $\nu_{n,m} = \nu_{p,q}$ and $(n,m) \neq (p,q)$. Then $n \neq p$ *and* $m \neq q$ (Why ?). Thus, $q^2 - m^2 \neq 0$, and so

$$(n/L)^2 + (m/M)^2 = (p/L)^2 + (q/M)^2 \quad \text{implies that} \quad \frac{n^2 - p^2}{q^2 - m^2} = \frac{L^2}{M^2} \,.$$

Since the left side of this last equation is rational, L^2/M^2 must be rational, and this contradicts the assumption. For the square, note that even if $n \neq m$, we have $\nu_{m,n} = \nu_{n,m}$. Thus, for a square, the (m,n)–th harmonic and the (n,m)–th harmonic have the same frequency. □

Summary 9.1

1. Two–dimensional heat equation : The solution of the initial/boundary–value problem

$$\text{D.E.} \quad u_t = k(u_{xx} + u_{yy}), \qquad 0 \le x \le L,\ 0 \le y \le M,\ t \ge 0$$

$$\text{B.C.} \begin{cases} u(x,0,t) = 0, & u(x,M,t) = 0 \\ u(0,y,t) = 0, & u(L,y,t) = 0 \end{cases} \tag{S1}$$

$$\text{I.C.} \quad u(x,y,0) = f(x,y),$$

where

$$f(x,y) = \sum_{n,m=1}^{N} b_{n,m} \sin\left(\frac{n\pi x}{L}\right)\sin\left(\frac{m\pi y}{M}\right), \tag{S2}$$

is given by

$$u(x,y,t) = \sum_{n,m=1}^{N} b_{n,m}\, e^{-\lambda_{n,m}kt} \sin\left(\frac{n\pi x}{L}\right)\sin\left(\frac{m\pi y}{M}\right),$$

where

$$\lambda_{n,m} = \pi^2[(n/L)^2 + (m/M)^2], \quad n, m = 1, 2, 3, \ldots .$$

2. Integral representation : In the above problem (S1), if $f(x,y)$ is continuous (but not necessarily of the form (S2)), on the closed rectangle, and is zero on the edges, then the unique continuous solution of (S1) is

$$u(x,y,t) = \frac{1}{4\pi kt} \int_{-\infty}^{\infty}\int_{-\infty}^{\infty} e^{-[(x-\bar{x})^2 + (y-\bar{y})^2]/(4kt)}\, \tilde{f}_{o,o}(\bar{x},\bar{y})\, d\bar{x}d\bar{y}, \quad t > 0,$$

(with $u(x,y,0) \equiv f(x,y)$), where $\tilde{f}_{o,o}(x,y)$ is the unique extension of $f(x,y)$ (to the whole plane) which is odd and periodic in x (of period $2L$) and odd and periodic in y (of period $2M$).

3. Double Fourier sine series : The double Fourier sine series of a function $f(x,y)$ on the rectangle $(0 \le x \le L,\ 0 \le y \le M)$ is the expression

$$\sum_{n,m=1}^{\infty} b_{n,m} \sin\left(\frac{n\pi x}{L}\right)\sin\left(\frac{m\pi y}{M}\right),$$

where

$$b_{n,m} = \frac{4}{LM}\int_0^M\int_0^L f(x,y)\sin\left(\frac{n\pi x}{L}\right)\sin\left(\frac{m\pi y}{M}\right) dxdy, \quad n, m = 1, 2, 3, \ldots,$$

provided these integrals exist.

4. Laplace's equation : The solution of the problem

$$\text{D.E.}\quad u_{xx} + u_{yy} + u_{zz} = 0 \qquad 0 < x < L,\ 0 < y < M\ ,\ 0 < z < P$$

$$\text{B.C.}\ \begin{cases} u(x,y,0) = f(x,y)\ ,\ u(x,y,P) = g(x,y) \\ u(x,y,z) = 0 \quad \text{on the faces} \quad x = 0,\ x = L,\ y = 0 \text{ and } y = M\ , \end{cases}$$

where

$$f(x,y) = \sum_{n,m=1}^{N} a_{n,m} \sin(\tfrac{n\pi x}{L})\sin(\tfrac{m\pi y}{M}) \quad \text{and} \quad g(x,y) = \sum_{n,m=1}^{N} b_{n,m} \sin(\tfrac{n\pi x}{L})\sin(\tfrac{m\pi y}{M})\ ,$$

is given by

$$u(x,y,z) = \sum_{n,m=1}^{N} Z_{n,m}(z)\sin(\tfrac{n\pi x}{L})\sin(\tfrac{m\pi y}{M})\ ,$$

where

$$Z_{n,m}(z) = \frac{1}{\sinh(P\sqrt{\lambda_{n,m}})}\left[a_{n,m}\sinh[(P-z)\sqrt{\lambda_{n,m}}] + b_{n,m}\sinh(z\sqrt{\lambda_{n,m}})\right]\ ,$$

and $\sqrt{\lambda_{n,m}} = \pi\sqrt{(n/L)^2+(m/M)^2}$.

5. Two–dimensional wave equation : The solution of the problem

$$\text{D.E.}\quad u_{tt} = a^2(u_{xx} + u_{yy}) \qquad 0 \le x \le L,\ 0 \le y \le M,\ -\infty < t < \infty$$

$$\text{B.C.}\ \begin{cases} u(x,0,t) = 0, \quad u(x,M,t) = 0 \\ u(0,y,t) = 0, \quad u(L,y,t) = 0 \end{cases}$$

$$\text{I.C.}\quad u(x,y,0) = f(x,y),\ u_t(x,y,0) = g(x,y)\ ,$$

where

$$f(x,y) = \sum_{n,m=1}^{N} A_{n,m} \sin(\tfrac{n\pi x}{L})\sin(\tfrac{m\pi y}{M})\ , \text{ and } \quad g(x,y) = \sum_{n,m=1}^{N} B_{n,m} \sin(\tfrac{n\pi x}{L})\sin(\tfrac{m\pi y}{M})\ ,$$

is

$$u(x,y,t) = \sum_{n,m=1}^{N} u_{n,m}(x,y,t), \quad \text{with} \quad a_{n,m} = A_{n,m} \text{ and } b_{n,m} = \frac{B_{n,m}}{a\sqrt{\lambda_{n,m}}}\ ,$$

where the (n,m)–th harmonic, $u_{n,m}(x,y,t)$, is

$$u_{n,m}(x,y,t) = \left[a_{n,m}\cos(at\sqrt{\lambda_{n,m}}) + b_{n,m}\sin(at\sqrt{\lambda_{n,m}})\right] \sin(\tfrac{n\pi x}{L})\sin(\tfrac{m\pi y}{M})\ ,$$

and $\sqrt{\lambda_{n,m}} = \pi\sqrt{(n/L)^2+(m/M)^2}$. The frequency of $u_{n,m}$ is $\frac{a}{2\pi}\sqrt{\lambda_{n,m}}$, $n, m = 1, 2, 3 \dots$.

Exercises 9.1

1. Solve the problem

D.E. $u_t = 6(u_{xx} + u_{yy})$, $0 \le x \le 2,\ 0 \le y \le 3,\ t \ge 0$

B.C. $\begin{cases} u(x,0,t) = 0, & u(x,3,t) = 0 \\ u(0,y,t) = 0, & u(2,y,t) = 0 \end{cases}$

I.C. $u(x,y,0) = 4\sin(3\pi x/2)\sin(\pi y) - 2\sin(\pi x)\sin(2\pi y/3)$.

2. Solve the problem

D.E. $u_t = u_{xx} + u_{yy}$, $0 \le x \le 1,\ 0 \le y \le 1,\ t \ge 0$

B.C. $\begin{cases} u(x,0,t) = 0, & u(x,1,t) = 0 \\ u_x(0,y,t) = 0, & u_x(1,y,t) = 0 \end{cases}$

I.C. $u(x,y,0) = 2\sin^2(2\pi x)\sin(\pi y)$.

3. Solve the problem

D.E. $u_t = u_{xx} + u_{yy}$, $0 \le x \le 1,\ 0 \le y \le 1,\ t \ge 0$

B.C. $\begin{cases} u(x,0,t) = 0, & u_y(x,1,t) = 0 \\ u_x(0,y,t) = 0, & u(1,y,t) = 0 \end{cases}$

I.C. . $u(x,y,0) = \sin(\pi(3x+y)/2) - \sin(3\pi x/2)\cos(\pi y/2)$.

4. A rectangular plate $0 \le x \le L$, $0 \le y \le M$ with heat diffusivity constant k is insulated on the edges $y = 0$ and $y = M$, and is maintained at temperature zero on the edges $x = 0$ and $x = L$.

(a) Write down the appropriate initial/boundary–value problem for two–dimensional heat flow in this plate, for a given continuous initial temperature distribution $u(x,y,0) = f(x,y)$.

(b) Using a double Fourier series, formally solve the problem found in part (a).

(c) Obtain an integral representation of the formal solution in the form (12), by using the method of images (i.e., by using a suitable extension of the initial temperature to the whole xy–plane ; cf. Section 7.5).

5. Let $f(x,t)$, $g(y,t)$, and $h(z,t)$ solve the respective heat equations $f_t = kf_{xx}$, $g_t = kg_{yy}$ and $h_t = kh_{zz}$. Show that $u(x,y,z,t) \equiv f(x,t)g(y,t)h(z,t)$ solves $u_t = k(u_{xx} + u_{yy} + u_{zz})$. Will the same construction work in the case of the wave equation ? Why or why not ?

6. (a) Use the technique of Problem 5 to show that one solution of the problem

$$\text{D.E.} \quad u_t = k(u_{xx} + u_{yy} + u_{zz}), \quad -\infty < x, y, z < \infty, \quad t > 0$$

$$\text{I.C.} \quad u(x,y,z,0^+) = F(x)G(y)H(z)$$

is

$$u(x,y,z,t) = \frac{1}{(4\pi kt)^{3/2}} \int_{-\infty}^{\infty}\int_{-\infty}^{\infty}\int_{-\infty}^{\infty} e^{-[(x-\bar{x})^2 + (y-\bar{y})^2 + (z-\bar{z})^2]/(4kt)} F(\bar{x})G(\bar{y})H(\bar{z})\, d\bar{x}d\bar{y}d\bar{z},$$

if $F(x)$, $G(y)$ and $H(z)$ are continuous and bounded.

(b) If we replace $F(x)G(y)H(z)$ in (a) by a finite linear combination of functions of this form (e.g., a finite triple Fourier series), will the given solution still be valid ? Explain.

(c) Why is the solution in part (a) *not* unique ? **Hint.** See Example 1 of Section 7.4 .

7. Consider the cube in Example 3.

(a) If the cube is insulated on one of its faces, then what is the temperature at $(\frac{1}{2},\frac{1}{2},\frac{1}{2})$ for large t.

(b) What if the cube is insulated on all but one face ?

(c) If the cube is to be insulated on two faces, should they be chosen adjacent, or chosen opposite, if the goal is to minimize the temperature drop at the center for large t ?

8. Solve

$$\text{D.E.} \quad u_{xx} + u_{yy} + u_{zz} = 0, \quad 0 < x < 3\pi,\ 0 < y < 2\pi,\ 0 < z < 1$$

$$\text{B.C.} \begin{cases} u(x,y,0) = \sin(x)\sin(y) \quad u(x,y,1) = 0 \\ u(x,y,z) = 0 \quad \text{on the faces } x = 0,\ y = 0,\ x = 3\pi,\ y = 2\pi . \end{cases}$$

9. Solve

$$\text{D.E.} \quad u_{xx} + u_{yy} + u_{zz} = 0, \quad 0 < x, y, z < \pi$$

$$\text{B.C.} \begin{cases} u_x(0,y,z) = u_x(\pi,y,z) = 0,\ u_y(x,0,z) = u_y(x,\pi,z) = 0 \\ u_z(x,y,0) = \cos(x)\sin^2(y),\ u_z(x,y,\pi) = 0 \end{cases}$$

10. (a) Consider the rectangular solid $0 \le x \le L$, $0 \le y \le M$, $0 \le z \le P$. Suppose that real numbers are assigned to each of the eight vertices (corners) of the solid. Show that there is a *unique* harmonic function of the form $u(x,y,z) = axyz + bxy + cyz + dxz + ex + fy + gz + h$ which has the given values at the eight vertices. **Hint.** Apply the two–dimensional result (cf. Example 1 of Section 6.2) to obtain harmonic functions $f(x,y)$ and $g(x,y)$ on the two faces $z = 0$ and $z = P$, with the correct values at the corners. Let $u(x,y,z) = [(P-z)f(x,y) + zg(x,y)]/P$.

(b) Suppose that we are given a continuous function on each of the twelve *edges* (not faces) of a rectangular solid in part (a), such that the functions agree whenever 3 edges meet at a corner. Show that there is a harmonic function u (i.e., $u_{xx} + u_{yy} + u_{zz} = 0$) defined inside the solid, which extends continuously to the boundary such that the extension is a harmonic function inside each face and is equal to the given continuous functions on the edges. You may assume that the Dirichlet problem can be solved on each face (cf. Section 6.2). **Hint.** First, use part (a) to reduce the problem to the case where the given functions on the edges are 0 at the eight vertices. Then solve the Dirichlet problem on each pair of opposite faces and form a (convex) linear combination, as in part(a). Add the results for the three pairs of opposite faces and divide by two (Why?).

11. Explain why the result of Problem 10 allows us to convert a Dirichlet problem for a rectangular solid to a related problem, where the given function on the boundary of the solid is zero on the edges. Explain why such a reduction is necessary, if we are to solve the problem through the use of double Fourier sine series on the faces, as was done with (23).

12. By assuming a solution of the form $u(x,y,z) = Ax^2 + By^2 + Cz^2 + Dx + Ey + Fz$, solve the Neumann problem on a solid cube, where the normal derivative is a given constant on each face :

$$\text{D.E.} \quad u_{xx} + u_{yy} + u_{zz} = 0 \quad (0 \le x,\ y,\ z \le 1)$$

$$\text{B.C.} \begin{cases} u_x(0,y,z) = a_0\ ,\ u_y(x,0,z) = b_0\ ,\ u_z(x,y,0) = c_0 \\ u_x(1,y,z) = a_1\ ,\ u_y(x,1,z) = b_1\ ,\ u_z(x,y,1) = c_1\ . \end{cases}$$

Why is there no solution of this problem unless $a_0 + b_0 + c_0 = a_1 + b_1 + c_1$? Would Fourier series or separation of variables be of any value in solving this problem ? Why or why not ?

13. Find a formal solution of the problem

$$\text{D.E.} \quad u_{tt} = a^2(u_{xx} + u_{yy}) \quad 0 \le x,\ y \le 1$$

$$\text{B.C.} \begin{cases} u(x,0,t) = 0\ ,\ \ u(x,1,t) = 0 \\ u(0,y,t) = 0\ ,\ \ u(1,y,t) = 0 \end{cases}$$

$$\text{I.C.} \quad u(x,y,0) = 0\ ,\ \ u_t(x,y,0) = x(x-1)y(y-1)\ .$$

14. The speed a of wave propagation in a rectangular drum is 600 feet/sec.. The lowest two frequencies of the drum are 300 and 400 cycles per second. What are the length and the width of the drum ?

9.2 The Eigenfunction Viewpoint

Here we introduce the concepts of eigenfunctions and eigenvalues of the Laplace operator. These concepts provide a unified viewpoint from which boundary–value problems for the heat and wave equations on arbitrary domains in any dimension can be understood. Moreover, these notions are independent of the type of coordinate system which is being used, and thus they provide a guiding principle when we study PDEs in spherical and cylindrical coordinates in subsequent sections. We also prove a convergence theorem for Fourier series in higher dimensions, and discuss higher–dimensional Fourier transforms.

Eigenfunctions and eigenvalues of Δ

Observe that while the functions

$$f_{n,m}(x,y) = \sin(\tfrac{n\pi x}{L})\sin(\tfrac{m\pi y}{M})$$

are not harmonic since they do not satisfy Laplace's equation, the Laplacian of $f_{n,m}$ is

$$\Delta f_{n,m} = (f_{n,m})_{xx} + (f_{n,m})_{yy} = -\pi^2[(n/L)^2 + (m/M)^2]\, f_{n,m} = -\lambda_{n,m} f_{n,m}\,. \qquad (1)$$

Thus, when the Laplace operator $\Delta = \dfrac{\partial^2}{\partial x^2} + \dfrac{\partial^2}{\partial y^2}$ operates on the function $f_{n,m}$, the result is a constant multiple of this function, namely $-\lambda_{n,m} f_{n,m}$.

In general, functions $g(x,y)$ which have the property that $\Delta g + \lambda g = 0$ for some constant λ, are known as **eigenfunctions** of Δ, provided $g(x,y)$ is not identically zero. The constant λ is called the **eigenvalue** associated with the eigenfunction $g(x,y)$.

Remark. "Eigen" is the German word for "self", and presumably the name comes from the fact that Δ sends such a function to itself, aside from a constant factor. Hence, (1) demonstrates that $f_{n,m}(x,y)$ is an eigenfunction of Δ with eigenvalue $\lambda_{n,m}$. Those who are familiar with the notion of eigenvalue from linear algebra should observe that $-\lambda_{n,m}$ is the eigenvalue of Δ associated with $f_{n,m}$ in the linear algebra sense ($\Delta f_{n,m} = -\lambda_{n,m} f_{n,m}$). However, the above definition is more convenient for the present setting, since it is easier to refer to the nonnegative quantities $\lambda_{n,m}$ (as opposed to $-\lambda_{n,m}$) as being the eigenvalues.

Example 1. Show that if $g(x,y)$ is any C^2 eigenfunction of Δ, with eigenvalue $\lambda \geq 0$, then

(a) $v(x,y,t) = be^{-\lambda kt}g(x,y)$ is a solution of the heat equation $v_t = k\Delta v$, and

(b) $u(x,y,t) = [b_1\cos(a\sqrt{\lambda}\ t) + b_2\sin(a\sqrt{\lambda}\ t)]g(x,y)$ solves the wave equation $u_{tt} = a^2\Delta u$, where b, b_1, and b_2 are arbitrary constants.

Solution. Note that $\Delta v = \Delta(be^{-\lambda kt}g) = be^{-\lambda kt}\Delta g = -be^{-\lambda kt}\lambda g = -\lambda v$, and similarly $\Delta u = -\lambda u$. Thus, $v_t = -\lambda bke^{-\lambda kt}g = -k\lambda v = k\Delta v$ and $u_{tt} = -(a\sqrt{\lambda})^2 u = -a^2\lambda u = a^2\Delta u$. □

This example shows that eigenfunctions of Δ can be used to construct solutions of the heat and wave equations. Indeed, all of the product solutions of these PDEs are of the form as in Example 1. Of course, solutions of Laplace's equation $\Delta u = 0$ are eigenfunctions with eigenvalue $\lambda = 0$, and the solutions constructed in Example 1 are then steady–state solutions. In a problem with time–independent B.C., the solutions constructed in Example 1 will satisfy these B.C., provided the eigenfunction $g(x,y)$ meets these B.C. . Note that all of these observations hold in any number of spatial dimensions. In particular, for dimension 1, the Laplace operator reduces to $\frac{d^2}{dx^2}$, whose eigenfunctions are of the form $c_1\sin(bx) + c_2\cos(bx)$ (with eigenvalue b^2) or $c_1e^{bx} + c_2e^{-bx}$ (with eigenvalue $-b^2$). The B.C. which we dealt with in Chapters 3 and 5 led us to eigenfunctions and eigenvalues of a particular form depending on the B.C. (e.g., $\sin(\frac{n\pi x}{L})$, when $u(0,t) = 0$ and $u(L,t) = 0$). For heat and wave problems with homogeneous D.E. and B.C., our past strategy could be phrased in terms of eigenfunctions, in the following steps.

1. Using separation of variables, find those eigenfunctions, say $g_n(x,y)$ with eigenvalues λ_n $(n = 1, 2, 3,...)$, which satisfy the B.C. . Then, as in Example 1, multiply $g_n(x,y)$ by the appropriate function of time to obtain a function $u_n(x,y,t)$ which satisfies the D.E. and B.C. .

2. Apply the superposition principle to form the more general solution $\sum c_n u_n(x,y,t)$ of the D.E. and B.C. .

3. Approximate each function in the I.C. (e.g., initial temperature, or position and velocity) by a linear combination of eigenfunctions, say by computing Fourier coefficients.

4. By substituting the sum $\sum c_n u_n(x,y,t)$ of product solutions into each I.C., determine the arbitrary constants in the time–dependent parts of the $u_n(x,y,t)$ by equating coefficients with the coefficients found in step 3.

Assuming that the eigenfunctions $g_n(x,y)$ in step 1 have been determined, the only potential difficulty remaining is Step 3. In broad terms, Fourier series (single, double or multiple) is the study of adequately representing functions in terms of linear combinations of eigenfunctions satisfying given linear, homogeneous B.C.. Indeed, Fourier series are often referred to as **eigenfunction expansions**. Although, there are theorems which prove the existence of eigenfunctions and the validity of eigenfunction expansions of reasonably nice functions, one major barrier to the approach is that it is difficult to find the eigenfunctions, when the region of a boundary–value problem is not of a standard shape, such as a rectangular box, a disk, a ball, a cylinder, etc. We will determine the eigenfunctions for boundary–value problems in the cases of balls and cylinders in Sections 9.3 – 9.5.

Remark. Fourier transforms may also be regarded as providing eigenfunction expansions when a *continuous* superposition (i.e., an integral) of eigenfunctions $e^{i\xi x}$ (with eigenvalues ξ^2) is needed to represent certain functions on infinite domains. For example, if $f(x)$ is a rapidly decreasing function with Fourier transform $\hat{f}(\xi)$, then the Inversion Theorem (cf. Section 7.3) states that $f(x)$ is given by

$$f(x) = \frac{1}{\sqrt{2\pi}} \int_{-\infty}^{\infty} \hat{f}(\xi)\, e^{i\xi x}\, d\xi\,, \tag{2}$$

which is a continuous superposition of the eigenfunctions $e^{i\xi x}$ of Δ. □

If one can obtain an eigenfunction expansion for the function in the I.C. for a heat problem with homogeneous B.C., then the solution of the problem is at hand. One just needs to multiply each eigenfunction in the expansion by the correct time–dependent factor in Example 1(a). For instance, if (2) is the "eigenfunction expansion" for an initial temperature $f(x)$, then the associated hypothetical solution of the heat equation $u_t = ku_{xx}$ is (noting that the eigenvalue for $e^{i\xi x}$ is ξ^2)

$$u(x,t) = \frac{1}{\sqrt{2\pi}} \int_{-\infty}^{\infty} \hat{f}(\xi)\, e^{-\xi^2 kt}\, e^{i\xi x}\, d\xi = \frac{1}{\sqrt{2\pi}} \int_{-\infty}^{\infty} \frac{1}{\sqrt{2\pi}} \int_{-\infty}^{\infty} f(w) e^{-i\xi w} dw\, e^{-\xi^2 kt} e^{i\xi x}\, d\xi$$

$$= \frac{1}{2\pi} \int_{-\infty}^{\infty} \left[\int_{-\infty}^{\infty} e^{i\xi(w-x)}\, e^{-\xi^2 kt}\, d\xi \right] f(w)\, dw = \frac{1}{\sqrt{4\pi kt}} \int_{-\infty}^{\infty} e^{-(w-x)^2/(4kt)}\, f(w)\, dw,$$

which is the solution found in Section 7.4. The same technique (but with multiple Fourier transforms) provides the higher–dimensional version of this solution, namely (12) in Section 9.1. We could also use this technique to find the higher–dimensional analogs of D'Alembert's formula for the wave equation on the plane or space, but there are easier ways (covered in Section 9.3) of obtaining these formulas.

Example 2. Find an eigenfunction of Δ which is zero on the boundary of a square, but which is not a product of the form $f(x)h(y)$.

Solution. Note that if $g_1(x,y)$ and $g_2(x,y)$ are two eigenfunctions of Δ with the same eigenvalue λ, then any linear combination $c_1g_1 + c_2g_2$ is also an eigenfunction with eigenvalue λ. Indeed, using the linearity of Δ, we have $\Delta(c_1g_1 + c_2g_2) = c_1\Delta g_1 + c_2\Delta g_2 = -\lambda(c_1g_1 + c_2g_2)$. Also observe that if g_1 and g_2 are zero on the boundary of the square, then $c_1g_1 + c_2g_2$ also satisfies this boundary condition. In particular, consider the square $0 \le x, y \le \pi$ and let $g_1(x,y) = \sin(nx)\sin(my)$ and let $g_2(x,y) = \sin(mx)\sin(ny)$ for unequal positive integers n and m. These are two eigenfunctions of Δ with common eigenvalue $(n^2 + m^2)$. Thus, any linear combination of them will also be an eigenfunction with eigenvalue $(n^2 + m^2)$. In particular let α be any constant in $[0,2\pi)$ and consider the linear combination

$$g(x,y) = \cos\alpha \sin(nx)\sin(my) + \sin\alpha \sin(mx)\sin(ny) . \tag{3}$$

Of course, when $\alpha = 0$, $\pi/2$, π or $3\pi/2$, this reduces to a product of the form $f(x)h(y)$. However, for every other value of α in $[0,2\pi)$, $g(x,y)$ cannot be put in this form (Why ?). For definiteness, consider the case when $n = 1$, $m = 2$, and $\alpha = 3\pi/4$. In this case, $g(x,y) = [-\sin(x)\sin(2y) + \sin(2x)\sin(y)]/\sqrt{2} = \sqrt{2}\sin(x)\sin(y)[-\cos(y) + \cos(x)]$, which is zero not only on the boundary of the square, but also on the diagonal $y = x$, and nowhere else. A function of the form $f(x)h(y)$ cannot vanish on this diagonal, unless it is identically 0 on the square (Why ?). □

Nodal curves, symmetry breaking and eigenspaces

The curves on which an eigenfunction $g(x,y)$ is zero, are known as **nodal curves.** If λ is the eigenvalue (i.e., $\Delta g + \lambda g = 0$), then we form an associated solution, say $\cos(at\sqrt{\lambda})g(x,y)$, of the wave equation $u_{tt} = a^2\Delta u$, as in Example 1. Note that as the membrane vibrates in this mode, the points on the nodal curves remain fixed (i.e., they do not move up and down as the membrane vibrates). If the ratio L^2/M^2 of the squares of the dimensions of a rectangular drum is irrational, then, by Example 5 of Section 9.1, no two harmonics have the same frequency (or equivalently, no two independent eigenfunctions have the same eigenvalue). In this case, the nodal curves of any eigenfunction consist of equally spaced horizontal and/or vertical line segments. If the ratio L^2/M^2 is rational (e.g., as with a square), then the nodal curves of various linear combinations of eigenfunctions with the same eigenvalue (e.g., (3)) can assume a wide variety of forms. For the linear combination (3) with $n = 1$ and $m = 2$, the nodal curves for various α are shown in Figure 1. The associated modes of vibration $u(x,y,t) = \cos(at\sqrt{5})g(x,y)$ are also graphed at a fixed time $t = \pi/(a\sqrt{5})$.

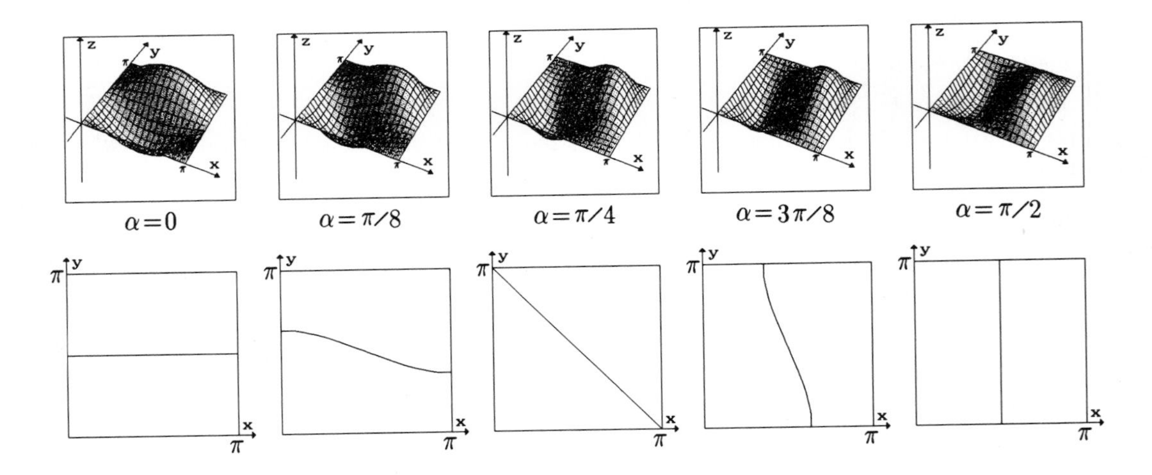

Figure 1

When $n = 1$ and $m = 3$, more complicated patterns arise, as is shown in Figure 2.

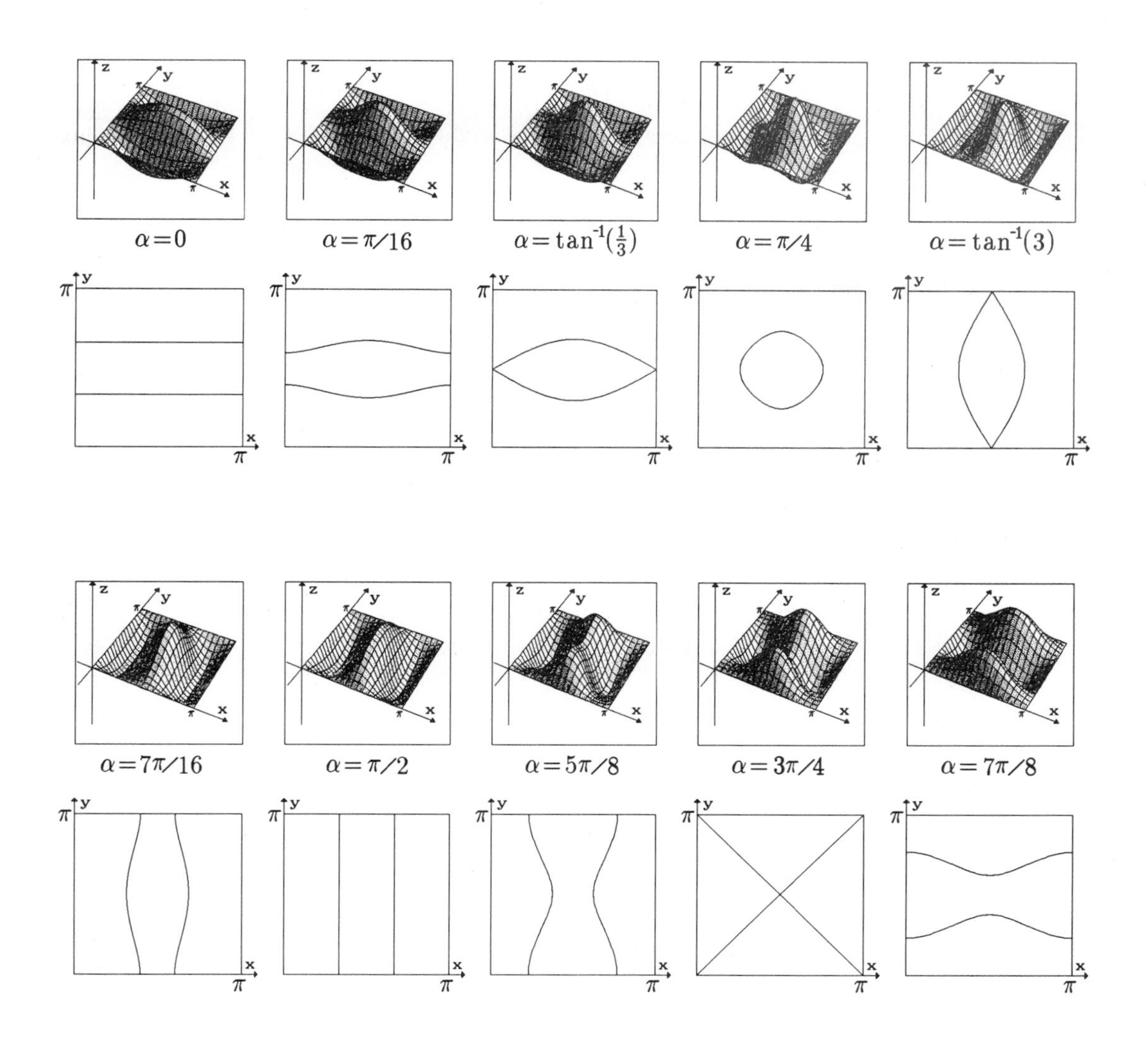

Figure 2

If a real physical square drum is vibrated at a frequency associated with several different modes, then it usually will settle into just one of the many possible modes, due to slight deviations from squareness, slight imperfections in the membrane, or some unknown cause. This is an example of the important notion of "spontaneous symmetry breaking". Even though a symmetry is theoretically precise, it is almost always broken in practice. Usually, if a domain has some symmetry, then one can conclude that many of the eigenvalues will have more than one eigenfunction, because under a symmetry operation which flips or rotates the domain without changing it, an eigenfunction is possibly converted to another eigenfunction with the same eigenvalue. For example, when the square is flipped about the diagonal $y = x$, the eigenfunction $\sin(nx)\sin(my)$ is converted into $\sin(ny)\sin(my)$, since flipping interchanges x and y. Since disks and balls have even higher degrees of symmetry, we expect to have even larger families of eigenfunctions (say vanishing on the boundary) which share a common eigenvalue of Δ. Each such family is known as the **eigenspace** associated with the eigenvalue. As we will see in Section 9.5, the dimensions of the eigenspaces of the Schrödinger operator $\Delta + 1/\rho$ (in three–dimensional space, where $\rho = (x^2+y^2+z^2)^{\frac{1}{2}}$ are largely responsible for the periodicity occurring in the periodic table of elements. Thus, the eigenspace concept is very important, and we will have much more to say about it in subsequent sections.

Multiple Fourier series and transforms

Here we give a concise treatment of double Fourier series of functions $f(x,y)$ on a rectangle R, given by $-L \le x \le L$, $-M \le y \le M$. Our discussion easily extends to the n–dimensional case. For two integrable complex–valued functions $f(x,y)$ and $g(x,y)$ on R, we define the inner product

$$\langle f,g\rangle = \int_{-M}^{M}\int_{-L}^{L} f(x,y)\overline{g(x,y)}\,dxdy\,. \tag{4}$$

For (x,y) in R, define

$$E_{n,m}(x,y) = e^{i\pi(nx/L + my/M)} = e^{in\pi x/L}\,e^{im\pi y/M}\,. \tag{5}$$

It is easy to check that, for any integers n, m, p, q, we have

$$\langle E_{n,m}, E_{p,q}\rangle = \begin{cases} 0 & \text{if } n \neq p \text{ or } m \neq q \\ 4LM & \text{if } n = p \text{ and } m = q \end{cases}, \tag{6}$$

i.e., the functions $E_{n,m}$ form an orthogonal family of functions of norm–square 4LM.

Definition. The (complex) **double Fourier series** of a function $f(x,y)$ defined on the rectangle R $(0 \le x \le L,\ 0 \le y \le M)$ is the expression

$$\text{FS } f(x,y) = \sum_{n,m=-\infty}^{\infty} c_{n,m}\,E_{n,m}(x,y) = \sum_{n,m=-\infty}^{\infty} c_{n,m}e^{i\pi(nx/L + my/M)}, \tag{7}$$

where

$$c_{n,m} = \frac{1}{4LM}\langle f,E_{n,m}\rangle = \frac{1}{4LM}\int_{-M}^{M}\int_{-L}^{L} f(x,y)e^{-i\pi(mx/L + ny/M)}\,dx\,dy\,, \tag{8}$$

and $E_{n,m}(x,y)$ is defined by (5), provided that all of these integrals exist.

Note that formula (8) for $c_{n,m}$ could be obtained formally by taking the inner product of both sides of (7) with $E_{n,m}$ and applying the orthogonality result (6). In the sum (7) the integers n and m run independently over all integers. The astute reader may wonder if the series (7) will converge and whether the same result is obtained regardless of the order in which the terms are added. Although the order of addition does not matter in finite sums, it can make a difference in infinite sums. For example, $\pi/4 = 1 - 1/3 + 1/5 - 1/7 \ldots$, but rewriting this sum in the form $(1 + 1/5) - 1/3 + (1/9 + 1/13) - 1/7 + (1/17 + 1/21) - 1/11 + \ldots$, we get a value larger than 13/15 which is greater than $\pi/4$. However, if the sum of the *absolute values* (or magnitudes, if the terms are complex numbers) of the terms of an infinite series is finite (i.e. the series is absolutely convergent), then the order in which the terms are added will not matter. The trouble with the series $1 - 1/3 + 1/5 - \ldots$ is that it is not absolutely convergent (i.e., $1 + 1/3 + 1/5 + \ldots = \infty$). The Fourier series (7) is **absolutely convergent**, if the sum $\sum |c_{n,m}|$ of the magnitudes of all of the coefficients is finite. In this case, the order in which the terms of the Fourier series are added will be immaterial. Thus, in the following theorem, we first give criteria for the function f(x,y), which guarantee that $\sum |c_{n,m}| < \infty$.

Theorem 1 (Convergence Theorem for Double Fourier Series). Let $f(x,y)$ **be a** C^k **function** $(k \geq 3)$ **on the rectangle** $-L \leq x \leq L$, $-M \leq y \leq M$. **Let** K **be the largest of the maxima of** $|\partial^k f/\partial x^k|$ **and** $|\partial^k f/\partial y^k|$ **on this rectangle. Assume that for each** $j = 0, 1, \ldots, k-1$, **we have**

$$\frac{\partial^j f}{\partial x^j}(-L,y) = \frac{\partial^j f}{\partial x^j}(L,y) \quad \text{and} \quad \frac{\partial^j f}{\partial y^j}(x,-M) = \frac{\partial^j f}{\partial y^j}(x,M) \,. \tag{9}$$

Then, for any positive integer N,

$$\sum_{\substack{|n| > N \\ \text{or} \\ |m| > N}} |c_{n,m}| \leq \frac{2\pi\ K\ (\sqrt{2}/\pi)^k\ \max(L^k,M^k)}{(k-2)\ N^{k-2}}\,, \tag{10}$$

where the $c_{n,m}$ **are the Fourier coefficients** $\frac{1}{4LM} <f,E_{n,m}>$. **In particular,** $\sum |c_{n,m}| < \infty$. **Also, for the** N–th **partial sum of the double Fourier series of** f, **obtained by summing** m **and** n **from** $-N$ **to** N, **we have**

$$\left| f(x,y) - \sum_{n,m=-N}^{N} c_{n,m}\, e^{i\pi(nx/L + my/M)} \right| \leq \frac{2\pi\ K\ (\sqrt{2}/\pi)^k\ \max(L^k,M^k)}{(k-2)N^{k-2}}\,, \tag{11}$$

(i.e., **these partial sums converge uniformly to the function** f(x,y) **on the rectangle**).

Proof. Integrating by parts repeatedly with respect to x and using the equations (9) to eliminate the endpoint evaluations, we have

$$c_{n,m} = \frac{1}{4LM} \int_{-M}^{M} \int_{-L}^{L} f(x,y)\, \overline{E_{n,m}(x,y)}\, dx\, dy = \dots$$

$$= \frac{1}{4LM} \left[\frac{L}{n\pi i} \right]^k \int_{-M}^{M} \int_{-L}^{L} \left[\frac{\partial^k f}{\partial x^k} \right] \overline{E_{n,m}(x,y)}\, dx\, dy\ , \quad n \neq 0\ . \tag{12}$$

A similar result holds, if we integrate by parts with respect to y. Thus, by the definition of K,

$$|c_{n,m}| \le K \left[\frac{L}{|n|\pi} \right]^k \quad \text{and} \quad |c_{n,m}| \le K \left[\frac{M}{|m|\pi} \right]^k , \quad n, m \neq 0\ . \tag{13}$$

We may replace the numerators L and M in (13) by the larger (i.e., by $\max(L,M)$), and then choose the stronger of the two resulting inequalities (i.e., the one with the larger denominator) to conclude that

$$|c_{n,m}| \le K \left[\frac{\max(L,M)}{\pi \max(|m|,|n|)} \right]^k \le K \left[\frac{\sqrt{2}}{\pi} \right]^k \max(L^k, M^k)\, [n^2 + m^2]^{-k/2}, \tag{14}$$

since $[(n^2 + m^2)/2]^{1/2} \le \max(|m|,|n|)$. Using a two–dimensional integral comparison, we have

$$\sum_{\substack{|n| > N \\ \text{or } |m| > N}} [n^2 + m^2]^{-k/2} \le \iint_{u^2 + v^2 > N^2} [u^2 + v^2]^{-k/2}\, du\, dv$$

$$= \int_0^{2\pi} \int_N^{\infty} r^{-k}\, r\, dr\, d\theta = 2\pi N^{2-k}/(k-2)\ . \tag{15}$$

Combining (14) and (15), we obtain the result (10).

Once we prove that FS $f(x,y) = f(x,y)$, the left side of (11) becomes

$$\left| \sum_{\substack{|n| > N \\ \text{or } |m| > N}} c_{n,m}\, e^{i\pi(nx/L + my/M)} \right| \le \sum_{\substack{|n| > N \\ \text{or } |m| > N}} |c_{n,m}|\ ,$$

and so (11) would then follow from (10). Thus, it remains to prove that FS $f(x,y) = f(x,y)$. Note that since we have assumed that $f(x,y)$ is C^k $(k \ge 3)$ with the endpoint conditions (9) holding, we certainly know that for each fixed y in $[-M,M]$, $f(x,y)$ (as a function of x) amply satisfies the conditions of any of the convergence theorems of Section 4.2. Thus we know that

$$f(x,y) = \sum_{|n|=0}^{\infty} g_n(y)\, e^{in\pi x/L}, \quad \text{where} \quad g_n(y) = \frac{1}{2L} \int_{-L}^{L} f(x,y) e^{-in\pi x/L}\, dx\ , \tag{16}$$

According to Leibniz's rule, $g_n(y)$ is certainly C^2 and, using (9), we have $g_n(-M) = g_n(M)$ and $g_n'(-M) = g_n'(M)$. Thus, $g_n(y)$ is also equal to its Fourier series, as a function of y, i.e.,

$$g_n(y) = \sum_{|m|=0}^{\infty} C_{n,m}\, e^{im\pi y/M}\,, \quad \text{where} \quad C_{n,m} = \frac{1}{2M}\int_{-M}^{M} g_n(y)\, e^{-im\pi y/M}\, dy = c_{n,m}\,.$$

Hence, substituting this expression for $g_n(y)$ into (16),

$$f(x,y) = \sum_{|n|=0}^{\infty}\Big[\sum_{|m|=0}^{\infty} c_{n,m}\, e^{im\pi y/M}\Big] e^{in\pi x/L} = \text{FS } f(x,y)\,,$$

where we have implicitly used the fact that $\sum |c_{n,m}| < \infty$ (i.e., in the last equality, we needed to know that FS f(x,y) is independent of any reordering of terms). □

Remarks. Through the use of Euler's formula

$$e^{i\pi(nx/L + my/M)} = e^{in\pi x/L}\, e^{im\pi y/M} = [\cos(\tfrac{n\pi x}{L}) + i\,\sin(\tfrac{n\pi x}{L})]\,[\cos(\tfrac{m\pi y}{M}) + i\,\sin(\tfrac{m\pi y}{M})]\,,$$

the complex Fourier series can be written in terms of the four possible products

$$\sin(\tfrac{n\pi x}{L})\sin(\tfrac{m\pi y}{M}),\ \ \cos(\tfrac{n\pi x}{L})\cos(\tfrac{m\pi y}{M}),\ \ \sin(\tfrac{n\pi x}{L})\cos(\tfrac{m\pi y}{M}),\ \ \cos(\tfrac{n\pi x}{L})\sin(\tfrac{m\pi y}{M})\,. \qquad (17)$$

Moreover, if the function f(x,y) (defined on the rectangle $-L \le x \le L$, $-M \le y \le M$) is real–valued, then all of the imaginary terms cancel in the complex Fourier series, leaving only a series involving the terms (17) with real coefficients. If the function f(x,y) is odd in x and odd in y, then only the terms $\sin(\frac{n\pi x}{L})\sin(\frac{m\pi y}{M})$ will be involved. Thus, if one desires to represent a function defined on the rectangle $0 \le x \le L$, $0 \le y \le M$ by a double sine series, one can extend the function oddly in x and oddly in y, to the larger rectangle $-L \le x \le L$, $-M \le y \le M$, and compute the complex Fourier series of this extension. This is entirely equivalent to computing the double Fourier sine series of the unextended function, using formula (10) of Section 9.1. In general, by using different types of extensions, all of the various double Fourier series are seen to be special cases of the complex Fourier series for suitably extended functions. All of the above treatment of double Fourier series can be easily modified to handle the case of triple Fourier series (or multiple Fourier series in any finite dimension). □

Double Fourier transforms

Suppose that f(x,y) is a function defined for all (x,y), and suppose that f(x,y) is absolutely integrable (i.e., $\int_{-\infty}^{\infty}\int_{-\infty}^{\infty} |f(x,y)|\, dx\, dy < \infty$). Then the **double Fourier transform** of f(x,y) is the function

$$\hat{f}(\xi,\eta) = \frac{1}{2\pi}\int_{-\infty}^{\infty}\int_{-\infty}^{\infty} f(x,y)\, e^{-i(\xi x + \eta y)}\, dx\, dy\, . \tag{18}$$

Formally, $\hat{f}(\xi,\eta)$ can also be obtained by taking the Fourier transform first in the variable x and then taking the Fourier transform of the result with respect to y, i.e.,

$$\hat{f}(\xi,\eta) = \frac{1}{\sqrt{2\pi}}\int_{-\infty}^{\infty} \hat{f}(\xi,y)\, e^{-i\eta y}\, dy, \quad \text{where} \quad \hat{f}(\xi,y) = \frac{1}{\sqrt{2\pi}}\int_{-\infty}^{\infty} f(x,y)\, e^{-i\xi x}\, dx\, .$$

In other words,

$$\hat{f}(\xi,\eta) = [\hat{f}(\xi,y)]\hat{\ }(\eta). \tag{19}$$

The **inverse double Fourier transform** of an absolutely integrable function $g(\xi,\eta)$ is the function

$$\check{g}(x,y) = \frac{1}{2\pi}\int_{-\infty}^{\infty}\int_{-\infty}^{\infty} g(\xi,\eta)\, e^{i(\xi x + \eta y)}\, d\xi\, d\eta\, . \tag{20}$$

Clearly, $\check{g}(x,y) = [\check{g}(x,\eta)]\check{\ }(y)$. By formally applying twice the Inversion Theorem (cf. Section 7.3) for one–dimensional Fourier transforms, we can obtain formally the Inversion Theorem for double Fourier transforms, namely

$$f(x,y) = [\hat{f}(\xi,\eta)]\check{\ }(x,y) = \frac{1}{2\pi}\int_{-\infty}^{\infty}\int_{-\infty}^{\infty} f(\xi,\eta)\, e^{-i(\xi x + \eta y)}\, d\xi\, d\eta\, . \tag{21}$$

One can also obtain formally the (double) Convolution Theorem from a two–fold application of the single variable Convolution Theorem ($[f*g]\hat{\ }(\xi) = \sqrt{2\pi}\, \hat{f}(\xi)\hat{g}(\xi)$), as follows. For "nice", suitably decaying functions f(x,y) and h(x,y), we have

$$[f*h](x,y) \equiv \int_{-\infty}^{\infty}\int_{-\infty}^{\infty} f(x-w,y-z)h(w,z)\, dw\, dz\, . \tag{22}$$

Then

$$[f*h]\hat{}\,(\xi,\eta) = \frac{1}{2\pi}\int_{-\infty}^{\infty}\int_{-\infty}^{\infty}\int_{-\infty}^{\infty}\int_{-\infty}^{\infty} f(x-w,y-z)h(w,z)\, e^{-i(\xi x + \eta y)}\, dw\, dz\, dx\, dy$$

$$= \frac{1}{\sqrt{2\pi}}\int_{-\infty}^{\infty}\int_{-\infty}^{\infty}\left[\frac{1}{\sqrt{2\pi}}\int_{-\infty}^{\infty}\left[\int_{-\infty}^{\infty} f(x-w,y-z)h(w,z)\, dw\right] e^{-i\xi x}\, dx\right] dz\right] e^{-i\eta y}\, dy$$

$$= \frac{1}{\sqrt{2\pi}}\int_{-\infty}^{\infty}\left[\int_{-\infty}^{\infty} \sqrt{2\pi}\, \hat{f}(\xi,y-z)\, \hat{h}(\xi,z)\, dz\right] e^{-i\eta y}\, dy$$

$$= 2\pi\, \hat{f}(\xi,\eta)\, \hat{h}(\xi,\eta),$$

where we have formally used Fubini's theorem (cf. Appendix A.3), the single–variable Convolution Theorem, and (19).

Note that all of these formal manipulations certainly hold in the case where all functions in sight are rapidly decreasing. In using Fourier transform methods to find hypothetical solutions of PDEs, there is no advantage in establishing the most general circumstances under which the Inversion and Convolution Theorems hold, since such solutions must be checked independently. Also, the above definitions and results readily extend to the case of Fourier transforms in an arbitrary finite dimension n.

Summary 9.2

1. **Eigenfunctions and eigenvalues** : Functions $g(x,y)$ such that $\Delta g + \lambda g = 0$, for some constant λ, are known as eigenfunctions of Δ, provided that $g(x,y)$ is not identically zero. The constant λ is called the eigenvalue associated with the eigenfunction $g(x,y)$.

2. **Multiple Fourier series** : The (complex) double Fourier series of a function $f(x,y)$ defined on the rectangle R, $-L \le x \le L$, $-M \le y \le M$, is the expression

$$\mathrm{FS}\, f(x,y) = \sum_{n,m=-\infty}^{\infty} c_{n,m} E_{n,m}(x,y) = \sum_{n,m=-\infty}^{\infty} c_{n,m} e^{i\pi(nx/L + my/M)},$$

where

$$c_{n,m} = \frac{1}{4LM} \langle f, E_{n,m} \rangle = \frac{1}{4LM} \int_{-M}^{M} \int_{-L}^{L} f(x,y) e^{-i\pi(mx/L + ny/M)}\, dx\, dy,$$

provided all of the integrals exist, where $E_{n,m}(x,y) = e^{i\pi(nx/L + my/M)}$. Theorem 1 gives criteria under which the series $\mathrm{FS}\, f(x,y)$ converges uniformly to $f(x,y)$ on R.

3. **Double Fourier transforms** : Suppose that $f(x,y)$ is a function defined for all (x,y), and suppose that $f(x,y)$ is absolutely integrable (i.e., $\int_{-\infty}^{\infty}\int_{-\infty}^{\infty} |f(x,y)|\, dx\, dy < \infty$). Then the double Fourier transform of $f(x,y)$ is the function

$$\hat{f}(\xi,\eta) = \frac{1}{2\pi} \int_{-\infty}^{\infty}\int_{-\infty}^{\infty} f(x,y)\, e^{-i(\xi x + \eta y)}\, dx\, dy.$$

The inverse double Fourier transform of an absolutely integrable function $g(\xi,\eta)$ is the function

$$\check{g}(x,y) = \frac{1}{2\pi} \int_{-\infty}^{\infty}\int_{-\infty}^{\infty} g(\xi,\eta)\, e^{i(\xi x + \eta y)}\, d\xi\, d\eta.$$

By formally applying twice the Inversion Theorem for one–dimensional Fourier transforms, we can obtain formally the Inversion Theorem for double Fourier transforms, namely

$$f(x,y) = [\hat{f}(\xi,\eta)]^{\vee}(x,y) = \frac{1}{2\pi} \int_{-\infty}^{\infty}\int_{-\infty}^{\infty} \hat{f}(\xi,\eta)\, e^{-i(\xi x + \eta y)}\, d\xi\, d\eta.$$

One can also obtain formally the following (double) Convolution Theorem from a two–fold application of the single variable Convolution Theorem.

$$\text{If } [f*h](x,y) \equiv \int_{-\infty}^{\infty}\int_{-\infty}^{\infty} f(x-w,y-z)h(w,z)\, dw\, dz, \text{ then } [f*h]^{\wedge}(\xi,\eta) = 2\pi\, \hat{f}(\xi,\eta)\, \hat{h}(\xi,\eta).$$

Exercises 9.2

1. (a) Find all of the eigenfunctions $f(x,y)$ of $\Delta = \partial^2/\partial x^2 + \partial^2/\partial y^2$ on the rectangle R, $0 \le x \le 3$, $0 \le y \le 2$, which satisfy the boundary conditions $f_x(0,y) = f_x(3,y) = 0$ $(0 \le y \le 2)$ and $f_y(x,0) = f_y(x,2) = 0$ $(0 \le x \le 3)$.

(b) For each of the eigenfunctions found in part (a), construct a corresponding solution of the heat equation $u_t = k\Delta u$ on R. State an initial/boundary–value problem whose solution is a linear combination of these constructed solutions.

(c) Do part (b) again, but now in relation to the wave equation $u_{tt} = a^2\Delta u$ on R.

2. (a) What is the relationship between the frequencies of the harmonics of a rectangular drum and the eigenvalues for the eigenfunctions of the Laplace operator, which are zero on the boundary of the rectangle.

(b) Let λ_1 and λ_2 be the smallest two *distinct* eigenvalues for the eigenfunctions in part (a). Show that the length and width of the drum are determined by λ_1 and λ_2. Prove that

$\frac{1}{4} < \lambda_1/\lambda_2 \le \frac{2}{5}$, with the upper bound $\frac{2}{5}$ only being achieved for the square drum.

Remark. In general, it is unknown whether the precise shape of a drum (not necessarily rectangular) can be determined, even if one knows all of the eigenvalues (or all the frequencies). For further reading on this subject, we recommend the article: Marc Kac, *Can one hear the shape of a drum ?* American Mathematical Monthly **73** (1966) pp. 1–23.

3. Give an example of a rectangular drum which is not a square drum, but which has two linearly independent (i.e., one is not a constant times the other) eigenfunctions with the same eigenvalue. **Hint**. See Example 5 of Section 9.1 . There are infinitely many possible examples.

4. Consider a two–dimensional heat flow in the unit square $0 \le x, y \le 1$ with insulated edges. Explain why the maximum of the temperature $u(x,y,t)$ with $u(x,y,0) = \cos(3\pi x)\cos(\pi y)$ will approach zero faster (as $t \to \infty$) than the maximum of the temperature $v(x,y,t)$ with $v(x,y,0) = \cos(2\pi x)\cos(2\pi y)$. In general, how does the size of the eigenvalue of an initial eigenfunction temperature distribution affect the relative rate of decline of the maximum temperature ?

5. Use Theorem 1 to state and prove a uniform convergence theorem for the double Fourier sine series of a suitable function $g(x,y)$ defined on the rectangle $0 \le x \le L$, $0 \le y \le M$. In particular, explain why one must assume $g_{xx}(0^+,y) = 0$ $(0 \le y \le M)$ in order to use Theorem 1.

6. State and prove the analog of the convergence theorem in this section, in the case of *triple* complex Fourier series.

7. By means of *formal* application of the properties of double Fourier transforms derive the hypothetical solution (12) of Section 9.1 for the heat problem

$$\begin{array}{ll} \text{D.E.} & u_t = k(u_{xx} + u_{yy}) \quad -\infty < x,y < \infty,\ t > 0 \\ \text{I.C.} & u(x,y,0) = F(x,y)\ , \end{array}$$

for a given absolutely integrable, continuous function $F(x,y)$.

8. (a) State a Maximum Principle for the heat equation on a rectangular solid.

(b) Why is it important to have such a result if the initial and/or boundary temperatures are only approximately known.

(c) Mimic the proof of the Maximum Principle for the one–dimensional heat equation in Section 3.2, in order to establish the Maximum Principle that you stated in part (a). Does essentially the same proof work for a heat flow in a rectangular solid ?

9. By means of formal calculations, verify Parseval's equality for double Fourier Transforms:

$$\int_{-\infty}^{\infty}\int_{-\infty}^{\infty} f(x,y)\overline{g(x,y)}\,dx\,dy = \int_{-\infty}^{\infty}\int_{-\infty}^{\infty} \hat{f}(\xi,\eta)\overline{\hat{g}(\xi,\eta)}\,d\xi\,d\eta\ .$$

Hint. You may apply Parseval's equality to each variable separately. Alternatively, write $f(x,y)$ and in terms of $\hat{f}(\xi,\eta)$ using the Inversion Theorem, and then interchange the order of integration.

10. (a) State and prove Bessel's inequality for double Fourier series on a rectangle.

(b) State and prove Parseval's equality for functions $f(x,y)$ as in Theorem 1.

Hint. For part (a), mimick the proof of Bessel's inequality in the single variable case (cf. Section 4.2), and for part (b) use the ideas in Problems 5 and 7 of Exercises 4.2.

9.3 PDEs in Spherical Coordinates

In Section 6.2, we have seen that Laplace's equation $u_{xx} + u_{yy} = 0$ in dimension 2 retains its form under a rotational change of coordinates. The same is true for Laplace's equation in space, since any rotation of space is the composition of rotations about the x, y and z axes, and each of these rotations preserves the form by the two–dimensional result. Thus, we expect that by viewing Laplace's equation $\Delta u \equiv u_{xx} + u_{yy} + u_{zz} = 0$ (or the three–dimensional heat equation $u_t = k\Delta u$ or the wave equation $u_{tt} = a^2\Delta u$) in terms of spherical coordinates, we can exploit this rotational symmetry. Also, many natural or synthetic objects have approximately spherical shapes (e.g., the earth, the sun, bubbles, eyeballs, tumors, tennis balls, balloons, drops, oranges, certain cells and viruses, atoms, etc.), and the use of spherical coordinates for boundary–value problems for these objects is indispensable.

The Laplace operator in spherical coordinates – A geometric construction

In Figure 1 below, P is some arbitrary point (other than the origin O), ρ (rho) is the distance of P to the origin, φ is the angle $(0 \le \varphi \le \pi)$ from the positive z–axis to the segment OP, and θ is the angle $(0 \le \theta < 2\pi)$ from the positive x–axis and the projection OP′ of OP onto the xy–plane.

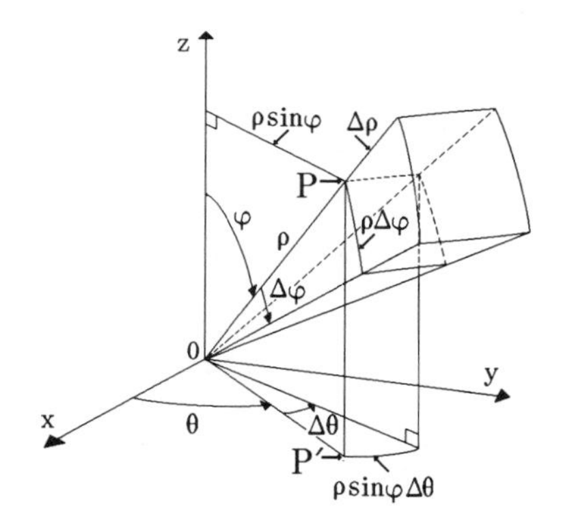

Figure 1

The **spherical coordinates** of P are (ρ,φ,θ). Observe that when P is on the xy–plane the spherical coordinates of P are $(r,\pi/2,\theta)$. Thus, spherical coordinates are a natural extension of polar coordinates. In many books in applied fields, θ is taken to be the angle from the z–axis (i.e., φ and θ are switched in Figure 1). It is good to be aware of this source of confusion. Indeed, many books use (r,θ) for polar coordinates and later define θ to be the angle from the z–axis, when spherical coordinates are introduced ! The transformation equations connecting spherical with cartesian coordinates are

$$x = \rho\sin\varphi\cos\theta \qquad \rho = (x^2 + y^2 + z^2)^{\frac{1}{2}}$$

$$y = \rho\sin\varphi\sin\theta \qquad \varphi = \arccos\left[\frac{z}{(x^2+y^2+z^2)^{\frac{1}{2}}}\right] \quad (x^2+y^2+z^2 \neq 0) \tag{1}$$

$$z = \rho\cos\varphi \qquad \theta = \begin{cases} \arccos\left[\dfrac{x}{(x^2+y^2)^{\frac{1}{2}}}\right], & y \geq 0 \quad (x^2+y^2 \neq 0) \\ \arccos\left[\dfrac{-x}{(x^2+y^2)^{\frac{1}{2}}}\right] + \pi, & y < 0 . \end{cases}$$

Consider a solid region of the form $\rho_1 \leq \rho \leq \rho_2$, $\varphi_1 \leq \varphi \leq \varphi_2$, $\theta_1 \leq \theta \leq \theta_2$. We call this a **spherical solid.** The volume of this solid is given by

$$\int_{\theta_1}^{\theta_2}\int_{\varphi_1}^{\varphi_2}\int_{\rho_1}^{\rho_2} \rho^2 \sin\varphi \, d\rho \, d\varphi \, d\theta = \tfrac{1}{3}(\rho_2{}^3 - \rho_1{}^3)[\cos(\varphi_1) - \cos(\varphi_2)](\theta_2 - \theta_1) .$$

In particular, we get $4\pi\rho_2{}^3/3$ for the sphere $0 \leq \rho \leq \rho_2$, $0 \leq \theta \leq 2\pi$, $0 \leq \varphi \leq \pi$. This integral is the result of summing up all of the infinitesimal volumes, $dV = (d\rho)(\rho\sin\varphi d\theta)(\rho d\varphi) = \rho^2\sin\varphi \, d\rho d\varphi d\theta$, of the small, nearly rectangular, spherical solids shown in Figure 1. The quantity $\rho^2\sin\varphi \, d\rho \, d\varphi \, d\theta$ is known as the **spherical volume element** and it must be included when computing triple integrals of functions in spherical coordinates (just as $r dr d\theta$ must be included when computing double integrals in polar coordinates). We now prove the formula which allows the computation of the Laplacian of a function in terms of spherical coordinates. While one can prove the following result by applying the chain rule, our proof provides some geometric insight which we will need in the next subsection.

Proposition 1 (The Laplacian in spherical coordinates). Let $u(x,y,z)$ **be a** C^2 **function defined in a region of** xyz–**space. If** u **is expressed in terms of spherical coordinates, say** $u(x,y,z) = U(\rho,\varphi,\theta)$, **then for** $\rho\sin\varphi \neq 0$,

$$\begin{aligned} u_{xx} + u_{yy} + u_{zz} &= \rho^{-2}\left[(\rho^2 U_\rho)_\rho + \frac{1}{\sin\varphi}(\sin\varphi \, U_\varphi)_\varphi + \frac{1}{\sin^2\varphi} U_{\theta\theta}\right] \\ &= U_{\rho\rho} + 2\rho^{-1}U_\rho + \rho^{-2}\left[U_{\varphi\varphi} + \cot\varphi \, U_\varphi + \frac{1}{\sin^2\varphi} U_{\theta\theta}\right] . \end{aligned} \tag{2}$$

Proof. The Laplacian of a function f at any point P can be computed by forming the sum of the second derivatives at P of the function along three *straight* lines, parametrized by arclength and intersecting orthogonally at P. The variable ρ parametrizes the radial line through P with respect to arclength, and indeed the first term in (2) is the second derivative of U along this line. For the other two lines, we choose the tangent lines to the the curves of constant latitude and longitude (parametrized by θ and φ, respectively). Let $(\rho(t),\theta(t),\varphi(t))$ be the spherical coordinates of the point at distance t (say, in the downward direction) along the straight line which is tangent to the curve of longitude through the point $(\rho_0,\theta_0,\varphi_0)$, $\varphi_0 \neq 0$ or π. We have $\rho(t)^2 = t^2 + \rho_0^2$, $\theta(t) \equiv \theta_0$, and $\sin(\varphi(t)-\varphi_0) = t/\rho(t)$. The second derivative of U along this line is then

$$\frac{d}{dt}\left[\frac{d}{dt}U(\rho(t),\varphi(t),\theta(t))\right] = \frac{d}{dt}\left[U_\rho\frac{d\rho}{dt} + U_\varphi\frac{d\varphi}{dt} + U_\theta\frac{d\theta}{dt}\right]$$
$$= U_{\rho\rho}\left[\frac{d\rho}{dt}\right]^2 + U_\rho\,\rho''(t) + U_{\varphi\varphi}\left[\frac{d\varphi}{dt}\right]^2 + U_\varphi\,\varphi''(t) + U_{\theta\theta}\left[\frac{d\theta}{dt}\right]^2 + U_\theta\,\theta''(t)\,. \tag{3}$$

Now $2\rho\rho' = 2t$ and $2\rho'^2 + 2\rho\rho'' = 2$. Thus, at $t = 0$, we have $\rho' = 0$ and $\rho'' = 1/\rho_0$. Also, $\cos(\varphi - \varphi_0)\,\varphi' = \rho^{-1} - t\rho^{-2}\rho'$ and $-\sin(\varphi - \varphi_0)\varphi'^2 + \cos(\varphi - \varphi_0)\varphi'' = -2\rho^{-2}\rho' + 2\rho^{-3}\rho'^2 - t\rho^{-2}\rho''$. Thus, at $t = 0$, we get $\varphi' = \rho_0^{-1}$ and $\varphi'' = 0$. Hence, (3) becomes

$$\rho_0^{-1}\,U_\rho + \rho_0^{-2}\,U_{\varphi\varphi}\,. \tag{4}$$

It remains to show that the second derivative of U along the straight line tangent to the curve of constant latitude through $(\rho_0,\theta_0,\varphi_0)$ is

$$\rho_0^{-1}\,U_\rho + \rho_0^{-2}\cot(\varphi_0)\,U_\varphi + \rho_0^{-2}\sin^{-2}(\varphi_0)\,U_{\theta\theta}\,. \tag{5}$$

For this line, we still have $\rho(t)^2 = \rho_0^2 + t^2$ and $\sin(\theta(t) - \theta_0) = t/[\rho(t)\sin(\varphi_0)]$ and (using (1)) $\cos(\varphi(t)) = \rho_0\cos(\varphi_0)/\rho(t)$. Thus, at $t = 0$, again $\rho' = 0$, $\rho'' = \rho_0^{-1}$, $\theta' = [\rho_0\sin(\theta_0)]^{-1}$, and $\theta'' = 0$, as before. To compute $\varphi'(0)$ and $\varphi''(0)$, we first note that $-\sin(\varphi)\varphi' = \rho_0\cos(\varphi_0)[-\rho^{-2}\rho']$ and $-\cos(\varphi)\varphi'^2 + -\sin(\varphi)\,\varphi'' = \rho_0\cos(\varphi_0)[2\rho^{-3}\rho'^2 - \rho^{-2}\rho'']$. Thus, at $t = 0$, we have $\varphi' = 0$ and $\varphi'' = \rho_0^{-2}\cot(\varphi_0)$. Putting these values into (3), we obtain (5), and adding (4) and (5) to $U_{\rho\rho}$, we obtain equation (2). □

Example 1. The function $u(x,y,z) = xyz$ is clearly harmonic (i.e., $\Delta u = 0$), but rewrite $u(x,y,z)$ as a function $U(\rho,\varphi,\theta)$ in spherical coordinates and verify that the right side of (2) is 0.

Solution. Using (1), $u(x,y,z) = U(\rho,\varphi,\theta) = \rho^3\sin^2\varphi\cos\varphi\cos\theta\sin\theta = \frac{1}{2}\rho^3\sin^2\varphi\cos\varphi\sin2\theta$. Thus,

$$U_{\rho\rho} + 2\rho^{-1}U_{\rho} + \rho^{-2}\Big[U_{\varphi\varphi} + \cot\varphi\, U_{\varphi} + \frac{1}{\sin^2\varphi} U_{\theta\theta}\Big]$$

$$= 6\rho\sin^2\varphi\cos\varphi\sin2\theta + \rho\Big[(2\sin\varphi\cos^2\varphi - \sin^3\varphi)_{\varphi} + \cot\varphi(2\sin\varphi\cos^2\varphi - \sin^3\varphi)\Big]\tfrac{1}{2}\sin2\theta - 2\rho\cos\varphi\sin2\theta$$

$$= 6\rho\sin^2\varphi\cos\varphi\sin2\theta + \rho\Big[(2\sin\varphi - 3\sin^3\varphi)_{\varphi} + \cot\varphi(2\sin\varphi\cos^2\varphi - \sin^3\varphi)\Big]\tfrac{1}{2}\sin2\theta - 2\rho\cos\varphi\sin2\theta.$$

$$= 6\rho\sin^2\varphi\cos\varphi\sin2\theta + \rho\Big[2\cos\varphi - 9\sin^2\varphi\cos\varphi + (2\cos^3\varphi - \cos\varphi\sin^2\varphi)\Big]\tfrac{1}{2}\sin2\theta - 2\rho\cos\varphi\sin2\theta$$

$$= 6\rho\sin^2\varphi\cos\varphi\sin2\theta + \rho\Big[-2\cos\varphi - 9\sin^2\varphi\cos\varphi + (2 - 3\sin^2\varphi)\cos\varphi\Big]\tfrac{1}{2}\sin2\theta$$

$$= 6\rho\sin^2\varphi\cos\varphi\sin2\theta - 6\rho\sin^2\varphi\cos\varphi\sin2\theta = 0. \quad \square$$

The Laplace operator Δ_s on the unit sphere

In the proof of Proposition 1, we saw that the Laplacian in spherical coordinates splits into three parts which correspond to the second derivatives of U along lines which are tangent to the coordinate curves parametrized by ρ, φ, and θ:

$$U_{\rho\rho} + \Big[\rho^{-2}U_{\varphi\varphi} + \rho^{-1}U_{\rho}\Big] + \Big[\rho^{-2}[\sin^{-2}(\varphi)\, U_{\theta\theta} + \cot(\varphi)U_{\varphi}] + \rho^{-1}U_{\rho}\Big]\,. \tag{6}$$

This provides a geometrical interpretation of the otherwise geometrically obscure terms in (2). If we have a function $f(\varphi,\theta)$ that is originally defined just on the unit sphere which is parametrized by φ and θ, then we can extend this function to all of space outside of the origin, by defining $\bar{f}(\rho,\varphi,\theta) \equiv f(\varphi,\theta)$ (i.e., extend the function constantly along the normal rays to the sphere). The definition of the Laplace operator Δ_s, for C^2 functions $f(\varphi,\theta)$ on the sphere $\rho = 1$, is

$$\Delta_s f(\varphi,\theta) = \Delta\bar{f}(\rho,\varphi,\theta)\Big|_{\rho=1} = \frac{1}{\sin\varphi}\,[\sin\varphi\, f_{\varphi}]_{\varphi} + \frac{1}{\sin^2\varphi} f_{\theta\theta}\,. \tag{7}$$

According to (6) with $U = \bar{f}$, the Laplacian $\Delta_s f$ at a point P on the sphere is the sum of the second derivatives of $\bar{f}$ along a pair of tangent lines intersecting orthogonally at P. If the lines are rotated, the sum of the second derivatives is invariant. Thus, despite appearances, the expression (7) will have a well–defined limit, even as φ approaches 0 or π (i.e., the Laplacian of a C^2 function on a sphere is continuous, since the Laplacian can be computed *geometrically* in the same way at each point). In other words, the apparent singularity of $\Delta_s f$ in (7) at $\varphi = 0$ and $\varphi = \pi$ is always removable if f is a C^2 function on the sphere (cf. Example 2 below). The somewhat complicated form (7) of the Laplacian is due to the fact that the circles of latitude have lengths $2\pi\sin\varphi$ which depend on φ. In other words, the asymmetry in formula (7) for Δ_s is necessary to counteract the asymmetry in the coordinates (φ,θ) which depend on a selection of north and south poles. □

Example 2. For (x,y,z) on the unit sphere $x^2 + y^2 + z^2 = 1$ (or $\rho = 1$), let $u(x,y,z) = xyz$. Write this function in terms of the coordinates (φ,θ) on the sphere, say $u(x,y,z) = f(\varphi,\theta)$. Compute $\Delta_s f$, and show that $\Delta_s f + 12f = 0$.

Solution. From Example 1, we know that for (x,y,z) unrestricted, $u(x,y,z) = U(\rho,\varphi,\theta) = \frac{1}{2}\rho^3\sin^2\varphi\cos\varphi\sin 2\theta$. Since $\rho = 1$ on the unit sphere, $f(\varphi,\theta) = \frac{1}{2}\sin^2\varphi\cos\varphi\sin 2\theta$. In order to compute $\Delta_s f$, note that (7) implies that we can simply set $\rho = 1$, and compute the Laplacian (in space) of $\tilde{f}(\rho,\varphi,\theta) = f(\varphi,\theta)$. The computation is exactly as in Example 1, except that $\rho = 1$ and the part $U_{\rho} + 2\rho^{-1}U_{\rho} = 6\rho\cos\varphi\sin^2\varphi\sin 2\theta$ no longer appears, because $\tilde{f}$ does not depend on ρ. Thus, the computation in Example 1 leads to the result $\Delta_s f = -6\sin^2\varphi\cos\varphi\sin 2\theta = -12f$. □

Definition. A C^2 function $f(\varphi,\theta) \not\equiv 0$ defined on the unit sphere, such that $\Delta_s f + \lambda f = 0$ for some constant λ, is known as a **spherical harmonic**.

In other words, the spherical harmonics are the eigenfunctions of the Laplace operator Δ_s on the unit sphere, and the associated constants λ are the eigenvalues of Δ_s. In Example 2, we found that $\frac{1}{2}\cos\varphi\sin^2\varphi\sin 2\theta$ is a spherical harmonic with eigenvalue 12.

Example 3. Let $f = f(\varphi,\theta)$ be a C^2 function defined on the sphere, which satisfies $\Delta_s f + \lambda f = 0$. If $\lambda < 0$, then show that $f \equiv 0$, and if $\lambda = 0$, show that f must be constant. In other words, show that the eigenvalues of Δ_s are nonnegative, and the only eigenfunctions (spherical harmonics) with eigenvalue 0 are nonzero constant functions.

Solution. By assumption, $\lambda f^2 = -f\Delta_s f$. If we integrate both sides over the sphere, we obtain

$$\lambda \int_0^{2\pi}\int_0^{\pi} f(\varphi,\theta)^2 \sin\varphi\, d\varphi\, d\theta = -\int_0^{2\pi}\int_0^{\pi} f(\varphi,\theta)\Big[\,[\sin\varphi\, f_{\varphi}]_{\varphi} + \frac{1}{\sin\varphi} f_{\theta\theta}\Big]\, d\varphi\, d\theta$$

$$= -\int_0^{2\pi}\int_0^{\pi} f(\varphi,\theta)\,\frac{\partial}{\partial\varphi}\Big[\sin\varphi\, f_{\varphi}\Big]\, d\varphi\, d\theta - \int_0^{\pi}\frac{1}{\sin\varphi}\Big[\int_0^{2\pi} f\, f_{\theta\theta}\, d\theta\Big]\, d\varphi$$

$$= -\,f\,\sin\varphi\, f_{\varphi}\Big|_{\varphi=0}^{\varphi=\pi} + \int_0^{2\pi}\!\!\int_0^{\pi} (f_{\varphi})^2 \sin\varphi\, d\varphi\, d\theta - \int_0^{\pi} \frac{1}{\sin\varphi}\left[(f\,f_{\theta})\Big|_{\theta=0}^{\theta=2\pi} - \int_0^{2\pi} (f_{\theta})^2\, d\theta\right] d\varphi$$

$$= \int_0^{\pi}\!\!\int_0^{2\pi} \left[(f_{\varphi})^2 + \frac{1}{\sin^2\varphi}(f_{\theta})^2\right] \sin\varphi\, d\theta\, d\varphi \;\geq\; 0\,,$$

where we have integrated by parts with respect to both φ and θ, and we have used the facts that $f\,f_{\theta}$ is a periodic function of θ and $\sin\varphi \geq 0$ on $[0,\pi]$. In order to justify switching the order of integration, we have used the fact that $\Delta_s f(\varphi,\theta)$ must extend continuously to $\varphi = 0$ and $\varphi = \pi$, since $\Delta_s f$ is a continuous function on the sphere for the C^2 function f. The quantity $(f_{\varphi})^2 + \sin^{-2}\varphi\,(f_{\theta})^2$ is the square of the length of the **gradient** $\nabla f \equiv f_{\varphi}\,\mathbf{e}_{\varphi} + (\sin\varphi)^{-1} f_{\theta}\,\mathbf{e}_{\theta}$, where $\mathbf{e}_{\varphi}$ and $\mathbf{e}_{\theta}$ are the unit vectors in the increasing φ and θ directions. In other words, writing $dA = \sin\varphi\, d\varphi\, d\theta$, and denoting the integral over the sphere by $\int_S$, we have shown

$$\lambda \int_S f^2\, dA = -\int_S f\,\Delta_s f\, dA = \int_S \|\nabla f\|^2\, dA \geq 0\,. \tag{8}$$

Thus, if λ is negative, then $f \equiv 0$ (Why ?). Moreover, if $\lambda = 0$, then $\|\nabla f\| = 0$, in which case f must be constant (Why ?). □

Remark. In Section 9.4, we will find all of the spherical harmonics. There are infinitely many spherical harmonics, and each of them is a polynomial in $\cos\varphi$, $\sin\varphi$, $\cos\theta$ and/or $\sin\theta$. Moreover, we show that not only are the eigenvalues of Δ_s nonnegative, but also they must be of the form $\lambda = n(n+1)$ for some $n = 0, 1, 2, 3, \ldots$. (Note that $\lambda = 12 = 3(3+1)$, for the spherical harmonic in Example 2.) These eigenvalues should be compared with the eigenvalues n^2 $(n = 0, 1, 2, \ldots)$ for the Laplace operator $\partial^2/\partial\theta^2$, on the unit circle, where the eigenfunctions are $\sin(n\theta)$ and $\cos(n\theta)$. The eigenvalues $n(n+1)$ of Δ_s arise in the quantum mechanics of the atom. Indeed, in quantum mechanics, the operator $-\hbar^2\Delta_s$ (where $\hbar = h/(2\pi)$ and $h \approx 6.6 \times 10^{-27}$ erg·sec is Planck's constant) is known as the square of the orbital angular momentum, and it operates on the Schrödinger wave function. The eigenvalues of $\hbar^2\Delta_s$ are usually denoted by $\hbar^2\ell(\ell+1)$, $\ell = 0, 1, 2, \ldots$, and they are interpreted as the possible discrete measurements of the square of the length of the orbital angular momentum vector for an electron in an atom. We consider the quantum mechanics of the electron in a hydrogen atom in more detail in Section 9.5. □

Eigenfunctions of the Laplace operator in space

We have seen in Section 9.2 that the eigenfunctions of the Laplace operator can be used to easily form solutions of the heat and wave equations. Here we seek eigenfunctions F (such that $\Delta F + cF = 0$ for some constant c) which are of the form $F(\rho,\varphi,\theta) = R(\rho)f(\theta,\varphi)$. According to equations (2) and (7), we have

$$\Delta F \equiv \rho^{-2}\left[(\rho^2 F_\rho)_\rho + \frac{1}{\sin\varphi}(\sin\varphi\, F_\varphi)_\varphi + \frac{1}{\sin^2\varphi} F_{\theta\theta}\right]$$

$$= \rho^{-2}\left[f(\theta,\varphi)\,(\rho^2 R'(\rho))' + R(\rho)\left[\frac{1}{\sin\varphi}(\sin\varphi\, f_\varphi)_\varphi + \frac{1}{\sin^2\varphi} f_{\theta\theta}\right]\right]$$

$$= \rho^{-2}\left[f(\theta,\varphi)\,(\rho^2 R'(\rho))' + R(\rho)\,\Delta_s f\right] .$$

Thus, the eigenfunction equation $\Delta F + cF = 0$ becomes

$$\rho^{-2}[(\rho^2 R'(\rho))' f(\varphi,\theta) + R(\rho)\Delta_s f(\varphi,\theta)] + cR(\rho)f(\varphi,\theta) = 0$$

or

$$\frac{(\rho^2 R'(\rho))'}{R(\rho)} + c\rho^2 = \frac{-\Delta_s f(\varphi,\theta)}{f(\varphi,\theta)} . \tag{9}$$

Each side of this equation must be a constant, say λ. Thus, we arrive at the two equations

$$(\rho^2 R'(\rho))' + (c\rho^2 - \lambda)R(\rho) = 0 \tag{10}$$

and

$$\Delta_s f(\varphi,\theta) + \lambda f(\varphi,\theta) = 0 . \tag{11}$$

In particular, this shows that the **angular part**, $f(\varphi,\theta)$, of the function $F = R(\rho)f(\varphi,\theta)$ must be an eigenfunction (with some eigenvalue λ) of the Laplace operator Δ_s on the unit sphere (i.e., $f(\varphi,\theta)$ must be a spherical harmonic). In summary, we have the following theorem.

Theorem 1. $F(\rho,\varphi,\theta) = R(\rho)f(\varphi,\theta)$ **is an eigenfunction of the Laplace operator** Δ (on space) **with eigenvalue** c (i.e., $\Delta F + cF = 0$) **if and only if** $f(\varphi,\theta)$ **is a spherical harmonic with eigenvalue** λ (i.e., $\Delta_s f + \lambda f = 0$), **and** $R(\rho)$ **is a solution of the equation** (10).

Remark. The eigenfunction equation $\Delta F + cF = 0$ is known as the **Helmholtz equation.** [German physiologist and physicist Hermann Ludwig Ferdinand von Helmholtz (1821–1894) made fundamental advances in electrical theory, optics and mathematics. He was the first to measure the speed of nerve impulses.] □

Recall that $1/\rho$ defines a harmonic function for $\rho > 0$. Thus, $1/\rho$ must be a solution of (10), when $c = 0$ and $\lambda = 0$. Thus, it is natural to try a solution of the form $R(\rho) = g(\rho)/\rho$, when c is arbitrary. Then, we have $(\rho^2 R'(\rho))' = (\rho g'(\rho) - g(\rho))' = \rho g''(\rho)$. Thus, (10) becomes $\rho g''(\rho) + (c\rho - \lambda/\rho)g(\rho) = 0$, and we have

$$R(\rho) = g(\rho)/\rho\,, \quad \text{if} \quad g''(\rho) + (c - \lambda\rho^{-2})g(\rho) = 0\,. \tag{12}$$

For now, we consider two cases, namely $\lambda = 0$ and $c = 0$. When $c = 0$, the form of the equation $g'' - \lambda\rho^{-2}g = 0$ suggests that we try a solution of the form $g(\rho) = \rho^m$. Then we obtain $[m(m-1) - \lambda]\rho^{m-2} = 0$. We will eventually find that the eigenvalues λ for Δ_s are of the form $(n+1)n$ for $n = 0, 1, 2, \ldots$. Thus, when $c = 0$, we take $m = n+1$ and $-n$, and

$$R_n(\rho) = a_n\rho^n + b_n\rho^{-(n+1)} \quad \text{for} \quad c = 0 \text{ and } \lambda_n = n(n+1),\ n = 0, 1, \ldots\,. \tag{13}$$

When $\lambda = 0$ and $c = \pm 1$, we obtain, from (12), that the functions

$$\begin{aligned} R_0^-(\rho) &= (a_0 e^{\rho} + b_0 e^{-\rho})/\rho & \lambda = 0,\ c = -1 \\ R_0^+(\rho) &= [a_0\cos(\rho) + b_0\sin(\rho)]/\rho = \frac{A}{\rho}\sin(\rho + \delta) & \lambda = 0,\ c = 1\,, \end{aligned} \tag{14}$$

satisfy (10), where a_0, b_0, A and δ are arbitrary constants. For $c \neq 0$, say $c = \pm b^2$ for $b > 0$, we have two solutions $R_0^+(b\rho)$ and $R_0^-(b\rho)$ of (10). Indeed, it is a general fact that if $R(\rho)f(\varphi,\theta)$ is an eigenfunction of Δ with eigenvalue ± 1, then $R(b\rho)f(\varphi,\theta)$ will be an eigenfunction with eigenvalue $\pm b^2$ (cf. Problem 4). The formula for the general solution of (10), in the general case ($\lambda = n(n+1)$, $n = 0, 1, 2, \ldots$, $c = \pm b^2 \neq 0$), is

$$R_n^{\pm}(b\rho), \quad \text{where} \quad R_n^{\pm}(\rho) \equiv \rho^n \left[\frac{1}{\rho}\frac{d}{d\rho}\right]^n [R_0^{\pm}(\rho)]\,, \quad b > 0\,. \tag{15}$$

We derive this formula in Section 9.4 (cf. Theorem 5).

Example 4 (**Radial heat and wave equations in space**). Find all solutions (valid for $\rho > 0$), of the three–dimensional heat and wave equations, which are of the form $R(\rho)T(t)$, and which are bounded as t and ρ approach $+\infty$.

Solution. Substituting $R(\rho)T(t)$ into the heat and wave equations $u_t = k\Delta u$ and $u_{tt} = a^2\Delta u$

respectively, and separating variables, we obtain

$$\frac{T'}{kT} = \Delta R/R = -c \qquad \text{and} \qquad \frac{T''}{a^2 T} = \Delta R/R = -c\,.$$

Thus, $R(\rho)$ must be an eigenfunction of Δ with eigenvalue c. By Theorem 1, $R(\rho)$ must be one of the functions $R_0^{\pm}(b\rho)$ with $c = \pm b^2 \neq 0$, or a steady–state solution $A\rho^{-1} + B$ when $c = 0$. In view of the boundedness condition, we obtain (with arbitrary constants A, γ and δ)

for the heat equation: $u = A\rho^{-1}\sin(b\rho+\delta)\, e^{-b^2kt}$ $\qquad c = b^2 > 0,$ (16)

and

$$\text{for the wave equation:} \quad u = \begin{cases} A\rho^{-1}\, e^{-b(\rho+at)} & c = -b^2 < 0 \quad (17) \\ A\rho^{-1}\sin(b\rho+\delta)\sin(bat+\gamma) & c = +b^2 > 0\,, \quad (18) \end{cases}$$

and for both equations, there is the steady–state solution when $c = 0$. The solutions (16) [or (18)], which are C^{∞} only when $\delta = 0$, can be used to solve heat [or wave] problems where the temperature [or wave amplitude] is assumed to be independent of φ and θ, say in a homogeneous ball whose spherical boundary is insulated or maintained at a constant temperature [or whose amplitude has vanishing normal derivative or is constant on the boundary] and whose initial temperature distribution [or initial amplitude and velocity distributions] depend(s) only on ρ (cf. Example 5 below and Problems 8 – 10). Initial data which depend on φ or θ will be handled through the use of spherical harmonics with $\lambda = n(n+1)$ and $n > 0$ (cf. Example 6 and Problems 17 – 20 of Section 9.4). It is interesting to note that at each time t, solution (17) is proportional to the famous **Yukawa potential** $\rho^{-1}e^{-b\rho}$ for the short range nuclear force, as opposed to the coulomb or gravitational potential ρ^{-1}. □

Remark. From the considerations leading to (12), we see that if we assume solutions of the form $U(\rho,t) = g(\rho,t)/\rho$, then the heat equation $U_t = k\rho^{-2}(\rho^2 U_\rho)_\rho$ becomes $g_t = kg_{\rho\rho}$, while the wave equation $U_{tt} = a^2\rho^{-2}(\rho^2 U_\rho)_\rho$ becomes $g_{tt} = a^2 g_{\rho\rho}$. Thus, if it is known that the solution of a certain heat or wave problem has *no* angular dependence, then such problems reduce to the one–dimensional cases considered in Chapters 3 and 5, as is illustrated in Example 5 below. However, if a problem has angular dependence in the B.C. or I.C., then this technique fails, and a more general approach involving spherical harmonics may be needed.

Example 5 (Temperature in a ball). Consider a solid, homogeneous ball of diameter 1, and with heat diffusivity constant k. The spherical surface of the ball is maintained at temperature 0 for $t > 0$, and it has a constant initial temperature distribution $U(\rho,0) = 100°\ C$. Find a formal series solution for the temperature $U(\rho,\varphi,\theta,t)$.

Solution. Since there is no angular dependence in the boundary condition or in the initial temperature, the formal solution $U(\rho,t)$ does not involve φ and θ and it formally solves

$$\begin{aligned}&\text{D.E.}\quad U_t = k\rho^{-2}(\rho^2 U_\rho)_\rho \qquad 0 < \rho \le \tfrac{1}{2},\ t \ge 0\\ &\text{B.C.}\quad U(\tfrac{1}{2},t) = 0\\ &\text{I.C.}\quad U(\rho,0) = 100\ .\end{aligned}$$

Writing $U(\rho,t) = g(\rho,t)/\rho$, according to the above remark, this problem is equivalent to

$$\begin{aligned}&\text{D.E.}\quad g_t = kg_{\rho\rho} \qquad 0 \le \rho \le 1/2,\ t \ge 0\\ &\text{B.C.}\quad g(0,t) = 0\ ,\ g(\tfrac{1}{2},t) = 0\\ &\text{I.C.}\quad g(\rho,0) = 100\rho\ .\end{aligned}$$

This new problem is a familiar one–dimensional problem. We expand $g(\rho,0) = 100\rho$ as a Fourier sine series on $[0,\frac{1}{2}]$ and insert the time–dependent exponential factors to formally obtain

$$U(\rho,t) = g(\rho,t)/\rho = \frac{100}{\pi\rho}\left[\ \sum_{n=1}^{\infty} (-1)^{n+1}\frac{1}{n}\, e^{-4n^2\pi^2kt}\sin(2n\pi\rho)\right].$$

Observe that for large t, the first term will be dominant and

$$U(\rho,t) \approx \frac{100}{\pi\rho}\, e^{-4\pi^2kt}\sin(2\pi\rho)\ ,\quad \text{as}\quad t \to \infty\ .$$

In particular, the temperature at the center of the ball is $\approx 200\, e^{-4\pi^2kt}$ (Why ?). Recall that the corresponding temperature at the center of the unit cube (cf. Example 3 of Section 9.1) is $6400\pi^{-3}e^{-3\pi^2kt} \approx 206.4\, e^{-3\pi^2kt}$. Since $-4\pi^2kt < -3\pi^2kt$, the sphere's center eventually cools more rapidly than the cubes's center. This is to be expected since a sphere of diameter 1 can be placed inside a unit cube. However, if the radius of the sphere is chosen to be $(4\pi/3)^{-1/3}$, so that the volume is the same as the volume of the unit cube, then one can check that the sphere's center eventually cools more slowly than the cube's center (cf. Problem 7). □

Example 6 (**D'Alembert's formula in space**). By means of formal computations, find the analog of D'Alembert's formula for the solution of the following wave problem in space.

$$\begin{aligned}&\text{D.E.}\quad u_{tt} = a^2(u_{xx} + u_{yy} + u_{zz}) \qquad -\infty < x,\ y,\ z,\ t < \infty\\ &\text{I.C.}\quad u(x,y,z,0) = f(x,y,z)\ ,\ \ u_t(x,y,z,0) = g(x,y,z)\ .\end{aligned} \tag{19}$$

Solution. We proceed formally, by applying Fourier transform methods. Taking the three–dimensional Fourier transform of both sides of the D.E., we obtain

$$\hat{u}_{tt}(\xi,\eta,\zeta,t) = -a^2\,(\xi^2 + \eta^2 + \zeta^2)\,\hat{u}(\xi,\eta,\zeta,t)\,,$$

and so

$$\hat{u}(\xi,\eta,\zeta,t) = \hat{u}(\xi,\eta,\zeta,0)\cos(at[\xi^2+\eta^2+\zeta^2]^{\frac{1}{2}}) + \hat{u}_t(\xi,\eta,\zeta,0)\,\frac{\sin(at[\xi^2+\eta^2+\zeta^2]^{\frac{1}{2}})}{a[\xi^2+\eta^2+\zeta^2]^{\frac{1}{2}}}\,.$$

Using $\hat{u}(\xi,\eta,\zeta,0) = \hat{f}(\xi,\eta,\zeta)$ and $\hat{u}_t(\xi,\eta,\zeta,0) = \hat{g}(\xi,\eta,\zeta)$, and formally applying the Inversion Theorem for Fourier transforms of functions on space (i.e., applying the one–dimensional Inversion Theorem three times, once for each spatial variable), we have

$$\begin{aligned} u(x,y,z,t) &= (2\pi)^{-\frac{3}{2}}\int_{-\infty}^{\infty}\int_{-\infty}^{\infty}\int_{-\infty}^{\infty}\left[\hat{f}(\xi,\eta,\zeta)\cos(at[\xi^2+\eta^2+\zeta^2]^{\frac{1}{2}})\right]e^{i(\xi x+\eta y+\zeta z)}\,d\xi\,d\eta\,d\zeta \\ &+ (2\pi)^{-\frac{3}{2}}\int_{-\infty}^{\infty}\int_{-\infty}^{\infty}\int_{-\infty}^{\infty}\left[\hat{g}(\xi,\eta,\zeta)\,\frac{\sin(at[\xi^2+\eta^2+\zeta^2]^{\frac{1}{2}})}{a[\xi^2+\eta^2+\zeta^2]^{\frac{1}{2}}}\right]e^{i(\xi x+\eta y+\zeta z)}\,d\xi\,d\eta\,d\zeta\,. \end{aligned} \tag{20}$$

Recall that D'Alembert's formula for the one–dimensional wave problem (cf. Section 5.2) did not involve the Fourier transforms of the initial data $f(x)$ and $g(x)$, and thus we wish to rid the formal solution (20) of Fourier transforms. Note that the first integral of (20) is just the time derivative of the second integral, but with f replaced by g. Hence, we will simplify the second integral, and take the time derivative of the result to obtain the first integral. We have

$$g(\bar{x},\bar{y},\bar{z}) = (2\pi)^{-\frac{3}{2}}\int_{-\infty}^{\infty}\int_{-\infty}^{\infty}\int_{-\infty}^{\infty}\hat{g}(\xi,\eta,\zeta)\,e^{-i(\xi\bar{x}+\eta\bar{y}+\zeta\bar{z})}\,d\xi\,d\eta\,d\zeta\,. \tag{21}$$

Now, it would be desirable to write the sine term in (20) in exponential form, and at the same time, eliminate the denominator. Let $\omega \equiv [\xi^2+\eta^2+\zeta^2]^{\frac{1}{2}}$. The function $\sin(at\omega)/(a\omega)$ in (20) is a spherically symmetric eigenfunction of $\tilde{\Delta} \equiv \partial^2/\partial\xi^2 + \partial^2/\partial\eta^2 + \partial^2/\partial\zeta^2$ with eigenvalue $(at)^2$, according to (14) and (15), with $n = 0$, $\delta = 0$ and $b = |at|$, and Theorem 1. For any fixed point $(\bar{x},\bar{y},\bar{z})$ with $(\bar{x}^2+\bar{y}^2+\bar{z}^2)^{\frac{1}{2}} = |at|$, we have that $\exp[-i(\bar{x}\xi + \bar{y}\eta + \bar{z}\zeta)]$ is another eigenfunction of $\tilde{\Delta}$ with eigenvalue $(at)^2$, which is *not* spherically symmetric (Why ?). However, by integrating over the sphere $\bar{\rho} = |at|$ in $(\bar{x},\bar{y},\bar{z})$–space (i.e., averaging over a sphere of directions), we obtain a spherically symmetric eigenfunction (cf. the first integral in (22) below) with eigenvalue $(at)^2$. Moreover, since any two spherically–symmetric nonsingular eigenfunctions with same eigenvalue have a constant ratio (by Theorem 1 and (14)), we must have (for some $C(t)$ independent of (ξ,η,ζ), but possibly depending on t)

$$\begin{aligned} \frac{\sin(at\omega)}{a\omega} &= C(t)\int_{\bar{\rho}=|at|} e^{-i(\bar{x}\xi+\bar{y}\eta+\bar{z}\zeta)}\,d\bar{A} \\ &\equiv C(t)\int_0^{2\pi}\int_0^{\pi}\exp[-i|at|(\xi\sin\bar{\varphi}\cos\bar{\theta} + \eta\sin\bar{\varphi}\sin\bar{\theta} + \zeta\cos\bar{\varphi})]\,(at)^2\sin\bar{\varphi}\,d\bar{\varphi}\,d\bar{\theta}\,. \end{aligned} \tag{22}$$

Then taking the limit of both sides of (22) as $\omega \to 0^+$, we must have $t = C(t)\,(at)^2 4\pi$, or $C(t) = (4\pi a^2 t)^{-1}$. Substituting this for $C(t)$ into (22), we obtain

$$\frac{\sin(at\omega)}{a\omega} = \frac{t}{4\pi(at)^2}\int_{\bar\rho=|at|} e^{-i(\bar x\xi+\bar y\eta+\bar z\zeta)}\, d\bar A\,. \tag{23}$$

Thus, the second integral, say $I(x,y,z,t)$, in (20) becomes

$$I(x,y,z,t) = (2\pi)^{-\frac{3}{2}}\int_{-\infty}^{\infty}\int_{-\infty}^{\infty}\int_{-\infty}^{\infty}\hat g(\xi,\eta,\zeta)\left[\frac{t}{4\pi(at)^2}\int_{\bar\rho=|at|} e^{-i(\bar x\xi+\bar y\eta+\bar z\zeta)}d\bar A\right]e^{i(\xi x+\eta y+\zeta z)}\, d\xi d\eta d\zeta$$

$$= t\,\frac{1}{4\pi(at)^2}\int_{\bar\rho=|at|}\left[(2\pi)^{-\frac{3}{2}}\int_{-\infty}^{\infty}\int_{-\infty}^{\infty}\int_{-\infty}^{\infty}\hat g(\xi,\eta,\zeta)\,\exp\Big[i[\xi(x-\bar x)+\eta(y-\bar y)+\zeta(z-\bar z)]\Big]\, d\xi\, d\eta\, d\zeta\right] d\bar A$$

$$= t\,\frac{1}{4\pi(at)^2}\int_{\bar\rho=|at|} g(x-\bar x,y-\bar y,z-\bar z)\, d\bar A \;\equiv\; t\,M_g(x,y,z;|at|)\,, \tag{24}$$

where we have formally interchanged the order of integration and applied the Inversion Theorem (i.e., (21)). The function $M_g(x,y,z;|at|)$, defined by the last equation, is the mean (average) of the function g over a sphere, centered at (x,y,z), of radius $|at|$. Initial disturbances at points on this sphere will reach the center (x,y,z) in time t. Thus, $I(x,y,z,t)$ is a superposition of all of the initial velocity sources at points which are of distance $|at|$ from (x,y,z). One might think that since the surface area of the sphere of radius $|at|$ is $4\pi a^2 t^2$, there should be a factor of t^2 multiplying $M_g(x,y,z;|at|)$, but observe from Example 4 that the amplitude of a spherically–symmetric wave originating at a point decreases inversely with the distance from the source. Thus, only one factor of t remains. As we mentioned, the first integral of (20) is simply the derivative of the second with g replaced with f. Thus, when $t > 0$, the formal solution of problem (19) is given by **D'Alembert's formula in space** :

$$u(x,y,z,t) = \frac{d}{dt}\Big[tM_f(x,y,z;|at|)\Big] + tM_g(x,y,z;|at|)\,. \tag{25}$$

If f is C^2 and g is C^1, it can be verified (cf. Problem 13) that (25) is indeed a C^2 solution of the problem (19) . □

Remark. There is a very important property that the solution (25) has, but which the one–dimensional D'Alembert's solution,

$$\tfrac{1}{2}[f(x+at) + f(x-at)] + \frac{1}{2a}\int_{x-at}^{x+at} g(r)dr\,, \tag{26}$$

does not. Observe that (25) only depends on the value of g on a spherical shell of points whose distance from (x,y,z) is exactly at. However, (26) depends on the values of g at all points whose distance from x is *less than* or equal to $|at|$ (i.e., points in the interval of dependence discussed in Section 5.2). In the one–dimensional case a velocity disturbance which is sent from a point P and received at a point P′ has an effect which lasts beyond the time of arrival, whereas in dimension 3, the effect is felt only upon the instant of arrival and there is no aftereffect (reverberation). Thus, sharp signals can be sent in dimension 3, but reverberation can happen in dimension 1. Actually the confusion due to reverberation in dimension 1 is not too serious because, once the transient part of the signal has passed, its lasting effect is constant (i.e., for fixed x, the integral in (26) is eventually constant, if g vanishes outside of a small interval), and this means that effective communication via wires is possible. However, in dimension 2, there is also reverberation, and the reverberation is not eventually constant. Indeed, the analog of **D'Alembert's solution in dimension 2** is (cf. Problem 14)

$$u(x,y,t) = \frac{1}{2\pi a}\frac{\partial}{\partial t}\left[\iint_{\bar{r}\leq|at|} \frac{f(x-\bar{x},y-\bar{y})}{\sqrt{a^2t^2-(\bar{r})^2}}\,d\bar{x}\,d\bar{y}\right] + \frac{1}{2\pi a}\iint_{\bar{r}\leq|at|} \frac{g(x-\bar{x},y-\bar{y})}{\sqrt{a^2t^2-(\bar{r})^2}}\,d\bar{x}\,d\bar{y}, \quad \bar{r} \equiv \sqrt{\bar{x}^2+\bar{y}^2}\,. \tag{27}$$

The fact that the denominator depends on t implies that the reverberation is not eventually constant in dimension 2, and thus the sharp transmission of signals (say electrical or acoustic) is impossible in this case. The statement that signals are transmitted on well–defined expanding shell–like wavefronts, is known as **Huygens' Principle.** Thus, Huygens' Principle holds for the wave equation in dimension 3, but it fails in dimension 2 (as well as in every *even* dimension, cf. [Courant and Hilbert, p. 690]). It should also be noted that formula (23), which was the key to the derivation of (25), is false in dimension 2 (Why ?). □

Summary 9.3

1. The Laplace operator in spherical coordinates : For $u(x,y,z) = U(\rho,\varphi,\theta)$, as in Proposition 1,

$$u_{xx} + u_{yy} + u_{zz} = \rho^{-2}\left[(\rho^2 U_\rho)_\rho + \frac{1}{\sin\varphi}(\sin\varphi\, U_\varphi)_\varphi + \frac{1}{\sin^2\varphi} U_{\theta\theta}\right].$$

2. The Laplace operator Δ_s on the unit sphere : For a C^2 function $f(\varphi,\theta)$ on the unit sphere,

$$\Delta_s f(\varphi,\theta) = \Delta\tilde{f}(\rho,\varphi,\theta)\Big|_{\rho=1} = \frac{1}{\sin\varphi}[\sin\varphi\, f_\varphi]_\varphi + \frac{1}{\sin^2\varphi} f_{\theta\theta},$$

where $\tilde{f}(\rho,\varphi,\theta) = f(\varphi,\theta)$ is the extension of f which is constant in the radial direction.

3. Eigenfunctions of the Laplace operator in space : According to Theorem 1, $F(\rho,\varphi,\theta) = R(\rho)f(\varphi,\theta)$ is an eigenfunction of the Laplace operator Δ (on space) with eigenvalue c (i.e., $\Delta F + cF = 0$) if and only if $f(\varphi,\theta)$ is a spherical harmonic with eigenvalue λ (i.e., $\Delta_s f + \lambda f = 0$), and $R(\rho)$ is a solution of the equation

$$(\rho^2 R'(\rho))' + (c\rho^2 - \lambda)R(\rho) = 0. \qquad (*)$$

As will be proven (cf. Theorem 1′ of Section 9.4), the eigenvalues λ are of the form $n(n+1)$, $n = 0, 1, 2, \ldots$, and for each such n there are $2n+1$ linearly independent spherical harmonics described in Section 9.4. Moreover, in Theorem 5 of Section 9.4 we prove that, if $\lambda = n(n+1)$ and $c = \pm b^2 \neq 0$, then the general solution of (*) is $R_n^\pm(b\rho)$, where

$$R_n^\pm(\rho) \equiv \rho^n\left[\frac{1}{\rho}\frac{\partial}{\partial\rho}\right]^n [R_0^\pm(\rho)], \quad \text{and} \quad \begin{array}{l} R_0^-(\rho) \equiv (a_0 e^\rho + b_0 e^{-\rho})/\rho \\ R_0^+(\rho) \equiv (a_0\cos(\rho) + b_0\sin(\rho))/\rho \end{array},$$

and where a_0 and b_0 are arbitrary constants. When $c = 0$, the general solution of (*) with $\lambda = n(n+1)$ is $a\rho^n + b\rho^{-(n+1)}$.

4. Applications : In Example 4, we found the product solutions for radial heat flow and wave propagation in space, and radial heat flow in a ball was covered in Example 5. D'Alembert's formula (25) for arbitrary wave propagation in space was derived in Example 6. This formula shows that Huygens' Principle holds in dimension 3.

Exercises 9.3

1. Express the following functions $u(x,y,z)$ in terms of spherical coordinates and verify that they are harmonic (i.e., $\Delta u = 0$) by computing the Laplacian in spherical coordinates using (2) in Proposition 1.

(a) $u(x,y,z) = (x^2 + y^2 + z^2)^{-\frac{1}{2}}$ (b) $u(x,y,z) = x(x^2 + y^2 + z^2)^{-\frac{3}{2}}$

(c) $u(x,y,z) = x^2 + y^2 - 2z^2$ (d) $u(x,y,z) = z^3 - 3zx^2$.

2. Suppose that (x,y,z) is restricted to the unit sphere $x^2 + y^2 + z^2 = 1$ (or $\rho = 1$). Then write the functions in Problem 1 in terms of the coordinates φ and θ on the unit sphere. In each case, verify that the resulting function $f(\varphi,\theta)$ is a spherical harmonic (i.e., $\Delta_s f + \lambda f = 0$ for some constant $\lambda \geq 0$). Is λ of the form $n(n+1)$, $n = 0, 1, 2, \ldots$?

3. Compute $\Delta_s(\sin\varphi)$. Is $\sin\varphi$ C^2 on the unit sphere ? Is it C^1 ? Is it continuous ?

4. (a) Show that if $f(x,y,z)$ is an eigenfunction of Δ with eigenvalue c [i.e., $f_{xx} + f_{yy} + f_{zz} + cf = 0$, with $f \neq 0$], then the function $g(x,y,x) = f(bx,by,bz)$ $(b \neq 0)$ is an eigenfunction of Δ with eigenvalue b^2c.

(b) Use part (a) and Theorem 1 to deduce that if $R(\rho)S(\varphi,\theta)$ is an eigenfunction of Δ, with eigenvalue ± 1 and if $\Delta_s S + \lambda S = 0$, then $r(\rho) \equiv R(b\rho)$ satisfies equation (10) with $c = \pm b^2$.

(c) Show directly that if $R(\rho)$ satisfies (10) with $c = \pm 1$, then $r(\rho)$ satisfies (10) with $c = \pm b^2$.

5. (a) Use the same idea as in Example 3 to prove that the nonconstant C^2 eigenfunctions $f(\theta)$ of $d^2/d\theta^2$ (i.e., $f'' + \lambda f = 0$) on a unit circle must have nonnegative eigenvalues.

(b) Show that the eigenvalues of $d^2/d\theta^2$ in part (a) must be of the form n^2, $n = 0, 1, 2, \ldots$.

6. It is known that on any smooth surface T without boundary (e.g., a sphere or doughnut surface), the Laplace operator on the surface is the unique operator Δ such that, for all C^2 functions f and g on T, $\int_T f\,\Delta g\,dA = \int_T \nabla f \cdot \nabla g\,dA$, where dA is the area element on T and ∇f denotes the gradient of f on T (cf. Problem 11 of Exercises 9.6 for a proof of this formula based on the Divergence Theorem). Show that the nonconstant eigenfunctions for Δ on T have positive eigenvalues.

7. In Example 5, suppose that the radius of the ball is changed to $(4\pi/3)^{-\frac{1}{3}}$, so that the volume is 1. Under the conditions in Example 5, show that the center of a unit cube will then cool more rapidly (for large t) than the sphere's center in this case.

8. Expressing $u(x,y,z,t)$ as $U(\rho,t)$, solve the heat equation $u_t = k\Delta u$ on the ball $\rho \le 1$ with initial temperature distribution $U(\rho,0) = \rho^{-1}\sin^3(\pi\rho)$ and B.C. $U(1,t) = 0$.

9. (a) Writing $U(\rho,t) = g(\rho,t)/\rho$, transform the following heat problem for the *insulated* unit ball to a one–dimensional heat problem for the unknown function $g(\rho,t)$.

$$\text{D.E.} \quad U_t = k\rho^{-2}(\rho^2 U_\rho)_\rho \qquad 0 < \rho < 1,\ t > 0$$

$$\text{B.C.} \quad U_\rho(1,t) = 0$$

$$\text{I.C.} \quad U(\rho,0) = f(\rho)\ .$$

(b) In the new problem (found in part (a)) for $g(\rho,t)$, find the Sturm–Liouville problem (cf. Section 4.4) which is satisfied by the radial part $R(\rho)$ of the product solutions $R(\rho)T(t)$ of the D.E. and B.C. of the problem for $g(\rho,t)$. What equation determines the eigenvalues and what are the eigenfunctions ? (Be sure to consider $\lambda \le 0$.) Assuming that $\rho f(\rho)$ can be approximated by a sum of eigenfunctions, how can we obtain a solution of the original problem in (a) ?

10. Redo Problem 9 in the case of the wave problem

$$\text{D.E.} \quad U_{tt} = a^2\rho^{-2}(\rho^2 U_\rho)_\rho \qquad 0 < \rho < 1,\ -\infty < t < \infty$$

$$\text{B.C.} \quad U_\rho(1,t) = 0$$

$$\text{I.C.} \quad U(\rho,0) = f(\rho)\ ,\ U_t(\rho,0) = 0.$$

11. (a) Suppose that the B.C. is dropped in part (a) of Problem 9, so that a radially–symmetric heat flow problem, on *all* of space, results. Use the integral solution of the one–dimensional heat equation for the semi–infinite rod to solve the related problem for $g(\rho,t)$.

(b) Solve the following radially symmetric wave problem on all of space

$$\text{D.E.} \quad U_{tt} = a^2\rho^{-2}(\rho^2 U_\rho)_\rho \qquad 0 < \rho < \infty,\ -\infty < t < \infty$$

$$\text{I.C.} \quad U(\rho,0) = f(\rho)\ ,\ U_t(\rho,0) = h(\rho).$$

by transforming the problem to a one–dimensional wave problem to which D'Alembert's formula may be applied. Check that your answer agrees with the the result obtained by directly using D'Alembert's formula in space (25).

12. (a) For any complex number c show that $U(\rho) \equiv e^{c\rho}/\rho$ $(\rho > 0)$ is an eigenfunction of Δ with eigenvalue $-c^2$ (i.e., $\Delta U - c^2 U = 0$).

(b) Deduce that $V(\rho,t) \equiv e^{kc^2t+c\rho}/\rho$ solves the heat equation $v_t = k\Delta v$.

(c) Set $c = \alpha(1+i)$ for some real α, and consider the real part of $V(\rho,t)$ in part (b). By replacing t by $t + \delta$ for some δ, find a solution $U(\rho,t)$ of the heat equation $u_t = k\Delta u$ for $\rho > \rho_0$, such that $U(\rho_0,t) = \cos(\omega t)$ and $U(\rho,t) \to 0$ as $\rho \to \infty$.

(d) By deleting the condition that $U(\rho,t) \to 0$ as $\rho \to \infty$ and forming a superposition, find (if possible) a solution $U(\rho,t)$ of $u_t = k\Delta u$, such that $U(\rho_0,t) = \cos(\omega t)$, where u is C^2 on the ball, for $\rho \leq \rho_0$, even at $\rho = 0$. For a fixed $\omega > 0$, does a solution exist for any ρ_0?

13. Verify directly that (25) does indeed solve the three–dimensional wave problem (19), when $f(x,y,z)$ is C^3 and $g(x,y,z)$ is C^2. **Hint.** Use spherical coordinates centered about (x,y,z).

14. By formal calculations (or use the hypotheses of Problem 13), obtain the solution (27) of the two–dimensional wave problem, by using the three–dimensional solution (25) in the case where f and g do not depend on z. This technique for obtaining solutions of lower–dimensional problems from solutions of higher–dimensional problems is known as **the method of descent.**

9.4 Spherical Harmonics, Laplace Series and Applications

In this section we will find all of the spherical harmonics (i.e., eigenfunctions of Δ_s) and prove that any C^2 function on the sphere can be written as a uniformly convergent series (the Laplace series) of spherical harmonics. This Laplace series for functions on a sphere is analogous to the Fourier series for functions on a circle. It is used in boundary–value problems involving spherical coordinates when angular dependence of solutions is evident, as we will illustrate in examples.

Example 1. Show that if $U(\rho,\varphi,\theta) = R(\rho)f(\varphi,\theta)$ is a solution of Laplace's equation $\Delta U = 0$ in space, then $f(\varphi,\theta)$ must be a spherical harmonic. Use this observation to find several spherical harmonics by writing some obvious solutions of Laplace's equation $u_{xx} + u_{yy} + u_{zz} = 0$ (say x, y, z; xy, xz, yz, $x^2 - y^2$, $y^2 - z^2$, $z^2 - x^2$), in terms of spherical coordinates. What are the eigenvalues of the spherical harmonics obtained in this way ?

Solution. If $U(\rho,\varphi,\theta) = R(\rho)f(\varphi,\theta)$ solves Laplace's equation $\Delta U = 0$, then U is an eigenfunction of Δ with eigenvalue $c = 0$ (i.e., $\Delta U + 0\cdot U = 0$). Thus, $f(\varphi,\theta)$ must be a spherical harmonic, by Theorem 1 of Section 9.3. Using equations (1) of Section 9.3, we have

$$x = \rho\sin\varphi\cos\theta\,, \quad y = \rho\sin\varphi\sin\theta \quad \text{and} \quad z = \rho\cos\varphi\,. \tag{1}$$

Thus, the functions $\sin\varphi\cos\theta$, $\sin\varphi\sin\theta$, $\cos\varphi$ are all spherical harmonics. As a check, we compute Δ_s of $\sin\varphi\cos\theta$. Using formula (7) of Section 9.3, we obtain

$$\begin{aligned}\Delta_s[\sin\varphi\cos\theta] &= \frac{1}{\sin\varphi}\Big[(\sin\varphi\cos\varphi\cos\theta)_\varphi\Big] + \frac{1}{\sin^2\varphi}\Big[-\sin\varphi\cos\theta\Big] \\ &= \frac{1}{\sin\varphi}(1 - 2\sin^2\varphi)\cos\theta - \frac{1}{\sin\varphi}\cos\theta = -2(\sin\varphi\cos\theta)\,.\end{aligned}$$

Thus, $\sin\varphi\cos\theta$ is indeed an eigenfunction of Δ_s and the eigenvalue is 2. One can also check that $\sin\varphi\sin\theta$ and $\cos\varphi$ also have eigenvalue 2, but we actually can deduce this from spherical symmetry. Indeed, a rotation of space about the line $x = y = z$ by 120° carries the point (x,y,z) to (y,z,x), and hence the function x becomes the function y, y becomes z, and z becomes x. Since the operators Δ and Δ_s are rotationally invariant, we deduce from the symmetry that each of these functions has the same eigenvalue, namely 2. There are infinitely many other eigenfunctions with eigenvalue 2, but, as we will eventually show, they are all obtained by forming linear combinations $c_1\sin\varphi\cos\theta + c_2\sin\varphi\sin\theta + c_3\cos\varphi$. For a given eigenvalue λ, the set of all solutions $f(\varphi,\theta)$ of $\Delta_s f + \lambda f = 0$ is known as the **eigenspace** of Δ_s associated with λ. Thus, the set of linear combinations above is the eigenspace of Δ_s associated with the eigenvalue 2, and this eigenspace is three–dimensional. We also know from Theorem 1 of Section 9.3, that the quadratic harmonic functions 2xy, 2yz, 2zx, $x^2 - y^2$ and $y^2 - z^2$, when restricted to the sphere $\rho = 1$, are all eigenfunctions of Δ_s. In terms of φ and θ, these harmonics are (using (1)), respectively,

$$2\sin^2\varphi\cos\theta\sin\theta,\quad 2\sin\varphi\cos\varphi\sin\theta,\quad 2\sin\varphi\cos\varphi\cos\theta,$$
$$\sin^2\varphi\,(\cos^2\theta-\sin^2\theta)\quad\text{and}\quad \sin^2\varphi\sin^2\theta-\cos^2\varphi. \tag{2}$$

These spherical harmonics are all carried by rotations into one another (e.g., note that $2xy$ is carried into x^2-y^2 by a rotation of 45° about the z–axis, and x, y, z may be permuted cyclically by rotations about the line $x = y = z$, as above). Thus, they all have the same eigenvalue. This eigenvalue could be found by computing Δ_s of any these functions. Instead, there is an easier way to find it. Note that $2xy = \rho^2 2\sin^2\varphi\cos\theta\sin\theta$, which is of the form $R(\rho)f(\varphi,\theta)$, where $R(\rho) = \rho^2$, and this is a special case of (13) in Section 9.3 with $n = 2$ and $\lambda = 2(2+1) = 6$. Thus, the eigenvalue for any of the spherical harmonics (2) obtained from harmonic polynomials of degree 2 (with no lower degree terms) is 6. We will also show that the five functions (2) generate (via linear combinations) the entire five–dimensional eigenspace of Δ_s for the eigenvalue 6. (Note that $x^2-z^2 = (x^2-y^2)+(y^2-z^2)$, and so this does not restrict to a sixth *independent* eigenfunction of Δ_s.) □

Definitions. An n–th degree polynomial $u(x,y,z)$ is called **homogeneous** if all of its terms have degree n (e.g., x^3-3xy^2+xyz). Moreover, a function $u(x,y,z)$ is called **harmonic** if $u(x,y,z)$ satisfies Laplace's equation $\Delta u \equiv u_{xx}+u_{yy}+u_{zz} = 0$ (e.g., x^3-3xy^2+xyz is also harmonic).

In the same way as in Example 1, one can prove the following extension of Example 1.

Theorem 1. Each n–th **degree, homogeneous, harmonic polynomial** $u(x,y,z)$ **can be written in the form** $U(\rho,\varphi,\theta)=\rho^n f(\varphi,\theta)$, **where** $f(\varphi,\theta)$ **is an eigenfunction of** Δ_s **(i.e., a spherical harmonic) with eigenvalue** $n(n+1)$.

The proof of the following deeper result will be carried out in the subsection on Laplace series.

Theorem 1′. For every (C^2) **spherical harmonic** $f(\varphi,\theta)$, **there is an integer** $n \ge 0$, **such that** $\rho^n f(\varphi,\theta)$ **is a harmonic polynomial when expressed in terms of** x, y, z. **The eigenvalue for** $f(\varphi,\theta)$ **is** $n(n+1)$, **and there are** $2n+1$ **linearly independent spherical harmonics which have eigenvalue** $n(n+1)$ **(i.e., the eigenspace of** Δ_s **corresponding to this eigenvalue has dimension** $2n+1$).

For small values of n, one might easily find the 2n+1 independent homogeneous harmonic polynomials just by trial and error. One can always find several of these polynomials just by considering the real and imaginary parts of $(x + iy)^n$, $(y + iz)^n$ and $(z + ix)^n$ (Why ?). However, we need to develop a systematic way of finding all of them. There are actually many different approaches. The most common approach is to search for harmonic polynomials of the form $\rho^n L(\varphi)M(\theta)$, when they are written in terms of spherical coordinates. One substitutes this form into (9) of Section 9.3, with $c = 0$, and obtains

$$n(n+1) = \frac{-1}{L(\varphi)M(\theta)}\left[\frac{1}{\sin\varphi}[\sin\varphi\, L'(\varphi)]'M(\theta) + \frac{1}{\sin^2(\varphi)} L(\varphi)M''(\theta)\right] . \qquad (3)$$

Separation of variables leads to the conclusion that M''/M is a constant. This constant must be be of the form $-m^2$ for integers $m = 0, \pm1, \pm2, \ldots$, for otherwise $M(\theta)$ cannot be periodic of period 2π (Why is this needed ?). Thus, $M(\theta)$ is of the form $c_1\cos(m\theta) + c_2\sin(m\theta)$, but it is customary to use the complex notation

$$M_m(\theta) = e^{im\theta} \qquad m = 0, \pm1, \pm2, \ldots . \qquad (4)$$

For each m, (3) yields, with $M(\theta) = M_m(\theta)$, the following ODE for $L(\varphi)$:

$$\sin\varphi\, [\sin\varphi\, L'(\varphi)]' + [n(n+1) \sin^2\varphi - m^2]L(\varphi) = 0 . \qquad (5)$$

When $m = 0$ this is known as **Legendre's differential equation**, and for $m \neq 0$ it is known as the **associated Legendre equation.** The usual method of solving this equation is first to make the substitution $w = \cos\varphi$ and define $h(w) = h(\cos\varphi) \equiv L(\varphi)$. Using the chain rule $\frac{d}{d\varphi} = \frac{dw}{d\varphi}\frac{d}{dw} = -\sin\varphi\frac{d}{dw}$, and $\sin^2\varphi = 1 - w^2$, we immediately obtain

$$(1-w^2)\frac{d}{dw}\left[(1-w^2)\frac{dh}{dw}\right] + \left[n(n+1)(1-w^2) - m^2\right]h(w) = 0 . \qquad (6)$$

When $m = 0$ (Legendre's differential equation), this ODE can be solved by assuming that a solution is in the form of a power series $\sum a_k w^k$. We will produce the desired nonsingular solutions in a more natural way (cf. Example 3). The power series approach shows that the only solutions which are well–behaved as $w = \cos\varphi$ nears ± 1 (i.e., as we approach the north and south poles of the sphere) are constant multiples of a polynomial of degree n in w, known as the **n–th Legendre polynomial** $P_n(w)$. Any second independent solution of the homogeneous second–order ODE (6) is singular at $w = \pm 1$, as is easily shown by examining the Wronskian (cf. Problem 8). The first five Legendre polynomials are

$$P_0(w) = 1,\ P_1(w) = w,\ P_2(w) = \tfrac{1}{2}(3w^2 - 1),\ P_3(w) = \tfrac{1}{2}(5w^3 - 3w),$$
$$P_4(w) = \tfrac{1}{8}(35w^4 - 30w^2 + 3),\ P_5(w) = \tfrac{1}{8}(63w^5 - 70w^3 + 15w)\,. \tag{7}$$

Should any higher Legendre polynomials be needed, the general formula is (cf. Example 2 below) :

$$P_n(w) = \frac{1}{2^n}\sum_{k=0}^{[n/2]} \frac{(2n-2k)!(-1)^k}{k!(n-k)!(n-2k)!}\, w^{n-2k}\,, \tag{8}$$

where $[n/2] = n/2$, if n is even, and $[n/2] = n/2 - 1/2$, if n is odd.

Remark. Alternatively, there is a beautifully symmetric formula

$$P_n(w) = \sum_{k=0}^{n} \left[\frac{n!}{k!(n-k)!}\right]^2 \left[\frac{w-1}{2}\right]^{n-k} \left[\frac{w+1}{2}\right]^{k} = \sum_{k=0}^{n} (-1)^{n-k}\left[\frac{n!}{k!(n-k)!}\sin^{(n-k)}(\varphi/2)\cos^{k}(\varphi/2)\right]^2.$$

□

Example 2. Find a harmonic polynomial of degree 3 in x, y and z, such that its restriction to the sphere $\rho = 1$, is a spherical harmonic $f(\varphi,\theta)$ which is independent of θ.

Solution. We will first find the appropriate spherical harmonic, and then construct the polynomial. The product spherical harmonics are of the form $L(\varphi)e^{im\theta}$, where $L(\varphi)$ is a solution of the associated Legendre equation (5). Since we desire a spherical harmonic which is independent of θ, we take $m = 0$ and deduce that $L(\varphi) = P_n(\cos\varphi)$. On the sphere, $z = \cos\varphi$. Thus, as we wish to obtain a polynomial of degree 3, we take $n = 3$, and a suitable harmonic polynomial is $2\rho^3 P_3(\cos\varphi) = \rho^3(5\cos^3\varphi - 3\cos\varphi) = 5z^3 - 3z\rho^2 = 5z^3 - 3z(x^2 + y^2 + z^2)$. □

Example 3. Consider the familiar harmonic potential $u(x,y,z) = [x^2 + y^2 + z^2]^{-\frac{1}{2}} = \rho^{-1}$, and let $u_c(x,y,z) \equiv u(x,y,z - c)$ be its translate along the z–axis. In terms of spherical coordinates, u_c is given by $U_c(\rho,\varphi,\theta) = [\rho^2 - 2c\rho\cos\varphi + c^2]^{-\frac{1}{2}}$. By expanding U_c into a power series in the variable c, show that

$$U_c(\rho,\varphi,\theta) = \sum_{n=0}^{\infty} P_n(\cos\varphi)\rho^{-(n+1)}\, c^n\,, \tag{9}$$

where $P_n(\cos\varphi)$ is given by (8). Deduce that $P_n(\cos\varphi)$ does indeed solve Legendre's differential equation (i.e., equation (5) with $m = 0$).

Solution. Note that $x^2 + y^2 + (z-c)^2 = \rho^2 - 2cz + c^2 = \rho^2 - 2c\rho\cos\varphi + c^2 = \rho^2(1 - 2\cos(\varphi)c/\rho + (c/\rho)^2)$. Thus, letting $\xi = c/\rho$ and $w = \cos\varphi$, we obtain

$$U_c(\rho,\varphi,\theta) = [\rho^2 - 2c\rho\cos\varphi + c^2]^{-\frac{1}{2}} = \rho^{-1}[1 - 2\xi\cos\varphi + \xi^2]^{-\frac{1}{2}} = \rho^{-1}[1 - 2\xi(w - \xi/2)]^{-\frac{1}{2}}.$$

Using the binomial expansion $[1 - 2b]^{-\frac{1}{2}} = \sum_{r=0}^{\infty} \frac{(2r)!}{2^r\,(r!)^2} b^r$ (convergent for $|b| < \frac{1}{2}$) with $b = \xi(w - \xi/2)$, we then obtain (for $|\xi(w - \xi/2)| < \frac{1}{2}$, say $|\xi| < \frac{1}{3}$ or $|c| < \rho/3$)

$$\begin{aligned} U_c(\rho,\varphi,\theta) &= \rho^{-1}\sum_{r=0}^{\infty} \frac{(2r)!}{2^r(r!)^2}\,[\xi(w - \xi/2)]^r \\ &= \rho^{-1}\sum_{r=0}^{\infty} \frac{(2r)!}{2^r(r!)^2}\left[\xi^r \sum_{s=0}^{r} (-1)^s \frac{r!}{s!(r-s)!}\, w^{r-s}(\xi/2)^s\right] \\ &= \rho^{-1}\sum_{r=0}^{\infty}\sum_{s=0}^{r} \frac{(2r)!(-1)^s}{2^{r+s}(r!)^2}\,\frac{r!}{s!(r-s)!}\, w^{r-s}\,\xi^{r+s} \\ &= \sum_{r=0}^{\infty}\sum_{s=0}^{r} \frac{(-1)^s(2r)!}{2^{r+s}r!(r-s)!s!}\, w^{r-s}\rho^{-(1+r+s)}c^{r+s}, \end{aligned}$$

for $|c| < \rho/3$. Since we want a series in terms of c^n, we make the change of indices $n = r+s$, $k = s$. Since $k = s \le r = n - s = n - k$, we have $2k \le n$, in which case k ranges from 0 to $[n/2]$. Thus, we have

$$U_c(\rho,\varphi,\theta) = \sum_{n=0}^{\infty}\left[\frac{1}{2^n}\sum_{k=0}^{[n/2]} \frac{(-1)^k(2n-2k)!}{k!(n-k)!(n-2k)!}(\cos\varphi)^{n-2k}\right]\rho^{-(n+1)}\,c^n, \quad \text{for } |c| < \rho/3,$$

which is the desired relation (9). In order to deduce that $P_n(\cos(\varphi))$ is a solution of the Legendre's differential equation, we need only to show that $P_n(\cos(\varphi))\rho^{-(n+1)}$ is harmonic. However, since (9) is the Taylor series of the function $f(c) = U_c(\rho,\varphi,\theta)$, we must have that

$$n!\,P_n(\cos(\varphi))\rho^{-(n+1)} = f^{(n)}(0)$$

$$= \frac{d^n}{dc^n}[x^2 + y^2 + (z-c)^2]^{-\frac{1}{2}}\bigg|_{c=0} = (-1)^n \frac{\partial^n}{\partial z^n}[x^2 + y^2 + z^2]^{-\frac{1}{2}}, \tag{10}$$

which is a partial derivative of a harmonic function and hence is harmonic itself (Why ?). □

We know that if $L(\varphi)$ is a solution of the associated Legendre equation (5) of order m, then $e^{im\theta}L(\varphi)$ is a spherical harmonic. So far, we have constructed the spherical harmonics when $m = 0$, namely $P_n(\cos\varphi)$. These spherical harmonics depend only on φ (not on θ), and are called **zonal spherical harmonics**. Appropriate solutions $P_{n,m}(w)$ of (5), for an arbitrary integer m, are given by

$$P_{n,m}(w) = (1-w^2)^{m/2} \frac{d^m}{dw^m}\left[P_n(w)\right] \quad \text{or} \quad P_{n,m}(\cos\varphi) = \sin^m\varphi \, P_n^{(m)}(\cos\varphi), \quad \text{for } m \geq 0,$$

and

$$P_{n,m}(w) = P_{n,-m}(w)\,, \qquad \text{for } m < 0. \tag{11}$$

Note that $P_{n,m}(w) = 0$, if $|m| > n$, since $P_n(w)$ is a polynomial of degree n. One way of checking that $P_{n,m}(w)$ satisfies equation (6) is by induction on $|m|$, using the fact that $P_n(w)$ solves the equation when $m = 0$. This does not explain how (11) was found in the first place. A derivation is provided in the last subsection (cf. Theorem 6).

Theorem 2. For any integer $n \geq 0$, **the following** $2n + 1$ **functions are eigenfunctions of** Δ_s **(i.e., spherical harmonics) with common eigenvalue** $n(n+1)$:

$$S_{n,m}(\varphi,\theta) \equiv e^{im\theta} P_{n,m}(\cos\varphi) = e^{im\theta}\sin^{|m|}(\varphi)\, P_n^{(|m|)}(\cos\varphi)\,, \quad m = -n, \ldots, n\,. \tag{12}$$

By taking the real and imaginary parts, we have the real–valued spherical harmonics

$$P_n(\cos\varphi), \quad \cos(m\theta)\sin^m\varphi\, P_n^{(m)}(\cos\varphi), \quad \sin(m\theta)\sin^m(\varphi)\, P_n^{(m)}(\cos\varphi), \quad m = 1, \ldots, n\,. \tag{12$'$}$$

Each of the functions (12′), **when multiplied by** ρ^n **becomes a harmonic polynomial of degree** n **when expressed in terms of** x, y **and** z.

Before giving the proof, in Figure 1 we provide the reader with a graphical representation of some of the spherical harmonics . The very first picture is the unit sphere. In the other pictures of Figure 1, to represent a given spherical harmonic, say $S(\varphi,\theta)$, we have pushed each point (φ,θ) on the unit sphere away from the origin by a distance of $\alpha \cdot S(\varphi,\theta)$, where α is a positive constant (usually .5 or less, so that the deformation of the sphere is not too severe). (If $S(\varphi,\theta) < 0$, then the point (φ,θ) is pushed *toward* the origin.) Note that the zonal spherical harmonics $P_n(\cos\varphi)$ are represented by surfaces of revolution about the z–axis, while the other spherical harmonics are not, because of the factors of $\sin(m\theta)$, although these other surfaces still have an m–fold discrete rotational symmetry about the z–axis. Two views of $P_3(\cos(\varphi))$ and $P_5(\cos(\varphi))$ are given, since these surfaces are not invariant under reflection in the xy–plane. Note that if $m+n$ is even, the surface $P_{n,m}(\cos\varphi)\sin(m\theta)$ is invariant under reflection in the xy–plane, since $P_{m,n}(\cos\varphi)$ is an even function about $\varphi = \pi/2$ in that case (Why?). When m is even, there is symmetry through the z–axis. If n is even and m is odd, the surface is symmetric through the origin (Why?). These observations allow us to "see" parts of the surfaces which do not show.

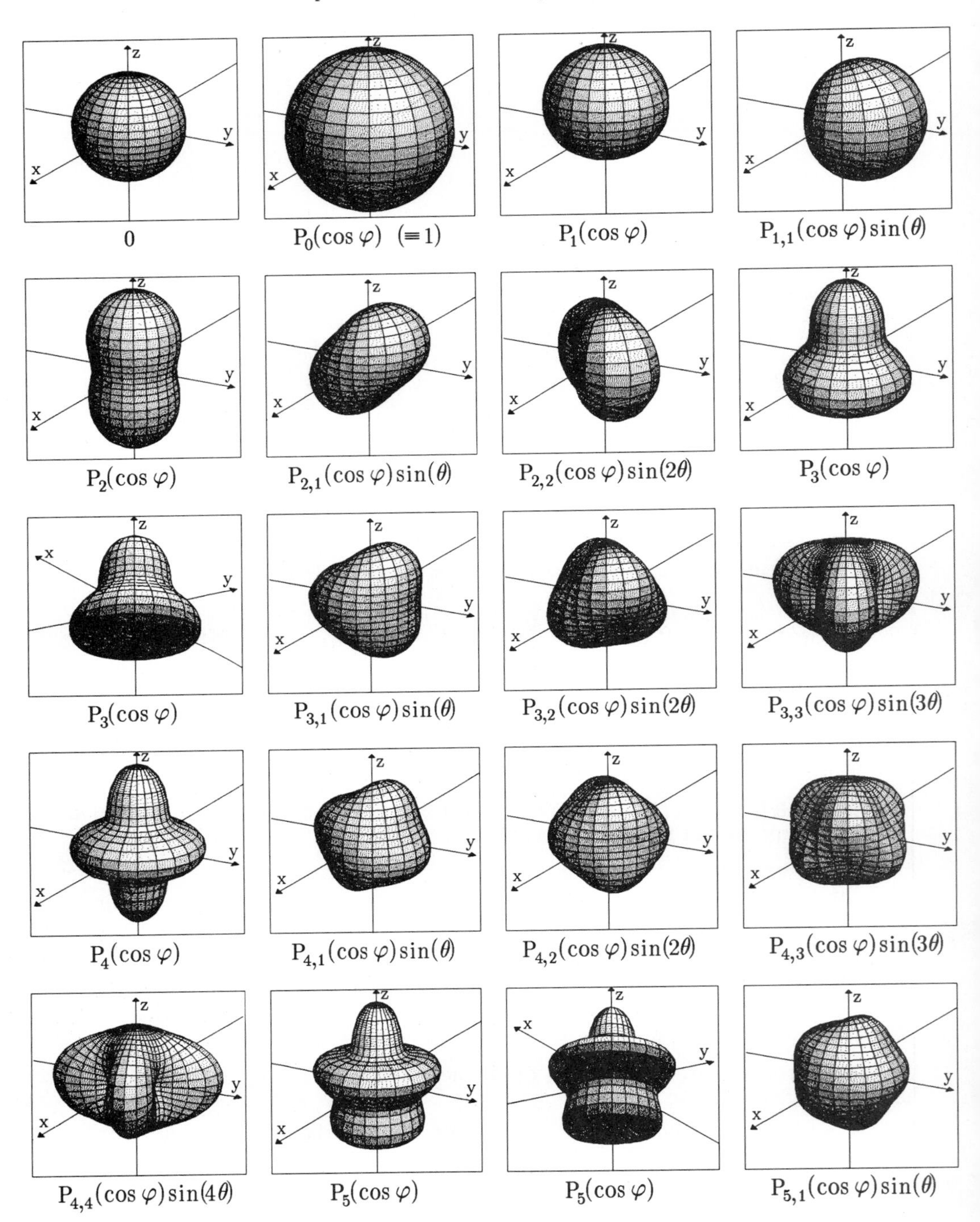
0
$P_0(\cos\varphi)$ $(\equiv 1)$
$P_1(\cos\varphi)$
$P_{1,1}(\cos\varphi)\sin(\theta)$
$P_2(\cos\varphi)$
$P_{2,1}(\cos\varphi)\sin(\theta)$
$P_{2,2}(\cos\varphi)\sin(2\theta)$
$P_3(\cos\varphi)$
$P_3(\cos\varphi)$
$P_{3,1}(\cos\varphi)\sin(\theta)$
$P_{3,2}(\cos\varphi)\sin(2\theta)$
$P_{3,3}(\cos\varphi)\sin(3\theta)$
$P_4(\cos\varphi)$
$P_{4,1}(\cos\varphi)\sin(\theta)$
$P_{4,2}(\cos\varphi)\sin(2\theta)$
$P_{4,3}(\cos\varphi)\sin(3\theta)$
$P_{4,4}(\cos\varphi)\sin(4\theta)$
$P_5(\cos\varphi)$
$P_5(\cos\varphi)$
$P_{5,1}(\cos\varphi)\sin(\theta)$

Figure 1

Proof of Theorem 2. Each function in (12) is a product of the function $M_m(\theta)$ (cf. (4)) and a solution of (5), and hence is a solution of $\Delta_s f = -n(n+1)f$ (cf. (3)). It easily follows that the real and imaginary parts are solutions of this linear homogeneous equation as well. Since ρ^n solves (10) of Section 9.3 with $\lambda = n(n+1)$ and $c = 0$, we know from Theorem 1 of Section 9.3 that when any of the functions in (12′) is multiplied by ρ^n, we obtain a harmonic function on space. It remains to show that these harmonic functions are polynomials of degree n, when expressed in terms of x, y and z. Let $r = (x^2+y^2)^{\frac{1}{2}} = \rho\sin\varphi$. Observe that $\cos(m\theta)\rho^m\sin^m(\varphi) = r^m\cos(m\theta)$ $= \text{Re}[(x+iy)^m]$, and $\sin(m\theta)\rho^m\sin^m(\varphi) = \ldots = \text{Im}[(x+iy)^m]$, which are polynomials. Moreover, $\rho^{n-m}\ P_n^{(m)}(\cos\varphi)$ has a leading term of the form Az^{n-m} (since $z = \rho\cos\varphi$), while the next term is of the form $Bz^{n-m-2}\rho^2$, and so forth. It is important to realize that the powers of w in $P_n^{(m)}(w)$ are all even or all odd (Why ?), so that only even powers of ρ will occur. Since $\rho^2 =$ $x^2 + y^2 + z^2$, these even powers of ρ are all polynomials (unlike ρ itself), and hence $\rho^{n-m}P_n^{(m)}(\cos\varphi)$ is in fact a polynomial. Thus, each of the harmonic functions obtained by multiplying the functions (12′) by ρ^n is indeed a polynomial. □

Remark. Problem 12 shows that every real harmonic polynomial of degree n is a linear combination of the harmonic polynomials obtained by multiplying the spherical harmonics (12') by ρ^n. None of these functions is a linear combination of the others, since they are orthogonal as functions of θ (and hence, linearly independent). More generally, we have the following result. □

Theorem 3 (Orthogonality of spherical harmonics). For all integers $n, n' \geq 0$ **and** $-n \leq m \leq n$ **and** $-n' \leq m' \leq n'$, **using the notation in** (12),

$$\int_0^{2\pi}\int_0^{\pi} S_{n,m}(\varphi,\theta)\,\overline{S_{n',m'}(\varphi,\theta)}\,\sin\varphi\,d\varphi\,d\theta = \begin{cases} 0\,, & \textbf{if } m \neq m' \textbf{ or } n \neq n' \\ \dfrac{(n+m)!}{(n-m)!}\dfrac{4\pi}{2n+1}\,, & \textbf{if } m = m' \textbf{ and } n = n'. \end{cases} \tag{13}$$

Proof. For continuous complex–valued functions $f(\varphi,\theta)$ and $g(\varphi,\theta)$ on the unit sphere, we define

$$\langle f,g\rangle = \int_0^{2\pi}\int_0^{\pi} f(\varphi,\theta)\overline{g(\varphi,\theta)}\,\sin\varphi\,d\varphi d\theta\,, \tag{14}$$

If f and g are C^2, then integration by parts with respect to φ and θ, as in Example 3 of Section 9.1, yields

$$<f,\Delta_s g> = \int_0^{2\pi}\int_0^{\pi} f(\varphi,\theta)\left[\frac{\partial}{\partial\varphi}(\sin\varphi\,\bar{g}_\varphi) + \frac{1}{\sin\varphi}\bar{g}_{\theta\theta}\right] d\varphi\, d\theta$$

$$= \int_0^{2\pi}\int_0^{\pi} (f_\varphi\,\bar{g}_\varphi + \frac{1}{\sin^2\varphi} f_\theta\,\bar{g}_\theta)\sin\varphi\, d\varphi\, d\theta\,, \tag{15}$$

where we have used the periodicity of g_θ f in θ and the fact that $\sin(0) = \sin(\pi) = 0$, to eliminate the endpoint differences. Taking the complex conjugate of (14) or (15) has the same effect as interchanging f and g. Thus, we have $\overline{<f,\Delta_s g>} = <g,\Delta_s f>$ or

Green's Formula on the unit sphere : $\quad <f,\Delta_s g> = \overline{<g,\Delta_s f>} = <\Delta_s f,g>$. (16)

Using the fact that $\Delta_s S_{n,m} = -n(n+1)S_{n,m}$, we then have

$$-[n(n+1) - n'(n'+1)]<S_{n,m},S_{n',m'}> = <\Delta_s S_{n,m}, S_{n',m'}> - <S_{n,m}, \Delta S_{n',m'}> = 0\,.$$

Hence, if $n \neq n'$, then we have $<S_{n,m}, S_{n',m'}> = 0$ (i.e., (13) holds, if $n \neq n'$). If $n = n'$ and $m \neq m'$, then we just use the orthogonality of $e^{im\theta}$ and $e^{im'\theta}$ on the θ–interval $[0,2\pi]$. In Problem 10, we assist the reader in verifying the result (13), when $n = n'$ and $m = m'$. □

Laplace series for functions on the sphere

Definition. The **Laplace series** of a function $f(\varphi,\theta)$ defined on the sphere is the expression

$$\text{LS } f(\varphi,\theta) = \sum_{n=0}^{\infty}\sum_{m=-n}^{n} c_{n,m}S_{n,m}(\varphi,\theta)\,, \tag{17}$$

where

$$c_{n,m} = \frac{(n-m)!}{(n+m)!}\frac{2n+1}{4\pi}\int_0^{2\pi}\int_0^{\pi} f(\varphi,\theta)e^{-im\theta}P_{n,m}(\cos\varphi)\sin\varphi\, d\varphi d\theta = \|S_{n,m}\|^{-2}<f,S_{n,m}>, \tag{18}$$

and $P_{n,m}$ and $S_{n,m}$ are defined by (11) and (12′) respectively, if the integrals (18) exist.

Remark. As with Fourier series on a circle or double Fourier series on a rectangle, the Laplace series of a function on a sphere is an eigenfunction expansion where the operator is the Laplace operator on the given domain of the function. The formula for the coefficients is obtained formally from the orthogonality relation (13). □

Theorem 4 (Uniform convergence of Laplace series). **Let $f(\varphi,\theta)$ be a C^2 function on the unit sphere. Then the Laplace series of f converges uniformly to f, in the sense that**

$$|f(\varphi,\theta) - \sum_{n=0}^{N} \sum_{m=-n}^{n} c_{n,m}\, S_{n,m}(\varphi,\theta)| \le \frac{M\sqrt{2}}{\sqrt{2N+1}}, \tag{19}$$

where M is the maximum of the continuous function $|\Delta_s f(\varphi,\theta)|$ over the sphere.

Proof. Let σ_N denote the double sum in (19), and let p be any point on the sphere. It can be proven (cf. Problem 13) that even though it appears that the value of $\sigma_N(p)$ might depend on the choice of the north pole (i.e., the choice of spherical coordinates), in fact it does not. Thus, to check the convergence of $\sigma_0(p)$, $\sigma_1(p)$, $\sigma_2(p)$,... at any point on the sphere, we may choose spherical coordinates with p at the north pole. Assuming that this choice has been made, we need only to check that (19) is valid at the pole $\varphi = 0$. At the point p, we then have

$$\begin{aligned}
\sigma_N(p) &= \sum_{n=0}^{N} \sum_{m=-n}^{n} c_{n,m} P_{n,m}(1)\, e^{im\theta} = \sum_{n=0}^{N} c_{n,0} \\
&= \frac{1}{4\pi} \int_0^{2\pi} \int_0^{\pi} \sum_{n=0}^{N} (2n+1) P_n(\cos\varphi) f(\varphi,\theta) \sin\varphi \, d\varphi \, d\theta \\
&= \int_{-1}^{1} \tfrac{1}{2}[P_0(w) + 3P_1(w) + \ldots + (2N+1)P_N(w)]\, F(w)\, dw,
\end{aligned}$$

where $F(w) = F(\cos\varphi) = \frac{1}{2\pi}\int_0^{2\pi} f(\varphi,\theta)\, d\theta$ is the average of f on the latitude $\varphi = \cos^{-1}(w)$. In Problem 9, we show that $P_0(w) + 3P_1(w) + \ldots + (2n+1)P_N(w) = P_N{}'(w) + P_{N+1}{}'(w)$. Thus,

$$\begin{aligned}
\sigma_N(p) &= \frac{1}{2}\int_{-1}^{1} [P_N{}'(w) + P_{N+1}{}'(w)]\, F(w)\, dw \\
&= \tfrac{1}{2}[P_N(w) + P_{N+1}(w)] F(w) \Big|_{-1}^{1} - \frac{1}{2}\int_{-1}^{1} F'(w)\, [P_N(w) + P_{N+1}(w)]\, dw \\
&= F(1) - \frac{1}{2}\int_{-1}^{1} F'(w)\, [P_N(w) + P_{N+1}(w)]\, dw,
\end{aligned} \tag{20}$$

where we have used the facts $P_n(1) = 1$ and $P_n(-1) = (-1)^n$ (cf. Problem 9). Since $F(1) = f(p)$, we have (by (20) and the Cauchy–Schwarz inequality ($|\langle f,g\rangle| \le \|f\|\,\|g\|$)

$$|f(p) - \sigma_N(p)| = \tfrac{1}{2}\,|<F',P_N+P_{N+1}>| \le \tfrac{1}{2}\|F'\|\,<P_N+P_{N+1},P_N+P_{N+1}>^{\frac{1}{2}}$$

$$= \tfrac{1}{2}\|F'\|\,[2(2N+1)^{-1} + 2(2N+3)^{-1}]^{\frac{1}{2}} \le \|F'\|\,(2N+1)^{-\frac{1}{2}}, \qquad (21)$$

where $\|F'\|^2 = \int_{-1}^{1} F'(w)^2\,dw$, and $\|P_N\|^2 = 2(2N+1)^{-1}$ by Problem 10. We next estimate $\|F'\|$ in terms of $M \equiv \max|\Delta_s f(\varphi,\theta)|$. Since $\frac{dw}{d\varphi} = -\sin\varphi$, we obtain, by Leibniz's rule

$$-2\pi\sin\varphi\, F'(w) = \sin\varphi\frac{d}{d\varphi}\int_0^{2\pi} f(\varphi,\theta)\,d\theta = \int_0^{2\pi}\int_0^{\varphi} (\sin\varphi\, f_\varphi(\varphi,\theta))_\varphi\, d\varphi\, d\theta$$

$$= \int_0^{\varphi}\int_0^{2\pi} \left[\frac{1}{\sin\varphi}(\sin\varphi\, f_\varphi)_\varphi + \frac{1}{\sin^2\varphi} f_{\theta\theta}\right] \sin\varphi\, d\theta\, d\varphi\,,$$

where we have used $\int_0^{2\pi} f_{\theta\theta}\,d\theta = 0$, by the periodicity of f_θ. Thus,

$$\sin\varphi|F'(w)| \le M\int_0^{\varphi} \sin\varphi\, d\varphi = M\,(1-\cos\varphi)\,.$$

For $0 \le \varphi \le \pi/2$, we have $\sin\varphi \ge \sin^2\varphi = (1-\cos^2\varphi) \ge 1-\cos\varphi$. Thus, we have $|F'(w)| \le M$ for $0 \le w \le 1$. Similarly, replacing $\int_0^{\varphi}$ by $-\int_{\varphi}^{\pi}$ in the above, we have $|F'(w)| \le M$, for $-1 \le w \le 0$. Consequently, $\int_{-1}^{1} F'(w)^2 dw \le 2M^2$, and $\|F'\| \le \sqrt{2}\,M$, and (21) yields (19). □

Proof of Theorem 1′. Suppose that there is a real–valued C^2 eigenfunction f of Δ_s which has eigenvalue λ. If this eigenvalue is not of the form $n(n+1)$, then $-(\lambda - n(n+1))\cdot\langle f,S_{n,m}\rangle = <\Delta_s f,S_{n,m}> - <f,\Delta_s S_{n,m}> = 0$, and so $<f,S_{n,m}> = 0$. Thus, LS $f(\varphi,\theta) \equiv 0$, and since f is C^2, by Theorem 4, we then have $f(\varphi,\theta) = \text{LS } f(\varphi,\theta) \equiv 0$. If λ is of the form $n(n+1)$, then f is in the eigenspace for the eigenvalue $n(n+1)$, and we still have $f(\varphi,\theta) = \text{LS } f(\varphi,\theta)$ which is a linear combination of the $S_{n,m}$, $m = -n,\dots,n$. Since f is real–valued, the imaginary parts of this linear combination must mutually cancel, meaning that f is a linear combination of the real eigenfunctions

$$P_n(\cos\varphi),\quad \cos(m\theta)P_{n,m}(\cos\varphi),\quad \sin(m\theta)P_{n,m}(\cos\varphi),\qquad m = 1,\dots,n. \qquad (22)$$

Hence, any C^2 eigenfunction of Δ_s must have its eigenvalue of the form $n(n+1)$, and any function in the eigenspace for $n(n+1)$ must be a linear combination of the functions (22) (i.e., the dimension of this eigenspace is $2n+1$). We know already from Theorem 2 that the functions (22) are restrictions (to the unit sphere) of harmonic polynomials of degree n. □

Examples and applications

Example 4 (Steady–state temperature in a ball). Solve the following Dirichlet problem for a ball.

$$\text{D.E.} \quad u_{xx} + u_{yy} + u_{zz} = 0 \qquad \rho < 2$$

$$\text{B.C.} \quad u(x,y,z) = x^2 + 2y^2 + 3z^2 \qquad \rho = 2\,. \tag{23}$$

Solution. We first write $x^2 + 2y^2 + 3z^2$ in terms of spherical coordinates with $\rho = 2$:

$$x^2 + 2y^2 + 3z^2 = 4\sin^2\varphi\cos^2\theta + 8\sin^2\varphi\sin^2\theta + 12\cos^2\varphi \equiv f(\varphi,\theta)\,. \tag{24}$$

We will express $f(\varphi,\theta)$ as a linear combination of spherical harmonics. Then, we need only to insert a factor of $(\rho/2)^n$ in front of each term with eigenvalue $n(n+1)$, and the result will solve the problem (Why ?). One could compute the Laplace series of $f(\varphi,\theta)$ by evaluating the integrals (18) for the Laplace coefficients. Since (in (24)) the maximum degree of the polynomial factors in $\sin\varphi$ and $\cos\varphi$ is two, we know that $c_{n,m} = 0$ for $n \geq 3$, but this still leaves 9 integrals to compute. Instead, we try to write $f(\varphi,\theta)$ in the form of a Laplace series, using the half–angle formulas (e.g., $\sin^2\theta = \frac{1}{2}[1 - \cos(2\theta)]$). First we write $f(\varphi,\theta)$ as a Fourier series in θ, with coefficients which are functions of φ :

$$f(\varphi,\theta) = (6\sin^2\varphi + 12\cos^2\varphi) - 2\sin^2\varphi\cos(2\theta) = (6\cos^2\varphi + 6) - 2\sin^2\varphi\cos(2\theta)\,,$$

where we have written the first term as a polynomial in $w = \cos\varphi$, since we want to express this term as a sum of Legendre polynomials. Since $P_2(w) = \frac{1}{2}(3w^2 - 1)$ and $P_0(w) = 1$, we have $6w^2 + 6 = (4P_2(w) + 2) + 6 = 4P_2(w) + 8P_0(w)$. We also want to express the factor $-2\sin^2\varphi$ as a linear combination of associated Legendre functions $\sum a_n P_{n,2}(w) = \sin^2\varphi \sum a_n P_n''(\cos\theta)$. Thus, we want $\sum a_n P_n''(w) = -2$, and clearly $a_n = 0$ except for $n = 2$, in which case $a_2 P_2''(w) = a_2 \cdot 3 = -2$ implies $a_2 = -2/3$. Thus, we have the following Laplace series for $f(\varphi,\theta)$:

$$f(\varphi,\theta) = 8P_0(\cos\varphi) + [4P_2(\cos\varphi) - \tfrac{2}{3}P_{2,2}(\cos\varphi)\cos(2\theta)] = 8 + [(6\cos^2\varphi - 2) - 2\sin^2\varphi\cos(2\theta)].$$

Since the term in brackets is a spherical harmonic with $n = 2$, we multiply this term by $(\rho/2)^2$, and we arrive at the solution

$$U(\rho,\varphi,\theta) = 8 + [(6\cos^2\varphi - 2) - 2\sin^2\varphi\cos(2\theta)]\,\rho^2/4 \quad \text{or}$$

$$u(x,y,z) = 8 + \tfrac{3}{2}z^2 - \tfrac{1}{2}(x^2+y^2+z^2) - \tfrac{1}{2}(x^2-y^2) = 8 - x^2 + z^2\,. \tag{25}$$

As a check, note that $u(x,y,z) = 8 - x^2 + z^2$ is clearly harmonic, and when $x^2+y^2+z^2 = 4$, we have $u(x,y,z) = 2(x^2+y^2+z^2) - x^2 + z^2 = x^2 + 2y^2 + 3z^2$, and so (25) is a solution of the problem. Actually, there is a maximum/minimum principle (with proof strictly analogous to the proof of the two–dimensional version in Section 6.4) for harmonic functions on bounded regions in space. As a consequence, solutions of Dirichlet problems for bounded regions are unique (e.g., (25) is the unique solution of (23)). □.

Remark. The same approach that was used in Example 4 can be used to solve the Dirichlet problem in any ball, where the function on the boundary sphere is an n–th degree polynomial, say $p(x,y,z)$. Indeed, the solution will be a harmonic polynomial $h(x,y,z)$ of degree $\le n$. We have $p(x,y,z) = h(x,y,z)$ at all points (x,y,z) on the sphere, but usually not elsewhere. □

Example 5 (A vibrating balloon). Let $v(\varphi,\theta,t)$ be the displacement in the radial (positive ρ) direction of the point (ρ_0,φ,θ) of a vibrating spherical balloon of equilibrium radius ρ_0. Assuming homogeneity, one can prove that the only linear, second–order equation for (undamped) v is of the form $v_{tt} = a^2\Delta_s v - \omega^2 v$, for some constants a and ω. Find a formal solution of

$$\begin{array}{ll} \text{D.E.} & v_{tt} = a^2\Delta_s v - \omega^2 v \qquad 0 \le \varphi,\ \theta/2 \le \pi,\ -\infty < t < \infty \\ \text{I.C.} & v(\varphi,\theta) = f(\varphi,\theta),\ v_t(\varphi,\theta) = g(\varphi,\theta)\,. \end{array} \tag{26}$$

Solution. Separation of variables, with $v = T(t)F(\varphi,\theta)$, yields

$$\frac{1}{a^2}\left[\frac{T''}{T} + \omega^2\right] = \frac{\Delta_s F}{F} = -\lambda\,.$$

Thus, $F(\varphi,\theta)$ must be some spherical harmonic with eigenvalue λ, necessarily of the form $\lambda_n = n(n+1)$, and $T(t) = c_1\cos[(\omega^2+a^2\lambda_n)^{\frac12}t] + c_2\sin[(\omega^2+a^2\lambda_n)^{\frac12}t]$. Hence, formally

$$v(\varphi,\theta,t) = \sum_{n=0}^{\infty}\sum_{m=-n}^{n}\left[a_{n,m}\cos[(\omega^2+a^2\lambda_n)^{\frac12}t] + b_{n,m}\sin[(\omega^2+a^2\lambda_n)^{\frac12}t]\right]S_{n,m}(\varphi,\theta)\,,$$

where

$$a_{n,m} = \|S_{n,m}\|^{-2}\langle f,S_{n,m}\rangle \quad\text{and}\quad b_{n,m} = (\omega^2+a^2\lambda_n)^{-\frac12}\|S_{n,m}\|^{-2}\langle g,S_{n,m}\rangle$$

and

$$\lambda_n = n(n+1),\quad S_{n,m}(\varphi,\theta) = e^{im\theta}P_{n,m}(\cos\varphi),\quad \|S_{n,m}\|^{-2} = \frac{(n-m)!}{(n+m)!}\frac{(2n+1)}{4\pi}\,.$$

If the initial displacement and velocity distributions f and g are C^2, then (according to

Theorem 4), we can approximate these functions to within any positive experimental error by a partial sum of their Laplace series, thereby obtaining a C^∞ solution of the D.E. which meets the I.C. within experimental error. We leave the proof of uniqueness and continuous dependence of solutions on initial data to the interested reader. □

Example 6 (A heat problem on a ball). Describe all of the eigenfunctions of the Laplace operator, which are C^2 throughout the interior of the ball $\rho < \rho_0$ and which vanish on the boundary sphere $\rho = \rho_0$. Use these eigenfunctions to formally solve the heat problem

$$\begin{array}{lll} \text{D.E.} & u_t = k(u_{xx} + u_{yy} + u_{zz}) & x^2 + y^2 + z^2 \le \rho_0^2,\ t \ge 0 \\ \text{B.C.} & U(\rho_0,\varphi,\theta,t) = 0 & 0 \le \varphi,\ \theta/2 \le \pi \\ \text{I.C.} & U(\rho,\varphi,\theta,0) = f(\rho,\varphi,\theta) & 0 \le \rho \le \rho_0. \end{array} \tag{27}$$

Solution. By Theorem 1, we know that if $R(\rho)f(\varphi,\theta)$ is an eigenfunction of Δ, then $f(\varphi,\theta)$ is a spherical harmonic, say with eigenvalue $n(n+1)$, and $R(\rho)$ solves the **radial equation**

$$(\rho^2 R'(\rho))' + (c\rho^2 - n(n+1))R(\rho) = 0. \tag{28}$$

For $c = 0$, the function $R(\rho)$ must be a multiple of ρ^n, since the other independent solution ρ^{-n-1} is singular at $\rho = 0$. However, ρ^n does not vanish at $\rho = \rho_0$, and thus we must take $c \neq 0$. When $c = -b^2$, we have (cf. Theorem 5 in the next subsection) the general solution $R_n^-(b\rho)$, where $R_n^-(\rho) = \rho^n(\rho^{-1}\frac{d}{d\rho})^n[R_0^-(\rho)]$, and $R_0^-(\rho) = (ae^\rho + be^{-\rho})/\rho$ (cf. (14) of Section 9.3). The only form for $R_n^-(b\rho)$ which is nonsingular at the origin is $A\cdot\sinh(b\rho)/\rho$, but this cannot be 0 when $\rho = \rho_0$, unless $A = 0$. When $c = +b^2$, we have (cf. Theorem 5) the general solution $R_n^+(b\rho)$, where $R_n^+(\rho) = \rho^n(\rho^{-1}\frac{d}{d\rho})^n[R_0^+(\rho)]$ and $R_0^+(\rho) = A\cdot\sin(\rho+\delta)/\rho$. Hence if $R_0^+(\rho)$ is nonsingular at $\rho = 0$ (i.e., $\delta = 0$), then $R_0^+(\rho)$ is a constant multiple of

$$j_0(\rho) \equiv \sin(\rho)/\rho = 1 - \rho^2/3! + \rho^4/5! - \dots, \tag{29}$$

which is a power series with an infinite radius of convergence. When the operator $\rho^{-1}\frac{d}{d\rho}$ is applied termwise to a convergent series of the form $c_0 + c_2\rho^2 + c_4\rho^4 + \ldots$(involving just even powers of ρ, as in (29)), the first term disappears and the exponents in the other terms drop by 2, yielding another series of this form. Applying this fact repeatedly to the series (29), it follows that some nonsingular solutions of (28) with $c = +b^2$ are constant multiples of

$$j_n(b\rho), \text{ where } j_n(\rho) \equiv (-1)^n\rho^n\left[\rho^{-1}\frac{d}{d\rho}\right]^n[j_0(\rho)], \quad n = 0, 1, 2, \ldots . \tag{30}$$

Moreover, the factor of ρ^n ensures that the first $n-1$ derivatives of $j_n(\rho)$ are zero at $\rho = 0$. If the operator $\rho^n\left[\rho^{-1}\frac{d}{d\rho}\right]^n$ is applied to a series of the form $c_{-1}\rho^{-1} + c_0\rho^0 + c_1\rho^1 + \ldots$ with $c_{-1} \neq 0$ (such as the series for $\cos(\rho)/\rho$ or $e^{\pm\rho}/\rho$), then the result is always singular (Why ?). Hence, constant multiples of $j_n(\rho)$ are the *only* nonsingular solutions of (28). Because of the B.C., we need to select those values for b, say $b_{n,q}$ $(q = 1, 2, 3,\ldots)$, such that $j_n(b_{n,q}\,\rho_0) = 0$. Let $\beta_{n,q}$ denote the q–th positive real number for which $j_n(\beta_{n,q}) = 0$ (i.e., the q–th place where the graph of $j_n(\beta)$ crosses the positive β–axis, or the q–th positive zero of j_n). Then

$$b_{n,q} = \beta_{n,q}/\rho_0\,, \quad \text{since} \quad j_n(b_{n,q}\rho_0) = j_n(\beta_{n,q}) = 0,\ q = 1, 2, 3,\ldots . \tag{31}$$

For $n = 0$, we have $\beta_{0,q} = q\pi$. When $n \geq 1$, there are also infinitely many $b_{n,q}$ (cf. (32) below and Example 8 of Section 4.4). However, for $n \geq 1$, the $\beta_{n,q}$ are more difficult to determine. For example, $j_1(\rho) = \frac{d}{d\rho}(\sin(\rho)/\rho) = \rho^{-2}[\rho\cos\rho - \sin\rho]$, and the $\beta_{1,q}$ are the positive solutions of the equation $\rho = \tan\rho$, which can found numerically to any degree of accuracy, say with the Newton–Raphson method (cf. Problem 12 of Section 8.2). However, there are tables which provide some of the $b_{n,q}$ for values of n and q which are not too large, and there are formulas which yield approximate values of $b_{n,q}$ for any n and *large* q. The functions $j_n(\rho)$ are known as the **spherical Bessel functions** and these are related to the **Bessel functions** $J_\nu(\rho)$ (which we will define in Section 9.4), via

$$j_n(\rho) = \left[\frac{\pi}{2\rho}\right]^{1/2} J_{n+\frac{1}{2}}(\rho)\,. \tag{32}$$

We have graphed $j_n(\rho)$ for n = 0, 1, and 2, in Figure 1 below.

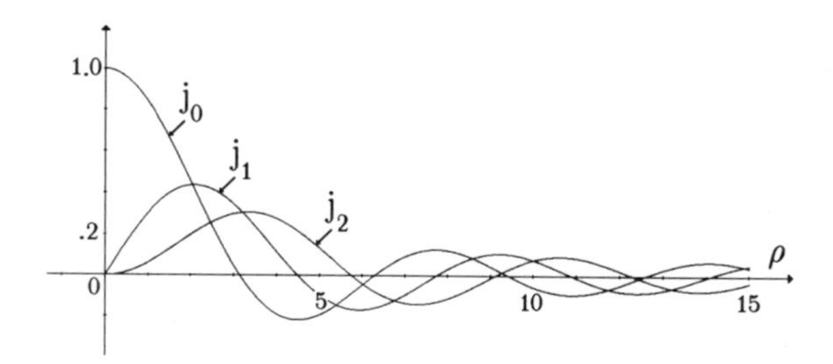

Figure 2

The eigenfunctions of Δ, which are zero on the sphere $\rho = \rho_0$, are constant multiples of the members of the family (where n = 0, 1, 2,...; q = 1, 2,...; and m = −n, −n+1, ..., n) :

$$E_{n,q,m}(\rho,\varphi,\theta) \equiv j_n(b_{n,q}\rho)S_{n,m}(\varphi,\theta) = j_n(b_{n,q}\rho)\, e^{im\theta} \sin^{|m|}(\varphi)\, P_n^{(|m|)}(\cos\varphi) . \tag{33}$$

Since the eigenvalue for $E_{n,q,m}$ is $b_{n,q}^2$, the product solutions of the D.E. which meet the B.C. are obtained by multiplying the $E_{n,q,m}$ by the time–dependent factors $\exp(-b_{n,q}^2 kt)$:

$$U_{n,q,m}(\rho,\varphi,\theta,t) = \exp(-b_{n,q}^2 kt)E_{n,q,m}(\rho,\varphi,\theta) . \tag{34}$$

We aim for a formal solution of (27) by taking an infinite superposition:

$$U(\rho,\varphi,\theta,t) = \sum_{q=1}^{\infty} \sum_{n=0}^{\infty} \exp(-b_{n,q}^2 kt) \sum_{m=-n}^{n} c_{n,q,m}\, E_{n,q,m}(\rho,\varphi,\theta) . \tag{35}$$

Setting t = 0, and equating the result with $f(\varphi,\theta)$ in the I.C., leads to the requirement

$$f(\rho,\varphi,\theta) \overset{?}{=} \sum_{q=1}^{\infty} \sum_{n=0}^{\infty} \sum_{m=-n}^{n} c_{n,q,m}\, E_{n,q,m}(\rho,\varphi,\theta) . \tag{36}$$

In the event that all but a finite number of terms in (36) are zero, then (35) is the exact solution of problem (27). For applications one would like to know how (or if) a sufficiently nice function $f(\rho,\varphi,\theta)$, vanishing for $\rho = \rho_0$, can be approximated to within some positive error by such a finite sum. If so, then a maximum principle which can be proved for solutions of the heat equation on a ball can be used to demonstrate that the solution that results from the approximation is close to the exact solution (if such exists). In Problem 15, one demonstrates the orthogonality of the

$E_{n,q,m}$ relative to <f,g> on the ball (i.e., integral of $f\,\overline{g}$ over the ball). The expression for the coefficients $c_{n,q,m}$ which results from the usual formal application of orthogonality is

$$c_{n,q,m} = \|E_{n,q,m}\|^{-2} \langle f,E_{n,q,m}\rangle , \tag{37}$$

where

$$\|E_{n,q,m}\|^2 = \langle E_{n,q,m},E_{n,q,m}\rangle = \langle S_{n,m},S_{n,m}\rangle \int_0^{\rho_0} [j_n(b_{n,q}\,\rho)]^2 \rho^2\, d\rho$$
$$= \frac{4\pi(n+m)!}{(2n+1)(n-m)!}\left[\tfrac{1}{2}\rho_0^3\,[j_n'(\beta_{n,q})]^2\right] . \tag{38}$$

Here we have used (13) to obtain $\|S_{n,m}\|^2$, and we have used the result of Problem 4 in Section 9.5 to compute the integral over ρ. As a consequence of Theorem 3 in Section 9.6 about the uniform convergence of eigenfunction expansions, the series (36) converges uniformly to $f(\rho,\varphi,\theta)$, if $f(\rho,\varphi,\theta)$ is C^2 on the ball $\rho \le \rho_0$ and zero on the sphere $\rho = \rho_0$. As we have seen, general estimates, for the number of terms which are needed to approximate f to within a certain error, are usually very conservative. Moreover, an integral comparison test typically requires at least a rough knowledge of the behavior of the coefficients $c_{n,q,m}$, which may not be easy to ascertain. Thus, in practice, one just truncates the series at a certain convenient number of terms, and checks the truncated series against the function $f(\rho,\varphi,\theta)$, hoping that no more coefficients need to be computed. □

The generation of eigenfunctions via the Cauchy–Riemann operator

We next derive formulas for all of the product eigenfunctions $R(\rho)S(\varphi,\theta)$ of the Laplace operator Δ on space. In particular, formula (11) for the nonsingular associated Legendre functions and formula (15) of Section 9.3 for the functions $R_n^{\pm}(\rho)$ is derived. The remarkable operator which generates all of the desired solutions is defined as follows.

Definition. The **Cauchy–Riemann operator** $\partial_x + i\partial_y$ is the first–order differential operator which assigns, to each C^1 complex–valued function $u(x,y,z) + iv(x,y,z)$, the function

$$(\partial_x + i\partial_y)(u + iv) = (u_x - v_y) + i(u_y + v_x) \qquad \left(\partial_x \equiv \frac{\partial}{\partial x} \right) . \tag{39}$$

Remark. In the case of functions $f = u + iv$ which only depend on (x,y), we see from (39) that the condition $(\partial_x + i\partial_y)f = 0$ is the same as the pair of Cauchy–Riemann equations which ensure that f is a complex–analytic function of $x + iy$. The Cauchy–Riemann operator can be regarded as the square–root of the Laplace operator in dimension two, in the sense that $(\partial_x + i\partial_y)\overline{(\partial_x + i\partial_y)} = \partial_x^{\,2} + \partial_y^{\,2}$. Such square roots for the Laplace operator exist in any dimension, and are constructed using matrix algebras. The physicist Paul Dirac (1902–1984), who shared the 1933 Nobel prize in physics with Schrödinger, discovered that the more precise wave equation for the electron is not Schrödinger's equation, but the Dirac equation which involves the square root of the wave operator $\partial_t^2 - \partial_x^2 - \partial_y^2 - \partial_z^2$, namely the **Dirac operator** $\gamma_0\partial_t + \gamma_1\partial_x + \gamma_2\partial_y + \gamma_3\partial_z$, whose coefficients γ_μ are certain 4×4 matrices, found in many books on quantum field theory (e.g., [Bjorken and Drell]). □

Proposition 1 (The Cauchy–Riemann operator in polar coordinates). **Let** $g(x,y)$ **be a** C^∞ **function on the punctured plane** $\mathbb{R}^2 - (0,0)$, **and let** $g(x,y) = G(r,\theta)$ **in terms of polar coordinates. Then**

$$(\partial_x + i\partial_y)[g] = e^{i\theta}(\partial_r + i\tfrac{1}{r}\partial_\theta)[G] = e^{i\theta}(G_r + i\tfrac{1}{r}G_\theta)\,. \tag{40}$$

Moreover, if $G_\theta = 0$ (i.e., G **only depends on** r), **then**

$$(\partial_x + i\partial_y)^m[g] = e^{im\theta}r^m\left[\frac{1}{r}\frac{d}{dr}\right]^m[G]\,, \quad m = 0, 1, 2, \ldots\,. \tag{41}$$

Proof. On the right side of (40), replace G_r by the equivalent expression $g_x x_r + g_y y_r = g_x\cos\theta + g_y\sin\theta$, and replace G_θ by $g_x x_\theta + g_y y_\theta = -g_x r\sin\theta + g_y r\cos\theta$. Then algebraic simplification of the result yields $(g_x + ig_y)$, proving (40). We know, from (40), that (41) holds when $m = 1$. By induction, it suffices to show that if (41) holds for $m = k$, then (41) holds for $m = k+1$. However,

$$(\partial_x + i\partial_y)^{k+1}[G] = e^{i\theta}(\partial_r + i\tfrac{1}{r}\partial_\theta)\left[e^{ik\theta}r^k\left[\frac{1}{r}\frac{d}{dr}\right]^k[G]\right]$$

$$= e^{i(k+1)\theta}\left[kr^{k-1}\left[\frac{1}{r}\frac{d}{dr}\right]^k[G] + r^k\frac{\partial}{\partial r}\left[\left[\frac{1}{r}\frac{d}{dr}\right]^k[G]\right] + ki^2r^{k-1}\left[\frac{1}{r}\frac{d}{dr}\right]^k[G]\right]$$

$$= e^{i(k+1)\theta}r^{k+1}\left[\frac{1}{r}\frac{d}{dr}\right]\left[\frac{1}{r}\frac{d}{dr}\right]^k[G]\,,$$

which proves the result for $m = k + 1$, as desired. □

Proposition 2. **If** $F(\rho)$ **is a** C^∞ **function depending only on** $\rho = (x^2+y^2+z^2)^{1/2}$, **then**

$$(\partial_x + i\partial_y)^m[F(\rho)] = e^{im\theta}(\rho\sin\varphi)^m\left[\frac{1}{\rho}\frac{d}{d\rho}\right]^m[F(\rho)]\,, \quad m = 0, 1, 2, \ldots\,. \tag{42}$$

Proof. Fix z, and consider the function $g(r) = F(\sqrt{r^2+z^2})$ with $r = \sqrt{x^2+y^2}$. For $\rho = \sqrt{r^2+z^2}$, we have $r^{-1}\partial_r g = r^{-1}F'(\rho)\rho_r = r^{-1}F'(\rho)\,\frac{1}{2}\rho^{-1}2r = \rho^{-1}\frac{d}{d\rho}F$, and repeating this calculation m times, we have $(r^{-1}\partial_r)^m[g] = (\rho^{-1}\frac{d}{d\rho})^m[F]$. Thus, applying (41) to the function $g(r)$, for each fixed z, we obtain (42). □

Theorem 5. For $n = 0$ **and** $c = \pm 1$, **let** $R_0^{\pm}(\rho)$ **be the nonzero solutions of**

$$\rho^2 R''(\rho) + 2\rho R'(\rho) + (c\rho^2 - n(n+1))R(\rho) = 0\,. \tag{43}$$

(cf. (14) in Section 9.3). **Then, for any** $n = 0, 1, 2, \ldots$,

$$(\partial_x + i\partial_y)^n[R_0^{\pm}(\rho)] = e^{in\theta}\sin^n(\varphi)\,\rho^n\left[\frac{1}{\rho}\frac{d}{d\rho}\right]^n[R_0^{\pm}(\rho)], \tag{44}$$

and (44) **is an eigenfunction of** Δ **with eigenvalue** ± 1. **Moreover,**

$$R_n^{\pm}(\rho) \equiv \rho^n\left[\frac{1}{\rho}\frac{d}{d\rho}\right]^n[R_0^{\pm}(\rho)], \qquad n = 0, 1, 2, \ldots\,. \tag{45}$$

solves (43), **for** $c = \pm 1$, **and** $R_n^{\pm}(b\rho)$ **solves** (43) **for** $c = \pm b^2$.

Proof. Equation (44) is immediate from (42). By Theorem 1 of Section 9.3, we know that $\Delta[R_0^{\pm}(\rho)] + \pm R_0^{\pm}(\rho) = 0$. Since the order of differentiation, with respect to x, y, and z, does not matter (i.e., the operators ∂_x, ∂_y, and ∂_z commute), we have

$$\Delta\Big[(\partial_x + i\partial_y)^n\,[R_0^{\pm}(\rho)]\Big] = (\partial_x + i\partial_y)^n\,\Delta[R_0^{\pm}(\rho)] = \pm(\partial_x + i\partial_y)^n[R_0^{\pm}(\rho)]\,,$$

which shows that (44) is an eigenfunction of Δ with eigenvalue ± 1 (The fact that (44) is not identically zero follows from the power series considerations in Example 6). By Theorem 1 of Section 9.3, we know that $e^{in\theta}\sin^n\varphi$ is a spherical harmonic (which has eigenvalue $n(n+1)$, since it is proportional to $S_{n,n}(\varphi,\theta) = e^{in\theta}\sin^n\varphi\,P_n^{(n)}(\cos\varphi)$). Then, also by Theorem 1 of Section 9.3, we know that $R_n^{\pm}(\rho)$ solves (43) with $c = \pm 1$, and $R_n^{\pm}(b\rho)$ solves (43) with $c = \pm b^2$ (cf. Problem 4 of Exercises 9.3). □

Theorem 6. **Let** $u_c(x,y,z) \equiv [x^2+y^2+(z-c)^2]^{-1/2} = [\rho^2-2c\rho\cos\varphi + c^2]^{-1/2} \equiv U_c(\rho,\varphi,\theta)$ **be the shift of the standard harmonic potential** $1/\rho$ **by** c **units in the** z **direction (cf. Example 3). Then, for** $m = 0, 1, 2, \ldots$,

$$(\partial_x + i\partial_y)^m\, u_c = \sum_{n=0}^{\infty} (-1)^m\, e^{im\theta} \sin^m\varphi\, P_n^{(m)}(\cos\varphi)\, \rho^{-n-1}\, c^{n-m}\,. \tag{46}$$

Moreover, $\sin^m\varphi\, P_n^{(m)}(\cos\varphi)$ **solves the associated Legendre equation** (5), **and for** $n \geq m$,

$$\frac{(-1)^n}{(n-m)!}(\partial_x + i\partial_y)^m\, \partial_z^{n-m}\left[\frac{1}{\rho}\right] = e^{im\theta}\sin^m\varphi\, P_n^{(m)}(\cos\varphi)\, \rho^{-n-1} = S_{n,m}(\varphi,\theta)\rho^{-n-1}\,. \tag{47}$$

Proof. Let T_c be the translation operator which assigns to each function $f(x,y,z)$ the new function $f(x,y,z-c)$ (i.e., $T_c[f](x,y,z) = f(x,y,z-c)$). This operator commutes with the operators ∂_x, ∂_y, and ∂_z, in the sense that $\partial_x(T_c[f]) = T_c[\partial_x f]$, and similarly for ∂_y and ∂_z (i.e., we get the same result whether we translate first and differentiate second or vice–versa). Thus,

$$(\partial_x + i\partial_y)^m\, [u_c] = (\partial_x + i\partial_y)^m\, [T_c[u]] = T_c[(\partial_x + i\partial_y)^m[u]]\,. \tag{48}$$

By (42), we have

$$\begin{aligned}(\partial_x + i\partial_y)^m\, (\rho^{-1}) &= e^{im\theta}\sin^m\varphi\,\rho^m\,(\rho^{-1}\partial_\rho)^m\,[\rho^{-1}] = \ldots \\ &= (-1)^m\, 1\cdot 3\cdot 5\cdot \ldots\cdot(2m+1)\rho^{-m-1}e^{im\theta}\sin^m\varphi\,.\end{aligned} \tag{49}$$

We apply T_c to this function in order to evaluate (46). We have $T_c(\rho) = (\rho^2 - 2c\rho\cos\varphi + c^2)^{\frac{1}{2}}$, and $T_c(\sin\varphi) = T_c(r/\rho) = r/T_c(\rho) = \rho\sin\varphi/T_c(\rho)$, since $r = \sqrt{x^2+y^2}$ is invariant under T_c. Thus, (49) yields

$$(\partial_x + i\partial_y)^m[u_c] = (-1)^m\, 1\cdot 3\cdot 5\cdot\ldots\cdot(2m+1)\, e^{im\theta}\sin^m\varphi\left[\frac{\rho^m}{(\rho^2 - 2c\rho\cos\varphi + c^2)^{m+\frac{1}{2}}}\right]. \tag{50}$$

The expression in brackets can essentially be obtained from $[\rho^2 - 2c\rho w + c^2]^{-\frac{1}{2}}$, by repeated differentiation with respect to $w = \cos\varphi$, as follows :

$$\frac{d^m}{dw^m}[\rho^2 - 2\rho cw + c^2]^{-1/2} = 1\cdot 3\cdot 5\cdot\ \ldots\ \cdot(2m+1)\rho^m c^m[\rho^2 - 2\rho cw + c^2]^{-m-\frac{1}{2}}\,. \tag{51}$$

Thus, from (50), (51), and (9), we obtain

$$
\begin{aligned}
(\partial_x + i\partial_y)^m[u_c] &= (-1)^m c^{-m} \frac{d^m}{dw^m}\left[[\rho^2 - 2\rho c w + c^2]^{-1/2} \right] \sin^m\varphi\, e^{im\theta} \\
&= (-1)^m c^{-m} \frac{d^m}{dw^m}\left[\sum_{n=0}^{\infty} P_n(w)\rho^{-n-1}c^n \right] \sin^m\varphi\, e^{im\theta} \\
&= \sum_{n=m}^{\infty} \left[(-1)^m e^{im\theta} \sin^m\varphi\, P_n^{(m)}(\cos\varphi)\rho^{-n-1} \right] c^{n-m} . \qquad (52)
\end{aligned}
$$

The differentiation under the summation can be justified without any appeal to uniform convergence theorems for derivatives. Indeed, for $\rho > 0$, $f(c) \equiv \frac{d^m}{dw^m}\left[[\rho^2 - 2\rho c w + c^2]^{-1/2} \right]$ has a valid power series expansion in c, and the Taylor coefficients of $f(c)$ may be computed just as well by differentiating $[\rho^2 - 2\rho c w + c^2]^{-1/2}$ with respect to c first, and then with respect to w (m times), as the order of differentiation does not matter. The result (47) follows immediately from the fact that the expression multiplying c^{n-m} in (52) must be

$$
\frac{1}{(n-m)!} \frac{d^{n-m}}{dc^{n-m}}\left[(\partial_x + i\partial_y)^m[u_c] \right]\Big|_{c=0} = \frac{(-1)^{n-m}}{(n-m)!} (\partial_x + i\partial_y)^m \partial_z^{n-m}[u] .
$$

Since the left side of (47) is harmonic, we know that the function $S_{n,m}(\varphi,\theta)$ on the right side is a spherical harmonic, and hence $\sin^m\varphi\, P_n^{(m)}(\cos\varphi)$ solves the associated Legendre equation (5). □

Summary 9.4

1. Spherical harmonics : According to Theorem 1, each n–th degree, homogeneous, harmonic polynomial $u(x,y,z)$ can be written in the form $U(\rho,\varphi,\theta) = \rho^n f(\varphi,\theta)$, where $f(\varphi,\theta)$ is an eigenfunction of Δ_s (i.e., a spherical harmonic) with eigenvalue $n(n+1)$. Conversely, by Theorem 1′, for every (C^2) spherical harmonic $f(\varphi,\theta)$, there is an integer $n \geq 0$, such that $\rho^n f(\varphi,\theta)$ is a harmonic polynomial when expressed in terms of x, y, z. The eigenvalue for $f(\varphi,\theta)$ is $n(n+1)$, and there are $2n+1$ linearly independent spherical harmonics which have eigenvalue $n(n+1)$ (i.e., the eigenspace of Δ_s corresponding to this eigenvalue has dimension $2n+1$). These independent spherical harmonics are given by the real–valued functions

$$P_n(\cos\varphi), \quad P_{n,m}(\cos\varphi)\cos(m\theta), \quad P_{n,m}(\cos\varphi)\sin(m\theta)\ , \quad m = 1, \ldots, n\ ,$$

where $P_n(\cos\varphi)$ and $P_{n,m}(\cos\varphi)$ are described in 2 below. Alternatively, allowing complex coefficients, the spherical harmonics with eigenvalue $n(n+1)$ are all linear combinations of

$$S_{n,m}(\varphi,\theta) \equiv e^{im\theta}\, P_{n,m}(\cos\theta) = e^{im\theta} \sin^{|m|}(\varphi)\, P_n^{(|m|)}(\cos\varphi)\ , \qquad m = -n, \ldots, n\ .$$

2. Legendre′s equations : When separation of variables is used to find product solutions $f(\varphi,\theta) = L(\varphi)M(\theta)$ of the equation $\Delta_s f = -n(n+1)f$, the ODEs which result are $M''(\theta) + m^2 M(\theta) = 0$ and

$$\sin\varphi[\sin\varphi\, L'(\varphi)]' + [n(n+1)\sin^2\varphi - m^2]L(\varphi) = 0, \qquad (*)$$

which is known as Legendre's differential equation when $m = 0$, and as Legendre's associated equation for nonzero integers m. In terms of $w = \cos\varphi$, the n–th Legendre polynomial is the solution of (*) given by

$$P_n(w) = \frac{1}{2^n} \sum_{k=0}^{[n/2]} \frac{(2n-2k)!(-1)^k}{k!(n-k)!(n-2k)!}\, w^{n-2k}\ .$$

The first five Legendre polynomials are

$$P_0(w) = 1,\ P_1(w) = w,\ P_2(w) = \tfrac{1}{2}(3w^2 - 1),\ P_3(w) = \tfrac{1}{2}(5w^3 - 3w),$$

$$P_4(w) = \tfrac{1}{8}(35w^4 - 30w^2 + 3),\ P_5(w) = \tfrac{1}{8}(63w^5 - 70w^3 + 15w)\ .$$

For any integer m, the only solutions of (*) which are C^1 on the sphere are constant multiples of

$$P_{n,m}(\cos\varphi) = P_{n,m}(w) \equiv (1-w^2)^{m/2} \frac{d^m}{dw^m}\Big[P_n(w)\Big] = \sin^m\varphi\, P^{(m)}(\cos\varphi), \quad \text{for } m > 0,$$

and

$$P_{n,m}(w) \equiv P_{n,-m}(w)\ , \qquad \text{for } m < 0.$$

3. Orthogonality : For continuous complex–valued functions $f(\varphi,\theta)$ and $g(\varphi,\theta)$ on S, we define

$$<f,g> = \int_0^{2\pi}\int_0^{\pi} f(\varphi,\theta)\overline{g(\varphi,\theta)} \sin\varphi \, d\varphi d\theta ,$$

If f and g are C^2, then integration by parts with respect to φ and θ, yields

Green's Formula on the unit sphere : $<f,\Delta_s g> = \overline{<g,\Delta_s f>} = <\Delta_s f,g>$,

which is used in proving the following orthogonality relation for the spherical harmonics :

$$\int_0^{2\pi}\int_0^{\pi} S_{n,m}(\varphi,\theta)\, \overline{S_{n',m'}(\varphi,\theta)} \sin\varphi \, d\varphi \, d\theta = \begin{cases} 0 , & \text{if } m \neq m' \text{ or } n \neq n' \\ \dfrac{(n+m)!}{(n-m)!}\dfrac{4\pi}{2n+1} , & \text{if } m = m' \text{ and } n = n'. \end{cases}$$

4. Laplace series for functions on the sphere : The Laplace series of a function $f(\varphi,\theta)$ defined on the sphere is the expression

$$\text{LS } f(\varphi,\theta) = \sum_{n=0}^{\infty} \sum_{m=-n}^{n} c_{n,m} S_{n,m}(\varphi,\theta) ,$$

where

$$c_{n,m} = \frac{(n-m)!}{(n+m)!}\frac{2n+1}{4\pi} \int_0^{2\pi}\int_0^{\pi} f(\varphi,\theta)e^{-im\theta}P_{n,m}(\cos\varphi) \sin\varphi \, d\varphi d\theta = \|S_{n,m}\|^{-2}<f,S_{n,m}> ,$$

and $S_{n,m}$ and $P_{n,m}$ are defined in 1 and 2 above, provided the integrals exist. If f is C^2, the uniform convergence of the Laplace series LS f to f is ensured by Theorem 5.

5. The Cauchy–Riemann operator : The Cauchy–Riemann operator $\partial_x + i\partial_y$ is the first–order differential operator which assigns, to each C^1 complex–valued function $u(x,y,z) + iv(x,y,z)$, the function

$$(\partial_x + i\partial_y)(u + iv) = (u_x - v_y) + i(u_y + v_x) .$$

This operator generates all the spherical harmonics via the formula (where $0 \le m \le n$)

$$\frac{(-1)^n}{(n-m)!} (\partial_x + i\partial_y)^m \, \partial_z^{n-m}\left[\frac{1}{\rho}\right] = e^{im\theta}\sin^m\varphi \, P_n^{(m)}(\cos\varphi) \, \rho^{-n-1} = S_{n,m}(\varphi,\theta)\rho^{-n-1},$$

which shows that the functions $\sin^m\varphi \, P^{(m)}(\cos\varphi)$ must solve the associated Legendre equation (cf. Theorem 6). Moreover, this operator is also used to obtain the functions $R_n^{\pm}(\rho)$ from the functions $R_0^{\pm}(\rho)$ (cf. Theorem 5).

Exercises 9.4

1. Choose one of the functions $f(\varphi,\theta)$ in (2) and verify by direct computation that it is an eigenfunction of Δ_s with eigenvalue 6.

2. Describe a rotation or sequence of rotations in space which

(a) carries $x^2 - y^2$ to $2xy$ (b) carries $2xy$ to $2zx$

(c) carries $x^2 - y^2$ to $y^2 - x^2$ (d) carries $2xy$ to $y^2 - z^2$.

3. For $n = 0, 1, 2, 3$, check that

(a) the Legendre polynomials $P_n(w)$ (cf. (7)) satisfy Legendre's differential equation (6), with $m = 0$, and

(b) the functions $P_n(\cos\varphi)$ satisfy (5).

4. (a) Find a harmonic polynomials of degree 4 in x, y and z, such that when it is restricted to the sphere $\rho = 1$, the resulting spherical harmonics is independent of θ (cf. Example 2).

(b) If a spherical harmonic is independent of φ, must it be constant ? Explain.

5. For $n = 2$ and $m = 1$ and 2, check that $P_{n,m}(\cos\varphi)$ in (11) satisfies the associated Legendre equation (5).

6. (a) Express each of the spherical harmonics (2) arising from quadratic polynomials as a linear combination of the spherical harmonics $S_{2,m}(\varphi,\theta) = e^{im\theta}\sin^{|m|}(\varphi)\, P_2^{(|m|)}(\cos\varphi)$ $(-2 \le m \le 2)$.

(b) Write the real and imaginary parts of $\rho^2 S_{2,m}(\varphi,\theta)$ $(-2 \le m \le 2)$ as harmonic quadratic polynomials in x, y and z.

7. Find seven harmonic polynomials $p(x,y,z)$ of degree 3, none of which is a linear combination of the others. Can anyone find eight ? Why not ?

8. Show that any solution of the associated Legendre equation (5) which is C^2 on $[0,\pi]$ must be a constant multiple of the solution $P_{n,m}(\cos\varphi)$.
Hint. Consider the Wronskian $W(\theta) \equiv y_1(\theta)y_2'(\theta) - y_1'(\theta)y_2(\theta)$ of two solutions, and use the fact (cf. Problem 18 of Chapter 1) that $W(\theta) = C\exp(-\int \cos\varphi/\sin\varphi \, d\varphi) = C/\sin\varphi$. Thus, at least one of two independent solutions is not C^1 at the poles $\varphi = 0$ or π (Why ?).

9. A function $G(w,c)$ is a generating function for a sequence of functions $f_0(w), f_1(w), f_2(w), \ldots,$ if $G(w,c) = \sum_{n=0}^{\infty} f_n(w)c^n$ for $|c| < \epsilon$, where ϵ is some positive constant.

(a) Using Example 3, show that $L(w,c) \equiv (1 - 2cw + c^2)^{-\frac{1}{2}}$ is a generating function for the sequence $P_0(w), P_1(w), P_2(w), \ldots$ of Legendre polynomials.

(b) By observing that $L(1,c) = (1-c)^{-1} = 1 + c + c^2 + \ldots$ (for $|c| < 1$), deduce that $P_n(1) = 1$. Similarly deduce that $P_n(-1) = (-1)^n$.

(c) Use the formula $L(w,c) = (1 - 2cw + c^2)^{-\frac{1}{2}}$ to show that $2c^2L_c + cL = c^2L_w - L_w$. By writing both sides as power series and equating coefficients of like powers of c, deduce that for $n \geq 1$, we have $(2n+1)P_n(w) = P_{n+1}'(w) - P_{n-1}'(w)$. Use this to obtain the fact (used in the proof of Theorem 4) that $P_0(w) + 3P_1(w) + \ldots + (2n+1)P_n(w) = P_{n+1}'(w) + P_n'(w)$.

(d) By reviewing the proof of Theorem 6, find a generating function for the sequence of associated Legendre functions $P_{0,m}(w), P_{1,m}(w), P_{2,m}(w), \ldots$, for a fixed positive integer m.

10. (a) Use the orthogonality result (13) in Theorem 3, when $m = m' = 0$ and $n \neq n'$, to show that $\int_{-1}^{1} P_n(w)P_{n'}(w)\,dw = 0$ (i.e., the Legendre polynomials are orthogonal on $[-1,1]$).

(b) Use the result of part (a) and Problem 9(a) to deduce that

$$\int_{-1}^{1} \frac{dw}{1 - 2wc + c^2} = \sum_{n=0}^{\infty} \int_{-1}^{1} P_n(w)^2\,dw\; c^{2n}. \qquad (*)$$

(c) By evaluating the integral in part (b) and expanding the result in terms of a power series in c, deduce that $\int_{-1}^{1} P_n(w)^2\,dw = \frac{2}{2n+1}$, by comparing the coefficients in $(*)$. Why does this result give us the orthogonality relation (13) in the case $m = m' = 0$ and $n = n'$?

(d) Show that $\int_{-1}^{1} P_{n,m}(w)^2 dw = \int_{-1}^{1} P_n^{(m)}(w)\left[(1-w^2)^m\, P_n^{(m)}(w)\right] dw = \ldots$ (via integration by parts m-times)... $= \int_{-1}^{1} P_n(w)Q_n(w)\,dw$, where $Q(w) = (-1)^m \frac{d^m}{dw^m}\left[(1-w^2)^m\, P_n^{(m)}(w)\right]$ is a polynomial of degree n with highest power term $\frac{(n+m)!}{(n-m)!}\, a_n w^n$, where $a_n w^n$ is the highest power term in $P_n(w)$.

(e) Explain why any polynomial of degree k can be written as a linear combination of Legendre polynomials of degree $\leq k$. Use this fact and part (d) to conclude that

$$Q_n(w) = \frac{(n+m)!}{(n-m)!}P_n(w) + c_{n-1}P_{n-1}(w) + \dots + c_1P_1(w) + c_0P_0(w), \text{ for constants } c_0, \dots, c_{n-1}.$$

(f) Use parts (a), (c), (d) and (e) to conclude that $\int_{-1}^{1} P_{n,m}(w)^2\, dw = \frac{(n+m)!}{(n-m)!}\frac{2}{2n+1}$, $0 \leq m \leq n$, and that the relation (13) holds.

11. (a) Find a solution $u(x,y,z)$ of Laplace's equation $\Delta u = 0$ in the ball $\rho < 1$, such that on the boundary sphere $\rho = 1$, we have $u(x,y,z) = x + y^2 + z^3$.

(b) Find a solution of $\Delta u = 0$ *outside* of the ball, which still satisfies the boundary condition in part (a), and which tends to zero as $\rho \to \infty$.

12. Let $f(\varphi,\theta)$ be an eigenfunction of Δ_s with eigenvalue $n(n+1)$.

(a) Let $f_m(\varphi,\theta) \equiv e^{im\theta}\int_0^{2\pi} f(\varphi,\omega)e^{-im\omega}d\omega$. Show that if $f(\varphi,\theta)$ is smooth enough to permit differentiation under the integral, then $f_m(\varphi,\theta)$ also satisfies $\Delta_s f_m = -n(n+1)f_m$.

(b) Show that $|\nabla f_m|^2 \geq \sin^{-2}\varphi \left|\frac{\partial}{\partial\theta} f_m\right|^2 = (\sin^{-2}\varphi)\, m^2\, |f_m|^2 \geq m^2\, |f_m|^2$.

(c) Use (a), (b) and $\int_S f_m\, \Delta_s f_m\, dA = -\int_S |\nabla f_m|^2 dA$, to prove that $f_m \equiv 0$, if $|m| > n$.

(d) Deduce from (c) and Problem 8, that f must be a linear combination of the standard spherical harmonics $S_{n,m}(\varphi,\theta) \equiv e^{im\theta} \sin^{|m|}\varphi\, P_n^{(|m|)}(\cos\varphi)$, $m = -n, \dots, n$.

(e) Deduce from (d) that any harmonic polynomial in (x,y,z), of degree n, must be a linear combination of the harmonic polynomials obtained from (12′) via multiplication by ρ^n.

13. Let H_n denote the set of all homogeneous harmonic polynomials in x, y, z of degree n, restricted to the sphere $\rho = 1$ (together with the zero function). Let F be a continuous function on the sphere. In the following parts, we demonstrate that there is a (unique) h in H_n such that $\|F-h\|^2 \equiv \langle F-h,F-h\rangle \equiv \int_S (F-h)\overline{(F-h)}\, dA$ is smaller than $\|F-k\|^2$ for any other k in H_n. Note that h is thus characterized in a way that is independent of any choice of spherical coordinate system. We will show that $h = h_n \equiv \Sigma_{m=-n}^{n}\, c_{n,m}S_{n,m}$, where $c_{n,m} \equiv \|S_{n,m}\|^{-2}\langle F,S_{n,m}\rangle$, whence even though the terms in this sum may depend on the choice of spherical coordinates, the entire sum does not.

(a) Let k be in H_n. We know from Problem 12(e) that $k = \Sigma_{m=-n}^{n} a_m S_{n,m}$ for some constants $a_{-n}, \ldots, a_n$ (Why ?). Show that $\langle h_n - k, F - h_n \rangle = 0$.

(b) Show that $\|F-k\|^2 = \langle (F-h_n) + (h_n-k), (F-h_n) + (h_n-k) \rangle = \|F-h_n\|^2 + \|h_n-k\|^2$, where one uses part (a) for the last equality.

(c) Conclude that $\|F-k\|^2 > \|F-h_n\|^2$, unless $k = h_n$ (i.e., $h = h_n$).

(d) Show that (c) implies that σ_N in the proof of Theorem 4 is indeed independent of the choice of north pole for the spherical coordinate system, as claimed. **Hint.** $\sigma_N = \Sigma_{n=0}^{N} h_n$.

14. Let σ_N be the double sum in (19) in Theorem 4. Use (19) to show that $\|f - \sigma_N\|^2 \to 0$ as $N \to \infty$, and use this to prove Parseval's equality $\|f\|^2 = \sum_{n=0}^{\infty} \frac{4\pi}{2n+1} \sum_{m=-n}^{n} \frac{(n+m)!}{(n-m)!} |c_{n,m}|^2$.
Hint. Show that $\|f - \sigma_N\|^2 = \|f\|^2 - 2\langle f, \sigma_N \rangle + \|\sigma_N\|^2 = \|f\|^2 - \|\sigma_N\|^2$.

15. (a) Let $f(\rho,\varphi,\theta)$ and $g(\rho,\varphi,\theta)$ be C^2 functions on the ball $\rho \le \rho_0$ such that f and g are zero on the boundary. Let $\langle f,g \rangle$ denote the integral of fg (or $f\bar{g}$, if f and g are allowed to be complex) over the ball, with respect to the volume element $\rho^2 \sin\varphi \, d\rho \, d\varphi \, d\theta$. Prove Green's formula $\langle \Delta f, g \rangle = \langle f, \Delta g \rangle$. **Hint.** The computation is quite easy if you write (cf. Proposition 1 of Section 9.3) $\Delta u = \rho^{-2}(\rho^2 U_\rho)_\rho + \rho^{-2} \Delta_s U$, and use the known result (cf. (16)) $\langle \Delta_s F, G \rangle = \langle F, \Delta_s G \rangle$ on a sphere.

(b) Let $E_{n,q,m}$ be the eigenfunctions defined by (33). Deduce from part (a) that $\langle E_{n,q,m}, E_{n',q',m'} \rangle = 0$ if $n \ne n'$, $q \ne q'$ or $m \ne m'$. **Hint.** We know that this result holds if $n \ne n'$ or $m \ne m'$ (Why ?). Hence, assume that $n = n'$ and $m = m'$ and $q \ne q'$, and use $\beta_{n,q} \ne \beta_{n,q'}$ (cf. (31)). Thus, we avoid the possibility that $\beta_{n,q} = \beta_{n',q'}$ for $n \ne n'$ and $q \ne q'$.

(c) Conclude from part (b) that $\int_0^{\rho_0} j_n(\beta_{n,q}\rho/\rho_0) j_n(\beta_{n,q'}\rho/\rho_0) \, \rho^2 \, d\rho = 0$, if $q \ne q'$. How might this result be proved directly ? Would a direct proof be easier ?

16. (a) Compute the spherical Bessel functions $j_1(\rho)$ and $j_2(\rho)$ and show that they satisfy the radial equation (28).
(b) Show that for small ρ, $j_1(\rho) \approx \rho/3$ and $j_2(\rho) \approx \rho^2/15$.

17. Solve D.E. $u_t = k\Delta u \quad \rho \le 1\ ,\ t \ge 0$

B.C. $U(1,\varphi,\theta,t) = 0$

I.C. $U(\rho,\varphi,\theta,0) = j_0(2\pi\rho) + j_1(\beta_{1,3}\rho)\cos\varphi + j_2(\beta_{2,1}\rho)\sin^2\varphi\sin(2\theta)$,

where $\beta_{n,q}$ are as in (31).

18. Suppose that the B.C. in (27) of Example 6 is replaced by $U_\rho(\rho_0,\varphi,\theta,t) = 0$ (i.e., the ball is insulated). Let $\gamma_{n,q}$ denote the q–th positive number where the *derivative* $j_n'(\rho)$ is zero. Express the formal solution of this new problem by using the numbers $\gamma_{n,q}$.

19. Change the D.E. in (27) of Example 6 to the wave equation $u_{tt} = a^2\Delta u$ and adjoin the I.C. $U_t(\rho,\varphi,\theta,0) = 0$. What is the formal solution of this new problem ?

20. (a) Argue that the heat equation for the temperature $u = U(\varphi,\theta,t)$, in a thin, homogeneous metallic unit sphere which is insulated on the inner and outer surfaces, is of the form $u_t = k\Delta_s u$.

(b) Solve this equation formally in the case where the initial temperature is $U(\varphi,\theta,0) = f(\varphi,\theta)$.

(c) Find an exact solution when $f(\varphi,\theta) = 2\cos\varphi\sin^2\varphi\cos^2\theta$.

(d) Prove that for any solution v of $v_t = k\Delta_s v$, we have $\frac{d}{dt}\int_S v^2\,dA = -2k\int_S |\nabla v|^2\,dA$, and conclude that the solution found in (c) is unique.

(e) State and prove a maximum principle for solutions of $u_t = k\Delta_s u$.

21. (a) Show that the Laplace series for the C^3 function $\sin^3\varphi$ has an infinite number of terms, whereas the Laplace series for $\sin^3\varphi\cos(3\theta)$ has only one term.

(b) Use Theorem 4 to estimate the number of terms of the Laplace series for $\sin^3\varphi$ that suffice to approximate $\sin^3\varphi$ to within an error of .01 . (The industrious reader may wish to compute the coefficients of the Laplace series to see if the convergence is really this slow.)

(c) Why can we not apply Theorem 4 to the function $\sin\varphi$?

22. Show that every homogeneous polynomial $p(x,y,z)$ (not necessarily harmonic) of degree N can be written in the form $p(x,y,z) = h_N(x,y,z) + \rho^2 h_{N-2}(x,y,z) + \rho^4 h_{N-4}(x,y,z) + \dots$, where h_k is a harmonic polynomial of degree k. Is this representation unique ? Why ? **Hint.** Consider the Laplace series of $p(x,y,z)$ on the unit sphere and apply Theorem 2.

9.5 Special Functions and Applications

The solutions of the ODEs which are encountered after separation of variables in various coordinate systems are of such fundamental importance, that they are given names and are the objects of much scrutiny. Such solutions fall under the category of special functions. The Legendre functions $P_{n,m}(\cos\varphi)$ and spherical Bessel functions $j_n(\rho)$ of Section 9.4 are examples of special functions. Here we will study some of the special functions (e.g., Bessel functions) which are indispensable in applications to boundary–value problems for cylinders and disks, and we consider the Laguerre polynomials and Hermite functions in connection with the quantum mechanics of the hydrogen atom and the harmonic oscillator.

Radial special functions

We have seen (cf. Example 1 of Section 9.2) that solutions u of the eigenfunction equation $\Delta u + cu = 0$ immediately lead to solutions $e^{-ckt}u$ of the heat equation $u_t = k\Delta u$, and to solutions $e^{\pm i\sqrt{c}at}u$ of the wave equation $u_{tt} = a^2\Delta u$. Moreover, equations of the more general form

$$\Delta u + [f(\|\mathbf{x}\|) + c]u = 0 , \tag{1}$$

(where f is a given function of the distance $\|\mathbf{x}\| = \sqrt{x_1^2 + \ldots + x_n^2}$ of the point $\mathbf{x} = (x_1, x_2, \ldots, x_n)$ to the origin in $\mathbb{R}^n$) arise in certain heat [or wave] problems where there is a temperature [or displacement] dependent source term. But most often such equations occur in quantum mechanics, which we will consider later in this section. Assuming a product solution of (1) of the form $R(r)H(\theta)$ in dimension 2 [using polar coordinates (r,θ)], and $R(\rho)S(\varphi,\theta)$ in dimension 3 [using spherical coordinates (ρ,φ,θ)], we obtain the respective separated equations :

$$H''(\theta) + m^2H(\theta) = 0, \qquad r^2R''(r) + rR'(r) + [cr^2 - m^2 + r^2f(r)]R(r) = 0, \tag{2}$$

$$\Delta_s S(\varphi,\theta) + n(n+1)S(\varphi,\theta) = 0, \quad \rho^2R''(\rho) + 2\rho R'(\rho) + [c\rho^2 - n(n+1) + \rho^2 f(\rho)]R(\rho) = 0 . \tag{3}$$

We refer to the equations for $R(r)$ and $R(\rho)$ as **radial equations**. In the case where $f(\rho) \equiv 0$, we saw already in Theorem 5 of Section 9.4 that the general solution of (3) for any nonzero $c = \pm b^2$ is given by

$$R_n^{\pm}(b\rho), \text{ where } R_n^{\pm}(\rho) \equiv \rho^n\left[\frac{1}{\rho}\frac{\partial}{\partial\rho}\right]^n[R_0^{\pm}(\rho)] , \text{ and } \quad \begin{aligned} R_0^{-}(\rho) &= (a_0e^{b\rho} + b_0e^{-b\rho})/\rho \\ R_0^{+}(\rho) &= [a_0\cos(\rho) + b_0\sin(\rho)]/\rho . \end{aligned} \tag{4}$$

When $c = 0$ and $f(r) \equiv 0$, the general solution of the radial equation in (2) is $K\log r + C$ when $m = 0$, and $c_1 r^m + c_2 r^{-m}$ for $m = \pm 1, \pm 2, \ldots$; while the general solution of (3) is $c_1 \rho^{-n-1} + c_2 \rho^n$, for $n = 0, 1, 2, \ldots$. Note that the solutions (4) of (3) can be expressed as simple combinations of elementary functions (i.e., x^k, $\sin(x)$, $\cos(x)$, e^x, $\log(x)$). However, this is not the case for solutions of (2), with $f(r) \equiv 0$ and $c \neq 0$, which we consider next.

Bessel functions

When $c = 1$ and $f(r) \equiv 0$ in (2), we obtain **Bessel's equation of order m** :

$$r^2 R''(r) + rR'(r) + [r^2 - m^2]R(r) = 0 . \tag{5}$$

[Friedrich Wilhelm Bessel (1784–1846) was a German astronomer who first determined the distance of a star (other than the sun) from the earth. He was the first to routinely use Bessel functions.]. For each $m = 0, 1, 2, \ldots$, it can be shown that one solution of (5), known as the **Bessel function of the first kind of order m**, can be written in a power series

$$J_m(r) = (r/2)^m \sum_{k=0}^{\infty} \frac{(-r^2/4)^k}{k!(m+k)!} , \tag{6}$$

which converges for all r (even if r is complex), but it is known that $J_m(r)$ cannot be expressed algebraically in terms of elementary functions. This is one reason why we considered the eigenvalue equation $\Delta u + cu = 0$ first in spherical coordinates instead of polar or cylindrical coordinates. As can be seen from the graphs (cf. Figure 1 below and Figure 2 of Section 9.4), the functions $J_m(r)$ are similar to the (elementary) spherical Bessel functions $j_n(\rho)$ which solve the "spherical Bessel equation" (3) with $c = 1$ and $f = 0$. Just as the $j_n(\rho)$ can be used to solve heat and wave problems in a solid ball, the $J_m(r)$ can be used to solve such problems in a disk or a cylinder. Indeed, Bessel functions are also known as **cylindrical functions**.

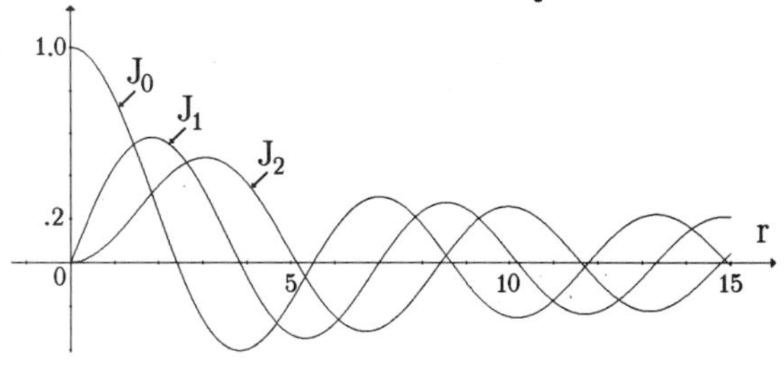

Figure 1

Example 1 **(The vibrating circular drum).** Determine the transverse amplitude of a vibrating circular drum head of radius $r_0 > 0$, by formally solving the problem

$$\begin{aligned} &\text{D.E.}\quad u_{tt} = a^2(u_{xx} + u_{yy}) \qquad x^2 + y^2 \le r_0^2\,,\ -\infty < t < \infty \\ &\text{B.C.}\quad U(r_0,\theta,t) = 0 \\ &\text{I.C.}\quad U(r,\theta,0) = f(r,\theta),\quad U_t(r,\theta,0) = g(r,\theta)\,. \end{aligned} \tag{7}$$

Solution. The product solutions $T(t)R(r)H(\theta)$ of the D.E. and B.C. are of the form

$$[A\cdot\cos(bat) + B\cdot\sin(bat)]\ R_m(br)\ [C\cdot\cos(m\theta) + D\cdot\sin(m\theta)]\,, \tag{8}$$

where $R_m(r)$ solves (2) with $c = b^2$, and $R_m(r_0) = 0$. For any two nonsingular solutions $y_1(r)$ and $y_2(r)$ of (2), the Wronskian $W = y_1y_2' - y_1'y_2$ is $C\exp(-\int 1/r\, dr) = C/r$, which shows that at most one of two independent solutions of (2) can be C^1 at $r = 0$. Thus, $R_m(r)$ must be a constant multiple of $J_m(br)$. By (6), for small r, $J_m(r) \approx [2^m m!]^{-1} r^m$, whence near the origin, $J_m(r)\cos(m\theta)$ and $J_m(r)\sin(m\theta)$ resemble $\mathrm{Re}[(x+iy)^m]$ and $\mathrm{Im}[(x+iy)^m]$, and the solutions (8) can be shown to be C^∞ everywhere. Let $j_{m,q}$ be the q–th positive value for r such that $J_m(r) = 0$. (As is suggested by the above graphs, it can be shown that J_m has infinitely many positive zeros (cf. Example 8 of Section 4.4). Then, the values for b, such that (8) meets the B.C. of (7), are of the form $j_{m,q}/r_0$. For a formal solution, we consider

$$U(r,\theta,t) = \sum_{q=1}^{\infty} \sum_{m=-\infty}^{\infty} [a_{m,q}\cos(aj_{m,q}t/r_0) + b_{m,q}\sin(aj_{m,q}t/r_0)]\ J_m(j_{m,q}r/r_0)\ e^{im\theta}, \tag{9}$$

where, for negative integers m, $J_m(r) \equiv (-1)^m J_{-m}(r)$. To meet the initial conditions, we need

$$f(r,\theta) = \sum_{q=1}^{\infty} \sum_{m=-\infty}^{\infty} a_{m,q} J_m(j_{m,q}r/r_0)\ e^{im\theta} \tag{10}$$

and

$$g(r,\theta) = \sum_{q=1}^{\infty} \sum_{m=-\infty}^{\infty} b_{m,q}a\left[j_{m,q}/r_0\right] J_m(j_{m,q}r/r_0)\ e^{im\theta}\,. \tag{11}$$

The (m,q)–th term in these series is an eigenfunction of the Laplace operator $\Delta = \partial_x^{\,2} + \partial_y^{\,2} = r^{-1}\partial_r(r\partial_r) + r^{-2}\partial_\theta^{\,2}$ with eigenvalue $(j_{m,q}/r_0)^2$ (Why ?). It is clear that terms with different values of m are orthogonal on the disk $r \le r_0$. Moreover, two terms with the same m, but different q, will have different eigenvalues, and hence will be orthogonal by virtue of Green's formula (cf. Problem 3(a)). Consequently, a formal application of orthogonality yields

$$a_{m,q} = \Omega_{m,q}^{-1} \int_0^{2\pi} \int_0^{r_0} f(r,\theta)\, J_m(j_{m,q} r/r_0)\, e^{-im\theta}\, r\, dr\, d\theta \tag{12}$$

and

$$b_{m,q} = [a j_{m,q} \Omega_{m,q}/r_0]^{-1} \int_0^{2\pi} \int_0^{r_0} g(r,\theta)\, J_m(j_{m,q} r/r_0)\, e^{-im\theta}\, r dr\, d\theta\,, \tag{13}$$

where

$$\Omega_{m,q} \equiv 2\pi \int_0^{r_0} J_m(j_{m,q} r/r_0)^2\, r\, dr = \pi\, r_0^2\, J_m'(j_{m,q})^2\,, \tag{14}$$

(cf. Problem 3(b) for the last equality). As a consequence of Theorem 3 (Uniform Convergence of Eigenfunction Expansions) in Section 9.6, the series (10) and (11) will converge uniformly, provided f and g are C^2 on the disk $r \le r_0$ and are 0 on the boundary circle $r = r_0$. Thus, in this case one can truncate the formal series solution (9) at sufficiently large values for q and m obtaining an exact solution of the D.E. and B.C. which meets the I.C. to within any given positive experimental error. □

Remark 1. In the case where the function f in (7) only depends on r, we have $a_{m,q} = 0$ in (10), if $m \ne 0$ (Why ?), and the series (10) is the **Fourier–Bessel series (of order 0)**

$$\text{FBS } f(r) \equiv \sum_{q=0}^{\infty} a_{0,q}\, J_0(j_{0,q} r/r_0)\,, \quad \text{where} \quad a_{0,q} = \Omega_{0,q}^{-1} \int_0^{r_0} f(r)\, J_0(j_{0,q} r/r_0)\, r\, dr\,, \tag{15}$$

and $\Omega_{0,q}$ is given by (14). If f(r) defines a C^2 function on the disk $r \le r_0$, then FBS f(r) converges uniformly to f(r), as a special case of Theorem 3 of Section 9.6 . □

Remark 2. The real and imaginary parts of the terms in the formal solution (9) are the **fundamental modes** of the drum. Some of these fundamental modes at a fixed time are illustrated in Figure 2 below. For convenience, we have chosen $r_0 = 1$. As time varies, one should imagine each of these surfaces oscillating between itself and its negative. From (9) we see that the frequency of oscillation for a mode, with given m and q, is $a j_{m,q}/(2\pi r_0)$. Thus, the frequencies are determined by the zeros of the Bessel functions. Many of these zeros have been computed numerically to many places and there is a table of some of the zeros in Appendix A.6. Using this table, we find that $j_{0,1} < j_{1,1} < j_{2,1} < j_{0,2} < j_{1,2} < j_{2,2} < j_{0,3} < j_{1,3} < j_{2,3} < j_{0,4} < j_{1,4} < j_{2,4}$. Thus, one can deduce the ordering according to frequency in Figure 2. For the modes shown in Figure 2, all of the frequencies in one column are lower than all the frequencies in the next column to the right, but this would not hold if we had added one more row. Indeed, $j_{3,1} > j_{0,2}$. Moreover, the modes shown do *not* have the twelve lowest possible frequencies. □

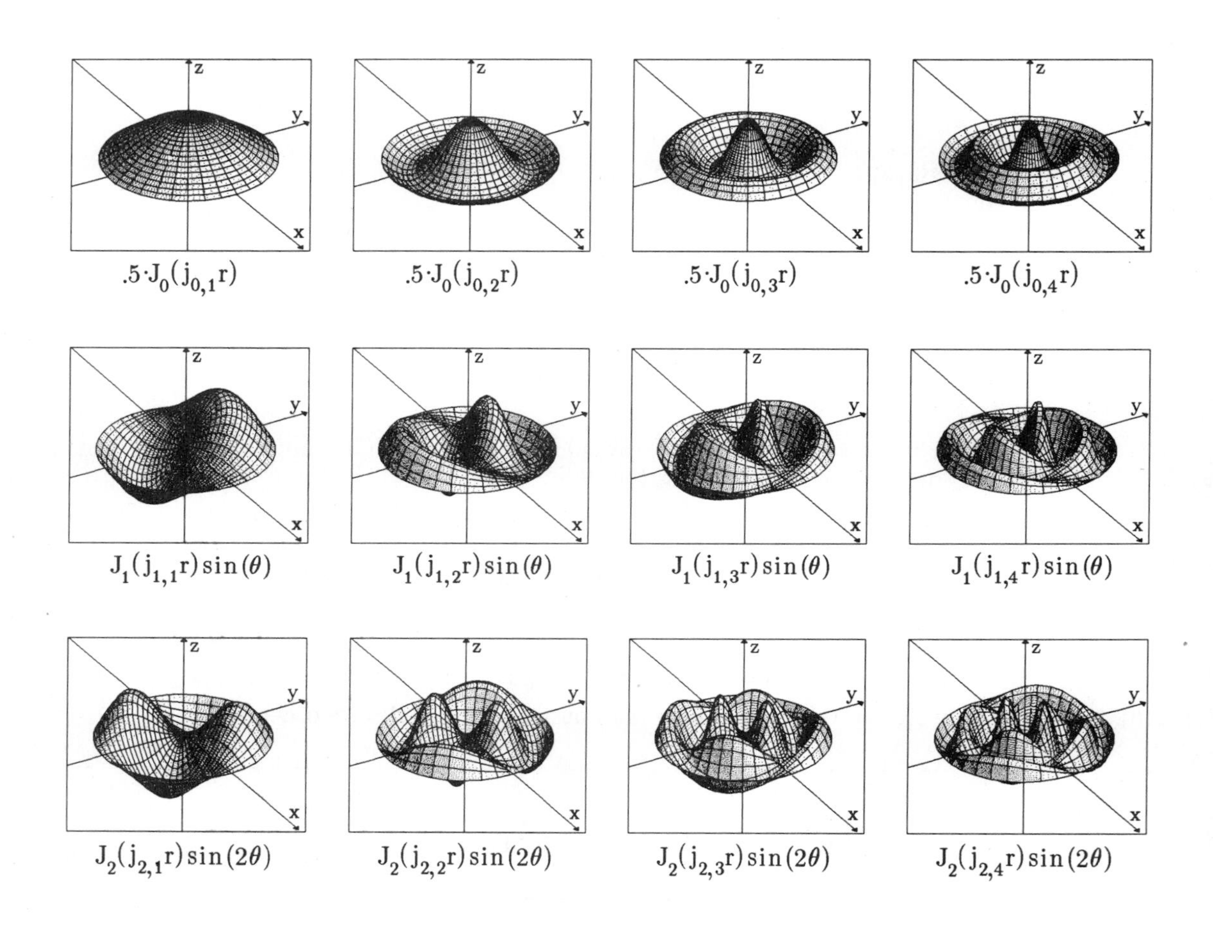

Figure 2

Remark 3. A second solution of Bessel's equation (5), which is linearly independent of $J_m(r)$, is the **Bessel function of the second kind**, denoted by $Y_m(r)$ (Hankel introduced this function in 1869). As noted in Example 1, these solutions must be singular at $r = 0$. Indeed, for *small* r, $Y_0(r) \approx \frac{2}{\pi}\log(r)$ and $Y_m(r) \approx -\frac{1}{\pi}(r/2)^{-m}/(m-1)!$, $m > 0$. Such solutions are used in the problem of vibrating annular drum head or for heat flow in an annulus. Recall that the Dirichlet and Neumann problems for steady–state temperatures (cf. Section 6.3) in an annulus required the use of the singular solutions $\log(r)$ and r^{-m} of the radial equation (2), in the case $c = 0$ which is appropriate for Laplace's equation. The precise formula for $Y_m(r)$ and the asymptotic behavior of $J_m(r)$ and $Y_m(r)$ for large r are summarized in Appendix A.6 . □

As the next example illustrates, there are applications of the equation (2) in the case where $c = -b^2 < 0$. In this case, the solutions are obtained from the solutions $R(br)$ of (2) with $c = b^2 > 0$, by replacing b by ib. Such solutions are known as **modified Bessel functions**. For example, replacing r in $J_m(r)$ by ir, and multiplying by $(-i)^m$ to ensure reality, we obtain the following modified Bessel function which solves (5) with $[r^2-m^2]$ replaced by $-[r^2+m^2]$.

$$I_m(r) \equiv (-i)^m J_m(ir) = (r/2)^m \sum_{k=0}^{\infty} \frac{(r^2/4)^k}{k!(m+k)!} . \tag{16}$$

The graphs of I_0, I_1 and I_2 are shown in Figure 3 below.

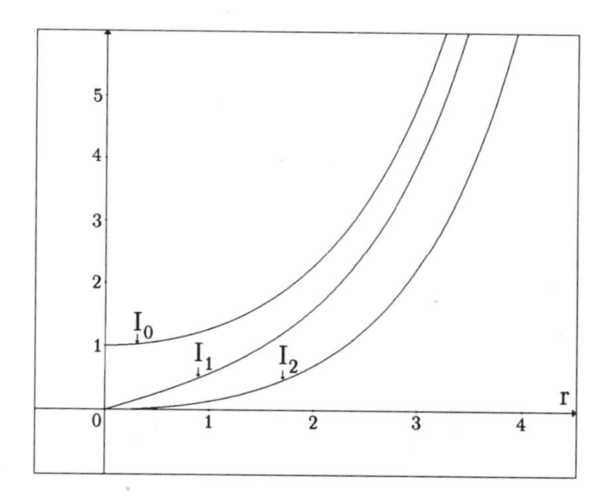

Figure 3

Example 2 (Steady–state temperature distribution in a cylinder). Consider the problem of determining the steady–state temperature distribution of a homogeneous solid cylinder which is insulated on the top and bottom faces with a prescribed temperature on the lateral surface:

$$\text{D.E.} \quad u_{xx} + u_{yy} + u_{zz} = 0 \qquad x^2 + y^2 \le r_0^2 , \quad 0 \le z \le z_0$$

$$\text{B.C.} \quad \begin{cases} U(r_0,\theta,z) = f(\theta, z/z_0) \\ U_z(r,\theta,0) = 0, \quad U_z(r,\theta,z_0) = 0 . \end{cases} \tag{17}$$

Discuss the behavior of the solution as $z_0 \to 0^+$, in relation to the Dirichlet problem for the disk.

Solution. Separation of variables for the product solution $U(r,\theta,z) = R(r)H(\theta)Z(z)$, leads to

$$r^2R''(r) + rR'(r) + (-p^2r^2 - m^2)R(r) = 0\,,\quad H'' + m^2H = 0 \quad\text{and}\quad Z'' + p^2Z = 0. \tag{18}$$

Note that, by the B.C., we must have $p = p_n = n\pi/z_0$ $(n = 0, 1, 2, \ldots)$, and $Z(z) = a_n\cos(n\pi z/z_0)$. Since $H(\theta)$ must be periodic in θ, we conclude that $H(\theta) = c_m e^{im\theta}$, $m = 0, \pm1, \pm2, \ldots$. In the case $n = 0$, we obtain the nonsingular solution r^m for $R(r)$. In the case $n \geq 1$, note that since $J_m(br)$ is a solution of the equation $r^2R''(r) + rR'(r) + (b^2r^2 - m^2)R(r) = 0$ (cf. Problem 1), we will obtain a solution for $R(r)$ in (18), if we choose $b = ip_n$. Thus, we take $R_{m,n}(r) = (-i)^mJ_m(ip_nr) = I_m(p_nr)$ (cf. (16)). For a formal solution, we consider

$$U(r,\theta,z) = \sum_{m=-\infty}^{\infty} \tfrac{1}{2}c_{m,0}r^{|m|}e^{im\theta} + \sum_{n=1}^{\infty}\sum_{m=-\infty}^{\infty} c_{m,n}\, I_m(n\pi r/z_0)e^{im\theta}\cos(n\pi z/z_0)\,. \tag{19}$$

The B.C. $U(r_0,\theta,z) = f(\theta,z/z_0)$ is satisfied if

$$\begin{aligned} f(\theta,z/z_0) &= U(r_0,\theta,z) \\ &= \sum_{m=-\infty}^{\infty} \tfrac{1}{2}c_{m,0}r_0^{|m|}\, e^{im\theta} + \sum_{n=1}^{\infty}\sum_{m=-\infty}^{\infty} c_{m,n}I_m(n\pi r_0/z_0)e^{im\theta}\cos(n\pi z/z_0). \end{aligned} \tag{20}$$

This is a double Fourier series for $f(\theta,z/z_0)$ on the "rectangle" $0 \leq \theta \leq 2\pi,\ 0 \leq z \leq z_0$, if

$$c_{m,0} = \frac{1}{\pi z_0}\int_{-\pi}^{\pi}\int_0^{z_0} f(\theta,z/z_0)e^{im\theta}\,dz\,d\theta = \frac{1}{\pi}\int_{-\pi}^{\pi}\int_0^1 f(\theta,\zeta)\, e^{im\theta}\, d\zeta\, d\theta \quad (\zeta \equiv z/z_0)$$

and

$$c_{m,n} = [I_m(n\pi r_0/z_0)\,\pi]^{-1}\int_{-\pi}^{\pi}\int_0^1 f(\theta,\zeta)\cos(n\pi\zeta)e^{im\theta}\, d\zeta\, d\theta\,,$$

for $m = 0, \pm1, \pm2, \ldots$ and $n = 1, 2, \ldots$. As a consequence of Theorem 1 of Section 9.2, if $f(\theta,z)$ is C^3 with $f_z(\theta,0) = 0$ and $f_z(\theta,1) = 0$, then the series (20) will converge uniformly to $f(\theta,z)$. By truncating the formal solution (19) at a sufficiently large value for m and n, we obtain a solution of the problem (17), within experimental error. As $z_0 \to 0^+$, we should obtain the solution for the Dirichlet problem on the disk with boundary function $f(\theta) \equiv \int_0^1 f(\theta,\zeta)\, d\zeta$ (i.e., the double sum in (19) should approach 0). In fact, it is known that for large r, $I_m(r) \approx e^r/\sqrt{2\pi r}$ (regardless of m; cf. [Whittaker and Watson, p. 373]). Thus, $I_m(n\pi r/z_0)/I_m(n\pi r_0/z_0) \approx \exp[-n\pi(r_0-r)/z_0]\sqrt{r_0/r}$. Hence, at least for $0 < \epsilon < r < r_0 - \epsilon$, the second sum is of the order

$\exp(-\pi\epsilon/z_0)$, which tends rapidly to zero as $z_0 \to 0^+$. For $0 \le r \le \epsilon$, we have $I_m(n\pi r/z_0)/I_m(n\pi r_0/z_0) \le I_m(n\pi\epsilon/z_0)/I_m(n\pi r_0/z_0) \approx \exp[-n\pi(r_0-\epsilon)/z_0]\sqrt{r_0/\epsilon}$ which also tends rapidly to zero as $z_0 \to 0^+$. □

For problems involving heat flow in a sector of a disk, with central angle α, it is necessary to find solutions of Bessel's equation (5) in the case where $m = n\pi/\alpha$ $(n = 1, 2, 3, \ldots)$. Note that m will not necessarily be an integer unless π/α is an integer. Thus, it is necessary in such problems to consider solutions of Bessel's equation when the order m is not an integer. In the definition (6), there appears the quantity $(m+k)!$, which makes sense only when m is an integer. Thus, when m is not an in integer, in (6) one replaces $(m+k)!$ by $\Gamma(m+k+1)$, where Γ is the **gamma function** which is defined by

$$\Gamma(s) = \int_0^\infty e^{-x} x^{s-1}\, dx\,, \quad \text{for real } s > 0\,. \tag{21}$$

Using repeated integration by parts (cf. Problem 6), one can easily show that, for a positive integer n, we have $\Gamma(n+1) = n!$. Thus, the gamma function can be used to extend the factorial function to all positive real numbers. In particular,

$$\Gamma\left[\tfrac{1}{2}\right] = \int_0^\infty e^{-x} x^{-1/2}\, dx = \int_0^\infty e^{-y^2} y^{-1}\, 2y\, dy = \sqrt{\pi}\,.$$

Moreover, integrating by parts, we find $\Gamma(s+1) = s\Gamma(s)$, from which we obtain

$$\Gamma\left[n + \tfrac{1}{2}\right] = \frac{1\cdot 3\cdot 5\cdot \ldots \cdot(2n-1)}{2^n}\sqrt{\pi}\,, \qquad n = 1, 2, 3, \ldots\,. \tag{22}$$

Example 3. Show that the spherical Bessel function j_n (cf. (29) and (30) of Section 9.4) is related to $J_{n+\frac{1}{2}}$ by the formula

$$j_n(\rho) = \sqrt{\tfrac{1}{2}\pi/\rho}\; J_{n+\frac{1}{2}}(\rho)\,. \tag{23}$$

Solution. First we show that $R(\rho)$ is a solution of the three–dimensional radial equation (3), if and only if $\sqrt{r}\,R(r)$ solves Bessel's equation (5) with $m = (n+\frac{1}{2})$:

$$r^2(\sqrt{r}\,R(r))'' + r(\sqrt{r}\,R(r))' + (r^2 - (n+\tfrac{1}{2})^2)\sqrt{r}\,R(r)$$
$$= \sqrt{r}\left[\, r^2R'' + rR' - \tfrac{1}{4}R + rR' + \tfrac{1}{2}R + [r^2 - (n+\tfrac{1}{2})^2]\,R \,\right]$$
$$= \sqrt{r}\left[\, r^2R'' + 2rR' + [r^2 - n(n+1)]\,R \,\right].$$

Thus, $\sqrt{r}\,j_n(r)$ is a solution of Bessel's equation of order $(n+\frac{1}{2})$. Bessel's equation cannot have two independent solutions which are C^1 at $r = 0$, since the Wronskian is Cr^{-1}. Thus, the ratio $\sqrt{r}\,j_n(r)/J_{n+\frac{1}{2}}(r)$ is constant. For small $r > 0$, we have (using (22) and (6))

$$J_{n+\frac{1}{2}}(r) \approx (r/2)^{n+\frac{1}{2}}/\Gamma(n+\tfrac{1}{2}+1) = \sqrt{2r/\pi}\,r^n/[1\cdot 3\cdot\ldots\cdot(2n+1)] \approx \sqrt{2r/\pi}\,j_n(r)\,,$$

where we have obtained the leading term for $j_n(r)$ from the definition

$$j_n(r) \equiv (-1)^n r^n \left[\frac{1}{r}\frac{\partial}{\partial r}\right]^n \sum_{m=0}^{\infty} \frac{(-1)^m}{(2m+1)!}\,r^{2m} = (-1)^n r^n \left[\frac{1}{r}\frac{\partial}{\partial r}\right]^{n-1} \sum_{m=1}^{\infty} \frac{(-1)^m 2m}{(2m+1)!}\,r^{2m-2}$$
$$= (-1)^n r^n \left[\frac{1}{r}\frac{\partial}{\partial r}\right]^{n-2} \sum_{m=2}^{\infty} \frac{(-1)^m 2m(2m-2)}{(2m+1)!}\,r^{2m-4} = \ldots = \frac{r^n}{1\cdot 3\cdot 5\cdot\ldots\cdot(2n+1)} + \ldots\,. \quad \square$$

Schrödinger's equation and quantum mechanics

Much of the behavior of very small objects, such as an electron in an atom, cannot be adequately described in terms of the classical mechanics of Newton, whereby an object, subject to a given force, moves along a trajectory which is determined by its initial position and velocity. Indeed, nature has placed a fundamental restriction on how well one can simultaneously measure an object's position and velocity – a restriction which no measuring instrument can overcome, regardless of how technologically advanced it is. More precisely, in any direction (say **i**), for a particle P of mass M, the error δx in the measurement of P's x–coordinate and the error δv_1 in the measurement of the **i**–component of P's velocity, must satisfy the **(Heisenberg) uncertainty relation**

$$\delta x\,\delta v_1 \ge \tfrac{1}{2}\hbar/M\,, \quad \text{where } \hbar = \frac{h}{2\pi} \text{ and } h \approx 6.626 \times 10^{-27}\ \mathrm{g\ cm^2\ s^{-1}}. \tag{24}$$

For an object with a mass of a few grams, the product of the errors is quite small. However, for an electron of mass $\approx 9.108 \times 10^{-28}$ g, the product $\delta x\,\delta v_1$ of errors must be larger than .5 $\mathrm{cm^2\ s^{-1}}$, regardless of advances in technology. Indeed, the notion that a particle can have a well–defined position and velocity is in dispute. Some people, who have thought deeply about such things, have come to the conclusion that a particle has no position or velocity which is independent of an act of observation. Because one cannot reliably use position and velocity as the description of the state of a small particle, a different characterization of the state of a particle must be adopted. Quantum mechanics provides a surprisingly accurate description of small–scale phenomena, where Newtonian mechanics demonstrably fails. According to quantum mechanics (in its simplest form), a **quantum state** of a particle at fixed time is a complex–valued function $\psi(x,y,z)$, defined on $\mathbb{R}^3$, which is known as a (Schrödinger) **wave function**. The state ψ is said to be **normalizable**, if the integral of the square of the modulus of ψ over $\mathbb{R}^3$, namely $\|\psi\|^2 \equiv \int_{\mathbb{R}^3} |\psi(x,y,z)|^2 \, dxdydz$, is finite and nonzero (the zero function is not a permissible state). Since two states are regarded as equivalent, if one is a constant multiple of the other, one can always multiply a normalizable wave function by a constant, in order to achieve $\|\psi\|^2 = 1$, in which case ψ is said to be **normalized**. For a normalized state ψ, the probability that the particle will be in any given region A in space is $\int_A |\psi|^2 \, dxdydz$, a number in $[0,1]$ (e.g., if $A = \mathbb{R}^3$, the integral is 1, meaning that the particle is certainly somewhere in $\mathbb{R}^3$). Thus, the wave function contains some information about the location of the particle, but only in a probabilistic sense. Indeed, $|\psi(x,y,z)|^2$ is said to give the density of the probability cloud of possible positions for the particle. The state ψ also provides some information about the velocity of the particle. In fact, the density of the probability cloud for the possible velocities (v_1,v_2,v_3) of the particle is the square modulus $|\hat{\psi}(Mv_1/\hbar,Mv_2/\hbar,Mv_3/\hbar)|^2$, where $\hat{\psi}$ is the Fourier transform of ψ. The uncertainty relation (24) is a consequence of this fact (cf. Problem 13).

Consider a particle of mass M which is subject to a force $\mathbf{F}(x,y,z)$ of the form $-\nabla V(x,y,z)$ for some function $V(x,y,z)$, known as a potential for $\mathbf{F}$. A quantum state with a definite energy E (classically, kinetic plus potential energy), satisfies the PDE

$$\frac{\hbar^2}{2M}\Delta\psi + V(x,y,z)\psi = E\psi, \qquad (25)$$

which is known as the (time–independent) **Schrödinger equation**. Note that equation (25) says that ψ is an eigenfunction of the operator $\frac{\hbar^2}{2M}\Delta + V$, and the energy E is the associated eigenvalue. Depending on the potential, it may happen that the eigenvalues of this operator are not arbitrary, but instead form a sequence of discrete values in some specified range. In other words, the possible energies of a particle, bound by a potential, may be limited to a certain discrete set of values. Spectral analysis reveals that atoms emit and absorb light of fairly definite wavelengths (or energies), and this is a consequence of the fact that the possible energies of the quantum states of electrons in atoms are discrete, in accordance with quantum mechanics, but in opposition to classical mechanics which incorrectly predicts a continuous range of energies.

Of particular importance is the case in which the potential V only depends on the distance ρ to the origin, say $V = V(\rho)$. In this case, Schrödinger's equation can be put in the form

$$\Delta\psi + [f(\rho) + c]\psi = 0\,, \quad \text{where} \quad f(\rho) = -2M\hbar^{-2}V(\rho) \quad \text{and} \quad c = 2M\hbar^{-2}E\,. \qquad (26)$$

The two most important forms for V are $V(\rho) = \frac{1}{2}k\rho^2$ (the **harmonic oscillator potential**; cf. Problem 9), and $V(\rho) = -e^2\rho^{-1}$, the **Coulomb potential** (due to a proton of charge e) for the electric force on an electron in a hydrogen atom, which we consider in the following example.

Example 4 (Electron states in the hydrogen atom). Using Schrödinger's equation, show that the possible negative energies E of the normalizable electron states ψ in a hydrogen atom, consisting of an electron of mass $m_e = 9.108\times10^{-28}$ g and a proton of mass $M_p \approx 1836m_e$, are of the form $E = E_n = -\dfrac{\mu e^4}{2\hbar^2 n^2}$, $n = 1, 2, \ldots$, where $\mu = m_e M_p/(M_p + m_e) \approx m_e$ is the "reduced mass", and e is the charge of the electron. Determine the states ψ having these energies, and show that there are n^2 independent states having energy E_n, ignoring spin.

Solution. Since the origin is customarily placed at the center of proton instead of the center of mass of the electron/proton system, one uses the μ, instead of m_e, for the mass in Schrödinger's equation, but it does not make much difference, since $M_p >> m_e$. The electrostatic potential energy between the electron and the proton is $V(\rho) = -e^2/\rho$. Thus, we obtain from (25),

$$\frac{\hbar^2}{2\mu}\Delta\psi + \frac{e^2}{\rho}\psi + E\psi = 0 . \tag{27}$$

Assuming a product solution $\psi = R(\rho)S(\varphi,\theta)$ of (27), the radial equation which results from separation of variables (according to (26) and (3)) is

$$\rho^2R''(\rho) + 2\rho R'(\rho) + [2\mu E\hbar^{-2}\rho^2 - \ell(\ell+1) + \rho^2 2\mu\hbar^{-2}e^2\rho^{-1}]R(\rho) = 0$$

or

$$\rho^2R''(\rho) + 2\rho R'(\rho) - [b^2\rho^2 + \ell(\ell+1) - k\rho]R(\rho) = 0, \quad b \equiv \hbar^{-1}\sqrt{-2\mu E}, \quad k \equiv 2\mu e^2/\hbar^2 , \tag{28}$$

and $S(\varphi,\theta) = S_{\ell,m}(\varphi,\theta) = e^{im\theta}\sin^{|m|}(\varphi)\, P_\ell^{(|m|)}(\cos\varphi)$, $\ell = 0, 1, 2, \ldots, -\ell \le m \le \ell$. The use of the symbol ℓ, instead of n, is conventional in quantum mechanics, since n is reserved for labeling the energy levels E_n. For large ρ, the term $b^2\rho^2$ dominates the terms $\ell(\ell+1)$ and $k\rho$. Thus for large ρ, we expect the general solution of (28) to resemble the solution $(c_1e^{b\rho} + c_2e^{-b\rho})/\rho$, when the terms $\ell(\ell+1)$ and $k\rho$ are absent. In order that ψ be normalizable in this approximation, we must have $c_1 = 0$, and we expect $R(\rho)$ to decay as $e^{-b\rho}$ (as $\rho \to \infty$). For small ρ, the terms $b\rho^2$ and $k\rho$ in (28) are negligible, and in this case the general solution should behave as $d_1\rho^{-\ell-1} + d_2\rho^\ell$. To ensure continuity at the origin, we must have $d_1 = 0$. Thus, in view of the suspected behavior of the solution of (28) for large and small ρ, it is natural to seek a solution of (28) of the form $R(\rho) = L(\rho)\rho^\ell e^{-b\rho}$. Substituting this form into (28), we obtain the following ODE for $L(\rho)$

$$\rho L'' + 2(\ell + 1 - b\rho)L' + [k - 2b(1+\ell)]L = 0 \,. \tag{29}$$

It can be shown [Schiff, 1968] that any solution of (29), which is not a polynomial in ρ, must grow as $e^{2b\rho}$, in which case $R(\rho) = \rho^\ell L(\rho)e^{-b\rho} \approx \rho^\ell e^{b\rho}$ will not yield a normalizable function ψ. Let $P_\nu(\rho)$ be a polynomial of degree ν, say with leading term ρ^ν. If we put $P(\rho)$ into (29), then the left side of (29) has a highest power term $[-2b\nu + k - 2b(\ell+1)]\rho^\nu$. Thus, letting $n \equiv (\nu+\ell+1)$ (a positive integer), we cannot obtain a polynomial solution $P_\nu(\rho)$ of degree ν, unless $[k - 2bn]\rho^\nu = 0$, or

$$b = b_n \equiv \frac{k}{2n} \quad \text{or} \quad \hbar^{-1}\sqrt{-2\mu E} = \frac{2\mu e^2/\hbar^2}{2n} \quad \text{or} \quad E = E_n = \frac{-\mu e^4}{2n^2\hbar^2}\,. \tag{30}$$

Hence, we have obtained the necessary form for the negative energies for the electron. However, if E is of the form (30), then one can successively determine the coefficients of the lower degree terms of $P_\nu(\rho)$ in order that (29) be satisfied. Thus, for each n and ℓ, with $0 \le \ell \le n - 1$ there is a polynomial solution of (29) of degree $\nu = n - 1 - \ell$, which we denote by $P_{n,\ell}(\rho)$ (cf. (33) below, for an explicit formula for $P_{n,\ell}(\rho)$). Recall from Section 9.4 that, for each ℓ, there are $2\ell + 1$ possible values for m, corresponding to the spherical harmonics $S_{\ell,m}(\varphi,\theta)$. Thus, we have a family of normalizable electron wave functions with fixed energy E_n $(n = 0, 1, 2, \ldots)$:

$$\psi_{n,\ell,m}(\rho,\varphi,\theta) \equiv P_{n,\ell}(\rho)\,\rho^\ell\, e^{-b_n\rho}\, S_{\ell,m}(\varphi,\theta) \qquad (\ell = 0, 1, \ldots, n-1\,;\; m = -\ell, \ldots, \ell). \tag{31}$$

For a given n, the number of these wave functions is $1 + 3 + 5 + \ldots + [2(n-1) + 1] = n^2$. With some work, it can be shown that there are no other linearly independent normalizable wave functions with energy E_n (How ?). □

Remark 1. In (31), the number n is known as the **total quantum number** of the state, while ℓ is the **orbital angular momentum** and m is the **magnetic quantum number**. For each of the wave functions, it turns out that there are two independent "spin" states. Thus, there are actually $2n^2$ independent states in the n–th energy shell. The sequence $2, 8, 18, \ldots, 2n^2, \ldots$ is manifested in the periodic table of elements. Indeed, elements with the same number of electrons in the outermost (highest) energy shell, have similar chemical properties. Actually, for atoms with Z protons in the nucleus, one must use the potential $-Ze^2/\rho$, and consequently E_n acquires a factor of Z^2, and in general, e^2 must be replaced by Ze^2 in all of our computations. A more detailed analysis [Schiff, 1968] shows that mutual interactions of electrons in atoms, along with the spin of the electrons and certain other effects, actually split the $2n^2$ states into a number of different energies clustered about E_n. □

Remark 2. For any integers $p, q \geq 0$, the **generalized Laguerre polynomial** $L_q^p(x)$ is the q–th degree polynomial solution of **Laguerre's ODE** $xL'' + (p + 1 - x)L' + qL = 0$, given by

$$L_q^p(x) = \sum_{s=0}^{q} \frac{(-1)^s}{s!} \frac{(p+q)!}{(p+s)!\,(q-s)!} x^s . \tag{32}$$

By setting $x = k\rho/n$, and $F(x) = P_{n,\ell}(\rho)$, one finds (from the definition of $P_{n,\ell}(\rho)$, (29) and (30)) that $xF'' + ([2\ell+1] + 1 - x)F' + (n-\ell-1)F = 0$, which is Laguerre's ODE with $p = 2\ell+1$ and $q = n-\ell-1$. Thus, in view of the fact that there is at most one independent polynomial solution of this ODE (Why ?), we have

$$P_{n,\ell}(\rho) = c_{n,\ell} L_{n-\ell-1}^{2\ell+1}(n\rho/k) = c_{n,\ell} \sum_{s=0}^{n-\ell-1} \frac{(-1)^s}{s!} \frac{(n+\ell)!}{(2\ell+1+s)!\,(n-\ell-1-s)!} \left(\frac{n\rho}{k}\right)^s , \tag{33}$$

for some constant $c_{n,\ell}$. □

Summary 9.5

1. Radial special functions : When the method of separation of variables is applied to the equation $\Delta u + [f(\|\mathbf{x}\|) + c]u = 0$, in terms of polar coordinates in dimension 2 and in terms of spherical coordinates in dimension 3, we obtain the following respective equations :

$$H''(\theta) + m^2H(\theta) = 0\,, \quad r^2R''(r) + rR'(r) + [cr^2 - m^2 + r^2f(r)]R(r) = 0, \tag{S1}$$

$$\Delta_s S(\varphi,\theta) + n(n+1)S(\varphi,\theta) = 0, \quad \rho^2R''(\rho) + 2\rho R'(\rho) + [c\rho^2 - n(n+1) + \rho^2 f(\rho)]R(\rho) = 0\,. \tag{S2}$$

The equations for $R(r)$ and $R(\rho)$ are known as radial equations. When $f(\rho) = 0$ and $c = \pm b^2 \neq 0$, the general solution of the equation (S2), is

$$R_n^{\pm}(b\rho), \text{ where } R_n^{\pm}(\rho) \equiv \left[\frac{1}{\rho}\frac{d}{d\rho}\right]^n [R_0^{\pm}(\rho)], \text{ and } \quad \begin{aligned} R_0^-(\rho) &= (a_0 e^{b\rho} + b_0 e^{-b\rho})/\rho \\ R_0^+(\rho) &= [a_0\cos(\rho) + b_0\sin(\rho)]/\rho\,, \end{aligned} \tag{S3}$$

as was shown in Section 9.3. However, the solutions of (S1) (with $f(r) \equiv 0$) cannot be expressed in terms of elementary functions, and we describe them next.

2. Bessel functions : When $f(r) \equiv 0$ and $c = 1$, (S1) is Bessel's equation of order m :

$$r^2R''(r) + rR'(r) + [r^2 - m^2]R(r) = 0\,. \tag{S4}$$

When m is a nonnegative integer, any solution of Bessel's equation which is C^1 for all r must be a constant multiple of the Bessel function $J_m(r)$, defined by

$$J_m(r) = (r/2)^m \sum_{k=0}^{\infty} \frac{(-r^2/4)^k}{k!(m+k)!}\,, \tag{S5}$$

When m is an arbitrary positive real number, $J_m(r)$ is defined by formula (S5), provided $(m+k)!$ is replaced by $\Gamma(m+k+1)$, where Γ is the gamma function defined by

$$\Gamma(s) = \int_0^{\infty} e^{-x} x^{s-1}\, dx\,, \text{ for real } s > 0\,.$$

The solutions $R_n^+(\rho)$ in (S3) are known a spherical Bessel functions, and they can be related to Bessel functions of half–integer order. For example, the spherical analog of $J_m(r)$ is

$$j_n(\rho) \equiv (-1)^n \left[\frac{1}{\rho}\frac{d}{d\rho}\right]^n [\rho^{-1}\sin\rho] = \sqrt{\tfrac{1}{2}\pi/\rho}\; J_{n+\frac{1}{2}}(\rho)\,.$$

There are other solutions of Bessel's equation which are singular at $r = 0$. The definitions and properties of these Bessel functions of the second kind are given in Appendix A.6 .

3. Applications of Bessel functions in solving PDEs : We used Bessel functions to solve the wave problem for a vibrating circular drum (Example 1), and we found that the modified Bessel function $I_m(r)$ appears in the solution of a steady–state temperature problem for a solid cylinder which is insulated on the top and bottom faces (Example 2). In general, Bessel functions arise in problems involving cylindrical regions or in problems with symmetry about an axis.

4. Schrödinger's equation and quantum mechanics : Heisenberg's uncertainty relation states that the product of the errors in measurement of a particle's position and velocity in any fixed direction cannot be smaller than $\frac{1}{2}\hbar/M$, where M is the mass of the particle. According to quantum mechanics, a quantum state of a particle, at a fixed time, is a complex–valued function $\psi(x,y,z)$, defined on $\mathbb{R}^3$, which is known as a (Schrödinger) wave function. The state is said to be normalized if $\|\psi\|^2 \equiv \int_{\mathbb{R}^3} |\psi(x,y,z|^2\, dxdydz = 1$, and for such a state, the probability that the particle will be found in a region A in $\mathbb{R}^3$ is $\int_A |\psi(x,y,z)|^2\, dxdydz$. A particle with energy E and which is subject to a force with potential $V(x,y,z)$, is described by a state ψ that is a solution of Schrödinger's (time–independent) equation,

$$\frac{\hbar^2}{2M}\Delta\psi + V(x,y,z)\psi = E\psi, \qquad \text{(S6)}$$

In Example 4, Schrödinger's equation is used to determine the possible energies and normalizable states for an electron bound by the Coulomb potential $e\rho^{-1}$ of a proton in a hydrogen atom. The corresponding problem for the harmonic oscillator potential $\frac{1}{2}k\rho^2$ is addressed in Problem 9.

Exercises 9.5

1. Show that if $F(r)$ solves Bessel's differential equation $r^2F''(r) + rF'(r) + [r^2-m^2]F(r) = 0$ $(b \neq 0)$, then $R(r) \equiv F(br)$ solves $r^2R''(r) + rR'(r) + [b^2r^2 - m^2]R(r) = 0$. Can this be done, by using the idea in Problem 4(b) of Exercises 9.3 ?

2. (a) Use integration by parts to establish (by means of formal calculations) Green's formula

$$\int_0^{2\pi}\int_0^{r_0} [g(r,\theta)\Delta f - f(r,\theta)\Delta g]\, rdrd\theta = \int_0^{2\pi} [g(r_0,\theta)f_r(r_0,\theta) - f(r_0,\theta)g_r(r_0,\theta)]\, r_0\, d\theta$$

for the disk $r \leq r_0$.

(b) Formulate and verify Green's formula for functions $f(\rho,\varphi,\theta)$ and $g(\rho,\varphi,\theta)$ defined on the solid ball $\rho \leq \rho_0$.

Hint. A special case is covered in Problem 15 of Section 9.4 (cf. also Example 3 of Section 9.3).

3. (a) Use Green's formula for a disk (cf. Problem 2 (a)) to prove that for any $m = 0, 1, 2, \ldots$,

$$(\beta^2 - b^2)\int_0^{r_0} J_m(br)\, J_m(\beta r)\, r\, dr = r_0[bJ_m'(br_0)J_m(\beta r_0) - \beta J_m(br_0)J_m'(\beta r_0)] \,. \qquad (*)$$

(b) Suppose that $J_m(\beta r_0) = 0$. Differentiate both sides of $(*)$ with respect to b and then set $b = \beta$, in order to deduce that $\int_0^{r_0} J_m(\beta r)^2\, r\, dr = r_0^2 J_m'(\beta r_0)^2/2$.

(c) Assuming that the formula in part (a) is still valid when b and β are complex numbers, prove that $J_m(z)$ cannot be zero for any z which is not real. **Hint.** If $J_m(z) = 0$, then show that $J_m(\bar{z}) = 0$, by examining the power series definition for J_m. In part (a), take $r_0 = 1, b = z$ and $\beta = \bar{z}$ $(r_0 = 1)$, and note that the case where z is purely imaginary is treated differently.

4. (a) Use Green's formula for a ball (cf. Problem 2(b)) to prove that for $n = 0, 1, 2, \ldots$,

$$(\beta^2 - b^2)\int_0^{\rho_0} j_n(b\rho)\, j_n(\beta\rho)\, \rho^2\, d\rho = \rho_0^2\, [bj_n'(b\rho_0)j_n(\beta\rho_0) - \beta j_n(b\rho_0)j_n'(\beta\rho_0)] \,, \qquad (**)$$

where $j_n(\rho) \equiv (-1)^n \left[\frac{1}{\rho}\frac{d}{d\rho}\right]^n [\rho^{-1}\sin(\rho)]$.

(b) Assume that $j_n(\beta r_0) = 0$. Differentiate both sides of $(**)$ with respect to b, and then set $b = \beta$, in order to deduce that $\int_0^{\rho_0} j_n(\beta\rho)^2\, \rho^2\, d\rho = \rho_0^3\, j_n'(\beta\rho_0)^2/2$ (cf. Problem 3(b)).

5. (a) Find a formal solution of the problem

$$\begin{array}{lll} \text{D.E.} & u_t = k(u_{xx} + u_{yy}) & x^2 + y^2 < r_0^2 \,,\ t \geq 0 \\ \text{B.C.} & U(r_0,\theta,t) = 0 \,, & -\pi \leq \theta \leq \pi \\ \text{I.C.} & U(r,\theta,0) = f(r,\theta) \,. & \end{array}$$

(b) Find a formal solution, in the case when the B.C. is replaced by $U_r(r_0,\theta) = 0$ (i.e., the rim of the disk is insulated).

6. Show that $\Gamma(n+1) = n!$ for $n = 1, 2, 3, \ldots$.

7. Find the eigenfunctions of Δ on a wedge $(0 \leq r \leq r_0,\ 0 \leq \theta \leq \alpha)$ which are zero on the boundary $\theta = 0$, $\theta = \alpha$, $r = r_0$. What kinds of heat and wave problems could be solved with these eigenfunctions ? Give some examples.

8. Suppose that the B.C. in Example 2 are replaced by $U(r_0,\theta,z) = 0$ $(0 \le z \le z_0)$, $U(r,\theta,0) = 0$, $U(r,\theta,z_0) = f(r,\theta)$ $(0 \le r \le r_0)$. Find the formal solution in this case.

9. Here we determine possible energies E and the associated normalizable wave functions (quantum states) of a particle of mass m subject to a a spring force $\mathbf{F} = -\nabla V = -k\mathbf{r}$, with potential $V(\rho) = \frac{1}{2}k\rho^2 = \frac{1}{2}k(x^2+y^2+z^2)$ (i.e., the harmonic oscillator potential).

(a) Show that the product solutions of Schrödinger's equation $-\frac{\hbar^2}{2m}\Delta\psi + \frac{1}{2}k\rho^2\psi = E\psi$ are of the form $X(x)Y(y)Z(z)$, where

$$-\frac{\hbar^2}{2m}X'' + \tfrac{1}{2}kx^2X = e_1X \qquad -\frac{\hbar^2}{2m}Y'' + \tfrac{1}{2}ky^2Y = e_2Y \qquad -\frac{\hbar^2}{2m}Z'' + \tfrac{1}{2}kz^2Z = e_3Z\,,$$

and e_1, e_2, e_3 are constants such that $e_1 + e_2 + e_3 = E$.

(b) Show that if we write $X(x) = F(\alpha x)$, then the equation for X is transformed into the ODE $\frac{1}{2}F''(\alpha x) + \frac{1}{2}(\alpha x)^2F(\alpha x) = \lambda_1F(\alpha x)$, by taking $\alpha = [km/\hbar^2]^{\frac{1}{4}}$ and $\lambda_1 = \hbar^{-1}\sqrt{m/k}\,e_1$. Deduce from Problem 16 of Exercises 7.2, that when $\lambda_1 = n+\frac{1}{2}$ for some integer $n \ge 0$, we have a rapidly decreasing solution $X_n(x) = H_n(\alpha x)e^{-\frac{1}{2}\alpha^2x^2}$, where H_n is the n–th degree Hermite polynomial.

(c) Conclude from (b) that when E is of the form $\left[N + \frac{3}{2}\right]\hbar\sqrt{k/m}$ (for an integer $N \ge 0$), we have a the following normalizable wave function satisfying Schrödinger's equation in part (a)

$$\psi_{n,m,p}(x,y,z) = H_n(\alpha x)H_m(\alpha y)H_p(\alpha z)e^{-\frac{1}{2}\alpha^2\rho^2}\,, \quad \text{provided} \quad n + m + p = N\,.$$

(d) With some work it can be shown that the only energies E for which there are normalizable solutions of Schrödinger's equation are of the form in (c), and any normalizable wave function for the N–th energy is a linear combination of the independent wave functions in (c). Show that there are $(N+1)(N+2)/2$ wave functions of the form in (c).

(e) Can the particle ever have energy zero ? What is the minimum energy ?

10. Suppose that the quantum state of a particle of mass M is given by a normalized, rapidly decreasing wave function $\psi(x,y,z)$. The mean x–coordinate of the possible position measurements of the particle is $\bar{x} \equiv \int_{\mathbb{R}^3} x|\psi(x,y,z)|^2\,dxdydz = \langle x\psi,\psi\rangle$ (i.e., the x–coordinate of the center of gravity of its probability cloud, whose density is $|\psi(x,y,z)|^2$). The standard deviation (i.e., the typical error) of the measurements of x is $\delta x = [\langle (x-\bar{x})^2\psi,\psi\rangle]^{\frac{1}{2}} = \|(x-\bar{x})\psi\|$. The density of the probability cloud of possible velocities (v_1,v_2,v_3) is (according to quantum mechanics) $\omega(v_1,v_2,v_3) \equiv (\hbar/M)^{\frac{3}{2}}\,\hat{\psi}(Mv_1/\hbar,\ Mv_2/\hbar,\ Mv_3/\hbar)$, where $\hat{\psi}$ is the Fourier transform of ψ.

Moreover, the mean measurement of the i–component of the particle's velocity is $\overline{v}_1 \equiv \langle v_1\omega,\omega\rangle$, and the standard deviation is $\delta v_1 \equiv [\langle (v_1-\overline{v}_1)^2\omega,\omega\rangle]^{\frac{1}{2}} = \|(v_1-\overline{v}_1)\omega\|$.

(a) Use $(\partial_x\psi)\hat{}\,(\xi) = i\xi\hat{\psi}(\xi)$ and Parseval's equality (cf. Section 7.3) to show that

$$\|(v_1-\overline{v}_1)\omega\| = (\hbar/M)\|-i\partial_x\psi-(M/\hbar)\overline{v}_1\psi)\|\ , \quad \text{where } \partial_x \equiv \frac{\partial}{\partial x}.$$

(b) Use the Cauchy–Schwarz inequality $|\langle f,g\rangle| \le \|f\|\ \|g\|$, to deduce that

$$\begin{aligned}\|(x-\overline{x})\psi\|\ \|(-i\partial_x\psi-(M/\hbar)\overline{v}_1\psi\| &\ge |\langle (x-\overline{x})\psi,-i\partial_x\psi-(M/\hbar)\overline{v}_1\psi\rangle| \\ &\ge \mathrm{Im}[\langle (x-\overline{x})\psi,-i\partial_x\psi-(M/\hbar)\overline{v}_1\psi\rangle]\ .\end{aligned}$$

(c) Show that the last expression in (b) is $\mathrm{Im}[\langle x\psi,-i\partial_x\psi\rangle]$.

(d) Show that $\mathrm{Im}[\langle x\psi,-i\partial_x\psi\rangle] = -\frac{1}{2}i[\langle x\psi,-i\partial_x\psi\rangle - \langle -i\partial_x\psi,x\psi\rangle] = \frac{1}{2}\langle \partial_x(x\psi)-x\partial_x\psi,\psi\rangle$ $= \frac{1}{2}\langle\psi,\psi\rangle$.

(e) Deduce from the above, that the uncertainty relation $\delta x\ \delta v_1 \ge \frac{1}{2}\hbar/M$ holds.

(f) Assume that $\overline{x} = \overline{y} = \overline{z} = 0$ and $\overline{v}_1 = \overline{v}_2 = \overline{v}_3 = 0$. Show that if $\delta x = \delta y = \delta z$ and $\delta x\ \delta v_1 = \delta y\ \delta v_2 = \delta z\ \delta v_3 = \frac{1}{2}\hbar/M$ (i.e., the uncertainty is minimal), then ψ must be the ground state (state of lowest energy) of a particle in a harmonic oscillator potential (cf. Problem 9). How is the spring constant k related to δx for this state ?

9.6 Solving PDEs on Manifolds

In this section we define the Laplacian of functions on fairly general k–dimensional subsets of $\mathbb{R}^n$ known as smooth k–manifolds with or without boundary. This is done in the same way as the Laplacian Δ_s was defined for the unit sphere (an example of a 2–manifold) in $\mathbb{R}^3$, by extending f to be constant in the normal direction and taking the usual Laplacian of the extended function. We then examine the general properties of eigenfunctions and eigenvalues for the Laplace operator on a "compact" k–manifold. A uniform convergence theorem for eigenfunction expansions for sufficiently differentiable functions on manifolds is stated, and references for the proof are given. These eigenfunction expansions are used to solve the heat, wave, and Poisson–Laplace equations on a manifold. We also show how to write the solutions in terms of integral formulas, by direct construction of Green's functions from the eigenfunctions and eigenvalues of the Laplace operator. We emphasize that much of the material previously covered on the heat, wave and Laplace equations is a collection of special cases of the general viewpoint of this section, which thus solidifies what we have learned already. The notion of manifolds has become a fundamental part of the description of the universe in modern theoretical physics, from the space–time continuum of general relativity and cosmology to the continuous symmetry groups which act on the wave functions of elementary particles. For this reason also, we believe it is good to expose the reader to manifolds. For a more thorough introduction to manifolds and related notions, we recommend [Abraham, Marsden and Ratiu].

The definition of a k–manifold in $\mathbb{R}^n$ and its Laplace operator

Let f_1, f_2 and f_3 be C^∞ functions defined on some open set B in $\mathbb{R}^3$. To each point $P = (x,y,z)$ in B, there is assigned a point $F(P) \equiv (f_1(P), f_2(P), f_3(P))$, also in $\mathbb{R}^3$. If the set of points F(P), as P varies throughout B, is an open set C, then we say that F is a **smooth mapping** from B onto C. If there is also a smooth mapping G of C onto B, such that $G(F(P)) = P$ for all points P in B, then F is called a **diffeomorphism** from B onto C (i.e., for open sets B and C, a diffeomorphism is a smooth mapping from B onto C which has a smooth inverse mapping from C onto B). Similarly, one can define the notion of a diffeomorphism from one open set of $\mathbb{R}^n$ to another open subset of $\mathbb{R}^n$, for any $n = 1, 2, 3, 4, \ldots$. Let k and n be integers with $1 \le k \le n$. The **standard k–dimensional subspace in $\mathbb{R}^n$** is the set, denoted by $\mathbb{R}^k_n$, of points in $\mathbb{R}^n$ of the form $(x_1, x_2, \ldots, x_k, \overbrace{0, \ldots, 0}^{\leftarrow n-k \rightarrow})$, where $x_1, x_2, \ldots, x_k$ are arbitrary real numbers (e.g., $\mathbb{R}^2_3$ is the xy–plane in $\mathbb{R}^3$, and $\mathbb{R}^1_3$ is the x–axis in $\mathbb{R}^3$). The **standard, closed, k–dimensional half–space in $\mathbb{R}^n$** is the subset H^k_n of all points in $\mathbb{R}^k_n$ which have a nonnegative first coordinate (i.e., $x_1 \ge 0$) (e.g., H^2_3 is the half–plane $x \ge 0$, $z = 0$ in $\mathbb{R}^3$).

Definition. A (smooth) **k–manifold in $\mathbb{R}^n$** is subset M of $\mathbb{R}^n$ which has the following property. For each point P in M, there is an open set B in $\mathbb{R}^n$ and a diffeomorphism F from B to an open set C containing P, say $P = F(P')$ for some P' in B, such that (cf. Figure 1) the intersection of B with H_n^k is mapped by F onto the entire intersection of C with M. If P' is on the edge of H_n^k, then the corresponding point P on M is called a **boundary point** of **M**. The set of all boundary points of M is called the **boundary** of M, and it is denoted by ∂M.

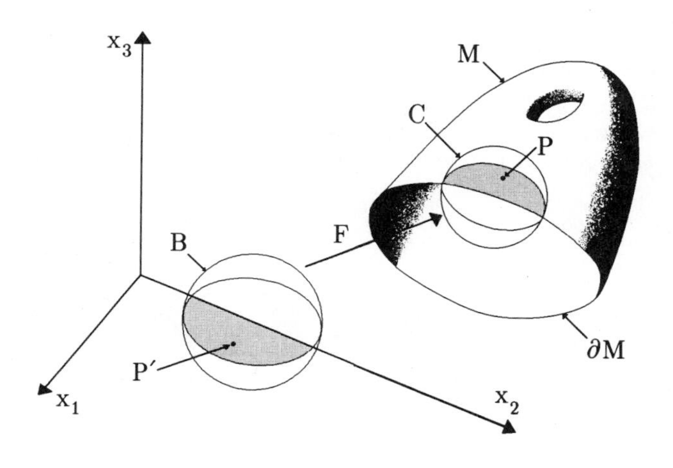

Figure 1

Example 1. Verify explicitly that the quarter sphere Q defined by $\rho = 1$, $z \geq 0$, $x > 0$ is a 2–manifold in $\mathbb{R}^3$, whose boundary is the half equator E defined by $\rho = 1$, $z \geq 0$, $x > 0$.

Solution. Let F be the mapping which assigns to each point (x,y,z) in the open rectangular solid B, $-\pi/2 < x < \pi/2$, $-\pi/2 < y < \pi/2$, $\frac{1}{2} < z < 2$, the point $F(x,y,z) \equiv (z\cos(x)\cos(y), z\cos(x)\sin(y), z\sin(x))$. This is a diffeomorphism from B onto the portion, say C, of the spherical shell, defined by $\frac{1}{2} < \rho < 2$, which lies in the half–space $x > 0$. Indeed, the inverse G of F is defined for any point (x,y,z) in C, by (cf. Problem 1)

$$G(x,y,z) = (\sin^{-1}(r/\rho), \sin^{-1}(y/r), \rho)\ , \text{ where } r = (x^2 + y^2)^{\frac{1}{2}} \text{ and } \rho = (r^2 + z^2)^{\frac{1}{2}}.$$

The intersection $B \cap H_3^2$ is carried by K onto the quarter sphere Q. Moreover, the edge $x = 0$ of $H_3^2 \cap R$ is mapped onto E, which is then ∂Q. □

Remark. The above example is simple in the sense that F(B) contained all of Q. It is often necessary to use several diffeomorphisms F_1, F_2, ..., defined on open sets B_1, B_2, ..., in order to ensure that each point of the prospective manifold is contained in some $F(B_i)$. For example, to show that the hemisphere $\rho = 1$, $z \geq 0$ is a 2–manifold with the equator as its boundary, we would need more than one diffeomorphism, because the entire circular equator cannot be the image of a portion of the edge line of H_3^2 under a single diffeomorphism (Why ?). One would also need two diffeomorphisms to exhibit the sphere $\rho = 1$ as a 2–manifold without boundary. □

Fortunately, with a little experience, it is usually quite easy to recognize a manifold when you see one, and only rarely does anyone bother to prove that a given subset of $\mathbb{R}^n$ is a manifold. Roughly speaking a k–manifold M in $\mathbb{R}^n$ is a "smooth" k–dimensional subset of $\mathbb{R}^n$ which, in a sufficiently small open n–dimensional ball about any of its points, say P, resembles either a possibly curved) k–dimensional disk (if P is not part of the boundary ∂M), or a k–dimensional half–disk with the (k–1)–dimensional "flat" part containing P (if P is part of ∂M). Some 2–manifolds in $\mathbb{R}^3$ are illustrated in Figure 2 below. Of course one should imagine these surfaces to be smooth, without the angles or corners which are a consequence of computer generation. Only the first of these surfaces has a boundary which consists of two circles.

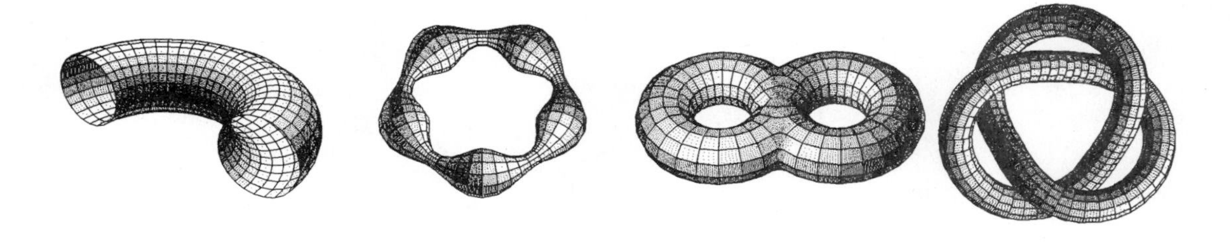

Figure 2

There are some fairly simple objects which are *not* (smooth) k–manifolds. For example, two planes in space which intersect in a line do not constitute a manifold, since this object does not locally resemble a single disk or half–disk in a neighborhood of any point on the line of intersection. The surface of a cube is not a smooth manifold, because corners cannot result from applying a diffeomorphism to a disk in H_3^2. For the same reason, a cone is not a smooth manifold, nor is a closed rectangular solid. The question, as to whether a physical object or system has a mathematical corner or not, is rather meaningless or at least irrelevant (e.g., consider the atoms near a "corner"). It seems reasonable to assert that most domains which arise in "real" applications can be approximated *within experimental error* by smooth k–manifolds. Considering the large number of variables that can enter into the description of even a fairly simple mechanical, electrical, or economic system, it does *not* suffice to restrict one's attention to the case $k \leq 3$ or even $k \leq 1000$.

Let f be a function defined on a k–manifold M with boundary ∂M (possibly empty) in $\mathbb{R}^n$ It can be proven (cf. [Lang, p. 96]) that f can be extended to a function $\bar{f}$ on some open set A of $\mathbb{R}^n$ containing M, in such a way that $\bar{f}$ is constant on each line segment which intersects

M orthogonally at a single point, and which lies in A. We say that f is C^2 on M, if $\tilde{f}$ is C^2 on A. Since $\tilde{f}$ is then a C^2 function on the open set A in $\mathbb{R}^n$, it makes sense to apply the Laplace operator $\Delta_n \equiv \partial^2/\partial x_1^2+...+\partial^2/\partial x_n^2$ in $\mathbb{R}^n$ to the function $\tilde{f}$.

Definition of Δ_M. We define the **Laplace operator** Δ_M on C^2 functions f on M by $\Delta_M f(P) \equiv \Delta_n \tilde{f}(P)$, for all points P in M, where $\tilde{f}$ is the above defined extension (of f) which is constant in the directions normal to M. Equivalently, $\Delta_M f(P)$ is the sum of the second derivatives at P of $\tilde{f}$ along k lines which are tangent to M and intersect orthogonally at P.

Remark. This is exactly how the Laplace operator Δ_s (on the unit sphere) was defined in (7) of Section 9.3 . While the above definition gives an immediate geometrical interpretation to Δ_M, there are other equivalent definitions which express $\Delta_M f$ in terms of the so–called "metric coefficients" of M relative to a coordinate system. Since we will not be solving the equations $\Delta_M u + \lambda u = 0$ explicitly on any manifolds not previously considered, we will not discuss metric coefficients or the formula for Δ_M in terms of them. Instead, we will concentrate on general properties of the eigenfunctions and eigenvalues without finding them explicitly, which is a formidable task, even for simple manifolds. □

Eigenfunctions and eigenvalues for Δ_M on k–manifolds M

In solving boundary–value problems for PDEs (such as the heat equation $u_t = \Delta_M u$) on k–manifolds M in $\mathbb{R}^n$, it is desirable to find the eigenfunctions of Δ_M which meet the boundary conditions. Of course, in the case of manifolds without boundary (e.g., a spherical surface) there is no boundary and hence no boundary condition. In order that we may apply the superposition principle, the boundary conditions need to be homogeneous and linear. We restrict ourselves to a standard type, which we describe as follows. Let f be a C^2 function on a k–manifold (i.e., f extends to a C^2 function $\tilde{f}$ described above). The **outward normal derivative of** f **at a point** p **on** ∂M is the directional derivative

$$\partial_\nu f(p) \equiv \nabla\tilde{f}\cdot\mathbf{n}_p \tag{1}$$

of $\tilde{f}$ in the direction of the unique unit vector $\mathbf{n}_p$ (based at p) which is normal to ∂M, tangent to M, and pointing away from the **interior** M^o (of M) which consists of all points in M but not in ∂M. We introduce a fairly general homogeneous boundary condition as follows. Let N be the (possibly empty) intersection of ∂M with some open subset of $\mathbb{R}^n$, and let D be the set of points in ∂M which are not in N (i.e., $D = \partial M - N$). Let u be a function which is continuous on M, and C^1 on $M^o \cup N$. Then it is meaningful to impose the boundary condition

$$\text{B.C.} \begin{cases} u \equiv 0 & \text{on } D \\ \partial_\nu u \equiv 0 & \text{on } N\,. \end{cases} \tag{2}$$

In the special case where N is empty, (2) is a Dirichlet condition, and when D is empty (2) is a Neumann condition. We say that a k–manifold in $\mathbb{R}^n$ is **compact**, if it is a closed subset of $\mathbb{R}^n$ and is contained in a sufficiently large n–dimensional ball (cf. Appendix A.4). Henceforth, all manifolds are assumed to be compact, and **connected**, in the sense that any two points in M can be joined by a continuous curve which lies in M. The following key result is essentially stated in [Chavel, 1984], but a complete proof (in a form which is comprehensible to undergraduates) does not appear to be located in a single place, but rather it is a consequence of results which are scattered over the literature. For a readable introduction to the literature, we recommend the article by Jozef Dodziuk, "*Eigenvalues of the Laplacian and the heat equation*" , Amer. Math. Monthly, vol. **88**, (1981), 686–695.

Theorem 1 (**The Dirichlet/Neumann eigenvalue problem**). **Let** M **be a compact, k–manifold in** $\mathbb{R}^n$, **with boundary** $\partial M = D \cup N$ (possibly empty). **For each real value of** λ, **the problem**

$$\begin{aligned} &\text{D.E.} \quad \Delta_M u + \lambda u = 0 \quad \text{on } M \\ &\text{B.C.} \begin{cases} u \equiv 0 & \text{on } D \\ \partial_\nu u \equiv 0 & \text{on } N\,, \end{cases} \end{aligned} \tag{3}$$

has at most a finite number of linearly independent solutions (i.e., **the eigenspace for a given eigenvalue** λ **is finite dimensional**). **The solutions are** C^∞ **on** M, **in the sense that they extend to** C^∞ **functions on an open subset** (of $\mathbb{R}^n$) **which contains** M. **The values of** λ **for which this problem has a nonzero solution form a sequence** $0 \le \lambda_0 \le \lambda_1 \le \lambda_2 \le \ldots$, **in which each eigenvalue is repeated a number of times equal to the dimension of its eigenspace.**

Remark. For large q, the behavior of the eigenvalues λ_q is given by **Weyl's formula**

$$\lambda_q \sim 4\pi^2[\omega_k \text{Vol}(M)]^{-2/k}\, q^{2/k}\,, \quad \text{as } q \to \infty\,, \tag{4}$$

where ω_k is the k–dimensional "volume" of the "solid" unit ball in $\mathbb{R}^k$ (i.e., $\omega_3 = 4\pi/3$, $\omega_2 = \pi$, $\omega_1 = 2$ and in general $\omega_k = \pi^{k/2}/\Gamma(1+\frac{1}{2}k)$, where Γ is the gamma function (cf. (21) and (22) of Section 9.5), and Vol(M) is the k–dimensional volume of the the k–manifold M (e.g., the surface area, when $k = 2$). The precise meaning of the symbol " $\sim$ " is that the ratio of the left and right sides of (4) approaches 1 as $q \to \infty$. [Hermann Weyl (1885–1955) was a German mathematical physicist who made profound contributions to relativity, quantum mechanics, and pure mathematics.] For a proof of Weyl's formula, see [Chavel, Chapters VI and VII]. □

Example 2. Show the validity of Weyl's formula directly, in the simple case when M is the 1–manifold consisting of the interval [0,L]. Consider all possible B.C. (2).

Solution. When M is an interval [0,L], and the B.C. are given by $u(0) = 0$ and $u(L) = 0$, we have the eigenfunctions $\sin(\frac{n\pi x}{L})$ of $\Delta \equiv \mathrm{Vol}(M) = L$, $\lambda_0 = (\pi/L)^2$ and in general $\lambda_q = [(q+1)\pi/L]^2$. Note that

$$4\pi^2 (\omega_1 \mathrm{Vol}(M))^{-2} q^{2/k} = 4\pi^2[2L]^{-2}q^2 = (q\pi/L)^2 \sim [(q+1)\pi/L]^2 = \lambda_q .$$

Similarly, for the B.C. $u'(0) = 0$, $u'(L) = 0$, we have $\lambda_0 = 0$ and $\lambda_q = (q\pi/L)^2$. While for $u'(0) = 0$, $u(L) = 0$, we have $\lambda_q = [(q+\frac{1}{2})\pi/L]^2$. Thus, Weyl's formula (4) is valid in all of these cases. □

One can define the integral of a continuous function on a compact k–manifold M in $\mathbb{R}^n$ in the following way. When $k = n$, the interior of M is open subset of $\mathbb{R}^n$ and the boundary is a smooth (n–1)–manifold. The integral is defined in the usual way, by introducing a rectangular subdivision, forming Riemann sum approximations, and letting the norm of the mesh go to zero. When $k < n$, we use the following construction. For each $\epsilon > 0$, let M_ϵ be the set in $\mathbb{R}^n$ consisting of all of the points in $\mathbb{R}^n$ that can be connected to M by a segment, of length $\leq \epsilon$, which meets M orthogonally. For small enough ϵ, each point of M_ϵ is connected to M by a unique shortest segment (of length less than ϵ). Let f be a continuous function on M and extend f to a function f_ϵ defined on M_ϵ, by taking f_ϵ to be constant on these shortest segments. Let I_ϵ be the integral of f_ϵ over M_ϵ , which is a "nice" n–dimensional closed, bounded set. Now, I_ϵ depends on ϵ, and $\lim_{\epsilon\to 0} I_\epsilon = 0$, since I_ϵ is nearly proportional to the volumes $\omega_{n-k}(\epsilon) \equiv \epsilon^{n-k}\pi^{(n-k)/2}/\Gamma(1+\frac{1}{2}k)$ of the normal (n–k)–dimensional disks of radius ϵ. We define

$$\int_M f(P)\, dM(P) \equiv \lim_{\epsilon\to 0} I_\epsilon/\omega_{n-k}(\epsilon) . \tag{5}$$

Although (5) defines the entire expression on the left, one regards the symbol "dM(P)" as the "volume element" for M (e.g. $dM(\varphi,\theta) = \sin\varphi \, d\varphi d\theta$ for the unit sphere).

Definition. For two continuous complex–valued functions f and g on a compact k–manifold M in $\mathbb{R}^n$, we define the inner product

$$<f,g> = \int_M f(P)\overline{g(P)} \, dM(P) . \tag{6}$$

Theorem 2 (Green's formula on a compact k–manifold). Let f and g be C^2 functions on a compact k–manifold in $\mathbb{R}^n$. Then

$$<\Delta_M f,g> - <f,\Delta_M g> = \int_{\partial M} g(Q)\partial_\nu f(Q) - f(Q)\partial_\nu g(Q) \, d\partial M(Q) , \tag{7}$$

where the right side is defined to be zero if ∂M is empty.

Remark. We omit the proof, which typically is carried out in with much technical machinery (cf. [Abraham, Marsden and Ratiu; Chapter 7]), but which (in the final analysis) reduces to the fundamental theorem of calculus, as in the one–dimensional case. For a relatively simple proof, one might first establish the Gauss Divergence Theorem for n–manifolds in $\mathbb{R}^n$, which can be done in essentially the same way as in advanced calculus books. Then, for $k < n$, one can apply this result to the solid M_ϵ above, to obtain the Gauss Divergence Theorem on M. Green's formula on M then follows easily from the Gauss Divergence Theorem on M (cf. Problem 10). □

Observe that if f and g satisfy the B.C. (2), then the right side of (7) vanishes, and it follows that two real–valued solutions u_i and u_j of the eigenvalue problem, with different eigenvalues, say λ_i and λ_j , must be orthogonal (i.e., $<u_i,u_j> = 0$). Even within a single eigenspace of dimension d, one can find d functions which are mutually orthogonal and of unit length. Consequently, there is a sequence, u_0, u_1, u_2, ..., of C^∞ real–valued eigenfunctions for the problem (3) such that

$$\Delta_M u_0 = -\lambda_0 u_0 \, , \; \Delta_M u_1 = -\lambda_1 u_1 , \; \Delta_M u_3 = -\lambda_3 u_3 , \; \dots \, ,$$

where (8)

$$<u_i,u_j> = \begin{cases} 0 & \text{for } i \neq j \\ 1 & \text{for } i = j \end{cases} \qquad i, j = 0, 1, 2, \dots .$$

Although we have taken the u_q to be real–valued, in some situations it convenient to consider complex–valued eigenfunctions (e.g., the spherical harmonics $S_{n,m}(\varphi,\theta)$ in Section 9.4). All of our considerations can be trivially modified in order to treat this case.

Definition. Let f be a function on M, such that for all functions u_q $(q = 0, 1, 2, \ldots)$ in (8), the integrals $<f,u_q>$ exist. Then the **eigenfunction expansion of f** (for $P \in M$) is the expression

$$E\,f(P) \equiv \sum_{q=0}^{\infty} c_q\, u_q(P)\,, \quad \text{where } c_n = \|u_n\|^{-2}<f,u_q> = <f,u_q>\,. \tag{9}$$

Remark. In essence, all Fourier series (ignoring the issue of corners), as well as the Laplace series on a sphere and the full Fourier–Bessel series on a disk, are special cases of (9). Although the following result is rarely stated in the form given here, it follows from simple modifications of known results in [Palais]. □

Theorem 3 (Uniform convergence of eigenfunction expansions). Let M be a compact k–manifold, and let f be a C^m function, where $m > k/2$, which satisfies the B.C. (2). **Let $u_0, u_1, u_2, \ldots$ be the sequence of eigenfunctions given by** (8). **Then $E\,f$ converges uniformly to f on M, in the sense that**

$$\lim_{N\to\infty}\ \max_{P\in M} |S_N(P) - f(P)| = 0, \tag{10}$$

where $S_N(P)$ is the N–th partial sum of $E\,f(P)$.

The standard initial boundary–value problems on manifolds

In Theorems 4 and 5, we provide the solutions of the standard heat and wave problems on a compact k–manifold M, when the functions in the initial conditions are *finite* linear combinations of the eigenfunctions u_0, u_1, u_2, ..., in (8), which (by definition) satisfy the B.C. in these problems. In other words, we assume that the initial data has been approximated within experimental error by such sums, which is reasonable in view of Theorem 3. Essentially, Theorems 4 and 5 include, as special cases, all previous results for heat and wave problems, on bounded domains (e.g., finite intervals, disks, spheres, balls, and rectangular regions, ignoring "corner technicalities"), in which the D.E. and B.C. (Dirichlet and/or Neumann) were homogeneous, and the function(s) in I.C. was (were) assumed to be finite linear combinations of the spatial parts of product solutions. The proofs consist simply of substituting the given solution into the D.E., B.C. and I.C., to verify that these equations are satisfied. (Remember that $\Delta_M u_q = -\lambda_q u_q$ for all $q = 0, 1, 2, \ldots$.). In practice, the difficult part of the solution process is finding explicit expressions or approximations for the u_q and λ_q, and adequately expressing the initial functions in terms of the u_q. Theorems 4 and 5 show that the rest of the solution process is easy.

Theorem 4 (The homogeneous heat problem). For a compact k–manifold M**, the solution of**

$$\text{D.E.} \quad u_t = \kappa\, \Delta_M u \quad \text{on } M \text{ for } t \geq 0$$

$$\text{B.C.} \begin{cases} u \equiv 0 & \text{on } D \\ \partial_\nu u \equiv 0 & \text{on } N \end{cases} \tag{11}$$

$$\text{I.C.} \quad u(P,0) = f(P) \equiv \sum_{q=0}^{Q} c_q u_q(P)$$

is

$$u(P,t) = \sum_{q=0}^{Q} c_q\, e^{-\lambda_q \kappa t} u_q(P)\,. \tag{12}$$

Theorem 5 (The homogeneous wave problem). For a compact k–manifold M**, the solution of**

$$\text{D.E.} \quad u_{tt} = a^2 \Delta_M u \quad \text{on } M \text{ for } -\infty < t < \infty$$

$$\text{B.C.} \begin{cases} u \equiv 0 & \text{on } D \\ \partial_\nu u \equiv 0 & \text{on } N \end{cases} \tag{13}$$

$$\text{I.C.} \quad u(P,0) = f(P) \equiv \sum_{q=0}^{Q} a_q u_q(P)\,, \quad u_t(P,0) = g(P) \equiv \sum_{q=0}^{Q} b_q u_q(P)$$

is

$$u(P,t) = \sum_{q=0}^{Q} \left[a_q \cos(\sqrt{\lambda_q}\, at) + (a\sqrt{\lambda_q}\,)^{-1}\, b_q \sin(\sqrt{\lambda_q}\, at)\right] u_q(P)\,, \tag{14}$$

where it is understood that the second term in the brackets is $b_q t$ **if** $\lambda_q = 0$.

The next two theorems show that eigenfunction expansions also yield solutions of the inhomogeneous heat and wave equations which arise when there are heat sources or forced vibrations, respectively. Again, the proofs consist of directly substituting the given solutions into the D.E., B.C., and I.C. to verify that they are satisfied. In each case, the given solution can be derived, by assuming that the solution is of the form $u(P,t) = \sum_{q=0}^{Q} \beta_q(t) u_q(P)$, substituting this form into the D.E. to obtain an ODE for each $\beta_q(t)$ ($q = 0, 1, 2, \ldots, Q$), and solving these ODEs subject to the initial condition(s) implied by the I.C. . Alternatively, one can use Duhamel's principle, by which solutions are constructed by forming a continuous superposition (i.e., an integral) of solutions of related family of initial–value problems (cf. Sections 3.4 and 5.3).

Theorem 6 (**The heat problem with a source**). **For a compact k–manifold M, the solution of**

$$\text{D.E.}\quad u_t(P,t) = \kappa\Delta_M u(P,t) + \varphi(P,t) \quad \text{on } M \text{ with } t \geq 0$$

$$\text{B.C.}\begin{cases} u \equiv 0 & \text{on } D \\ \partial_\nu u \equiv 0 & \text{on } N \end{cases} \tag{15}$$

$$\text{I.C.}\quad u(P,0) = 0,$$

with heat source

$$\varphi(P,t) = \sum_{q=0}^{Q} \alpha_q(t) u_q(P),$$

is

$$u(P,t) = \sum_{q=0}^{Q} \left[\left[e^{-\lambda_q \kappa t} \int_0^t e^{\lambda_q \kappa s} \alpha_q(s)\, ds \right] u_q(P) \right]. \tag{16}$$

Theorem 7 (**The wave problem with a source**). **For a compact k–manifold M, the solution of**

$$\text{D.E.}\quad u_{tt}(P,t) = a^2\Delta_M u(P,t) + \varphi(P,t)$$

$$\text{B.C.}\begin{cases} u \equiv 0 & \text{on } D \\ \partial_\nu u \equiv 0 & \text{on } N \end{cases} \tag{17}$$

$$\text{I.C.}\quad u(P,0) = 0,\ u_t(P,0) = 0,$$

with source

$$\varphi(P,t) = \sum_{q=0}^{Q} \alpha_q(t) u_q(P),$$

is

$$u(P,t) = \sum_{q=0}^{Q} \left[\left[\int_0^t \frac{\sin[a\sqrt{\lambda_q}\,(t-s)]}{a\sqrt{\lambda_q}} \alpha_q(s)\, ds \right] u_q(P) \right], \tag{18}$$

(We use the convention in Theorem 5, when $\lambda_q = 0$.)

Green's functions

The solutions in the above theorems can be formally rewritten in terms of integrals by convolving the given functions in the I.C. or inhomogeneous term in the D.E. with a certain kernel or Green's function for the problem, which is constructed from the eigenfunctions. We illustrate this construction in the next examples.

Example 3 (The kernel for the homogeneous heat problem). Formally, find integral formulas for the solutions of the heat problem (11).

Solution. In the solution (12) of problem (11), we write $c_q = \langle f,u_q \rangle$ (cf. (8) and (9)) in terms of an integral, and we interchange the sum and integrals in the following computation

$$\begin{aligned} u(P,t) &= \sum_{q=0}^{Q} c_q e^{-\lambda_q \kappa t} u_q(P) = \sum_{q=0}^{Q} \int_M f(P')u_q(P')\,dM(P')\,e^{-\lambda_q \kappa t} u_q(P) \\ &= \int_M f(P')\left[\sum_{q=0}^{Q} e^{-\lambda_q \kappa t} u_q(P)u_q(P')\right] dM(P'). \end{aligned} \tag{19}$$

The given form for f implies that f is orthogonal to the u_q with $q > Q$. Thus, Q in (19) may be replaced by any larger finite value. It is tempting to simply replace Q by ∞, but there is no guarantee that this sum will even converge. Indeed, the sum does not converge when $t = 0$ and $P = P'$. It can be proven (cf. [Chavel, Chapters VI and VII]) that for $t > 0$, the sum does converge to a C^∞ function on the set $\mathbb{R}^+ \times M \times M$ of triples $(t,P;P')$, $t > 0$. We denote this sum by

$$H(t,P;P') = \sum_{q=0}^{\infty} e^{-\lambda_q \kappa t} u_q(P)u_q(P'), \qquad \text{for } (t,P;P') \text{ in } \mathbb{R}^+ \times M \times M. \tag{20}$$

Moreover, one can show that when the initial temperature f is continuous on M and not necessarily a finite linear combination of eigenfunctions, we have the following solution of the D.E. and B.C. for $t > 0$ (cf. [Chavel, Chapters VI and VII]),

$$u(P,t) = \int_M H(t,P;P')\,f(P')\,dM(P') \qquad t > 0 \tag{21}$$

and

$$\lim_{t \to 0^+} u(P,t) = f(P). \tag{22}$$

Remark. The function $H(t,P;P')$ is known as the **heat kernel** or **fundamental solution of the heat equation on** M (for the given B.C.). The interpretation of $H(t,P;P')$ is that it represents the temperature at P, at time t, due to an initial heat source concentrated at P' at $t = 0$. In the remark after the next example, we will show that $H(t,P;P')$ arises also in the integral formula for the solution of the problem (15) for the inhomogeneous heat equation (cf. (30)). □

Example 4 (The Green's function for the heat problem with a source). Express the solution (16) of problem (15) in terms of an integral formula.

Solution. The solution (16) can be written in the form

$$u(P,t) = \sum_{q=0}^{Q} e^{-\lambda_q \kappa t} \int_0^t \alpha_q(s)\, e^{\lambda_q \kappa s}\, ds\; u_q(P)$$

$$= \int_M \int_0^t \Big[\sum_{q=0}^{Q} e^{-\lambda_q \kappa (t-s)} u_q(P) u_q(P') \Big]\, \varphi(P',s)\, ds\, dM(P') \,, \tag{23}$$

We take $Q = \infty$, and formally define

$$G(t,P;t',P') = \epsilon(t-t') \sum_{q=0}^{\infty} e^{-\lambda_q \kappa (t-t')} u_q(P) u_q(P') \,, \tag{24}$$

where

$$\epsilon(s) \equiv \begin{cases} 1 & \text{for } s \geq 0 \\ 0 & \text{for } s < 0 \,. \end{cases} \tag{25}$$

The solution (23) of problem (15) can then be written as

$$u(P,t) = \int_M \int_0^\infty G(t,P;t',P') \varphi(P',t')\, dt'\, dM(P') \,. \quad \square \tag{26}$$

Remark. The function G on $\mathbb{R}^+ \times M \times \mathbb{R}^+ \times M$ is known as the **Green's function** for the problem (15). If we apply the heat operator $\partial_t - \kappa \Delta_M$ to the equation (26), and formally differentiate under the integral, we obtain the following equation involving the source term $\varphi(P,t)$ in (15) :

$$\varphi(P,t) = u_t - \kappa \Delta_M u = \int_M \int_0^\infty \Big[(\partial_t - \kappa \Delta_M)\, G(t,P;t',P') \Big]\, \varphi(P',t')\, dM(P') \,. \tag{27}$$

The expression in parentheses cannot be an ordinary function, but instead it is a "generalized function" or a "distribution" which is known as the **Dirac delta** at (t,P) (cf. (33) of Section 7.4), which may be loosely defined as the generalized function $\delta(t,P;t',P')$ which has the property that

$$\varphi(P,t) = \int_M \int_0^\infty \delta(t,P;t',P')\, \varphi(P',t')\, dt'\, dM(P') \,. \tag{28}$$

Thus,

$$(\partial_t - \kappa \Delta_M)[G(t,P;t',P')] = \delta(t,P;t',P') \,. \tag{29}$$

While we will not go into the theory of distributions, by which precise sense can be made of the relation (29), it can be said that ultimately (29) states that the integral operator $I[\varphi]$, defined by the right side of (26), is the inverse of the differential operator $\partial_t - \kappa \Delta_M$ on the set of sufficiently smooth functions which satisfy the B.C. and I.C. of (15). Intuitively, one can think of

$G(t,P;t',P')$ as the temperature at (P,t) due to a concentrated point source at (P',t'). Recalling the interpretation for $H(t,P;P')$ in the remark following Example 3, we should have $H(t,P;P') = G(t,P;0,P')$, and checking the above definitions, this is the case. More generally,

$$G(t,P;t',P') = H(t-t',P;P') , \tag{30}$$

which says that the temperature at P at time t, due to a source at P' at time t', is the same as that for an "initial source", if time is measured from the starting time t'. Thus, (30) can be regarded as a reformulation of Duhamel's principle. □

For small values of t and for P and P' in the interior $M^o \equiv M - \partial M$, we expect that $H(t,P,P')$ will be close the heat kernel for $\mathbb{R}^k$. In other words, for t small,

$$H(t,P;P') \sim [4\pi\kappa t]^{-k/2} \exp[-d(P,P')^2/4\kappa t], \qquad P,\ P' \text{ in } M^o , \tag{31}$$

where the distance $d(P,P')$, between P and P' in M^o, is small. Essentially, (31) is true because, at P, the contribution due to a nearby source, at P', at a small time t later, should be very much like the effect in $\mathbb{R}^k$, since locally M^o looks like $\mathbb{R}^k$. If we set $P' = P$ in relation (20) and integrate $H(t,P;P)$ with respect to P over M, then using orthogonality (cf. (8)),

$$\begin{aligned} \int_M H(t,P,P)\, dM(P) &= \sum_{q=0}^{\infty} e^{-\kappa\lambda_q t} \int_M u_q(P)^2\, dM(P) \\ &= \sum_{q=0}^{\infty} e^{-\kappa\lambda_q t} \langle u_q,u_q\rangle = \sum_{q=0}^{\infty} e^{-\kappa\lambda_q t} . \end{aligned} \tag{32}$$

On the other hand, integrating (31) over M (with $P = P'$), we obtain

$$\int_M H(t,P,P)\, dM(P) \sim [4\pi\kappa t]^{-k/2}\, \mathrm{Vol}(M) . \tag{33}$$

By comparing (32) and (33), we arrive at

$$\sum_{q=0}^{\infty} e^{-\kappa\lambda_q t} \sim [4\pi\kappa t]^{-k/2}\, \mathrm{Vol}(M) . \tag{34}$$

In other words, the volume of M can be determined from the eigenvalues of Δ_M. Actually, we already know this from Weyl's formula (4). Indeed, it is not too hard to show that (4) and (34) are equivalent, in the sense that one can be derived directly from the other (use the Karamata Tauberian Theorem in [Feller, p. 466]). There is a subtle correction to (34) due to the curvature or general shape of M. Indeed, in the case of a compact 2–manifold without boundary (i.e., ∂M is empty), such as a possibly deformed sphere, with a number (say h) of holes (e.g., a doughnut surface has one hole), we have the more accurate estimate (cf. [Chavel, p. 155])

$$\sum_{q=0}^{\infty} e^{-\kappa\lambda_q t} \sim \frac{\mathrm{Vol(M)}}{4\pi\kappa t} + \frac{1}{3}(1-h) \quad \text{or} \quad h = 1 + 3\lim_{t\to 0^+}\left[\frac{\mathrm{Vol(M)}}{4\pi t} - \sum_{q=0}^{\infty} e^{-\lambda_q t}\right]. \tag{35}$$

Thus, the number of holes can be determined from the eigenvalues. Since the eigenvalues can be thought of as the negative squares of the frequencies of fundamental modes of vibration, one might say that both the surface area and the number of holes of a vibrating surface can be "heard" (cf. Problem 2 of Section 9.2), assuming that one can hear arbitrarily high frequencies. □

Example 5 (**Green's function for a wave problem**). Find the Green's function for the wave problem (17), and use it to find an integral formula for the solution of the initial–value problem (13).

Solution. As with the heat problem in Example 4, we write the solution (18) as

$$u(P,t) = \int_M \int_0^t \left[\sum_{q=0}^{Q} \frac{\sin[a\sqrt{\lambda_q}\,(t-t')]}{a\sqrt{\lambda_q}}\, u_q(P)u_q(P')\right] \varphi(P',t')\, dM(P'). \tag{36}$$

Thus, formally taking $Q = \infty$, we have a Green's function (cf. (25) for the definition of ϵ)

$$G(t,P;t',P') = \epsilon(t-t')\sum_{q=0}^{\infty} \frac{\sin[a\sqrt{\lambda_q}\,(t-t')]}{a\sqrt{\lambda_q}}\, u_q(P)u_q(P'), \tag{37}$$

which (in a distributional sense) satisfies the equation,

$$(\partial_{tt} - a^2\,\Delta_M)\, G(t,P;t',P') = \delta(t,P;t',P'), \tag{38}$$

and hence $G(t,P;t',P')$ may be regarded as the amplitude at P at time t, due to a certain source disturbance concentrated at P' at time t'. Formally allowing $Q = \infty$, the solution (16) for the initial–value problem (13) with homogeneous D.E. can be written in the form

$$\int_M \left[\sum_{q=0}^{\infty} \cos(a\sqrt{\lambda_q}\,t)\, u_q(P)u_q(P')\right] f(P') + \left[\sum_{q=0}^{\infty} \frac{\sin(a\sqrt{\lambda_q}\,t)}{a\sqrt{\lambda_q}}\, u_q(P)u_q(P')\right] g(P')\, dM(P')$$

$$= \int_M -\partial_{t'} G(t,P;0,P')\, f(P') + G(t,P;0,P')\, g(P')\, dM(P')$$

$$= \int_M \left[G(t,P;0,P')\, u_{t'}(P',0) - G_{t'}(t,P;0,P')\, u(P',0)\right] dM(P'). \tag{39}$$

Thus, the solution of the initial–value problem (13) can be elegantly written in terms of the Green's function, and indeed the Green's function (37) is essentially the kernel of the wave problem where the initial position $u(P,0)$ is zero and the initial velocity profile is arbitrary. Hence, the Green's function $G(t,P;t',P')$ can be interpreted as the displacement at P at time t due to a concentrated velocity source at P' at time t'. □

Example 6 (**A Poisson kernel**). Formally find an integral solution for the Poisson problem

$$\text{D.E.} \quad \Delta_M u(P) = h(P) \quad \text{on } M$$

$$\text{B.C.} \begin{cases} u \equiv 0 & \text{on } D \\ \partial_\nu u \equiv 0 & \text{on } N \,. \end{cases} \tag{40}$$

Solution. Taking the inner product of both sides of the D.E. with u_q, we obtain

$$\langle h,u_q \rangle = \langle \Delta_M u,u_q \rangle = \langle u,\Delta_M u_q \rangle = -\langle u,\lambda_q u_q \rangle = -\lambda_q \langle u,u_q \rangle \,.$$

Thus, in order that there be a solution, it is necessary that

$$\langle h,u_q \rangle = 0, \text{ if } \lambda_q = 0 \,. \tag{41}$$

Actually, λ_q can be zero only in the case where $q = 0$ and $u_0 \equiv$ constant $\neq 0$, which will occur only when D in the B.C. is empty (i.e., either the B.C. are purely Neumann conditions, or ∂M is empty; cf. Problem 11). In any case, we obtain the following formal solution, assuming that (41) is met :

$$\begin{aligned} u(P) &= \sum_{\lambda_q \neq 0} \langle u,u_q \rangle u_q(P) = \sum_{\lambda_q \neq 0} -\lambda_q^{-1} \langle h,u_q \rangle \, u_q(P) \\ &= \int_M G(P,P') \, h(P') \, dM(P') \,, \end{aligned} \tag{42}$$

where the **Poisson kernel** $G(P,P')$ is given by

$$G(P,P') \equiv \sum_{\lambda_q \neq 0} -\lambda_q^{-1} \, u_q(P) u_q(P') \,. \quad \square \tag{43}$$

Example 7 (**Steady–state heat problem**). By means of formal manipulations and Green's formula (7), find a hypothetical solution of the steady–state heat problem

$$\text{D.E.} \quad \Delta_M u = 0 \quad \text{on } M$$

$$\text{B.C.} \begin{cases} u = f & \text{on } D \\ \partial_\nu u \equiv g & \text{on } N \,. \end{cases} \tag{44}$$

Solution. We consider first the case $\lambda_0 \neq 0$. According to (42), we formally have

$$h(P) = \Delta_M \int_M G(P,P')\, h(P')\, dM(P') = \int_M \Delta_M G(P,P')\, h(P')\, dM(P') . \tag{45}$$

We apply (45) in the case where h(P) is a solution (presumed to exist) u of the problem (44) :

$$\begin{aligned} u(P) &= \int_M \Delta_M G(P,P')u(P') - G(P,P')\Delta_M u(P')\, dM(P') \\ &= \int_{\partial M} \partial_\nu G(P,Q)u(Q) - G(P,Q)\partial_\nu u(Q)\, d\partial M(Q) , \end{aligned} \tag{46}$$

or in view of the B.C.,

$$u(P) = \int_{\partial M} \partial_\nu G(P,Q)\, f(Q) - G(P,Q)\, g(Q)\, d\partial M(Q) . \tag{47}$$

Thus, in the case where $\lambda_0 \neq 0$ (i.e., D nonempty), (47) provides us with an integral formula for a hypothetical solution of (44). In particular, if $D = \partial M$ and ∂M is nonempty, we see that $\partial_\nu G(P,Q)$ serves as a hypothetical Poisson kernel for the Dirichlet problem for M. With considerable effort (cf. [Cordes, p. 82]), it can be proved that $G(P,P')$ is a C^∞ function on M×M, except along the "diagonal", where $P = P'$. Moreover, u defined by (47) is harmonic in the interior M^o (i.e., $\Delta_M u = 0$ on M^o), and if the given boundary function f is continuous, then u extends continuously to f on the boundary. We now consider the case where D is empty. If ∂M is also empty, then there is no boundary condition to be satisfied and any constant function will solve the problem (44). Assuming that M is connected, the constant functions are the only solutions of (44) (cf. Problem 14). If D is empty and ∂M is not empty, then (44) is a pure Neumann problem on M, and there is a compatibility condition which must hold in order that a solution exist. Indeed, using Green's formula, we have the **compatibility condition** (cf. Sections 6.1–6.3) :

$$0 = \int_M 1\, \Delta_M u - u\Delta_M 1\, dM = \int_{\partial M} 1\partial_\nu u - u\partial_\nu 1\, d\partial M = \int_{\partial M} \partial_\nu u\, d\partial M = \int_{\partial M} g\, d\partial M , \tag{48}$$

i.e., physically, the net prescribed temperature flux on the boundary must be 0, in order that a steady–state temperature distribution exist. Now, suppose that u is a solution of (44). Then, by subtracting the average $\bar{u} \equiv \mathrm{Vol}(M)^{-1}\int_M u\, dM$ from u, we obtain a function $u - \bar{u}$ which satisfies condition (48) (i.e., $\langle u - \bar{u}, u_0\rangle = 0$). Consequently, (47) applies with $f \equiv u - \bar{u}$, and

$$u(P) - \bar{u} = \int_{\partial M} -G(P,Q)\, g(Q)\, d\partial M(Q) . \tag{49}$$

Of course, adding a constant to any solution of a pure Neumann problem, yields another solution, whence we have the the following formal general solution of the pure Neumann problem

$$u(P) = \int_{\partial M} -G(P,Q)\, g(Q)\, d\partial M(Q) + c , \tag{50}$$

where c is an arbitrary constant, provided the compatibility condition (48) is met. □

Summary 9.6

1. Manifolds : Although a rigorous definition of manifold was provided, intuitively a k–manifold in $\mathbb{R}^n$ $(k \leq n)$ is a smooth k–dimensional subset, say M, of $\mathbb{R}^n$ which may have a smooth (k–1)–dimensional boundary ∂M.

2. The Laplace operator Δ_M : A function defined on a k–manifold M in $\mathbb{R}^n$ is C^2, if it can be extended constantly in the normal directions (to M) to a C^2 function $\bar{f}$ on defined on an open subset A (of $\mathbb{R}^n$) which essentially "thickens" M to an n–dimensional object. The Laplace operator Δ_M, for C^2 functions f on M, is then defined by $\Delta_M f(P) = \Delta\bar{f}(P)$, for all points P in M, where $\Delta = \partial^2_{x_1} + \ldots + \partial^2_{x_n}$ is the Laplace operator for $\mathbb{R}^n$. We then can consider the standard heat equation $(u_t = k\Delta_M u)$ and wave equation $(u_{tt} = a^2\Delta_M u)$ on M, subject to initial conditions and certain standard boundary conditions, say

$$\text{B.C.} \begin{cases} u \equiv 0 & \text{on } D \\ \partial_\nu u \equiv 0 & \text{on } N \end{cases}, \tag{S1}$$

where $\partial_\nu u$ denotes the outward normal derivative of u at points in ∂M, N is the intersection of ∂M with an open subset of $\mathbb{R}^n$, and $D \equiv \partial M - N$ is the set of points in ∂M but not in N.

3. Eigenfunctions and eigenvalues of Δ_M : Of primary importance in the solution of the heat and wave problems on a k–manifold M in $\mathbb{R}^n$, is the determination of the eigenfunctions of Δ_M which satisfy the B.C. (S1). We restrict ourselves to compact M (i.e. M is a closed and bounded subset of $\mathbb{R}^n$). In this case, there is a sequence $u_0, u_1, u_2, \ldots$ of real–valued eigenfunctions of Δ_M, meeting the B.C. (S1), and associated eigenvalues $0 \leq \lambda_0 \leq \lambda_1 \leq \lambda_2 \leq \ldots$, with

$$\Delta_M u_0 = -\lambda_0 u_0,\ \Delta_M u_1 = -\lambda_1 u_1,\ \Delta_M u_3 = -\lambda_3 u_3, \ldots,$$

where

$$\langle u_i,u_j\rangle = \begin{cases} 0 & \text{for } i \neq j \\ 1 & \text{for } i = j \end{cases} \qquad i, j = 0, 1, 2, \ldots . \tag{S2}$$

Here $\langle f,g\rangle \equiv \int_M fg\, dM$ denotes the inner product of functions on M. Here, each eigenvalue is repeated a number of times equal to the dimension of its eigenspace. The fact that eigenfunctions with different eigenvalues are orthogonal is a direct consequence of

Green's Formula: $\langle\Delta_M f,g\rangle - \langle f,\Delta_M g\rangle = \int_{\partial M} g(Q)\partial_\nu f(Q) - f(Q)\partial_\nu g(Q)\, d\partial M(Q)$.

4. Eigenfunction expansions : Let f be a function on M, such that for all functions u_q (q = 0, 1, 2, ...) in (S2), the integrals $<f,u_q>$ exist. Then

$$\mathrm{E}\, f(P) \equiv \sum_{q=0}^{\infty} c_q\, u_q(P)\,, \quad \text{where } c_n = \|u_n\|^{-2}<f,u_q> = <f,u_q>\,. \tag{S3}$$

If f is C^m where $m > k/2$ (where k is the dimension of M) and if f satisfies the B.C. (S1), then E f converges uniformly to f (cf. Theorem 3).

5. Heat and wave problems on manifolds : Solutions of the initial–value problems for the heat and wave equations on a compact k–manifold M with B.C. (S1) are given in Theorems 4 and 5, in the *practical* case when the initial functions are finite eigenfunction expansions, say within experimental error. Solutions to the inhomogeneous heat and wave equations with the B.C. (S1), are provided by Theorems 6 and 7, when the source terms have finite eigenfunctions with time–dependent coefficients.

6. Green's functions : The solutions of the inhomogeneous heat, wave and Poisson equations with the B.C. (S1) can expressed in terms of integral formulas where the inhomogeneous term is convolved with a Green's function for the given problem. In each case, the Green's function was constructed from the eigenfunctions u_q and eigenvalues λ_q in (S2). For details, see Examples 3 through 7.

7. Weyl's formula : The behavior of the eigenvalues λ_q for large q is influenced by the dimension k and the k–dimensional volume [Vol(M)], of a compact, connected k–manifold. This is evident from

$$\text{Weyl's formula :} \quad \lambda_q \sim 4\pi^2[\omega_k \mathrm{Vol}(M)]^{-2/k}\, q^{2/k}\,, \text{ as } q \to \infty\,, \tag{S4}$$

or equivalently,

$$\sum_{q=0}^{\infty} e^{-\kappa\lambda_q t} \sim [4\pi\kappa t]^{-k/2}\mathrm{Vol}(M)\,, \text{ as } t \to 0^+. \tag{S5}$$

Refinements of the formula (S5) show that certain geometric (or "topological") properties of a k–manifold in $\mathbb{R}^n$ are also determined by the behavior of the eigenvalues λ_q for large q. For example, the number h of holes of a compact surface (2–manifold) without boundary is determined by

$$\sum_{q=0}^{\infty} e^{-\lambda_q t} \sim \frac{\mathrm{Vol}(M)}{4\pi t} + \frac{1}{3}(1-h) \quad \text{or} \quad h = 1 + 3 \lim_{t \to 0^+} \left[\frac{\mathrm{Vol}(M)}{4\pi t} - \sum_{q=0}^{\infty} e^{-\lambda_q t} \right].$$

Exercises 9.6

1. Show that the mapping F in Example 1 is a diffeomorphism between the sets B and C , by showing that every point P in C is $F(P')$ for some point P' in B, and that $G(F(P')) = P'$.

2. (a) Let $f(x,y)$ be a C^∞ function, defined for all (x,y) in the xy–plane. Find a diffeomorphism from $\mathbb{R}^3$ onto $\mathbb{R}^3$ such that the xy–plane is carried onto the graph of $f(x,y)$.

(b) How can the diffeomorphism in (a) be used to show that the graph of $f(x,y)$ above the region $x - y \geq 0$ is a 2–manifold in $\mathbb{R}^3$ with boundary consisting of points of the form $(x,x,f(x,x))$.
Hint. You may have to modify the diffeomorphism in (a) by a rotation of space.

3. Prove that the circle $x^2 + y^2 = 1$ is a 1–manifold without boundary in $\mathbb{R}^2$.

4. Prove that the square $|x| + |y| = 1$ is not a (smooth) 1–manifold in $\mathbb{R}^2$.
Hint. Demonstrate that a line segment cannot be mapped to a "corner" by a diffeomorphism.

5. Check the validity of the Weyl's formula (4) in the case of

(a) a circle of radius r (b) a sphere of radius ρ (c) an $L \times M$ rectangle .

Hint. It is crucial to recall that each eigenvalue is repeated a number of times equal to the dimension of the associated eigenspace.

6. Verify by direct substitution, that the claimed solutions in Theorems 4 and 5 do in fact solve the D.E., B.C., and I.C. .

7. Do Problem 6 for Theorems 6 and 7.

8. For a compact manifold M with boundary ∂M, convert the following problem to a problem with homogeneous B.C. .

D.E. $u_t = \kappa \Delta_M u$ on M, for $t > 0$

B.C. $u(Q,t) = f(Q,t)$ for all Q on ∂M for $t > 0$

I.C. $u(P,0) = 0$ for all P in M.

How might the resulting "related homogeneous problem" be formally solved ?

Hint. To get a particular function satisfying the B.C., solve the Dirichlet problem with the given B.C. at each time $t > 0$.

9. Suppose that the B.C. in Problem 8 is changed to $\partial_\nu u(Q,t) = f(Q,t)$. What difficulty is there with finding a function which satisfies the B.C. by solving the Neumann problem at each time t? How might this difficulty be overcome through first solving the Poisson D.E. $\Delta w = c(t)$, at each time t, for $c(t) \equiv \mathrm{Vol}(M)^{-1}\int_{\partial M} f(Q,t)\, d\partial M(Q)$?

Hint. Use $\int_M \Delta_M w(P)\, dM(P) = \int_{\partial M} \partial_\nu w(Q)\, d\partial M(Q)$, as a special case of Green's formula, and consider the related problem for $v = u - w$.

10. A vector field on a compact k–manifold M in $\mathbb{R}^n$ assigns to each point P in M a vector $\mathbf{V}(P)$ which is tangent to M. Let $\bar{\mathbf{V}}$ be the extension of $\mathbf{V}$ to an open subset A of M in $\mathbb{R}^n$, which is obtained by translating $\mathbf{V}$ along the normal lines to M. We define the **divergence** (denoted $\mathrm{div}(\mathbf{V})$) of $\mathbf{V}$ at P to be the divergence of $\bar{\mathbf{V}}$ at P, namely $\nabla \cdot \bar{\mathbf{V}} \equiv \partial \bar{V}_1/\partial x_1 + \cdots + \partial \bar{V}_n/\partial x_n$ (at P), where $\bar{V}_i$ is the i–th component of $\bar{\mathbf{V}}$. The **Gauss divergence theorem** for the compact k–manifold M states that for any C^1 vector field $\mathbf{V}$ on M,

$$\int_M \mathrm{div}(\mathbf{V})\, dM = \int_{\partial M} \mathbf{V}(Q)\cdot \mathbf{n}(Q)\, d\partial M(Q),$$

where $\mathbf{n}(Q)$ is the outward pointing normal to ∂M at Q in ∂M, and the right side is defined to be zero, if ∂M is empty.

(a) Show that for a C^2 function f on M, we have $\Delta_M f = \mathrm{div}(\nabla f)$ where ∇f is the vector field on M defined by $\nabla f \equiv \nabla \bar{f} = (\bar{f}_{x_1}, \ldots, \bar{f}_{x_n})$, where $\bar{f}$ is the extension of f which is constant on normal lines to M.

(b) For C^2 functions f and g on M, show that $\mathrm{div}(f\,\nabla g) = f\Delta_M g + \nabla f \cdot \nabla g$, and deduce that $f\Delta_M g - g\Delta_M f = \mathrm{div}(f\nabla g - g\nabla f)$.

(c) From parts (a) and (b) and the Gauss divergence theorem, deduce Green's formula (7).

$$\langle \Delta f, g\rangle - \langle f, \Delta g\rangle = \int_{\partial M} g(Q)\partial_\nu f(Q) - f(Q)\partial_\nu g(Q)\, d\partial M(Q).$$

11. Let M be a compact k–manifold with or without boundary.

(a) Use Problem 10 to show that $\int_M v\Delta_M w\, dM = -\int_M \nabla v \cdot \nabla w\, dM + \int_{\partial M} v\partial_\nu w\, d\partial M$, for any C^1 function v and any C^2 function w.

(b) Use part (a) to show that the eigenvalues λ_n of Δ_M with the standard B.C. ($u = 0$ on $D \subset \partial M$, and $\partial_\nu u = 0$ on $N \subset \partial M$ (cf. Theorem 1)) are nonnegative, and that $\lambda_0 = 0$ only in the case $N = \partial M$ (i.e., the B.C. is purely Neumann or ∂M is empty).

(c) Conclude from part (b) that the Poisson problem in Example 6 is at least formally solvable for all "nice" functions h, in the case when D is not empty (i.e., condition (41) is then vacuous).

12. Show that the usual proof (cf. end of Section 3.2) of the Maximum Principle for the heat equation, still works in the case when the spatial domain is a compact k–manifold M in $\mathbb{R}^n$.

13. Assuming that there is a C^2 function v on the compact manifold M with boundary ∂M, such that $\Delta_M v \equiv 1$ (or more generally, $\Delta_M v > 0$), prove the Maximum Principle for harmonic functions u on M^o (i.e., $\Delta_M u = 0$ on the interior M^o of M) which extend continuously to ∂M (cf. Theorem 1 of Section 6.4) . (Note that the existence of such a function v in the case where ∂M is not empty is made plausible by part (c) of Problem 11.) **Hint.** Consider the Laplacian at a maximum of $u + \epsilon v$, for sufficiently small $\epsilon > 0$.

14. Assume that M is a compact k–manifold *without* boundary. Use Green's formula to show that there is no C^2 function v on M such that $\Delta_M v \geq 0$, unless v is constant.

15. Let M be a compact k–manifold with or without boundary.

(a) Use part (a) of Problem 11 to show that for any C^2 solution of the heat problem (11) in Theorem 4, we must have $\frac{d}{dt}\int_M u(P,t)^2\, dM(P) \leq -2\kappa \int_M \|\nabla u\|^2\, dM$.

(b) Use this fact in part (a) to prove the uniqueness of C^2 solutions of the heat problem (11).

16. Let u be a solution of the wave equation $u_t = a^2\Delta_M u$ on the compact manifold M in $\mathbb{R}^n$. Consider the energy integral $H(t) = \frac{1}{2}\int_M u_t^{\,2} + a^2\|\nabla u\|^2\, dM$ (cf. (12) Section 5.1).

(a) Show that $H'(t) = a^2\int_M u_t\Delta_M u + \nabla u_t \cdot \nabla u\, dM = a^2\int_M \operatorname{div}(u_t\nabla u)\, dM$

$= a^2\int_{\partial M} u_t\nabla u \cdot \mathbf{n}\, d\partial M$ (cf. Part (b) of Problem 10).

(b) Use part (a) to prove a uniqueness theorem for the wave problem

$$\text{D.E.}\quad u_{tt} = a^2 \Delta_M u \quad \text{on } M \text{ with } -\infty < t < \infty$$

$$\text{B.C.}\quad \begin{cases} u \equiv 0 & \text{on } D \\ \partial_\nu u \equiv 0 & \text{on } N \end{cases}$$

$$\text{I.C.}\quad u(P,0) = f(P)\ ,\ u_t(P,0) = g(P) \quad \text{for all } P \text{ in } M\ .$$

17. Prove that there is at most one C^∞ function $H(t,P;P')$ (for $t > 0$), such that the problem

$$\text{D.E.}\quad u_t = \Delta_M u \quad \text{on } M \text{ (a compact manifold)}$$

$$\text{B.C.}\quad \begin{cases} u \equiv 0 & \text{on } D \\ \partial_\nu u \equiv 0 & \text{on } N \end{cases}$$

$$\text{I.C.}\quad u(P,0) = f(P)$$

(cf. (15)) has the solution $u(P,t) = \int_M H(t,P;P')\ f(P')\ dM(P')$, for every initial C^∞ temperature $f(P)$ (i.e., the heat kernel is unique).
Hint. Let $f(P) = u_n(P)$, $n = 0, 1, 2, \ldots$, and use uniqueness of solutions (cf. Problem 15), in order to find the eigenfunction expansion of $H(t,P;P')$ as a function of P' for fixed t and P.

18. If $f(x)$ is defined on $[-\pi,\pi]$ and has a C^∞ periodic extension $\tilde{f}(x)$, the solution of

$$\text{D.E.}\quad u_t = u_{xx} \quad -\pi < x < \pi\ ,\ t > 0$$

$$\text{B.C.}\quad u(-\pi,t) = u(\pi,t)\ ,\ u_x(-\pi,t) = u_x(\pi,t)$$

$$\text{I.C.}\quad u(x,0) = f(x)$$

is given by $u(x,t) = \dfrac{1}{\sqrt{4\pi t}} \int_{-\infty}^{\infty} e^{-(x-y)^2/(4t)}\ \tilde{f}(y)\ dy$ (cf. Section 7.5).

(a) Show (at least formally) that $u(x,t) = \int_{-\pi}^{\pi} \left[\sum_{m=-\infty}^{\infty} \dfrac{1}{\sqrt{4\pi t}}\, e^{-(x-z-2m\pi)^2/(4t)} \right] f(z)\ dz$.

(b) Deduce from (a) and Problem 17 that (cf. Problem 20 of Section 7.2)

$$\sum_{m=-\infty}^{\infty} e^{-m^2 t}\, e^{im(x-z)} = \frac{1}{\sqrt{4\pi t}} \sum_{m=-\infty}^{\infty} e^{-(x-z-2m\pi)^2/(4t)}\ .$$

19. Use Problems 11–13 of Exercises 6.4, Example 1 of Section 9.5 and (42) to formally show

$$\frac{1}{\pi}\sum_{k=1}^{\infty}\sum_{m=-\infty}^{\infty} -\beta_{m,k}^{-2}\, J_m'(\beta_{m,k})^{-2} J_m(r\beta_{m,k}) J_m(r'\beta_{m,k}) e^{im(\theta-\theta')} = \frac{1}{\pi}\log\left|\frac{1-rr'e^{i(\theta-\theta')}}{re^{i\theta}-r'e^{i\theta'}}\right| ,$$

for $re^{i\theta} \neq r'e^{i\theta'}$, $r < 1$ and $r < 1$.

20. For $\mathbf{r} = (x,y,z)$ and $\mathbf{r}' = (x',y',z')$, here we show that the Green's function for

$$\begin{aligned} &\text{D.E.}\quad \Delta u = h\,, \quad \rho^2 = x^2 + y^2 + z^2 \leq 1 \\ &\text{B.C.}\quad u(x,y,z) = 0 \quad\ \rho = 1\,, \end{aligned} \tag{$*$}$$

can be written in the following form, where $\rho' \equiv \|\mathbf{r}'\|$:

$$G(\mathbf{r},\mathbf{r}') = -\frac{1}{4\pi}\left[\frac{1}{\|\mathbf{r}-\mathbf{r}'\|} - \frac{\rho'^{-1}}{\|\mathbf{r}-\rho'^{-2}\,\mathbf{r}'\|}\right]. \tag{$**$}$$

(a) Show that $\dfrac{\rho'^{-1}}{\|\mathbf{r}-\rho'^{-2}\,\mathbf{r}'\|} = \dfrac{\rho^{-1}}{\|\mathbf{r}'-\rho^{-2}\,\mathbf{r}\|}$, and that $G(\mathbf{r},\mathbf{r}') = 0$ when $\rho = 1$.

Hint. Verify that $\rho'^2\|\mathbf{r}-\rho'^{-2}\mathbf{r}'\|^2 = \rho^2\rho'^2 - 2\mathbf{r}\cdot\mathbf{r}' + 1$.

(b) If h in $(*)$ is a C^2 function on space which vanishes outside of some ball, show that $u(\mathbf{r}) = \int_{\rho'\leq 1} G(\mathbf{r},\mathbf{r}')h(\mathbf{r}')\, dx'dy'dz'$ solves the problem $(*)$. You may use part (a) and take the limit as $\rho \to 1$ under the integral to formally check the B.C. .

Hint. Observe that if $\rho' < 1$, then the second term of $G(\mathbf{r},\mathbf{r}')$ in $(**)$ is a harmonic function of $\mathbf{r}$ in the ball $\rho < 1$, whereas the first term has a singularity at $\mathbf{r} = \mathbf{r}'$. Thus, only the first term of $G(\mathbf{r},\mathbf{r}')$ contributes to $\Delta u(\mathbf{r})$. To evaluate $\Delta u(\mathbf{r})$, use spherical coordinates in $x'y'z'$–space, but with $\mathbf{r}$ as the origin. The computation is similar to the one carried out in Problem 9 of Section 6.4.

21. Express the Green's function in Problem 20 in terms of the spherical Bessel functions and spherical harmonics, as was done in Problem 19 in the case of the two–dimensional unit disk.

APPENDIX

A.1 The Classification Theorem

Here we prove that, by a change of variables, any second–order linear PDE (for an unknown function of two variables) with constant coefficients (i.e., of the form (1) below) can be transformed into one of the four standard types, namely the generalized wave, Poisson or heat equation or the standard degenerate equation, depending on whether the original PDE is hyperbolic, elliptic, parabolic or degenerate, respectively (cf. Section 1.3). Thus, the Classification Theorem reduces the study of (1) to the study of the four standard types, in that, by reversing the change of variables (2), any solution of the appropriate standard PDE can be transformed back to a solution of the original PDE (1). We pay particular attention to deriving the quantities which distinguish between the parabolic and the degenerate cases, namely $2cd - be$ and $2ae - bd$, as these quantities are not commonly discussed in the literature.

The idea behind the proof of the Classification Theorem comes from analytic geometry, where one uses rotations and translations of coordinates in order to put the quadratic equation

$$ax^2 + bxy + cy^2 + dx + ey = f$$

into the standard form of the of the equation of a hyperbola (or a pair of lines intersecting at a point), an ellipse (or a point or the empty set), a parabola, or a pair of (possibly coincident) parallel lines. A crucial step in the proof is that, under rotations of coordinates, the **discriminant** $b^2 - 4ac$ is invariant (i.e., unchanged).

Theorem 1 (The Classification Theorem). Consider the second–order, linear PDE

$$aU_{\xi\xi} + bU_{\xi\tau} + cU_{\tau\tau} + dU_{\xi} + eU_{\tau} + kU = F(\xi,\tau), \quad (a^2 + b^2 + c^2 \neq 0) \tag{1}$$

where the unknown function $U = U(\xi,\tau)$ **is** C^2**, and** a, b,c,d,e **and** k **are given real constants and** $F(\xi,\tau)$ **is a given continuous function. Then there is a change of variables of the form**

$$\begin{cases} x = \alpha\xi + \beta\tau \\ t = -\beta\xi + \alpha\tau \\ u(x,t) = \rho^{-1}\exp(\gamma\xi + \delta\tau)U(\xi,\tau)\,, \end{cases} \tag{2}$$

where α, β, γ, δ **and** ρ $(\rho \neq 0)$ **are real constants with** $\alpha^2 + \beta^2 = 1$**, such that (1) is transformed into exactly one of the following forms (where** A **and** K **are real constants and** $A > 0$**) :**

(i) $-A^2 u_{xx} + u_{tt} + Ku = f(x,t)$, (3)

if $b^2 - 4ac > 0$ (**the hyperbolic case**);

(ii) $A^2 u_{xx} + u_{tt} + Ku = f(x,t)$, (4)

if $b^2 - 4ac < 0$ (**the elliptic case**);

(iii) $A^2 u_{xx} + u_t + Ku = f(x,t)$, (5)

if $b^2 - 4ac = 0$, **and** $2cd \neq be$ **or** $2ae \neq bd$, (**the parabolic case**);

(iv) $u_{xx} + Ku = f(x,t)$, (6)

if $b^2 - 4ac = 0$, **and** $2cd = be$ **and** $2ae = bd$, (**the degenerate case**),

where $f(x,t) = \exp[\gamma\xi + \delta\tau]F(\xi,\tau)$, $\xi = \alpha x - \beta t$ **and** $\tau = \beta x + \alpha t$.

Proof. Our strategy is to write the left side of (1) in terms of u_{xx}, u_{xt}, u_{tt}, u_x, u_t and u, and choose the constants α, β, γ, δ and ρ, such that the left side becomes (after multiplication by $\exp[\gamma\xi + \delta\tau]$) one of the desired forms. From this argument, it will be apparent that only one of the forms (3) – (6) can be achieved for the given equation (1). We write

$$U(\xi,\tau) = \rho e^{\eta} u(x,t) \text{ , where } \quad \eta \equiv -(\gamma\xi + \delta\tau).$$

Then

$$U_\xi = \rho\left[e^{\eta}\eta_\xi u + e^{\eta}\eta(u_x x_\xi + u_t t_\xi)\right] = \rho e^{\eta}\left[\alpha u_x - \beta u_t - \gamma u\right] ,$$

and

$$\begin{aligned} U_{\xi\xi} &= \rho e^{\eta}\eta_\xi\left[\alpha u_x - \beta u_t - \gamma u\right] \\ &\quad + \rho e^{\eta}\left[\alpha(u_{xx}x_\xi + u_{xt}t_\xi) - \beta(u_{tx}x_\xi + u_{tt}t_\xi) - \gamma(u_x x_\xi + u_t t_\xi)\right] \\ &= \rho e^{\rho}\left[\alpha^2 u_{xx} - 2\alpha\beta u_{xt} + \beta^2 u_{tt} - 2\alpha\gamma u_x + 2\beta\gamma u_t + \gamma^2 u\right] . \end{aligned}$$

Similarly,

$$U_\tau = \rho e^{\eta}\left[\beta u_x + \alpha u_t - \delta u\right]$$

and

$$U_{\tau\tau} = \rho e^{\rho}\left[\beta^2 u_{xx} + 2\alpha\beta u_{xt} + \alpha^2 u_{tt} - 2\beta\delta u_x - 2\alpha\delta u_t + \delta^2 u\right] .$$

Moreover,

$$U_{\xi\tau} = \rho e^{\eta}\left[\alpha\beta u_{xx} + (\alpha^2 - \beta^2)u_{xt} - \alpha\beta u_{tt} - (\alpha\delta + \beta\gamma)u_x + (\beta\delta - \gamma\alpha)u_t + \delta\gamma u\right] .$$

Substituting the above expressions into (1), we obtain

$$\rho e^{\eta}\Big[Au_{xx} + Bu_{xt} + Cu_{tt} + Du_x + Eu_t + Ku\Big] = F(\xi,\tau)\ , \tag{7}$$

where

$$A = a\alpha^2 + b\alpha\beta + c\beta^2, \tag{8}$$

$$B = 2(c-a)\alpha\beta + b(\alpha^2 - \beta^2), \tag{9}$$

$$C = a\beta^2 - b\alpha\beta + c\alpha^2, \tag{10}$$

$$D = -a2\alpha\gamma - b(\gamma\beta + \delta\alpha) - c2\beta\delta + d\alpha + e\beta\ , \tag{11}$$

$$E = a2\beta\gamma + b(\beta\delta - \alpha\gamma) - c2\alpha\delta - d\beta + e\alpha\ , \tag{12}$$

$$K = a\gamma^2 + b\delta\gamma + c\delta^2 - d\gamma - e\delta + k\ . \tag{13}$$

Using the assumption $\alpha^2 + \beta^2 = 1$, a calculation shows that $B^2 - 4AC = (b^2 - 4ac)(\alpha^2 + \beta^2) = (b^2 - 4ac)$ (i.e., the discriminant is invariant). We can choose α and β, such that $B = 0$. Indeed, since $\alpha^2 + \beta^2 = 1$, there is an angle θ such that $\alpha = \cos\theta$ and $\beta = \sin\theta$. Then, $\alpha^2 - \beta^2 = \cos(2\theta)$ and $2\alpha\beta = \sin(2\theta)$, and we need only to choose θ, such that the unit vector $(\cos(2\theta),\sin(2\theta))$ is orthogonal to the vector $(b,a-c)$, in order that $B = 0$. Then, we have $-4AC = B^2 - 4AC = b^2 - 4ac$. We consider the cases, when $b^2 - 4ac$ is positive, negative or zero.

Case 1 ($-4AC = b^2 - 4ac > 0$). Here A and C must be nonzero and of opposite sign. We choose $\rho = C^{-1}$. Then, dividing (7) by e^{η}, we achieve the form (3) (where $A^2 = -A/C > 0$ and $K = K/C$), provided that we can show that it is possible to choose γ and δ such that $D = 0$ and $E = 0$, or in other words (cf. (11) and (12)),

$$\begin{aligned} (2a\alpha + b\beta)\gamma + (2c\beta + b\alpha)\delta &= d\alpha + e\beta\ , \\ (2a\beta - b\alpha)\gamma + (b\beta - 2c\alpha)\delta &= d\beta - e\alpha\ . \end{aligned} \tag{14}$$

We can solve this system for γ and δ, if the determinant of the coefficient matrix is nonzero. However, using the fact that $\alpha^2 + \beta^2 = 1$, we have

$$\begin{vmatrix} 2a\alpha + b\beta & 2c\beta + b\alpha \\ 2a\beta - b\alpha & b\beta - 2c\alpha \end{vmatrix} = b^2 - 4ac\ . \tag{15}$$

Thus, in the case at hand, we can (uniquely) choose the constants γ and δ, so that $D = E = 0$:

$$\gamma = \frac{1}{b^2 - 4ac}\begin{vmatrix} d\alpha + e\beta & 2c\beta + b\alpha \\ d\beta - e\alpha & b\beta - 2c\alpha \end{vmatrix} = \frac{be - 2cd}{b^2 - 4ac}$$

and

$$\delta = \frac{1}{b^2 - 4ac}\begin{vmatrix} 2a\alpha + b\beta & d\alpha + e\beta \\ 2a\beta - b\alpha & d\beta - e\alpha \end{vmatrix} = \frac{bd - 2ae}{b^2 - 4ac}\ . \tag{16}$$

Case 2 ($-4AC = b^2 - 4ac < 0$). Here A and C are of the same sign. Since $b^2 - 4ac \neq 0$, we may select the constants γ and δ as in (16), so that $D = E = 0$. We set $\rho = C^{-1}$, and divide (7) by e^{η}, thereby achieving the form (4).

Case 3 ($-4AC = b^2 - 4ac = 0$). In this case, we must have $A = 0$ or $C = 0$. Note that A and C cannot both be zero. Indeed, according to equations (8) and (10), we have $0 = A + C = (a + c)(\alpha^2 + \beta^2)$. Then $a = -c$ yields $0 = b^2 - 4ac = b^2 + 4a^2 = b^2 + 2a^2 + 2c^2$ or $a = b = c = 0$, which contradicts the assumption that equation (1) is of order 2. Observe that in formulas (8)–(10), if we replace α by β and β by $-\alpha$, then B changes to $-B$, A changes to C, and C changes to A. Thus, in the present case, if necessary, we can alter the values of α and β, so that $A \neq 0$ and $C = 0$, while $B = 0$ still. Since $b^2 - 4ac = 0$, we are no longer guaranteed that the equations (14) can be solved simultaneously for γ and δ, so as to obtain $D = 0$ and $E = 0$. However, in the first equation of (14), if

$$(2a\alpha + b\beta) \neq 0 \quad \text{or} \quad (2c\beta + b\alpha) \neq 0\ , \tag{17}$$

then we can at least choose values for γ and δ, such that this first equation holds (i.e., $D = 0$). In order to deduce that (17) holds, we note that if $2a\alpha + b\beta = 0$ and $2c\beta + b\alpha = 0$, then we arrive at the contradiction that $A = 0$. Indeed, by (8),

$$2A = 2a\alpha^2 + 2b\alpha\beta + 2c\beta^2 = \alpha(2a\alpha + b\beta) + \beta(2c\beta + b\alpha).$$

Hence, we can find γ and δ such that $D = 0$. Since the determinant (15) vanishes in the case at hand, there is a simultaneous solution (γ,δ) of the system (14) if and only if the numerators in (16) both vanish, i.e.,

$$be = 2cd \quad \text{and} \quad bd = 2ae\ . \tag{18}$$

Thus, it possible to choose γ and δ such that $D = E = 0$ if and only if equations (18) both hold. In that case, we achieve the form (6) by setting $\rho = A^{-1}$ and dividing (7) by e^{η}. In the event that one or both of the equations (18) fails to hold, we can still choose γ and δ, so that $D = 0$. We can also ensure that A and E have the same sign. Indeed, observe that if we replace α by $-\alpha$ and β by $-\beta$, then (12) shows that E changes to $-E$, while A is invariant according to (8) [also note that B, C and D remain 0]. Then, we choose $\rho = E^{-1}$ in order to obtain the form (5) with $A^2 = A/E$. □

Remark. If the coefficients of the PDE (1) are allowed to be functions of ξ and τ, then the type of the equation may vary. For example, the PDE $U_{\xi\xi} - \xi U_{\tau\tau} = 0$ has discriminant $b^2 - 4ac = 4\xi$. Thus, this PDE is hyperbolic for $\xi > 0$, elliptic for $\xi < 0$, and parabolic for $\xi = 0$. □

A.2 Fubini's Theorem

In the text, we have frequently referred to Leibniz's rule to justify differentiating under an integral, and occasionally we have used Fubini's theorem (after the Italian mathematician Guido Fubini (1879–1943), who proved the result around 1910) to interchange the order of integration in multiple integrals. Since Fubini's theorem seems conceptually simpler than Leibniz's rule, we cover Fubini's theorem first. In order to show that some care is needed when interchanging the order of integration, we consider a few examples.

Example 1. Let

$$f(x,y) = \begin{cases} \dfrac{x^2 - y^2}{(x^2 + y^2)^2} & \text{if } (x,y) \neq (0,0) \\ 0 & \text{if } (x,y) = (0,0)\,. \end{cases} \tag{1}$$

We show that

$$\int_0^1 \int_0^1 f(x,y)\,dy\,dx \neq \int_0^1 \int_0^1 f(x,y)\,dx\,dy\,. \tag{2}$$

The left side of (2) is

$$\int_0^1 \left[\frac{y}{x^2 + y^2}\right]_{y=0}^{y=1} dx = \int_0^1 \frac{1}{x^2 + 1}\,dx = \arctan(1) = \frac{\pi}{4}\,.$$

However, the right side of (2) is

$$\int_0^1 \left[\frac{-x}{x^2 + y^2}\right]_{x=0}^{x=1} dy = \int_0^1 \frac{-1}{y^2 + 1}\,dy = -\arctan(1) = -\frac{\pi}{4}\,.$$

Actually, from the fact that $f(y,x) = -f(x,y)$, we can deduce that the two sides of (2) have opposite signs, and so it would have been sufficient to demonstrate that one side is nonzero. □

Example 2. Let $f(x,y) = (2xy - x^2y^2)e^{-xy}$. This function is C^∞ throughout the xy–plane, and yet we will show that

$$\int_0^1 \int_0^\infty f(x,y)\,dy\,dx \neq \int_0^\infty \int_0^1 f(x,y)\,dx\,dy\,. \tag{3}$$

For $x > 0$, we have

$$\int_0^\infty f(x,y)\,dy = \lim_{R\to\infty} \int_0^R (2xy - x^2y^2)e^{-xy}dy = \lim_{R\to\infty} \left[xy^2e^{-xy}\right]\Big|_{y=0}^{y=R} = 0\,.$$

Since $f(0,y) = 0$, we also get the same result when $x = 0$. Thus, the left side of (3) is 0. However, the right side of (3) is

$$\int_0^\infty \int_0^1 (2xy - x^2y^2)e^{-xy}\, dx\, dy = \int_0^\infty \left[x^2ye^{-xy} \right]_{x=0}^{x=1} dy$$

$$= \int_0^\infty ye^{-y}\, dy = \lim_{R\to\infty} \left[-e^{-y} - ye^{-y} \right]_{y=0}^{y=R} = 1 . \quad \square$$

Although these examples show that in general the order of integration can make a difference, the following version of Fubini's theorem gives criteria under which the order of integration can be switched without altering the result. This is not the most general version of Fubini's theorem, but it suffices for all of our applications. □

Fubini's Theorem. Let $-\infty \le a < b \le \infty$ **and** $-\infty \le c < d \le \infty$. **Let** $f(x,y)$ **be a function which is continuous on the region** $a < x < b,\ c < y < d$, **except possibly on a set which is the union of a finite number of lines. Suppose that**

$$\int_c^d \int_a^b |f(x,y)|\, dx\, dy < \infty \quad \textbf{or} \quad \int_a^b \int_c^d |f(x,y|\, dy\, dx < \infty . \tag{4}$$

Then,

$$\int_c^d \int_a^b f(x,y)\, dx\, dy = \int_a^b \int_c^d f(x,y)\, dy\, dx . \tag{5}$$

Remarks. If $f(x,y) \ge 0$, then (5) holds, even if (4) does not hold. Indeed, if $f(x,y) \ge 0$ and (4) does not hold, then both sides of (5) are $+\infty$, since $|f(x,y)| = f(x,y)$. Of course, since (5) fails in Examples 1 and 2, in spite of the fact that both sides of (5) are finite, it must be the case that condition (4) is violated by the functions $f(x,y)$ in those examples, since the continuity condition of Fubini's theorem is met by these functions. We leave the verification that (4) is violated in these examples to the interested reader. With some additional hypotheses, we will give a proof of Fubini's theorem based in part on Leibniz's rule at the end of Appendix A.3. For now, we offer the following intuitive explanation. Both sides of (5) ought to represent the net volume between the graph of the function and the xy–plane, where the volume counts negatively when the graph is below the xy–plane. The two sides merely represent two ways of summing up the volume elements $f(x,y)\, dxdy$. If the total volume, say V^+ above the xy–plane is finite and the total volume below the xy–plane, say V^-, is finite (i.e., if either one of the integrals (4) is finite.), then the two methods of computing the net volume should give the same result, namely $V^+ - V^-$ (i.e., (5) should hold). However, if V^+ and V^- are both infinite, then $V^+ - V^-$ is $\infty - \infty$ which is indeterminate (e.g. $\lim_{x\to\infty} (x + 1) = \infty$ and $\lim_{x\to\infty} (x - 1) = \infty$, but $\lim_{x\to\infty} [(x + 1) - (x - 1)] = 2$ while $\lim_{x\to\infty} [(x - 1) - (x + 1)] = -2$). Indeed, one method of summing up the volume may draw from V^+ more, as it proceeds, than it draws from V^-, while another method may draw more from V^- than it does from V^+. This occurs in Example 1. □

A.3 Leibniz's Rule

Before stating and proving Leibniz's rule, we give some examples which show that taking a derivative under an integral does not always yield the correct result.

Example 1. For each positive value of x, we have

$$I(x) \equiv \int_0^\infty \frac{\sin(xy)}{y}\, dy \equiv \lim_{R\to\infty} \int_0^R \frac{\sin(xy)}{y}\, dy = \frac{\pi}{2}. \tag{1}$$

Indeed, the change of variables from y to $z = xy$ shows that $I(x)$ is independent of the positive value chosen for x, and the result (1) when $x = 1$ is shown in the Supplement at the end of Chapter 7. Thus, we have

$$\frac{d}{dx}\int_0^\infty \frac{\sin(xy)}{y}\, dy = I'(x) = 0. \tag{2}$$

However, if we differentiate under the integral before integrating, we obtain

$$\int_0^\infty \frac{\partial}{\partial x}\left[\frac{\sin(xy)}{y}\right] dy = \int_0^\infty \cos(xy)\, dy = \lim_{R\to\infty} \frac{\sin(xR)}{x}, \tag{3}$$

which does not exist. Since 0 exists, (2) and (3) are unequal, i.e.,

$$\frac{d}{dx}\int_0^\infty \frac{\sin(xy)}{y}\, dy \neq \int_0^\infty \frac{\partial}{\partial x}\left[\frac{\sin(xy)}{y}\right] dy \quad (x > 0).$$

Thus, we do not get the same result, when differentiating under the integral. Indeed, we do not get anything. □

Example 2. Let $f(x,y) = \begin{cases} \dfrac{x-y}{|x-y|} & \text{if } x \neq y \\ 0 & \text{if } x = y \end{cases}$,

(i.e., $f(x,y) = 1$, if $x > y$; $f(x,y) = -1$, if $x < y$; and $f(x,y) = 0$, if $x = y$). The graph of $f(x,y)$ over the region $x > y$ is a horizontal half–plane. Thus, $f_x(x,y) = 0$ for $x > y$. Similarly, $f_x(x,y) = 0$, for $x < y$. When $y = x$, f_x does not exist. Recall that a definite integral of a function (of a single variable) which is redefined or undefined at a finite number of points is unaffected by this change. In particular, for any fixed x, the function $f_x(x,y)$ of the variable y is zero except at a single point (namely, at x) where it is undefined, and consequently its integral with respect to y over any interval is zero. Hence, $\int_0^1 \frac{\partial}{\partial x} f(x,y)\, dy = 0$.

On the other hand,

$$\int_0^1 f(x,y)\,dy = \int_0^x -1\,dy + \int_x^1 1\,dy = -x + (1-x) = 1 - 2x.$$

Thus,

$$\frac{d}{dx}\int_0^1 f(x,y)\,dy = -2 \neq 0 = \int_0^1 \frac{\partial}{\partial x} f(x,y)\,dy\,.$$

Hence, we cannot differentiate under the integral without changing the result. □

Example 3. Let $f(0,0) = 0$ and $f(x,y) = \dfrac{x}{x^2 + y^2}$, for $(x,y) \neq (0,0)$. We have

$$\int_0^1 f(x,y)\,dy = \int_0^1 \frac{x}{x^2 + y^2}\,dy = \int_0^{1/x} \frac{1}{1 + z^2}\,dz = \arctan(\tfrac{1}{x})$$

Thus,

$$\frac{d}{dx}\int_0^1 f(x,y)\,dy = \frac{1}{1 + x^{-2}}(-x^{-2}) = \frac{-1}{1 + x^2}\,.$$

However,

$$\int_0^1 \frac{\partial}{\partial x} f(x,y)\,dy = \int_0^1 \frac{y^2 - x^2}{x^2 + y^2}\,dy = \frac{y}{x^2 + y^2}\Bigg|_{y=0}^{y=1}$$

$$= \frac{1}{x^2 + 1} \neq \frac{d}{dx}\int_0^1 f(x,y)\,dy. \quad □$$

Although the above examples show that differentiation under an integral is unjustified in general, Leibniz's rule gives criteria under which this operation is permissible. Our proof depends on the following elementary version of the Dominated Convergence Theorem for Riemann integrals which gives a criterion under which one can take a limit under an integral. The proof was inspired by [Lewin, 1986, 1988].

Dominated Convergence Theorem. Let $f(x)$ **and** $f_n(x)$ $(n = 1, 2, 3, \ldots)$ **be piecewise continuous functions which are defined on some nonempty interval** I **(which may be open or closed, finite or infinite) of real numbers. Suppose that for each** x, $f(x) = \lim_{n\to\infty} f_n(x)$. **Assume that there is a piecewise continuous function** $g(x) \geq 0$, **such that** $|f_n(x)| \leq g(x)$, **and the integral of** $g(x)$ **over** I, **denoted by** $\int_I g(x)\,dx$, **is finite (We say that the sequence of functions** $f_n(x)$ **is *dominated* by the integrable function** $g(x)$.**) Then,**

$$\lim_{n\to\infty}\int_I f_n(x)\,dx = \int_I \lim_{n\to\infty} f_n(x)\,dx = \int_I f(x)\,dx\,. \tag{4}$$

Proof. Note that

$$\left| \int_I f_n(x)\,dx - \int_I f(x)\,dx \right| \le \int_I |f_n(x) - f(x)|\,dx . \tag{5}$$

Thus, to show (4), it suffices to show that the right–hand side of (5) can be made arbitrarily small, by choosing n sufficiently large. Now by (5) we need only to prove that the theorem is true in the case where $f_n(x) \ge 0$ and $f(x) \equiv 0$ (i.e., replace $f_n(x)$ by $|f_n(x) - f(x)|$, and $g(x)$ by $2g(x)$). Henceforth we assume that $f_n(x) \ge 0$ and $f(x) \equiv 0$. Let L be any positive real number, and let J be the set of points x in I such that $|x| > L$. Since $\int_I g(x)\,dx < \infty$, we can make $\int_J g(x)\,dx$ arbitrarily small by choosing L large enough. Moreover, since $0 \le f_n(x) \le g(x)$, $\int_J f_n(x)\,dx$ can also be made arbitrarily small (for all n simultaneously) by choosing L large enough. Let K be the set of points in I which are not in J. Thus, K is a finite interval contained in $[-L,L]$. It remains to prove that $\lim_{n\to\infty} \int_K f_n(x)\,dx = 0$. A **step function** defined on K, is a function which is constant on each of a finite number of subintervals of K and is zero elsewhere in K. Since piecewise continuous functions on a closed interval are Riemann integrable, we know that $\int_K f_n(x)\,dx$ may be approximated arbitrarily closely by a lower sum. In other words, there is a step function $s_n(x)$, such that $0 \le s_n(x) \le f_n(x)$ for all x in K, and $0 \le \int_K f_n(x)\,dx \le \int_K s_n(x)\,dx + n^{-1}$. Thus, it suffices to show that $\lim_{n\to\infty} \int_K s_n(x)\,dx = 0$. Let $G(x)$ be a step function, defined on K, such that $G(x) \ge g(x)$ for all x in K, and $\int_K G(x)\,dx \le \int_K g(x)\,dx + 1 < \infty$. For any fixed $\epsilon > 0$, let

$$E_n \equiv \{\, x \in K : s_n(x) > \epsilon \cdot G(x) \,\} \quad \text{and} \quad T_n \equiv \{\, x \in K : s_n(x) \le \epsilon \cdot G(x) \,\} .$$

Since s_n and G are step functions, E_n and T_n are unions of finitely many intervals. Then

$$\begin{aligned} \int_K s_n(x)\,dx &= \int_{E_n} s_n(x)\,dx + \int_{T_n} s_n(x)\,dx \le \int_{E_n} s_n(x)\,dx + \int_{T_n} \epsilon \cdot G(x)\,dx \\ &\le \int_{E_n} G(x)\,dx + \epsilon \cdot \left[\int_K g(x)\,dx + 1 \right] . \end{aligned}$$

Thus,

$$\int_K s_n(x)\,dx \le \mu(E_n) + \epsilon \left[\int_K g(x)\,dx + 1 \right] , \tag{6}$$

where $\mu(A) \equiv \int_A G(x)\,dx$, for any subset A of K, which is a union of finitely many intervals.

We call such subsets A **elementary**. Since ϵ is arbitrary, it suffices to show $\lim_{n\to\infty} \mu(E_n) = 0$.

Let $\quad F_n \equiv \cup_{m\geq n} E_m = \{x \in K : s_m(x) > \epsilon \cdot G(x)$ for at least one $m \geq n\}$. $\quad$ (7)

Note that $F_1 \supset F_2 \supset F_3 \supset \ldots$ (i.e. the sets F_n "shrink" as n increases). There is no point which is in all of the sets F_n, n = 1, 2, 3, ... , because $f_n(x)$ converges to 0 and $f_n(x) \geq s_n(x)$ on K. (Even if $\epsilon \cdot G(x) = 0$ for some x, we still have $0 \leq s_n(x) \leq f_n(x) \leq g(x) \leq G(x) = 0$, i.e. $s_n(x) = 0$ for all n, and thus x is in none of the sets F_n.) Recall that we want to show that $\lim_{n\to\infty} \mu(E_n) = 0$. Suppose, on the contrary, that there is a constant c, such that $\mu(E_n) > c > 0$ for infinitely many values of n. For each n = 1, 2, 3, ... , let H_n be a subset of F_n, such that H_n consists of a finite number of *closed* intervals and

$$\mu(H_n) > \mu(A) - c2^{-n}, \quad (8)$$

for every elementary subset A of F_n. Note that $\mu(A) < \mu(K) < \infty$, so that H_n exits. Let $Z_n \equiv \cap_{m\leq n} H_m = H_1 \cap H_2 \cap \ldots \cap H_n$. Then the Z_n form a shrinking sequence of closed, bounded sets and they must have a point of intersection if they are all nonempty. [For each n, select a point p_n from Z_n and note that p_n will have a subsequence which converges to a point in each of the closed bounded sets Z_n by the Bolzano–Weierstrass Theorem of Appendix A.4 .] However, there can be no point in every Z_n, since $Z_n \subset H_n \subset F_n$ and the sets F_n have no common point. Thus, Z_{n_0} must be empty for some n_0, and hence Z_n is empty for all $n \geq n_0$. However, for $n \geq n_0$ and $0 \leq i \leq n$, E_n is a subset of F_i, and $H_i \cup E_n$ is a subset of F_i consisting of a finite number of intervals. By (6), $\mu(H_i) \geq \mu(E_n \cup H_i) - c2^{-i} = \mu(E_n - H_i) + \mu(H_i) - c2^{-i}$ or $\mu(E_n - H_i) \leq c2^{-i}$. Then

$$\mu(E_n) = \mu(E_n - Z_n) \leq \mu(E_n - H_1) + \ldots + \mu(E_n - H_n) \leq (2^{-1} + \ldots + 2^{-n})c < c. \quad (9)$$

for $n > n_0$. Now (9) contradicts our assumption that $\mu(E_n) > c$ for infinitely many n. Thus, our assumption that $\lim_{n\to\infty} \mu(E_n) \neq 0$ is false, and $\lim_{n\to\infty} \mu(E_n) = 0$ is true. Hence, by (6), we can make $\int_K s_n(x)\,dx$ as small as desired, by first choosing ϵ small and then choosing n sufficiently large. □

In many situations it is convenient to have the following "continuous" version of the Dominated Convergence Theorem, which follows rather easily from the above "discrete" version.

Corollary (Dominated Convergence Theorem – continuous version). For each real number h in some interval H containing h_0, let $F(x,h)$ be a piecewise continuous function of x defined on some nonempty interval I (which may be open or closed, finite or infinite) of real numbers. Suppose that $f(x) \equiv \lim_{h\to h_0} F(x,h)$ is piecewise continuous on I. Assume that there is a piecewise continuous function $g(x) \geq 0$, such that $|F(x,h)| \leq g(x)$ for all $x \in I$ and $h \in H$, and the integral of $g(x)$ over I, denoted by $\int_I g(x)\,dx$, is finite. Then,

$$\lim_{h\to h_0} \int_I F(x,h)\,dx = \int_I \lim_{h\to h_0} F(x,h)\,dx = \int_I f(x)\,dx . \tag{4'}$$

Proof. Let $h_1, h_2, \ldots$ be any sequence of points in H, such that $\lim_{n\to\infty} h_n = h_0$. Let $f_n(x) = F(x,h_n)$. Since the $f(x)$ and $f_n(x)$ satisfy the hypotheses of the Dominated Convergence Theorem and the sequence $h_1, h_2, \ldots$ approaching h_0 is arbitrary, we have

$$\lim_{h\to h_0} \int_I F(x,h)\,dx = \lim_{n\to\infty} \int_I f_n(x)\,dx = \int_I f(x)\,dx = \int_I \lim_{h\to h_0} F(x,h)\,dx . \quad \square$$

Theorem (Leibniz's rule). Let R be the region $-\infty \leq a < x < b \leq +\infty$, $-\infty \leq c < t < d \leq +\infty$ in the xt–plane. Let $f(x,t)$ be a continuous function defined on R, such that $f_t(x,t)$ is also continuous on R. Moreover, assume that $\int_c^d |f(x,t)|\,dx < \infty$ for each x in (a,b). Suppose that there is a piecewise continuous function $g(x)$, such that for all (x,t) in R,

$$|f_t(x,t)| \leq g(x) \quad \text{and} \quad \int_c^d g(x)\,dx < \infty . \tag{10}$$

Then,

$$\frac{d}{dt}\int_c^d f(x,t)\,dx = \int_c^d \frac{\partial}{\partial t} f(x,t)\,dx \quad (a < t < b). \tag{11}$$

If c and d are finite, and $f(x,t)$ and $f_t(x,t)$ are continuous on the region $c \leq x \leq d$ and $a < t < b$ (*including the edges* $x = c$ and $x = d$), then we do not need (10) in order to get (11).

Proof. From the definition of derivative,

$$\frac{d}{dt}\int_c^d f(x,t)\,dx = \lim_{h\to 0}\frac{1}{h}\left[\int_c^d f(x,t+h)\,dx - \int_c^d f(x,t)\,dx\right] = \lim_{h\to 0}\int_c^d \frac{f(x,t+h)-f(x,t)}{h}\,dx .$$

We justify taking the limit under the integral, as follows. Using the Fundamental Theorem of Calculus, and the assumption, $|f_t(x,t)| \le g(x)$, we have

$$|f(x,t+h) - f(x,t)| = \left|\int_t^{t+h} f_t(x,s)\,ds\right| \le \int_t^{t+h} |f_t(x,s)|\,ds \le g(x)\cdot h$$

or

$$\left|\frac{f(x,t+h) - f(x,t)}{h}\right| \le g(x) .$$

Thus, we may apply (4′) (cf. the above corollary) to obtain

$$\frac{d}{dt}\int_c^d f(x,t)\,dx = \lim_{h\to 0}\int_c^d \frac{f(x,t+h)-f(x,t)}{h}\,dx = \int_c^d \lim_{h\to 0}\frac{f(x,t+h)-f(x,t)}{h}\,dx = \int_c^d \frac{\partial}{\partial t}f(x,t)\,dx .$$

We now verify the last assertion of the theorem. For any $t \in (a,b)$, select α and β such that $a < \alpha \le t \le \beta < b$. If c and d are *finite* and $f_t(x,t)$ is continuous on the closed rectangular region $c \le x \le d$, $a < \alpha \le t \le \beta < b$, then we can take the function $g(x)$ (required in (10), but now with $t \in (\alpha,\beta)$) to be constant function, equal to the maximum of $|f_t(x,t)|$ on this closed rectangular region. The existence of this maximum is established independently in Appendix A.4, where it is proved that a continuous function (e.g., $|f_t|$) on a closed rectangle has a maximum. Letting the maximum be M, we have $\int_c^d g(x)\,dx = M(d-c) < \infty$. Thus, the assumptions in (10) are satisfied in this case, and the last assertion of the theorem is proved. □

Remark (Generalizations). There is a more general version of Leibniz's rule which states that if $f(x,t)$ satisfies the assumptions in the above theorem, and if $c(t)$ and $d(t)$ are C^1 functions with $c < c(t)$ and $d(t) < d$, for $a < t < b$, then

$$\frac{d}{dt}\int_{c(t)}^{d(t)} f(x,t)\,dx = f(d(t),t)d'(t) - f(c(t),t)c'(t) + \int_{c(t)}^{d(t)} f_t(x,t)\,dx . \tag{12}$$

The proof of this result proceeds just as the proof of the Lemma preceding Theorem 1 (Duhamel's principle) in Section 3.4. In other words, one considers a function $H(w,z,t) = \int_w^z f(x,t)\,dx$, and (12) follows at once from the chain rule $\frac{d}{dt}\left[H(c(t),d(t),t)\right] = H_w\,c'(t) + H_z\,d'(t) + H_t$, the Fundamental Theorem of Calculus, and Leibniz's rule. We also mention that the Dominated Convergence Theorem and Leibniz's rule extend without much further difficulty to the case of multiple integrals. Moreover, the above proof of Leibniz's rule is still valid when $f(x,t)$ is continuous in t, but only piecewise continuous in x, and $f_t(x,t)$ is piecewise continuous in t. As a consequence, we easily prove the following useful result concerning termwise differentiation of an infinite series of functions. □

Theorem (Termwise differentiation of series). **Let** $f_1(t), f_2(t), \ldots$ **be a sequence of** C^1 **functions. Suppose that there are constants** M_n, **such that** $|f_n'(t)| \leq M_n$ **and** $\sum_{n=1}^{\infty} M_n < \infty$. **Then**

$$\frac{d}{dt}\left[\sum_{n=1}^{\infty} f_n(t)\right] = \sum_{n=1}^{\infty} f_n'(t) . \tag{13}$$

Proof. We define $f(x,t) = f_n(t)$ if $n-1 \leq x < n$, $n = 1, 2, \ldots$. Observe that $f_t(x,t)$ is continuous in t and piecewise continuous in x (cf. the end of the above Remark). Also, let $g(x) = M_n$ if $n-1 \leq x < n$, $n = 1, 2, \ldots$. Then $|f_t(x,t)| \leq g(x)$ and $\int_0^{\infty} g(x)\, dx = \sum_{n=1}^{\infty} M_n < \infty$. Thus, Leibniz's rule yields,

$$\frac{d}{dt}\left[\sum_{n=1}^{\infty} f_n(t)\right] = \frac{d}{dt}\left[\int_0^{\infty} f(x,t)\, dx\right] = \int_0^{\infty} f_t(x,t)\, dx = \sum_{n=1}^{\infty} f_n'(t) . \quad \square$$

Leibniz's rule is also used in the following proof of Fubini's theorem :

Proof of Fubini's theorem (cf. the statement in Appendix A.2) with some new assumptions. In addition to the assumptions in Fubini's theorem of Appendix A.2, here we assume that :

1. $f(x,y)$ is bounded on any closed subrectangle of $a < x < b$, $c < y < d$.
2. The function $f(x,y)$ is piecewise continuous in x and in y.
3. For all α and β with $a \leq \alpha < \beta \leq b$, the integrals $\int_{\alpha}^{\beta} f(x,y)\, dx$ and $\int_{\alpha}^{\beta} |f(x,y)|\, dx$ are piecewise continuous functions of y.
4. For all γ and δ with $c \leq \gamma < \delta \leq d$, the integrals $\int_{\gamma}^{\delta} f(x,y)\, dy$ and $\int_{\gamma}^{\delta} |f(x,y)|\, dy$ are piecewise continuous functions of x.

Suppose that $\int_c^d \int_a^b |f(x,y)|\, dx\, dy < \infty$. We prove that $\int_c^d \int_a^b f(x,y)\, dx\, dy = \int_a^b \int_c^d f(x,y)\, dy\, dx$ (The other case follows by symmetry). For $c < y < d$, let $G(x,y) = \int_c^y f(x,z)\, dz$, and for $a < \alpha < \beta < b$, let $H(y) = \int_{\alpha}^{\beta} G(x,y)\, dx$. For any value x for which $f(x,y)$ is a continuous function of y at $y = y_0$, we have $G_y(x,y_0) = f(x,y_0)$ by the Fundamental Theorem of Calculus. Thus, by the assumptions on $f(x,y)$, $G_y(x,y)$ is a piecewise continuous function of x and of y, which is bounded on any closed subrectangle. Moreover, by using the boundedness of $f(x,y)$ on closed subrectangles, we see that $G(x,y)$ is a continuous function of y for each x (Why ?). Thus, by Leibniz's rule (cf. also the end of the above Remark),

$$H'(y) = \int_\alpha^\beta G_y(x,y)\,dx = \int_\alpha^\beta f(x,y)\,dx\,.$$

The Fundamental Theorem of Calculus then yields,

$$\int_\alpha^\beta \int_c^d f(x,y)\,dy\,dx = H(d) = \int_c^d H'(y)\,dy = \int_c^d \int_\alpha^\beta G_y(x,y)\,dx\,dy = \int_c^d \int_\alpha^\beta f(x,y)\,dx\,dy\,.$$

Now,

$$\int_\alpha^b \int_c^d f(x,y)\,dy\,dx = \lim_{\beta \to b} \int_\alpha^\beta \int_c^d f(x,y)\,dy\,dx = \lim_{\beta \to b} \int_c^d \int_\alpha^\beta f(x,y)\,dx\,dy\,.$$

Let $F(\beta,y) = \int_\alpha^\beta f(x,y)\,dx$, for $c < y < d$, $\alpha < \beta < b$. Then $|F(\beta,y)| \leq g(y) \equiv \int_a^b |f(x,y)|\,dx$ and $\int_c^d g(y)\,dx = \int_c^d \int_a^b |f(x,y)|\,dx\,dy < \infty$, by assumption. Thus, the Dominated Convergence Theorem yields

$$\lim_{\beta \to b} \int_c^d \int_\alpha^\beta f(x,y)\,dx\,dy = \lim_{\beta \to b} \int_c^d F(\beta,y)\,dy = \int_c^d \lim_{\beta \to b} F(\beta,y)\,dy = \int_c^d \int_\alpha^b f(x,y)\,dx\,dy\,.$$

Using a similar argument for the limit $a \to \alpha$, the desired result follows. □

A.4 The Maximum/Minimum Theorem

Not every function has a maximum or minimum. For example, the function $f(x) = x^3$ (for $-\infty < x < \infty$) has no maximum or minimum. Also, even though the function $f(x) = \arctan(x)$ has values which are abitrarily close to $\pi/2$, this value is never achieved for any real value of x. Our goal in this section is to find criteria which ensure that a function of several variables achieves maximum and minimum values at points in a prescribed region. For concreteness, we deal with subsets of the xy–plane (i.e., subsets of $\mathbb{R}^2$) and functions f(x,y), but all of our considerations can easily be generalized to the case of n–variables, n = 1, 2, 3,... . A subset C of $\mathbb{R}^2$ is called **closed**, if every point, which is the limit of a convergent sequence of points in C, is itself in C. For example, a rectangular region is closed, only if the edges are included in the region. A subset of $\mathbb{R}^2$ is **bounded**, if it can enclosed in a sufficiently large (but finite) square. For example, a disk is bounded, but a half–plane is not bounded. A function, defined on a subset C of $\mathbb{R}^2$, is **continuous,** if, for every sequence of points $p_n = (x_n,y_n)$ (n = 1, 2, 3,...) in C, which converges to some point $p = (x,y)$ in C, we have that the sequence of values $f(p_n)$ converges to the value f(p). (There are other equivalent definitions of continuity, but this one is convenient for our purposes.) Our main goal is to prove the Maximum/Minimum Theorem which says that a function, which is defined and continuous on a closed, bounded subset, has a maximum and a minimum value. For the proof, we need the following result.

Theorem 1 (The Bolzano–Weierstrass Theorem for $\mathbb{R}^2$). Let $p_n = (x_n,y_n)$ (n = 1, 2, 3, ...) **be a sequence of points which form a bounded subset** B **of** $\mathbb{R}^2$. **Then the sequence** $p_1, p_2, p_3, \ldots$ **has a convergent subsequence.**

Proof. For convenience, we uniformly shrink and/or translate the bounded set B, so that it lies in the unit square $\{ (x,y) : 0 < x, y < 1\}$. Let $D \equiv \{0, 1, 2,\ldots,9\}$ be the set of digits. There are digits a_1 and b_1 in D, such that for infinitely many values of n, the decimal expansion of x_n begins with $.a_1$ and the decimal expansion of y_n begins with $.b_1$. Let N_1 be the set of all such values of n. There are digits a_2 and b_2 such that for infinitely many values of n in N_1, the decimal expansion of x_n and y_n begin with $.a_1a_2$ and $.b_1b_2$, respectively. Let N_2 be the (infinite) subset of N_1 consisting of such values of n. Similarly, we can construct infinite sets $N_3, N_4, N_5,\ldots$(each contained in the previous one). Let $q_1 = p_{n_1}$ for some n_1 in N_1, and let $q_2 = p_{n_2}$ for some n_2 (greater than n_1) in N_2, and define $q_3, q_4,\ldots$ similarly. Let $q = (.a_1a_2a_3\ldots , .b_1b_2b_3\ldots)$. Since the distance from q_n to q is $\leq [(10^{-n})^2 + (10^{-n})^2]^{\frac{1}{2}} = \sqrt{2}\, 10^{-n}$, the subsequence $q_1, q_2, q_3, \ldots$ converges to q. □

The Maximum/Minimum Theorem. **Let** $f(x,y)$ **be a continuous function defined on a closed, bounded subset** C **of** $\mathbb{R}^2$. **Then there are points** $(\underline{x},\underline{y})$ **and** $(\overline{x},\overline{y})$ **in** C, **such that**

$$f(\underline{x},\underline{y}) \leq f(x,y) \leq f(\overline{x},\overline{y})$$

for all (x,y) **in** C. **In particular,** $|f(x,y)| \leq K$ **for some constant** K.

Proof. We first prove that there is a constant M such that $f(x,y) < M$. If M does not exist, then we can find a sequence of points $p_1, p_2, p_3, \ldots$ such that $f(p_n) > n$. By Theorem 1, there is a subsequence $q_1, q_2, q_3, \ldots$, which converges to some point q in the closed, bounded set C. However, since f is assumed to be continuous, we have that $f(q) = \lim_{n\to\infty} f(q_n)$ which does not exist. This contradicts the assumption that f is defined at all points of C. Let S be the set of all values $f(x,y)$ as (x,y) varies throughout C. For any real number r and positive integer n, let $r[n]$ be the truncation of r at the n–th decimal place (i.e., $\sqrt{2}\,[3] = 1.414$). Let s_n be a number in S, such that $s_n[n]$ is as large as possible. The existence of s_n is ensured by the existence of the constant M which is larger than all members of S. Let s be the real number such that $s[n] = s_n[n]$ for all $n = 1, 2, 3, \ldots$. By construction, all numbers in S are less than or equal to s. Thus, if we can show that there is a point $(\overline{x},\overline{y})$ in C, such that $f(\overline{x},\overline{y}) = s$, then $f(x,y) \leq f(\overline{x},\overline{y})$, for all (x,y) in C. Let $P_n = (x_n,y_n)$ be a point in C, such that $f(x_n,y_n) = s_n$, defined above. By Theorem 1, the sequence $P_1, P_2, P_3, \ldots$ has a subsequence $Q_1, Q_2, Q_3, \ldots$ (say $Q_i = P_{n_i}$, $i = 1, 2, \ldots$) which converges to some point Q in the closed, bounded set C. By the continuity of f, we have $f(Q) = \lim_{i\to\infty} f(Q_i) = \lim_{i\to\infty} s_{n_i} = s$. Thus, Q serves as a point $(\overline{x},\overline{y})$, such that $f(x,y) \leq f(\overline{x},\overline{y})$. Since $-f(x,y)$ is also continuous on C, we now know that there is also some point $(\underline{x},\underline{y})$ such that $-f(x,y) \leq -f(\underline{x},\underline{y})$ (or $f(\underline{x},\underline{y}) \leq f(x,y)$) for all (x,y) in C. □

Remark. A more common proof of the Maximum/Minimum Theorem is as follows. Let M_0 be the smallest constant which is greater than or equal to all of the values $f(x,y)$ as (x,y) ranges over C. Since we have shown that $f(x,y) < M$ for some M, the existence of M_0 follows from the so–called **least upper bound axiom**, which we purposely avoided in the above "constructive" proof. We need only to show that $M_0 = f(x_0,y_0)$ for some (x_0,y_0) in C. If this is not the case, then $g(x,y) = [M_0 - f(x,y)]^{-1}$ is continuous on C (Why ?). Thus, $g(x,y) < K$, for some positive constant K. Hence $M_0 - f(x,y) = g(x,y)^{-1} > K^{-1}$ or $f(x,y) < M_0 - K^{-1}$, which contradicts the definition of M_0. □

A.5 A Table of Fourier Transforms*

$$\hat{f}(\xi) = \frac{1}{\sqrt{2\pi}} \int_{-\infty}^{\infty} f(x) e^{-i\xi x}\, dx$$

f(x)	$\hat{f}(\xi)$
1. $f'(x)$	$i\xi \hat{f}(\xi)$
2. $f^{(m)}(x)$ (m = 0, 1, 2, ...)	$i^m \xi^m \hat{f}(\xi)$
3. $xf(x)$	$i\frac{d}{d\xi}\hat{f}(\xi)$
4. $x^m f(x)$ (m = 0, 1, 2, ...)	$i^m \frac{d^m}{d\xi^m}\hat{f}(\xi)$
5. $f(x+a)$ (a real)	$e^{ia\xi}\hat{f}(\xi)$
6. $e^{ibx}f(ax)$ (a, b real, $a \neq 0$)	$\frac{1}{a}\hat{f}(\frac{\xi-b}{a})$
7. $(f*g)(x) \equiv \int_{-\infty}^{\infty} f(y-x)g(y)\, dy$	$\sqrt{2\pi}\,\hat{f}(\xi)\hat{g}(\xi)$
8. $f(x) = \begin{cases} 1 & \text{if } \lvert x\rvert \leq L \\ 0 & \text{if } \lvert x\rvert > L \end{cases}$	$\frac{1}{\sqrt{2\pi}}\frac{2\sin(\xi L)}{\xi}$
9. $f(x) = \begin{cases} L - \lvert x\rvert & \text{if } \lvert x\rvert \leq L \\ 0 & \text{if } \lvert x\rvert \geq L \end{cases}$	$\sqrt{2/\pi}\,\frac{1 - \cos(\xi L)}{\xi^2}$
10. $e^{-\frac{1}{2}ax^2}$ (a > 0)	$\frac{1}{\sqrt{a}} e^{-\frac{1}{2}\xi^2/a}$
11. $e^{-a\lvert x\rvert}$ (a > 0)	$\frac{1}{\sqrt{2\pi}}\frac{2a}{a^2+\xi^2}$
12. $\frac{1}{a^2+x^2}$ (a > 0)	$\frac{\sqrt{2\pi}}{2a} e^{-a\lvert\xi\rvert}$
13. $e^{-a\lvert x\rvert}\cos(bx)$ (a,b > 0)	$\frac{a}{\sqrt{2\pi}}\left[\frac{1}{[a^2+(b-\xi)^2]} + \frac{1}{[a^2+(b+\xi)^2]}\right]$
14. $e^{-\frac{1}{2}ax^2}\sin(bx)$ (a,b > 0)	$\frac{1}{2i\sqrt{a}}\left[e^{-\frac{1}{2}(\xi-b)^2/a} - e^{-\frac{1}{2}(\xi+b)^2/a}\right]$
15. $\delta(x-c)$ (c real)	$e^{-ic\xi}/\sqrt{2\pi}$

* The results tabulated here are derived in Chapter 7.

A.6 Bessel Functions

In the text, Bessel functions are discussed in Sections 9.2, 9.3 and 4.4 . A classical reference for the detailed treatment of Bessel functions is G. N. Watson, Bessel Functions, 2nd ed., Cambridge, Cambridge University Press, 1958 (see also E.T. Whittaker and G.N. Watson, 4th ed. , A Course of Modern Analysis Cambridge, Cambridge University Press, 1952). The following is a collection of some of the basic definitions and properties of Bessel functions.

I. **Bessel's (differential) equation of real order** ν is

$$x^2y'' + xy' + (x^2 - \nu^2)y = 0 . \tag{1}$$

The **Bessel function of the first kind of order** ν, satisfies (1) and is defined by

$$J_\nu(x) = (\tfrac{1}{2}x)^\nu \sum_{k=0}^{\infty} \frac{(-\frac{1}{4}x^2)^k}{k!\Gamma(1+\nu+k)} , \tag{2}$$

where Γ is the **Gamma function,** $\Gamma(s) \equiv \int_0^\infty e^{-x}x^{s-1}\,dx$, $s > 0$, and $\Gamma(1+n) = n!$, if $n = 1, 2, \dots$. The reciprocal of the Gamma function uniquely extends to an analytic function "$1/\Gamma(z)$" which is defined for all complex z, and $1/\Gamma(z)$ is zero only for $z = 0, -1, -2, \dots$. This fact can be used to make sense of (2), even when ν is negative.

When ν is *not* an integer, the functions $J_{-\nu}(x)$ and $J_\nu(x)$ are two independent solutions of (1), and any other solution will be a linear combination of them. One standard linear combination is the **Bessel function** $Y_\nu(x)$ **of the second kind,** defined by

$$Y_\nu(x) = \frac{J_\nu(x)\cos(\nu\pi) - J_{-\nu}(x)}{\sin(\nu\pi)} , \qquad \nu \neq 0, \pm1, \pm2, \dots . \tag{3}$$

For integers $m, k \geq 0$, recall from above that $1/\Gamma(1-m+k) = 0$ for $1-m+k \leq 0$ or $k \leq m-1$. Thus, with $\nu = -m$, (1) yields $J_{-m}(x) = (-1)^m J_m(x)$. Thus, $J_m(x)$ and $J_{-m}(x)$ are *not* linearly independent, if m is an integer. However, a solution of (1) with $\nu = m$, which is linearly independent of $J_m(x)$, is the **Bessel function** $Y_m(x)$ **of the second kind of integer order** m, defined as the limit of $Y_\nu(x)$ in (3) as $\nu \to m$. In all cases, $c_1J_\nu(x) + c_2Y_\nu(x)$ is the general solution of (1). For $m = 0, 1, 2, \dots$,

$$Y_m(x) = \frac{2}{\pi}\log(\tfrac{1}{2}x)\, J_m(x) - \frac{1}{\pi}(\tfrac{1}{2}x)^{-m} \sum_{k=0}^{m-1} \frac{(m-k-1)!}{k!}(\tfrac{1}{4}x^2)^k$$
$$-\frac{1}{\pi}(\tfrac{1}{2}x)^m \sum_{k=0}^{\infty} \frac{(-\frac{1}{4}x^2)^k}{k!(m+k)!}\Big[\psi(k+1) + \psi(m+k+1)\Big], \qquad (4)$$

where the first sum is suppressed if $m = 0$, and $\psi(n) \equiv -\gamma + \sum_{q=1}^{n-1} \frac{1}{q}$ (for $n = 2, 3, \ldots$), where $\gamma = 0.57721566\ldots$ is **Euler's constant** $[\ \gamma \equiv \lim_{n\to\infty} (\sum_{q=1}^{n-1} \frac{1}{q} - \log n)\ ]$ and $\psi(1) \equiv -\gamma$. For integers $m \le 0$, it follows from (2) that $Y_{-m}(x) = (-1)^m Y_m(x)$, upon taking a limit as $\nu \to -m$.

II. Some basic properties of Bessel functions of integer order m

1. Differential recurrence relations :

$$\frac{d}{dx}\Big[x^m J_m(x)\Big] = x^m J_{m-1}(x) \quad \text{and} \quad \frac{d}{dx}\Big[x^m Y_m(x)\Big] = x^m Y_{m-1}(x) \qquad (m \ge 1)\,. \qquad (5)$$

2. Pure recurrence relations :

$$2m\, J_m(x) = x[J_{m-1}(x) + J_{m+1}(x)]$$

and

$$2m\, Y_m(x) = x[Y_{m-1}(x) + Y_{m+1}(x)]\,. \qquad (m > 1) \qquad (6)$$

3. Bessel's integral for $J_m(x)$ $(m = 0, 1, 2, \ldots)$:

$$J_m(x) = \frac{1}{\pi}\int_0^{\pi} \cos(mt - x \sin t)\, dt\,. \qquad (7)$$

4. A generating function for $J_m(x)$:

$$f(x,t) \equiv \exp\Big[\frac{x}{2}(t - \frac{1}{t})\Big] = \sum_{m=-\infty}^{\infty} J_m(x) t^m \qquad (t \ne 0)\,. \qquad (8)$$

Since $f(x,t)f(-x,t) \equiv 1$ $(t \ne 0)$, it follows that

$$J_0(x)^2 + 2\sum_{m=0}^{\infty} J_m(x)^2 = 1\,, \qquad (9)$$

and hence $|J_0(x)| \le 1$ and $|J_m(x)| \le \sqrt{\tfrac{1}{2}}$ $(m = 1, 2, 3, \ldots;\ x \text{ real})$. (10)

5. Asymptotic bevavior : For fixed m, as $x \to \infty$, we have

$$J_m(x) = \left[\frac{2}{\pi x}\right]^{\frac{1}{2}} \cos(x - \tfrac{1}{2}\pi m - \tfrac{1}{4}\pi) + O(x^{-\frac{3}{2}}) \quad (x \to \infty) \tag{11}$$

$$Y_m(x) = \left[\frac{2}{\pi x}\right]^{\frac{1}{2}} \sin(x - \tfrac{1}{2}\pi m - \tfrac{1}{4}\pi) + O(x^{-\frac{3}{2}}) \quad (x \to \infty). \tag{12}$$

III. Half–order Bessel functions of the first kind

$$J_{\frac{1}{2}}(x) = \left[\frac{2}{\pi x}\right]^{\frac{1}{2}} \sin x , \qquad J_{-\frac{1}{2}}(x) = \left[\frac{2}{\pi x}\right]^{\frac{1}{2}} \cos x , \tag{13}$$

$$J_{m+\frac{3}{2}}(x) = -\, x^{m+\frac{1}{2}} \frac{d}{dx}\left[x^{-(m+\frac{1}{2})} J_{m+\frac{1}{2}}(x)\right] \tag{14}$$

$$(m = 0, 1, 2, ...)$$

$$J_{m-\frac{1}{2}}(x) = x^{-(m+\frac{1}{2})} \frac{d}{dx}\left[x^{(m+\frac{1}{2})} J_{m+\frac{1}{2}}(x)\right] . \tag{15}$$

IV. A table of zeros of $J_m(x)$ (m = 0, 1, 2, 3, 4)*

$J_m(x)$ (m = 0, 1, 2,...) has an infinite number of zeros, all of which are real. If $j_{m,k}$ denotes the k–th positive zero of $J_m(x)$, arranged in increasing order, then (cf. (11))

$$j_{m,k} \approx (k + \tfrac{1}{2}m - \tfrac{1}{4})\pi , \quad \text{for} \quad k >> m . \tag{16}$$

m	$j_{m,1}$	$j_{m,2}$	$j_{m,3}$	$j_{m,4}$	$j_{m,5}$
0	2.404825...	5.520078...	8.653727...	11.791534...	14.930917...
1	3.831706...	7.015586...	10.173468...	13.323691...	16.470630...
2	5.135622...	8.417244...	11.619841...	14.795951...	17.959819...
3	6.380161...	9.761023...	13.015200...	16.223464...	19.409414...
4	7.588342...	11.064709...	14.372536...	17.615966...	20.826933...

A table of some zeros of $J_m(x)$ for m = 0, 1, 2, 3, 4 .

*We used Macsyma on a VAX/785 to obtain the zeros tabulated here.

REFERENCES

REFERENCES

A

Abbott, M. B., *An introduction to the method of characteristics*, American Elsevier Pub. Co., New York, 1966.

Abraham, R., Marsden, J.E., and Ratiu, T., *Manifolds, Tensor Analysis, and Applications*, 2nd ed., Springer-Verlag, New York, 1988.

Agmon, S., *Lectures on elliptic boundary value problems*, Van Nostrand, Princeton, N.J., 1965.

Ahlfors, L. V., *Complex analysis ; an introduction to the theory of analytic functions of one complex variable*, 2nd ed., McGraw-Hill, New York, 1966.

Ames, W. F., *Numerical methods for partial differential equations*, Nelson, London, 1969.

Ames, W. F., *Nonlinear partial differential equations in engineering*, Academic Press, New York, 1972.

Ames, W. F. and Rogers, C., *Nonlinear boundary value problems in science and engineering*, Academic Press, Boston, 1989.

Andrews, L. C., *Elementary partial differential equations with boundary value problems*, Academic Press, Orlando, 1986.

Andrews, L. C., *Special functions for engineers and applied mathematicians*, MacMillan, New York, 1985.

Antman, S. S., The equations for large vibrations of strings, Amer. Math. Monthly **87** (1980), 359-370.

Apostol, T. M., *Calculus*, vols. 1 and 2, 2nd ed., Blaisdell Pub. Co., Waltham, Mass., 1969.

Apostol, T. M., *Mathematical analysis; a modern approach to advanced calculus*, Addison-Wesley, Reading, Mass., 1957.

Avery, J., *Hyperspherical harmonics : applications in quantum theory*, Kluwer Academic, Boston, 1989.

Axelsson, O. and Barker, V. A., *Finite element solution of boundary value problems : theory and computation*, Academic Press, Orlando, 1984.

Ayres, F., *Schaum's outline of theory and problems of differential equations*, McGraw-Hill, New York, 1952.

B

Bari, N. K., *A treatise on trigonometric series*, Pergamon Press, New York, 1964.

Bateman, H., *Partial differential equations of mathematical physics*, Dover, New York, 1944.

Bathe, K. and Wilson E. L., *Numerical methods in finite element analysis*, Prentice-Hall, Englewood Cliffs, N.J., 1976.

Beckenbach, E. F., *Modern mathematics for the engineer*, vols. 1 and 2., McGraw-Hill, New York, 1956-1961.

Beckenbach, E. F. and Bellman, R. E., *Inequalities*, Springer-Verlag, Berlin, 1965.

Bellman, R. E., *Stability theory of differential equations*, McGraw-Hill, New York, 1953.

Bellman, R. E. and Adomian, G., *Partial differential equations : new methods for their treatment and solution*, Reidel Pub. Co., Hingham, Mass., 1985.

Bellman, R. E. and Cooke, K. L., *Modern elementary differential equations*, 2nd ed., Addison-Wesley, Reading, Mass., 1971.

Bellman, R. E., Kalaba, R. E. and Lockett, J., *Numerical inversion of the Laplace transform : application to biology, economics, engineering, and physics*, American Elsevier Pub. Co., New York, 1966.

Bérard, P. H., *Spectral geometry : direct and inverse problems*, Springer-Verlag, New York, 1986.

Berg, P. W. and McGregor, J. L., *Elementary partial differential equations*, Holden-Day, San Francisco, 1966.

Berman, A. and Plemmons, R. J., *Nonnegative matrices in the mathematical sciences*, Academic Press, New York, 1979.

Birkhoff, G. and Lynch, R. E., *Numerical solution of elliptic problems*, SIAM, Philadelphia, 1984.

Birkhoff, G. and Rota, G. C., *Ordinary differential equations*, 2nd ed., Blaisdell, Waltham, Mass., 1969.

Birkhoff, G. and Schoenstadt, A. L., *Elliptic problem solvers II*, Academic Press, Orlando, 1984.

Bjorck, A., Plemmons, R. J. and Schneider, H., *Large scale matrix problems*, North-Holland, New York, 1981.

Bjorken, J. D. and Drell, S. D., *Relativistic quantum fields*, McGraw-Hill, New York, 1965.

Bland, D. R., *Wave theory and applications*, Oxford University Press, Oxford, 1988.

Bleecker, D., *Gauge theory and variational principles*, Addison-Wesley, Reading, Mass., 1981.

Bleistein, N., *Mathematical methods for wave phenomena*, Academic Press, New York, 1984.

Boas, R. P., Inversion of Fourier and Laplace transforms, Amer. Math. Monthly **69** (1962), 955-960.

Boccardo, L. and Tesei, A., *Nonlinear parabolic equations : qualitative properties of solutions*, Longman Scientific & Technical, Harlow, 1987.

Bowman, F., *Introduction to Bessel functions*, Dover, New York, 1958.

Bowman, F. and Gerard, F. A., *Higher Calculus*, Cambridge Univ. Press, London, 1967.

Botha, J. F. and Pinder, G. F., *Fundamental concepts in the numerical solution of differential equations*, Wiley, New York, 1983.

Boyce, W. E. and Di Prima, R. C., *Elementary differential equations and boundary value problems*, Wiley, New York, 1965.

Brand, L., *Advanced calculus; an introduction to classical analysis*, Wiley, New York, 1955.

Brand, L., *Differential and difference equations*, Wiley, New York, 1966.

Braun, M., *Differential equations and their applications: an introduction to applied mathematics*, 3rd ed., Springer-Verlag, New York, 1983.

Broman, A. E., *Introduction to partial differential equations ; from Fourier series to boundary-value problems*, Addison-Wesley, Reading, Mass., 1970.

Buck, R. C. and Buck, E. F., *Advanced calculus*, 2nd ed., McGraw-Hill, New York, 1965.

Burkill, J. C., *The theory of ordinary differential equations*, 2nd ed., Interscience Publishers, New York, 1962.

Byerly, W. E., *An elementary treatise on Fourier's series, and spherical, cylindrical, and ellipsoidal harmonics, with applications to problems in mathematical physics*, Dover, New York, 1959.

C

Cajori, F., The early history of partial differential equations and of partial differentiation and integration, Amer. Math. Monthly **35** (1928), 459-467.

Cannon, J. R., *The one-dimensional heat equation*, Cambridge Univ. Press, New York, 1984.

Carasso, A., Stone, A. P. and John, F., *Improperly posed boundary value problems*, Pitman, San Francisco, 1975.

Carnahan, B., Luther, H. A. and Wilkes, J. O., *Applied numerical methods*, Wiley, New York, 1969.

Carslaw, H. S., *Introduction to the theory of Fourier's series and integrals*, 3rd ed., Dover, New York, 1930.

Carslaw, H. S. and Jaeger, J. C., *Conduction of heat in solids*, 2nd ed., Clarendon Press, Oxford, 1959.

Cercignani, C., *The Boltzmann equation and its applications*, Springer-Verlag, New York, 1988.

Ceschino, F. and Kuntzmann, J., *Numerical solution of initial value problems*, Prentice-Hall, Englewood Cliffs, N.J., 1966.

Chester, C. R., *Techniques in partial differential equations*, McGraw-Hill, New York, 1971.

Chavel, I., *Eigenvalues in Riemannian geometry*, Academic Press, New York, 1984.

Churchill, R. V. and Brown, J. W., *Fourier series and boundary value problems*, 4th ed., McGraw-Hill, New York, 1987.

Ciarlet, P., Kesavan, S., Ranjan, A. and Vanninathan, M., *Lectures on the finite element method*, Tata Institute of Fundamental Research, Bombay, 1975.

Coddington, E. A., *An introduction to ordinary differential equations*, Prentice-Hall, Englewood Cliffs, N.J., 1961.

Coddington, E. A. and Levinson, N., *Theory of ordinary differential equations*, McGraw-Hill, New York, 1955.

Collatz, L., *The numerical treatment of differential equations*, 3rd ed., Springer-Verlag, Berlin, 1960.

Collatz, L., *Differential equations : an introduction with applications*, Wiley, New York, 1986.

Colton, D. L., *Partial differential equations in the complex domain*, Pitman, San Francisco, 1976.

Coppel, W. A., J.B. Fourier — on the occasion of his two-hundredth birthday, Amer. Math. Monthly **76** (1969), 468-483.

Copson, E. T., *Partial differential equations*, Cambridge Univ. Press, New York, 1975.

Cordes, H. O., *Spectral theory of linear differential operators and comparison algebras*, Cambridge Univ. Press, New York, 1987.

Courant, R. and Hilbert, D., *Methods of mathematical physics*, vols. 1 and 2, Interscience Publishers, New York, 1953.

Craig, I. J. and Brown, J. C., *Inverse problems in astronomy : a guide to inversion strategies for remotely sensed data*, Adams Hilger Ltd, Bristol and Boston, 1986.

Cryer, C. W., *Numerical functional analysis*, Clarendon Press, Oxford, 1982.

Csordas, G. and Varga, R. S., Comparisons of regular splittings of matrices, Numer. Math. vol. **44** (1984), 23-35.

D

Davies, E. B., *Heat kernels and spectral theory*, Cambridge Univ. Press, New York, 1989.

Davis, H. F., *Fourier series and othogonal functions*, Allyn and Bacon, Boston, 1963.

Davis, J. L., *Wave propagation in solids and fluids*, Springer-Verlag, New York, 1988.

Deans, S. R., *The radon transform and some of its applications*, Wiley, New York, 1983.

Debnath, L., *Nonlinear waves*, Cambridge Univ. Press, New York, 1983.

De Luca, L. J. and Sedlock, J. T., *Calculus: a first course*, Prentice-Hall, Englewood Cliffs, N.J., 1973.

Dennemeyer, R., *Introduction to partial differential equations and boundary value problems*, McGraw-Hill, New York, 1968.

Dezin, A. A., *Partial differential equations : an introduction to a general theory of linear boundary value problems*, Springer-Verlag, New York, 1987.

Diaz, J. I., *Nonlinear partial differential equations and free boundaries*, Pitman, Boston, 1985.

Dodziuk, J., Eigenvalues of the Laplacin and the heat equation, Amer. Math. Monthly **88** (1981), 686-695.

Doniach, S. and Sondheimer, E. H., *Green's functions for solid state physicists*, W. A. Benjamin, Reading, Mass., 1974.

Downing, H.H., Sums of sines converted into numerical sums, Amer. Math. Monthly **56** (1949), 630-631.

Driver, R. D., *Introduction to ordinary differential equations*, Harper & Row, New York, 1978.

Du Chateau, P. and Zachmann, D., *Applied partial differential equations*, Harper & Row, New York, 1989.

Du Chateau, P. and Zachmann, D., *Schaum's outline of theory and problems of partial differential equations*, McGraw-Hill, New York, 1986.

Duff, G. F. D., *Partial differential equations*, Toronto Univ. Press, Toronto, 1956.

Duff, G. F. D. and Naylor, D., *Differential equations of applied mathematics*, Wiley, New York, 1966.

Dunford, N. and Schwartz J. *T*, *Linear operators*, Interscience Publishers, New York, 1958.

Dym, H. and McKean, H. P., *Fourier series and integrals*, Academic Press, New York, 1972.

E

Ecomomou, E. N., *Green's functions in quantum physics*, 2nd ed., Springer-Verlag, New York, 1983.

Edwards, C. H., *Advanced calculus of several variables*, Academic Press, New York, 1973.

Edwards, R. E., *Fourier series, a modern introduction*, 2nd ed., Springer-Verlag, New York, 1979.

Eggermont, P. P. B., Noncentral difference quotients and the derivative, Amer. Math. Monthly **95** (1988), 551-553.

Epstein, B., *Partial differential equations*, McGraw-Hill, New York, 1962.

Evans, L. C., *Weak convergence methods for nonlinear partial differential equations*,Amer. Math. Soc., Providence, R. I., 1990.

F

Fattorini, H. O., *Second order linear differential equations in Banach spaces*, North-Holland, New York, 1985.

Farlow, S. J., *Partial differential equations for scientists and engineers*, Wiley, New York, 1982.

Feineman, G. , Garrett, S. J. and Karaus, A. D., *Applied differential equations*, Spartan Books, Washington, 1965.

Feller, W., *An introduction to probability theory and its application*, vol II, 3rd ed., Wiley, New York, 1967.

Forsythe, G. E., *Numerical analysis and partial differential equations*, Wiley, New York, 1958.

Forsythe, G. E. and Wasow, W. R., *Finite-difference methods for partial differential equations*, Wiley, New York, 1960.

Fox, L., *An introduction to numerical linear algebra, with exercises*, Oxford Univ. Press, New York, 1965.

Franklin, P., *An introduction to Fourier methods and the Laplace transformation*, Dover, New York, 1958.

Friedman, A., *Partial differential equations of parabolic type*, Prentice-Hall, Englewood Cliffs, N.J., 1964.

Friedman, A., *Variational principles and free-boundary problems*, Wiley, New York, 1982.

Friedman, A., *Partial differential equations of parabolic type*, Prentice-Hall, Englewood Cliffs, N.J., 1964.

Fulks, W., *Advanced calculus*, 2nd ed., Wiley, New York, 1969.

G

Garabedian, P., *Partial differential equations*, Wiley, New York, 1964.

Gilbarg, D. and Trudinger, N. S., *Elliptic partial differential equations of second order*, Springer-Verlag, New York, 1977.

Gilbert, R. P., *Constructive methods for elliptic equations*, Springer-Verlag, New York, 1974.

Gilbert, R. P., *Function theoretic methods in partial differential equations*, Academic Press, New York, 1969.

Gilkey, P. B., *The index theorem and the heat equation*, Publish or Perish, Inc., Boston, 1974.

Glowinski, R., *Numerical methods for nonlinear variational problems*, Springer-Verlag, New York, 1984.

Glowinski, R., Lions J. L. and Tremolieres, R., *Numerical analysis of variational inequalities*, North-Holland, New York, 1981.

Goffman, C. and Waterman, D., Some aspects of Fourier series, Amer. Math. Monthly **77** (1970), 119-133.

Golomb, M. and Shanks, M. E., *Elements of ordinary differential equations*, 2nd ed, McGraw-Hill, New York, 1965.

Golub, G. H. and Van Loan, C. F., *Matrix computations*, Johns Hopkins Univ. Press, 1983.

Good, I. J., Analogues of Poisson's summation formula, Amer. Math. Monthly **69** (1962), 259-266.

Graham, A., *Nonnegative matrices and applicable topics in linear algebra*, Halsted Press, New York, 1987.

Greenberg, M. D., *Foundations of applied mathematics*, Prentice-Hall, Englewood Cliffs, N.J., 1978.

Greenberg, M. D., *Advanced engineering mathematics*, Prentice-Hall, Englewood Cliffs, N.J., 1988.

Groetsch, C.W., *Generalized inverses of linear operators : representation and approximation*, Dekker, New York, 1977.

Grosswald, E., *Bessel polynomials*, Springer-Verlag, New York, 1978.

Guenther, R.B., *Partial differential equations of mathematical physics and integral equations*, Prentice-Hall, Englewood-Cliffs, N.J., 1988.

Gustafson, K. E., *Introduction to partial differential equations and Hilbert space methods*, 2nd ed., Wiley, New York, 1987.

H

Haberman, R., *Mathematical models : mechanical vibrations, population dynamics, and traffic flow : an introduction to applied mathematics*, Prentice-Hall, Englewood Cliffs, N.J., 1977.

Haberman, R., *Elementary applied partial differential equations : with Fourier series and boundary value problems*, Prentice-Hall, Englewood Cliffs, N.J., 1983.

Hackbusch, W., *Multi-grid methods and applications*, Springer-Verlag, New York, 1985.

Hagin, F., *A first course in differential equations*, Prentice-Hall, Engelwood Cliffs, N.J., 1975.

Hardy, G. H., A Theorem concerning Fourier transforms, J. London Math. Soc., **8** (1933), 227-231.

Hardy, G. H. and Rogosinski, W. W., *Fourier series*, Cambridge Tracts no. 38, Cambridge Univ. Press, New York, 1950.

Harris, S., *An introduction to the theory of the Boltzman equation*, Holt, Rinehart and Winston, New York, 1971.

Hartman, P., *Ordinary differential equations*, 2nd ed., Birkhauser, Boston, 1982.

Hellwig, G., *Partial differential equations : an introduction*, 2nd ed., Blaisdell Pub. Co., New York, 1977.

Henrici, P., *Applied and computational complex analysis*, vols. 1, 2 and 3, Wiley, New York, 1974.

Henrici, P., *Error propagation for difference methods*, Wiley, New York, 1963.

Henrici, P., *Elements of numerical analysis*, Wiley, New York, 1964.

Higgins, J. R., Five short stories about the cardinal series, Bull. Amer. Math. Soc. (New Series), **12**, (1985), 45-89.

Hidebrand, F. B., *Advanced calculus for applications*, Prentice-Hall, Englewood Cliffs, N.J., 1962.

Hildebrand, F. B., *Introduction to numerical analysis*, McGraw-Hill, New York, 1956.

Hlavacek, I., *Solution of variational inequalities in mechanics*, Springer-Verlag, New York, 1988.

Hochstadt,

Hodge, P. G., On the method of characteristics, Amer. Math. Monthly **57** (1950), 621-623.

Hoffman, S. P., *Advanced calculus*, Prentice-Hall, Englewood Cliffs, N.J., 1970.

Holzapfel, R., *Geometry and arithmetic around Euler partial differential equations*, Kluwer Academic, Boston, 1986.

Holbrook, J. G., *Laplace transforms for electrical engineers*, 2nd rev. ed., Pergamon Press, New York, 1966.

Hormander, L., *The analysis of linear partial differential equations*, Springer-Verlag, New York, 1983.

Hornbeck, R. W., *Numerical methods*, Quantum Publishers, New York, 1975.

Householder, A. S., *Principles of numerical analysis*, McGraw-Hill, New York, 1953.

Huntley, I. and Johnson, R. M., *Linear and nonlinear differential equations*, Halsted Press, New York, 1983.

Hunzeker, H. L., The separation of partial differential equations with mixed derivatives, Amer. Math. Monthly **68** (1961), 131-134.

Hurewicz, W., *Lectures on ordinary differential equations*, M.I.T. Press, Cambridge, Mass., 1958.

I

Ince, E. L. and Sneddon, I. N., *The solution of ordinary differential equations*, 2nd ed., Wiley, New York, 1987.

Isaacson, E. and Keller, H. B., *Analysis of numerical methods*, Wiley, New York, 1966.

J

John, F., *Lectures on advanced numerical analysis*, Gordon and Breach, New York, 1967.

John, F., *Partial differential equations*, 4th ed., Springer-Verlag, New York, 1982.

Jordan, D. W. and Smith, P., *Nonlinear ordinary differential equations*, 2nd ed., Oxford Univ. Press, Oxford, 1987.

Jordan, K., *Calculus of finite differences*, 3rd ed., Chelsea Pub. Co., New York, 1965.

K

Kac, M., Can one hear the shape of a drum, Amer. Math. Monthly **73S** (1966), 1-23.

Kaplan, W., *Advanced calculus*, Addison-Wesley, Reading, Mass., 1952.

Keller, H. B., *Numerical methods for two-point boundary-value problems*, Blaisdell Pub. Co., Waltham, Mass., 1968.

Kellner, R., On a theorem of Pólya, Amer. Math. Monthly **73** (1966), 856-858.

Kelman, R. B., Regularity of certain explicit solutions to Laplace's equation at artificial interfaces, Amer. Math. Monthly **73** (1966), 1073-1078.

Kober, H., *Dictionary of conformal representations*, 2nd ed., Dover, New York,1957.

Konig, M., *Schauder's estimates and boundary value problems for quasilinear partial differential equations*, Presses de l'Universite de Montreal, Montreal, 1985.

Kovach, L. D., *Boundary-value problems*, Addison-Wesley, Reading, Mass., 1984.

Krein, M. G. and Gohberg, I., *Topics in differential and integral equations and operator theory*, Birkhauser Verlag, Boston, 1983.

Krylov, N. V., *Nonlinear elliptic and parabolic equations of the second order*, Kluwer Academic, Boston, 1987.

Krzyzanski, M., *Partial differential equations of second order*, Polish Scientific Publishers, Warszawa PWN, 1971.

Kunz, K. S., *Numerical analysis*, McGraw-Hill, New York, 1957.

L

Ladyzhenskaia, O., it Linear and quasi-linear equations of parabolic type, Translations of mathematical monographs, v. 23, American Mathematical Society, Providence, R.I., 1968.

Laforgia, A. and Muldoon, M. E., Some consequences of the Sturm comparison theorem, Amer. Math. Monthly **93** (1986), 89-94.

Lang, S., *Differential manifolds*, Addison-Wesley, Reading, Mass., 1972.

Lapidus, L. and Schiesser, W. E., *Numerical methods for differential systems : recent developments in algorithms, software, and applications*, Academic Press, New York, 1976.

Lax, P. D. and Phillips, R. S., *Scattering theory*, rev. ed., Academic Press, Boston, 1989.

Lax, P. D., Numerical solution of partial differential equations, Amer. Math. Monthly **72S** (1965), 74-84.

Lebedev, N. N. and Silverman, R. A., *Special functions and their applications*, Rev. Engish ed., Prentice-Hall, Englewood Cliffs, N.J., 1965.

Leighton, W., *Ordinary differential equations*, 2nd ed., Wadsworth, Belmont, Calif., 1966.

Leis, R., *Partial differential equations of applied mathematics*, Wiley, New York, 1983.

Leis, R., *Initial value boundary problems in mathematical physics*, Wiley, New York, 1986.

Le Page, W. R., *Complex variables and the Laplace transform for engineers*, McGraw-Hill, New York, 1961.

Leung, A. W., it Systems of nonlinear partial differential equations : applications to biology and engineering, Kluwer Academic, Boston, 1989.

Levitan, B. M. and Sargsjan, I. S., *Sturm-Liouville and Dirac operators*, Kluwer Academic, Boston, 1989.

Lewin, J. W., A truly elementary approach to the bounded convergence theorem, Amer. Math. Monthly **93** (1986), 395-397.

Lewin, J. W., Some applications of the bounded convergence theorem for an introductory course in analysis, Amer. Math. Monthly **94** (1987), 988-993.

Lewin, J. W., *An introduction to mathematical analysis*, McGraw-Hill, New York, 1988.

Lewis, R. W. , Morgan, K. and Zienkiewicz, O. C., *Numerical methods in heat transfer*, Wiley, New York, 1981.

Lieberstein, H. M., *Theory of partial differential equations*, Academic Press, New York, 1972.

Lions, J. L. and Magenes, E., *Non-Homogeneous boundary-value problems and applicatons*, Springer-Verlag, New York, 1972.

Lomen, D. and Mark, J., *Ordinary differential equations with linear algebra*, Prentice-Hall, Englewood Cliffs, N.J., 1986.

Loomis, L. H. and Sternberg, S., *Advanced calculus*, Addison-Wesley, Reading, Mass., 1968.

M

Magnus, W. and Winkler, S., *Hill's Equation*, Interscience Publishers, New York, 1966.

Martin, W. T. and Reissner, E., *Elementary differential equations*, 2nd ed., Addison-Wesley, Reading, Mass., 1961.

Mathews, J. H., *Basic complex variables for mathematics and engineering*, Allyn and Bacon, Boston, 1982.

Mathews, J. and Walker, R. L., *Mathematical methods of physics*, W. A. Benjamin, New York, 1964.

McShane, E. J., The Fourier transform and mean convergence, Amer. Math. Monthly **68** (1961), 205-211.

Michel, A. N., *Ordinary differential equations*, Academic Press, New York, 1982.

Mikhlin, S. G., *Linear equations of mathematical physics*, Holt, Rinehart and Winston, New York, 1967.

Miles, E.P. and Williams, E., A basic set of polynomial solutions for the Euler-Poisson-Darboux and Beltrami equations, Amer. Math. Monthly **63** (1956), 401-404.

Miller, K. S., *Advanced complex calculus*, Harper, New York, 1960.

Miller, K. S., *Partial differential equations in engineering problems*, Prentice-Hall, New York, 1953.

Miller, R. K. and Michel, A. N., *Ordinary differential equations*, Academic Press, New York, 1982.

Milne, W. E., *Numerical calculus: approximations, interpolation, finite differences, numerical integration and curve fitting*, Princeton University Press, Princeton, 1949.

Milne, W. E., *Numerical solution of differential equations*, Wiley, New York, 1953.

Mitchell, A. and Wait, R., *The finite element method in partial differential equations*, Wiley, New York, 1977.

Mitchell, A. R., *Computational methods in partial differential equations*, Wiley, New York, 1969.

Moon, P. H. and Spencer, D. E., *Partial differential equations*, D. C. Heath, Lexington, Mass., 1969.

Myint-U, T. and Debnath, L., *Partial differential equations for scientists and engineers*, North-Holland, New York, 1987.

N

Nirenberg, L., *Lectures on linear partial differential equations*, Regional conference series in mathematics, no. 17, American Mathematical Society, Providence, R.I., 1972.

Noble, B. and Daniel, J. W., *Applied linear algebra*, 3rd ed., Prentice-Hall, Englewood Cliffs, N.J., 1988.

Noye, J., *Computational techniques for differential equations* (North-Holland Mathematics Studies 83), North-Holland, New York, 1984.

O

Olmsted, J. M. H., *Advanced calculus*, Appleton-Century-Crofts, New York, 1961.

Ortega, J. M., *Numerical analysis; a second course*, Academic Press, New York, 1972.

Ortega, J. M. and Rheinboldt, W. C., *Iterative solutions of nonlinear equations in several variables*, Academic Press, New York, 1970.

Osserman, R., The isoperimetric inequality, Bull. Amer. Math. Soc., **84** (1978), 1182-1238.

Osserman, R. and Weinstein, A., eds., *Geometry of the Laplace operator*, Proceedings of symposia in pure mathematics, *v.* XXXVI, American Mathematical Society, Providence, R.I., 1980.

P

Palais, R. S., *Seminar on the Atiyah-Singer References theorem*, Princeton Univ. Press, Princeton, 1965.

Papoulis, A., *The Fourier integral and its applications*, McGraw-Hill, New York, 1962.

Paris, R. B. and Wood, A. D., *Asymptotics of high order differential equations*, Wiley, New York, 1986.

Payne, L. E., *Improperly posed problems in partial differential equations*, Regional conference series in applied mathematics, 22, Society for Industrial and Applied Mathematics, Philadelphia, 1975.

Pearson, C. E., *Numerical methods in engineering and science*, Van Nostrand Reinhold Co, New York, 1986.

Peek, R. L., Flow of heat in a disk heated by a gas stream, Amer. Math. Monthly **39** (1932), 276-280.

Pinsky, M. A., *Introduction to partial differential equations with applications*, McGraw-Hill, New York, 1984.

Pólya, G. and Szegö, G., *Isoperimetric inequalities in mathematical physics*, Princeton University Press, Princeton, 1951.

Prenter, P. M., *Splines and variational methods*, Wiley, New York, 1975.

Protter, H. M., The characteristic initial value problem for the wave equation and Riemann's method, Amer. Math. Monthly **61** (1954), 702-705.

Protter, H. M. and Morrey, C. B., *College calculus with analytic geometry*, 3rd ed., Addison-Wesley, Reading, Mass., 1977.

Protter, M. H. and Weinberger, H. F., *Maximum principles in differential equations*, Prentice-Hall, Englewood Cliffs, N.J., 1967.

R

Rabenstein, A. L., *Introduction to ordinary differential equations*, Academic Press, New York, 1972.

Rainville, E. D., *Intermediate differential equations*, 2d ed., Macmillan, New York, 1964.

Rainville, E. D., *Special functions*, Macmillan, New York, 1960.

Rainville, E. D., *The Laplace transform : an introduction*, Macmillan, New York, 1963.

Rainville, E. D. and Bedient, P. E., *Elementary differential equations*, 4th ed., Macmillan, 1969.

Reid, W. T., *Ordinary differential equations*, Wiley, New York, 1971.

Reissig, R., Sansone, G. and Conti, R., *Non-linear differential equations of higher order*, Noordhoff International Publishing, Leyden, 1974.

Rheinboldt, W. C., *Numerical analysis of parameterized nonlinear equations*, Wiley, New York, 1986.

Rhoades, B. E., On forming partial differential equations, Amer. Math. Monthly **66** (1959), 473-476.

Richtmyer, R. D., *Difference methods for initial-value problems*, Interscience Publishers, New York, 1957.

Roach, G. F., *Green's functions*, 2nd ed., Cambridge Univ. Press, New York, 1982.

Rogers, C. and Shadwick, W. F., *Backlund transformations and their applications*, Academic Press, New York, 1982.

Roseau, M., *Asymptotic wave theory*, American Elsevier Pub. Co., New York, 1976.

Rosenberg, von, D. U., *Methods for the numerical solution of partial differential equations*, American Elsevier Pub. Co., New York, 1969.

Ross, S.L., *Differential equations*, 1st ed., Blaisdell, New York, 1964.

Royden, H. L., *Real analysis*, 2nd ed., Macmillan, New York, 1968.

Rozhdestvenskii, B. L. and IAnenko, N. N., *Systems of quasilinear equations and their application to gas dynamics*, 2nd ed., American Mathematical Society, Providence, R.I., 1983.

Rubinstein, Z., *A course in ordinary and partial differential equations*, Academic Press, New York, 1969.

Rudin, W., *Functional Analysis*, 3rd ed.,McGraw-Hill, New York, 1973.

Rudin, W., *Principles of mathematical analysis*, 3rd. ed., McGraw-Hill, New York, 1976.

Rudin, W., *Real and complex analysis*, McGraw-Hill, New York, 1987.

S

Sagan, H., *Boundary and eigenvalue problems in mathematical physics*, Wiley, New York, 1961.

Sakamoto, R., *Hyperbolic boundary value problems*, Cambridge Univ. Press, New York, 1982.

Salas, S. L., Hille, E. and Anderson, J. T., *Calculus : one and several variables, with analytic geometry*, 5th ed., Wiley, New York, 1986.

Salkover, M., The homogeneous linear differential equation with constant coefficients, Amer. Math. Monthly **37** (1930), 524-529.

Schechter, M., *Operator methods in quantum mechanics*, North-Holland, New York, !981.

Schechter, M., *Modern methods in partial differential equations : an introduction*, 2nd ed., McGraw-Hill, New York, 1977.

Scheid, F. J., *Schaum's outline of theory and problems of numerical analysis*, McGraw-Hill, New York, 1968.

Schiff, L. I., *Quantum mechanics*, 3rd ed., McGraw-Hill, New York, 1968.

Schultz, M., *Spline analysis*, Prentice-Hall, Englewood Cliffs, N. J., 1973.

Scott, E. J., Determination of the Riemann function, Amer. Math. Monthly **80** (1973), 906-909.

Seeley, R. T., *An introduction to Fourier series and integrals*, W. A. Benjamin, 1966.

Seeley, R. T., *Calculus of Several variables*, Scott, Foresman, Glenview, Ill., 1970.

Sewell, G., *The numerical solution of ordinary and partial differential equations*, Academic Press, Boston, 1988.

Silverman, R. A., *Modern calculus and analytic geometry*, Macmillan, New York, 1969.

Simmons, G. F., *Differential equations, with applications and historical notes*, McGraw-Hill, New York, 1972.

Smith, G. D., *Numerical solution of partial differential equations : finite difference equations*, 3rd ed., Oxford University Press, Oxford, 1985.

Smoller, J. A., ed., *Nonlinear partial differential equations*, Contemporary mathematics, v. 17, American Mathematical Society, Providence, R.I., 1983.

Sneddon, I. N., *Fourier transforms*, McGraw-Hill, New York, 1951.

Sneddon, I. N., *Elements of partial differential equations*, McGraw-Hill, New York, 1957.

Sneddon, I. N., *Special functions of mathematical physics and chemistry*, 3rd ed., Longman, London, 1980.

Sobolev, S. L., *Partial differential equations of mathematical physics*, Addison-Wesley, Reading, Mass., 1964.

Sod, G. A., *Numerical methods in fluid dynamics : initial and initial boundary-value problems*, Cambridge University Press, New York, 1985.,

Spain, B., *Ordinary differential equations*, Van Nostrand Reinhold, New York, 1969.

Sperb, R. P., *Maximum principles and their applications*, Academic Press, New York, 1981.

Spivak, M., *Calculus on manifolds : a modern approach to classical theorems of advanced calculus*, Benjamin and Cummings Pub. Co., Menlo Park, Calif., 1965.

Stakgold, I., *Green's functions and boundary value problems*, Wiley, New York, 1979.

Sternberg, W. J. and Smith, T. L., *The theory of potential and spherical harmonics*, 2nd ed., University of Toronto, Toronto, 1952.

Strang, G., *Linear algebra and its applications*, 2nd ed., Academic Press, New York, 1980.

T

Taira, K., *Diffusion processes and partial differential equations*, Academic Press, Orlando, 1988.

Taylor, A. E., *Calculus, with analytic geometry*, Prentice-Hall, Englewood Cliffs, N.J., 1959.

Taylor, A. E. and Mann, W. R., *Advanced calculus*, 2nd ed., Xerox College Pub., Lexington, Mass., 1972.

Temam, R., *Navier-Stokes equations : theory and numerical analysis,* Rev. ed., North-Holland, New York, 1979.

Thomas, G. B. and Finney, R. L., *Calculus and analytic geometry*, 6th ed., Addison-Wesley, Reading, Mass., 1984.

Titchmarsh, E.C., *Eigenfunction expansions associated with second-order differential equations*, 2nd ed., Clarendon Press, Oxford, 1962.

Titchmarsh, E. C., *Introduction to the theory of Fourier integrals*, 2nd ed., Clarendon Press, Oxford, 1967.

Titchmarsh, E. C., *The theory of functions*, 2nd ed., Oxford Univ. Press, 1939.

Treves, F., *Basic linear partial differential equations*, Academic Press, New York, 1975.

Treves, F., Applications of distributions to pde theory, Amer. Math. Monthly **77** (1970), 241-248.

V

Varga, R. S., *Matrix iterative analysis*, Prentice-Hall, Englewood Cliffs, N.J., 1962.

Vemuri, V. and Karplus, Walter J., *Digital computer treatment of partial differential equations*, Englewood Cliffs, N.J., 1981.

Vichnevetsky, R., *Computer methods for partial differential equations*, Prentice-Hall, Englewood Cliffs, N.J., 1981.

W

Warner, F., *Foundations of differentiable manifolds and Lie groups*, Scott, Foresman and Co., Glenview, Ill., 1971.

Watson, E. J., *Laplace transforms and applications*, Van Noststrand Reinhold, New York, 1981.

Watson, G. N., *A treatise on the theory of Bessel functions*, 2nd ed., Cambridge Univ. Press, Cambridge, 1966.

Watson, N. A., *Parabolic equations on an infinite strip*, Dekker, New York, 1989.

Webster, A. G., *Partial differential equations of mathematical physics*, 2nd corr. ed., Dover, New York, 1966.

Weinberger, H. F. , *A first course in partial differential equations with complex variables and transform methods*, 1st ed., Blaisdell, New York, 1965.

Wells, C. P., Separability conditions for some self-adjoint partial differential equations, Amer. Math. Monthly **66** (1959), 684-689.

Whittaker, E. T., *A course of modern analysis*, 4th ed., Cambridge Univ. Press, Cambridge, 1963.

Widder, D. V., *Advanced calculus*, 2nd ed., Prentice-Hall, Englewood Cliffs, N.J., 1961.

Widder, D. V., *The Laplace transform*, Princeton University Press, Princeton, 1941.

Widder, D. V., *An introduction to transform theory*, Academic Press, series of monographs and textbooks, *v.* 42, New York, 1971.

Widder, D. V., *The heat equation*, Academic Press, series of monographs and textbooks, *v.* 67, New York, 1975.

Wilf, H. S., *Mathematics for the physical sciences*, Wiley, New York, 1962.

Wilker, J. B., The nth derivative as a limit, Amer. Math. Monthly **94** (1987), 354-356.

Williams, W. E., *Partial differential equations*, Clarendon Press, New York, 1980.

Williamson, R. E., *Introduction to differential equations : ODE, PDE, and series*, Prentice-Hall, Englewood Cliffs, N.J., 1986.

Wimp, J., *Sequence transformations and their applications*, Academic Press, New York, 1981.

Witten, M., *Hyperbolic partial differential equations*, Pergamon Press, New York, 1983.

Wloka, J. *Partial differential equations*, Cambridge Univ. Press, New York, 1987.

Y

Young, E. C., *Partial differential equations; an introduction*, Allyn and Bacon, Boston, 1972.

Young, D. M., *Iterative solution of large linear systems*, Academic Press, New York, 1971.

Z

Zachmanoglou, E. C. and Thoe, D. W., *Introduction to partial differential equations with applications*, The Williams and Wilkins Co., Baltimore, 1976.

Zauderer, E., *Partial differential equations of applied mathematics*, Wiley, New York, 1983.

Zuily, C., *Uniqueness and non-uniqueness in the Cauchy problem*, Birkhauser Verlag, Boston, 1983.

Zygmund, A., *Trigonometric series*, 2nd ed., Cambridge Univ. Press, New York, 1959.

SELECTED ANSWERS

SELECTED ANSWERS

Answers 1.1

1. (a) $y = Ce^{x^2/2}$ (c) $\arctan(y) = \frac{1}{3}(x^3 - 3x) + C$

(e) $Cx - x\cos(t) = 1$ (g) $(t-1)e^t + e^{-x} = C$ (i) $T(t) = Ce^{-3t}$.

3. 2 minutes.

4. (a) $y = \frac{1}{3}e^{-2x}\left[e^{3x} + 2\right]$ (c) $y = 2\sin(x)\log|\sin(x)|$

(e) $y = (2x+1)e^{-x^2}$ (g) $x = 100e^{-\sin(t)}$ (i) $x = (2t + 100)^{-\frac{3}{2}}$.

6 (a) $y = c_1x + c_2$ (c) $y = c_1\cos(x\sqrt{3}) + c_2\sin(x\sqrt{3})$

(e) $y = c_1e^{3x} + c_2$ (g) $y = c_1e^{-x/2} + c_2e^{-2x}$ (i) $y = (c_1 + c_2x)e^{2x}$.

7. (a) $y = e^{2t}$ (c) $y = a\cos(t) + b\sin(t)$ (e) $y = (5/3)e^{-(4/5)t}\sin(3t/5)$.

13. $\omega_R = 1/\sqrt{LC}$.

15. $x(t) = e^t\cos(t)$, $y(t) = -e^t\sin(t)$. The graph spirals away from the origin in the clockwise sense.

17. (a) $x(t) = (1-2t)e^t$ $y(t) = (1-t)e^t$
(c) $x(t) = (4/\sqrt{33})e^{5t/2}\sinh(t\sqrt{33}/2)$ $y(t) = (3/\sqrt{33})e^{5t/2}[\sinh(t\sqrt{33}/2) + (\sqrt{33}/3)\cosh(t\sqrt{33}/2)]$.

Answers 1.2

5. (a) third–order , linear, homogeneous .
(b) first–order, nonlinear.
(c) fourth–order, linear, inhomogeneous.
(d) second–order, nonlinear.
(e) second–order, linear, homogeneous.

7. (a) $u(x,y) = x^2 + 2xy - y^2$.

8. (b) $n - 1$.
(c) For larger n, the hot and cold regions are in closer proximity and the temperature gradient between these regions is greater.

9. (c) $(m+1)(n+1)(p+1)$ compartments.

11. (b) $u(x,t) = \cos(\pi at)\sin(\pi x) + (2/3\pi a)\sin(3\pi at)\sin(3\pi x)$
$v(x,t) = \cos(\pi at)\sin\pi x) + (1/3\pi a)\sin(3\pi at)\sin(3\pi x)$.

13. (a) $u(x,y) = \frac{1}{2}Cx^2 + \frac{1}{6}Ax^3 + \frac{1}{6}By^3$.
(b) Let $v(x,y) = u(x,y) + h(x,y)$, where $h(x,y)$ is any of the infinitely many solutions of Laplace's equation.

18. $A^{-1} = \frac{d}{dx} + p(x)$.

Answers 1.3

1. (a) $u(x,y) = x^3 + y^2x + f(y)$, f in C^1
(c) $u(x,y,z) = f(x,y) + g(y,z) + h(x,z)$, f, g, h in C^3 .

2. (a) $u(x,y) = f(y)e^{2x}$, f in C^1
(c) $u(x,y) = e^{-x^2}(2ye^{x^2} + f(y))$, f in C^1
(e) $u(x,y) = f(x)\exp[xy] + g(x)\exp[-xy]$, f, g in C^2 .

3. (a) $u(x,y) = ye^{2x}$
(c) $u(x,y) = 2y(1 - e^{y^2-x^2})$
(e) $u(x,y) = \cosh(xy)$.

4. The following solutions are not the only possible ones.
(a) $u(x,t) = e^{-2\lambda^2 t}(c_1\cos(\lambda x) + c_2\sin(\lambda x))$
(c) $u(x,t) = (c_1e^{\lambda x} + c_2e^{-\lambda x})(d_1e^{4\lambda t} + d_2e^{-4\lambda t})$
(e) $u(x,y,z) = \exp[z(a^2 + b^2)^{\frac{1}{2}}]\cos(ax)\cos(by)$.

5. If $\lambda = 0$, $u(x,t) = (c_1x + c_2)e^{-t}$.

If $\lambda > 0$, $u(x,t) = (c_1e^{x\sqrt{\lambda}} + c_2e^{-x\sqrt{\lambda}})\exp[(\lambda - 1)t]$.
If $\lambda = \mu^2$, then $u(x,t) = (c_1e^{x\mu} + c_2e^{-x\mu})\exp[(\mu^2 - 1)t]$.

6. (a) $u(x,y) = \exp[x + r(2y - 3x)]$.
(c) $u(x,y) = \exp[(x/rs) + ry + sz]$, $rs \neq 0$.
(e) $u(x,y) = \exp[rx + y(1 - r^2)^{\frac{1}{2}}]$.

8. Only the ratio a/b is uniquely determined. Thus, the answers below are not unique.
(a) $u(x,y) = f(2x - y)$
(c) $u(x,y) = f(dx - cy)$.

9. (a) $u(x,y) = x - \frac{1}{2}y$.

(c) $u(x,y) = (1/49)[(3y + 4x)^2 - 7(3y + 4x)]$.

Answers 2.1

1. In each solution below, f is an arbitrary C^1 function.
(a) $u(x,y) = f(3x+2y) + \frac{1}{4}x^2$ (c) $u(x,y) = -e^{x+y} + e^{2y}f(2x-y)$ (e) $v(w,z) = w^3 + f(3z-w)$.

2. (a) $u(x,y) = -e^{x+y} + [e^{x-\frac{1}{2}y} + \sin((x-\frac{1}{2}y)^2)]e^{2y}$

(c) $u(x,y) = -e^{x+y} + [\frac{2}{3}(x-\frac{1}{2}y) + 1]e^{\frac{4}{3}(x-\frac{1}{2}y)}e^{2y}$.

4. $g(x)$ must be of the form $-1+ke^x$ for some constant k. There are infinitely many solutions of the problem if $g(x)$ has this form. Any function of the form $u(x,y) = -1 + C(y-3x)e^x$ will be a solution, as long as $C(0) = k$ and C is C^1.

5. There are infinitely many solutions. Two of them are

$$u(x,y) = -1 + 2\cos(y-3x)e^x \quad \text{and} \quad u(x,y) = -1 + 2\exp(y-3x)e^x = -1 + 2e^{y-2x} .$$

6. $u(x,y) = (y + 2x)^2$.

7. Both are correct. Note that $e^{-cy/b} = e^{-cx/a}e^{c(bx-ay)/ab}$.

12. (a) By (26), $P_\infty(y) = C\cdot\exp(-\int_0^y .1\, dy) = Ce^{-y/10}$.

(b) The number of avocados on the shelf in the long run is $\int_0^\infty P_\infty(y)\, dy = 10\cdot C$. Thus, $C = 30$. Since about 30 are lost each day, naturally about 30 need to be acquired daily.

Answers 2.2

1. (In each of the following, C is an arbitrary C^1 function.)
(a) $u(x,y) = C(x^2/y)$ (c) $u(x,y) = xC(ye^x)$.

2. (a) $u(x,y) = [x^2/y]^{\frac{1}{3}}$, $y > 0$ (c) $u(x,y) = xye^x$.

3. (a) $X(s,t) = s \cdot e^t$, $Y(s,t) = e^{2t-s}$, $U(s,t) = \sin(s)$

(c) $X(s,t) = s^2 \cdot e^t$, $Y(s,t) = s \cdot \exp[s^2 \cdot (1-e^t)]$, $U(s,t) = s^3 \cdot e^t$.

7. Let $u(x,y) = 0$ for $x \leq 0$, and let $u(x,y) = yx^2$ for $x \geq 0$. Note that $u(x,y) \neq C(yx^2)$, since $u(-x,y) \neq u(x,y)$ for x and $y \neq 0$.

8. (d) $u(x,y) = f_n((-1)^n y \sin(x))$, for $n\pi \leq x \leq n\pi + \pi$, $n = 0, \pm 1, \pm 2, \ldots$.

Answers 2.3

1. $u(x,y,z) = [x^2 + y^2 + 2z^2 - 2yz - 2xz]e^z$.

2. (a) $y = -x + \alpha$, $z = \beta$, $\bar{x} = x + y$, $\bar{y} = z$, $\bar{z} = x$ $\quad \bar{u}_{\bar{z}} + u = \bar{y} \qquad \bar{u} = \bar{y} + C(\bar{x},\bar{y})e^{-\bar{z}}$

$u = z + C(x+y,z)e^{-x} \quad u(0,y,z) = y^2e^z = z + C(y,z) \quad C(y,z) = y^2e^z - z$

$u(x,y,z) = z + ((x+y)^2e^z - z)e^{-x}$.

(c) $u(x,y,x+y) = (x+y) + e^y \Rightarrow x+y + C(x+y,x+y)e^{-x} = (x+y) + e^y \Rightarrow C(x+y,x+y) = e^{x+y}$. Thus, we need only to choose C such that $C(r,r) = e^r$. Consider $C(r,s) = e^r f(s-r)$, where f is any C^1 function with $f(0) = 1$.

3. (a) $u(x,y,z,t) = f(x + t, y + 2t, z - t)$.

(b) $u(x,y,z,t) = (x + t)^2 + (y + 2t)^2 + (z - t)^2$.

(c) $(-t,-2t,t)$.

4. (a) $u(x,y,z) = C(x \cos(z) + y \sin(z), y \cos(z) - x \sin(z))$, where C is an arbitrary C^1 function of two variables.

6. (a) $C(x-u^2, y-u^2) = 0$ (Other equivalent answers are possible.)

(b) $u(x,y) = \sqrt{2x-y+1}$, for $2x-y+1 > 0$.

7. (a) $C(xy, u^2-x^2) = 0$.

(b) $u(x,y) = x(x^2y^4+2y^2+1)^{\frac{1}{2}}$.

12. (b) $t = \tau$, $x = [V(s) + sV'(s)]\tau + s$, $\rho = f(s)$.

15. (a) $x(t) = -c_0[\tfrac{1}{2}(\gamma-1)\alpha]^{-1}(\alpha t + 1) + C(1 + \tfrac{1}{2}(\gamma+1)\alpha t)^{2/(\gamma+1)}$; $C = x(0) - c_0[\tfrac{1}{2}(\gamma-1)\alpha]^{-1}$.

Answers 3.1

3. (a) $u(x,t) = 4e^{-(2\pi/3)^2 2t}\sin(2\pi x/3) - e^{-(5\pi/3)^2 2t}\sin(5\pi x/3)$

(c) $u(x,t) = \frac{3}{4}e^{-2(\pi/3)^2 t}\sin(\pi x/3) - \frac{1}{4}e^{-2\pi^2 t}\sin(\pi x)$

(e) $u(x,t) = 6e^{-(128\pi^2 t/9)}\sin(8\pi x/3) + e^{-50\pi^2 t}\sin(5\pi x)$.

6. (a) $u(x,t) = 5e^{-t}\cos(x) + 3e^{-64t}\sin(8x)$

(c) $u(x,t) = \frac{1}{2}(9 + e^{-36t}\cos(6x))$

(e) $u(x,t) = \frac{1}{2}(5 + 3e^{-4t}\cos(2x) + 4e^{-4t}\sin(2x))$.

9. (c) $u(x,t) = e^{-ht}\sum_{n=1}^{N} b_n e^{-(n\pi/L)^2 kt}\sin(n\pi x/L)$.

Answers 3.2

4. Let $f(x) = 5\sin(3x) - 3\sin(5x)$. Then $f'(x) = 15\cos(3x) - 15\cos(5x) = 30\cdot\sin(4x)\cdot\sin(x) = 0$ when $x = 0, \pi/4, \pi/2, 3\pi/4, \pi$. Now, $f(0) = f(\pi) = 0$, $f(\pi/4) = f(3\pi/4) = 5\sqrt{2}/2 + 3\sqrt{2}/2 = 4\sqrt{2}$, while $f(\pi/2) = -8$. Thus, the Maximum/Minimum Principles yield $-8 \le u(x,t) \le 4\sqrt{2}$.

Answers 3.3.

3. (b) $u(x,t) = \sum_{n=0}^{N} c_n \exp[-(n+\frac{1}{2})^2\pi^2 kt/L^2]\cdot\sin[(n+\frac{1}{2})\pi x/L]$.

4. $u(x,t) = u_p(x,t) + v(x,t) = x - 1 + e^{-9\pi^2 t/2}\sin(3\pi x/2)$.

5. $u(x,t) = u_p(x,t)+v(x,t) = \frac{1}{20}\cdot x^2 + 2x + \frac{1}{2}t + e^{-5\pi^2 t}\cos(\pi x)$.

7. $u(x,t) = 4 - 2\pi + 2x + 7e^{-9t/4}\cos(3x/2)$.

Answers 3.4

1. (b) $u(x,t) = \frac{1}{18}(1 - e^{-18t})\sin(3x)$.

3. $u(x,t) = e^{-9t}\sin(3x) + \sin(t)e^{-4t}\sin(2x)$.

5. (a) $w(x,t) = \frac{(b-a)x^2}{2L} + ax + \frac{(b-a)k}{L}t + \int h(t)\,dt$.

7. $u(x,t) = (e^t - t^2)x/\pi + t^2 + e^{-4t}\sin(2x) + \sin(x)\,(t - 1 + e^{-t})$

9. $u(x,t) = t[x - x^2 + e^{-4\pi^2 t}\cos(2\pi x)]$.

Answers 4.1

2. (a) FS $f(x) = \frac{1}{8} - \frac{1}{8}\cos(4\pi x)$.

(b) FS $f(x) = \frac{1}{2}\cos(x) + \sin(x) - \frac{1}{2}\cos(3x)$.

4. (a) FS $f(x) = \frac{1}{2} + \frac{1}{\pi}\sum_{n=1}^{\infty} \frac{1}{n}\cdot [1 - (-1)^n]\sin(n\pi x/L)$.

7. (a) FS $f(x) = \frac{1}{\pi}(e^{\pi} - e^{-\pi})\left[\frac{1}{2} + \sum_{n=1}^{\infty} \frac{(-1)^n}{1 + n^2}[\cos(nx) - n\cdot\sin(nx)]\right]$.

8. FS $f(x) = \frac{1}{\pi} + \frac{1}{2}\sin(x) - \frac{2}{\pi}[\cos(2x)/(1\cdot 3) + \cos(4x)/(3\cdot 5) + \ldots]$.

9. (a) FS $f(x) = \frac{8}{15} - \frac{48}{\pi^4}\sum_{n=1}^{\infty} \frac{(-1)^n}{n^4}\cos(n\pi x)$.

(b) $N = 3243$.

(c) $N = 6$.

Answers 4.2

2. $f(x) = a(x^3 - L^2 x) + d$ for any constants $a \neq 0$ and d .

Answers 4.3

4. (a) FCS $f(x) = \frac{L}{2} - \frac{4L}{\pi^2}\sum_{k=1}^{\infty} \frac{1}{(2k+1)^2}\cos[(2k+1)\pi x/L]$.

(b) FSS $f(x) = \frac{2L}{\pi}\sum_{n=1}^{\infty} (-1)^{n+1}\frac{1}{n}\sin(n\pi x/L)$.

5. (a) FCS $f(x) = 1$ and FSS $f(x) = \frac{4}{\pi}\sum_{k=1}^{\infty} \frac{1}{2k+1}\sin[(2k+1)\pi x/L]$

(b) FSS $f(x) = \frac{8}{\pi}\sum_{k=1}^{\infty} \frac{k}{4k^2-1}\sin(2kx)$ and FCS $f(x) = \cos(x)$.

8. (a) $u(x,t) = \frac{\pi^2}{3} + 4\sum_{n=1}^{\infty} \frac{(-1)^n}{n^2} e^{-n^2kt}\cos(nx)$.

9. (a) $u(x,t) = \frac{4}{\pi^2}\sum_{m=0}^{\infty} \frac{(-1)^m}{(2m+1)^2} e^{-(2m+1)^2\pi^2 t}\sin((2m+1)\pi x)$.

10. (a) The formal solution is $u(x,t) = \frac{2}{\pi}\sum_{n=0}^{\infty} \frac{\cos([n+\frac{1}{2}]\pi/2)}{n+\frac{1}{2}} e^{-[n+\frac{1}{2}]^2kt}\sin[(n+\frac{1}{2})x]$.

11. $u(x,t) = x - 1 + 4\sum_{n=0}^{\infty} \left[\frac{1}{(2n+1)\pi} - \frac{2(-1)^n}{(2n+1)^2\pi^2}\right] e^{-(n+\frac{1}{2})^2\pi^2kt}\sin[(n+\frac{1}{2})\pi x]$.

12. $u(x,t) = \frac{kt}{10} + \frac{x^2}{20} + 2x - \frac{35}{3} + 20\cdot\sum_{n=1}^{\infty} \frac{(2 - 3(-1)^n)}{\pi^2 n^2} e^{-(n\pi/10)^2kt} \cos(\frac{n\pi x}{10})$.

Answers 4.4

1. (a) $\lambda_n = \frac{(n-\frac{1}{2})^2\pi^2}{L^2}$, $y_n(x) = A_n\sin\left[\frac{(n-\frac{1}{2})\pi x}{L}\right]$, $n = 1, 2, 3, \ldots$

(c) $\lambda_n = \frac{(n-1)^2\pi^2}{L^2}$, $y_n(x) = B_n\cos\left[\frac{(n-1)\pi x}{L}\right]$, $n = 1, 2, 3, \ldots$.

2. (a) $\lambda_1 = -1$, $y_1(x) = A_1e^x$

$\lambda_n = (n-1)^2$, $y_n(x) = a_n[\sin((n-1)x) + (n-1)\cos((n-1)x)]$, $n = 2, 3, 4, \ldots$

(c) $\lambda_n = (2n-1)^2\pi^2$, $y_n(x) = A_n\sin((2n-1)\pi x) + B_n\cos((2n-1)\pi x)$, $n = 1, 2, 3, \ldots$.

5. (b) $L \neq n\pi$, $n = 1, 2, 3, \ldots$.

7. (a) $(xy')' + \left[\frac{x^2 - m^2}{x}\right]y = 0$, $x > 0$.

(c) $((1-x^2)^{\frac{1}{2}}y')' + (1-x^2)^{-\frac{1}{2}}m^2y = 0$, $-1 < x < 1$.

14. (b) $y_n(x) = B_n\cos(n\pi x/L)$, $z_n(x) = A_n\sin(n\pi x/L)$, A_n, $B_n \neq 0$, $n \geq 1$.

Answers 5.1

1. (a) $u(x,t) = 3\cos(\frac{\pi at}{L})\sin(\frac{\pi x}{L}) + \frac{L}{4\pi a}\sin(\frac{2\pi at}{L})\sin(\frac{2\pi x}{L}) - \cos(\frac{4\pi at}{L})\sin(\frac{4\pi x}{L})$.

(c) $u(x,t) = \frac{L}{4\pi a}\sin(\frac{\pi at}{L})\sin(\frac{\pi x}{L}) + \frac{L}{12\pi a}\sin(\frac{3\pi at}{L})\sin(\frac{3\pi x}{L})$.

3. (b) $u(x,t) = U(x) = -\frac{g}{2a^2}x(L-x)$, which represents a hanging string.

6. (a) $u(x,t) = \frac{8}{\pi}\sum_{k=0}^{N}(2k+1)^{-3}\cos((2k+1)t)\sin((2k+1)x)$, where $N \geq 13$.

Answers 5.2

1. (a) $\frac{1}{2}(1 + \frac{1}{2a})(x + at)^2 + \frac{1}{2}(1 - \frac{1}{2a})(x - at)^2$.

(c) t.

(e) $\sin(x + at)$.

4. $u(x,t) = \frac{1}{2}|x - at|^3 + \frac{1}{2}|x + at|^3$.

Answers 5.3

2. (a) $u(x,t) = \frac{1}{2} + \frac{1}{2}t + \frac{1}{2}[\cos(2at) - \frac{1}{2a}\sin(2at)]\cos(2x)$.

(b) $u(x,t) = \frac{1}{2}[\cos^2(x+at) + \cos^2(x-at)] + \frac{1}{2a}[2at + \frac{1}{4}\sin(2(x+at)) - \frac{1}{4}\sin(2(x-at))]$.

4. (a) $u(x,1) = 0,\quad u(x,2) = -f(x),\quad u(x,3) = 0,\quad u(x,4) = f(x)$.

5. (b) $a/4L$, which is half the lowest frequency $a/2L$ in the case where both ends are fixed.

6. $u(x,t) = \frac{4t^2}{\pi} + \frac{x^2}{\pi} - x + 2\cos(6t)\cos(3x) + \frac{1}{2}\sin(2t)\cos(x)$.

7. $u(x,t) = \frac{1}{1+a^2}\Big[e^{-t}\cos(x) + \frac{1}{2a}[\sin(x+at)-\sin(x-at)] - \frac{1}{2}[\cos(x+at)+\cos(x-at)]\Big]$.

9. $u(x,t) = \begin{cases} (9a^2 - \omega^2)^{-1}[\cos(\omega t) - \cos(3at)]\sin(3x), & \omega \neq 3a, \\ (t/6a)\sin(3at)\sin(3x), & \omega = 3a \text{ (resonance)}. \end{cases}$

Answers 6.1

2. (c) Among the many possibilities, consider x and $2x$. More generally, consider x and any harmonic function $u(x,y)$ which is not of the form $ay + b$ (cf. part (b)).

3. For arbitrary constants a, b, c, A and B, we have the product solutions $(ax + b)(Ay + B)$,

$$(ae^{cx} + be^{-cx})(A\sin(cy) + B\cos(cy)) \quad \text{and} \quad (ae^{cy} + be^{-cy})(A\sin(cx) + B\cos(cx)).$$

The factor $(ae^{cx} + be^{-cx})$ may be replaced by $(\alpha\cosh(cx) + \beta\sinh(cx))$ which is often more useful, and similarly for $(ae^{cy} + be^{-cy})$.

6. In the Dirichlet problem, the value of the steady–state temperature is prescribed at each point on the boundary of the plate, whereas in the Neumann problem the flux of the heat energy through the boundary of the plate is specified at all points along the boundary.

7. (a) $q(x,y)$ is proportional to the rate of heat energy production per unit area at the point (x,y).

(b) In order that there be a steady–state temperature distribution $u(x,y)$, the rate of heat energy production inside the plate must equal the rate of heat loss through the boundary.

Answers 6.2

2. (a) $u(x,y) = 9[\sinh(8\pi M/L)]^{-1}\sin(8\pi x/L)\sinh[8\pi(M-y)/L]$

(b) $u(x,y) = [\sinh(\pi M/L)]^{-1}\sin(\pi x/L)\sinh(\pi y/L)$.

3. $u(x,y) = (\sinh\pi)^{-1}[\sinh(\pi-y)\sin x + \sinh y \sin x + \sinh(\pi-x)\sin y + \sinh x \sin y]$.

4. (a) $U(x,y) = x - y + 2xy$

(b) $u(x,y) = 3[\sinh(\pi)]^{-1}\sin(\pi x)\sinh(\pi-\pi y) + [\sinh(2\pi)]^{-1}\sin(2\pi y)\sinh(2\pi-2\pi x)] + U(x,y)$.

5. (b) two terms (cf. Example 5)

(c) Yes, by applying the Maximum/Minimum Principle to the difference.

7. (b) $u(x,y) = \dfrac{-\cos x\ \cosh(\pi-y)}{\sinh\pi} - \dfrac{\cos(2x)\ \cosh[2(\pi-y)]}{2\sinh(2\pi)} + \coth\pi + \frac{1}{2}\coth(2\pi)$.

Answers 6.3

1. (a) $U(r,\theta) = \log(r)/\log(2) + ([2/(2-\frac{1}{2})]r + [-2/(2-\frac{1}{2})]r^{-1})\cos\theta$

$+ ([1/(4-\frac{1}{4})]r^2 + [-1/(4-\frac{1}{4})]r^{-2})\cos(2\theta) + ([-\frac{1}{4}/(4-\frac{1}{4})]r^2 + [-\frac{1}{4}/(4-\frac{1}{4})]r^{-2})\sin(2\theta)$

$= \log(r)/\log(2) + \frac{4}{3}(r - r^{-1})\cos\theta + \frac{4}{15}(r^2 - r^{-2})\cos(2\theta) + \frac{1}{15}([-r^2 + 16r^{-2})\sin(2\theta)$.

(c) $U(r,\theta) = b + (a - b)\log(r)/\log(2)$

(d) $U(r,\theta) = [(a - \frac{1}{8}b)/(8 - \frac{1}{8})]r^3 + [(8b - a)/(8 - \frac{1}{8})]r^{-3}]\sin(3\theta)$

$= \{[(8a - b)/63]r^3 + [(64b - 8a)/63]r^{-3}\}\sin(3\theta)$.

3. (a) $U(r,\theta) = 1 + r\cos\theta + \frac{1}{4}r^2\cos(2\theta)$ (c) $U(r,\theta) = a$ (d) $U(r,\theta) = \frac{1}{8}\, a\, r^3\sin(3\theta)$.

We cannot use the solutions in Problem 1 because they are singular at the origin.

4. (a) $U(r,\theta) = \frac{1}{2\pi}\int_{-\pi}^{\pi} \frac{(4 - r^2)t^2}{4 - 4r\cos(\theta - t) + r^2}\, dt$, by Theorem 1.

(b) $\text{FS } f(\theta) = \frac{1}{3}\pi^2 + 4\left[\sum_{n=1}^{\infty} \frac{(-1)^n}{n^2}\cos(n\theta)\right]$.

Thus, $U(r,\theta) = \frac{1}{3}\pi^2 + 4\left[\sum_{n=1}^{\infty} \frac{(-1)^n}{n^2}(r/2)^n\cos(n\theta)\right]$, which is exact, according to Theorem 4 and the fact that FS $f(\theta) = f(\theta)$, since the periodic extension of $f(\theta)$ is continuous and piecewise C^1 (i.e., the series converges to $f(\theta)$ when $r = 2$).

(c) By the Maximum Principle (Theorem 1 of Section 6.4), the maximum of the difference between a truncation, say $U_N(r,\theta)$, of the infinite series solution and the exact solution $U(r,\theta)$ is achieved on the boundary. Thus, we need only to estimate the number of terms of FS $f(\theta)$ needed to approximate $f(\theta)$ to within .01 . We use an integral comparison :

$$|f(\theta) - S_N f(\theta)| \le 4\int_0^{\infty} x^{-2}\, dx = 4/N \text{ . Thus, } N = 401 \text{ will suffice.}$$

5. (a) $\frac{2\pi}{(1-a^2)^{\frac{1}{2}}}$ (b) (i) $2\pi r^n/(1-a^2)^{\frac{1}{2}}$, where $r = [1 - (1-a^2)^{\frac{1}{2}}]/a$, for $n = 0, 1, 2, \ldots$.

9. By the Mean–Value Theorem, the steady–state temperature (i.e., the harmonic function with the given boundary values) is the average $\frac{1}{2\pi}\int_{-\pi}^{\pi} 10\,\theta^2\, d\theta = \frac{1}{2\pi}\, 20\,\pi^3/3 = 10\,\pi^2/3$.

Answers 6.4

3. Note that $u(x,y) = c\sinh(y)\sin(x)$ solves the problem, for any constant c. This does not contradict Theorem 2, because the domain $0 < x < \pi$, $0 < y < \infty$ is not bounded.

4. The functions u_1 and u_2 are singular at the origin and thus are not harmonic everywhere in the disk $x^2 + y^2 \le 1$.

7. (a) If $u(x,y)$ is a solution, then so is $u(x,y) + y$.

Answers 6.5

2. The function $\log r$ is not defined, and hence not harmonic, at the origin.

10. (d) The streamlines result from a fluid flow through the slit from -1 to 1 on the v–axis. Electrostically, the hyperbolas represent the equipotential curves formed when a large positively charged plate is placed edge–to–edge with large negatively charged plate, so that there is a slit between them. If the plates are maintained at different constant temperatures, then the hyperbolas will be the isotherms of the resulting steady–state temperature distribution.

11. (a) The image of $(-\infty, -1] + i\,\epsilon$ slightly above the u–axis under $g(w) = (1 - w^2)^{\frac{1}{2}}$ is slightly to the right of the positive y–axis. The image of $[-1, 1] + i\,\epsilon$ slightly above the u–axis runs slightly above the segment $[0,1]$ on the x–axis and then runs slightly below this segment back toward the origin. The image of $[1,+\infty) + i\epsilon$ is slightly to the right of the negative y–axis. Thus, the streamlines flow around a segment perpendicular to the y–axis.

(b) The streamlines are (cross–sectional) isotherms of a steady–state heat distribution about a large flat plate with a perpendicular strip attached, or the equipotential curves of the potential created by a charged conductor which has the same shape.

Answers 7.1

1. The sums in (a), (b) and (c) below run over all nonzero integers.

(a) $\mathrm{FS_c}\, f(x) = L/(i\pi) \sum \S(-1)^{m+1}/m\ddagger\, e^{im\pi x/L}$

(b) $L^2/3 + \sum (-1)^m \S 2L^2/(m\pi)^2\ddagger\, e^{im\pi x/L}$,

(c) $L/2 + L/\pi^2 \sum \lambda(1 - (-1)^m)/m^2\ddagger\, e^{im\pi x/L}$,

3. (a) $\hat{f}(\xi) = (1/\xi^2)\sqrt{2/\pi}\,[1 - \cos(\xi L)]$,

(c) $\hat{f}(\xi) = 1/(2i\sqrt{a})\left[e^{-(\xi-b)^2/2a} - e^{-(\xi+b)^2/2a}\right]$,

11. $f(x) = x$, among many other possibilities.

Answers 7.2

2. (a) $\hat{f}(\xi) = (1 - \xi^2)\, e^{-\xi^2/2}$

7. $\hat{f}(\xi) = \sqrt{2/\pi}\,(1 + i\xi)^{-3}$.

19. (a) $\sum_{m=1}^{\infty} \sin(am)/m = \frac{1}{2}(\pi - a)$.

(c) $\frac{1}{a^2} + 2\sum_{m=1}^{\infty} 1/(a^2 + m^2) = \frac{\pi}{a}\coth(\pi a) \quad (a \neq 0)$.

Answers 7.3

1. (a) $\check{g}(x) = e^{-\frac{1}{2}ax^2}$ (c) $\check{g}(x) = \begin{cases} 0 & |x| > 1 \\ \frac{1}{2} & x = \pm 1 \\ 1 & |x| < 1 \end{cases}$.

3. $\hat{h}(\xi) = \sqrt{2\pi}\, i^{n+1}\, \xi\, \hat{f}(\xi)\hat{f}^{(n)}(\xi)$.

5. (a) $\hat{f}_n(\xi) = \frac{n!}{\sqrt{2\pi}}\left[\frac{1 - i\xi}{1 + \xi^2}\right]^{n+1}$

6. (a) $g(x) = \frac{b}{\sqrt{2\pi(b-a)}} \exp[-\frac{1}{2}abx^2/(b-a)]$.

7. (a) $\frac{1}{2}\pi a^3$.

(c) $\pi a^{-2}[1 - e^{-ab}]$.

10. $f(x) = 1 - |x|$ if $|x| < 1$, and $f(x) = 0$ if $|x| \geq 1$.

Answers 7.4

2. $u(x,t) = \sqrt{t/2k}\, e^{-x^2/4kt}$.

11. $u(x,t) = \frac{\sin(x+at)}{1 + (x+at)^2} + \frac{\sin(x-at)}{1 + (x-at)^2}$.

14. $u(x,y) = \frac{1}{\pi}\int_{-\infty}^{\infty} \frac{y f(s)}{y^2 + (x-s)^2}\, ds = \frac{1}{\pi}\int_{-\infty}^{0} \frac{-y}{y^2 + (x-s)^2}\, ds + \frac{1}{\pi}\int_{0}^{\infty} \frac{y}{y^2 + (x-s)^2}\, ds$

$= \frac{1}{\pi}\arctan[(x-s)/y]\Big|_{-\infty}^{0} - \frac{1}{\pi}\arctan[(x-s)/y]\Big|_{0}^{\infty} = \frac{2}{\pi}\arctan(x/y)$.

Answers 7.5

2. $u(x,t) = -\frac{1}{\sqrt{\pi}}\int_0^t (t-s)^{-\frac{1}{2}} e^{-x^2/4(t-s)}\, h(s)\, ds$.

3. $u(x,t) = \frac{1}{\sqrt{4\pi kt}}\int_0^\infty \left[e^{-(x+y)^2/4kt} + e^{-(x-y)^2/4kt}\right] f(y)\, dy - \frac{1}{\sqrt{\pi}}\int_0^t (t-s)^{-\frac{1}{2}} e^{-x^2/4(t-s)}\, h(s)\, ds.$

5. $u(x,y) = \frac{1}{\pi}\int_0^\infty y\left[\frac{1}{y^2 + (x-s)^2} - \frac{1}{y^2 + (x+s)^2}\right] f(s)\, ds$,

8. (a) $u(x,y) = \frac{1}{\pi}\int_0^\infty \left[y\left[\frac{1}{y^2+(x-s)^2} - \frac{1}{y^2+(x+s)^2}\right]f(s) + x\left[\frac{1}{x^2+(y-s)^2} - \frac{1}{x^2+(y+s)^2}\right]g(s)\right] ds$

Note that other solutions may be obtained by simply adding multiples of xy or $\mathrm{Im}[(x+iy)^{2n}]$, n = 1, 2, 3, ...

Answers 8.1

1. (a) $|\sin(x)| \leq 1$

(c) $|e^x - (1 + x + x^2/2!)| = \left|\sum_{n=3}^\infty \frac{1}{n!} x^n\right| \leq |x|^3 \left|\sum_{n=3}^\infty \frac{1}{n!} x^{n-3}\right| \leq e\cdot|x|^3$ for $|x| \leq 1$.

5. (b) If $f(x) = x$ and $g(x) = x^2$, then $f(x) + g(x) \neq O(x^2)$. In general, choose any $f(x)$ such that $f(x) = O(x)$, but $f(x) \neq O(x^2)$. Then $f(x) + g(x) \neq O(x^2)$ for otherwise $f(x) = [f(x) + g(x)] - g(x) = O(x^2)$.

10. $\lim_{\Delta x\to 0} \frac{f(\Delta x) - f(-\Delta x)}{2\Delta x} = \lim_{\Delta x\to 0} \frac{1}{2}\left[\frac{f(\Delta x) - f(0)}{\Delta x} + \frac{f(-\Delta x) - f(0)}{-\Delta x}\right] = \frac{1}{2}(f'(0) + f'(0)).$

If $f(x) = |x|$, then $\lim_{\Delta x\to 0} \frac{f(\Delta x) - f(-\Delta x)}{2\Delta x} = 0$, but $f'(0)$ does not exist.

Answers 8.2

1. (a) $v_{1,1} = 3.5$, $u(0.5, 1) = 3.5$,

   (b) (i) $v_{1,32} = 3.625$, $v_{2,32} = 3.5$, $v_{3,32} = 3.625$,

2. (a) $v_{1,1} = 5$, $v_{2,1} = 8$, $v_{3,1} = 13$, $v_{4,1} = 20$,

   (b) (i) $v_{1,4} = 5$, $v_{2,4} = 8$, $v_{3,4} = 13$, $v_{4,4} = 20$,

3. (d) $u_4(0.1,\ 0.005) = 0.984...$, while $v_{1,2} = 0.9875$.

6. $|w_{i,j}| \le 0.0016276...$. Note that $u(.5,.5) - v_{4,128} = .500073831... - .500071505... = .00000232...$, which is much less than the upper bound $.0016276...$.

13. $\beta_1 = .653271...$, $\beta_2 = 3.292310...$, $\beta_3 = 6.361620...$, $\beta_4 = 9.477485...$.

Answers 8.3

2. (a) $y_n = c_1 2^n + c_2 3^n$ (b) $y_n = c_1 + c_2\, n$ (c) $y_n = 2^{n/2}(c_1 \cos(\frac{n\pi}{4}) + c_2 \sin(\frac{n\pi}{4}))$.

6. (b) Using (30), $v_{\frac{1}{2}M,N} = [1 - 4\cdot\frac{1}{6}\sin^2(\frac{\pi}{2M})]^N \sin(\frac{\pi}{2}) = [1 - \frac{2}{3}\sin^2(\frac{\pi}{2M})]^{3M^2}$, $N = 3M^2$.

(d) If $r > 0$, $M = \pi\, e^{-\frac{1}{12}\pi^2}\, (3240)^{-\frac{1}{6}}\, r^{-\frac{1}{6}} \approx (0.3588...)\, r^{-\frac{1}{6}}$. If $r < 0$, then $M = \pi e^{-\frac{1}{12}\pi^2}(1620)^{-\frac{1}{6}}\, |r|^{-\frac{1}{6}} \approx (0.4027...)\, |r|^{-\frac{1}{6}}$.

(e) With $r = -10^{-7}$, theoretically the error is smallest when $M \approx (0.4027...)\, |10|^{\frac{7}{6}} = 5.91...$. Numerical calculation by computer yields $u(.5,.5) = .0071918...$. The values $\bar{v}_{M,3M^2}$ computed via the explicit difference method with a systematic round off error introduced by dropping all but the first 6 decimal places are given by $\bar{v}_{2,12} = .007706$, $\bar{v}_{4,48} = .007213$, $\bar{v}_{6,108} = .007184$, $\bar{v}_{8,192} = .007170$, $\bar{v}_{10,300} = .007156$, $\bar{v}_{12,432} = .007141$. Note that the error $|u(.5,.5) - \bar{v}_{M,3M^2}|$ is smallest when $M = 6$, as predicted.

Answers 9.1

1. $u(x,y,t) = 4\exp\left[-\pi^2[(\frac{3}{2})^2+1^2]6t\right]\sin(3\pi x/2)\sin(\pi y) - 2\exp\left[-\pi^2[1^2+(\frac{2}{3})^2]6t\right]\sin(\pi x)\sin(2\pi y/3)$.

3. $u(x,y,t) = \exp\left[-\pi^2[(\frac{3}{2})^2+(\frac{1}{2})^2]t\right]\cos(3\pi x/2)\sin(\pi y/2)$.

7. For the justification of the following approximations, see Example 3.

(a) $u(\frac{1}{2},\frac{1}{2},\frac{1}{2},t) \approx 6400\pi^{-3}\cdot\frac{1}{2}\sqrt{2}\, e^{-9\pi^2 kt/4}$. Note that $\sin(\pi/4) = \frac{1}{2}\sqrt{2}$ and $1^2+1^2+(\frac{1}{2})^2 = 9/4$.

(b) $u(\frac{1}{2},\frac{1}{2},\frac{1}{2},t) \approx 400\pi^{-1}\cdot\frac{1}{2}\sqrt{2}\, e^{-\pi^2 kt/4}$.

(c) For opposite insulated faces, $u(\frac{1}{2},\frac{1}{2},\frac{1}{2},t) \approx 1600\pi^{-2}e^{-2\pi^2 kt}$.

For adjacent insulated faces, $u(\frac{1}{2},\frac{1}{2},\frac{1}{2},t) \approx 6400\pi^{-3}\frac{1}{2}e^{-3\pi^2 kt/2}$. Thus, the temperature drop in the middle is smallest for *adjacent* insulated faces, when t is large, because $\frac{3}{2} < 2$.

8. $u(x,y,z) = \sin(x)\sin(y)\sinh(\sqrt{2}(1-z))/\sinh(\sqrt{2})$.

13. $u(x,y,t) = \sum_{n,m=1}^{\infty} c_n c_m \,[\pi a(n^2+m^2)^{\frac{1}{2}}]^{-1} \sin[\pi a(n^2+m^2)^{\frac{1}{2}}t]\, \sin(n\pi x)\sin(m\pi y)$.

where $c_k = 2\int_0^1 z(z-1)\sin(k\pi z)\,dz = 2 <s_k, z(z-1)> = -2\,(k\pi)^{-2} <s_k'', z(z-1)>$

$= -2\,(k\pi)^{-2}\left[\{s_k'(z)z(z-1) - s_k(z)(2z-1)\}\Big|_0^1 + <s_k, 2>\right] = -2\,(k\pi)^{-2} <s_k, 2>$

$= 2\,(k\pi)^{-3}\, 2\cos(k\pi z)\Big|_0^1 = 2\,(k\pi)^{-3}\, 2[\cos(k\pi) - 1] = 2\,(k\pi)^{-3}\, 2[(-1)^k - 1]$

$= -\left[\frac{2}{k\pi}\right]^3$ if k is odd, and 0 otherwise.

Thus, $u(x,y,t) = \sum_{n,m = 1\ (\text{odd})}^{\infty} \left[\frac{4}{nm\pi\pi}\right]^3 [\pi a(n^2+m^2)^{\frac{1}{2}}]^{-1} \sin[\pi a(n^2+m^2)^{\frac{1}{2}}t]\, \sin(n\pi x)\sin(m\pi y)$.

Answers 9.2

1. (a) $f_{n,m}(x,y) = \cos(n\pi x/3)\cos(m\pi y/2)$ $m, n = 0, 1, 2, \ldots$.

(b) The associated solutions of $u_t = k\Delta u$ are of the form

$u_{n,m}(x,y,t) = \exp\left[-k\pi^2(m^2/3^2 + n^2/2^2)t\right]\cos(n\pi x/3)\cos(m\pi y/2)$.

Note that $u(x,t) = \sum_{n=0}^{N}\sum_{m=0}^{M} u_{n,m}(x,y,t)$ solves the problem

D.E. $u_t = k\Delta u$ on R with $t > 0$

B.C. $u_y(x,0,t) = u_y(x,2,t) = 0$ $(0 \le x \le 3)$ and $u_x(0,y) = u_x(3,y) = 0$ $(0 \le y \le 2)$

I.C. $u(x,y,0) = \sum_{n=0}^{N}\sum_{m=0}^{M} a_{n,m}\cos(n\pi x/3)\cos(m\pi y/2)$.

4. The two solutions are of the form $u_1(x,y,t) = \exp(-\pi^2(3^2+1^2)kt)\cos(3\pi x)\cos(\pi y)$ and $u_2(x,y,t) = \exp(-\pi^2(2^2+2^2)kt)\cos(2\pi x)\cos(2\pi y)$. The solution u_1 decays more rapidly, since $3^2 + 1^2 > 2^2 + 2^2$. In general, the larger the eigenvalue, the more rapid the decrease.

Answers 9.3

1. (a) $U(\rho,\varphi,\theta) = \rho^{-1}$ (c) $U(\rho,\varphi,\theta) = \rho^2\sin^2(\varphi) - 2\rho^2\cos^2(\varphi) = \rho^2(1 - 3\cos^2\varphi)$.

3. $\Delta_s(\sin\varphi) = (\sin\varphi)^{-1}\cos^2\varphi - \sin\varphi$; $\sin\varphi$ is not C^2 (nor C^1) on the unit sphere.

8. $U(\rho,t) = \rho^{-1}\left[\frac{3}{4}e^{-\pi^2kt}\sin(\pi\rho) - \frac{1}{4}e^{-9\pi^2kt}\sin(3\pi\rho)\right]$.

11. (a) $U(\rho,t) = \frac{1}{\rho}\frac{1}{\sqrt{4\pi kt}}\int_0^\infty\left[e^{-(\rho-\bar\rho)^2/4kt} - e^{-(\rho+\bar\rho)^2/4kt}\right]\bar\rho f(\bar\rho)\,d\bar\rho$.

Answers 9.4

2. The following answers are not unique

(a) $(x,y,z) \mapsto (\sqrt{\tfrac{1}{2}}(y+x),\sqrt{\tfrac{1}{2}}(y-x),z)$; rotation by $-45°$ about the third axis.

(c) $(x,y,z) \mapsto (-y,x,z)$; rotation by $90°$ about the third axis.

4. (a) $8\rho^4P_4(\cos\varphi) = \rho^4(35\cos^4\varphi - 30\cos^2\varphi + 3) = 35z^4 - 30\rho^2z^2 + 3\rho^4$

$= 35z^4 - 30(x^2 + y^2 + z^2)z^2 + 3(x^2 + y^2 + z^2)^2$.

(b) Yes, since $f(\varphi,\theta) = f(0,\theta)$, which is the constant value of f at the north pole.

6. (a) $2\sin^2\varphi\cos\theta\sin\theta = \frac{1}{6}S_{2,2}(\varphi,\theta) - \frac{1}{6}S_{2,-2}(\varphi,\theta)$.

$2\sin\varphi\cos\varphi\sin\theta = \frac{1}{6}S_{2,1}(\varphi,\theta) - \frac{1}{6}S_{2,-1}(\varphi,\theta)$.

$2\sin\varphi\cos\varphi\cos\theta = \frac{1}{6}S_{2,1}(\varphi,\theta) + \frac{1}{6}S_{2,-1}(\varphi,\theta)$.

$\sin^2\varphi(\cos^2\theta - \sin^2\theta) = \frac{1}{6}S_{2,2}(\varphi,\theta) + \frac{1}{6}S_{2,-2}(\varphi,\theta)$.

$\sin^2\varphi\sin^2\theta - \cos^2\varphi = -\frac{1}{2}\left[\frac{1}{6}S_{2,2}(\varphi,\theta) + \frac{1}{6}S_{2,-2}(\varphi,\theta)\right] - S_{2,0}(\varphi,\theta)$.

11. (a) $u(x,y,z) = \frac{1}{3} + x + \frac{6}{5}z - \frac{1}{3}x^2 + \frac{2}{3}y^2 - \frac{1}{3}z^2 + z^3 - \frac{6}{5}z(x^2+y^2+z^2)$.

(b) $U(\rho,\varphi,\theta) = x\rho^{-3} + \frac{1}{3}\rho^{-1} - \frac{1}{2}(z^2 - \frac{1}{3}\rho^2)\rho^{-5} - \frac{1}{2}(x^2-y^2)\rho^{-5} + z^3\rho^{-7} - \frac{6}{5}z\rho^{-5} + \frac{6}{5}\rho^{-3}$.

17. $U(\rho,\varphi,\theta,t) = \exp[-4\pi^2kt]\,j_0(2\pi\rho) + \exp[-(\beta_{1,3})^2kt]j_1(\beta_{1,3}\,\rho)\cos\varphi$

$+ \exp[-(\beta_{2,1})^2kt]\,j_2(\beta_{2,1}\,\rho)\sin^2\varphi\sin(2\theta)$.

Answers 9.5

7. Let $f_{\nu,q}(r,\theta) = J_\nu(j_{\nu,q} r/r_o)\sin(n\pi\theta/\alpha)$, $\nu = n/\alpha$, $n = 1, 2, 3, \ldots$, where $J_\nu(r)$ is defined as in (6), but with m replaced by ν and $(m+k)!$ replaced by $\Gamma(\nu+k+1)$, and $j_{\nu,q}$ is the q–th positive zero of $J_\nu(r)$. The eigenfunctions $f_{\nu,q}(r,\theta)$ (with eigenvalues $j_{\nu,q}/r_o$) can be used to solve for heat flow in a thin wedge insulated on the top and bottom surfaces, and maintained at 0 on the edges. Moreover, they can be used for describing the vibrations of a wedge–shaped drum.

8. $$U(r,\theta,z) = \sum_{m=-\infty}^{\infty} \sum_{q=1}^{\infty} c_{m,q}\, J_m(j_{m,q} r/r_o) \sinh(j_{m,q} z/r_o)\, e^{im\theta},$$

where $$c_{m,q} = \left[\pi\, r_o^2 J_m'(j_{m,q})^2 \sinh(j_{m,q} z_o/r_o)\right]^{-1} \int_0^{2\pi} \int_0^{r_o} f(r,\theta)\, J_m(j_{m,q} r/r_o) e^{im\theta}\, dr d\theta .$$

9. (e) No, the minimum energy is $\frac{3}{2}\hbar\sqrt{k/m}$, which is obtained when $N = 0$.

Answers 9.6

8. Let $U(P,t) \equiv \int_{\partial M} \partial_\nu G(P,Q)\, f(Q,t)\, d\partial M(Q)$. From (47), $U(P,t)$ solves the D.E. $\Delta_M U = 0$, with B.C. $U(Q,t) = f(Q,t)$ for $Q \in \partial M$. Let $v(P,t) \equiv u(P,t) - U(P,t)$. Then v solves the related problem with homogeneous B.C. and with the I.C. $v(P,0) = -U(P,0)$. This related problem has the formal solution $v(P,t) = \sum_{q=1}^{\infty} c_q\, e^{-\lambda_q \kappa t} u_q(P)$, where $c_q = -\langle U(\cdot,0),u_q\rangle$.

Thus, $$u(P,t) = \int_{\partial M} \partial_\nu G(P,Q)\, f(Q,t)\, d\partial M(Q) + \sum_{q=1}^{\infty} c_q\, e^{-\lambda_q \kappa t} u_q(P) .$$

21. $$G(\mathbf{r},\mathbf{r}') = \sum_{n=0}^{\infty} \sum_{q=1}^{\infty} \sum_{m=-n}^{n} -(\beta_{n,q})^{-1} \|E_{n,q,m}\|^{-2} j_n(\beta_{n,q}\rho) S_{n,m}(\varphi,\theta) j_n(\beta_{n,q}\rho') S_{n,m}(\varphi',\theta'),$$

where (ρ,θ,φ) [resp. (ρ',θ',φ')] are the spherical coordinates of $\mathbf{r}$ [resp. $\mathbf{r}'$].

INDEX OF NOTATION

INDEX OF NOTATION

INDEX

INDEX

(See also Index of Notation)